Lecture Notes
in Control and Information Sciences   258

Editor: M. Thoma

T0138119

Lecture Notes
in Control and Information Sciences   258

Springer
London
Berlin
Heidelberg
New York
Barcelona
Hong Kong
Milan
Paris
Singapore
Tokyo

Alberto Isidori, Françoise Lamnabhi-Lagarrigue, and Witold Respondek (Eds)

Nonlinear Control in the Year 2000

Volume 1

With 86 Figures

 Springer

Editors

Alberto Isidori, Professor
Dipartimento di Informatica e Sistemistica, Università di Roma "La Sapienza",
00184 Rome Italy

Françoise Lamnabhi-Lagarrigue, Docteur D'état
Laboratoire des Signaux et Systems, CNRS SUPELEC,
91192 Gif-sur-Yvette, France

Witold Respondek, Professor
Laboratoire de Mathématique et Informatique, INSA de Rouen,
76131 Mont Saint Aignan, France

ISBN 1-85233-363-4 Springer-Verlag London Berlin Heidelberg

British Library Cataloguing in Publication Data
A catalog record for this book is available from the British Library

Library of Congress Cataloging-in-Publication Data
Nonlinear control in the year 2000 / Alberto Isidori, Françoise Lamnabhi-Lagarrigue,
and Witold Respondek.
        p. cm. -- (Lecture notes in control and information sciences, ISSN 0170-8643 ; 258-259)
    Includes bibliographical references.
    ISBN 1-85233-363-4 (v. 1 : acid-free paper) -- ISBN 1-85233-364-2 (v.2 : acid-free paper)
    1. Nonlinear control theory. I. Isidori, Alberto. II. Lamnabhi-Lagarrigue, F.
    (Françoise), 1953- III. Respondek, W. IV. Series.
    QA402.35 .3N66 2001
    003.5--dc21   00-045600

Typesetting: Camera ready by editors
Printed and bound at the Athenæum Press Ltd., Gateshead, Tyne & Wear
69/3830-543210 Printed on acid-free paper SPIN 10768731

Preface

These two volumes contain papers based on talks delivered at the 2nd Workshop of the Nonlinear Control Network (http://www.supelec.fr/lss/NCN), held in Paris, June 5-9, 2000. The Authors of the presented papers, as well as the Editors of these two volumes, hope that "Nonlinear Control in the Year 2000" is not only one more book containing proceedings of a workshop. Two main reasons justifying our hope are that, firstly, the end of the century is a natural moment to think about past developments in nonlinear control and about its perspectives in the twenty-first century; and, secondly, we believe that nonlinear control has reached an age of maturity which enables the community to sketch a state-of-the-art of the field. We hope that papers contained in the two volumes serve to fulfill these goals: many of them have their roots and their origins in nonlinear control theories which have been developed in past three decades and which, by now, form a basis of canonical results in the field. Such papers form a bridge between the actual theory and its future developments. Many other papers contained in the book present completely new ideas and suggest new directions and, in this sense, they are directed towards the future of the nonlinear control.

We would like to emphasize one peculiarity of our field: nonlinear control is an example of a theory situated at a crossroad between mathematics and engineering science. Due to this position, nonlinear control has its roots in both fields and, as we deeply believe, can bring new ideas and new results for both domains. The book reflects very well this "double character" of nonlinear control theory: the reader will find in it results which cover a wide variety of problems: starting from pure mathematics, through its applications to nonlinear feedback design, all way to recent industrial advances.

Eight papers contained in the book are based on invited talks delivered at the Workshop by:

> Alessandro Astolfi,   John Baras,   Christopher Byrnes,
> Bronisław Jakubczyk, Anders Rantzer, Kurt Schlacher,
> Eduardo Sontag,   Hector Sussmann.

Altogether the book contains 80 papers and therefore it is impossible to mention all discussed topics: to give a flavour of the presented material let us mention a few of them. For many theoretical papers a common factor is optimal control, for example sub-Riemannian geometry (in particular fundamental results on (non)subanalyticity of small balls and of the distance function), the use of generalized differentials in generalized Maximum Principle, singular and constrained optimal control problems. Another subdomain of nonlinear control attracting many contributions to the book is stability and asymptotic behavior. In this area, stemming from traditional Lyapunov techniques, the methods based on the concept of input-to-state stability have established a leading role in analysis and design, and new ideas have emerged, such as those aiming at the analysis of non-equilibrium steady-state behaviors, or at the evaluation of asymptotic convergence for almost all initial conditions via dual Lyapunov analysis. Applications of these ideas in nonlinear control are widespread: stabilization of nonlinear systems, trajectory tracking, adaptive control. Other subdomains of nonlinear control attracting a lot of attention, and represented in the book, are that of observability and observers for nonlinear systems, sliding mode control, theory of nonlinear feedback (invariants, classification, normal forms, flatness and dynamic feedback), recursive design (backstepping and feedforwarding). The papers present in the two volumes cover various aspects of all just mentioned topics. Moreover, the book contains also papers discussing main results of a plenary Session devoted to industrial applications.

We wish to thank all invited speakers and all contributors to the 2nd Nonlinear Control Network Workshop for making this conference an outstanding intellectual celebration of the area of Nonlinear Control at the turn of the century. The Editors are grateful to all the chairpersons:

Dirk Aeyels,	*Andrei Agrachev,*	*Andrea Bacciotti,*
Alfonso Baños,	*Georges Bastin,*	*Antonio Bicchi,*
Fritz Colonius,	*Emmanuel Delaleau,*	*Michel Fliess,*
Halina Frankowska,	*Jean-Paul Gauthier,*	*Henri Huijberts,*
Bronisław Jakubczyk,	*Philippe Jouan,*	*Ioan Landau,*
Jean Lévine,	*Antonio Loria,*	*Riccardo Marino,*
Frédéric Mazenc,	*Gérard Montseny,*	*Claude Moog,*
Philippe Mullhaupt,	*Henk Nijmeijer,*	*Romeo Ortega,*
Elena Panteley,	*Laurent Praly,*	*Anders Rantzer,*
Pierre Rouchon,	*Joachim Rudolph,*	*Andrey Sarychev,*
Arjan van der Schaft,	*Jacquelien Scherpen,*	*Rodolphe Sepulchre,*
Fatima Silva Leite,	*Hebertt Sira-Ramirez,*	*Fabian Wirth,*
Alan Zinober.		

They have excellently played their role during the presentations and also have helped us in reviewing the contributed papers accepted for publication in this book. We would like to thank the TMR Program for the financial support which, in particular, helped numerous young researchers in attending the Workshop. We express our thanks to *Christiane Bernard* (Mathematics and Information Sciences Project Officer at the European Community) and *Radhakisan Baheti* (Program Director for Control at NSF) for their participation at the Session on Founding to Academic Research on Control. We also thank the CNRS staff as well as PhD students and Postdocs at L2S in Gif-sur-Yvette for their help in the organization of the Workshop.

Françoise Lamnabhi-Lagarrigue   *Alberto Isidori*   *Witold Respondek*

Gif-sur-Yvette   Rome   Rouen

Contents

Volume 1

Volume 2

Subanalyticity of Distance and Spheres in Sub-Riemannian Geometry

Andrei Agrachev[1] and Jean-Paul Gauthier[2]

[1] S.I.S.S.A., Via Beirut 2-4
34013 Trieste, Italy
and Steklov Mathematical Institute, ul. Gubkina 8
117966 Moscow, Russia
agrachev@sissa.it
[2] Laboratoire d'analyse appliquée et optimisation
Universit de Bourgogne
Département de Mathématiques, B.P. 47870
21078 Dijon, France
gauthier@u-bourgogne.fr

1 Introduction

Let M be a C^∞ Riemannian manifold, $\dim M = n$. A distribution on M is a smooth linear subbundle Δ of the tangent bundle TM. The number $k = \dim \Delta_q$ is the *rank* of the distribution. The restriction of the Riemannian structure to Δ is a *sub-Riemannian structure*.

We fix a point $q_0 \in M$ and study only admissible paths starting from this point, i.e. meeting the initial condition $q(0) = q_0$. Sections of the linear bundle Δ are smooth vector fields; we set

$$\bar{\Delta} = \{X \in \text{Vec}\, M : X(q) \in \Delta_q,\ q \in M\},$$

the space of sections of Δ. Iterated Lie brackets of the fields in $\bar{\Delta}$ define a flag

$$\Delta_{q_0} \subset \Delta_{q_0}^2 \subset \cdots \subset \Delta_{q_0}^m \cdots \subset T_q M$$

in the following way:

$$\Delta_{q_0}^m = span\{[X_1, [X_2, [\ldots, X_m]\ldots](q_0) : X_i \in \bar{\Delta},\ i = 1, \ldots, m\}.$$

A distribution Δ is *bracket generating* at q_0 if $\Delta_{q_0}^m = T_{q_0}M$ for some $m > 0$. If Δ is bracket generating, and if q_0 and q_1 are close, the Riemannian length of the shortest path, connecting q_0 to q_1, is the *sub-Riemannian distance* or *Carnot—Caratheodory distance* between q_0 and q_1.

In the remainder of the paper we assume that Δ is bracket generating at a given initial point q_0. We denote by $\rho(q)$ the sub-Riemannian distance between q_0 and q. ρ is a continuous function defined on a neighborhood of q_0. Moreover, ρ is Hölder-continuous with the Hölder exponent $\frac{1}{m}$, where $\Delta_{q_0}^m = T_{q_0}M$.

We study mainly the case of real-analytic M and Δ. The germ at q_0 of a Riemannian distance is the square root of an analytic germ. This is not true for a sub-Riemannian distance function ρ. Moreover, ρ is never smooth in a punctured neighborhood of q_0 (i.e. in a neighborhood without the pole q_0). It may happen that ρ is not even subanalytic. Main results of the paper concern subanalyticity properties of ρ in the case of a generic real-analytic Δ.

We prove that, generically, the germ of ρ at q_0 is subanalytic if $n \leq (k-1)k+1$ (Th. 5) and is not subanalytic if $n \geq (k-1)\left(\frac{k^2}{3} + \frac{5k}{6} + 1\right)$ (Th. 8). The balls $\rho^{-1}([0, r])$ of small enough radius are subanalytic if $n > k \geq 3$ (Th. 7). This statement about the balls is valid not only generically, but up to a set of distributions of codimension ∞.

In particular, if $k \geq 3$, $n \geq (k-1)\left(\frac{k^2}{3} + \frac{5k}{6} + 1\right)$, then (generically!) the balls $\rho^{-1}([0, r])$ are subanalytic but ρ is not! The results are just stated herein, and the complete proofs can be found in [16].

This paper is a new step in a rather long research line, see [1,5,6,9,10,12,13,15]. The main tools are the nilpotent approximation, Morse-type indices of geodesics, both in the normal and abnormal cases, and transversality techniques.

2   Nilpotentization, endpoint mapping and geodesics

Nilpotentization or nilpotent approximation is a fundamental operation in the geometric control theory and sub-Riemannian geometry; this is a real nonholonomic analog of the usual linearization (see [2,3,7,8]). Here we refer specially to [7].

We are working in a small neighborhood O_{q_0} of $q_0 \in M$, where we fix an orthonormal frame $X_1, \ldots, X_k \in \text{Vec}\, O_{q_0}$ of the sub-Riemannian structure under consideration. Admissible paths are thus solutions of the Cauchy problem:

$$\dot{q} = \sum_{i=1}^{k} u_i(t) X_i(q), \quad q \in O_{q_0}, \quad q(0) = q_0, \tag{1}$$

where $u = (u_1(\cdot), \ldots, u_k(\cdot)) \in L_2^k[0, 1]$. The nilpotentization of (1) has the following equation :

$$\dot{x} = \sum_{i=1}^{k} u_i(t) \chi_* \hat{X}_i(x), \quad x \in \mathbb{R}^n, \quad x(0) = 0, \tag{$\hat{1}$}$$

where $\chi : O_{q_0} \to \mathbb{R}^n$ is an "adapted coordinate map", and $\hat{X}_i(x)$ is the "nilpotentization" of X_i, $i = 1, \cdots, k$. In adapted coordinates :

$$\chi(q_0) = 0, \quad \chi_*\big|_{T_{q_0}M}(\Delta_{q_0}^i) = \mathbb{R}^{k_1} \oplus \cdots \oplus \mathbb{R}^{k_i}, \quad 1 \leq i \leq l,$$

$$\mathbb{R}^n = \mathbb{R}^{k_1} \oplus \cdots \oplus \mathbb{R}^{k_l},$$

$$\dim(\Delta_{q_0}^i/\Delta_{q_0}^{i-1}) = k_i, i = 1, \ldots, l.$$

We introduce a dilation $\delta_t : \mathbb{R}^n \to \mathbb{R}^n$, $t \in \mathbb{R}$, by the formula:

$$\delta_t(x_1, x_2, \ldots, x_l) = (tx_1, t^2 x_2, \ldots, t^l x_l). \tag{2}$$

Below $\|u\| = \left(\int\limits_0^1 \sum\limits_{i=0}^k u_i^2(t) \, dt \right)^{\frac{1}{2}}$ is the norm in $L_2^k[0,1]$. Let

$$U_r = \{u \in L_2^k[0,1] : \|u\| = r\},$$

be the sphere of radius r in $L_2^k[0,1]$. We consider the *endpoint mapping* f : $u \mapsto q(1)$. It is a well-defined smooth mapping of a neighborhood of the origin of $L_2^k[0,1]$ into M. Clearly, $\rho(q) = \min\{\|u\| : u \in L_2^r[0,1], \ f(u) = q\}$.

We define also the mapping $\hat{f} : L_2^k[0,1] \to \mathbb{R}^n$ by the rule $\hat{f} : u(\cdot) \mapsto x(1)$, where $x(\cdot) = x(\cdot; u)$ is the solution of $(\hat{1})$.

Proposition 1. *Let* $\chi = (\chi_1, \ldots, \chi_l)$, $\chi_j : O_{q_0} \to \mathbb{R}^{k_j}$, $j = 1, \ldots, l$. *Then the following identities hold for any* $u(\cdot) \in L_2^k[0,1]$, $\varepsilon \in \mathbb{R}$:

$$\hat{f}(u(\cdot)) = \left(\int\limits_0^1 \sum\limits_{i=1}^k u_i(t)\hat{X}_i\chi_1(q_0) \, dt, \ldots, \right.$$

$$\left. \int\limits_{0 \le t_1 \le \cdots \le t_l \le 1} \cdots \int \sum\limits_{i_j=1}^k u_{i_1}(t_1) \cdots u_{i_l}(t_l)\hat{X}_{i_1} \circ \cdots \circ \hat{X}_{i_l}\chi_l(q_0) \, dt_1 \cdots dt_l \right);$$

$\hat{f}(\varepsilon u(\cdot)) = \delta_\varepsilon \hat{f}(u(\cdot))$, *where* δ_ε *is the dilation (2).*

We set $f_\varepsilon(u)) = \delta_{\frac{1}{\varepsilon}}\chi(f(\varepsilon u))$. Then f_ε is a smooth mapping from a neighborhood of 0 in $L_2^k[0,1]$ to $\mathbb{R}^n$. Moreover, any bounded subset of $L_2^k[0,1]$ is contained in the domain of f_ε for ε small enough.

Theorem 1. $f_\varepsilon \longrightarrow \hat{f}$ *as* $\varepsilon \longrightarrow 0$ *in the* C^∞ *topology of the uniform convergence of the mappings and all their derivatives on the balls in* $L_2^k[0,1]$.

Recall that $\rho(q) = \min\{\|u\| : f(u) = q, \ u \in L_2^k[0,1]\}$ is the sub-Riemannian distance function. We set:

$$\rho_\varepsilon(x) = \min\{\|u\| : f_\varepsilon(u) = x, \ u \in L_2^k[0,1]\} = \frac{1}{\varepsilon}\rho\left(\chi^{-1}(\delta_\varepsilon x)\right)$$

and:

$$\hat{\rho}(x) = \min\{\|u\| : \hat{f}(u) = x, \ u \in L_2^k[0,1]\}.$$

Thus $\hat{\rho}$ is the sub-Riemannian distance for the nilpotentization of the original system.

Lemma 1. *The family of functions* $\rho_\varepsilon\big|_K$ *is equicontinuous for any compact* $K \subset \mathbb{R}^n$.

Theorem 2. $\rho_\varepsilon \longrightarrow \hat{\rho}$ *uniformly on compact subsets of* $\mathbb{R}^n$ *as* $\varepsilon \longrightarrow 0$.

The following proposition is a modification of a result by Jacquet [13].

Proposition 2. *Let* $\mathcal{M}_r = \{u \in U_r : \exists \alpha \in (0,1] \text{ s.t. } \alpha u \text{ is minimal for } (1)\}$. *Then* $\overline{\mathcal{M}}_r$ *is a compact subset of the Hilbert sphere* U_r *and* $\hat{f}(\overline{\mathcal{M}}_r \setminus \mathcal{M}_r) \subset \hat{\rho}^{-1}(r)$; *in particular, any element of* $\overline{\mathcal{M}}_r \setminus \mathcal{M}_r$ *is a minimizing control for system* $(\hat{1})$.

Subriemannian minimizers (geodesics) can be of two types: normal or abnormal. Normal geodesics are projections $\pi\psi(t)$ on M of trajectories on T^*M of the Hamiltonian vector field $\overrightarrow{h}$ associated with the Hamiltonian h, $h(\psi) = \frac{1}{2}\sum_{i=1}^{k} h_i^2(\psi)$, $\forall q \in M$, $\psi \in T_q^*M$, where h_i is the Hamiltonian lift of X_i.

Abnormal geodesics are solutions of:

$$\dot{\psi} = \sum_{i=1}^{k} u_i(t)h_i(\psi), \quad h_i(\psi(t)) = 0. \tag{3}$$

There is the following fundamental result:

Proposition 3. *(Goh-Legendre condition) If* $u(\cdot)$ *is a strictly abnormal minimizer, it does exist a* ψ *satisfying (3) with:*

$$\{h_i, h_j\}(\psi(t)) = 0 \quad \forall i,j \in \{1,\ldots,k\}, \tag{4}$$

$$\sum_{i,j=1}^{k} \{h_i, \{h_j, \sum_{i=1}^{k} u_i(t)h_i\}\}v_iv_j \le 0 \quad \forall (v_1,\ldots,v_k) \in \mathbb{R}^k, \tag{5}$$

for almost all $t \in [0,1]$, *where* $\{a,b\} = ab$ *is the Poisson bracket of the Hamiltonians* a, b. □

Remark. Identity (4) is called the Goh condition while Inequality (5) is the generalized Legendre condition. It is easy to see that both conditions are actually intrinsic: Identity (4) doesn't depend on the choice of the orthonormal frame $X_1,\ldots,X_k$ since $h_i(\psi(t))$, $i = 1,\ldots,k$, vanish anyway. Inequality (5) doesn't depend on the choice of the orthonormal frame provided that (4) is satisfied.

We say that $u(\cdot)$ is a *Goh control* if (4) is satisfied for an appropriate $\psi(\cdot)$; it is a *Goh–Legendre control* if both (4) and (5) are satisfied.

We come to the first subanalyticity result, which is a consequence of theorem 2 and of general facts about subanalytic sets:

Theorem 3. *If the germ of* ρ *at* q_0 *is subanalytic, then* $\hat{\rho}$ *is subanalytic.*

3 Subanalyticity and nilpotentization

In this section, we assume that everything is analytic.

Theorem 4. *If the nilpotent system (2̂) does not admit nonzero Goh–Legendre abnormal controls, then the germ of ρ at q_0 is subanalytic.* □

The system (2) is said to be *medium fat* if:

$$T_{q_0} M = \Delta_{q_0}^2 + span\{[X, [X_i, X_j]](q_0) : i, j = 1, \ldots, k\}$$

for any $X \in \bar{\Delta}$, $X(q_0) \neq 0$ (see [5]). Medium fat systems do not admit nontrivial Goh controls. It follows directly from the definitions that a system is medium fat if and only if its nilpotentization is. We come to the following:

Corollary 1. *If the system (2) is medium fat, then the germ of ρ at q_0 is subanalytic.*

It is proved in [5] that generic germs of distributions are medium fat for $n \leq (k-1)k + 1$. This gives the following general result.

Theorem 5. *Assume that $n \leq (k-1)k + 1$. Then the germ of the sub-Riemannian distance function associated with a generic germ of a rank k distribution on an n-dimensional real-analytic Riemannian manifold is subanalytic.*

4 Exclusivity of Goh Controls for Rank > 2 Distributions

First we'll make precise the term exclusivity. Rank k distributions on M are smooth sections of the "Grassmannization" $H_k TM$ of the tangent bundle TM. The space of sections is endowed with the C^∞ Whitney topology and is denoted by $\overline{H_k TM}$. Smooth families of distributions parametrized by the manifold N are sections of the bundle $p_*^N H_k TM$ over $N \times M$ induced by the standard projection $p^N : N \times M \to M$. Let $\mathcal{A} \subset \overline{H_k TM}$ be a set of distributions. We say that $\mathcal{A}$ has codimension ∞ in $\overline{H_k TM}$ if the subset:

$$\{D \in \overline{p_*^N H_k TM} : D\big|_{x \times M} \notin \mathcal{A}, \ \forall x \in N\},$$

is everywhere dense in $\overline{p_*^N H_k TM}$, $\forall N$.

We will also use a real-analytic version of the definition, just given. The only difference with the smooth case is that the manifolds and the sections are assumed to be real-analytic, while the topology remains the same Whitney topology.

Theorem 6. *For any $k \geq 3$, the distributions admitting nonzero Goh controls form a subset of codimension ∞ in the space of all smooth rank k distributions on M.*

It was proved in [1, Cor.4] that the small sub-Riemannian balls are subanalytic for any real-analytic sub-Riemannian structure without nontrivial Goh controls. Combining this fact with Theorem 6 we obtain the folowing result. Recall that all over the paper we keep the notation $\rho(q)$, $q \in M$, for the sub-Riemannian distance between q and the fixed point q_0. The sub-Riemannian distance is defined by a given distribution Δ on the Riemannian manifold M.

Theorem 7. *Suppose that M is real-analytic and $k \geq 3$. There exists a subset $\mathcal{A}$ of codimension ∞ in the space of rank k real-analytic distributions on M such that the relation $\Delta \notin \mathcal{A}$ implies the subanalyticity of the sub-Riemannian balls $\rho^{-1}([0, r])$ for all r, small enough.*

5   Nilpotent Systems

The system:

$$\dot{x} = \sum_{i=1}^{k} u_i(t) Y_i(x), \quad x \in \mathbb{R}^n, \quad x(0) = 0, \tag{6}$$

is called *nilpotent* if it coincides with its own nilpotentization expressed in adapted coordinates.

In other words, $\mathbb{R}^n$ is presented as a direct sum $\mathbb{R}^n = \mathbb{R}^{k_1} \oplus \cdots \oplus \mathbb{R}^{k_l}$, $k_1 = k$, so that any vector $x \in \mathbb{R}^n$ takes the form $x = (x_1, \ldots, x_l)$, $x_i = (x_{i1}, \ldots, x_{ik_i}) \in \mathbb{R}^{k_i}$, $i = 1, \ldots, l$. The vector fields Y_i, $i = 1, \ldots, k$, are polynomial and quasi-homogeneous. More precisely, they are homogeneous of weight -1 with respect to the dilation:

$$\delta_t : (x_1, x_2, \ldots, x_l) \mapsto (tx_1, t^2 x_2, \ldots, t^l x_l), \quad t \in \mathbb{R};$$

$\delta_{t*} Y_i = t Y_i$, $i = 1, \ldots, k$.

We keep the notation $\hat{f} : L_2^k[0, 1] \to \mathbb{R}^n$ for the endpoint mapping $u \mapsto x(1; u)$, where $x(\cdot; u)$ is the solution of (6), $u = (u_1(\cdot), \ldots, u_k(\cdot))$, and the notation $\hat{\rho} : \mathbb{R}^n \to \mathbb{R}_+$ for the sub-Riemannian distance, $\hat{\rho}(x) = \min\{\|u\| : \hat{f}(u) = x\}$. A special case of the system (6) with $n = l = 3$, $k_1 = 2$, $k_2 = 0$, $k_3 = 1$, is called "the flat Martinet system". We will use the special notation $\rho^m : \mathbb{R}^n \to \mathbb{R}_+$ for the sub-Riemannian distance in this case, which plays an important role below.

Proposition 4. *Assume that $k = 2, k_3 \neq 0$. Then there exists a polynomial submersion $\Phi : \mathbb{R}^n \to \mathbb{R}^3$ such that $(\rho^m))^{-1}([0, r]) = \Phi\left(\hat{\rho}^{-1}([0, r])\right)$, $\forall r \geq 0$.*

Corollary 2. *Under the conditions of proposition 4 the sub-Riemannian balls $\hat{\rho}([0, r])$, $r > 0$, are not subanalytic.*

Now consider nilpotent distributions of rank greater than 2, i.e. $k = k_1 > 2$. We restrict ourselves to the case of maximal possible k_2, k_3. It means

$$k_2 = \min\{n - k, \frac{k(k-1)}{2}\} \quad k_3 = \min\{n - \frac{k(k+1)}{2}, \frac{(k+1)k(k-1)}{3}\}.$$

Remark. Generic germs of distributions and their nilpotentizations have the maximal possible growth vector and, in particular, the maximal possible k_2, k_3.

Proposition 5. *Assume that $n \geq (k-1)\left(\frac{k^2}{3} + \frac{5k}{6} + 1\right)$ and k_2, k_3 are maximal possible. Then there exists a polynomial submersion $\Phi : \mathbb{R}^n \to \mathbb{R}^3$ such that $(\rho^m))^{-1}(r) = \Phi\left(\hat{\rho}^{-1}(r)\right)$, $\forall r \geq 0$.*

Corollary 3. *Under the conditions of Proposition 5, the sub-Riemannian balls $\hat{\rho}([0, r])$, $r > 0$, are not sub-analytic.*

Let now Δ be an arbitrary (not necessarily nilpotent) germ of a bracket generating distribution at $q_0 \in M$, and let ρ be the germ of the associated sub-Riemannian distance function. Combining corollaries 2, 3, and Theorem 3 we obtain the following:

Theorem 8. *Assume that either $k = 2$ and $\Delta_{q_0}^3 \neq \Delta_{q_0}^2$ or*

$$\dim M \geq (k-1)\left(\frac{k^2}{3} + \frac{5k}{6} + 1\right)$$

and the segment $(k, \dim \Delta_{q_0}^2, \dim \Delta_{q_0}^3)$ of the growth vector is maximal. Then ρ is not subanalytic. In particular, generic germs are such that ρ is not subanalytic.

Finally, combining Theorem 8 with Theorem 7 we come to the following surprising result.

Corollary 4. *Let ρ be a germ of sub-Riemannian distance function associated with a generic germ of real-analytic distribution of rank $k \geq 3$, on a n-dimensional manifold, $n \geq (k-1)\left(\frac{k^2}{3} + \frac{5k}{6} + 1\right)$. Then the balls $\rho^{-1}([0, r])$ are subanalytic for all small enough r, but the function ρ is not subanalytic!*

8   Andrei Agrachev and Jean–Paul Gauthier

References

1. A. A. Agrachev, *Compactness for sub-Riemannian length-minimizers and sub-analyticity*. Rend. Semin. Mat. Torino, 1998, v.56
2. A. A. Agrachev, R. V. Gamkrelidze, A. V. Sarychev, *Local invariants of smooth control systems*. Acta Applicandae Mathematicae, 1989, v.14, 191–237.
3. A. A. Agrachev, A. V. Sarychev, *Filtrations of a Lie algebra of vector fields and nilpotent approximation of control systems*. Dokl. Akad. Nauk SSSR, 1987, v.295, 777–781; English transl. in Soviet Math. Dokl., 1988, v.36, 104–108.
4. A. A. Agrachev, A. V. Sarychev, *Abnormal sub-Riemannian geodesics: Morse index and rigidity*. Annales de l'Institut Henri Poincaré—Analyse non linéaire, 1996, v.13, 635–690.
5. A. A. Agrachev, A. V. Sarychev, *Sub-Riemannian metrics: minimality of abnormal geodesics versus subanalyticity*. J. ESAIM: Control, Optimisation and Calculus of Variations, 1999, v.4, 377–403.
6. A. A. Agrachev, B. Bonnard, M. Chyba, I. Kupka, *Sub-Riemannian sphere in Martinet flat case*. J. ESAIM: Control, Optimisation and Calculus of Variations, 1997, v.2, 377–448.
7. A. Bellaïche, *The tangent space in sub-Riemannian geometry*. In the book: "Sub-Riemannian geometry", Birkhäuser, 1996, 1–78.
8. R. M. Bianchini, G. Stefani, *Graded approximations and controllability along a trajectory*. SIAM J. Control Optim., 1990, v.28, 903–924.
9. B. Bonnard, M. Chyba, *Méthodes géométriques et analytique pour étudier l'application exponentiele, la sphère et le front d'onde en géometrie SR dans le cas Martinet*. J. ESAIM: Control, Optimisation and Calculus of Variations, submitted.
10. B. Bonnard, G. Launay, E. Trélat, *The transcendence we need to compute the sphere and the wave front in Martinet SR-geometry*. Proceed. Int. Confer. Dedicated to Pontryagin, Moscow, Sept.'98, to appear.
11. J.-P. Gauthier, I. Kupka, *Observability for systems with more outputs than inputs and asymptotic observers*. Mathem. Zeitshrift, 1996, v.223, 47–78.
12. Zhong Ge, *Horizontal path space and Carnot-Caratheodory metric*, Pacific J. Mathem., 1993, v.161, 255–286.
13. S. Jacquet, *Subanalyticity of the sub-Riemannian distance*. J. Dynamical and Control Systems, 1999, v.5
14. P. K. Rashevsky, *About connecting two points of a completely nonholonomic space by admissible curve*. Uch. Zapiski Ped. Inst. Libknechta, 1938, n.2, 83-94.
15. H. J. Sussmann, *Optimal control and piecewise analyticity of the distance function*. In: A. Ioffe, S. Reich, Eds., Pitman Research Notes in Mathematics, Longman Publishers, 1992, 298–310.
16. A. Agrachev, J-P. Gauthier, *On Subanaliticity of Carnot-Caratheodory Distances*. March 2000. Submited to Annales de l'Institut Henri Poincaré.

Principal Invariants of Jacobi Curves

Andrei Agrachev[1] and Igor Zelenko[2]

[1] S.I.S.S.A.
Via Beirut 2-4
34013 Trieste, Italy
and
Steklov Mathematical Institute
ul. Gubkina 8
117966 Moscow, Russia
agrachev@sissa.it
[2] Department of Mathematics, Technion-Israel Institute of Technology
Haifa 32000, Israel
zigor@techunix.technion.ac.il

Abstract. Jacobi curves are far going generalizations of the spaces of "Jacobi fields" along Riemannian geodesics. Actually, Jacobi curves are curves in the Lagrange Grassmannians. Differential geometry of these curves provides basic feedback or gauge invariants for a wide class of smooth control systems and geometric structures. In the present paper we mainly discuss two principal invariants: the *generalized Ricci curvature*, which is an invariant of the parametrized curve in the Lagrange Grassmannian providing the curve with a natural projective structure, and a *fundamental form*, which is a 4-order differential on the curve. This paper is a continuation of the works [1, 2], where Jacobi curves were defined, although it can be read independently.

1 Introduction

Suppose M is a smooth manifold and $\pi : T^*M \to M$ is the cotangent bundle to M. Let H be a codimension 1 submanifold in T^*M such that H is transversal to T_q^*M, $\forall q \in M$; then $H_q = H \cap T_q M$ is a smooth hypersurface in $T_q M$. Let ς be the canonical Liouville form on $T_q^* M$, $\varsigma_\lambda = \lambda \circ \pi_*$, $\lambda \in T^*M$, and $\sigma = d\varsigma$ be the standard symplectic structure on T^*M; then $\sigma|_H$ is a corank 1 closed 2-form. The kernels of $(\sigma|_H)_\lambda$, $\lambda \in H$ are transversal to $T_q^* M$, $q \in M$; these kernels form a line distribution in H and define a *characteristic 1-foliation* $\mathcal{C}$ of H. Leaves of this foliation are *characteristic curves* of $\sigma|_H$.

Suppose γ is a segment of a characteristic curve and O_γ is a neighborhood of γ such that $N = O_\gamma/(\mathcal{C}|_{O_\gamma})$ is a well-defined smooth manifold. The quotient manifold N is in fact a symplectic manifold endowed with a symplectic structure $\bar{\sigma}$ induced by $\sigma|_H$. Let $\phi : O_\gamma \to N$ be the canonical factorization; then $\phi(H_q \cap O_\gamma)$, $q \in M$, are Lagrangian submanifolds in N. Let

$L(T_\gamma N)$ be the Lagrange Grassmannian of the symplectic space $T_\gamma N$, i.e.
$L(T_\gamma N) = \{\Lambda \subset T_\gamma N : \Lambda^{\angle} = \Lambda\}$, where $D^{\angle} = \{e \in T_\gamma N : \bar{\sigma}(e, D) = 0\}$,
$\forall D \subset T_\gamma N$. Jacobi curve is the mapping

$$\lambda \mapsto \phi_*(T_\lambda H_{\pi(\lambda)}), \quad \lambda \in \gamma,$$

from γ to $L(T_\gamma N)$.

Jacobi curves are curves in the Lagrange Grassmannians. They are invariants
of the hypersurface H in the cotangent bundle. In particular, any differential
invariant of the curves in the Lagrange Grassmannian by the action of the
linear Symplectic Group produces a well-defined function on H.

Set $W = T_\gamma N$ and note that the tangent space $T_\Lambda L(W)$ to the Lagrange
Grassmannian at the point Λ can be naturally identified with the space of
quadratic forms on the linear space $\Lambda \subset W$. Namely, take a curve $\Lambda(t) \in$
$L(W)$ with $\Lambda(0) = \Lambda$. Given some vector $l \in \Lambda$, take a curve $l(\cdot)$ in W such
that $l(t) \in \Lambda(t)$ for all t and $l(0) = l$. Define the quadratic form $q_{\Lambda(\cdot)}(l) =$
$\bar{\sigma}(\frac{d}{dt}l(0), l)$. Using the fact that the spaces $\Lambda(t)$ are Lagrangian, i.e. $\Lambda(t)^{\angle} =$
$\Lambda(t)$, it is easy to see that the form $q_{\Lambda(\cdot)}(l)$ depends only on $\frac{d}{dt}\Lambda(0)$. So, we
have the map from $T_\Lambda L(W)$ to the space of quadratic forms on Λ. A simple
counting of dimension shows that this mapping is a bijection.

Proposition 1. *Tangent vectors to the Jacobi curve J_γ at a point $J_\gamma(\lambda)$,
$\lambda \in \gamma$, are equivalent (under linear substitutions of variables in the correspondent quadratic forms) to the "second fundamental form" of the hypersurface
$H_{\pi(\lambda)} \subset T^*_{\pi(\lambda)}M$ at the point λ.*

In particular, the velocity of J_γ at λ is a sign-definite quadratic form if and
only if the hypersurface $H_{\pi(\lambda)}$ is strongly convex at λ.

A similar construction can be done for a submanifold of codimension 2 in
T^*M. In the codimension 2 case characteristic curves do not fill the whole
submanifold; they are concentrated in the characteristic variety consisting of
the points, where the restriction of σ to the submanifold is degenerate.

We are mainly interested in submanifolds that are dual objects to smooth
control systems. Here we call a smooth control system any submanifold $V \subset$
TM, transversal to fibers. Let $V_q = V \cap T_q M$; The "dual" *normal variety* H^1
and *abnormal variety* H^0 are defined as follows:

$$H^1 = \bigcup_{q \in M} \{\lambda \in T^*_q M : \exists v \in V_q, \ \langle \lambda, v \rangle = 1, \ \langle \lambda, T_v V_q \rangle = 0\},$$

$$H^0 = \bigcup_{q \in M} \{\lambda \in T^*_q M \setminus 0 : \exists v \in V_q, \ \langle \lambda, v \rangle = \langle \lambda, T_v V_q \rangle = 0\}.$$

These varieties are not, in general, smooth manifolds; they may have singularities, which we do not discuss here. Anyway, one can obtain a lot of
information on the original system just studying smooth parts of H^1, H^0.

Characteristic curves of $\sigma|_H^1$ $(\sigma|_H^0)$ are associated with normal (abnormal) extremals of the control system V. The corresponding Jacobi curves admit a purely variational construction in terms of the original control system and in a very general setting (singularities included), see [1, 2, 3].

One of the varieties H^1, H^0 can be empty. In particular, if $V_q = \partial W_q$, where W_q is a convex set and $0 \in int W_q$, then $H^0 = \emptyset$. Moreover, in this case the Liouville form never vanishes on the tangent lines to the characteristic curves of $\sigma|_{H^1}$, and any characteristic curve γ has a canonical parametrization by the rule $\langle \varsigma, \dot{\gamma} \rangle = 1$.

If subsets $V_q \subset T_q M$ are conical, $\alpha V_q = V_q$, $\forall \alpha > 0$, then, in contrast to the previous case, $H^1 = \emptyset$ and ς vanishes on the tangent lines to the characteristic curves of $\sigma|_{H^0}$. The characteristic curves are actually unparametrized.

If V_q are compact, then H^1 has codimension 1 in $T^* M$, while H^0 has codimension ≥ 2 in all nontrivial cases.

The rank of the "second fundamental form" of the submanifolds H_q^1 and H_q^0 of $T_q^* M$ at any point is no greater than $\dim V_q$. Indeed, let $\lambda \in H_q^1$; then $\lambda \in (T_v V_q)^\perp$, $\langle \lambda, v \rangle = 1$, for some $v \in V_q$. We have $\lambda + (T_v V_q + \mathbb{R}v)^\perp \subset H_q^1$. So λ belongs to an affine subspace of dimension $n - \dim V_q - 1$, which is contained in H_q^1. For $\lambda \in H_q^0$, $\exists v \in V_q$ such that $\lambda \in (T_v V_q)^\perp$, $\langle \lambda, v \rangle = 0$. Then the affine subspace $\lambda + (T_v V_q + \mathbb{R}v)^\perp$ is contained in H_q^0.

Suppose that H^1 has codimension 1 in $T^* M$ and γ is a characteristic curve of $\sigma|_{H^1}$. Then the velocity of the Jacobi curve $\lambda \mapsto J_\gamma(\lambda)$, $\lambda \in \gamma$, has rank no greater than $\dim V_{\pi(\lambda)}$ (see proposition 1). The same is true for the Jacobi curves associated with characteristic curves of $\sigma|_{H^0}$, if H^0 has codimension 2.

Dimension of V_q is the number of inputs or control parameters in the control system. Less inputs means more "nonholonomic constraints" on the system. It happens that the rank of velocity of any Jacobi curve generated by the system never exceeds the number of inputs.

2   Derivative Curve

Let Λ be a Lagrangian subspace of W, i.e. $\Lambda \in L(W)$. For any $w \in \Lambda$, the linear form $\bar{\sigma}(\cdot, w)$ vanishes on Λ and thus defines a linear form on W/Λ. The nondegeneracy of $\bar{\sigma}$ implies that the relation $w \mapsto \sigma(\cdot, w)$, $w \in \Lambda$, induces a canonical isomorphism $\Lambda \cong (W/\Lambda)^*$ and, by the conjugation, $\Lambda^* \cong W/\Lambda$.

We set $\Lambda^\pitchfork = \{\Gamma \in L(W) : \Gamma \cap \Lambda = 0\}$, an open everywhere dense subset of $L(W)$. Let $Sym^2(\Lambda)$ be the space of self-adjoint linear mappings from Λ^* to Λ; this notation reflects the fact that $Sym^2(\Lambda)$ is the space of quadratic forms on Λ^* that is the symmetric square of Λ. $\Lambda^\pitchfork$ possesses a canonical structure of an affine space over the linear space $Sym^2(\Lambda) = Sym^2((W/\Lambda)^*)$. Indeed, for any $\Delta \in \Lambda^\pitchfork$ and coset $(w + \Lambda) \in W/\Lambda$, the intersection $\Delta \cap (w + \Lambda)$ of the linear subspace Δ and the affine subspace $w + \Lambda$ in W consists of exactly one point. To a pair $\Gamma, \Delta \in \Lambda^{\pitchfork}$ there corresponds a mapping $(\Gamma - \Delta) : W/\Lambda \to \Lambda$, where

$$(\Gamma - \Delta)(w + \Lambda) \stackrel{def}{=} \Gamma \cap (w + \Lambda) - \Delta \cap (w + \Lambda).$$

It is easy to check that the identification $W/\Lambda = \Lambda^*$ makes $(\Gamma - \Delta)$ a self-adjoint mapping from Λ^* to Λ. Moreover, given $\Delta \in \Lambda^{\pitchfork}$, the correspondence $\Gamma \mapsto (\Gamma - \Delta)$ is a one-to-one mapping of $\Lambda^{\pitchfork}$ onto $Sym^2(\Lambda)$ and the axioms of the affine space are obviously satisfied.

Fixing $\Delta \in \Lambda^{\pitchfork}$ one obtains a canonical identification $\Delta \cong W/\Lambda = \Lambda^*$. In particular, $(\Gamma - \Delta) \in Sym^2(\Lambda)$ turns into the mapping from Δ to Λ. For the last linear mapping we will use the notation $\langle \Delta, \Gamma, \Lambda \rangle : \Delta \to \Lambda$. In fact, this mapping has a much more straightforward description. Namely, the relations $W = \Delta \oplus \Lambda$, $\Gamma \cap \Lambda = 0$, imply that Γ is the graph of a linear mapping from Δ to Λ. Actually, it is the graph of the mapping $\langle \Delta, \Gamma, \Lambda \rangle$. In particular, $\ker \langle \Delta, \Gamma, \Lambda \rangle = \Delta \cap \Gamma$. If $\Delta \cap \Gamma = 0$, then $\langle \Lambda, \Gamma, \Delta \rangle = \langle \Delta, \Gamma, \Lambda \rangle^{-1}$.

Let us give coordinate representations of the introduced objects. We may assume that

$$W = \mathbb{R}^m \oplus \mathbb{R}^m = \{(x, y) : x, y \in \mathbb{R}^m\},$$

$$\bar{\sigma}((x_1, y_1), (x_2, y_2)) = \langle x_1, y_2 \rangle - \langle x_2, y_1 \rangle, \quad \Lambda = \mathbb{R}^m \oplus 0, \quad \Delta = 0 \oplus \mathbb{R}^m.$$

Then any $\Gamma \in \Delta^{\pitchfork}$ takes the form $\Gamma = \{(x, Sx) : x \in \mathbb{R}^n\}$, where S is a symmetric $m \times m$ matrix. The operator $\langle \Lambda, \Gamma, \Delta \rangle : \Lambda \to \Delta$ is represented by the matrix S, while the operator $\langle \Delta, \Gamma, \Lambda \rangle$ is represented by the matrix S^{-1}.

The coordinates in Λ induce the identification of $Sym^2\Lambda$ with the space of symmetric $m \times m$ matrices. $\Lambda^{\pitchfork}$ is an affine subspace over $Sym^2\Lambda$; we fix Δ as the origin in this affine subspace and thus obtain a coordinatization of $\Lambda^{\pitchfork}$ by symmetric $m \times m$ matrices. In particular, the "point" $\Gamma = \{(x, Sx) : x \in \mathbb{R}^n\}$ in $\Lambda^{\pitchfork}$ is represented by the matrix S^{-1}.

A subspace $\Gamma_0 = \{(x, S_0 x) : x \in \mathbb{R}^n\}$ is transversal to Γ if and only if $det(S - S_0) \neq 0$. Let us pick coordinates $\{x\}$ in Γ_0 and fix Δ as the origin in the affine space $\Gamma_0^{\pitchfork}$. In the induced coordinatization of $\Gamma_0^{\pitchfork}$ the "point" Γ is represented by the matrix $(S - S_0)^{-1}$.

Let $t \mapsto \Lambda(t)$ be a smooth curve in $L(W)$. We say that the curve $\Lambda(\cdot)$ is *ample* at τ if $\exists k > 0$ such that for any representative $\Lambda_\tau^k(\cdot)$ of the k-jet of $\Lambda(\cdot)$ at τ, $\exists t$ such that $\Lambda_\tau^k(t) \cap \Lambda(\tau) = 0$. The curve $\Lambda(\cdot)$ is called *ample* if it is ample at any point.

We have given an intrinsic definition of an ample curve. In coordinates it takes the following form: the curve $t \mapsto \{(x, S_t x) : x \in \mathbb{R}^n\}$ is ample at τ if and only if the function $t \mapsto det(S_t - S_\tau)$ has a root of *finite order* at τ.

Assume that $\Lambda(\cdot)$ is ample at τ. Then $\Lambda(t) \in \Lambda(\tau)^{\pitchfork}$ for all t from a punctured neighborhood of τ. We obtain the curve $t \mapsto \Lambda(t) \in \Lambda(\tau)^{\pitchfork}$ in the affine space $\Lambda(\tau)^{\pitchfork}$ with the pole at τ. Fixing an "origin" in $\Lambda(\tau)^{\pitchfork}$ we make $\Lambda(\cdot)$ a vector function with values in $Sym^2(\Lambda)$ and with the pole at τ. Such a vector function admits the expansion in the Laurent series at τ. Obviously, only free term in the Laurent expansion depends on the choice of the "origin" we did to identify the affine space with the linear one. More precisely, the addition of a vector to the "origin" results in the addition of the same vector to the free term in the Laurent expansion. In other words, for the Laurent expansion of a curve in an affine space, the free term of the expansion is a point of this affine space while all other terms are elements of the corresponding linear space. In particular,

$$\Lambda(t) \approx \Lambda_0(\tau) + \sum_{\substack{i=-l \\ i \neq 0}}^{\infty} (t - \tau)^i Q_i(\tau), \tag{1}$$

where $\Lambda_0(\tau) \in \Lambda(\tau)^{\pitchfork}$, $Q_i(\tau) \in Sym^2 \Lambda(\tau)$.

Assume that the curve $\Lambda(\cdot)$ is ample. Then $\Lambda_0(\tau) \in \Lambda(\tau)^{\pitchfork}$ is defined for all τ. The curve $\tau \mapsto \Lambda_0(\tau)$ is called the *derivative curve* of $\Lambda(\cdot)$.

Another characterization of $\Lambda_0(\tau)$ can be done in terms of the curves $t \mapsto \langle \Delta, \Lambda(t), \Lambda(\tau) \rangle$ in the linear space $Hom(\Delta, \Lambda(\tau))$, $\Delta \in \Lambda(\tau)^{\pitchfork}$. These curves have poles at τ. The Laurent expansion at $t = \tau$ of the vector function $t \mapsto \langle \Delta, \Lambda(t), \Lambda(\tau) \rangle$ has zero free term if and only if $\Delta = \Lambda_0(\tau)$.

The coordinate version of the series (2.1) is the Laurent expansion of the matrix-valued function $t \mapsto (S_t - S_\tau)^{-1}$ at $t = \tau$, where $\Lambda(t) = \{(x, S_t x) : x \in \mathbb{R}^n\}$.

3   Curvature operator and regular curves.

Using derivative curve one can construct an operator invariant of the curve $\Lambda(t)$ at any its point. Namely, take velocities $\dot{\Lambda}(t)$ and $\dot{\Lambda}_0(t)$ of $\Lambda(t)$ and its derivative curve $\Lambda_0(t)$. Note that $\dot{\Lambda}(t)$ is linear operator from $\Lambda(t)$ to $\Lambda(t)^*$ and $\dot{\Lambda}_0(t)$ is linear operator from $\Lambda_0(t)$ to $\Lambda_0(t)^*$. Since the form σ defines the canonical isomorphism between $\Lambda_0(t)$ and $\Lambda(t)^*$, the following operator $R(t) : \Lambda(t) \to \Lambda(t)$ can be defined:

$$R(t) = -\dot{\Lambda}_0(t) \circ \dot{\Lambda}(t) \tag{2}$$

This operator is called *curvature operator* of Λ at t. Note that in the case of Riemannian geometry the operator $R(t)$ is similar to the so-called Ricci operator $v \to R^\nabla(\dot{\gamma}(t), v)\dot{\gamma}(t)$, which appears in the classical Jacobi equation $\nabla_{\dot{\gamma}(t)} \nabla_{\dot{\gamma}(t)} V + R^\nabla(\dot{\gamma}(t), V)\dot{\gamma}(t) = 0$ for Jacobi vector fields V along the geodesic $\gamma(t)$ (here R^∇ is curvature tensor of Levi-Civita connection ∇), see [1]. This is the reason for the sign "$-$" in (2).

The curvature operator can be effectively used in the case of so-called regular curves. The curve $\Lambda(t)$ in Lagrange Grassmannian is called *regular*, if the quadratic form $\dot{\Lambda}(t)$ is nondegenerated for all t. Suppose that the curve $\Lambda(\cdot)$ is regular and has a coordinate representation $\Lambda(t) = \{(x, S_t x) : x \in \mathbb{R}^n\}$, $S_\tau = 0$. Then the function $t \mapsto S_t^{-1}$ has a simple pole at $t = \tau$ and one can get the following formula for the curvature operator (see [1]):

$$R(t) = ((2S_t')^{-1} S_t'')' - ((2S_t')^{-1} S_t'')^2 \tag{3}$$

Note that the right-hand side of (3) is a matrix analog of so-called Schwarz derivative or Schwarzian . Let us recall that the differential operator:

$$\mathbb{S} : \varphi \mapsto \frac{1}{3}\left(\frac{d}{dt}\left(\frac{\varphi''}{2\,\varphi'}\right) - \left(\frac{\varphi''}{2\,\varphi'}\right)^2\right) = \frac{1}{6}\frac{\varphi^{(3)}}{\varphi'} - \frac{1}{4}\left(\frac{\varphi''}{\varphi'}\right)^2, \tag{4}$$

acting on scalar function φ is called *Schwarzian*. The operator $\mathbb{S}$ is characterized by the following remarkable property: General solution of the equation $\mathbb{S}\varphi = \rho$ w.r.t φ is a Möbius transformation (with constant coefficients) of some particular solution of this equation. The matrix analog of this operator has similar property, concerning "matrix Möbius transformation" of the type $(AS + B)(CS + D)^{-1}$. It implies that in the regular case the curvature operator $R(t)$ determines the curve completely up to a symplectic transformation.

4   Expansion of the cross-ratio and Ricci curvature.

For the nonregular curve $\Lambda(t) = \{(x, S_t x) : x \in \mathbb{R}^n\}$, the function $t \mapsto (S_t - S_\tau)^{-1}$ has a pole of order greater than 1 at τ and it is much more difficult to compute its Laurent expansion. In particular, as we will see later in the nonregular case the curvature operator does not determine the curve up to a symplectic transformation. However, using the notion of cross-ratio it is possible to construct numerical invariants for a very nonrestrictive class of curves.

Suppose that Λ_0, Λ_1, Λ_2, and Λ_3 are Lagrangian subspaces of W and Λ_0 and $\Lambda_1 \cap \Lambda_2 = \Lambda_3 \cap \Lambda_0 = 0$. We have $\langle \Lambda_0, \Lambda_1, \Lambda_2 \rangle : \Lambda_0 \to \Lambda_2$, $\langle \Lambda_2, \Lambda_3, \Lambda_0 \rangle : \Lambda_2 \to \Lambda_0$. The cross-ratio $\left[\Lambda_0, \Lambda_1, \Lambda_2, \Lambda_3\right]$ of four "points" Λ_0, Λ_1, Λ_2, and Λ_3 in the Lagrange Grassmannian is, by definition, the following linear operator in Λ_2:

$$\left[\Lambda_0, \Lambda_1, \Lambda_2, \Lambda_3\right] = \langle \Lambda_0, \Lambda_1, \Lambda_2 \rangle \langle \Lambda_2, \Lambda_3, \Lambda_0 \rangle. \tag{5}$$

This notion is a "matrix" analog of the classical cross-ratio of four points in the projective line. Indeed, let $\Lambda_i = \{(x, S_i x) : x \in \mathbb{R}^n\}$, then, in coordinates $\{x\}$, the cross-ratio takes the form:

$$\left[\Lambda_0, \Lambda_1, \Lambda_2, \Lambda_3\right] = (S_2 - S_1)^{-1}(S_1 - S_0)(S_0 - S_3)^{-1}(S_3 - S_2) \tag{6}$$

By construction, all coefficients of the characteristic polynomial of $\left[\Lambda_0, \Lambda_1, \Lambda_2, \Lambda_3\right]$ are invariants of four subspaces $\Lambda_0, \Lambda_1, \Lambda_2$, and Λ_3.

Now we are going to show how to use the cross-ratio in order to construct invariants of the curve $\Lambda(t)$ in the Lagrange Grassmannian. Let, as before, $t \mapsto \{(x, S_t x) : x \in \mathbb{R}^n\}$ be the coordinate representation of a germ of the curve $\Lambda(\cdot)$.

Assumption 1 *For all parameters t_1 the functions $t \to det(S_t - S_{t_1})$ have at $t = t_1$ zero of the same finite order k.*

By the above the function $(t_0, t_1, t_2, t_3) \to det\left[\Lambda(t_0), \Lambda(t_1), \Lambda(t_2), \Lambda(t_3)\right]$ is symplectic invariant of the curve $\Lambda(t)$. Using this fact, let us try to find symplectic invariants of $\Lambda(t)$ that are functions of t. For this it is very convenient to introduce the following function

$$\mathcal{G}(t_0, t_1, t_2, t_3) = \ln \left| \frac{det\left[\Lambda(t_0), \Lambda(t_1), \Lambda(t_2), \Lambda(t_3)\right]}{\left[t_0, t_1, t_2, t_3\right]^k} \right|, \qquad (7)$$

where $\left[t_0, t_1, t_2, t_3\right] = \frac{(t_0 - t_1)(t_2 - t_3)}{(t_1 - t_2)(t_3 - t_0)}$ is the usual cross-ratio of four numbers t_0, t_1, t_2, and t_3. The function $\mathcal{G}(t_0, t_1, t_2, t_3)$ is also a symplectic invariant of $\Lambda(t)$. It can be easily expanded in formal Taylor series at any "diagonal" point (t, t, t, t) and the coefficients of this expansion are invariants of the germ of $\Lambda(\cdot)$ at t.

Indeed, by Assumption 1, we have:

$$det(S_{t_0} - S_{t_1}) = (t_0 - t_1)^k X(t_0, t_1), \quad X(t, t) \neq 0 \qquad (8)$$

for any t. It follows that $X(t_0, t_1)$ is a symmetric function (changing the order in (8) we obtain that X can be symmetric or antisymmetric, but the last case is impossible by the fact that $X(t, t) \neq 0$). Define also the following function

$$f(t_0, t_1) = ln|X(t_0, t_1)| \qquad (9)$$

This function is also symmetric, so it can be expanded in the formal Taylor series at the point (t, t) in the following way:

$$f(t_0, t_1) \approx \sum_{i,j=0}^{\infty} \alpha_{i,j}(t)(t_0 - t)^i (t_1 - t)^j, \quad \alpha_{i,j}(t) = \alpha_{j,i}(t) \qquad (10)$$

One can easily obtain the following lemma, using the fact that

$$\mathcal{G}(t_0, t_1, t_2, t_3) = f(t_0, t_1) - f(t_1, t_2) + f(t_2, t_3) - f(t_3, t_0)$$

Lemma 1. *For any t the function $\mathcal{G}(t_0, t_1, t_2, t_3)$ has the following formal Taylor expansion at the point (t, t, t, t):*

$$\mathcal{G}(t_0, t_1, t_2, t_3) \approx \sum_{i,j=1}^{\infty} \alpha_{i,j}(t)(\xi_0^i - \xi_2^i)(\xi_1^j - \xi_3^j), \tag{11}$$

where $\xi_l = t_l - t$, $l = 0, 1, 2, 3$.

From (11) it follows that all coefficients $\alpha_{i,j}(t)$, $i, j \geq 1$, are symplectic invariants of the curve $\Lambda(t)$.

Definition 1. The first appearing coefficient $\alpha_{1,1}(t)$ is called Ricci curvature of $\Lambda(t)$.

In the sequel the Ricci curvature is denoted by $\rho(t)$. Note that, by a direct computation, one can get the following relation between $\rho(\tau)$ and curvature operator $R(\tau)$ for the regular curve: $\rho(\tau) = \frac{1}{3} tr R(\tau)$. Actually, this relation justifies the name Ricci curvature for the invariant $\rho(t)$.

In some cases we are interested in symplectic invariants of unparametrized curves in Lagrange Grassmannian (i.e., of one-dimensional submanifolds in Lagrange Grassmannian). For example, so-called abnormal extremals of vector distributions and consequently their Jacobi curves a priori have no special parametrizations.

Now we want to show how, using the Ricci curvature, one can define a canonical projective structure on the unparametrized curve $\Lambda(\cdot)$. For this let us check how the Ricci curvature is transformed by a reparametrization of the curve $\Lambda(t)$.

Let $t = \varphi(\tau)$ be a reparametrization and let $\bar{\Lambda}(\tau) = \Lambda(\varphi(\tau))$. Denote by $\bar{\mathcal{G}}$ the function playing for $\bar{\Lambda}(\tau)$ the same role as the function $\mathcal{G}$ for $\Lambda(t)$. Then from (7) it follows that

$$\bar{\mathcal{G}}(\tau_0, \tau_1, \tau_2, \tau_3) = \mathcal{G}(t_0, t_1, t_2, t_3) + k \ln\left(\frac{[\varphi(\tau_0), \varphi(\tau_1), \varphi(\tau_2), \varphi(\tau_3)]}{[\tau_0, \tau_1, \tau_2, \tau_3]}\right), \tag{12}$$

where $t_i = \varphi(\tau_i)$, $i = 0, 1, 2, 3$. By direct computation it can be shown that the function $(\tau_0, \tau_1, \tau_2, \tau_3) \mapsto \ln\left(\frac{[\varphi(\tau_0), \varphi(\tau_1), \varphi(\tau_2), \varphi(\tau_3)]}{[\tau_0, \tau_1, \tau_2, \tau_3]}\right)$ has the following Taylor expansion up to the order two at the point (τ, τ, τ, τ):

$$\ln\left(\frac{[\varphi(\tau_0), \varphi(\tau_1), \varphi(\tau_2), \varphi(\tau_3)]}{[\tau_0, \tau_1, \tau_2, \tau_3]}\right) = \mathbb{S}\varphi(\tau)(\eta_0 - \eta_2)(\eta_1 - \eta_3) + \ldots, \tag{13}$$

where $\mathbb{S}\varphi$ is Schwarzian defined by (4) and $\eta_i = \tau_i - \tau$, $i = 0, 1, 2, 3$.

Suppose for simplicity that in the original parameter t the Ricci curvature $\rho(t) \equiv 0$ and denote by $\bar{\rho}(\tau)$ the Ricci curvature of the curve $\bar{\Lambda}(\tau)$. Then from (11), (12), and (13) it follows easily that:

$$\bar{\rho}(\tau) = k \mathbb{S}\varphi(\tau). \tag{14}$$

Conversely, if the Ricci curvature $\rho(t)$ of the curve $\Lambda(t)$ is not identically zero we can find at least locally (i.e., in a neighbourhood of given point) a reparametrization $t = \varphi(\tau)$ such that $\bar{\rho}(\tau) \equiv 0$ (from (14) it follows that in this case $\varphi(\tau)$ has to satisfy the equation $\mathbb{S}(\varphi^{-1})(t) = \frac{\rho(t)}{k}$)

The set of all parametrization of $\Lambda(\cdot)$ with Ricci curvature identically equal to zero defines a projective structure on $\Lambda(\cdot)$ (any two parametrization from this set are transformed one to another by Möbius transformation). We call it *the canonical projective structure* of the curve $\Lambda(\cdot)$. The parameters of the canonical projective structure will be called *projective parameters*.

5   Fundamental form of the unparametrized curve.

The Ricci curvature $\rho(\cdot)$ is the first coefficient in the Taylor expansion of the function $\mathcal{G}$ at the point (t, t, t, t). The analysis of the next terms of this expansion gives the way to find other invariants of the curve $\Lambda(\cdot)$ that do not depend on $\rho(t)$. In this section we show how to find candidates for the "second" invariant of $\Lambda(\cdot)$ and then we construct a special form on unparametrized curve $\Lambda(\cdot)$ (namely, the differential of order four on $\Lambda(\cdot)$), which we call the fundamental form of the curve $\Lambda(\cdot)$.

First note that analyzing the expansion (10) one can easily obtain the following lemma

Lemma 2. *Let* $\alpha_{i,j}(t)$ *be as in expansion (10). Then the following relation holds*

$$\alpha'_{i,j}(t) = (i+1)\alpha_{i+1,j}(t) + (j+1)\alpha_{i,j+1}(t) \tag{15}$$

In particular, from (15) it follows easily that

$$\alpha_{2,1}(t) = \frac{1}{4}\rho'(t), \quad \alpha_{2,2}(t) = \frac{1}{8}\rho''(t) - \frac{3}{2}\alpha_{3,1}(t) \tag{16}$$

These relations imply that the function $\alpha_{3,1}(t)$ is a candidate for the second invariant (as well as the function $\alpha_{2,2}(t)$).

Now let t be a projective parameter on $\Lambda(\cdot)$. Then by definition $\rho(t) \equiv 0$, and by (16) $\alpha_{2,1}(t) \equiv 0$ and $\alpha_{2,2}(t) = -\frac{3}{2}\alpha_{3,1}(t)$. This together with (11) and the fact that $\alpha_{3,1}(t) = \alpha_{1,3}(t)$ implies that the function $\mathcal{G}(t_0, t_1, t_2, t_3)$ has the following Taylor expansion up to the order four at the point (t, t, t, t):

$$\mathcal{G}(t_0, t_1, t_2, t_3) = \alpha_{3,1}(t)p_4(\xi_0, \xi_1, \xi_2, \xi_3) + \cdots, \tag{17}$$

where $\xi_i = t_i - t$, $i = 0, 1, 2, 3$, and $p_4(\xi_0, \xi_1, \xi_2, \xi_3)$ is a homogeneous polynomial of degree four (more precisely, $p_4(\xi_0, \xi_1, \xi_2, \xi_3) = (\xi_0^3 - \xi_2^3)(\xi_1 - \xi_3) + (\xi_0 - \xi_2)(\xi_1^3 - \xi_3^3) - \frac{3}{2}(\xi_0^2 - \xi_2^2)(\xi_1^2 - \xi_3^2))$.

Let τ be another projective parameter on $\Lambda(\cdot)$ (i.e., $t = \varphi(\tau) = \frac{a\tau+b}{c\tau+d}$) and denote by $\bar{\alpha}_{3,1}(\tau)$ the function that plays the same role for the curve $\Lambda(\varphi(\tau))$ as the function $\alpha_{3,1}(t)$ for $\Lambda(t)$. Then from (12), (17), and the fact that the cross-ratio is preserved by Möbius transformations it follows that

$$\bar{\alpha}_{3,1}(\tau)(d\tau)^4 = \alpha_{3,1}(t)(dt)^4 \tag{18}$$

It means that the form $\alpha_{3,1}(t)(dt)^4$ does not depend on the choice of the projective parameter t. We will call this form *a fundamental form* of the curve $\Lambda(\cdot)$.

If t is an arbitrary (not necessarily projective) parameter on the curve $\Lambda(\cdot)$, then the fundamental form in this parameter has to be of the form $A(t)(dt)^4$, where $A(t)$ is a smooth function (the "density" of the fundamental form). For projective parameter $A(t) = \alpha_{3,1}(t)$. For arbitrary parameter it can be shown, using (11), (12), that $A(t) = \alpha_{3,1}(t) - \frac{1}{5k}\rho(t)^2 - \frac{1}{20}\rho''(t)$.

If $A(t)$ does not change sign, then the canonical length element $|A(t)|^{\frac{1}{4}}dt$ is defined on $\Lambda(\cdot)$. The corresponding parameter τ (i.e., length with respect to this length element) is called *a normal parameter* (in particular, it implies that abnormal extremals of vector distribution may have canonical (normal) parametrization). Calculating the Ricci curvature $\rho_n(\tau)$ of $\Lambda(\cdot)$ in the normal parameter, we obtain a functional invariant of the unparametrized curve. We will call it *projective curvature* of the unparametrized curve $\Lambda(\cdot)$. If $t = \varphi(\tau)$ is the transition function between a projective parameter t and the normal parameter τ, then by (14) it follows that $\rho_n(\tau) = k\,\mathbb{S}\varphi(\tau)$.

Note that all constructions of this section can be done for the curve in the Grassmannian $G(m, 2m)$ (the set of all m-dimensional subspaces in the $2m$-dimensional linear space) instead of Lagrange Grassmannian by the action of the group $GL(2m)$ instead of Symplectic Group.

6   The method of moving frame.

In this section we consider nonregular curves having two functional invariants and prove that the above defined invariants ρ and A constitute a complete system of symplectic invariants for these curves (completeness of the system of invariants means that the system defines the curve uniquely up to a symplectic transformation)

Assume that dimension of the symplectic space W is four and consider ample curves $\Lambda(t)$ in $L(W)$ such that for any t the velocity $\dot{\Lambda}(t)$ is a quadratic form of rank 1. Without loss of generality we can assume that $\dot{\Lambda}(t)$ is nonnegative definite for any t. Let us fix some parameter τ and let $\{(x, S_t x) : x \in \mathbb{R}^n\}$ be a coordinate representation of $\Lambda(t)$ such that $S_\tau = 0$. Since the curve $\Lambda(t)$ is ample, the curve S_t^{-1} has a pole at τ. The velocity $\frac{d}{dt}S_t^{-1} : \Lambda(\tau)^* \to \Lambda(\tau)$ is a well defined self-adjoint operator. Moreover, by our assumptions, $\frac{d}{dt}S_t^{-1}$ is a nonpositive self-adjoint operator of rank 1. So for $t \neq \tau$ there exists unique, up to the sign, vector $w(t, \tau) \in \Lambda(\tau)$ such that for any $v \in \Lambda(\tau)^*$

$$\langle v, \frac{d}{dt} S_t^{-1} v \rangle = -\langle v, w(t, \tau) \rangle^2 \tag{19}$$

It is clear that the curve $t \to w(t, \tau)$ also has the pole at τ. Suppose that the order of the pole is equal to l. Denote by $u(t, \tau)$ the normalized curve $t \to u(t, \tau) = (t - \tau)^l w(t, \tau)$ and define the following vectors in $\Lambda(\tau)$: $e_j(\tau) = \frac{\partial^{j-1}}{\partial t^{j-1}} u(t, \tau) \big|_{t=\tau}$.

It is not hard to show that the order of pole of $t \to w(t, \tau)$ at $t = \tau$ is equal to l if and only if l is the minimal positive integer such that the vectors $e_1(\tau)$ and $e_l(\tau)$ are linear independent (in particular, $e_1(\tau)$ and $e_2(\tau)$ are linear independent if and only if $l = 2$). It implies easily that the set of points τ, where the vectors $e_1(\tau)$ and $e_2(\tau)$ are linear dependent, is a set of isolated points in $\mathbb{R}$.

Assumption 2 $\Lambda(t)$ is a curve in $L(W)$ with $\dim W = 4$ such that for any t the velocity $\dot{\Lambda}(t)$ is a quadratic form of rank 1 and $e_1(t)$, $e_2(t)$ are linear independent.

By the above it is easy to see that if $\Lambda(\cdot)$ satisfies the previous assumption, then it satisfies Assumption 1 with $k = 4$. So, the invariants $\rho(\cdot)$ and $A(\cdot)$ are defined for $\Lambda(\cdot)$. Note that the curve $\Lambda(\cdot)$ can be described by the curve $t \to w(t, \tau)$ of the vectors on the plane, i.e. $\Lambda(\cdot)$ can be described by two functions. The natural question is whether $(\rho(\cdot), A(\cdot))$ is a complete system of symplectic invariants of $\Lambda(\cdot)$.

Since vector $w(t, \tau)$ is defined up to the sign, the vector $e_1(\tau)$ is also defined up to the sign. So, for any τ one can take $(e_1(\tau), e_2(\tau))$ or $(-e_1(\tau), -e_2(\tau))$ as the canonical bases on the plane $\Lambda(\tau)$. Recall that by constructions of the section 2 for the curve $\Lambda(\cdot)$ the derivative curve $\Lambda_0(\cdot)$ is defined and for any τ the subspaces $\Lambda(\tau)$ and $\Lambda_0(\tau)$ are transversal. So, in addition to the vectors $e_1(\tau), e_2(\tau)$ on the plane $\Lambda(\tau)$, one can choose two vectors $f_1(\tau)$ and $f_2(\tau)$ on the plane $\Lambda_0(\tau)$ such that four vectors $\left(e_1(\tau), e_2(t), f_1(\tau), f_2(\tau) \right)$ constitute symplectic basis (or Darboux basis) of W (it means that $\sigma(f_i(\tau), e_j(\tau)) = \delta_{i,j}$). So, the curve $\Lambda(\cdot)$ defines a moving frame $\left(e_1(\tau), e_2(\tau), f_1(\tau), f_2(\tau) \right)$ and one can derive the structural equation for this frame:

Proposition 2. *The frame* $\left(e_1(\tau), e_2(\tau), f_1(\tau), f_2(\tau)\right)$ *satisfies the following structural equation:*

$$
\begin{cases}
\dot{e}_1 = 3e_2 \\[2mm]
\dot{e}_2 = \frac{1}{4}\rho e_1 + 4f_2 \\[2mm]
\dot{f}_1 = -(\frac{35}{12}A - \frac{1}{8}\rho^2 + \frac{1}{16}\rho'')e_1 - \frac{7}{16}\rho' e_2 - \frac{1}{4}\rho f_2 \\[2mm]
\dot{f}_2 = -\frac{7}{16}\rho' e_1 - \frac{9}{4}\rho e_2 - 3f_1
\end{cases}
\tag{20}
$$

Note that the coefficients in the equation (20) depend only on ρ and A and any symplectic basis can be taken as an initial condition of (20). It implies the following:

Theorem 1. *The curve $\Lambda(\cdot)$ satisfying Assumption 2 is determined by its invariants $(\rho(\cdot), A(\cdot))$ uniquely up to the symplectic transformation of W.*

Remark 1. It can be shown by a direct calculation that the curvature operator $R(\tau) : \Lambda(\tau) \to \Lambda(\tau)$ of the curve $\Lambda(\cdot)$ satisfying Assumption 2 has the following matrix in the basis $(e_1(\tau), e_2(\tau))$: $R(\tau) = \begin{pmatrix} 0 & \frac{7}{4}\rho'(\tau) \\ 0 & 9\rho(\tau) \end{pmatrix}$, i.e., R depends only on ρ. This means that in contrast to the regular case, the curvature operator does not determine the curve $\Lambda(\cdot)$ uniquely up to a symplectic transformation.

Theorem 1 implies the following result on unparametrized curves:

Theorem 2. *Assume that the curve $\Lambda(\cdot)$ satisfies Assumption 2 for some parametrization and its fundamental form $A(t)(dt)^4$ does not vanish. Then the sign of $A(t)$ and the projective curvature $\rho_n(\cdot)$ determine $\Lambda(\cdot)$ uniquely up to a symplectic transformation of W and a reparametrization.*

7   Flat curves.

The following definition is natural.

Definition 2. The curve $\Lambda(t)$, satisfying Assumption 2, is called flat if $\rho(t) \equiv 0$, $A(t) \equiv 0$.

As a consequence of Theorem 1, expansion (11), and structural equation (20) one can obtain the following characterization of the flat curve:

Theorem 3. *The curve $\Lambda(t)$, satisfying Assumption 2, is flat if and only if one of the following condition holds:*

1) all coefficients $Q_i(t)$ with $i > 0$ in the Laurent expansion (1) are equal to zero;

2) the derivative curve $\Lambda_0(t)$ is constant, i.e., $\dot{\Lambda}_0(t) \equiv 0$;

3) for any t_0, t_1, t_2, t_3

$$det\Big([\Lambda(t_0), \Lambda(t_1), \Lambda(t_2), \Lambda(t_3)]\Big) = [t_0, t_1, t_2, t_3]^4. \tag{21}$$

The conditions 1), 2), and 3) are also equivalent for regular curves in $L(W)$ with symplectic space W of arbitrary dimension (we only need to replace the power 4 in (21) by $\dim W$). In the last case these conditions are also equivalent to the fact that curvature operator $R(t)$ is identically equal to zero.

Conjecture. *Suppose that $\Lambda(t)$ satisfies Assumption 1. Then the conditions 1), 2), and 3), with the power 4 replaced by k in relation (21), are equivalent.*

If the previous conjecture is true, then one of the conditions 1), 2), or 3) can be taken as a definition of the flat curve.

Now let us discuss the notion of flatness for unparametrized curves.

Definition 3. An unparametrized curve $\Lambda(\cdot)$, satisfying Assumption 2 for some parametrization, is called flat, if its fundamental form is identically zero.

It happens that, up to symplectic transformations and reparametrizations, there exists exactly one maximal flat curve.

Theorem 4. *There is an embedding of the real projective line $\mathbb{RP}^1$ into $L(W)$ as a flat closed curve endowed with the canonical projective structure; Maslov index of this curve equals 2. All other flat curves are images under symplectic transformations of $L(W)$ of the segments of this unique one.*

References

1. A.A. Agrachev, R.V. Gamkrelidze, Feedback-invariant optimal control theory - I. Regular extremals, J. Dynamical and Control Systems, **3**,1997, No. 3, 343-389.
2. A.A. Agrachev, Feedback-invariant optimal control theory - II. Jacobi Curves for Singular Extremals, J. Dynamical and Control Systems, 4(1998), No. 4 , 583-604.
3. I. Zelenko, Nonregular abnormal extremals of 2-distribution: existence, second variation and rigidity, J. Dynamical and Control Systems, 5(1999), No. 3, 347-383.

The De Casteljau Algorithm on SE(3)

Claudio Altafini

Optimization and Systems Theory
Royal Institute of Technology
SE-10044, Stockholm, Sweden
altafini@math.kth.se

Abstract. Smooth closed-form curves on the Lie group of rigid body motions are constructed via the De Casteljau algorithm. Due to the lack of a bi-invariant metric on $SE(3)$, the resulting curve depends on the choice of the metric tensor. The two most common cases are analyzed.

1 Introduction

The group of rigid body transformation $SE(3)$ arises naturally as the configuration space of many robotic systems like aerial [8] and underwater vehicles [9] or robotic manipulators [10,15]. Motivated by motion planning purposes for such systems, the search for methods for the generation of smooth trajectories on $SE(3)$ has given rise to a rich field of literature. We mention among others [5,7,11,12,16]. In particular, geometric techniques seem to appear naturally when one wants to construct trajectories in an invariant and coordinate-free way. For example, one would like to have a notion of distance that does not change with the way it is measured, but rather that it represents as much as possible an intrinsic property of a system. It is straightforward then to describe the problem using tools from Riemannian geometry and for $SE(3)$ this is done is a series of papers [1,14,13,17].

Working in $\mathbb{R}^n$, the natural way to generate trajectories is to solve some optimization problem, for example minimize energy or acceleration. Given a certain number of boundary conditions, the resulting optimal curves are normally polynomial splines. The same trajectories can also be generated in an analytic way by considering the polygonal support of the curve, using constructions like the Bézier polynomials or schemes like the De Casteljau algorithm [6]. When one tries to extend such methods from $\mathbb{R}^n$ to a noncommutative Lie group, the situation becomes more complicated, as the different constructions no longer coincide. The use of variational techniques for trajectory generation on matrix Lie groups is investigated for example in [3,16]. In [12,4], the generalization of closed form methods like the De Casteljau algorithm is investigated for compact Lie groups for which a "natural" (i.e. completely frame-independent) Riemannian structure exists. A known negative result reported in the above mentioned literature is that $SE(3)$ cannot be endowed with a bi-invariant Riemannian metric i.e. it does not admit a natural, univocal concept of distance. This complicates things further as the choice of the metric tensor is task (or designer) biased. In the above mentioned works (see also [2]) a couple of preferred metric structures emerges. We will call them *Ad-invariant pseudo-Riemannian* and *double-geodesic*. The first consists in choosing an inner product which is nondegenerate but that can assume both negative and positive values. This corresponds to have curves with both negative and positive energy and gives as geodesics the so-called screw motions. In the second case, instead, the group structure of $SE(3)$ is discarded in favor of a cartesian product of two distinct groups (rotations and translations). Both choices have advantages and disadvantages according to the task in mind.

The scope of the present paper is to extend to these two structures of $SE(3)$ the De Casteljau method. The algorithm used here is taken from [4] and produces a C^2 curve connecting two poses in $SE(3)$ with given boundary velocities. The advantage of such an algorithm with respect to the variational approach is that it gives a curve in closed form, function only of the boundary data (and of the metric structure) so that it can be useful in applications in which a (noncausal) trajectory exactly matching the data is required. On the other hand, the obtained trajectories do not seem to be the optimum of any variational problem [4]. For sake of simplicity, we treat here only the symmetric case as it avoids us the computation of covariant derivatives.

2   Basic properties of $SE(3)$

Rotations in 3-dimensional space. Consider the set of rotational matrices i.e. the *Special Orthogonal Group* $SO(3)$ defined as:

$$SO(3) = \left\{ R \in Gl_3(\mathbb{R}) \ \text{s.t.} \ RR^T = I_3 \ \text{and} \ \det R = +1 \right\}$$

$SO(3)$ is a matrix Lie group with respect to the matrix product. The tangent space at the identity I, called $\mathfrak{so}(3)$ the *Lie algebra of* $SO(3)$, is the vector space of skew-symmetric matrices:

$$\mathfrak{so}(3) = \left\{ \widehat{\omega} \in M_3(\mathbb{R}) \ \text{s.t.} \ \widehat{\omega}^T = -\widehat{\omega} \right\}$$

where the operator $\widehat{\cdot}$ is called the wedge product and constitutes the linear representation of the vector product in $\mathbb{R}^3$ i.e.

$$\widehat{\cdot} : \quad \mathbb{R}^3 \quad \rightarrow \quad \mathfrak{so}(3)$$

$$\omega = \begin{bmatrix} \omega_1 \\ \omega_2 \\ \omega_3 \end{bmatrix} \mapsto \widehat{\omega} = \begin{bmatrix} 0 & -\omega_3 & \omega_2 \\ \omega_3 & 0 & -\omega_1 \\ -\omega_2 & \omega_1 & 0 \end{bmatrix}$$

corresponding to

$$\widehat{\omega}\sigma = \omega \times \sigma \quad \forall \ \sigma \in \mathbb{R}^3$$

Rigid body motion in 3-dimensional space. Similarly, rotations and translations form the *Special Euclidean Group SE*(3) whose matrix representation utilizes the homogeneous coordinates:

$$SE(3) = \left\{ g \in Gl_4(\mathbb{R}), \ g = \begin{bmatrix} R & p \\ 0 & 1 \end{bmatrix} \text{ s.t. } R \in SO(3) \text{ and } p \in \mathbb{R}^3 \right\}.$$

The Lie algebra of $SE(3)$ is

$$\mathfrak{se}(3) = \left\{ V \in M_4(\mathbb{R}), \text{ s.t. } V = \begin{bmatrix} \widehat{\omega} & v \\ 0 & 0 \end{bmatrix} \text{ with } \widehat{\omega} \in \mathfrak{so}(3) \text{ and } v \in \mathbb{R}^3 \right\}$$

The elements of $\mathfrak{se}(3)$ have the physical interpretation of velocities with respect to a choice of frame. In particular, deriving $g \in SE(3)$, the kinematic equations can assume two useful forms:

$$\dot{g} = V^s g \quad \text{and} \quad \dot{g} = g V^b \qquad V^s, V^b \in \mathfrak{se}(3) \tag{1}$$

called respectively *right* and *left invariant representations*. In the right invariant representation, the infinitesimal generator V^s is called the *spatial velocity*

$$V^s = \begin{bmatrix} \widehat{\omega}^s & v^s \\ 0 & 0 \end{bmatrix}$$

because it represents the velocity of g translated to the identity and expressed in an inertial frame. Invariance here is with respect to a matrix multiplication from the right and means invariance to the choice of the body fixed frame. Considering the rotation and translation components of the kinematic equations, the right invariant representation looks like:

$$\begin{cases} \dot{R} = \widehat{\omega}^s R \\ \dot{p} = \widehat{\omega}^s p + v^s \end{cases}$$

Similarly, the left invariant representation expresses invariance to change of the inertial frame and $V^b \in \mathfrak{se}(3)$

$$V^b = \begin{bmatrix} \widehat{\omega}^b & v^b \\ 0 & 0 \end{bmatrix}$$

is called *body velocity*. The first order kinematic equations are then:

$$\begin{cases} \dot{R} = R\widehat{\omega}^b \\ \dot{p} = R v^b \end{cases} \tag{2}$$

Adjoint map. The relation between spatial and body velocity at $g \in SE(3)$ is expressed by the *adjoint map*

$$\mathrm{Ad}_g(Y) = gYg^{-1} \qquad \forall \, Y \in \mathfrak{se}(3)$$

that gives the change of basis on the Lie algebra:

$$V^s = \mathrm{Ad}_g(V^b) = gV^b g^{-1} \qquad\qquad V^b = \mathrm{Ad}_{g^{-1}}(V^s)$$

The derivation of the adjoint map with respect to g gives the *Lie bracket*

$$\mathrm{ad}_V(Y) = [V, Y] = VY - YV$$

i.e. the bilinear form defining the Lie algebra.

In the matrix representation of Lie groups and algebras, the adjoint map looks like bilinear. If instead we extract a minimal representation of the elements of the Lie algebra (i.e. we consider a vector of length equal to the dimension of the Lie algebra), then we obtain a linear representation of the operators $\mathrm{Ad}_g(\cdot)$ and $\mathrm{ad}_V(\cdot)$:

$$\mathrm{Ad}_g = \begin{bmatrix} R & 0 \\ \widehat{p}R & R \end{bmatrix} \qquad \mathrm{ad}_V = \begin{bmatrix} \widehat{\omega} & 0 \\ \widehat{v} & \widehat{\omega} \end{bmatrix}$$

For $Y \in \mathfrak{se}(3)$, if $Y \simeq \begin{bmatrix} y_\omega \\ y_v \end{bmatrix} \in \mathbb{R}^6$, then

$$\mathrm{Ad}_g(Y) = \mathrm{Ad}_g \begin{bmatrix} y_\omega \\ y_v \end{bmatrix} \qquad \mathrm{ad}_V(Y) = \mathrm{ad}_V \begin{bmatrix} y_\omega \\ y_v \end{bmatrix}$$

Exponential and logarithmic maps in $SE(3)$. For constant velocities V^s and V^b, the kinematic equations (1) "looks like" a linear system of 1st order differential equations. Its solution is explicitly described in terms of the *exponential map*. For $SO(3)$ and $SE(3)$ the following formulae are used:

- exp map in $SO(3)$: *Rodriguez' formula*

$$\exp_{SO(3)} : \quad \mathfrak{so}(3) \quad \to \quad SO(3)$$
$$\widehat{\omega} \quad \mapsto \quad I + \frac{\sin \|\omega\|}{\|\omega\|}\widehat{\omega} + \frac{1 - \cos \|\omega\|}{\|\omega\|^2}\widehat{\omega}^2$$

- exp map in $SE(3)$

$$\exp_{SE(3)} : \quad \mathfrak{se}(3) \quad \to \quad SE(3)$$
$$V = \begin{bmatrix} \widehat{\omega} & v \\ 0_{3\times 1} & 0 \end{bmatrix} \mapsto \begin{bmatrix} \exp_{SO(3)}(\widehat{\omega}) & A(\widehat{\omega})p \\ 0 & 1 \end{bmatrix}$$

where

$$A(\widehat{\omega}) = I + \frac{1 - \cos \|\omega\|}{\|\omega\|^2}\widehat{\omega} + \frac{\|\omega\| - \sin \|\omega\|}{\|\omega\|^3}\widehat{\omega}^2$$

The exponential map gives the one-parameter curves corresponding to constant generators in $\mathfrak{se}(3)$ i.e. to the orbits of (complete) constant vector fields and their left/right translations. Both exponential maps are onto but not one-to-one. In fact, the corresponding logarithmic maps, which also have closed form expressions, have to be restricted to their principal values.

- logarithmic map in $SO(3)$

$$\log_{SO(3)} : SO(3) \rightarrow \mathfrak{so}(3)$$

$$R \mapsto \widehat{\omega} = \frac{\phi}{2\sin\phi}(R - R^T)$$

   where ϕ s.t. $\cos\phi = \frac{tr(R)-1}{2}$, $|\phi| < \pi$ defined for $tr(R) \neq -1$
- logarithmic map in $SE(3)$

$$\log_{SE(3)} : \quad SE(3) \quad \rightarrow \quad \mathfrak{se}(3)$$

$$g = \begin{bmatrix} R & p \\ 0_{3\times 1} & 1 \end{bmatrix} \mapsto V = \begin{bmatrix} \widehat{\omega} & A^{-1}(\widehat{\omega}) \\ 0_{3\times 1} & 0 \end{bmatrix}$$

where

$$A^{-1}(\widehat{\omega}) = I - \frac{1}{2}\widehat{\omega} + \frac{2\sin\|\omega\| - \|\omega\|(1 + \cos\|\omega\|)}{2\|\omega\|^2 \sin\|\omega\|}\widehat{\omega}^2$$

Exp and log in $SE(3)$: twists and screws. In $SE(3)$, the exponential map being onto means that every two elements can be connected by a one-parameter curve called *screw* (Chasles Theorem). Its (normalized) constant infinitesimal generator is called *twist* and corresponds to the axis of the rigid body rototranslation.

2.1   Metric structure for $SE(3)$

The construction of a polynomial from boundary data in the Euclidean case is essentially a geometric construction. In $\mathbb{R}^n$, defining a cubic polynomial curve requires the notion of straight line and of osculating plane which are fixed implicitly by the Euclidean structure. In order to generate a trajectory in $SE(3)$, we need to introduce some corresponding tools. For a Lie group, the equivalent of straight line is given by a geodesic arc i.e. by the minimum energy curve associated to some non-degenerate quadratic form on the Lie algebra. The osculating plane is related to the notion of curvature of the manifold which is also chosen by selecting a metric tensor.

The following properties are known from the literature and are collected here from the original sources [1,2,13,17] for sake of convenience.

Metric properties of $SO(3)$. On $SO(3)$, by Cartan 2nd criterion, the *Killing form*

$$K \; : \; \mathfrak{so}(3) \times \mathfrak{so}(3) \to \mathbb{R}$$
$$(X, \, Y) \quad \mapsto K(X, \, Y) = \mathrm{tr}(\mathrm{ad}_X \cdot \mathrm{ad}_Y)$$

is symmetric, negative definite and *Ad-invariant* (or bi-invariant):

$$K(X, \, Y) = K\,(\mathrm{Ad}_g X, \, \mathrm{Ad}_g Y) \quad \forall \, g \in SO(3)$$

Therefore, by choosing as inner product $\langle \cdot, \, \cdot \rangle = \alpha \, K(\cdot, \, \cdot)$ on $\mathfrak{so}(3)$ and propagating it to the whole tangent bundle $TSO(3)$ by left or right invariant translation, $SO(3)$ is endowed with the structure of a Riemannian manifold. For a bi-invariant metric, the geodesics correspond to one-parameter subgroups and to their left/right translations, inducing a natural concept of distance on the group through the logarithmic function.

Metric properties of $SE(3)$. The Levi decomposition for $SE(3)$ gives the following semidirect product:

$$SE(3) = SO(3) \otimes_R \mathbb{R}^3$$

with $SO(3)$ semisimple and $\mathbb{R}^3$ abelian. Semidirect product means that $SE(3)$ as a manifold can be considered the direct product $SO(3) \times \mathbb{R}^3$ but its group structure includes the action of $SO(3)$ on $\mathbb{R}^3$ by isometries.

It is known that in $SE(3)$ there is no Ad-invariant Riemannian metric which implies that there is no natural way of transporting vector fields between points of $SE(3)$ and therefore no natural concept of distance on $SE(3)$. The two most common approaches to tackle this obstruction are:

1. *Ad-invariant pseudo-Riemannian structure*
2. *double geodesic*

Approach 1: Ad-invariant pseudo-Riemannian structure. It consists in insisting on the notion of one-parameter subgroups and on using the logarithmic map for a pseudo-Riemannian metric structure i.e. a metric tensor that is nondegenerate but not necessarily positive definite.

The Killing form in $SE(3)$ is only semidefinite negative: if we rewrite $X, Y \in \mathfrak{se}(3)$ as $X \simeq \begin{bmatrix} x_\omega \\ x_v \end{bmatrix} \in \mathbb{R}^6$ and $Y \simeq \begin{bmatrix} y_\omega \\ y_v \end{bmatrix} \in \mathbb{R}^6$, then a symmetric bilinear form $\begin{bmatrix} x_\omega \\ x_v \end{bmatrix}^T W \begin{bmatrix} y_\omega \\ y_v \end{bmatrix}$ based only on the Killing form corresponds to the semidefinite quadratic matrix

$$W = \begin{bmatrix} \alpha I & 0 \\ 0 & 0 \end{bmatrix}$$

Combining this with a *Klein form*

$$\begin{bmatrix} x_\omega \\ x_v \end{bmatrix}^T \begin{bmatrix} 0 & \beta I \\ \beta I & 0 \end{bmatrix} \begin{bmatrix} y_\omega \\ y_v \end{bmatrix}$$

we get the most general Ad-invariant quadratic form in $SE(3)$

$$W = \begin{bmatrix} \alpha I & \beta I \\ \beta I & 0 \end{bmatrix}$$

whose eigenvalues are nondegenerate but can be either positive or negative according to the values of α and β, which says that the geodesic curves can give either positive or negative energy. Such geodesics are the *screw motions* mentioned above. The expression for the inner product is

$$\langle X, Y \rangle_{Ad-inv} = \alpha \operatorname{tr}(\operatorname{ad}_{x_\omega} \cdot \operatorname{ad}_{y_\omega}) + \beta \langle x_v, y_\omega \rangle_{\mathbb{R}^3} + \beta \langle y_v, x_\omega \rangle_{\mathbb{R}^3}$$

where $\langle \cdot, \cdot \rangle_{\mathbb{R}^3}$ is the standard inner product in Euclidean spaces.

Approach 2: double geodesic. The second method relies on the so-called double geodesic approach i.e. in forgetting about the group structure of $SE(3)$ and considering separately the bi-invariant metric of $SO(3)$ and the Euclidean metric of $\mathbb{R}^3$. The corresponding quadratic form

$$W = \begin{bmatrix} \beta I & 0 \\ 0 & \alpha I \end{bmatrix}$$

was first proposed by Brockett [1]. The inner product of two vectors $X, Y \in \mathfrak{se}(3)$ is then:

$$\langle X, Y \rangle_{d-geo} = \alpha \operatorname{tr}(\operatorname{ad}_{x_\omega} \cdot \operatorname{ad}_{y_\omega}) + \beta \langle x_v, y_v \rangle_{\mathbb{R}^3}$$

Although such a representation is neither right nor left invariant, the right (or left) invariance of the metric is preserved. Consider for example the left-invariant representation of the system.

Proposition 1. *Two different representations are obtained considering or discarding the group structure in $SE(3)$. The former is left-invariant while the latter, corresponding to the double geodesic structure, is neither left nor right invariant with respect to coordinate change. As the action of $SO(3)$ on $\mathbb{R}^3$ is by isometries, the two structures give the same left-invariant metric.*

Proof. Discarding the group structure, we have that the left-invariant equation for $g = (R, p)$ are:

$$\begin{cases} \dot{R} = R \widehat{\omega}_b \\ \dot{p} = v_b \end{cases} \tag{3}$$

Changing inertial frame from g to $g_0 g$, where $g_0 = (R_0, p_0)$ we get $g_0 g = (R_0 R, R_0 p + p_0)$. Deriving the two components,

$$\frac{d}{dt}(R_0 R, R_0 p + p_0) = (R_0 R \widehat{\omega}_b, R_0 v_b)$$

we reobtain (3). Applying the same change of inertial frame in the group structure, left invariance gives

$$\frac{d}{dt}(g_0 g) = g_0 g V_b$$

or eq. (2) in components. The only difference between (3) and (2) is an arbitrary rotation in the $\mathbb{R}^3$ part which does not modify lengths. □

3   Algorithm for a closed-form C^2 trajectory in $SE(3)$

The idea is to find a closed form curve $g(\cdot) : [0, 1] \mapsto SE(3)$ satisfying the following boundary conditions:

$$\gamma(0) = g_0, \quad \left. \frac{d\gamma}{dt} \right|_{t=0} = \dot{g}_0, \quad \gamma(1) = g_f, \quad \left. \frac{d\gamma}{dt} \right|_{t=1} = \dot{g}_f$$

Here, choosing $t_f = 1$ as final time is meant to simplify the calculations. If $t_f \neq 1$, the time axis can be rescaled appropriately afterwards. For this case, it was shown in [4] how to generate an interpolating curve in an analytic way through a combination of exponentials. The method generalizes to Lie groups a classical construction for Euclidean spaces called the *De Casteljau algorithm* [6]. The basic idea in $\mathbb{R}^n$ is to transform the boundary conditions on the velocity into intermediate points (called "control points"). The combination of the straight line segments connecting the extreme points to the control points gives the desired polynomial. The generalization consists in substituting the line segments used for the construction in the Euclidean version with geodesic arcs. Likewise, a couple of iterated combinations of the geodesics gives a C^2 curve which shares the same boundary conditions with the original patching of geodesic arcs.

A sketch of the (left-invariant version of the) algorithm is as follows:

- *transform the 1st order boundary values in infinitesimal generators V_0^1 and V_2^1 (see [4] for the details)*

$$\left. \frac{d\gamma}{dt} \right|_{t=0} = 3 g_0 V_0^1 \quad \text{and} \quad \left. \frac{d\gamma}{dt} \right|_{t=1} = 3 g_f V_2^1$$

- *get the "control points"*

$$g_1 = g_0 e^{V_0^1}$$
$$g_2 = g_f e^{-V_2^1}$$

i.e. the points reached by the time-one one-parameter arcs from the extremes.

- using the logarithmic map, find the velocity V_1^1 s.t.

$$g_2 = g_1 e^{V_1^1}$$

The three velocities obtained so far V_0^1, V_1^1, V_2^1 $\in \mathfrak{se}(3)$ are constant. Their combination (through the exponential) gives rise to curves which not anymore correspond to one-parameter subgroups, but keep the same boundary values as the C^0 patch of geodesic arcs.

- construct $V_0^2(t)$ and $V_1^2(t)$ s.t.

$$e^{V_0^2(t)} = e^{(1-t)V_0^1} e^{tV_1^1} \qquad t \in [0, 1]$$
$$e^{V_1^2(t)} = e^{(1-t)V_1^1} e^{tV_2^1}$$

- construct and $V_0^3(t)$ s.t.

$$e^{V_0^3(t)} = e^{(1-t)V_0^2(t)} e^{tV_1^2(t)}$$

The velocities $V_0^2(t)$, $V_1^2(t)$, $V_0^3(t)$ are not constant but correspond to "polynomial" generators

- the interpolating C^2 curve is

$$\gamma(t) = g_0 e^{tV_0^1(t)} e^{tV_0^2(t)} e^{tV_0^3}$$

As in $SO(3)$ and $SE(3)$ the exponential and logarithmic maps have closed form expressions, and as long as the data are given in a symmetric fashion (i.e. we do not have to compute covariant derivatives), the procedure above requires only linear algebra tools plus exp and log maps.

4   Simulation results

The two cases above for $SE(3)$ lead to different curves because the geodesics are different. In the pseudo-Riemannian case, the velocities V_0^1, V_1^1 and V_2^1 correspond to twists in $\mathfrak{se}(3)$ and are obtained through the maps $\exp_{SE(3)}$ and $\log_{SE(3)}$. In the double-geodesic case, instead, we need to split all the data: $SE(3) \ni g_i \mapsto (R_i, p_i) \in SO(3) \times \mathbb{R}^3$, use the metrics in $SO(3)$ and $\mathbb{R}^3$ to construct the various $\widehat{\omega}_j^i \in \mathfrak{so}(3)$ and $v_j^i \in \mathbb{R}^3$ and then recombine them in homogeneous representations

$$(\widehat{\omega}_j^i, v_j^i) \mapsto \begin{bmatrix} \widehat{\omega}_j^i & v_j^i \\ 0 & 0 \end{bmatrix} \in \mathfrak{se}(3)$$

The curves one obtain in the two cases are different. In particular, in the double geodesic case, it is possible to maintain the idea of straight like independent of the amount of the rotation, provided that the difference of the two end positions p_0 and p_f is aligned with the direction of both boundary tangent vectors of the translational part, see Fig. 1 (a). The price to pay is that the curve is not left-invariant with respect to a coordinate transformation since we "forget" about a rotation in the Euclidean part (Prop. 1). In general, the more consistent is the rotation component of the desired motion (with respect to the translational part) the more the two curves will look different (compare Fig. 2 and Fig. 3).

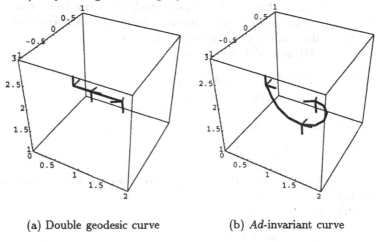

(a) Double geodesic curve   (b) *Ad*-invariant curve

Fig. 1. The two curve generated by the De Casteljau algorithm in the special situation in which the translational components of the boundary data lie on the same direction of the translational boundary velocities.

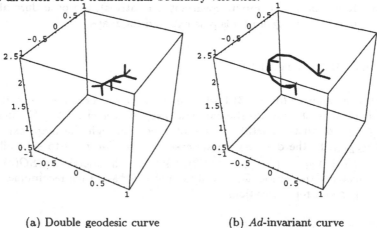

(a) Double geodesic curve   (b) *Ad*-invariant curve

Fig. 2. The two curves generated by the De Casteljau algorithm in a generic case.

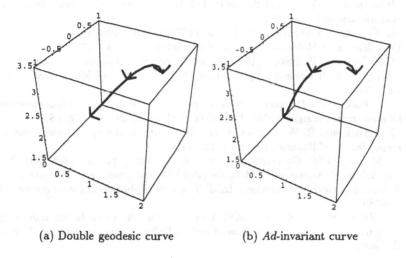

(a) Double geodesic curve   (b) *Ad*-invariant curve

Fig. 3. The two curves generated by the De Casteljau algorithm in another generic case.

References

1. R.W. Brockett (1990) Some mathematical aspects of Robotics. In J. Baillieul ed. *Robotics* Proceedings of Symposia in Applied Mathematics, American Mathematical Society, Providence.
2. F. Bullo and R. Murray (1995) Proportional Derivative (PD) Control on the Euclidean Group. Proc. of the 3rd European Control conference, Rome, Italy, 1091-1097.
3. M. Camarinha, F. Silva Leite and P.E. Crouch (1996) Second order optimality conditions for an higher order variational problem on a Riemannian manifold. Proc. 35th Conf. on Decision and Control, Kobe, Japan, 1636-1641.
4. P.E. Crouch, F. Silva Leite and G. Kun (1999) Geometric Splines. Proc. 14th IFAC World Congress, Beijing, China, vol.D, 533-538.
5. P.E. Crouch and F. Silva Leite (1995) The dynamic interpolation problem on Riemannian manifolds, Lie groups and symmetric spaces. Journal of Dynamical and Control Systems 1, 177-202.
6. G. Farin (1997) Curves and surfaces for computer-aided geometric design: a practical guide, 4th ed., Academic Press.
7. Q.J. Ge and B. Ravani (1994) Geometric construction of Bézier motions, ASME Journal of Mechanical Design, **116**, 749-755.
8. T.J. Koo and S. Sastry (1998) Output tracking control design of an helicopter based on approximate linearization. Proc. of the 37th Conference on Decision and Control, Tampa, FL.
9. N.E. Leonard and P.S. Krishnaprasad (1995) Motion control of drift-free left-invariant systems on Lie groups, IEEE Trans. on Automatic Control, **40**, 1539-1554.

10. R.M. Murray, Z. Li and S. Sastry (1994) A Mathematical Introduction to Robotic Manipulation, CRC Press.
11. L. Noakes, G. Heinzinger and B. Paden (1989) Cubic splines on curved spaces, IMA Journal of Mathematics, Control and Information, **6**, 465-473.
12. F.C. Park and B. Ravani (1995) Bézier curves on Riemannian manifolds and Lie groups with kinematic applications, ASME Journal of Mechanical Design **117**, 36-40.
13. F.C. Park (1995) Distance metrics on the rigid body motions with applications to mechanisms design, ASME Journal of Mechanical Design **117**, 48-54.
14. F.C. Park and R. W. Brockett (1994) Kinematic desterity of robotic mechanisms, Int. J. of Robotics Research, **12**, 1-15.
15. J. M. Selig (1996) Geometrical methods in Robotics, Springer, New York, NY.
16. M. Zefran, V. Kumar and C.B. Croke (1998) On the generation of smooth three-dimensional rigid body motions, IEEE Trans. on Robotics and Automation **12**, 576-589.
17. M. Zefran and V. Kumar (1998) Two methods for interpolating rigid body motions, Proc. of the 1998 IEEE Int. Conf. on Robotics and Automation, Leuven, Belgium.

Trajectory Tracking by Cascaded Backstepping Control for a Second-order Nonholonomic Mechanical System

Nnaedozie P.I. Aneke, Henk Nijmeijer, and Abraham G. de Jager

Eindhoven University of Technology
Faculty of Mechanical Engineering, Systems and Control Group
P.O. Box 513, 5600 MB Eindhoven, The Netherlands
Phone: +31 40 247 2611 / 2784, Fax : +31 40 246 1418
edo@wfw.wtb.tue.nl, a.g.de.jager@wfw.wtb.tue.nl, h.nijmeijer@tue.nl

Abstract. A design methodology is presented for tracking control of a class of second-order nonholonomic systems. The method consists of three steps. In the first step we transform the system into an extended chained-form system. This extended chained-form system is in cascade form and we apply a linear feedback to the first subsystem. In the second step, the second subsystem is exponentially stabilized by applying a backstepping procedure. In the third and final step it is shown that the closed-loop tracking dynamics of the extended chained-form system are globally exponentially stable under a persistence of excitation condition on the reference trajectory. The control design methodology is applied to an underactuated planar manipulator with two translational and one rotational joint (PPR). The simulation results show a good performance of the tracking controller.

1 Introduction

In the past few years there has been an increasing interest in the control of nonholonomic control systems [1]. The studies were primarily limited to the stabilization of first-order nonholonomic systems satisfying non-integrable kinematic or velocity constraints. Nonholonomic systems have a structural obstruction to the existence of smooth time-invariant stabilizing feedback laws and the feedback stabilization problem can only be solved by smooth time-varying feedback, non-smooth time-varying homogeneous feedback or discontinuous feedback [2].

In this contribution we consider tracking control of second-order nonholonomic systems. This class of systems arises from control problems related to mechanical systems with non-integrable acceleration constraints [3]. The tracking control problem for second-order nonholonomic systems has hardly received any attention. It is known that the time-varying Jacobian linearization around a trajectory can be controllable and therefore the tracking-error dynamics are locally smooth stabilizable. However, this only holds locally and it remains a challenging problem to develop control laws that achieve semi-global or even global trajectory tracking.

We try to solve the trajectory tracking problem by using a combined cascade and backstepping approach. Cascaded nonlinear systems are nonlinear systems that can be written as two subsystems, where the first subsystem is perturbed by the state of the second subsystem [4]. Backstepping is a well-known control design technique for nonlinear systems and the approach followed in this contribution is similar to that in [5]. The combined cascade and backstepping approach consists of three steps. In the first step, we transform the dynamic equations of motion of the mechanical system into an extended chained-form system with second-order derivatives (see [6]). This extended chained-form system is in cascade form and we apply a linear stabilizing feedback to the first subsystem. In the second step, the second subsystem is exponentially stabilized by applying a backstepping procedure. In the third and final step, we make conclusions on the exponential stability of the closed-loop tracking dynamics of the extended chained-form system by using a result for cascade systems from [7].

In the simulations, we focus on a subclass of the second-order nonholonomic control systems, namely underactuated robot manipulators. These systems are characterized by the fact that there are more degrees of freedom than actuators. We investigate the performance of the derived tracking controllers by simulation of a 3-link underactuated planar manipulator where the first two translational joints are actuated and the third revolute joint is not actuated [8].

In Sect. 2 we formulate the trajectory tracking problem for second-order nonholonomic mechanical systems. In Sect. 3 we present the control design methodology. In Sect. 4 we perform a simulation study with a 3-link planar underactuated manipulator and, finally, in Sect. 5 we draw conclusions and give recommendations for further research.

2   Problem Formulation

Consider a 3-link underactuated planar manipulator where the first two translational joints are actuated and the third revolute joint is not actuated. Fig. 1 shows the PPR planar manipulator. This system is a special case of an underactuated manipulator with second-order nonholonomic constraints. Oriolo and Nakamura [10] showed that planar underactuated manipulators have a structural obstruction to the existence of smooth time-invariant stabilizing feedback laws; they do not meet Brockett's well-known necessary condition for smooth feedback stabilization [2].

The tracking control problem for such underactuated manipulators has received little attention. However, underactuated manipulators are a challenging example of underactuated systems. The inclusion of the dynamics is mandatory and results in the presence of a drift term in the state-space model which makes the control more difficult [11].

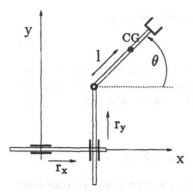

Fig. 1. A planar PPR manipulator

It should be noted that the linear approximation of this system is not controllable since the dynamics are not influenced by gravity. In the case of underactuated systems in the vertical plane, e.g. an inverted pendulum and the Acrobot [12], the linear approximation around an equilibrium state is controllable due to gravity [13]. Nevertheless, in [14] the 3-link underactuated planar manipulator was shown to be controllable by constructing an input to generate a trajectory from any given initial state to any given final state in finite time. The trajectories consist of simple translations and rotations together with combined translational and rotational movements. Also a particular feedback controller along this trajectory was presented [15].

2.1 Equations of Motion

Define $q = [r_x, r_y, \theta]$, where (r_x, r_y) denotes the displacement of the third joint from the origin $(x, y) = 0$ and θ denotes the orientation of the third link with respect to the positive x-axis, see Fig. 1. Let τ_1, τ_2 denote the inputs to the prismatic actuated joints. The mass and inertia of the links are denoted by m_i and I_i respectively and l denotes the distance between the center of mass of the third link and the third joint. The dynamic equations of motion can be written as

$$m_x \ddot{r}_x - m_3 l \sin(\theta)\ddot{\theta} - m_3 l \cos(\theta)\dot{\theta}^2 = \tau_1$$
$$m_y \ddot{r}_y + m_3 l \cos(\theta)\ddot{\theta} - m_3 l \sin(\theta)\dot{\theta}^2 = \tau_2 \qquad (1)$$
$$I\ddot{\theta} - m_3 l \sin(\theta)\ddot{r}_x + m_3 l \cos(\theta)\ddot{r}_y = 0,$$

where the configuration variables lie in the to be defined configuration-space, i.e $q \in \mathcal{C}$. We also introduced the parameters $m_x = (m_1 + m_2 + m_3)$, $m_y = m_2 + m_3$ and $I = I_3 + m_3 l^2$, see [8,16]. As we mentioned earlier, this system was shown to be controllable by constructing an input to generate a trajectory from any given initial state to any given final state in finite time [14]. The nonholonomic constraint of the system (1) can be written as

$$\lambda\ddot{\theta}(t) - \ddot{r}_x(t)\sin\theta(t) + \ddot{r}_y(t)\cos\theta(t) = 0, \tag{2}$$

where $\lambda = I/(m_3 l)$ equals the effective pendulum length when the third link is treated as a rigid-body pendulum suspended under the passive joint. In the next section we will show that the system can be transformed into a more simple form.

2.2   Transformation into Extended Chained-Form

In [8] a coordinate and feedback transformation was proposed to transform the system (1) into an extended chained form, see [6]. Let the transformed state $\xi = 0$ correspond to the equilibrium $q = 0$. The local coordinate transformation and input transformation are then given by

$$\begin{bmatrix} \xi_1 \\ \xi_2 \\ \xi_3 \end{bmatrix} = \begin{bmatrix} r_x + \lambda(\cos\theta - 1) \\ \tan(\theta) \\ r_y + \lambda\sin\theta \end{bmatrix} \tag{3}$$

$$\begin{bmatrix} \tau_1 \\ \tau_2 \end{bmatrix} = \begin{bmatrix} -m_3 l\cos\theta\dot{\theta}^2 + \left(m_x - \dfrac{I}{\lambda^2}\sin^2\theta\right)v_x + \left(\dfrac{I}{\lambda^2}\sin\theta\cos\theta\right)v_y \\ -m_3 l\sin\theta\dot{\theta}^2 + \left(\dfrac{I}{\lambda^2}\sin\theta\cos\theta\right)v_x + \left(m_y - \dfrac{I}{\lambda^2}\cos^2\theta\right)v_y \end{bmatrix}$$

where v_x and v_y are given by

$$\begin{bmatrix} v_x \\ v_y \end{bmatrix} = \begin{bmatrix} \cos\theta & \sin\theta \\ \sin\theta & -\cos\theta \end{bmatrix} \begin{bmatrix} \dfrac{u_1}{\cos(\theta)} + \lambda\dot{\theta}^2 \\ \lambda\left(u_2\cos^2\theta - 2\dot{\theta}^2\tan\theta\right) \end{bmatrix}.$$

When $\theta = \pi/2 \pm k\pi, k \in \mathbb{N}$, this coordinate transformation is not well-defined. We therefore restrict the configuration-space of the configuration coordinates q to the subspace $\mathcal{C}$ given by

$$\mathcal{C} = \left\{ q = (r_x, r_y, \theta) \in \mathbb{R}^3 \mid \theta \in (-\pi/2, \pi/2) \right\} \tag{4}$$

The dynamics of the underactuated manipulator are transformed into the extended chained-form system

$$\begin{aligned} \ddot{\xi}_1 &= u_1 \\ \ddot{\xi}_2 &= u_2 \\ \ddot{\xi}_3 &= \xi_2 u_1. \end{aligned} \tag{5}$$

The nonholonomic constraint (2) is transformed into the constraint $\ddot{\xi}_3 = \xi_2\ddot{\xi}_1$. Therefore, the state $q(t)$ corresponding to a solution $\xi(t)$ of (5) satisfies the nonholonomic constraint.

2.3   The State-Feedback Tracking Control Problem

Consider the extended chained-form system (5). Suppose that we want the third link to describe a predefined motion in time. We therefore consider a reference trajectory $q_d \in C$ for the system (1), i.e. we want the state $(q, \dot{q})$ of the system to follow a prescribed path $(q_d, \dot{q}_d)$. After transformation of this reference trajectory q_d we obtain a reference trajectory ξ_d for the extended chained-form system (5) satisfying a system of equations given by

$$
\begin{aligned}
\dot{\xi}_{11}^d &= \xi_{12}^d \quad \dot{\xi}_{12}^d = u_{1d} \\
\dot{\xi}_{21}^d &= \xi_{22}^d \quad \dot{\xi}_{22}^d = u_{2d} \\
\dot{\xi}_{31}^d &= \xi_{32}^d \quad \dot{\xi}_{32}^d = \xi_{21} u_{1d}.
\end{aligned}
\tag{6}
$$

Assumption 1. We assume that the reference trajectory $q_d(t) \in C$ $\forall t$, see (4), is continuously differentiable, i.e $q_d \in C^2$. Therefore the reference trajectory q_d is continuously differentiable and the corresponding inputs (u_{1d}, u_{2d}) in (6) are continuous.

The dynamics of the tracking-error $x = \xi - \xi_d$ in state-space form are given by

$$
\begin{aligned}
\dot{x}_{11} &= x_{12} \quad \dot{x}_{12} = u_1 - u_{1d} \\
\dot{x}_{21} &= x_{22} \quad \dot{x}_{22} = u_2 - u_{2d} \\
\dot{x}_{31} &= x_{32} \quad \dot{x}_{32} = x_{21} u_{1d} + \xi_{21}(u_1 - u_{1d}).
\end{aligned}
\tag{7}
$$

Problem 1. (**State-feedback tracking control problem**) The tracking control problem is said to be solvable if we can design appropriate Lipschitz continuous time-varying state-feedback controllers of the form

$$
u_1 = u_1(t, x, \bar{u}_d), \qquad u_2 = u_2(t, x, \bar{u}_d)
\tag{8}
$$

such that the closed-loop system (7,8) is globally uniformly asymptotically stable. The vector $\bar{u}_d$ contains $u_d = [u_{1d}, u_{2d}]$ and its higher order derivatives, i.e $\bar{u}_d = (u_d, \dot{u}_d, \ldots, u_d^{(3)}, \ldots)$.

Remark 1. We say that the tracking control problem has been solved when the tracking dynamics of the extended chained-form system (5) are GUAS. On the other hand, the tracking dynamics of the original mechanical system (1) are not globally asymptotically stable, but are only asymptotically stable when the coordinate-transform (3) is well-defined, i.e when $q_d(t) \in C, \forall t > 0$.

In the following sections we will investigate whether the error dynamics can be asymptotically or even exponentially stabilized. To that end we apply some of the results presented in [17,18] together with some theory for linear time-varying systems from [19].

3   Cascaded Backstepping Control

In this section we apply a cascade design to stabilize the error dynamics (7). We start by rewriting the tracking dynamics into a more convenient form.

$$
\Delta_1 \begin{cases} \dot{x}_{31} = x_{32} \\ \dot{x}_{32} = x_{21}u_{1d} + \xi_{21}(u_1 - u_{1d}) \end{cases} \quad \Delta_2 \begin{cases} \dot{x}_{21} = x_{22} \\ \dot{x}_{22} = u_2 - u_{2d} \end{cases}
$$

$$
\Pi \begin{cases} \dot{x}_{11} = x_{12} \\ \dot{x}_{12} = u_1 - u_{1d} \end{cases}
\tag{9}
$$

Suppose that the Π subsystem is stabilized by the input $u_1(u_{1d}, x_{11}, x_{12})$ and therefore $u_1 = u_{1d}$. Then we design the remaining input u_2 such that the remaining subsystem (Δ_1, Δ_2) is stabilized for $u_1 = u_{1d}$. In order to make conclusions on the exponential stability of the complete closed-loop system we will use a result from [17,18] for cascaded systems, which was based on a result in [4]. For a survey on the stability of cascaded systems see also chapter 5 of [9]. Consider the time-varying nonlinear system given by

$$
\begin{aligned}
\dot{z}_1 &= f_1(t, z_1) + g(t, z_1, z_2)z_2, \\
\dot{z}_2 &= f_2(t, z_2),
\end{aligned}
\tag{10}
$$

where $z_1 \in \mathrm{R}^n$, $z_2 \in \mathrm{R}^m$, $f_1(t, z_1)$ continuously differentiable in (t, z_1) and $f_2(t, z_2), g(t, z_1, z_2)$ continuous in their arguments and locally Lipschitz in z_2 and (z_1, z_2) respectively. The total system (10) is a system Σ_1 with state z_1 that is perturbed by the state z_2 of the system Σ_2, where

$$
\Sigma_1: \quad \dot{z}_1 = f_1(t, z_1) \quad \Sigma_2: \quad \dot{z}_2 = f_2(t, z_2),
\tag{11}
$$

and the perturbation term is given by $g(t, z_1, z_2)z_2$. We state the following result from [4,18].

Theorem 1. *The cascaded system* (10) *is globally uniform asymptotically stable (GUAS) if the following three assumptions hold:*

(1) Σ_1 : subsystem: *The subsystem* $\dot{z}_1 = f_1(t, z_1)$ *is GUAS and there exists a continuously differentiable function* $V(t, z_1) : \mathrm{R}^+ \times \mathrm{R}^n \to \mathrm{R}$ *and a positive definite proper function* $W(z_1)$ *such that*

(i) $V(t, z_1) \geq W(z_1)$

(ii) $\dfrac{\partial V}{\partial t} + \dfrac{\partial V}{\partial z_1} f_1(t, z_1) \leq 0, \qquad \forall \, \|z_1\| \geq \eta,$

(iii) $\left\| \dfrac{\partial V}{\partial z_1} \right\| \|z_1\| \leq \zeta V(t, z_1), \qquad \forall \, \|z_1\| \geq \eta,$

$\qquad\qquad$ (12)

where $\zeta > 0$ *and* $\eta > 0$ *are constants.*

(2) interconnection: *The function $g(t, z_1, z_2)$ satisfies*

$$\|g(t, z_1, z_2)\| \leq \kappa_1(\|z_2\|) + \kappa_2(\|z_2\|)\|z_1\|, \qquad \forall\, t \geq t_0, \tag{13}$$

where $\kappa_1, \kappa_2 : \mathrm{R}^+ \to \mathrm{R}^+$ are continuous functions.

(3) Σ_2 : subsystem: *The subsystem $\dot{z}_2 = f_2(t, z_2)$ is GUAS and satisfies*

$$\int_{t_0}^{\infty} \|z_2(t_0, t, z_2(t_0))\| dt \leq \zeta(\|z_2(t_0)\|), \qquad \forall\, t \geq t_0, \tag{14}$$

where the function $\zeta(\cdot)$ is a class $\mathcal{K}$ function.

Lemma 1. (see [4]) *If in addition to the assumptions in Theorem 1 both Σ_1 and Σ_2 are globally $\mathcal{K}$-exponentially stable, then the cascaded system (10) is globally $\mathcal{K}$-exponentially stable.*

Remark 2. In the system (9) the perturbation term $g(t, z_1, z_2)z_2$ is given by $\xi_{21}(u_1 - u_{1d})$ and depends on the, to be designed, feedback $u_1(t, x)$. Therefore, when considering Π as the unperturbed system Σ_2, in order to satisfy condition (2) in Theorem 1 the resulting perturbation matrix $g(t, z_1, z_2)$ has to be linear with respect to the variable $z_2 = (x_{11}, x_{12})$. This is the case when choosing the feedback u_1 as $u_1 = u_{1d} + k(t, x_{11}, x_{12})$, with $k : \mathrm{R}_+ \times \mathrm{R}^2 \to \mathrm{R}$ a linear function in (x_{11}, x_{12}).

3.1 Stabilization of the Subsystem Δ_1

Suppose that the Π subsystem has been stabilized by choosing

$$u_1 = u_{1d} - k_1 x_{11} - k_2 x_{12}, \qquad k_1 > 0, k_2 > 0, \tag{15}$$

where the polynomial $p(\lambda) = \lambda^2 + k_1\lambda + k_2$ is Hurwitz. Consider the time-varying subsystem Δ_1. For $u_1 = u_{1d}$ this subsystem can be written as the time-varying linear system

$$\begin{aligned} \dot{x}_{31} &= x_{32} \\ \dot{x}_{32} &= x_{21} u_{1d}, \end{aligned} \tag{16}$$

where x_{21} denotes a virtual input for the backstepping procedure. We aim at designing a stabilizing (virtual) feedback x_{21} for the subsystem Δ_1. Consider the first equation $\dot{x}_{31} = x_{32}$ and assume that x_{32} is the virtual input. A stabilizing function $x_{32} = \alpha_1(x_{31})$ for the x_{31}-subsystem is

$$\alpha_1 = -c_1 u_{1d}^{2d_1+2} x_{31}, \qquad c_1 > 0.$$

Define $\bar{x}_{32} = x_{32} - \alpha_1(x_{31}) = x_{32} + c_1 u_{1d}^{2d_1+2} x_{31}$ and consider the $\bar{x}_{32}$-subsystem

$$\dot{\bar{x}}_{32} = x_{21} u_{1d} + c_1 u_{1d}^{2d_1+2} x_{32} + c_1(2d_1 + 2) u_{1d}^{2d_1+1} \dot{u}_{1d} x_{31}.$$

Suppose that x_{21} is the virtual input, then a stabilizing function $x_{21} = \alpha_2(\bar{u}_{1d}, x_{31}, x_{32})$ for the $\bar{x}_{32}$-subsystem is given by

$$\alpha_2 = -\left(c_1 c_2 u_{1d}^{2d_1+2d_2+3} + c_1(2d_1+2)u_{1d}^{2d_1}\dot{u}_{1d}\right)x_{31} \tag{17}$$
$$- \left(c_1 u_{1d}^{2d_1+1} + c_2 u_{1d}^{2d_2+1}\right)x_{32}$$

where $c_2 > 0$ and we substituted $\bar{x}_{32} = x_{32} + c_1 u_{1d}^{2d_1+2}x_{31}$. Define $\bar{x}_{31} = x_{31}$. The closed-loop system becomes

$$\begin{aligned}
\dot{\bar{x}}_{31} &= -c_1 u_{1d}^{2d_1+2}\bar{x}_{31} + \bar{x}_{32} \\
\dot{\bar{x}}_{32} &= -c_2 u_{1d}^{2d_2+2}\bar{x}_{32}
\end{aligned} \tag{18}$$

We show that the resulting closed-loop system (18) is globally exponentially stable (GES) under the following assumption.

Assumption 2. The function $u_{1d}(t)$ is persistently exciting, i.e. there exist $\delta, \varepsilon_1, \varepsilon_2 > 0$ such that for all integers $r \geq 0$

$$\varepsilon_1 \leq \int_t^{t+\delta} u_{1d}^{2r+2}(\tau)d\tau \leq \varepsilon_2, \qquad \forall t > 0. \tag{19}$$

Under Assumption 2 the closed-loop system (18) is globally exponentially stable. The system (18) can be written as (10), and by applying Theorem 1 and Lemma 1 globally exponentially stability follows. The result is given in Proposition 1, which we state without proof. The proof will be given in a forthcoming paper.

Proposition 1. *The equilibrium $x = 0$ of (18) is globally exponentially stable (GES) for all reference inputs $u_{1d}(t)$ satisfying the persistence of excitation condition (19).*

The system (16), i.e the Δ_1-subsystem for $u_1 = u_{1d}$, in closed-loop with the linear time-varying feedback (17) is GES. By Theorem 3.12 in [9], we can prove the existence of a suitable Lyapunov function.

3.2   Stabilization of the Subsystem (Δ_1, Δ_2)

In Sect. 3.1 we saw that the Δ_1-subsystem can be exponentially stabilized by the virtual input $x_{21} = \alpha_2(t, u_{1d}(t), x_{31}, x_{32})$, given by (17). The virtual input x_{21} is a state of the system Δ_2 and we now continue the backstepping procedure for the Δ_2-subsystem as follows. Denote the state x of the Δ_1-subsystem by $x = [x_{31}, x_{32}]$. Define $\bar{x}_{21} = x_{21} - \alpha_2(t, u_{1d}(t), x)$ and consider the $\bar{x}_{21}$ sub-system $\dot{\bar{x}}_{21} = x_{22} - \mathrm{d}/\mathrm{dt}\,[\alpha_2(t, u_{1d}(t), x)]$, where x_{22} is the virtual input. A stabilizing function $\alpha_3(t, u_{1d}(t), x)$ for the $\bar{x}_{21}$-subsystem is given by

$$\alpha_3(t, u_{1d}(t), x) = \frac{\mathrm{d}}{\mathrm{dt}}\,[\alpha_2(t, u_{1d}(t), x)] - c_3\bar{x}_{21}^{2d_3+1}. \tag{20}$$

We define a new state as $\bar{x}_{22} = x_{22} - \alpha_3(t, u_{1d}(t), x)$ and consider the $\bar{x}_{22}$-subsystem $\dot{\bar{x}}_{22} = (u_2 - u_{2d}) - \mathrm{d}/\mathrm{dt}\,[\alpha_3(t, u_{1d}(t), x)]$. This subsystem can be stabilized by choosing the input u_2 such that

$$u_2 - u_{2d} = -c_4\bar{x}_{22}^{2d_4+1} + \frac{\mathrm{d}}{\mathrm{dt}}\,[\alpha_3(t, u_{1d}(t), x)]. \tag{21}$$

According to Theorem 3.12 in [9] there exists a Lyapunov function $V_1(t, x)$ for the Δ_1-subsystem such that $\dot{V}_1(t, x) \leq -\beta_3\|x\|^2$. Now consider the Lyapunov function given by $V(x_{21}, x_{22}, x) = V_1(t, x) + (\bar{x}_{21}^2 + \bar{x}_{22}^2)/2$. Its time-derivative satisfies

$$\dot{V}(x_{21}, x_{22}, x) \leq -\beta_3\|x\|^2 - (c_3\bar{x}_{21}^{2d_3+2} + c_4\bar{x}_{22}^{2d_4+2}),$$

showing that we have successfully stabilized the (Δ_1, Δ_2) subsystem. After substitution of α_3 and $\bar{x}_{21}$ respectively, we can write (21) as

$$u_2 - u_{2d} = -c_4\bar{x}_{22}^{2d_4+1} - c_3\frac{\mathrm{d}}{\mathrm{dt}}\left[\bar{x}_{21}^{2d_3+1}\right] + \frac{\mathrm{d}^2}{\mathrm{dt}^2}\,[\alpha_2(t, u_{1d}(t), x)] \tag{22}$$

where $\alpha_2(u_{1d}(t), x)$ and $\alpha_3(u_{1d}(t), x)$ are given by (17) and (20) respectively. This input $u_2 - u_{2d}$ stabilizes the (Δ_1, Δ_2)-subsystem.

3.3 Stability of the Complete Tracking Dynamics

In this section we show that the complete tracking dynamics are globally exponentially stable. In (9) we have stabilized the (Δ_1, Δ_2)-subsystem when $u_1 = u_{1d}$ and the Π subsystem. We can now use Theorem 1 to investigate the stability properties of the complete system. The result is stated in the following proposition. The proof will be given in a forthcoming paper.

Proposition 2. *Suppose that the reference trajectory $u_{1d}(t)$ satisfies Assumption 2. Consider the system (9) in closed-loop with the controller u_2 given by (22) and u_1 given by*

$$u_1 = u_{1d} - k_1x_{11} - k_2x_{12}, \qquad p(s) = s^2 + k_2s + k_1 \text{ is Hurwitz}, \tag{23}$$

with $k_1 > 0$, $k_2 > 0$. If the signal $\xi_{21}^d(t)$, the reference input u_{1d} and its derivative $\dot{u}_{1d}$, cf. (6), are uniformly bounded in t, then the closed-loop system is globally $\mathcal{K}$-exponentially stable.

Summarizing, we have exponentially stabilized the (Δ_1, Δ_2) and Π subsystems separately. We then concluded by Theorem 1 and Lemma 1 that the combined system is $\mathcal{K}$-exponentially stable when the reference input u_{1d} and its derivative $\dot{u}_{1d}$ are uniformly bounded over t.

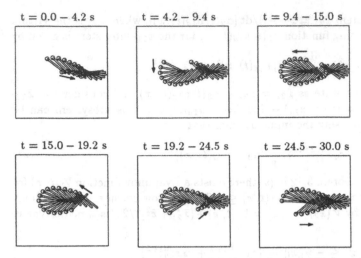

Fig. 2. Motion of the third joint and link in simulation of Fig. 3 ($\Delta t = 0.25$ s)

4   Simulation Study

In this section we apply the previous control design methodology to the 3-link planar underactuated manipulator. We evaluate the controller (22,23) by tracking a smooth trajectory (ξ_d, u_d), where the reference input u_{1d} and its derivative $\dot{u}_{1d}$ are uniformly bounded in t and the persistence of excitation condition (19) is satisfied. Therefore, we conclude by Proposition 2 that the tracking-error dynamics (7) are uniformly exponentially stable.

Consider the extended chained-form system (6) subject to the sinusoidal inputs $u_{1d} = -r_1 a^2 \sin(at)$ and $u_{2d} = -r_2 a^2 \cos(at)$. With initial condition $\xi_d(0) = [\xi_1, \dot{\xi}_1, \xi_2, \dot{\xi}_2, \xi_3, \dot{\xi}_3]$ given by $\xi_d(0) = [\xi_{1d}(0), r_1 a, r_2, 0, \xi_{3d}(0), a r_1 r_2 / 4]$ where $\xi_{1d}(0)$ and $\xi_{3d}(0)$ are still to be designed parameters, the closed-form solution $\xi_d \in \chi$ of the reference trajectory is given by

$$\xi_{1d}(t) = \xi_{1d}(0) + r_1 \sin(at), \quad \xi_{2d}(t) = r_2 \cos(at)$$
$$\xi_{3d}(t) = \xi_{3d}(0) + \frac{r_1 r_2}{8} \sin(2at). \tag{24}$$

The resulting reference trajectory q_d for the mechanical system (1) is given by

$$r_{xd}(t) = \xi_{1d}(0) + r_1 \sin(at) - \lambda \left(\cos(\arctan(r_2 \cos(at))) - 1 \right),$$
$$r_{yd}(t) = \xi_{3d}(0) + \frac{r_1 r_2}{8} \sin(2at) - \lambda \sin(\arctan(r_2 \cos(at))), \tag{25}$$
$$\theta_d(t) = \arctan(r_2 \cos(at)).$$

We consider a trajectory around the origin $(r_x, r_y) = 0$ and select $\xi_{1d}(0) = 0$, $\xi_{3d}(0) = 0$ and $r_1 = r_2 = a = 1$ in the reference trajectory (25). The controller

u_1 is given by (23). We choose $d_1 = d_2 = 0$ in the virtual input x_{21} (17) and $d_3 = d_4 = 0$ in the controller u_2 (22). We choose the control parameters as

$$k_1 = 4, \ k_2 = 2\sqrt{2}, \ c_1 = 2, \ c_2 = 2, \ c_3 = 4, \ c_4 = 4.$$

The perturbation term in (9) depends on $u_1 - u_{1d}$. Therefore, in order obtain fast convergence to the reference trajectory u_1 must converge to the reference value u_{1d} fast, which motivates our choice for k_1, k_2. The parameters c_1, c_2 determine the convergence of the chained state ξ_3 and its corresponding mechanical state r_y. The parameters c_3, c_4 mainly determine the convergence of the chained states ξ_2 and its corresponding mechanical state θ.

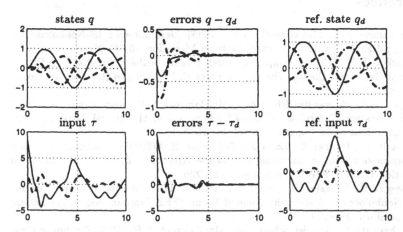

Fig. 3. Tracking of the trajectory (25) by the planar PPR manipulator; coordinates of system (1), r_x (*solid*), r_y (*dashed*), θ (*dash-dotted*), inputs τ_1 (*solid*), τ_2 (*dashed*)

The result of tracking the trajectory with initial condition $q(0) = 0$ is shown in Fig. 2. The state variables of the mechanical system are shown in Fig. 3. From the simulation it is clear that it takes about 5 seconds before the trajectory is successfully tracked. There still remains some freedom of choice in the control parameters, and when considering more complex trajectories this freedom of choice can be exploited to improve the transient response of the underactuated manipulator.

5 Conclusions

We have presented a control design method for exponential tracking of a second-order non-holonomic system. This control design method consisted of a combination of a cascade approach and a backstepping approach. It follows that the closed-loop tracking dynamics of the extended chained-form system are globally exponentially stable under a persistence of excitation condition on the reference trajectory. Furthermore, the tracking dynamics of the mechanical system are only exponentially stable on a subspace of the state-space where the coordinate transformation is well-defined.

The simulations show a good performance of the developed controller. However, in practice, there will be disturbances and mechanical friction (in the free rotational link) that can considerably deteriorate the performance. Our goal for future research is to extend the results to experiments with an underactuated RRR-robot [20].

References

1. Kolmanovsky I., McClamroch N.H. (1995) Developments in Nonholonomic Control Problems. IEEE Control Systems Magazine 15, 20–36
2. Brockett R.W. (1983) Asymptotic Stability and Feedback Stabilization. Differential Geometric Control Theory (Brockett R. W., Milman R. S., Sussmann H. J. (Eds.)) 181–191
3. Reyhanoglu M., van der Schaft A.J., McClamroch N.H., Kolmanovsky I. (1999) Dynamics and Control of a Class of Underactuated Mechanical Systems. IEEE Transactions on Automatic Control 44, 1663–1671
4. Pantely E., Lefeber E., Loría A., Nijmeijer H. (1998) Exponential Tracking Control of a Mobile Car using a Cascade Approach. In: IFAC Workshop on Motion Control, France, September, 221–226
5. Jiang Z. P., Nijmeijer H. (1999) A Recursive Technique for Tracking Control of Nonholonomic Systems in Chained Form. IEEE Transactions on Automatic Control 44, 265–279
6. Reyhanoglu M., van der Schaft A.J., McClamroch N.H. (1996) Nonlinear Control of a Class of Underactuated Systems. In: 35th IEEE Conference on Decision and Control, Japan, December, 2, 1682–1687
7. Panteley E. and Loría A., On Global Uniform Asymptotic Stability of Nonlinear Time-Varying Systems in Cascade. Systems and Control Letters 33, 131–138
8. Imura J., Kobayashi K., Yoshikawa T. (1996) Nonholonomic Control of a 3 Link Planar Manipulator with a Free Joint. In: 35th IEEE Conference on Decision and Control, Japan, December, 2, 1435–1436
9. Khalil H.K. (1996) Nonlinear Systems, 2nd edn. Prentice Hall, New York
10. Oriolo G., Nakamura Y. (1991) Free-Joint Manipulators: Motion Control under Second-Order Nonholonomic Constraints. In: IEEE/RSJ International Workshop on Intelligent Robots and Systems '91. Intelligence for Mechanical Systems, Japan, November, 3, 1248–1253
11. Oriolo G., Nakamura Y. (1991) Control of Mechanical Systems with Second-Order Nonholonomic Constraints: Underactuated Manipulators. In: 30th Conference on Decision and Control, England, December, 3, 2398–2403
12. Spong M.W. (1995) The Swing Up Control Problem for the Acrobot. IEEE Control Systems Magazine 15, 49–55
13. Arai H., Tanie K., Shiroma N. (1998) Nonholonomic Control of a Three-DOF Planar Underactuated Manipulator. IEEE Transactions on Robotics and Automation 14, 681–695

14. Arai H. (1996) Controllability of a 3-DOF Manipulator with a Passive Joint under a Nonholonomic Constraint. In: IEEE International Conference on Robotics and Automation, USA, April, 4, 3707–3713
15. Arai H., Tanie K., Shiroma N. (1997) Feedback Control of a 3-DOF Planar Underactuated Manipulator. In: IEEE International Conference on Robotics and Automation 1, 703–709
16. De Luca A., Mattone R., Oriolo G. (1995) Dynamic Mobility of Redundant Robots under End-Effector Commands. In: Nonlinear Control and Robotics Preprints, Universitá di Roma "La Sapienza", Italy
17. Lefeber E., Robertsson A., Nijmeijer H. (1999) Linear Controllers for Tracking Chained-Form Systems. Lecture Notes in Control and Information Sciences, Stability and Stabilization of Nonlinear Systems (Aeyels D., Lamnabhi-Lagarrigue F., van der Schaft A.J. (Eds.)), Springer, Heidelberg, 246, 183–199
18. Lefeber E., Robertsson A., Nijmeijer H. (1999) Output Feedback Tracking of Nonholonomic Systems in Chained Form. In: 5th European Control Conference, Germany, September, cd-rom, Paper 772 (AP-4)
19. Rugh W.J. (1993) Linear System Theory (Kailath T. (Ed.)), Information and System Sciences Series, 1st edn. Prentice-Hall, New Jersey
20. van Beek B., de Jager B. (1999) An Experimental Facility for Nonlinear Robot Control. In: IEEE International Conference on Control Applications, Hawai'i, August, 1, 668–673

Static Output Feedback Stabilization: from Linear to Nonlinear and Back

Alessandro Astolfi[1,2] and Patrizio Colaneri[1]

[1] Dip. di Elettronica e Informazione
Politecnico di Milano
Piazza Leonardo da Vinci 32
20133 Milano, Italy
[2] Dept. of Electrical and Electronic Engineering
Imperial College
Exhibition Road
London SW7 2BT, England
a.astolfi@ic.ac.uk

Abstract. The static output feedback stabilization problem for linear and nonlinear (affine) systems is discussed. A novel necessary and sufficient condition for linear systems is proposed. For nonlinear systems a sufficient condition is established and a (partial) converse is also discussed. The nonlinear formulation is used to derive a simple characterization of stabilizing static output feedback control laws for linear systems in terms of the intersection of two convex sets and a (generically) nonconvex set. This characterization is used to establish a series of simple obstructions to the solvability of the problem for linear SISO systems. A few worked out examples complete the chapter.

1 Introduction

The static output feedback (SOF) stabilization problem is probably one of the most known puzzle in systems and control. The simple statement of the problem is as follows: find a static output feedback control such that the closed-loop system is asymptotically stable. This problem is important in its own right, since static output feedback controllers are less expensive to be implemented and more reliable in practice. In particular, static output feedback controllers can be employed as *back-up* controllers, *i.e.* as controllers which are not active during the regular operation of the plant, but which are used in case of faults. For this reason back-up controllers have to be as simple as possible. Finally, it is possible to show that in many cases the design of a dynamical output feedback controller boils down to the solution of a static output feedback control problem [17,15].

Strictly speaking, static feedback controllers include the family of time-varying output gains. In most cases, the time-varying nature of the output gain is chosen in the narrower set of piecewise constant functions or time-periodic functions. The interest of such controllers rests upon the fact that they might stabilize a linear time-invariant system which is not stabilizable by constant static output feedback control. Unfortunately, a definite theory on the benefits of time-varying static output feedback control is still missing, and one has just to comment on simple particular examples and applications, see [7,1,18] for continuous time systems and [3] for discrete-time systems.

A simple way to stabilize a (dynamically output stabilizable) linear time-invariant system is that of using a generalized sampled-data hold output feedback with a suitable sampling period. It has been shown that the use of such control laws enables one to deal with problems otherwise unsolvable with time-invariant output controllers, such as pole assignment, simultaneous stabilization of a finite number of plants and gain margin improvement, see [13]. However, such control laws cannot be considered as static output feedback controllers. Indeed they are at all effect dynamical systems. Moreover, they introduce higher frequency components centered at multiples of the sampling frequency, which can possibly produce undesired ripples in the input and output responses. The study of benefits and inherent limitations of generalized sampled-data control systems are on their way and some aspects are still to be understood [9].

Turning back to static output feedback controllers, we can say that, despite the simplicity of its formulation, the fundamental question of the existence (in the general case) of a stabilizing static output feedback control law is still open. Many attempts have been made in the last years, so that at this stage we can count several nontrivial contributions to the problem, both numerical and speculative, see the recent paper [19], where the state of the art is presented and the existing methods are surveyed and compared. Several necessary conditions have been worked out. Among them, the parity interlacing property [22] and its extended version [20] give particularly interesting insight. Simple testable sufficient conditions have also been introduced, see e.g. [10]. Among the various technical design approaches to output feedback stabilization, it is worth mentioning the technique related to the solution of coupled linear matrix inequalities, which also provide a parameterization of all SOF gains [12]. A few algorithms have been proposed to get through the coupling condition. Among them the most interesting seem to be the min/max procedure proposed in [8], the cone complementary algorithm of [6], and the ILMI approach of [21].

All the contributions confirm that (generically) the SOF stabilization problem for linear time-invariant systems is intrinsically nonlinear, i.e. the linearity of the system to be stabilized does not yield any special advantage in finding analytic solutions or systematic procedures. This observation motivates the present paper, where the SOF stabilization problem is consider for affine nonlinear systems. Precisely, a sufficient condition is provided based on the solution of a suitable Hamilton-Jacobi inequality, and a partial converse is also proved. The nonlinear analysis is useful to formulate a novel necessary and sufficient condition for SOF stabilization of linear systems, given in terms of the coupled solution to a linear matrix inequality and a linear equality. From this equality, in the case of single-input single-output systems, it is possible to go deeply into the structure of the problem, establishing a *nested* series of testable necessary conditions (obstructions).

This chapter is organized as follows. In Section 2 the SOF stabilization problem for linear systems is recalled and a preliminary result is stated. This can be seen as a slight modification of the necessary and sufficient condition given in [14]. The SOF stabilization problem for nonlinear systems is tackled in Section 3, where the sufficient and the necessary conditions are provided in terms of the solution of a constrained Hamilton-Jacobi equation along with a rank condition. This sets the basis for the introduction of the novel necessary and sufficient condition provided in Section 4. Further results are proposed in Section 5, whereas a few interesting issues, some obstructions, and a discussion on the convexity properties for SISO systems are given in Section 6. Finally, Section 7 and Section 8 contain a few examples, some comments and hints for future research.

2   Preliminary results

Consider the continuous-time linear system

$$\dot{x} = Ax + Bu \qquad\qquad (1)$$
$$y = Cx \qquad\qquad (2)$$

where $x \in I\!\!R^n$, $u \in I\!\!R^m$, $y \in I\!\!R^p$ are the state, input and output vectors, respectively, and A, B, C are matrices with constant real coefficients and appropriate dimensions.

The *static output feedback stabilization* problem for system (1)-(2) consists in finding, if possible, a static control law described by

$$u = Fy \qquad\qquad (3)$$

such that the closed-loop system is asymptotically stable, *i.e.* the matrix $A + BFC$ has all its eigenvalues with negative real parts. If such an output feedback does exist, we say that the system (1)-(2) is *output stabilizable* and that F is a solution of the problem[1].

In what follows, whenever we deal with the linear system (1)-(2) we make the following standing assumptions.

(A1) The pair $\{A, B\}$ is controllable and B has full column rank.
(A2) The pair $\{A, C\}$ is observable and C has full row rank.
(A3) $m \leq p$, *i.e.* the transfer function $W(s) = C(sI - A)^{-1}B$ is *fat*.

---

[1] It is obvious that, if a solution exists, this is not unique.

The above assumptions are without loss of generality. First of all, the solvability of the problem depends only upon the controllable and observable part of the system, *i.e.* on the transfer function $W(s) = C(sI - A)^{-1}B$. Then, the rank assumptions can be always enforced, by elimination of redundant control inputs or measured outputs. Finally, if $m > p$, it is possible to consider the system

$$\dot{\lambda} = A'\lambda + C'v \tag{4}$$
$$\eta = B'\lambda \tag{5}$$

for which Assumption (A3) holds, and observe that the static output feedback stabilization problem for system (4)-(5) is solvable if and only if it is solvable for system (1)-(2). Moreover, if Ω (resp. F) is a solution of the static output feedback stabilization problem for system (4)-(5) (resp. (1)-(2)) then $F = \Omega'$ (resp. $\Omega = F'$) is a solution of the static output feedback stabilization problem for system (1)-(2) (resp. (4)-(5)).

A simple necessary and sufficient condition for the system to be output stabilizable is stated in the following result, whose proof is reported here for the sake of completeness, even though, in slight different forms, it can be found here and there in the existing literature, see *e.g.* [14].

Theorem 1. *Consider the system (1)-(2) with Assumptions (A1), (A2) and (A3). The system is output feedback stabilizable if and only if there exist a symmetric positive semidefinite matrix $P \in \mathbb{R}^{n \times n}$ and a matrix $G \in \mathbb{R}^{m \times n}$ such that*

$$0 = A'P + PA - PBB'P + C'C + G'G \tag{6}$$
$$0 = V(A'P + PA)V, \tag{7}$$

where[2]

$$V = I - C'(CC')^{-1}C.$$

Proof. (Only if.) Assume that the system (1)-(2) is output stabilizable and let F be such that $A + BFC$ is asymptotically stable. Then the Lyapunov Lemma implies the existence of a unique positive semidefinite solution of the Lyapunov equation

$$0 = (A + BFC)'P + P(A + BFC) + C'C + C'F'FC.$$

Setting $G = FC + B'P$, the above equation can be rewritten as

$$0 = A'P + PA - PBB'P + C'C + G'G.$$

Moreover, notice that $GV = B'PV$ and then, multiplying both sides of the above Riccati equation by V, it follows $V(A'P + PA)V = 0$.

---

[2] Observe that by Assumption (A2) the matrix V is well-defined.

(*If.*) Assume that there exist $P \geq 0$ and G satisfying equations (6) and (7). As a result

$$VPBB'PV = VG'GV.$$

This implies that there exists an orthogonal[3] matrix T such that $GV = TB'PV$. Defining

$$F = (T'G - B'P)C'(CC')^{-1}$$

it follows

$$FC = (T'G - B'P)(I - V) = (T'G - B'P),$$

hence

$$G'G = G'TT'G = (FC + B'P)'(FC + B'P).$$

Finally, substituting this last equation in equation (6) one has

$$0 = (A + BFC)'P + P(A + BFC) + C'C + C'F'FC.$$

Observe now that the observability of the pair $\{A, C\}$ is equivalent to the observability of the pair $\{(A+BFC), C'C+C'F'FC\}$; hence by the Lyapunov Lemma, the existence of $P \geq 0$ implies the stability of $(A+BFC)$. Note finally that, by observability of the pair $\{A, C\}$, $P > 0$.

Theorem 1 provides also a parameterization of all stabilizing static output feedback control laws, as expressed in the following statement, whose proof is easily obtained from the proof Theorem 1.

Corollary 1. *Consider the system (1)-(2). The family of all output feedback gains F such that the matrix $A + BFC$ is stable is given by*

$$F = (T'G - B'P)C'(CC')^{-1}$$

where $P = P' \geq 0$ and G solve (6) and (7) and T is any orthogonal matrix.

It is worth noting that, despite their simplicity, the conditions in Theorem 1 cannot be recast (to the best of the authors knowledge) in an LMI framework or in a simple computational scheme, because in general the static output feedback stabilization problem is nonconvex. Moreover, to obtain simpler conditions, *e.g.* conditions involving only the matrix P, it would be tempting to replace equation (6) with the matrix inequality

$$0 \geq A'P + PA - PBB'P + C'C = \mathcal{R}(P). \tag{8}$$

Unfortunately, using simple linear arguments, it is only possible to prove that the conditions expressed by equations (8) and (7) are necessary for static output feedback stabilizability. To recover also the sufficient part of the statement, one has to add a further rank condition on $\mathcal{R}(P)$, *i.e.* $\text{rank}(\mathcal{R}(P)) \leq m$.

---

[3] A matrix T is said to be orthogonal if $T'T = TT' = I$. In the same way, a (nonconstant) matrix $T(x)$ is said to be orthogonal if it is orthogonal for each fixed x.

However, using the more general formulation and solution of the problem given in the next section, it is possible to prove that the equations (8) and (7) provide a necessary and sufficient condition for the solvability of the static output feedback stabilization problem.

These conditions are obviously simpler than conditions (6) and (7), as they involve only one unknown, *i.e.* the matrix P. Moreover, it is fairly standard to recognize that equation (8) can be given an equivalent LMI formulation in the unknown P^{-1}.

3   Static output feedback stabilization for nonlinear systems

In this section we consider a nonlinear system described by equations of the form

$$\dot{x} = f(x) + g(x)u \tag{9}$$
$$y = h(x) \tag{10}$$

where $x \in \mathbb{R}^n$ denotes the state of the system, $u \in \mathbb{R}^m$ the control input, $y \in \mathbb{R}^p$ the measured output, and the mappings $f(x)$, $g(x)$ and $h(x)$ are smooth mappings defined in a neighborhood of the origin of $\mathbb{R}^n$. Moreover, we also assume that $x = 0$ is an equilibrium point, *i.e.* $f(0) = 0$ and $h(0) = 0$. In order to describe the main results of the section we need the following definitions.

Definition 1. The pair $\{f, h\}$ is said to be locally detectable (observable) if there exists a neighborhood U of the point $x = 0$ such that, if $x(t)$ is any integral curve of $\dot{x} = f(x)$ satisfying $x(0) \in U$, then $h(x(t))$ is defined for all $t \geq 0$ and $h(x(t)) = 0$ for all $t \geq 0$ implies $\lim_{t \to \infty} x(t) = 0$ ($x(t) = 0$ for all $t \geq 0$).

Definition 2. Given a smooth mapping $y = h(x)$, we denote with[4] $\ker(h)$ the set of all x such that $y = 0$.

We are now ready to present the main result of this section, which is a (partial) nonlinear counterpart of Theorem 1.

Theorem 2. *Consider the system (9)-(10) and assume that the pair $\{f, h\}$ is locally detectable and that* $\operatorname{rank}(dh(0)) = p$. *Suppose moreover that there exist a scalar function $V(x) \in C^1$, positive definite in a neighborhood of the origin, and a $m \times 1$ matrix function $G(x) \in C^1$ such that*

$$0 = V_x(x)f(x) - \frac{1}{4}V_x(x)g(x)g'(x)V_x'(x) + h'(x)h(x) + G'(x)G(x) \tag{11}$$
$$0 = V_x f(x), \qquad \forall\, x \in Ker(h). \tag{12}$$

---
[4] This is also denoted with $h^{-1}(0)$, see *e.g.* [11].

Then there exists a orthogonal matrix $T(x) \in C^1$ such that the function

$$\eta(x) = T(x)G(x) - \frac{1}{2}g'(x)V_x'(x) \tag{13}$$

is such that

(i) for all $x \in Ker(h)$

$$\eta(x) = 0; \tag{14}$$

(ii) the system $\dot{x} = f(x) + g(x)\eta(x)$ is locally asymptotically stable;

(iii) the trajectories $x(t)$ of the system $\dot{x} = f(x) + g(x)\eta(x)$ starting close to the origin are such that the output $y(t) = h(x(t))$ and the control $u(t) = \eta(x(t))$ are square integrable signals.

Moreover, if

$$p \geq k = rank(\frac{\partial \eta(x)}{\partial x}) \quad \forall x \in \Omega \tag{15}$$

for some constant k and some neighborhood Ω of $x = 0$, then

(iv) in a neighborhood of the origin, $\eta(x)$ is a function of y, i.e. $\eta(x) = \phi(y)$ for some smooth function $\phi(\cdot)$.

Proof. Point (i). By equations (11) and (12) it follows

$$\frac{1}{4}V_x(x)g(x)g'(x)V_x'(x) = G'(x)G(x), \quad \forall \, x \in Ker(h).$$

This means that there exists a smooth orthogonal matrix $T(x)$ such that

$$T(x)G(x) = \frac{1}{2}g'(x)V_x'(x), \quad \forall \, x \in Ker(h).$$

Point (ii). By equation (13), the Hamilton-Jacobi equation (11) can be rewritten as

$$0 = V_x(x)(f(x) + g(x)\eta(x)) + h'(x)h(x) + \eta'(x)\eta(x) \tag{16}$$

which, by detectability of the pair $\{f, h\}$, implies claim (ii).

Point (iii). Equation (16) can be rewritten as

$$\dot{V} = -h'(x(t))h(x(t)) - \eta'(x(t))\eta(x(t)).$$

Hence, by local asymptotic stability of the closed-loop system, one has

$$V(x(0)) = \int_0^\infty \|h(x(t))\|^2 + \|\eta(x(t))\|^2 dt,$$

which establishes the claim.

Point (iv). This claim follows from the assumptions, equation (15) and the rank theorem, see [11, Appendix 3].

The sufficient conditions in Theorem 2 are the nonlinear equivalent of the sufficient conditions in Theorem 1. They are obviously more involved, as they require the solution of a Hamilton-Jacobi equation. Moreover, in the linear case, for any fixed matrix G there exists a (unique) matrix $P = P' \geq 0$ solving the Riccati equation (6), whereas this fact is not in general true for the Hamilton-Jacobi equation (11). However, the nonlinear formulation allows to replace the Hamilton-Jacobi equation (11) with a Hamilton-Jacobi inequality involving only one unknown, as detailed in the following statement.

Corollary 2. *Assume that there exists a scalar function $W(x) \in C^1$, positive definite in a neighborhood of the origin, such that*

$$0 \geq W_x(x)f(x) - \frac{1}{4}W_x(x)g(x)g'(x)W_x'(x) + h'(x)h(x) \qquad (17)$$

$$0 = W_x f(x), \quad \forall \, x \in Ker(h). \qquad (18)$$

Then, there exist a (nonunique) $m \times 1$ matrix function $G(x) \in C^1$ and a scalar function $V(x) \in C^1$, positive definite in a neighborhood of the origin, such that equations (11) and (12) are satisfied.

Proof. The claim is trivially obtained setting $V(x) = W(x)$ and $G(x)$ such that

$$G'(x)G(x) = -V_x(x)f(x) + \frac{1}{4}V_x(x)g(x)g'(x)V_x'(x) - h'(x)h(x) \geq 0.$$

Observe that Theorem 2 admits the following partial converse.

Theorem 3. *Consider the system (9)-(10) and assume that the pair $\{f, h\}$ is locally observable. Assume moreover that there exists a continuous function $\phi(y)$, with $\phi(0) = 0$, such that*

(a) $\dot{x} = f(x) + g(x)\phi(y)$ is locally asymptotically stable;
(b) the trajectories $x(t)$ of the system $\dot{x} = f(x) + g(x)\phi(y)$ starting close to the origin are such that the output $y(t) = h(x(t))$ and the control $u(t) = \phi(y(t))$ are square integrable signals.

Then there exist a scalar function $V(x) \in C^1$, positive definite in a neighborhood Ω of the origin, a $m \times 1$ continuous function $G(x)$, and an orthogonal matrix $T(x) \in C^1$ such that

(i) $0 = V_x(x)f(x) - \frac{1}{4}V_x(x)g(x)g(x)'V_x(x) + h'(x)h(x) + G'(x)G(x)$

(ii) $0 = V_x(x)f(x), \quad \forall \, x \in Ker(h)$

(iii) $0 = T(x)G(x) - \frac{1}{2}g'(x)V_x(x), \quad \forall \, x \in Ker(h)$

(iv) $p \geq rank(\dfrac{\partial \phi}{\partial x}), \qquad \forall x \in \Omega.$

Proof. Given an initial condition $\bar{x} \in \Omega$, let $x(t)$ be the corresponding state trajectory of the system $\dot{x} = f(x) + g(x)\phi(y)$ and define

$$V(\bar{x}) = \int_0^\infty \|h(x(t))\|^2 + \|\phi(y(t))\|^2 \, dt.$$

By assumption (b) and the observability of the pair $\{f, h\}$, we conclude that $V(\bar{x})$ is positive definite and differentiable in Ω. Moreover, it satisfies equation (16) with $\eta(x)$ replaced by $\phi(h(x))$. Set now $G(x) = \phi(h(x)) + \frac{1}{2}g'(x)V_x'$ and $T(x) = I$ and observe that points (i)-(iii) are trivially satisfied. Finally, claim (iv) follows directly from $\phi(y)$ being a function of y only.

4   A new characterization for linear systems

In this section we exploit the results established in Section 3 to derive a new necessary and sufficient condition for the solvability of the output feedback stabilizability problem for system (1)-(2). This characterization is given in term of the intersection of two convex sets subject to a (generically) noncon-vex coupling condition. It is worth noting that the proposed condition heavily stems from the nonlinear control design methods established in the previous section. Moreover, in general, the resulting output feedback controller is non-linear, but (under simple regularity assumptions) a linear feedback can be computed.

Theorem 4. *Consider the system (1)-(2) with Assumptions (A1), (A2) and (A3). The system is output feedback stabilizable if and only if there exist two symmetric positive definite matrices X and P such that*

$$0 \le \begin{bmatrix} -XA' - AX + BB' & XC' \\ CX & I \end{bmatrix} \tag{19}$$

$$0 = V(A'P + PA)V \tag{20}$$

$$I = PX. \tag{21}$$

Proof. (*Only if*) Assume that the output feedback stabilization problem for system (1)-(2) is solvable. Then, by the necessity part of Theorem 1 there exist a symmetric positive semidefinite matrix P and a matrix G such that such that equations (6) and (7) are satisfied. Moreover, $P > 0$ by Assumption (A2) and observe that equations (7) and (20) are the same. Then, setting $X = P^{-1}$ (*i.e.* equation (21)), one can easily show that equation (6) implies the inequality (19).

(*If*) Assume now that there exist $P > 0$ and $X > 0$ such that (19), (20) and (21) hold. Then conditions (17) and (18) are satisfied with $W(x) = x'Px$,

$f(x) = Ax$, $g(x) = B$ and $h(x) = Cx$, and by Corollary 2 there exist a positive definite function $V(x) \in C^1$ and a matrix $G(x) \in C^1$ such that conditions (11) and (12) are also satisfied. Moreover, by Theorem 2 there exists an orthogonal matrix $T(x) \in C^1$ such that the function

$$\eta(x) = T(x)G(x) - \frac{1}{2}g'(x)V_x'(x) = T(x)G(x) - B'Px$$

satisfies conditions (i), (ii) and (iii) in Theorem (2).

Consider now the the $(m \times n)$-dimensional matrix

$$D\eta(\bar{x}) = \left.\frac{\partial \eta}{\partial x}\right|_{x=\bar{x}},$$

and let

$$\kappa^* = \max_{x \in \Omega} \mathrm{rank}(D\eta(x))$$

where Ω is a neighborhood of the origin. Note that, by Assumption (A3),

$$\kappa^* \le m \le p,$$

and there exists a point x_ϵ arbitrarily close to $x = 0$ such that[5]

(a) $\eta(x_\epsilon) = 0$;
(b) $\mathrm{rank}(D\eta(x)) = \kappa < \kappa^*$ for all x in a neighborhood B_ϵ of x_ϵ.

By the Rank Theorem (see [11]) and condition (i) of Theorem (2), in a neighborhood $\tilde{B}_\epsilon \subset B_\epsilon$ of x_ϵ the function $\eta(x)$ is such that

$$\eta(x) = \phi(Cx) = \phi(y). \tag{22}$$

To conclude the proof observe that by condition (ii) of Theorem (2) and equation (16) the function $\eta(x)$ is such that

$$\dot{x} = Ax + B\eta(x)$$

is locally exponentially stable. As a result the matrix $A + BD\eta(0)$ has all eigenvalues with negative real, and by a simple continuity argument also the matrix

$$A + BD\eta(x_\epsilon)$$

has all eigenvalues with negative real, for any x_ϵ in $\tilde{B}_\epsilon \cap \ker(Cx)$. However, for any point x in $\tilde{B}_\epsilon \cap \ker(Cx)$ one has

$$D\eta(x) = \left.\frac{\partial \phi}{\partial y}\right|_{y=0} C$$

---

[5] As a matter of fact, it is possible that all points x_ϵ close to zero and such that $\eta(x_\epsilon) = 0$ are singular points of $\eta(x)$, i.e. the rank of $\eta(x)$ is nonconstant in any neighborhood of such points. In this case it is still possible to draw the same conclusions but with more involved considerations.

where $\phi(y)$ is defined in equation (22), which proves that

$$F = \left.\frac{\partial \phi}{\partial y}\right|_{y=0}$$

is a solution of the static output feedback stabilization problem.

Remark 1. Notice that the set of all X satisfying the condition (19) in Theorem 4 is convex, and it is also convex the set of all P satisfying condition (20). Unfortunately, the coupling condition (21) is not convex. We conclude that the problem is in general nonconvex.

Remark 2. The result summarized in Theorem 4 can be easily extended to the case where not only stability but also an H_∞ performance bound is taken into account. For, consider a system described by the equations

$$\dot{x} = Ax + Bu + \bar{B}w \tag{23}$$

$$z = \begin{bmatrix} Cx \\ u \end{bmatrix} \tag{24}$$

and such that Assumptions (A1), (A2) and (A3) hold. Then it is possible to extend the results in Theorems 1 and 4 to prove that there exists a static output feedback control law $u = Fy$ such that the closed-loop system is asymptotically stable and the H_∞ norm from the input w to the output z is less then a prespecified level γ if and only if there exist two symmetric positive definite matrices X and P such that

$$0 \leq \begin{bmatrix} -XA' - AX + BB' - \dfrac{\bar{B}\bar{B}'}{\gamma^2} & XC' \\ CX & I \end{bmatrix} \tag{25}$$

$$0 = V(A'P + PA + \frac{P\bar{B}\bar{B}'P}{\gamma^2})V \tag{26}$$

$$I = PX. \tag{27}$$

5 Further results and special cases

In this section we focus on linear systems and propose further interesting characterizations of the SOF stabilization problem. Moreover, the SOF stabilization problem for a class of systems with special structure is addressed.

5.1 SOF stabilization, state feedback and output injection

In the introduction it has been noted that the problem of dynamic output feedback stabilization for system (1)-(2) can be recast as a SOF stabilization problem for an extended dynamical system. Moreover, the separation principle implies that dynamic output feedback stabilization problems can be solved if and only if stabilization through state feedback and through output injection can be achieved. The connection between static and dynamic output feedback stabilization can be strengthened, as discussed in the following statement.

Theorem 5. *Consider the system (1)-(2).*

The SOF stabilization problem is solvable if and only if there exist matrices (of appropriate dimensions) K, L and $P = P' > 0$ such that

$$\begin{aligned}
(A + BK)'P + P(A + BK) &< 0 \\
(A + LC)'P + P(A + LC) &< 0.
\end{aligned} \tag{28}$$

Remark 3. The conditions in Theorem 5 can be interpreted as follows. There exists a state feedback gain K and an output injections gain L such that the systems $\dot{x} = (A + BK)x$ and $\dot{\xi} = (A + LC)\xi$ are asymptotically stable and possess a common Lyapunov function. This is equivalent to say the SOF stabilization problem is solvable if and only if it is possible to solve a state feedback stabilization problem and an output injection stabilization problem in a way that the corresponding closed loop systems have a common Lyapunov function.

Proof. (Only if) Assume that the system (1)-(2) is static output feedback stabilizable and F is a solution to the problem. Then there exists $X = X' > 0$ such that

$$(A + BFC)'X + X(A + BFC) < 0.$$

As a result, $K = FC$, $L = BF$ $P = X$ are such that equations (28) hold.

(If) Assume equations (28) hold. Then, the result follows trivially from Theorem 3.8 in [19].

5.2 The *dual* of Theorem 4

Theorem 4 has been derived from Theorem 1 substituting the first equality, namely equation (6), with an inequality and removing the *unknown* G. As already remarked, the necessity part of Theorem 4 is a straightforward consequence of Theorem 1, whereas to prove the sufficiency we had to consider the (more general) nonlinear problem discussed in Section 3. However, if the idea of removing the unknown G from equation (6) and substituting the equality with an inequality is very natural, and indeed proves to be of interest, also the *dual* approach, namely the substitution of equation (7) with an inequality might be appealing in some applications. In particular, whenever the system to be stabilized is such that there exists a positive definite matrix P with the property that

$$A'P + PA \leq 0,$$

i.e. the system is marginally stable, condition (7) is automatically fulfilled if the equality sign is changed into an inequality. To make use of this *a priori* knowledge on the system it is necessary to derive a new characterization for the SOF stabilization problem, as discussed in the next statement.

Theorem 6. *Consider the system (1)-(2) with Assumptions (A1), (A2) and (A3). The system is output feedback stabilizable if and only if there exist a symmetric positive semidefinite matrix $P \in \mathbb{R}^{n \times n}$ and a matrix $G \in \mathbb{R}^{m \times n}$ such that*

$$0 = A'P + PA - PBB'P + C'C + G'G \tag{29}$$
$$0 \geq V(A'P + PA)V, \tag{30}$$

where V is as in Theorem 1.

5.3   Minimum phase systems with relative degree one

Consider the class of single-input single-output linear systems with relative degree one and asymptotically stable zero dynamics. For this class of systems a simple root locus reasoning shows that asymptotic stability can be achieved by means of high gain static output feedback. Goal of this section is to show how it is possible to recover this result from the theory presented in this work.

To begin with, note that the system matrices of a single-input single-output linear system with relative degree one can be written, in a proper set of coordinates and after a preliminary static output feedback, as

$$A = \begin{bmatrix} F & g \\ h & 0 \end{bmatrix} \qquad B = \begin{bmatrix} 0 \\ 1 \end{bmatrix} \qquad C = \begin{bmatrix} 0 & 1 \end{bmatrix}. \tag{31}$$

Moreover, by asymptotic stability of the zero dynamics $\sigma(F) \subset \mathbb{C}^-$. We are now ready to state the main result of this section.

Proposition 1. *Consider the system (1)-(2) and assume that $p = m = 1$ and that A, B and C are as in (31). Suppose moreover that the system has an asymptotically stable zero dynamics. Then the system is (obviously) asymptotic stabilizable by static output feedback.*

Moreover, let

$$P_0 = \begin{bmatrix} \tilde{P}_0 & h' \\ h & 0 \end{bmatrix}$$

with $\tilde{P}_0 = \tilde{P}_0' > 0$ *such that*

$$F'\tilde{P}_0 + \tilde{P}_0 F + h'h = 0,$$

and

$$P_1 = \begin{bmatrix} 0 & 0 \\ 0 & 1/2 \end{bmatrix}$$

then conditions (6) and (7) of Theorem 1 hold for some positive definite matrix P such that

$$\lim_{\alpha \to 0} P - P_0 - \frac{P_1}{\alpha} = 0$$

and some matrix G such that

$$\lim_{\alpha \to 0} G - \frac{1}{\alpha} [0 \quad 1/2] = 0.$$

Finally, a stabilizing output feedback gain is, for α sufficiently small

$$F = -\frac{1}{\alpha}$$

and an estimate of the minimum value of $\frac{1}{\alpha}$ required to achieve closed-loop stability is

$$\frac{1}{\alpha^\star} = 2\sigma_{\max}(h\tilde{P}_0^{-1}h').$$

We omit the proof of the above result which is conceptually simple. However, we point out that the result expressed in Proposition 1, which strongly rely on a particular selection of coordinates, can be given an equivalent coordinates free formulation. Furthermore, a similar result can be obtained for systems with asymptotically stable zero dynamics and relative degree two and for multivariable *square* systems. The extension to nonlinear systems is currently under investigation, but it must be noted that some interesting results in this direction can be found in [4,5].

6   SISO linear systems, obstructions and convexity properties

In this section we restrict our interest to linear systems described by equations of the form (1)-(2) with $p = m = 1$ and satisfying Assumption (A1) and (A2). Without loss of generality it is possible to write the system in the observability canonical form, *i.e.* with

$$A = \begin{bmatrix} 0 & 0 & \cdots & 0 & -a_n \\ 1 & 0 & \cdots & 0 & -a_{n-1} \\ \vdots & \vdots & \ddots & \vdots & \vdots \\ 0 & 0 & \cdots & 0 & -a_2 \\ 0 & 0 & \cdots & 1 & -a_1 \end{bmatrix}, \tag{32}$$

and

$$C = [\,0\ 0\ \cdots\ 0\ 1\,].$$ (33)

This special form allows to get more insight into the structure of the problem, as described in the following statement, where, for convenience, we define the column vector e_i as the $i - th$ vector of the identity matrix of dimension n and the set

$$\mathcal{X} = \{X = X' \mid X_{ij} = 0 \text{ if } i + j \text{ is odd}\}.$$

Theorem 7. *Consider the system (1)-(2) with A and C as in equations (32) and (33). The system is output feedback stabilizable only if there exists a positive definite matrix X satisfying the following conditions.*

(i)

$$0 \le \begin{bmatrix} -XA' - AX + BB' & XC' \\ CX & I \end{bmatrix};$$ (34)

(ii)

$$e_i'Xe_n = 0, \quad i = n-1,\ n-3,\ n-5,\ \cdots$$ (35)

(iii) $X \in \mathcal{X}$;
(iv) the polynomial

$$e_n'Xe_n\lambda^{n-1} + e_{n-2}'Xe_n\lambda^{n-3} + e_{n-4}'Xe_n\lambda^{n-5} + \cdots$$ (36)

has distinct roots all on the imaginary axis.

Proof. The proof is technical and it is omitted for brevity.

Remark 4. Observe that condition (iii) of Theorem 7 implies condition (ii). They have been both included since they express obstructions of diverse complexity and nature.

Theorem 7 provides an obstruction to the solvability of the static output feedback stabilization problem. It can be used in steps of increasing complexity. First condition (i) is checked. If it is feasible, then conditions (i) and (ii) can be simultaneously checked. If they are feasible, condition (iii) can be added, and finally conditions (i) to (iv) has to be checked together. If this procedure fails, then no solution exists. However, if the procedure works, no conclusion can be drawn.

It is interesting to note that conditions (i) to (iii) in Theorem 6 are convex, whereas condition (iv) is in general nonconvex. However, for low order systems the obstruction expressed in Theorem 7 can be easily checked, as formalized in the following statement.

Corollary 3. *Consider the system (1)-(2) with the matrices A and C as in equations (32) and (33).*

If $n \leq 2$ the set of all $X > 0$ satisfying conditions (i) to (iv) of Theorem 7 is convex and conditions (i) to (iv) of Theorem 7 are also sufficient.

If $n \leq 3$ and $B = C'$ the set of all $X > 0$ satisfying conditions (i) to (iv) of Theorem 7 is convex and conditions (i) to (iv) of Theorem 7 are also sufficient.

If $n \leq 4$ the set of all $X > 0$ satisfying conditions (i) to (iv) of Theorem 7 is convex.

If $n = 5$ or $n = 6$ the conditions (i) to (iv) of Theorem 7 can be recast in a convex optimization problem, i.e. as a quadratic optimization with linear constraints.

7   A few illustrative examples

In this section we discuss a few simple examples. Aim of the section is to show how the general results developed so far can be easily used in deciding the solvability of SOF stabilization problems for linear and nonlinear systems. In particular, for linear low dimensional systems, Theorem 7 provides a very simple test that can be also performed with *pen and paper*. For nonlinear systems, the sufficient conditions expressed in Theorem 2 is well suited to study conservative systems, *i.e.* systems which are Lyapunov stable and/or possess conserved quantities.

Example 1. Consider a linear 2-dimensional system in observability canonical form with

$$A = \begin{bmatrix} 0 & -3 \\ 1 & 4 \end{bmatrix}, \quad B = \begin{bmatrix} -5 \\ 1 \end{bmatrix}, \quad C = \begin{bmatrix} 0 & 1 \end{bmatrix}. \tag{37}$$

and observe that

$$V = \begin{bmatrix} 1 & 0 \\ 0 & 0 \end{bmatrix}.$$

This system satisfies the Parity Interlacing Property (PIP) [19], *i.e.* the property that guarantees the existence of a stable stabilizer, and satisfies also the Extended PIP [19], *i.e.* the property that guarantees the existence of a stable stabilizer with a stable inverse. Therefore, both the PIP and the EPIP do not give an obstruction to the solvability of the SOF stabilization problem.

A simple application of the necessary conditions in Theorem 7 yields the following conclusions. The system passes the first test, *i.e.* there exists a positive definite matrix X such that condition (34) holds. This is not surprising, as the test expressed by condition (34) is passed by any controllable and observable system. However, the system does not pass the second test, which requires the matrix X to be diagonal, $i.e.$ there exists no positive definite diagonal matrix X such that condition (34) holds. We conclude that the system is not stabilizable by static output feedback.

Remark 5. It is worth noting that, as the system satisfies both the PIP and the EPIP there exist stable stabilizers and also stable stabilizers with stable inverse. However, the order of these stabilizers may be large, in general. For the considered example it is easy to prove that there exists a stable stabilizer of order one, yet there is no stable stabilizer with stable inverse of order less then or equal to two.

Example 2. Consider a linear 2-dimensional system in observability canonical form with

$$A = \begin{bmatrix} 0 & 4 \\ 1 & -3 \end{bmatrix}, \quad B = \begin{bmatrix} -4 \\ 1 \end{bmatrix}, \quad C = \begin{bmatrix} 0 & 1 \end{bmatrix}. \tag{38}$$

Observe that

$$V = \begin{bmatrix} 1 & 0 \\ 0 & 0 \end{bmatrix},$$

and that the necessary conditions of Theorem 7 require the matrix X to be diagonal, $i.e.$

$$X = \begin{bmatrix} X_1 & 0 \\ 0 & X_3 \end{bmatrix}.$$

This implies that the sufficient conditions of Theorem 4 has to be tested with X and P diagonal. Simple calculations show that the set of all positive definite

$$P = \begin{bmatrix} P_1 & 0 \\ 0 & P_3 \end{bmatrix} = \begin{bmatrix} 1/X_1 & 0 \\ 0 & 1/X_3 \end{bmatrix}$$

satisfying condition (19) is described by

$$0 < 64P_1^2 P_3 - 32P_1^2 - P_3^2 - 8P_1 P_3 - 8P_1 P_3^2$$
$$0 < P_1$$
$$0 < P_3,$$

whereas condition (20) is trivially fulfilled if P is a diagonal matrix. The set of all feasible P, $i.e.$ the set of all P fulfilling condition (19) (with $X = P^{-1}$) and condition (20), is illustrated in Figure 1, which also shows that this set is convex.

To construct the stabilizing gain F we follow the procedure outlined in the proof of Theorem 2. For, we select an admissible pair $\{P_1, P_3\}$, $e.g.$ $\{1, 1\}$, construct the function $G(x)$, and the scalar $T(x)$. As a result, we obtain the nonlinear *state feedback* control law

$$\eta(x) = G(x) - 4x_1 + x_2$$

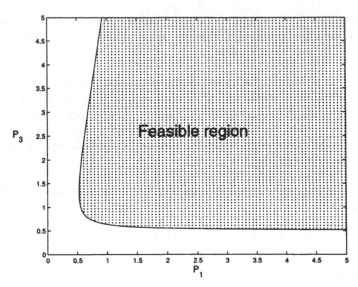

Fig. 1. The set of all pair $\{P_1, P_3\}$ satisfying condition (19) (with $X = P^{-1}$) and condition (20).

yielding local asymptotic stability for the closed-loop system. To compute the output feedback gain we have to differentiate the function $\eta(x)$ at some point close to the origin and such that $Cx = 0$. If the pair $\{P_1, P_3\} = (1, 1)$ is selected the corresponding output feedback gain is $F = 1.45$, which is indeed a stabilizing gain.

In this example it is possible to perform the calculations of the stabilizing gain in a parametric form, *i.e.* without selecting a priori a value for the pair $\{P_1, P_3\}$. This general procedure yields a family of stabilizing gains, which is described by the equation

$$F = F(P_3) = \sqrt{6P_3 - 1 + P_3^2} - P_3,$$

where, as can be seen from Figure 1 and verified with very simple calculations, $P_3 \in (1/2, \infty)$. For $P_3 \in (1/2, \infty)$, the function $F(P_3)$, depicted in Figure 2, takes value in the set[6] $(1, 3)$, and a very simple root locus analysis (see Figure 3) reveals that these are indeed all the static output feedback gains yielding closed-loop stability.

Example 3. Consider a rigid body in an inertial reference frame and let x_1, x_2 and x_3 denote the angular momentum components along a body fixed reference frame having the origin at the center of gravity and consisting of

---

[6] Note that $\lim_{P_3 \to \infty} F(P_3) = 3$.

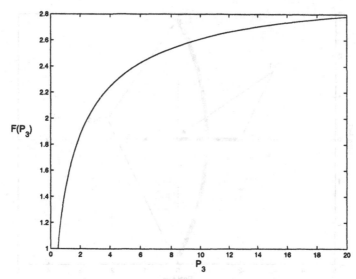

Fig. 2. The static output feedback gain $F = F(P_3)$ as a function of P_3, for $P_3 \in (1/2, 20)$.

the three principal axes. The Euler's equations for the rigid body with two independent controls are

$$\dot{x}_1 = \left(\frac{1}{I_3} - \frac{1}{I_2} \right) x_2 x_3 + u_1$$

$$\dot{x}_2 = \left(\frac{1}{I_1} - \frac{1}{I_3} \right) x_3 x_1 + u_2 \qquad (39)$$

$$\dot{x}_3 = \left(\frac{1}{I_2} - \frac{1}{I_1} \right) x_1 x_2 + \alpha u_1 + \beta u_2$$

where $I_1 > 0$, $I_2 > 0$ and $I_3 > 0$ denote the principal moments of inertia, $u = \mathrm{col}(u_1, u_2)$ the control torques, and α and β are two constant numbers depending on the location of the actuators. The only variables available for feedback are

$$\begin{aligned} y_1 &= c_{11}x_1 + c_{12}x_2 + c_{13}x_3 \\ y_2 &= c_{21}x_1 + c_{22}x_2 + c_{23}x_3 \end{aligned} \qquad (40)$$

where the c_{ij} are constant numbers which reflect the position of the sensing devices.

The problem of stabilization using dynamic feedback (either observer based or not) has been addressed in [2,16]. We now focus on the static output feedback stabilization problem and we show how it is possible to make use

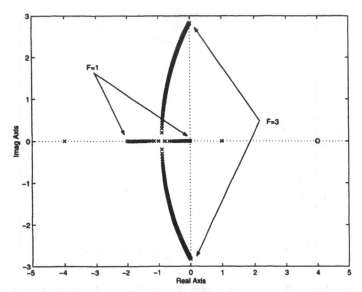

Fig. 3. The root locus of the system (38) for $F \in (1,3)$. Observe that for $F = 1$ and $F = 3$ the eigenvalues of the closed-loop system are on the imaginary axis.

of the result in Theorem 2. The point of departure is the consideration that for system (39) there exist two positive definite functions which are natural candidates to test the conditions in Theorem 2, namely the kinetic energy, *i.e.*

$$K(x) = \frac{1}{2} \left(\frac{x_1^2}{I_1} + \frac{x_2^2}{I_2} + \frac{x_3^2}{I_3} \right)$$

and the modulo of the angular momentum vector, *i.e.*

$$P(x) = \frac{1}{2} \left(x_1^2 + x_2^2 + x_3^2 \right).$$

In fact, simple calculations (which can be avoided by anyone with a basic knowledge of Newton's laws) show that along the trajectories of the system (39) with $u_1 = u_2 = 0$ one has

$$\dot{K} = \dot{P} = 0.$$

As a result the function

$$V(x) = \lambda P(x) + \mu K(x)$$

satisfies equation (12) for any λ and μ.

The next step is to verify if it is possible to find constants λ and μ such that the function $V(x)$ is positive definite and equation (11) in Theorem

2 is satisfied, for some $G(x)$. For the sake of brevity we omit the details of the discussion and present only the main result, the proof of which is a consequence of the above considerations.

Proposition 2. *Consider the system (39) with output (40). Suppose that*

$$\operatorname{rank} \begin{bmatrix} c_{11} & c_{12} & c_{13} \\ c_{21} & c_{22} & c_{23} \end{bmatrix} = 2.$$

The system is asymptotically stabilizable via static output feedback if

$$\{(\lambda, \mu) \in \operatorname{Ker} J\} \cap_{i=1,2,3} \{(\lambda, \mu) \mid \lambda + \frac{\mu}{I_i} > 0\} \neq \emptyset,$$

where

$$J = \begin{bmatrix} (\beta C_{12} - C_{13}) I_2 I_3 & \beta C_{12} I_2 - C_{13} I_3 \\ (\alpha C_{12} + C_{23}) I_1 I_3 & \alpha C_{12} I_1 + C_{23} I_3 \end{bmatrix}$$

and

$$C_{ij} = \det \begin{bmatrix} c_{1i} & c_{1j} \\ c_{2i} & c_{2j} \end{bmatrix}.$$

Remark 6. The condition expressed in Proposition 2 can be given a simple geometrical interpretation, as shown if Figure 4. In the space (λ, μ) the set

$$E = \cap_{i=1,2,3} \{(\lambda, \mu) \mid \lambda + \frac{\mu}{I_i} > 0\}$$

is a convex set, *i.e.* the shaded area in Figure 4, whereas the set

$$\mathcal{J} = \{(\lambda, \mu) \in \operatorname{Ker} J\}$$

is either a straight line, if $\det J = 0$, or the singleton $(\lambda, \mu) = (0, 0)$, if $\det J \neq 0$. As a result, the sufficient condition in Proposition 2 could be easily tested.

Remark 7. It is worth noting that, if $\det J = 0$ and $\operatorname{rank} C = 2$, the condition expressed in Proposition 2 is necessary and sufficient, *i.e.* the static output feedback stabilization problem for system (39) with output (40) is solvable if and only if

$$\mathcal{J} \cap E \neq \emptyset.$$

The sufficient part is a consequence of Proposition (2), whereas to prove the necessity two cases has to be considered, *i.e.* the case[7] $\mathcal{J} \cap (\partial E / \{0, 0\}) \neq \emptyset$ and the case $\mathcal{J} \cap E = \{0, 0\}$. In the former, the system can be nonasymptotically stabilized by static output feedback, whereas in the latter a simple application of classical instability theorems shows that for any (static) output feedback gain the closed-loop system is unstable.

---

[7] The notation ∂E is used to denote the boundary of the set E.

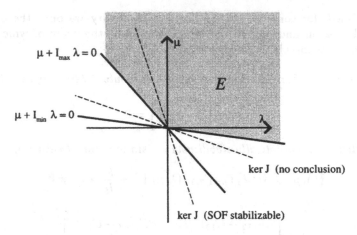

Fig. 4. Graphical representation of the sets E and $\mathcal{J}$. If $\mathcal{J} = \ker J$ intersects the shaded region then the stabilization problem is solvable, otherwise the sufficient condition does not yield any conclusion.

8   Concluding remarks

The problem of static output feedback stabilization for linear and nonlinear systems has been studied. For linear systems a new necessary and sufficient condition has been proposed. Moreover, for single-input single-output linear systems a set of simple to test necessary conditions have been derived. These necessary conditions, which can be recast in terms of *convex* conditions for low dimensional systems, can be exploited to decide the unsolvability of the problem.

A sufficient condition for nonlinear control affine systems has also been developed, together with a partial converse. It is worth noting that the proof of the sufficient condition for nonlinear systems is instrumental to develop the new characterization for linear systems.

The theoretical part is complemented with a few worked out examples. Future work will be directed toward the study of multi-input multi-output linear systems, to establish to what extent the necessary condition in Theorem 7 is also sufficient, to provide a description of all stabilizing gains, and to work out further nonlinear physically motivated examples.

References

1. B.D.O. Anderson and J.B. Moore, "Time-varying feedback laws for decentralized control" *IEEE Trans. Autom. Control*, Vol. 26, pp. 1133-1139, 1981.

2. A. Astolfi, "Output feedback control of the angular velocity of a rigid body", *Systems and Control Letters*, Vol. 36, pp. 181-192, 1999.
3. D. Aeyels and J.L. Willems, "Pole assignment for linear time-invariant systems by periodic memoryless output feedback", *Automatica*, Vol. 28, pp. 1159-1168, 1992.
4. C.I. Byrnes and A. Isidori, "Asymptotic stabilization of minimum phase nonlinear systems", *IEEE Trans. Autom. Control*, Vol. 36, pp. 1122-1137, 1991.
5. C.I. Byrnes, A. Isidori and J.C. Willems, "Passivity, feedback equivalence, and the global stabilization of minimum phase nonlinear systems", *IEEE Trans. Autom. Control*, Vol. 36, pp. 1228-1240, 1991.
6. L. El Ghaoui, F. Oustry and M. AitRami, "A cone complementarity linearization algorithm for static output-feedback and related problems", *IEEE Transactions on Automatic Control*, Vol. 42, pp. 1171-1176, 1997.
7. M.I. Freedman, "L_2 stability of time-varying systems - construction of multipliers with prescribed phase characteristics", *Int. J. Control*, Vol. 6, N. 4, pp. 559-578, 1968.
8. J.C. Geromel, C.C. de Souza and R.E. Skelton, "Static output feedback controllers: stability and convexity", *IEEE Transactions on Automatic Control*, Vol. 43, pp. 120-125, 1998.
9. A. Feuer and G.A. Goodwin, "Generalized sampled-data functions: frequency domain analysis and robustness, sensitivity, and intersample difficulties", *IEEE Trans. Autom. Control*, Vol. 39, pp. 1042-1047, 1994.
10. G. Gu, "Stabilizability conditions of multivariable uncertain systems via output feedback", *IEEE Trans. Autom. Control*, Vol. 35, pp. 925-927, 1990.
11. A. Isidori, *Nonlinear Control Systems*, Springer Verlag, 1995.
12. T. Iwasaki and R. Skelton, "Parameterization of all stabilizing controllers via quadratic Lyapunov functions", *Journal Optim. Theory Applic.*, Vol. 77, pp. 291-307.
13. P.T. Kabamba, "Control of linear systems using generalized sampled-data functions", *IEEE Trans. Autom. Control.*, Vol. 32, pp. 772-783, 1987.
14. V. Kucera and C.C. De Souza, "A necessary and sufficient condition for output feedback stabilizability", *Automatica* Vol. 31, pp. 1357-1359, 1995.
15. B. Martensson, "The order of any stabilizing regulator is sufficient a priori information for adaptive stabilization", *Systems and Control Letters*, Vol. 6, pp. 87-91.
16. F. Mazenc and A. Astolfi, "Robust output feedback stabilization of the angular velocity of a rigid body", *Systems and Control Letters*, Vol. 39, pp. 203-210, 2000.
17. C. Nett, D. Bernstein and W. Haddad, "Minimal complexity control law synthesis, Part. I: problem formulation and reduction to optimal static output feedback", *Proc. American Control Conference*, Pittsburgh, (USA), pp. 2056-2064.
18. A.W. Olbrot, "Robust stabilization of uncertain systems by periodic feedback", *Int. J. Control*, Vol. 45, N. 3, pp. 747-758, 1987.
19. V.L. Syrmos, C.T. Abdallah, P. Dorato and K. Grigoriadis, "Static output feedback - A survey", *Automatica*, Vol. 33, 2, pp. 125-137, 1997.
20. K. Wei, "Stabilization of a linear plant via a stable compensator having no real unstable zeros", *Systems and Control Letters*, Vol. 15, pp. 259-264, 1990.
21. Y.-Y. Cao, J. Lam and Y.-X. Sun, "Static output feedback stabilization: an ILMI approach", *Automatica*, Vol. 34, pp. 1641-1645, 1998.
22. D. Youla, J. Bongiorno and C. Lu, "Single loop stabilization of linear multivariable dynamical systems", *Automatica*, Vol. 10, pp. 159-173.

Semi-linear Diffusive Representations for Nonlinear Fractional Differential Systems

Jacques Audounet[1], Denis Matignon[2*], and Gérard Montseny[3]

[1] MIP / CNRS
   Université Paul Sabatier
   118, route de Narbonne.
   F-31068 Toulouse Cedex 4, France
   audounet@mip.ups-tlse.fr
[2] Ecole Normale Suprieure des Tlcommunications
   TSI Dept. & CNRS, URA 820
   46, rue Barrault
   F-75634 Paris Cedex 13, France
   matignon@tsi.enst.fr - http://www.tsi.enst.fr/~matignon/
[3] LAAS / CNRS
   7, avenue du Colonel Roche
   F-31077 Toulouse Cedex 4, France
   montseny@laas.fr - http://www.laas.fr/gt-opd/

Abstract. The stability of non-linear fractional differential equations is studied. A sufficient stability condition on the non-linearity is given for the input-output stability, thanks to many different reformulations of the system using diffusive representations of dissipative pseudo-differential operators. The problem of asymptotic internal stability is analyzed by a more involved functional analytic method. Finally, a *fractional* version of the classical Hartman–Grobman theorem for hyperbolic dynamical systems of order 1 is conjectured and reformulated, based upon known necessary and sufficient stability conditions for linear fractional differential equations.

1   Statement of the problem

We are interested in the following problem involving a non-linear dynamics f and a state $x \in \mathbb{R}$ (for simplicity sake):

$$d^\alpha x(t) = f(x(t)) + u(t); \qquad x(0) = x_0, \tag{1}$$

where d^α is the Caputo regularized version of the so-called Riemann-Liouville fractional derivative, with $0 < \alpha < 1$; meaning $d^\alpha x = I^{1-\alpha}\dot{x} = Y_{1-\alpha} \star \dot{x}$, with causal kernel $Y_\beta(t) = \frac{1}{\Gamma(\beta)}t_+^{\beta-1}$ for the fractional integral operator I^β of order β.

---

* The author would like to thank Raphaël Krikorian, from Centre de Mathématiques at École Polytechnique for helpful discussions and useful references.

Problem (1) can be advantageously reformulated in the equivalent Abel–Volterra equation:

$$x(t) = x_0 + I^\alpha \left[f(x(t)) + u(t) \right].$$ (2)

Then, using $Y_\alpha \star Y_{1-\alpha} = Y_1$ the Heaviside unit step, (2) can also be written as:

$$x(t) = I^\alpha \left[f(x(t)) + u(t) + x_0 Y_{1-\alpha}(t) \right];$$ (3)

alternatively, with the help of the new variable $z(t) = x(t) - x_0$, and of the new function $\tilde{f}(z) = f(z + x_0)$, (2) can also be written as:

$$z(t) = I^\alpha \left[\tilde{f}(z(t)) + u(t) \right].$$ (4)

As far as stability is concerned, trying to use geometrical or standard analytical techniques (such as those used in the integer case, see [11]), i.e. trying to extend them to the *fractional* differential case is of little help, unfortunately; for the main reason that quadratic forms prove hard to fractionally differentiate, since the fractional derivative is intrinsically a *non-local* pseudo-differential operator.

On the contrary, using diffusive representations of pseudo-differential operators (see [8,6,2]) proves useful, in so far as the problem can first be reformulated into one (or many equivalent) way(s) that is classical, namely a first order in time diffusion equation, on an infinite-dimensional state-space endowed with an appropriate Hilbert structure. Quite standard energy methods (Lyapunov functionals, LaSalle invariance principle) can therefore be used.

The paper is organized as follows:

- in section 2, the problem is reformulated in equivalent ways with many advantages for the analysis; in particular stability properties are more easily examined in this context; a main comparison result is established;
- in section 3, the problem is examined with *null* initial condition $x_0 = 0$, it requires LaSalle invariance principle, and gives strong stability of the internal state;
- in section 4, the problem of the initial condition alone is addressed: it requires more specific analytical tools pertaining to the properties of the heat equation, the use of which will be sketched as closely as possible;
- finally in section 5 we will indicate some natural extensions of the results, either straightforward ($x \in \mathbb{C}$ or $x \in \mathbb{C}^n$, other diffusive pseudo-differential operators that are dissipative: $\mu > 0$), or that seem to be within reach but still need to be fully developped.

2   Diffusive formulations

In subsection 2.1, system (1) is transformed into a *diagonal* infinite-dimensional system with an extra variable $\xi > 0$, and a state $\psi(\xi, t)$. The *heat equation* formulation can be recovered as follows: first let $\xi = 4\pi^2 \eta^2$ with $\eta \in \mathbb{R}$, then perform the inverse Fourier transform in the space of tempered distributions, a *heat* equation is then obtained with an extra space variable y and a state $\varphi(y,t)$ in subsection 2.2.

2.1 Diagonal diffusive formulations

Output form In the scalar case, problem (4) is *equivalent* to (see [10]):

$$\partial_t \psi(\xi,t) = -\xi \, \psi(\xi,t) + \widetilde{f}(z(t)) + u(t) ; \quad \psi(.,0) = 0 \quad \xi > 0, \tag{5a}$$

$$z(t) = \int_0^\infty \mu_\alpha(\xi) \, \psi(\xi,t) \, d\xi ; \tag{5b}$$

where μ_α stands for *the* diffusive representation of the fractional integral operator I^α, that is: $\mu_\alpha(\xi) = \frac{\sin \alpha \pi}{\pi} \xi^{-\alpha}$.

The energy associated to this equation is:

$$E_\alpha(t) = \frac{1}{2} \int_0^\infty \mu_\alpha(\xi) \, |\psi(\xi,t)|^2 \, d\xi , \tag{6}$$

for which it is easily proved that the following equality holds:

$$\frac{dE_\alpha}{dt}(t) = -\int_0^\infty \xi \, \mu_\alpha(\xi) \, |\psi(\xi,t)|^2 \, d\xi + z(t) \, \widetilde{f}(z(t)) + z(t) \, u(t) . \tag{7}$$

The functional spaces to be used are: $\mathcal{H}_\alpha = L^2_{\mu_\alpha}(\mathbb{R}^+)$, $\mathcal{V}_\alpha = L^2_{(1+\xi)\,\mu_\alpha}(\mathbb{R}^+)$ and $\mathcal{V}_\alpha' = L^2_{(1+\xi)^{-1}\,\mu_\alpha}(\mathbb{R}^+)$, and $\mathcal{V}_\alpha \hookrightarrow \mathcal{H}_\alpha \hookrightarrow \mathcal{V}_\alpha'$ with continuous and dense injections.

Balanced form Let us denote $\nu_\alpha(\xi) = \sqrt{\mu_\alpha(\xi)}$, which is meaningful thanks to $\mu > 0$ *only*; then by a straightforward change on ψ, and a slight abuse of notations, we get:

$$\partial_t \psi(\xi,t) = -\xi \, \psi(\xi,t) + \nu_\alpha(\xi) \left[\widetilde{f}(z(t)) + u(t) \right] ; \quad \psi(.,0) = 0 \quad \xi > 0 \tag{8a}$$

$$z(t) = \int_0^\infty \nu_\alpha(\xi) \, \psi(\xi,t) \, d\xi . \tag{8b}$$

The energy associated to this equation is:

$$E(t) = \frac{1}{2} \int_0^\infty |\psi(\xi,t)|^2 \, d\xi , \tag{9}$$

The functional spaces to be used are: $\mathcal{H} = L^2(\mathbb{R}^+)$, $\mathcal{V} = L^2_{(1+\xi)}(\mathbb{R}^+)$ and $\mathcal{V}' = L^2_{(1+\xi)^{-1}}(\mathbb{R}^+)$. They are *independent* of α.

2.2   Heat equation formulations

Now, tempered distributions will be used: $M_\alpha(y)$, with Fourier transform $m_\alpha(\eta) = 2 \sin(\alpha\pi) |2\pi\,\eta|^{1-2\alpha}$ for the output form, and $N_\alpha(y)$, with Fourier transform $n_\alpha(\eta) = \sqrt{m_\alpha(\eta)}$ for the balanced form. It is clear that, for $\frac{1}{2} < \alpha < 1$ both $M_\alpha(y) \propto |y|^{-2(1-\alpha)}$ and $N_\alpha(y) \propto |y|^{-(\frac{3}{2}-\alpha)}$ are regular L^1_{loc} functions; for $\alpha = \frac{1}{2}$, they are proportional to the Dirac measure δ, and for $0 < \alpha < \frac{1}{2}$ they are distributions of order 1 involving only *finite parts*: hence, integral terms such as $\int_{\mathbb{R}} M_\alpha(y)\,\varphi(y,t)\,dy$ have to be understood in the sense of *duality* brackets $< M_\alpha, \varphi(t) > = < 1, \varphi(t) >_{\mathcal{V}_\alpha', \mathcal{V}_\alpha}$.

Output form System (5a)-(5b) is *equivalent* to:

$$\partial_t\varphi(y,t) = \partial_y^2\varphi(y,t) + \left[\widetilde{f}(z(t)) + u(t)\right]\delta(y)\,; \quad \varphi(.,0) = 0\,, \tag{10a}$$

$$z(t) = \int_{\mathbb{R}} M_\alpha(y)\,\varphi(y,t)\,dy = < M_\alpha, \varphi(t) > . \tag{10b}$$

Balanced form System (8a)-(8b) is *equivalent* to:

$$\partial_t\varphi(y,t) = \partial_y^2\varphi(y,t) + \left[\widetilde{f}(z(t)) + u(t)\right]N_\alpha(y)\,; \quad \varphi(.,0) = 0\,, \tag{11a}$$

$$z(t) = \int_{\mathbb{R}} N_\alpha(y)\,\varphi(y,t)\,dy = < N_\alpha, \varphi(t) > . \tag{11b}$$

The energy associated to this equation is:

$$E(t) = \frac{1}{2}\int_{\mathbb{R}} |\varphi(y,t)|^2\,dy\,, \tag{12}$$

for which it is easily proved that the following equality holds:

$$\frac{dE}{dt}(t) = -\int_{\mathbb{R}} |\partial_y\varphi(y,t)|^2\,dy + z(t)\,\widetilde{f}(z(t)) + z(t)\,u(t)\,. \tag{13}$$

The functional spaces to be used are: $\mathcal{H} = L^2(\mathbb{R})$, $\mathcal{V} = H^1(\mathbb{R})$ and $\mathcal{V}' = H^{-1}(\mathbb{R})$. They are *independent* of α.

Remark 1. Note that these equivalent reformulations are interesting results on their own, for the following reasons:

- the system is local in time,
- a natural energy functional E is provided on an energy space $\mathcal{H}$, which helps prove that the system is dissipative under some specific conditions on the non-linearity f,
- a classical $(\mathcal{V}, \mathcal{H}, \mathcal{V}')$ functional analytic framework is being used, in which regularity results can be more easily obtained,

- on the heat equation formulations, the (weak or strong) maximum principle can be used, especially for comparison results,
- numerical approximation of diagonal diffusive formulations is straightforward, using standard schemes of numerical analysis (see [10]).

These features can *not* be captured on the original system (1) nor on any of the Abel-Volterra forms (2)-(4).

2.3 A comparison result

On formulation (11a)-(11b) with an extra forced term denoted by $g(t)$, the following quadratic a priori estimate will be useful in the sequel:

$$\frac{1}{2} \partial_t \|\varphi\|^2 + \|\partial_y \varphi\|^2 = f(t, < N_\alpha, \varphi(t) >) < N_\alpha, \varphi(t) > + < g(t), \varphi(t) >$$

The following theorem is an extension to the case $\alpha \neq \frac{1}{2}$ of a result of [1].

Theorem 1. *Suppose $f(t,.)$ is strictly decreasing on $\mathbb{R}$, let us consider φ_1, φ_2 solutions of:*

$$\partial_t \varphi_j - \partial_y^2 \varphi_j = f(t, < N_\alpha, \varphi_j >) \otimes N_\alpha + g_j \tag{14}$$

such that $t \mapsto z_j(t) = < N_\alpha, \varphi_j(t) > = < n_\alpha, \widehat{\varphi_j}(t) >$ be of class C^1 on $[0, T]$. If $g_1 \geq g_2$ on $[0, T]$, then $\varphi_1 \geq \varphi_2$ and $z_1 \geq z_2$ on $[0, T]$.

Proof (Sketch of). Function $\Phi = \varphi_1 - \varphi_2$ is the solution of:

$$\partial_t \Phi - \partial_y^2 \Phi = [f(t, < N_\alpha, \varphi_1 >) - f(t, < N_\alpha, \varphi_2 >)] \otimes N_\alpha + g_1 - g_2 ; \qquad \Phi_0 = 0 .$$

Multiplying this equation by Φ_- (where $\Phi = \Phi_+ - \Phi_-$ and $\Phi_+ \Phi_- = 0$) and integrating over $\mathbb{R}$ leads to:

$$\frac{1}{2} \partial_t \|\Phi_-\|^2 + \|\partial_x \Phi_-\|^2 =$$
$$- [f(t, < N_\alpha, \varphi_1 >) - f(t, < N_\alpha, \varphi_2 >)] < N_\alpha, \Phi_- > - < g_1 - g_2, \Phi_- >$$

Then, thanks to f strictly decreasing, $- [f(t, z_1) - f(t, z_2)] (z_1 - z_2)_- \leq 0$, with $z_j = < N_\alpha, \varphi_j >$. Hence, together with $g_1 - g_2 \geq 0$, we get:

$$\frac{1}{2} \partial_t \|\Phi_-\|^2 + \|\partial_x \Phi_-\|^2 \leq 0 + 0$$

Then function $\|\Phi_-\|$ is positive decreasing, with initial value 0, thus null a.e. It follows that $\varphi_1 \geq \varphi_2$ a.e. and $z_1 \geq z_2$ on $[0, T]$. $\square$

3   Analysis of the case $x_0 = 0$

First, we get a main theorem, the corollary of which is the stability of system (1) subject to specific conditions. Note that the proof needs to be performed on one of the four equivalent diffusive formulations only.

Theorem 2. *As soon as the input u has stopped, and provided f is strictly decreasing with $x f(x) < 0$, we get:* $\|\psi(.,t)\|_{\mathcal{H}} \to 0$ *and* $x(t) \to 0$ *as* $t \to \infty$.

Proof (Sketch of). The goal is to apply LaSalle invariance principle (see e.g. [3]); to this end, we proceed in six steps:

1. system (5a)-(5b) is dissipative: from (7), $\dot{E}_\alpha(t) \le 0$ thanks to $x f(x) < 0$,
2. moreover $E_\alpha = 0$ if and only if $\psi(.) = 0$ $\mu_\alpha -$ a.e.,
3. for any $\psi_0 \in \mathcal{H}_\alpha$, the trajectory $\{\psi(.,t)\}_{t \ge 0}$ is precompact (see [9]),
4. $\psi \to 0$ in $\mathcal{H}_\alpha$ strongly as $t \to \infty$,
5. $\psi \rightharpoonup 0$ in $\mathcal{V}_\alpha$ weakly,
6. hence, $x(t) = <1, \psi(.,t) >_{\mathcal{V}_\alpha ', \mathcal{V}_\alpha} \to 0$ as $t \to \infty$. □

Remark 2. In the case when f is *linear*, that is $f(x) = \lambda x$, the condition on f reads $\lambda < 0$; by a simple extension to the complex-valued case, one easily gets $\Re e(\lambda) < 0$, which happens to be sufficient but not necessary, since the *optimal* stability result for the linear case is: $|\arg(\lambda)| > \alpha \frac{\pi}{2}$ (see [5,6]).

Remark 3. Function f can be *discontinuous* at 0, in which case the results are still valid, though care must be taken that f is a *multivalued* function with $f(0) \ni 0$, and the following *differential inclusion*:

$$\partial_t \varphi(y,t) - \partial_y^2 \varphi(y,t) - \left[\tilde{f}(z(t)) + u(t)\right] \otimes N_\alpha \ni 0; \quad \varphi(.,0) = 0, \quad (15a)$$

$$z(t) = <N_\alpha, \varphi(t) >, \quad (15b)$$

which is nothing but a diffusive representation for the non-linear fractional differential inclusion:

$$d^\alpha x(t) - f(x(t)) - u(t) \ni 0; \qquad x(0) = x_0. \quad (16)$$

Once the well-posedness nature of problem (15a)-(15b) has been established, the solution $x = x_0 + z$ of (16) is uniquely determined as an output.

4   Analysis of the case $x_0 \ne 0$

This seems to be a more difficult problem than the previous one, mostly because of the long-memory behaviour.

4.1   Formulation through an extra forced term

From reformulation (3), the *pseudo*-initial condition x_0 in (1) can be taken into account by an extra *input* v instead of u in *any* of the equivalent diffusive formulations of section 2 with f, and can therefore be interpreted as a forced term, namely: $v(t) = x_0 \frac{1}{\Gamma(1-\alpha)} t_+^{-\alpha} + u(t)$.

Unfortunately, we cannot expect to use the stability result above (theorem 2), for it is clear that the extra input will never stop: the everlasting behaviour of the extra input comes from the hereditary aspect of the problem.

4.2   Formulation by a change of function and variable

From reformulation (4), the *pseudo*-initial condition x_0 in (1) can be taken into account by a change of variable $z = x - x_0$ and a change of function $\tilde{f}(z) = f(x_0 + z)$; we then use the heat equation formulation in balanced form (11a)-(11b). Suppose $u = 0$ from $t = 0$ on (the extension to $u = 0$ from $t = t_0$ will be addressed at the end of the section). Let $x_0 < 0^1$, then $\tilde{f}(0) = f(x_0) > 0$.

Lemma 1. *φ et z are increasing functions of t.*

Proof. With $g(t) = -N_\alpha \otimes \tilde{f}(0) \, \mathbf{1}_{[0,T]}$, φ_g is a solution of:

$$\partial_t \varphi_g - \partial_y^2 \varphi_g = N_\alpha \otimes \tilde{f}(z) + g(t); \qquad \varphi_{g0} = 0 ;$$

it is identically zero on $[0, T]$;[2] hence, thanks to the comparison result (theorem 1) with $g(t) \le 0$, we get $\varphi_g(.,t) = \varphi(.,t-T) \le \varphi(.,t)$, thus as $N_\alpha \ge 0$, $< N_\alpha, \varphi(.,t-T) > \le < N_\alpha, \varphi(.,t) >$. $\square$

Lemma 2. $\lim z \le -x_0$.

Proof. Otherwise, by continuity, $\exists t_0$ such that $z(t_0) = -x_0 \Rightarrow \tilde{f}(z(t_0)) = 0$, because φ is increasing; from what we deduce that:

- either $\varphi = cte$, that is an equilibrium state, implying φ is constant $\forall t > t_0$, hence $z(t)$ is constant.
- or $\varphi \ne cte$, in which case the concavity is of constant sign and negative, which contradicts φ increasing and $\varphi_0 = 0$. $\square$

Lemma 3. *There exists a unique equilibrium state $\varphi_\infty = cte$, and $z_\infty = -x_0$.*

---

[1] the case $x_0 < 0$ is treated similarly.
[2] Here g has been computed in such a way as to delay the start of the diffusion process by T.

Proof. At the equilibrium, $\partial_y^2 \varphi_\infty = -N_\alpha \otimes \tilde{f}(z_\infty)$. N_α being positive, the concavity of φ_∞ is of constant sign and negative, which is contradictory with φ increasing and $\varphi_0 = 0$, except if φ_∞ is constant. Thus, $\tilde{f}(z_\infty) = 0$ necessarily and $z_\infty = -x_0$ for f is injective. $\square$

Lemma 4. $z \to -x_0$.

Proof. $\lim z$ exists and $\leq -x_0$. If $z^* = \lim z < -x_0$, then $\tilde{f}(z) \geq k > 0$, which implies that on any compact subset $[-Y, Y]$, $\partial_t \varphi > \partial_x^2 \varphi + N_\alpha \otimes k$. From which we can easily deduce that $\varphi \to +\infty$; more precisely:

$$\forall K, Y, \exists t_0, \qquad \varphi(y) \geq K \text{ for } y \in [-Y, Y],$$

hence $z = <N_\alpha, \varphi> \geq K \int_{-Y}^{Y} N_\alpha \, dy > z^*$ for K large enough, which is contradictory. $\square$

Corollary 1. $x \to 0$ *as* $t \to \infty$.

Corollary 2. *It can be shown that the equilibrium state φ_∞ is asymptotically reached, in the following sense: $\varphi \to \varphi_\infty$ uniformly on any compact subset, that is in the weak-$*$ topology of $L^\infty(\mathbb{R})$.*

The previous analysis amounts to the *maximum principle* for the case $\alpha \neq \frac{1}{2}$. The formulation by a heat equation (namely, heat equation formulations) can *not* be overcome; it gives valuable information thanks to the evolution of an internal state of infinite dimension, from which the long-memory behaviour stems: this aspect is rather well controlled (from a functional analytic point of view) thanks to the specific properties of the heat equation. This is certainly one of the most *remarkable* applications of DR of PDOs; these techniques provide not only straightforward numerical schemes for the approximation, but also very sharp estimates for the analysis of the problem (and especially for *asymptotic* analysis).

Remark 4. It is noteworthy that $<N_\alpha, \varphi(., t)>$ tends to $-x_0$ as $t \to \infty$, but $<N_\alpha, \varphi_\infty>$ is *not* properly defined, because N_α and φ_∞ do *not* belong to dual spaces (except in the case $\alpha = \frac{1}{2}$). To some extent, the maximum principle *forces* this limit to exist without degenerating, but $\varphi(t)$ *diverges* in the energy space $L^2(\mathbb{R})$ (weak-$*$ convergence in $L^\infty(\mathbb{R})$).

When $u = 0$ for $t > t_0$, the state φ is initialized by $\varphi_0(y) = \varphi(t_0, y)$ at time t_0, with null input but $x_0 = x(t_0) \neq 0$ in general. Then, as the autonomous dynamics generated by $\varphi_0 \neq 0$ is stable (diffusion), it is the stability/unstability generated by x_0 which will play the major role and enable to conclude in a similar way.

5   Further extensions

The conditions can easily be extended to the complex-valued case, namely $\Re e(x^* f(x)) < 0$, and also to the vector-valued case, as $\Re e(x^H f(x)) < 0$; and the monotonicity of f must be translated in an appropriate way.

Moreover, the whole set of results obtained thanks to diffusive formulations can be extended to any other diffusive pseudo-differential operator of *dissipative* nature, that is $\mu > 0$.

Finally, in order to extend the sufficient stability condition, a more accurate result can be *conjectured*, as a *fractional* version of the Hartman–Grobman theorem, namely:

Theorem 3 (Conjecture). *The local stability of the equilibrium $x^* = 0$ of the non-linear fractional differential system $d^\alpha x = f(x)$ is governed by the global stability of the linearized system near the equilibrium $d^\alpha x = \lambda x$, where $\lambda = f'(0) \in \mathbb{C}$, namely:*

- *$x^* = 0$ is locally asymptotically stable if $|\arg(\lambda)| > \alpha\frac{\pi}{2}$,*
- *$x^* = 0$ is not locally stable if $|\arg(\lambda)| < \alpha\frac{\pi}{2}$.*

Note that nothing can be said if $|\arg(\lambda)| = \alpha\frac{\pi}{2}$, in which case the linearized system is asymptotically oscillating.

The idea is to use a semi-linear diffusive reformulation of the system, and then an infinite-dimensional version of the Hartman-Grobman theorem; more precisely:

$$\partial_t \varphi = \partial_y^2 \varphi + \tilde{f}(< N_\alpha, \varphi >) \otimes N_\alpha, \qquad \varphi_0 = 0, \tag{17a}$$

$$x(t) = x_0 + < N_\alpha, \varphi(t) >, \tag{17b}$$

is of the form $\partial_t \varphi = F(\varphi)$ with F linearizable in a weak sense (unbounded operators), $F = L + B$ with $L = \partial_y^2 + l$ the linear part and B a non-linear term of lower differential order; the solution and stability of (17a)-(17b) is known exactly when F reduces to L. Care must be taken that the equilibrium state φ_∞ does *not* belong to the energy space: specific methods from functional analysis and semi-linear diffusion PDEs must be investigated in order to tackle the problem properly.

References

1. J. Audounet, V. Giovangigli, and J.-M. Roquejoffre, *A threshold phenomenon in the propagation of a point-source initiated flame*, Physica D, (1998), pp. 295–316.
2. J. Audounet, D. Matignon, and G. Montseny, *Diffusive representations of fractional and pseudo-differential operators*, in Research Trends in Science and Technology, Beirut, Lebanon, March 2000, Lebanese American University, 10 pages.

3. A. Haraux, *Systèmes dynamiques dissipatifs et applications*, vol. 17 of Recherche en Mathématiques Appliquées, Masson, 1991.
4. S. O. Londen, *The qualitative behavior of the solution of nonlinear Volterra equations*, Michigan Math. Journal, 18 (1971), pp. 321–330.
5. D. Matignon, *Stability results for fractional differential equations with applications to control processing*, in Computational Engineering in Systems Applications, vol. 2, Lille, France, July 1996, IEEE-SMC, pp. 963–968.
6. ———, *Stability properties for generalized fractional differential systems*, ESAIM: Proceedings, 5 (1998), pp. 145–158.   URL: http://www.emath.fr/Maths/Proc/Vol.5/.
7. D. MATIGNON AND G. MONTSENY, eds., *Fractional Differential Systems: models, methods and applications*, vol. 5 of ESAIM: Proceedings, URL: http://www.emath.fr/Maths/Proc/Vol.5/, December 1998, SMAI.
8. G. Montseny, *Diffusive representation of pseudo-differential time-operators*, ESAIM: Proceedings, 5 (1998), pp. 159–175.   URL: http://www.emath.fr/Maths/Proc/Vol.5/.
9. ———, *Précompacité des trajectoires de systèmes pseudo-différentiels dissipatifs sous formulation diffusive*, Internal Report, June 2000, LAAS.
10. G. Montseny, J. Audounet, and D. Matignon, *Diffusive representation for pseudo-differentially damped non-linear systems*, in *Nonlinear Control in the Year 2000*, Springer Verlag, 2000.
11. J. Palis and W. de Melo, *Geometric theory of dynamical systems: an introduction*, Springer Verlag, 1982.
12. R. Temam, *Infinite dimensional dynamical systems in mechanics and physics*, vol. 68 of Applied Mathematical Sciences, Springer Verlag, 1988.

Controllability Properties of a Class of Control Systems on Lie Groups

Victor Ayala[1]* and Luiz A. B. San Martin[2]**

[1] Departamento de Matemática Universidad Católica del Norte
   Casilla 1280
   Antofagasta, Chile
   vayala@socompa.ucn.cl
[2] Instituto de Matemática
   Universidade Estadual de Campinas
   Cx. Postal 6065
   13081-970 Campinas SP, Brasil

Abstract. Linear control systems on Lie groups were introduced by Markus [3] and also studied by Ayala and Tirao in [1]. For this class of control systems we establish controllability results in the compact case and also in the semi-simple non-compact case. Wealso show that the forward orbit from the identity is not in general a semigroup.

1 Introduction

In this paper we study controllability properties of linear control systems on Lie groups. This class of control systems was introduced by Markus [3] and also studied by Ayala and Tirao [1]. A linear control system on a connected Lie group G is determined by the family of differential equations on G, parametrized by the class U of the piecewise constant addmisible control:

$$\dot{x}(t) = X\left(x(t)\right) + \sum_{i=1}^{m} u_i(t), Y_i\left(x(t)\right) \tag{1}$$

Here $x(t) \in G$, and X stand for an infinitesimal automorphism of G, i.e., the flow $(X_t)_{t \in \mathbb{R}}$ is a 1-parameter group of $\text{Aut}(G)$, the group of autmorphism of G. The control vectors Y_i are elements of the Lie algebra $\mathfrak{g}$ of G, considered as right invariant vector fields on G. First of all, we discuss the group of diffeomorphism generated by the system. We denote by $\text{Af}(G)$ the *affine group* of G i.,e., the semi-direct product $\text{Af}(G) = \text{Aut}(G) \times_s G$. If G is simple connected then it is well known that the Lie algebra $\mathfrak{aut}(G)$ of $\text{Aut}(G)$ is $\text{Der}(\mathfrak{g})$. On the other hand, we show that for a connected Lie group G, $\mathfrak{aut}(G)$ is a

* Partially supported by Proyecto FONDECYT-CONICYT n° 1990360 and Programa de Cooperacin Cientfica Internacional CNPq-CONICYT, Folio n° 97001
** Research partially supported by CNPq grant n° 301060/94-0

subalgebra of Der ($\mathfrak{g}$). We denote by $\mathfrak{g}_\Sigma$ the subalgebra of $\mathfrak{af}\,(G)$ generated by the vector fields of the system, and by G_Σ the connected subgroup of Af (G) whose Lie algebra is $\mathfrak{g}_\Sigma$. Of course G_Σ is exactly the group of diffeomorphisms generated by the control system, that is the systems group. The vector fields of the system are then right invariant vector fields in G_Σ. The original system lifts to a right invariant control system on G_Σ, whose forward orbit from the identity is a semigroup which we denote by S_Σ. And the (1) system itself is induced on G by the invariant system on G_Σ. We assume that the system satisfies the Lie algebra rank condition. This amounts to suppose that G has codimension one in the systems group G_Σ. By construction G identifies with a homogeneous space of G_Σ, and the system (1) is controllable if and only if S_Σ is transitive in this homogenous space. On the other hand, If G is a semi-simple non-compact Lie group denote by $\pi : G_\Sigma \rightarrow G$ the canonical homomorphism onto G_Σ/Z, induced by the decomposition of the Lie algebra $\mathfrak{g}_\Sigma = \mathfrak{z} \oplus \mathfrak{g}$, where $\mathfrak{z}$ is the one-dimensional center of $\mathfrak{g}_\Sigma$ and $Z = \exp(\mathfrak{z})$. Then, the invariant system on G_Σ defines through π an induced invariant control system on G. In [1] the authors extend the well known Kalman rank condition for a linear control systems on $\mathbb{R}^n$. In fact, they proved that the Lie algebra rank condition characterize controllability for abelian Lie groups. For a connected Lie group G they also prove local controllability from the identity element of G if the dimension of the subspace

$$V = \mathrm{Span}\left\{Y_i, ad^{(j)}(X)(Y_i) : i = 1, 2, ..., m, j \geq 1\right\} \subset \mathfrak{g}$$

is the dimension of G. We consider the global controllability property for a linear control system on a connected Lie group. According to the following cases, we prove:

1. In a compact and connected Lie group G, the linear system (1) is controllable if and only if it satisfies the Lie algebra rank condition.
2. The induced invariant system on G is controllable if the linear system (1) is controllable from the identity.

To get the last result we use the notion of left reversible semigroup and the following theorem proved in [5]: If L is semi-simple noncompact group and T is a semigroup with nonvoid interior then T is not left reversible unless $T = L$. This paper is organized as follows: In Section 2 we study the group of diffeomorphisms generated by the linear control system 1. Section 3 contains the main results about controllability: the compact case and the semisimple non-compact case. Section 4 contains one example showing a system which is locally controllable from the identity but not controllable from the same point on Sl $(2, \mathbb{R})$. In particular, this example shows that the forward orbit of (1) from the identity is not in general a semigroup.

2 Group of the system

In this section we discuss the group of diffeomorphisms generated the vector fields of the system (1). We start with some brief comments regarding the automorphisms of a group: intod. By the *affine group* Af (G) of a group G we understand the semi-direct product Af $(G) = $ Aut $(G) \times_s G$, where Aut (G) stands for the automorphism group of G. The multiplication in Af (G) is given by

$$(g, x) \cdot (h, y) = (gh, xg\,(y)) ,$$

and that the mapping $x \mapsto (1, x)$ – where 1 is the identity automorphism – embeds G into Af (G) as a normal subgroup. Through this embbeding left multiplication of elements of Aut (G) defines the action or Af (G) in G, $(g, x) \cdot y \mapsto xg\,(y)$. This action is transitive, and since the istropy at the identity is Aut (G), the group G becomes identified with the homogenous space Af $(G)/$Aut (G) of Af (G). Now, suppose that G is a Lie group and denote by $\mathfrak{g}$ its Lie algebra, and Aut $(\mathfrak{g})$ the automorphism group of $\mathfrak{g}$. Recall that Aut $(\mathfrak{g})$ is a Lie group whose Lie algebra is Der $(\mathfrak{g})$, the Lie algebra of derivations of $\mathfrak{g}$. In case G is connected and simply-connected the automorphism groups Aut (G) and Aut $(\mathfrak{g})$ are isomorphic. In fact, the isomorphism is given by the mapping which assigns to $g \in$ Aut (G) its differential dg_1 at the identity. Since any $g \in$ Aut $(\mathfrak{g})$ extends to an automorphism of G, it follows that this map is indeed an isomorphism of groups. Therefore for a simply connected group G the Lie algebra of Aut (G) is Der $(\mathfrak{g})$, hence the Lie algebra of Af (G), which we denote by $\mathfrak{af}\,(G)$, is the semi-direct product Der $(\mathfrak{g}) \times_s \mathfrak{g}$. The Lie bracket in this Lie algebra is given by

$$[(D_1, X_1) , (D_2, X_2)] = ([D_1, D_2], D_1 X_2 - D_2 X_1 + [X_1, X_2])$$

where inside the first coordinate the bracket is that of Der $(\mathfrak{g})$, while in the second that of $\mathfrak{g}$. From this description of the Lie algebras and the action of Aut (G) on G, it follows that any derivation $D \in$ Der $(\mathfrak{g})$ induces a vector field, say $\tilde{D}$, on G, whose flow is $\exp\,(tD) \in$ Aut (G). On the other hand if G is connected then $G = \tilde{G}/\Gamma$, where $\tilde{G}$ is the simply connected covering of G and $\Gamma \subset \tilde{G}$ is a discrete central subgroup. So that Aut (G) is the subgroup of Aut $\left(\tilde{G}\right)$ leaving Γ invariant, and since Γ is discrete the Lie algebra of Aut (G) is the subalgebra $\mathfrak{aut}\,(G)$ of Der $(\mathfrak{g})$ given by

$$\mathfrak{aut}\,(G) = \{D \in \text{Der}\,(\mathfrak{g}) : \tilde{D}\,(x) = 0 \text{ for all } x \in \Gamma\}.$$

In particular any $D \in \mathfrak{aut}\,(G)$ induces a vector field, also denoted by $\tilde{D}$, on G. Of course, if G is simply-connected then $\mathfrak{aut}\,(G) = $ Der $(\mathfrak{g})$. Moreover, $\mathfrak{af}\,(G)$ is the semi-direct product $\mathfrak{aut}\,(G) \times_s \mathfrak{g}$.

We turn now to the control system (1). In first place let X be an infinitesimal automorphism of the Lie group G, that is, X is a vector field on G such that its flow X_t is an automorphism for all t. Then X_t is an one-parameter subgroup of Aut (G), which implies that $X = \tilde{D}$ for some derivation $D \in \mathit{aut}\,(G)$. Therefore the vector fields of the system are elements of $\mathit{af}\,(G)$, so that the system itself lifts to a right invariant control system in the Lie group Af (G). We denote by $\mathfrak{g}_\Sigma$ the subalgebra of $\mathit{af}\,(G)$ generated by the vector fields of the system, and let G_Σ stand for the connected subgroup of Af (G) whose Lie algebra is $\mathfrak{g}_\Sigma$. Of course G_Σ is exactly the group of diffeomorphisms generated by the control system, that is the systems group.

Now, let the drift vector field of (1) be given by $X = \tilde{D}$ where $D \in \mathit{aut}\,(G)$, and denote by $\mathfrak{h}_\Sigma$ the smallest subalgebra of $\mathfrak{g}$ invariant under D and containing the control vectors Y_i, $i = 1, \ldots, m$. Clearly $(\mathbb{R}D) \times_s \mathfrak{h}_\Sigma$ is a subalgebra of $\mathit{af}\,(G)$. An easy inspection shows that $\mathfrak{g}_\Sigma = (\mathbb{R}D) \times_s \mathfrak{h}_\Sigma$. The connected subgroup of Aut (G) with Lie algebra $\mathfrak{g}_\Sigma$ is the semi-direct product

$$\exp{(\mathbb{R}D)} \times_s H_\Sigma,$$

where H_Σ is the connected subgroup of G whose Lie algebra is $\mathfrak{h}_\Sigma$, and $\exp{(\mathbb{R}D)}$ is the one-parameter subgroup generated by D. In general $\exp{(\mathbb{R}D)}$ may be diffeomorphic to $\mathbb{R}$ or to S^1. In order to simplify matters we shall consider as the systems group the group

$$G_\Sigma = \mathbb{R} \times_s H_\Sigma \tag{2}$$

with multiplication

$$(t_1, x_1) \cdot (t_2, x_2) = (t_1 + t_2, x_1 \,(\exp{(tD)}\, x_1)). \tag{3}$$

This group acts on G by $(t, x) \cdot y = x\,(\exp{(tD)}\,(y))$. The vector fields of the system are then right invariant vector fields in G_Σ, and the (1) system itself is induced on G by the invariant system on G_Σ. These descriptions of $\mathfrak{g}_\Sigma$ and G_Σ imply at once that (1) satisfies the Lie algebra rank condition if and only if $\mathfrak{h}_\Sigma = \mathfrak{g}$. In this case $\mathfrak{g}$ is an ideal of codimension one in $\mathfrak{g}_\Sigma$, and if G is connected it identifies with a homogenous space of G_Σ. We conclude this section proving a simple criteria for the existence of an one dimensional ideal complementing $\mathfrak{h}_\Sigma$. In order to state it we note that the restriction of D to $\mathfrak{h}_\Sigma$ is a derivation of this Lie algebra.

Proposition 21 *Let $\mathfrak{l}$ be a Lie algebra and $\mathfrak{m} \subset \mathfrak{l}$ be an ideal of codimension one. Take $B \in \mathfrak{l} \setminus \mathfrak{m}$ and consider the derivarion $D(X) = [B, X]$ of $\mathfrak{m}$. A necessary and sufficient condition for the existence of $A \in \mathfrak{l}$ such that $[A, \mathfrak{m}] = 0$ and $\mathfrak{l} = \mathbb{R}A \oplus \mathfrak{m}$ is that the derivation D is inner, i.e., $D = \mathrm{ad}\,(C)$ for some $C \in \mathfrak{m}$.*

Proof. Suppose that $[B, X] = [C, X]$ for some $C \in \mathfrak{m}$ and all $X \in \mathfrak{m}$. Then $A = B - C$ is the required vector complementing $\mathfrak{m}$.

Reciprocally, if A satisfies the given conditions then $A = x\,(B - C)$ for some $x \neq 0$ and $C \in \mathfrak{m}$. Then $(1/x)\,A + C$ so that $[B, X] = [C, X]$ for all $X \in \mathfrak{m}$, showing the derivation is inner.

3   Controllability

In this section we derive some results about the controllability of the linear system (1) on a Lie group G. From now on we assume that G is connected and the system satisfies the Lie algebra rank condition. This amounts to suppose that G has codimension one in the systems group G_Σ. The original system lifts to a right invariant control system in G_Σ, whose forward orbit from the identity is a semigroup which we denote by S_Σ. From the Lie algebra rank condition in G_Σ we know that the interior of S_Σ is dense in S_Σ.

By contruction G identifies with a homogeneous space of G_Σ, and the system (1) is controllable if and only if S_Σ is transitive in this homogenous space. We consider the controllability property according to the different types of groups.

3.1   Compact groups

If G is compact then its Lie algebra decomposes as $\mathfrak{g} = \mathfrak{z} \oplus \mathfrak{k}$ where $\mathfrak{z}$ is the center of $\mathfrak{g}$ and $\mathfrak{k}$ is a compact semi-simple Lie algebra. In order to discuss the Lie algebra $aut\,(G)$ we must consider the universal covering group $\tilde{G}$ of G. This is the only simply-connected group having Lie algebra $\mathfrak{g}$. Therefore, it is the direct product $\tilde{G} = Z \times K$, where Z is abelian simply-connected (isomorphic to the additive group of $\mathfrak{z}$) and K is compact with Lie algebra $\mathfrak{k}$. Let Γ be the central subgroup such that $G = \tilde{G}/\Gamma$. Since the connected subgroup of G having Lie algebra $\mathfrak{z}$ is compact, it follows that $Z/(Z \cap \Gamma)$ is a compact subgroup, so that Γ is a lattice in Z. Now, let D be a derivation of $\mathfrak{g}$. Then $D(\mathfrak{z}) \subset \mathfrak{z}$. In order that D induces an infinitesimal automorphism of G it is necessary that $\exp{(tD)}$ fixes every point of Γ. In particular the points of $Z \cap \Gamma$ are invariant under this group. This implies that, if we identify Z with $\mathfrak{z}$, then $D(X) = 0$ for all $X \in \Gamma$. But Γ is a lattice in $\mathfrak{z}$, hence spans the vector space $\mathfrak{z}$. Therefore D must be identically zero on $\mathfrak{z}$. This reduces the Lie algebra $aut\,(G)$ to a subalgebra of $\mathrm{Der}\,(\mathfrak{k})$.

On the other hand, let D be a derivation of $\mathfrak{k}$. The automorphisms $\exp{(tD)}$ can be extended to $Z \times K$ by putting $\exp{(tD)}\,(x, y) = (x, \exp{(tD)}\,(y))$ for $(x, y) \in Z \times K$. Now, pick $(x, y) \in \Gamma$. Then y belongs to the center $Z(K)$ of K. Since $\exp{(tD)}\,(y) \in Z(K)$ and this center is discrete, it follows that $\exp{(tD)}\,(y) = y$ for all $t \in \mathbb{R}$. Therefore the extension of $\exp{(tD)}$ to $Z \times K$ fixes every element of Γ so that D induces an infinitesimal automorphism of G.

Summarizing, we have that $aut\,(G)$ is isomorphic to $\mathrm{Der}\,(\mathfrak{k})$, which in turn is isomorphic to $\mathfrak{k}$, because every derivation of $\mathfrak{k}$ is inner.

Concerning the controllability of a linear control system in G, we observe that the group G_Σ is generated by left translations of G and inner automorphisms, that is, automorphisms of the form $y \mapsto xyx^{-1}$. Now, the Haar measure in G is bi-invariant, so that the flows of the vector fields of the system are measure preserving. Since the Haar measure is finite it follows that the system is controllable if and only if it satisfies the Lie algebra rank condition (see Lobry [2] and San Martin [4]). Thus we have proved the

Theorem 31 *In a compact and connected Lie group G, a linear system is controllable if and only if it satisfies the Lie algebra rank condition.*

3.2   Semi-simple noncompact groups

The center of any semi-simple group is a discrete subgroup implying that the elements of the center are fixed under a one-parameter group of automorphisms. Therefore, if G is a semi-simple Lie group a derivation D of its Lie algebra $\mathfrak{g}$ induces in G an infinitesimal automorphism, so that the algebra $aut\,(G)$ is just Der $(\mathfrak{g})$, which is isomorphic to $\mathfrak{g}$.

Now, assume that G is connected, and consider a linear system which satisfies the Lie algebra rank condition. According to the previous discussion the group of the system is $G_\Sigma = \mathbb{R} \times_s G$ with the multiplication given by (3), where D is the derivation defining the drift vector field. This group acts on G by the semi-direct product formula.

Appart from this action there is another one which comes from a homomorphism of G_Σ onto G, which is constructed as follows: By Proposition 21 we have that the Lie algebra $\mathfrak{g}_\Sigma$ associated with the system is a direct sum

$$\mathfrak{g}_\Sigma = \mathfrak{z} \oplus \mathfrak{g}$$

where $\mathfrak{z}$ is the one-dimensional center of $\mathfrak{g}_\Sigma$. Consider the subgroup $Z = \exp(\mathfrak{z})$. It is a closed subgroup of G_Σ. In fact, its closure $\mathrm{cl}\,(Z)$ is a connected Lie subgroup contained in the center of G. Hence its Lie algebra is contained in the center of $\mathfrak{g}_\Sigma$, and thus in $\mathfrak{z}$. Therefore the Lie algebra of $\mathrm{cl}\,(Z)$ is $\mathfrak{z}$, so that $\mathrm{cl}\,(Z) = Z$. We can then perform the coset space G_Σ/Z, which is a Lie group because Z is a normal subroup. Clearly the Lie algebra of G_Σ/Z is $\mathfrak{g}$. We claim that G_Σ/Z is isomorphic to G. To see this it is enough to check that $G \cap Z = \{1\}$. For this intersection we recall from Proposition 21 that a generator A of $\mathfrak{z}$ is given by $(D, -B)$ where $D = \mathrm{ad}\,(B)$, $B \in \mathfrak{g}$. Since A is in the center of $\mathfrak{g}_\Sigma$ it follows that $\exp tA = (\exp tD) \cdot (\exp(-tB))$. So that when we write $G_\Sigma = \mathbb{R} \times_s G$, we get

$$\exp(tA) = (t, \exp(-tB))\,.$$

¿From this expression it follows at once that $Z \cap G = \{1\}$, proving that $G \approx G_\Sigma/Z$ as claimed. We denote by $\pi : G_\Sigma \to G$ canonical homomorphism onto G_Σ/Z.

Thus we are laid to three control systems. First the invariant system on G_Σ:

$$\dot{x} = D(x) + \sum_{i=1}^{m} u_i Y_i(x) \qquad x \in G_\Sigma, \tag{4}$$

where D is viewed as a right invariant vector field in G_Σ. The linear system

$$\dot{x} = \tilde{D}(x) + \sum_{i=1}^{m} u_i Y_i(x) \qquad x \in G, \tag{5}$$

where $\tilde{D}$ is the infinitesimal automorphism of G induced by D, and finally, the induced invariant control system on G, given by

$$\dot{x} = \pi(D)(x) + \sum_{i=1}^{m} u_i Y_i(x) \qquad x \in G, \tag{6}$$

Our next objective is to prove that controllability of the linear system (5) from the identity implies the controllability of the invariant system (6). For this we recall first the following fact: A subsemigroup T of a group L is said to be left reversible in L if $TT^{-1} = L$. The following statement was proved in [5].

Proposition 32 *If L is semi-simple noncompact and T has nonvoid interior then T is not left reversible unless $T = L$.*

On the other hand we can show that the semigroup of certain controllable systems are reversible:

Proposition 33 *Suppose that L is a group acting transitively on a space M, and denote by H_x the isotropy group at $x \in M$. Let T be a subsemigroup of L and assume that $Tx = M$, and $T \cap H_x$ is a left reversible semigroup in H_x. Then T is left reversible in L.*

Proof. Take $h \in L$. We are required to show that $h \in TT^{-1}$, or equivalently, that there exists $g \in T$ such that $g^{-1}h \in T^{-1}$. Since $Tx = M$, there exists $g_1 \in T$ such that $g_1 x = hx$, so that $h^{-1}g_1 \in H_x$. Now, the left reversibility of $T \cap H_x$ ensures the existence of $g_2 \in T \cap H_x$ such that $h^{-1}g_1 g_2 \in T \cap H_x$. Taking $g = g_1 g_2 \in T$ it follows that $h^{-1}g \in T$ concluding the proof.

Joining together these two facts about reversible semigroups we obtain at once a necessary condition for the controllability of (5).

Theorem 34 *The invariant system (6) is controllable if the linear system (5) is controllable from the identity.*

Proof. The semigroup of the system contains the one-parameter semigroup $\{\exp(tD) : t \geq 0\}$. Since this semigroup is left reversible in the isotropy group of the identity, the previous proposition together with the controllability from the identiy imply that S_Σ is left reversible in G_Σ. This implies that $\pi(S_\Sigma)$ is left reversible in G, because $\pi(S_\Sigma)\pi(S_\Sigma)^{-1} = \pi(S_\Sigma S_\Sigma^{-1})$. Hence by Proposition 32, $\pi(S_\Sigma) = G$, showing that the invariant system (6) is controllable.

We observe that the assumption in this proposition is controllability from a specific point, namely the identity of G. This point is required in order that its isotropy subgroup meets S_Σ in a left reversible semigroup. Outside the identity this property may not hold.

We do not know whether controllability of the invariant projected system implies the controllability of (5), even from the identity. The following comments may shed some light at this problem: As discussed above $D = A + X$ where A is a generator of the center of $\mathfrak{g}_\Sigma$, and $X \in \mathfrak{g}$. Taking exponentials we have $\exp(tD) = \exp(tA)\exp(tX)$. With this equality in mind we can describe S_Σ as follows. Through the mapping

$$(t, g) \mapsto (\exp(tA), g)$$

the group G_Σ becomes isomorphic to direct product $\mathbb{R} \times G$. Let $\rho : \mathbb{R} \times G \to \mathbb{R}$ be the projection into the first coordinate. It is a homomorphism of groups. Since $\rho_*(D) = \rho_*(A)$ it follows that

$$S_\Sigma = \bigcup_{t \geq 0} \{t\} \times \mathcal{A}_t(1)$$

where $\mathcal{A}_t(1)$ stands for the accessible set, at time exactly t, from the identity for the invariant system in G. Now, suppose we can prove the converse of Proposition 34, namely that $\pi(S_\Sigma) = G$ implies that the linear system (5) is controllable from the identity. Then, by reversing time, we would have also that (5) is controllable *to* the identity, and hence that system is controllable from every point. However, a necessary condition for S_Σ to be transitive is that its interior meets all isotropy subgroups. For (5) the isotropy at a point h is $\{\exp(tD^h) : t \in \mathbb{R}\}$, where $D^h = A + X^h$ and $X^h = \mathrm{Ad}(h)X$. Hence, taking into account the above description of S_Σ, it follows that controllability of (5) is equivalent to have $\exp(tX^h) \in \mathrm{int}\mathcal{A}_t(1)$ for all $h \in G$ and some $t \in \mathbb{R}$. This seems to be a very strong property to get solely from the controllability of the invariant system (6).

We end this section with the following remark about the algebraic properties of the forward orbit from the identy of a linear system in G. Based on the examples of the classical linear system and of an invariant system it is natural to ask if that forward orbit is a semigroup of G. The example below shows that in semi-simple Lie groups the forward orbit is not in general a semigroup. In the next section we show that if G is nilpotent then the forward orbit from the identiy is a semigroup only in the trivial case of controllability.

4 Example

Let G be the simple group Sl$(2, \mathbb{R})$. Its Lie algebra is the subspace sl $(2, \mathbb{R})$ of trace zero 2×2 matrices. Consider the derivation $D = \mathrm{ad}\,(X)$, where

$$X = \begin{pmatrix} 1 & 0 \\ 0 & -1 \end{pmatrix},$$

and form the linear system $\dot{x} = \tilde{D}(x) + uY(x)$, where Y is the right invariant vector field given by the matrix

$$Y = \begin{pmatrix} 1 & 1 \\ 1 & -1 \end{pmatrix}.$$

An easy computation of brackets shows that $\{Y, [X, Y], [X, [X, Y]]\}$ span sl $(2, \mathbb{R})$. This implies that the linearized system at the identity is controllable, and hence that our linear system is locally controllable from $1 \in G$.

The group G_Σ is isomorphic to $\mathbb{R} \times$ Sl$(2, \mathbb{R})$ and hence to Gl$^+$ $(2, \mathbb{R})$, the group of matrices with $\det > 0$. Its Lie algebra is gl $(2, \mathbb{R})$. The centre of gl $(2, \mathbb{R})$ is spanned by the identity matrix. Hence we can normalize the isomorphism so that D, viewed as an element of $\mathfrak{g}_\Sigma = gl$ $(2, \mathbb{R})$, becomes the matrix

$$D = 1 + X = \begin{pmatrix} 2 & 0 \\ 0 & 0 \end{pmatrix}.$$

Thus S_Σ is the semigroup generated by

$$\{\exp t \begin{pmatrix} 2+u & u \\ u & -u \end{pmatrix} : t, u \in \mathbb{R}\}.$$

The projection $\pi(S_\Sigma)$ of S_Σ into Sl$(2, \mathbb{R})$ is the semigroup generated by $\exp(X + uY)$, which is the semigroup of the bilinear system $\dot{x} = Xx + uYx$. This bilinear system is not controllable because both matrices are symmetric. Hence $\pi(S_\Sigma)$ is proper so that the invariant system is not controllable.

Summarizing, we have got a linear control system which is locally controllable from the identity but not controllable from the same point. Since Sl$(2, \mathbb{R})$ is connected, it follows that the forward orbit from the identity is not a semigroup.

References

1. Ayala, V. and Tirao, J. (1999) Linear Control Systems on Lie Groups and Controllability. Proceedings of Symposia in Pure Mathematics (AMS) **64**, 47–64
2. Lobry, C. (1974) Controllability of Non Linear Systems on Compact Manifolds. SIAM J. Control **12**, 1–4

3. Markus, L. (1980) Controllability of multi-trajectories on Lie groups. Proceedings of Dynamical Systems and Turbulence, Warwick, Lecture Notes in Mathematics **898**, 250–265
4. San Martin, L.A.B. (1987) Controllability of families of measure preserving vector fields. Systems & Control Letters **8**, 459–462
5. San Martin, L. A. B. and Tonelli, P. A. (1995) Semigroup actions on homogeneous spaces. Semigroup Forum **50**, 59–88

Stability Analysis to Parametric Uncertainty: Extension to the Multivariable Case

Miguel Ayala Botto[1], Ton van den Boom[2], and José Sá da Costa[1]

[1] Technical University of Lisbon
   Instituto Superior Técnico
   Department of Mechanical Engineering, GCAR
   Avenida Rovisco Pais
   1049-001 Lisboa, Portugal
   Phone: +351-21-8419028 - Fax: +351-21-8498097 - migbotto@dem.ist.utl.pt
[2] Delft University of Technology
   Faculty of Information Technology and Systems, Control Laboratory
   P.O. Box 5031
   2600 GA Delft, The Netherlands
   Phone: +31-15-2784052 - Fax: +31-15-2786679 - t.j.j.vdboom@its.tudelft.nl

Abstract. This paper is concerned with stability robustness of nonlinear multivariable systems under input-output feedback linearization. A procedure is presented that allows plant uncertainty to be propagated through the control design, yielding an uncertainty description of the closed-loop in polytopic form. As feedback linearization aims for a linear closed-loop system, plant uncertainty in the nonlinear (open-loop) system causes the parameters of the resulting linear system to be uncertain. Due to the nonlinearity of the process under control, these closed-loop uncertainties will turn out to be nonlinear and state dependent. It is outlined, how, with a numerical procedure, these uncertainties can be bounded within intervals, thus allowing the construction of a polytopic uncertainty description. Stability robustness can then be verified with the aid of linear matrix inequalities (LMIs).

1 Introduction

The basic principle behind any feedback linearization control strategy is to provide a state feedback control law to cancel the system nonlinearities while simultaneously imposing some desired linear dynamics [1]. The success of this control scheme is strongly dependent on the exact model description of the system under consideration. However, since the exact model can never be obtained, there is a demand for a feedback linearization control scheme that can take plant uncertainty into account, while retaining the closed-loop performance and stability.

Plant uncertainty can be generally classified into two different types: parametric uncertainty, which represents imprecision of parameters within the model, and unstructured uncertainty, which represents unmodeled dynamics. Over the last decade, research has been focused on control design synthesis to address the problem of robustness of feedback linearization under bounded parametric uncertainty. There has been some research done on the problem of robust stability to parametric uncertainties under feedback linearization, based on sliding mode theory [2], or adaptive control [3], although in either cases an additional state feedback is required in order to guarantee stability robustness.

In this paper it is assumed that plant uncertainty is only due to parametric uncertainty. The system under control will be modeled with a nonlinear autoregressive with exogeneous input (NARX) model, which parameters are to be estimated according to a nonlinear optimization of a quadratic criterion. If certain conditions are met [4,5], statistical properties of such an estimation will provide confidence intervals for the estimated parameters. One of those conditions is that the system is in the model set, which can be easily checked with tools described in [6,7]. In order to analyze the influence of parametric uncertainty on the stability properties of the closed-loop system, it is necessary to transport the plant uncertainty through the control design. Developing such a procedure will be the main topic of this paper.

This paper is divided into 5 sections. In Section 2 the stability robustness analysis problem is formulated and its solution is pointed for a particular system, identical to the closed-loop system which will result from the application of the feedback linearization scheme. Section 3 describes the multivariable NARX model assumptions by focusing in the characterization of its parametric uncertainty, and further presents a possible formulation of the feedback linearization control scheme, which can easily take these uncertainties into account. The result will be then used in Section 4, where a state space uncertainty description of the closed-loop system is derived. This description is linked with the initial system configuration previously presented in Section 2, and so enabling a straightforward application of the robust stability analysis tools. In Section 5 some conclusions are drawn.

2   Problem statement

Consider the following state space description of a multivariable discrete-time system:

$$
\begin{aligned}
x_{k+1} &= (A + \delta A(x_k))x_k + (B + \delta B(x_k))v_k \\
y_k &= Cx_k
\end{aligned}
\tag{1}
$$

where $x_k \in \mathbf{X} \subset \Re^n$ is the state vector, $v_k \in \mathbf{V} \subset \Re^m$ the input vector and $y_k \in \mathbf{Y} \subset \Re^p$ the output vector of the system. It will be assumed throughout that the system is square, i.e., $m = p$, while the nominal system matrices A, B and C have appropriate dimensions and represent a stable linear system. A particular configuration is adopted where the uncertainties of the system are captured in nonlinear and state dependent terms, $\delta A(x_k)$ and $\delta B(x_k)$.

If $\|\delta B(0)\| < \infty$, without loss of generality, the following expansion can be given for $\delta B(x_k)$:

$$\delta B(x_k) = \sum_{i=1}^{n} \delta B_i^*(x_k)[x_k]_i + \delta B(0) \tag{2}$$

where $\delta B_i^*(x_k)$ is a $[n \times p]$ matrix. From (2), construct the $[n \times n]$ matrices $\delta \bar{B}_j(x_k)$ for which the i-th column is equal to the j-th column of $\delta B_i^*(x_k)$. In this way, the original system (1) can be described as:

$$\begin{aligned} x_{k+1} &= (A + \delta A(x_k) + \sum_{j=1}^{p} \delta \bar{B}_j(x_k)[v_k]_j)x_k + (B + \delta B(0))v_k \\ y_k &= Cx_k \end{aligned} \tag{3}$$

The stability of the system can then be easily checked, for every bounded input v_k and for any arbitrary initial state x_0, if a maximum bound for each element of the uncertainty terms $\delta A(x_k)$ and $\delta \bar{B}_j(x_k)$ is known in advance. As will be shown in the outcome of this paper, under some mild assumptions a quantitative measure for these bounds can be found by an off-line search procedure which spans x_k over the relevant regions of the system operating range. This procedure will provide the basic ingredients for a complete stability robustness analysis of system (3), since the uncertainty in matrix A can then be captured in a polytope defined according to the following convex hull, Co:

$$A + \delta A(x_k) + \sum_{j=1}^{p} \delta \bar{B}_j(x_k)[v_k]_j \in \text{Co}(A_1 \ldots, A_L) \tag{4}$$

which means that for each x_k and v_k there exist parameters $\lambda_1, \ldots, \lambda_L \geq 0$ with $\sum_{i=1}^{L} \lambda_i = 1$, such that the dynamics of the system can be recasted as:

$$A + \delta A(x_k) + \sum_{j=1}^{p} \delta \bar{B}_j(x_k)[v_k]_j = \sum_{i=1}^{L} \lambda_i A_i \tag{5}$$

According to Lyapunov's stability theory, a sufficient condition for the asymptotic stability of the autonomous system is proven if a positive definite matrix P is found, such that:

$$\begin{aligned} A_i^T P A_i - P &< 0 \\ P &> 0 \end{aligned} \quad \forall i = 1, \ldots, L \tag{6}$$

Moreover, expression (6) can be re-written as a Linear Matrix Inequality (LMI) by using $Q^{-1} = P$, resulting in:

$$\begin{aligned} A_i^T Q^{-1} A_i - Q^{-1} &< 0 \\ Q^{-1} &> 0 \end{aligned} \quad \forall i = 1, \ldots, L \tag{7}$$

According to the Schur's complement condition, expression (7) is equivalent to the following LMI:

$$\begin{bmatrix} Q & (A_iQ)^T \\ A_iQ & Q \end{bmatrix} > 0 \quad \forall i = 1, \ldots, L \tag{8}$$

which can be solved through computationally efficient algorithms [8]. Moreover, from the assymptotic stability of the autonomous system, together with the assumption of having a bounded input v_k, will follow that system (3) is also bounded input bounded output (BIBO) stable.

3   Feedback linearization with parametric uncertainties

Input-output feedback linearization control strategy aims at finding a state feedback control law, Ψ, which cancels the system nonlinearities while simultaneously imposing some desired closed-loop input-output dynamics [1]. In general terms, the feedback control law describes a nonlinear and state dependent dynamic mapping between the process inputs u_k and the new external inputs v_k, according to:

$$u_k = \Psi(x_k, x_k^l, v_k) \tag{9}$$

where x_k and x_k^l represent the state vector of the process and of the desired resulting linear closed-loop system, respectively. The success of feedback linearization strongly relies on the exact cancellation of the system nonlinear dynamics via feedback. Therefore, taking into account the plant uncertainties in this control loop is the first step towards a robust feedback linearization control scheme.

3.1   Model assumptions

One way to account for plant uncertainty is to assume that all plant uncertainty is captured in the model parameters, which underlies the assumption that both the model and the plant are structurally identical. Under these assumptions, consider the following NARX description of a given square MIMO plant:

$$y_{k+1} = f(x_k, u_k, \theta) \tag{10}$$

where the regression vector is partitioned as follows:

$$x_k = [y_k^T \quad \bar{y}_{k-1}^T \quad u_{k-1}^T \quad \bar{u}_{k-2}^T]^T \tag{11}$$

where y_k corresponds to the vector with the system output measurements at time instant k, u_{k-1} represents the system inputs at time instant $k-1$, where:

$$\bar{y}_{k-1} = [y_{k-1}^T \cdots y_{k-n_y}^T]^T \tag{12}$$

$$\bar{u}_{k-2} = [u_{k-2}^T \cdots u_{k-n_u}^T]^T \tag{13}$$

$$\theta = [\theta_1 \ldots \theta_t]^T \tag{14}$$

while $\theta \in \Theta \subset \Re^t$ represents the exact parameter set of the plant. This paper assumes that bounded confidence regions $\delta\theta$ for each nominal model parameter, $\hat{\theta}$, can be found, and so the following parametric uncertainty description can be adopted as a good measure for capturing the plant uncertainty:

$$\theta_i \in [\hat{\theta}_i - \delta\theta_i \quad \hat{\theta}_i + \delta\theta_i] \quad \forall i = 1, \ldots, t \tag{15}$$

where the probability that $\hat{\theta}$ deviates from θ more than $\delta\theta$, is smaller than the $(1-\alpha)$-level of the normal distribution, available in standard statistical tables. This model assumption, although possibly conservative for some parameters, is shown to be easily integrated in the approximate feedback linearization scheme, and so providing the means to obtain a robust stability analysis.

3.2 Approximate feedback linearization

Approximate feedback linearization assumes that the first order of Taylor's expansion of the nonlinear system (10), computed around the previous operating point, (x_{k-1}, u_{k-1}), is an accurate representation of the original nonlinear dynamics, as the higher order terms of this series expansion are considered to be negligible [9]. Following this procedure, the following model will be used throughout as representing the nonlinear plant dynamics:

$$\Delta y_{k+1} = F(x_{k-1}, u_{k-1}, \theta)\Delta x_k + E(x_{k-1}, u_{k-1}, \theta)\Delta u_k \tag{16}$$

with matrices $F(x_{k-1}, u_{k-1}, \theta)$ and $E(x_{k-1}, u_{k-1}, \theta)$ containing the partial derivatives of (10) with respect to x_k and u_k, respectively, both evaluated at the operating point, while $\Delta x_k = x_k - x_{k-1}$ and $\Delta u_k = u_k - u_{k-1}$. In order to have a model description which explicitly accounts for the parametric model uncertainties, present in $\delta\theta$, a second Taylor's expansion around the nominal model parameters $\hat{\theta}$ is now performed to the system description (16), resulting in the following expression:

$$\Delta y_{k+1} = [F(x_{k-1}, u_{k-1}, \hat{\theta}) + \delta F(x_{k-1}, u_{k-1}, \hat{\theta}, \delta\theta)]\Delta x_k +$$
$$+ [E(x_{k-1}, u_{k-1}, \hat{\theta}) + \delta E(x_{k-1}, u_{k-1}, \hat{\theta}, \delta\theta)]\Delta u_k \tag{17}$$

with $\delta F(x_{k-1}, u_{k-1}, \hat{\theta}, \delta\theta)$ and $\delta E(x_{k-1}, u_{k-1}, \hat{\theta}, \delta\theta)$ computed according to the following expressions:

$$\delta F(x_{k-1}, u_{k-1}, \hat{\theta}, \delta\theta) = \sum_{i=1}^{t} \bar{F}_i(x_{k-1}, u_{k-1}, \hat{\theta})[\delta\theta]_i \tag{18}$$

$$\delta E(x_{k-1}, u_{k-1}, \hat{\theta}, \delta\theta) = \sum_{i=1}^{t} \bar{E}_i(x_{k-1}, u_{k-1}, \hat{\theta})[\delta\theta]_i \qquad (19)$$

while $\bar{F}_i(x_{k-1}, u_{k-1}, \hat{\theta})$ and $\bar{E}_i(x_{k-1}, u_{k-1}, \hat{\theta})$ being respectively given by:

$$\bar{F}_i(x_{k-1}, u_{k-1}, \hat{\theta}) = \frac{\partial}{\partial\theta_i}[\tilde{F}(x_{k-1}, u_{k-1}, \theta)]_{\theta=\hat{\theta}} \qquad (20)$$

$$\bar{E}_i(x_{k-1}, u_{k-1}, \hat{\theta}) = \frac{\partial}{\partial\theta_i}[E(x_{k-1}, u_{k-1}, \theta)]_{\theta=\hat{\theta}} \qquad (21)$$

Therefore, provided that matrix $E(x_{k-1}, u_{k-1}, \hat{\theta})$ is invertible for all admissible pairs (x_{k-1}, u_{k-1}), the feedback law will be given by (letting fall arguments for a clear notation):

$$\Delta u_k = E^{-1}(-F\Delta x_k + CA\Delta x_k^l + CB\Delta v_k) \qquad (22)$$

where Δx_k^l is the linear state vector, Δv_k the newly created input signal, while A, B and C are appropriate choices for the state space matrices representing the desired imposed linear dynamics. The application of (22) to the model description with uncertainties found in (17), results in the following closed-loop description:

$$\Delta y_{k+1} = (CA + \delta EE^{-1}CA)\Delta x_k^l + (\delta F - \delta EE^{-1}F)\Delta x_k +$$
$$+(CB + \delta EE^{-1}CB)\Delta v_k \qquad (23)$$

Notice that in the case where the nominal model parameters exactly map the plant parameters, i.e., $\hat{\theta}=\theta$, which means no plant uncertainty, then the confidence regions will be given by $[\delta\theta]_i = 0$, for $i = 1, \ldots, t$ (see Section 3.1 for details). Further, terms $\delta F(x_{k-1}, u_{k-1}, \hat{\theta}, \delta\theta)$ and $\delta E(x_{k-1}, u_{k-1}, \hat{\theta}, \delta\theta)$ will vanish according to expressions (18) and (19), respectively, resuming the closed-loop description given in (23) to the following expression:

$$\Delta y_{k+1} = CA\Delta x_k^l + CB\Delta v_k \qquad (24)$$

which corresponds to a linear dynamic system having the following state space description:

$$\begin{aligned} \Delta x_{k+1}^l &= A\Delta x_k^l + B\Delta v_k \\ \Delta y_k &= C\Delta x_k^l \end{aligned} \qquad (25)$$

However, since plant uncertainty is not likely to be zero, a more general state space description which takes into account for the model parametric uncertainty will be presented in the next section.

4 Towards a state space uncertainty description

According to the partition of the regression vector given in (11), the prediction model with uncertainties described through expression (17) can be re-writen as:

$$\Delta y_{k+1} = (F_1 + \delta F_1)\Delta y_k + (F_2 + \delta F_2)\Delta \bar{y}_{k-1} + (F_3 + \delta F_3)\Delta u_{k-1} +$$
$$+(F_4 + \delta F_4)\Delta \bar{u}_{k-2} + (E + \delta E)\Delta u_k \tag{26}$$

In the same way, the feedback law (22) can be also be expanded according to:

$$\Delta u_k = E^{-1}\left[(-F_1 + A_{11})\Delta y_k + (-F_2 + A_{12})\Delta \bar{y}_{k-1} - F_3\Delta u_{k-1} - \right.$$
$$\left. -F_4\Delta \bar{u}_{k-2} + A_{15}\Delta v_{k-1} + A_{16}\Delta \bar{v}_{k-2} + B\Delta v_k\right] \tag{27}$$

where matrices A_{11}, A_{12}, A_{15} and A_{16} can be chosen to represent any linear time-invariant desired behaviour. The state space uncertainty description is then obtained by combining expressions (26) and (27) into a single state space description, considering the new extended linear state vector to be given by:

$$\Delta x_k^l = [\Delta y_k^T \quad \Delta \bar{y}_{k-1}^T \quad \Delta u_{k-1}^T \quad \Delta \bar{u}_{k-2}^T \quad \Delta v_{k-1}^T \quad \Delta \bar{v}_{k-2}^T]^T \tag{28}$$

resulting the state space uncertainty description given by:

$$\begin{aligned}
\Delta x_{k+1}^l &= (\bar{A} + \delta \bar{A}(\Delta x_k^l))\Delta x_k^l + (\bar{B} + \delta \bar{B}(\Delta x_k^l))\Delta v_k \\
\Delta y_k &= \bar{C}\Delta x_k^l
\end{aligned} \tag{29}$$

where matrices $\bar{A}$, $\bar{B}$ and $\bar{C}$ are given by the following expanded forms:

$$\bar{A} = \begin{bmatrix}
A_{11} & A_{12} & 0 & 0 & A_{15} & A_{16} \\
A_{21} & A_{22} & 0 & 0 & 0 & 0 \\
0 & 0 & 0 & 0 & 0 & 0 \\
0 & 0 & A_{43} & A_{44} & 0 & 0 \\
0 & 0 & 0 & 0 & 0 & 0 \\
0 & 0 & 0 & 0 & A_{65} & A_{66}
\end{bmatrix} \tag{30}$$

$$\bar{B} = [B \quad 0 \quad 0 \quad 0 \quad I \quad 0]^T \tag{31}$$

$$\bar{C} = [I \quad 0 \quad 0 \quad 0 \quad 0 \quad 0]^T \tag{32}$$

with matrices A_{21}, A_{22}, A_{43}, A_{44}, A_{65} and A_{66} respectively given by:

$$A_{21} = \begin{bmatrix} I_p \\ 0 \end{bmatrix} \in \Re^{(n_y+1)p \times p} \qquad A_{22} = \begin{bmatrix} 0 & 0 \\ I_{n_yp} & 0 \end{bmatrix} \in \Re^{(n_y+1)p \times (n_y+1)p} \tag{33}$$

$$A_{43} = \begin{bmatrix} I_p \\ 0 \end{bmatrix} \in \Re^{n_u p \times p} \quad A_{44} = \begin{bmatrix} 0 & 0 \\ I_{(n_u-1)p} & 0 \end{bmatrix} \in \Re^{n_u p \times n_u p} \tag{34}$$

$$A_{65} = \begin{bmatrix} I_p \\ 0 \end{bmatrix} \in \Re^{n_u p \times p} \quad A_{66} = \begin{bmatrix} 0 & 0 \\ I_{(n_u-1)p} & 0 \end{bmatrix} \in \Re^{n_u p \times n_u p} \tag{35}$$

while the uncertainty matrices $\delta \bar{A}$ and $\delta \bar{B}$ in (29) are respectively given by:

$$\delta \bar{A} = \begin{bmatrix} \delta F_1 - \delta E E^{-1}(F_1 - A_{11}) & \delta F_2 - \delta E E^{-1}(F_2 - A_{12}) & \cdots \\ 0 & 0 & \cdots \\ E^{-1}(-F_1 + A_{11}) & E^{-1}(-F_2 + A_{12}) & \cdots \\ 0 & 0 & \cdots \\ 0 & 0 & \cdots \\ 0 & 0 & \cdots \end{bmatrix}$$

$$\begin{bmatrix} \cdots & \delta F_3 - \delta E E^{-1} F_3 & \delta F_4 - \delta E E^{-1} F_4 & \delta E E^{-1} A_{15} & \delta E E^{-1} A_{16} \\ \cdots & 0 & 0 & 0 & 0 \\ \cdots & -E^{-1} F_3 & -E^{-1} F_4 & E^{-1} A_{15} & E^{-1} A_{16} \\ \cdots & 0 & 0 & 0 & 0 \\ \cdots & 0 & 0 & 0 & 0 \\ \cdots & 0 & 0 & 0 & 0 \end{bmatrix}^T \tag{36}$$

$$\delta \bar{B} = \begin{bmatrix} \delta E E^{-1} B \\ 0 \\ E^{-1} B \\ 0 \\ 0 \\ 0 \end{bmatrix} \tag{37}$$

At this stage, it should be noticed that the state space uncertainty description (29) has the same structure as the general system (1).

4.1   Robust feedback linearization

Since the uncertainty matrix $\delta \bar{B}$ in (37) is state dependent, a similar procedure as the one presented in Section 2 can be applied in order to incorporate this uncertainty term in the system dynamics (see expression (3) for details). Therefore, in order to analyze stability robustness to parametric uncertainty of the overall closed-loop system, a polytopic uncertainty description has to be constructed based on the maximum bounds for each element of matrices $\delta \bar{A}$ and $\delta \bar{B}$, described in (36) and (37), respectively. Finding these quantitative measures requires an off-line procedure to be applied based on a numerical search over all admissible operating points. The adopted strategy which assures that only relevant regions of the operating trajectory are considered, consists of designing a reference trajectory Δv_k and apply it to the feedback linearization of the process model. Then, as the model output travels along this trajectory, bounds on the uncertainty entries of matrices $\delta \bar{A}$ and $\delta \bar{B}$ can be collected such that the maximum values encountered along that trajectory are used to construct the polytopic description. One of the advantages of this approach is that uncertainty is considered only in relevant regions of the space spanned by the Δx_k vector during operation. Moreover, while this procedure is being performed, it can be automatically checked whether matrix $E(x_{k-1}, u_{k-1}, \hat{\theta})$ in (22), is invertible over all encountered operating points, a basic requirement for the success of the proposed control scheme.

5 Conclusions

In this paper a procedure was presented which allows for parametric plant uncertainty of a general nonlinear MIMO system to be propagated through an approximate feedback linearization control scheme, resulting in a linear closed-loop system with nonlinear and state dependent parametric uncertainties. By means of a numerical procedure these uncertainties could be bounded within intervals, allowing the construction of a polytopic uncertainty description. This particular configuration enables a straightforward application of robust stability analysis tools based on Lyapunov's stability concepts, as the overall stability problem could be formulated with the aid of linear matrix inequalities (LMIs).

Acknowledgements
This research is partially supported by Nonlinear Control Network, EC-TMR #ERBFMRXCT970137, and by PRAXIS XXI-BDP/20183/99.

References

1. Nijmeijer, H. van der Schaft, A. J. (1990) *Nonlinear Dynamical Control Systems*, Springer Verlag, New York
2. Utkin, Vadim I. (1992) *Sliding modes in control and optimization*, Springer, Communications and Control Engineering Series, Berlin
3. Slotine, Jean-Jacques E., Li, Weiping (1991) *Applied Nonlinear Control*, Prentice-Hall, New Jersey, USA
4. Seber, G.A.F., Wild, C.J. (1989) *Nonlinear regression*, John Wiley and Sons, Inc.
5. Ljung, Lennart (1987) *System Identification: Theory for the User*, Prentice-Hall, New Jersey, USA
6. Sjöberg, Jonas (1995) Nonlinear black-box modeling in system identification: a unified overview, *Automatica* **31**(12), 1691–1724
7. Billings, S.A., Voon, W.S.F. (1986) Correlation based model validity test for non-linear models, *International Journal of Control* 4(1), 235–244

8. Boyd, S., El Gahoui, L., Feron, E., Balakrishnan, V. (1994) *Linear matrix inequalities in system and control*, Vol. 15 of Studies in applied mathematics (SIAM)
9. te Braake, H.A.B., Ayala Botto, Miguel, van Can, H.J.L., Sá da Costa, José, Verbruggen, H.B. (1999) Linear predictive control based on approximate input-output feedback linearisation, *IEE Proceedings - Control Theory and Applications* 146(4), 295–300

External Stability and Continuous Liapunov Functions

Andrea Bacciotti

Dipartimento di Matematica del Politecnico
Torino, 10129 Italy
bacciotti@polito.it

Abstract. It is well known that external stability of nonlinear input systems can be investigated by means of a suitable extension of the Liapunov functions method. We prove that a complete characterization by means of continuous Liapunov functions is actually possible, provided that the definition of external stability is appropriately strengthened.

1 Introduction

A finite dimensional autonomous nonlinear system

$$\dot{x} = f(x, u) , \qquad x \in \mathbf{R}^n, \ u \in \mathbf{R}^m \tag{1}$$

is said to be *bounded input bounded state stable* (in short, BIBS stable) if for each initial state and each bounded input $u(t) : [0, +\infty) \to \mathbf{R}^m$ the corresponding solution is bounded for $t \geq 0$ (see [1] for a formal definition and comments). In the recent paper [3], uniform BIBS stability has been characterized by means of certain upper semi-continuous Liapunov functions. In fact, it is known that continuous Liapunov functions may not to exist for BIBS stable systems of the form (1).

The situation is exactly the same as in the theory of stability for equilibrium positions of systems without inputs (see [2], [4]). In this note we prove that the analogy can be further pursued. We extend to systems with inputs the theory developed in [2]. We show in particular that the existence of continuous Liapunov functions with suitable properties is equivalent to a type of external stability which is more restrictive than uniform BIBS stability.

In the next section we recall the basic notions (prolongations and prolongational sets associated to a dynamical system). Then we show how they generalize to systems with inputs. In Section 3 we introduce the definition of absolute bounded input bounded state stability (our strengthened form of external stability) and state the main result. The last section contains the proof.

2 Prerequisites

As already mentioned, for a locally stable equilibrium of a system without inputs

$$\dot{x} = f(x) , \qquad f \in C^1 \tag{2}$$

not even the existence of a continuous Liapunov function can be given for sure. In 1964, Auslander and Seibert ([2]) discovered that the existence of a continuous generalized Liapunov function is actually equivalent to a stronger form of stability. In order to illustrate the idea, it is convenient to begin with some intuitive considerations. Roughly speaking, stability is a way to describe the behavior of the system in presence of small perturbations of the initial state. More generally, let us assume that perturbations are allowed also at arbitrary positive times: under the effect of such perturbations, the system may jump from the present trajectory to a nearby one. Now, it may happens that an unfortunate superposition of these jumps results in an unstable behavior even if the system is stable and the amplitude of the perturbations tends to zero.

This phenomenon is technically described by the notion of *prolongation*, due to T. Ura and deeply studied in [2]. The existence of a continuous Liapunov function actually prevents the unstable behavior of the prolongational sets. On the other hand, the possibility of taking under control the growth of the prolongational sets leads to the desired strengthened notion of stability.

We proceed now formally to precise what we means for prolongation. First of all, we recall that from a topological point of view, very useful tools for stability analysis are provided by certain sets associated to the given system. These sets depend in general on the initial state. Thus, they can be reviewed as set valued maps. The simplest examples are the *positive trajectory* issuing from a point x_0

$$\Gamma^+(x_0) = \{y \in \mathbf{R}^n : y = x(t; x_0) \text{ for some } t \geq 0\} \tag{3}$$

where $x(\cdot; x_0)$ represents the solution of (2) such that $x(0; x_0) = x_0$, and the *positive limit set*.

We adopt the following agreements about notation. The open ball of center x_0 and radius $r > 0$ is denoted by $B(x_0, r)$. If $x_0 = 0$, we simply write B_r instead of $B(0, r)$. For $M \subset \mathbf{R}^n$, we denote $|M| = \sup_{x \in M} |x|$. Let $Q(x)$ be a set valued map from $\mathbf{R}^n$ to $\mathbf{R}^n$. For $M \subseteq \mathbf{R}^n$, we denote $Q(M) = \cup_{x \in M} Q(x)$. Powers of Q will be defined iteratively:

$$Q^0(x) = Q(x) \quad \text{and} \quad Q^k(x) = Q(Q^{k-1}(x))$$

for $k = 1, 2, \ldots$. Next, we introduce two operators, denoted by $\mathcal{D}$ and $\mathcal{I}$, acting on set valued maps. They are defined according to

$$(\mathcal{D}Q)(x) = \cap_{\delta > 0}\overline{Q(B(x, \delta))}$$

$$(\mathcal{I}Q)(x) = \cup_{k=0,1,2,\ldots}Q^k(x) \ .$$

The following characterizations are straightforward.

Proposition 1 *a)* $y \in (\mathcal{D}Q)(x)$ *if and only if there exist sequences* $x_k \to x$ *and* $y_k \to y$ *such that* $y_k \in Q(x_k)$ *for each* $k = 1, 2, \ldots$.

b) $y \in (\mathcal{I}Q)(x)$ *if and only if there exist a finite sequence of points* $x_0, \ldots, x_K$ *such that* $x_0 = x$, $y = x_K$ *and* $x_k \in Q(x_{k-1})$ *for* $k = 1, 2, \ldots, K$.

The operators $\mathcal{D}$ and $\mathcal{I}$ are idempotent. Moreover, for every set valued map Q, $\mathcal{D}Q$ has a closed graph, so that for every x the set $(\mathcal{D}Q)(x)$ is closed. However, $(\mathcal{I}Q)(x)$ is not closed in general, not even if $Q(x)$ is closed for each x.

When $\mathcal{I}Q = Q$ we say that Q is *transitive*. The positive trajectory is an example of a transitive map. In general, $\mathcal{D}Q$ is not transitive, not even if Q is transitive. In conclusion, we see that the construction

$$(\mathcal{D}(\ldots(\mathcal{I}(\mathcal{D}(\mathcal{I}(\mathcal{D}Q))))) \ldots))(x) \tag{4}$$

gives rise in general to larger and larger sets.

Definition 1 *A* prolongation *associated to system (2) is a set valued map* $Q(x)$ *which fulfils the following properties:*

(i) for each $x \in \mathbf{R}^n$, $\Gamma^+(x) \subseteq Q(x)$

(ii) $(\mathcal{D}Q)(x) = Q(x)$

(iii) If K *is a compact subset of* $\mathbf{R}^n$ *and* $x \in K$, *then either* $Q(x) \subset K$, *or* $Q(x) \cap \partial K \neq \emptyset$.

If Q is a prolongation and it is transitive, it is called a *transitive prolongation*. The following proposition will be used later (see [2]).

Proposition 2 *Let* K *be a compact subset of* $\mathbf{R}^n$ *and let* Q *be a transitive prolongation. Then* $Q(K) = K$ *if and only if* K *possesses a fundamental system of compact neighborhoods* $\{K_i\}$ *such that* $Q(K_i) = K_i$.

Starting from the map Γ^+ and using repeatedly the operators $\mathcal{D}$ and $\mathcal{I}$, we can construct several prolongational sets associated to (2). For instance, it is not difficult to see that

$$D_1(x) := (\mathcal{D}\Gamma^+)(x)$$

is a prolongation, the so called *first prolongation* of (2). The first prolongation characterizes stability. Indeed, it is possible to prove that an equilibrium x_0 of (1) is stable if and only if $D_1(x_0) = \{x_0\}$. The first prolongation in general is not transitive.

The intuitive construction (4) can be formalized by means of transfinite induction. This allows us to speak about higher order prolongations. More precisely, let α be an ordinal number and assume that the prolongation $D_\beta(x)$ of order β has been defined for each ordinal number $\beta < \alpha$. Then, we set

$$D_\alpha(x) = (\mathcal{D}(\cup_{\beta < \alpha}(\mathcal{I}D_\beta)))(x) \ .$$

The procedure saturates when $\alpha = \gamma$, the first uncountable ordinal number. Indeed, it is possible to prove that $\mathcal{I}D_\gamma = D_\gamma$, which obviously implies $D_\alpha(x) = D_\gamma(x)$ for each $\alpha \geq \gamma$.

Since, as already mentioned, (2) is stable at an equilibrium x_0 if and only if $D_1(x_0) = \{x_0\}$, it is natural to give the following definition.

Definition 2 *Let α be an ordinal number. The equilibrium x_0 is stable of order α (or α-stable) if $D_\alpha(x_0) = \{x_0\}$. The equilibrium x_0 is said to be absolutely stable when it is γ-stable.*

The main result in the Auslander and Seibert paper [2] is as follows.

Theorem 1 *The equilibrium x_0 is absolutely stable for system (2) if and only if there exists a generalized Liapunov function which is continuous in a whole neighborhood of the origin.*

3   Systems with input

The notion of prolongation applies also to systems with inputs ([5]). Let us adopt the following agreement: throughout this note an *admissible input* is any piecewise constant function $u(\cdot) : [0, +\infty) \to U$, where U is a preassigned constraint set of $\mathbf{R}^m$. In other words, for each admissible input there are sequences $\{t_k\}$ and $\{u_k\}$ such that

$$0 = t_0 < t_1 < t_2 < \ldots < t_k \ldots$$

and $u(t) \equiv u_k \in U$ for $t \in [t_{k-1}, t_k)$. Assume that for each $u \in U$, the vector field $f(\cdot, u)$ is of class C^1. A *solution* of (1) corresponding to an admissible input $u(\cdot)$ and an initial state x_0 is a continuous curve $x(\cdot; x_0, u(\cdot))$ such that $x(0; x_0, u(\cdot)) = x_0$ and coinciding with an integral curve of the vector field $f(\cdot, u_k)$ on the interval (t_{k-1}, t_k). The *reachable set* $A(x_0, U)$ relative to the system (1) and the constraint set U, is the set of all points lying on solutions corresponding to the initial state x_0 and any admissible input.

Reachable sets are the most natural candidate to play the role of the positive trajectories (3) in the case of systems with inputs. More precisely, let R be a positive real number, and let $U = \overline{B_R}$. We adopt the simplified notation $A^R(x_0) = A(x_0, \overline{B_R})$, and introduce the prolongations

$$D_1^R(x_0) = (\mathcal{D}A^R)(x_0) , \quad D_2^R(x_0) = (\mathcal{D}(\mathcal{I}(\mathcal{D}A^R)))(x_0) \text{ and so on.}$$

Definition 3 *We say that the system (1) is* absolutely bounded input bounded state *stable (in short ABIBS stable) if for each $R > 0$, there exists $S > 0$ such that*

$$|x_0| \le R , \implies |D_\gamma^R(x_0)| \le S ,$$

$\forall t \ge 0$.

The following characterization is easy. The proof is omitted.

Proposition 3 *System (1) is ABIBS-stable if and only if for each $R > 0$ there exists a compact set $K \subset \mathbf{R}^n$ such that $B_R \subset K$ and $D_\gamma^R(K) = K$.*

Definition 4 *A (generalized) ABIBS-Liapunov function for (1) is an everywhere continuous, radially unbounded function $V : \mathbf{R}^n \to \mathbf{R}$ which enjoys the following monotonicity property:*

(MP) *for all $R > 0$, there exists $\rho > 0$ such that for each admissible input $u(\cdot) : [0, +\infty) \to \overline{B_R}$ and each solution $x(\cdot)$ of (1) defined on an interval I and corresponding to $u(\cdot)$, one has that the composite map $t \mapsto V(x(t))$ is non-increasing on I, provided that $|x(t)| \ge \rho$ for each $t \in I$.*

We are now ready to state our main result.

Theorem 2 *System (1) is ABIBS-stable if and only if there exists an ABIBS-Liapunov function.*

The proof of Theorem 2 is given in the following section. We conclude by the remark that in general an ABIBS stable system does not admit ABIBS-Liapunov functions of class C^1. As an example, consider a system of the form (2) for which there exists a continuous function $V(x)$ which is radially unbounded and non-increasing along solutions, but not a C^1 function with the same properties. It is proved in [4] that such systems exist, even with $f \in C^\infty$. Of course, $f(x)$ can be thought of as a function of x and u, constant with respect to u. Any $V(x)$ which is radially unbounded and non-increasing along solutions, can be reinterpreted as an ABIBS Liapunov function.

4 The proof

Sufficient part

Assume that there exists a function $V(x)$ with the required properties. In what follows, we adopt the notation

$$W_\lambda = \{x \in \mathbf{R}^n : V(x) \le \lambda\} .$$

Fix $R_0 = 1$. According to (**MP**) we can associate to R_0 a number ρ_0. In fact, without loss of generality we can take $\rho_0 > R_0$. Let $m_0 = \max_{|y| \le \rho_0} V(y)$, and pick any $\lambda > m_0$. We note that

$$|x| \le \rho_0 \Longrightarrow V(x) \le m_0 \Longrightarrow x \in W_\lambda$$

that is, $B_{\rho_0} \subset W_\lambda$. In fact, there exist some $\eta > 0$ such that $B_{\rho_0+\eta} \subset W_\lambda$.

Lemma 1 *For each $\lambda > m_0$, we have $D_\gamma^{R_0}(W_\lambda) = W_\lambda$.*

Proof Of course, it is sufficient to prove that $D_\gamma^{R_0}(W_\lambda) \subseteq W_\lambda$.

<u>Step 1.</u> *For each $\lambda > m_0$ we have $A^{R_0}(W_\lambda) \subseteq W_\lambda$.*

Indeed, in the opposite case we could find $\bar\lambda > m_0$, $\bar x \in W_{\bar\lambda}$, $\bar y \notin W_{\bar\lambda}$, an admissible input $u(\cdot)$ with values in B_{R_0}, and a positive time T such that $x(T; x_0, u(\cdot)) = \bar y$. Set for simplicity $x(t) = x(t; x_0, u(\cdot))$. Let $\tau \in (0, T)$ such that $x(\tau) \in W_{\bar\lambda}$, while $x(t) \notin W_{\bar\lambda}$ for $t \in (\tau, T]$. Such a τ exists since the solutions are continuous. By construction, $V(x(\tau)) = \bar\lambda < V(\bar y)$. On the other hand, $|x(t)| \ge \rho_0$ on the interval $[\tau, T]$, so that $V(x(t))$ is non-increasing on this interval. A contradiction.

<u>Step 2.</u> *For each $\lambda > m_0$ we have $(\mathcal{D}A^{R_0})(W_\lambda) \subseteq W_\lambda$.*

Even in this case we proceed by contradiction. Assume that it is possible to find $\bar\lambda > m_0$, $\bar x \in W_{\bar\lambda}$, and $\bar y \in (\mathcal{D}A^{R_0})(W_{\bar\lambda})$ but $\bar y \notin W_{\bar\lambda}$. This means $V(\bar x) \le \bar\lambda < V(\bar y)$. Let $\varepsilon > 0$ be such that $\bar\lambda + 3\varepsilon \le V(\bar y)$. Since V is continuous, there exists $\delta > 0$ such that

$$V(x) \le \bar\lambda + \varepsilon < \bar\lambda + 2\varepsilon < V(y)$$

for all $x \in B(\bar x, \delta)$ and $y \in B(\bar y, \delta)$. By the definition of the operator $\mathcal{D}$, we can now take $\tilde x \in B(\bar x, \delta)$ and $\tilde y \in B(\bar y, \delta)$ in such a way that $\tilde y \in A^{R_0}(\tilde x)$. This is a contradiction to Step 1: indeed, since $\tilde x \in W_{\bar\lambda+\varepsilon}$, we should have $\tilde y \in W_{\bar\lambda+\varepsilon}$, as well. On the contrary, the fact that $\bar\lambda + 2\varepsilon < V(\tilde y)$ implies $\tilde y \notin W_{\bar\lambda+\varepsilon}$.

Thus, we have shown that $D_1^{R_0}(W_\lambda) = W_\lambda$ for each $\lambda > m_0$. To end the proof, we need to make use of transfinite induction. Let α be an ordinal number, and assume that the statement

$$D_\beta^{R_0}(W_\lambda) = W_\lambda \quad \text{for each} \ \ \lambda > m_0$$

holds for every ordinal number $\beta < \alpha$. It is not difficult to infer that also

$$(\mathcal{I}(D_\beta^{R_0}))(W_\lambda) = W_\lambda \quad \text{for each} \ \ \lambda > m_0$$

and, hence,

$$\cup_{\beta < \alpha}(\mathcal{I}(D_\beta^{R_0}))(W_\lambda) = W_\lambda \quad \text{for each} \ \ \lambda > m_0 \ . \tag{5}$$

For sake of convenience, let us set $E_\alpha^{R_0} = \cup_{\beta < \alpha}(\mathcal{I}(D_\beta^{R_0}))$. The final step is to prove that

$$(\mathcal{D}E_\alpha^{R_0})(W_\lambda) \subseteq W_\lambda \quad \text{for each} \ \ \lambda > m_0 \ .$$

Assume that there are $\bar\lambda > m_0$, $\bar x \in W_{\bar\lambda}$, and $\bar y \in (\mathcal{D}E_\alpha^{R_0})(W_{\bar\lambda})$ but $\bar y \notin W_{\bar\lambda}$. As before, we have $V(\bar x) \leq \bar\lambda < V(\bar y)$ and, by continuity, for sufficiently small ε we can find δ such that

$$V(x) \leq \bar\lambda + \varepsilon < \bar\lambda + 2\varepsilon < V(y)$$

for all $x \in B(\bar x, \delta)$ and $y \in B(\bar y, \delta)$. Let us choose $\tilde x$ and $\tilde y$ satisfying this last conditions, and such that $\tilde y \in E_\alpha^{R_0}(\tilde x)$. This is possible because of the definition of $\mathcal{D}$. In conclusion, we have $\tilde x \in W_{\bar\lambda + \varepsilon}$, $\tilde y \notin W_{\bar\lambda + \varepsilon}$, and $\tilde y \in E_\alpha^{R_0}(\tilde x)$. A contradiction to (5). The proof of the lemma is complete.

We are finally able to prove the sufficient part of Theorem 2. Fix $\lambda_0 > m_0$. Note that W_{λ_0} is closed (since V is continuous) and bounded (since V is radially unbounded). Hence, $W_{\lambda_0} \subset B_{R_1}$ for some $R_1 > \rho_0 > R_0 = 1$. In addition, it is not restrictive to take $R_1 \geq 2$. Using the properties of V, we find $\rho_1 > R_1$ and define $m_1 = \max_{|x| \geq \rho_1} V(x) \geq m_0$.

By repeating the previous arguments, we conclude that $D_\gamma^{R_1}(W_\lambda) = W_\lambda$ for each $\lambda > m_1$.

Fix $\lambda_1 > m_1$, and iterate again the procedure. We arrive to define a sequence of compact sets $\{W_{\lambda_i}\}$ such that $B_{R_i} \subset W_{\lambda_i}$, with $R_i \to +\infty$, and $D_\gamma^{R_i}(W_{\lambda_i}) = W_{\lambda_i}$.

Let finally R be an arbitrary positive number, and let R_i be the smallest number of the sequence $\{R_i\}$ such that $R \leq R_i$. Set $K = W_{\lambda_i}$. We clearly have

$$B_R \subset B_{R_i} \subseteq K \quad \text{and} \quad D_\gamma^R(K) \subseteq D_\gamma^{R_i}(K) = K \ .$$

The proof of the sufficient part is complete, by virtue of Proposition 3.

Necessary part

The idea is to construct a Liapunov function V by assigning its level sets for all numbers of the form

$$\frac{2^k}{j} \qquad j = 1, \ldots, 2^k, \quad k = 0, 1, 2, \ldots \tag{6}$$

namely, the reciprocals of the so called dyadic rationals. Note that they are dense in $[1, +\infty)$.

Let us start by setting $R_0 = 1$. According to Proposition 3, we can find a compact set denoted by W_{2^0} such that $B_{R_0} \subset W_{2^0}$ and $D_\gamma^{R_0}(W_{2^0}) = W_{2^0}$. Let $R_1 \geq \max\{2, |W_{2^0}|\}$. Using again Proposition 3, we find a compact set W_{2^1} such that $B_{R_1} \subset W_{2^1}$ and $D_\gamma^{R_1}(W_{2^1}) = W_{2^1}$. This procedure can be iterated. Assuming that W_{2^k} has been defined, we take $R_{k+1} \geq \max\{k + 2, |W_{2^k}|\}$ and the compact set $W_{2^{k+1}}$ in such a way that $B_{R_{k+1}} \subset W_{2^{k+1}}$ and $D_\gamma^{R_{k+1}}(W_{2^{k+1}}) = W_{2^{k+1}}$. The sequence $\{W_{2^k}\}$ satisfies the conditions

$$W_{2^k} \subset B_{R_{k+1}} \subset W_{2^{k+1}} \quad \text{and} \quad \cup_k W_{2^k} = \mathbf{R}^n .$$

We have so assigned a set to any dyadic reciprocals $\frac{2^k}{j}$ with $j = 1$, $k = 0, 1, 2, \ldots$. Next, consider pairs k, j such that $k \geq 1$ and $2^{k-1} \leq j \leq 2^k$, that is all dyadic reciprocals such that $2^0 \leq \frac{2^k}{j} \leq 2^1$.

By virtue of Proposition 2, there exists a compact neighborhood K of W_{2^0} such that K is properly contained in W_{2^1} and $D_\gamma^{R_0}(K) = K$. Call it $W_{4/3}$. Note that (beside the endpoints 1 and 2) 4/3 is the unique dyadic reciprocal with $k = 2$ included in the interval $[1, 2]$. Using again Proposition 2 applied to W_{2^0} and $W_{4/3}$ we define two new sets

$$W_{2^0} \subset W_{8/7} \subset W_{4/3} \subset W_{8/5} \subset W_{2^1}$$

such that $D_\gamma^{R_0}(W_{8/7}) = W_{8/7}$ and $D_\gamma^{R_0}(W_{8/5}) = W_{8/5}$. By repeating the procedure, we arrive to assign a compact set W_λ to any dyadic reciprocal $\lambda = \frac{2^k}{j}$ with $k \geq 1$ and $2^{k-1} \leq j \leq 2^k$, in such a way that $D_\gamma^{R_0}(W_\lambda) = W_\lambda$ and

$$W_{2^0} \subset W_\lambda \subset W_\mu \subset W_{2^1}$$

if $\lambda < \mu$. Then we turn our attention to dyadic reciprocals $\frac{2^k}{j}$ with $k \geq 2$ and $2^{k-2} \leq j \leq 2^{k-1}$, that is $2 \leq \frac{2^k}{j} \leq 4$. We proceed as above. This time, we obtain sets W_λ such that $D_\gamma^{R_1}(W_\lambda) = W_\lambda$ and

$$W_{2^1} \subset W_\lambda \subset W_\mu \subset W_{2^4}$$

if $\lambda < \mu$. This construction can be repeated for all k and j. We finally obtain an increasing family of compact sets $\{W_\lambda\}$ with the property that if $2^k \le \lambda < 2^{k+1}$ then $D_\gamma^{R_k}(W_\lambda) = W_\lambda$.

We are now ready to define the Liapunov function $V(x)$ for all $x \in \mathbf{R}^n$ as

$$V(x) = \inf\{\lambda : x \in W_\lambda\} \ .$$

Claim A. For each R there exists ρ such that if $|x| \ge \rho$ and $y \in A^R(x)$ then $V(y) \le V(x)$.

Let R be given and pick the integer k in such a way that $R_k < R \le R_{k+1}$. We prove that the choice $\rho = R_{k+2}$ works.

First of all, we remark that if $|x| \ge \rho$ then $x \notin W_{2^{k+1}}$, so that $V(x) > 2^{k+1}$. Let the integer p be such that $2^{k+1} \le 2^p \le V(x) < 2^{p+1}$, and let λ be a dyadic reciprocal such that $V(x) < \lambda < 2^{p+1}$. Of course $x \in W_\lambda$, and hence

$$A^R(x) \subseteq D_\gamma^R(x) \subseteq D_\gamma^{R_k+1}(x) \subseteq D_\gamma^{R_p}(x) \subseteq W_\lambda \ .$$

It follows that if $y \in A^R(x)$, then $V(y) \le \lambda$. Since λ can be taken arbitrarily close to $V(x)$, Claim A is proved.

Note that Claim A implies property (MP).

Claim B. $V(x)$ is radially unbounded.

Let $N > 0$, and let k be an integer such that $N < 2^k$. For $|x| > R_{2^k+1}$, we have $x \notin W_{2^k}$, that is $V(x) \ge 2^k$.

Claim C. $V(x)$ is continuous.

First, we remark that by construction, V is locally bounded. Assume that we can find a point $\bar{x} \in \mathbf{R}^n$ and a sequence $x_\nu \to \bar{x}$ such that $V(x_\nu)$ does not converge to $V(\bar{x})$. By possibly taking a subsequence, we have

$$\lim_\nu V(x_\nu) = l \ne V(\bar{x}) \ . \tag{7}$$

Assume first that $V(\bar{x}) < l$ and pick a dyadic reciprocal λ in such a way that $V(\bar{x}) < \lambda < l$ and $\bar{x} \in \mathrm{Int}\, W_\lambda$. For all sufficiently large ν, we should have $x_\nu \in W_\lambda$ as well. But then, $V(x_\nu) < \lambda$, and this is a contradiction to (7).

The case $V(\bar{x}) > l$ is treated in a similar way.

Also the proof of the necessary part is now complete.

References

1. Andriano V., Bacciotti A. and Beccari G. (1997) *Global Stability and External Stability of Dynamical Systems*, Journal of Nonlinear Analysis, Theory, Methods and Applications **28**, 1167-1185

112   Andrea Bacciotti

2. Auslander J. and Seibert P. (1964) *Prolongations and Stability in Dynamical Systems*, Annales Institut Fourier, Grenoble **14**, 237-268
3. Bacciotti A. and Mazzi L., *A Necessary and Sufficient Condition for Bounded Input Bounded State Stability of Nonlinear Systems*, SIAM Journal Control and Optimization, to appear
4. Bacciotti A. and Rosier L., *Regularity of Liapunov Functions for Stable Systems*, Systems and Control Letters, to appear
5. Tsinias J., Kalouptsidis N. and Bacciotti A. (1987) *Lyapunov Functions and Stability of Dynamical Polysystems*, Mathematical Systems Theory **19**, 333-354

Optimal Control with Harmonic Rejection of Induction Machine

Iyad Balloul and Mazen Alamir

Laboratoire d'Automatique de Grenoble
ENSIEG - BP 46
38402 Saint-Martin d'Hères, France
Iyad.Balloul@inpg.fr - Mazen.Alamir@inpg.fr

Abstract. In this paper a new algorithm is used to regulate flux and torque in induction machine. The aim is to show the possibility of attenuating the harmonics as early as the phase of the design of the control law while preserving good regulation performances. The control law results from an optimisation algorithm associated to a cost function. By adding some harmonic weighting factor to the regulation term in the cost function, the algorithm carries out a trade-off between regulation performances and harmonic attenuation.

Index terms: Optimisation algorithm, induction machine, optimal control, harmonic attenuation.

1 Introduction

Certainly, robustness and low cost are the key merits of induction machines. However, its structural simplicity hides a great functional complexity; due mainly to its nonlinear dynamics and to the variation of some parameters (such as the rotoric resistance or the load torque). Nevertheless, the induction machine becomes the favorable choice for a lot of applications ranging from railroad tracking to domestic appliances. The power range extends from tens of Watts to tens of Megawatts.
The required performances depend on the nature of the application. Beside the classical features (rise time, static error, *etc.*), a great interest is dedicated to new trends concerning robustness and harmonic rejection.
Harmonic rejection is used to be studied *a posteriori* when analysing the control law. The analysis is often employed to compare the harmonics related to the studied control law with the harmonics related to some classical control scheme (PWM for example).
This introduction is not intended to fully describe the state of the art of induction motor control. However, the interested reader is invited to consult [4] as an example of minimal energy approaches, and [5] as an example of passivity approaches.
In general, two classes of techniques are widely applied to control induction machines, namely:

- Field oriented Control (F.O.C.): where electrical vectors are considered in coordinates related to the rotating field. More precisely, one of the axes is directed in the sense of the rotoric or the statoric flux, so that the instantaneous torque has a scalar expression. This expression allows a dynamic regulation of the torque [6].

- Direct torque Control (D.T.C.): where commutations of the power inverter are directly controlled [2], resulting in a hybrid closed loop system. These approaches seem to better match the discontinuous nature of power machines which work typically on commutation (see §3.2 hereafter). But the poor theoretical foundation (caused mainly by the hybrid system framework) is still the common drawback of these approaches.

This paper shows the use of an optimal control algorithm to regulate the flux and the torque of the induction machine. This algorithm was first proposed to solve min-max non-convex optimisation problems [1]. The algorithm constitutes a tool for handling the constrained robust optimisation of general systems (that are not necessarily affine in control). But certainly, the most relevant property of the algorithm in the context of power machines is its ability to handle the robust optimisation problem with *non-convex* admissible control sets.

In this work, we investigate the application of the deterministic version of the algorithm to control induction machines. First, the cost function is constructed to generate the input profile that yields a minimal regulation error. Then, the cost function is modified to realise a compromise between regulation error and the spectral power at some frequency (i.e. the harmonic to be attenuated).

The proposed method does not only give a theoretically well-based reference for D.T.C. approaches, but also provides a rigorous method to treat new performance requirements:

- Harmonic attenuation problem can be treated when the control law is in the design phase.
- Model uncertainty and disturbances can be considered by the algorithm yielding a robust optimal control law.

In this paper only the first point (harmonic attenuation) is presented. Section (2) shows the formulation of the basic optimisation problem. Section (3) shows the application of the algorithm: first to simply achieve a regulation goal, and then to achieve both regulation and harmonic attenuation.

2   Basic result

Consider the following system defined on $[t_0, t_f]$:

$$\Sigma \begin{cases} \dot{x} = f(x, u, t) \; ; \; x(t_0) = x_0 \\ z = q(x, u, t) \end{cases} \tag{1}$$

where $x \in I\!\!R^n$ is the state, $u \in I\!\!R^m$ is the control input, and $z \in I\!\!R^s$ is the controlled output. The control u satisfies the following constraint:

$$u(t) \in U \; \forall t \in [t_0, t_f] \tag{2}$$

with U some compact set in $I\!\!R^m$.

In the sequel, $X(t; t_0; u)$ and $Z(t; t_0; u)$ stand for the solutions of (1) starting from

t_0 with the control u. The notation $U^{[t_1,t_2]}$ stands for the set of all functions defined on $[t_1, t_2]$ with value in U. Let:

$$J(t_0, t_f, x_0, u) = \Upsilon\big(x(t_f)\big) + \int_{t_0}^{t_f} z^T(t)z(t)dt \qquad (3)$$

be the criterion to be optimized. $\Upsilon(x(t_f))$ indicates the final desired performances, $x_0 = x(t_0)$ is the initial condition, and $z(t) := Z(t; t_0; u)$.
The associated optimisation problem is defined as follows:

$$P(t_0, t_f, x_0) : \min_{u \in U^{[t_0,t_f]}} J(t_0, t_f, x_0, u) \qquad (4)$$

or:

$$P(t_0, t_f, x_0) : \min_{u \in U^{[t_0,t_f]}} \left[\Upsilon\big(x(t_f)\big) + \int_{t_0}^{t_f} z^T(t)z(t)dt \right] \qquad (5)$$

with $x(t_f) = X(t_f; t_0; u)$ and $z(t) = Z(t; t_0; u)$
Since the initial state x_0 is fix, this problem can be written as:

$$P(t_0, t_f, x_0) = \min_{u \in U^{[t_0,t_f]}} \left[\int_{t_0}^{t_f} \left[\Upsilon_x\big(x(t)\big) \cdot f(x(t), u(t), t) + z^T(t)z(t) \right] dt \right] \qquad (6)$$

Using the minimum principle, the deterministic version of the algorithm [1] gives a solution $u \in U^{[t_0,t_f]}$ to the above problem (6). The efficiency of the solution is demonstrated via suitable simulations presented in next section.

3   Application on the induction machine

3.1   System definition

First, the used notations are defined:

- I, V, Φ and Γ designate respectively, currents, voltages, the flux and the torque.
- R, L and M indicate respectively, resistances, self and mutual inductances.
- The index r categorises rotoric variables, and s categorises statoric ones.
- p is the number of poles and Ω is the angular velocity.
- The intermediate coefficients are given by:

$$q = L_s L_r - M^2$$
$$\sigma = 1 - \frac{M^2}{L_s L_r} = \frac{q}{L_s L_r}$$
$$T_s = \frac{L_s}{R_s}$$
$$T_r = \frac{L_r}{R_r}$$

The equations describing the induction machine in the coordinates (α, β) are given by [8,7]:

$$\dot{x} = A(\Omega)x + Bu \\ y = h(x) \tag{7}$$

with $x = \begin{pmatrix} I_s \\ \Phi_s \end{pmatrix}$, $A(\Omega) = \left(\begin{array}{cc|cc} -(\frac{1}{\sigma T_r} + \frac{1}{\sigma T_s}) & -p\Omega & \frac{R_r}{q} & \frac{p\Omega L_r}{q} \\ p\Omega & -(\frac{1}{\sigma T_r} + \frac{1}{\sigma T_s}) & -\frac{p\Omega L_r}{q} & \frac{R_r}{q} \\ \hline & -R_s Id_2 & & 0 \end{array} \right)$

$B = \begin{pmatrix} \frac{L_r}{q} Id_2 \\ Id_2 \end{pmatrix}$ with Id_2 denoting the 2×2 identity matrix, and $u = \begin{pmatrix} V_{s\alpha} \\ V_{s\beta} \end{pmatrix}$

$$h(x) = \begin{pmatrix} h_1(x) \\ h_2(x) \end{pmatrix} = \begin{pmatrix} \Psi \\ \Gamma \end{pmatrix} = \begin{pmatrix} x_3^2 + x_4^2 \\ p(x_2 x_3 - x_1 x_4) \end{pmatrix}$$

where Ψ and Γ stand for flux and torque to be regulated. The numerical values considered in this paper are summarised in table (1).

L_s	L_r	M	R_s	R_r	p	Ω
0.031747 H	0.0323 H	0.031 H	0.07 Ω	0.052 Ω	2	$2\pi100$ rad/s

Table 1. Numerical values of the different parameters of the motor

3.2   Admissible control set

By construction, power machines work on commutation bases. The tension delivered to the machine by the power inverter can be represented in the plane (α, β) as shown in Fig. (1). Therefore, the set of admissible control can be defined by:

$$U = \{V_0, V_1, V_2, V_3, V_4, V_5, V_6, V_7\}$$

Note that V_0 and V_7 are identical; this is simply due to the definition of commutation cycle. The other vectors in the set has the same modulus (here equal to $2000V$), and the phase between two successive vectors is $\frac{\pi}{3}$.

Note also that when the inverter is powering the machine with multilevel control, the set U defined above can be simply extended to include the corresponding extra vectors.

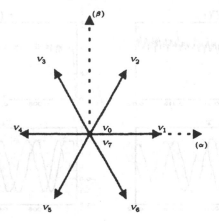

Fig. 1. Tension vectors in the plane (α, β)

3.3 Flux and torque regulation

Let $h_r(t) = \begin{pmatrix} \Psi_r \\ \Gamma_r \end{pmatrix}$ be the output reference (flux and torque). Define the criterion as:

$$J(t_0, t_f, x_0, u, w) = \int_{t_0}^{t_f} \|h(x) - h_r(t)\|_Q^2 dt \tag{8}$$

where $Q := \begin{pmatrix} 1 & 0 \\ 0 & \rho_r \end{pmatrix}$ is some weighting matrix, $\rho_r > 0$

Figure (2) shows the evolution of the machine for the step reference $\Psi_r = 1$ et $\Gamma_r = 100$.

Time step is $\delta = 50 \mu s$. The horizon of control is $t_f - t_0 = 0.2s$. An interesting feature to quote is the convergence of the algorithm after only one iteration.

3.4 Flux and torque regulation with harmonic attenuation

The structure of the optimisation scheme gives the opportunity to treat the harmonics as early as the design phase. Indeed, by including some harmonic-dependant penalty term in the cost function, the algorithm is compelled to select an input profile that attenuates the best this harmonic while preserving good regulation performances.

As an illustration, consider that a current harmonic at $\omega_0 = 2\pi f_0$ is to be attenuated. Then, by adding the following states to the previous model:

$$\dot{x}_5 = x_1 \cos \omega_0 t \qquad x_5(0) = 0 \tag{9}$$
$$\dot{x}_6 = x_1 \sin \omega_0 t \qquad x_6(0) = 0 \tag{10}$$
$$\dot{x}_7 = x_2 \cos \omega_0 t \qquad x_7(0) = 0 \tag{11}$$
$$\dot{x}_8 = x_2 \sin \omega_0 t \qquad x_8(0) = 0 \tag{12}$$

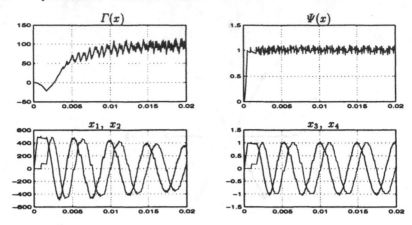

Fig. 2. Motor behavior in open-loop: $\Psi_r = 1$ et $\Gamma_r = 100$, without harmonic rejection

the new state matrix becomes:

$$
A_a(t) = \left(
\begin{array}{c|c}
A(\Omega) & \begin{matrix} 0\,0\,0\,0 \\ 0\,0\,0\,0 \\ 0\,0\,0\,0 \\ 0\,0\,0\,0 \end{matrix} \\
\hline
\begin{matrix} \cos\omega_0 t & 0 & 0\,0 \\ \sin\omega_0 t & 0 & 0\,0 \\ 0 & \cos\omega_0 t\,0\,0 \\ 0 & \sin\omega_0 t\,0\,0 \end{matrix} & 0
\end{array}
\right)
$$

Now, consider the cost function given by:

$$
J(t_0, t_f, x_0, u, w) = \Upsilon\big(x(t_f)\big) + \int_{t_0}^{t_f} \|h(x) - h_r(t)\|_Q^2 dt \tag{13}
$$

and the final performance function (suppose that $t_0 = 0$ to simplify the notation):

$$
\Upsilon(x(t_f)) = \frac{1}{t_f^2}\left(x_5^2(t_f) + x_6^2(t_f) + x_7^2(t_f) + x_8^2(t_f)\right) \tag{14}
$$

This function gives an approximation of the power of the harmonic corresponding to ω_0 in the time interval $[0, t_f]$.

From (14):

$$T_x^T(x(t)) = \frac{2}{t_f^2} \begin{pmatrix} 0 \\ 0 \\ 0 \\ 0 \\ x_5 \\ x_6 \\ x_7 \\ x_8 \end{pmatrix} \tag{15}$$

Although, the results presented in this section are preliminary, they show the efficiency of the algorithm in tackling the question of harmonic attenuation.
We start by showing the current power spectrum without harmonic rejection in Fig. (3).

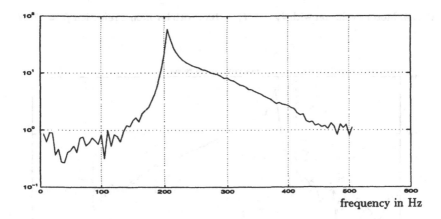

frequency in Hz

Fig. 3. Current power spectrum (logarithmical scale on y) without harmonic rejection

Now, consider that the unwanted harmonic is at $f_0 = \dfrac{\omega_0}{2\pi} = 170Hz$. The algorithm provides the corresponding control profile. Fig. (4) shows the evolution of the motor under this input. Figure (5) shows a comparison between the spectrum in this case and the previous one.
The parameters f and g in the figure are defined as follows:

- the frequency of the harmonic to be attenuated: $f = f_0$,
- the relative gain:

$$g = \frac{X(\omega_0) - X_n(\omega_0)}{X_n(\omega_0)}$$

where X is the current spectral power with harmonic attenuation. X_n is the current spectral power without harmonic attenuation.

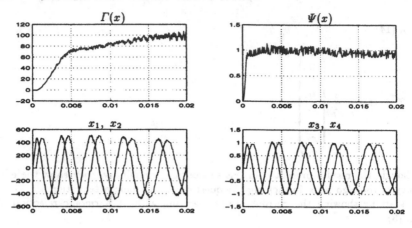

Fig. 4. Motor evolution with $\Psi_r = 1$, $\Gamma_r = 100$, with harmonic rejection $f_0 = 170Hz$

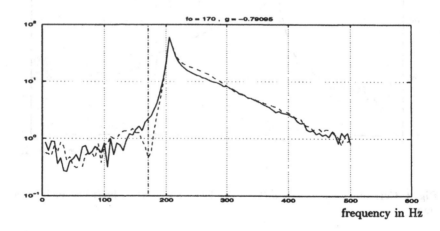

Fig. 5. Current power spectrum (logarithmical scale on y): ($-$) without harmonic rejection, (- -) with harmonic rejection

Remark 1.

The figure shows that the good attenuation of the harmonic at the selected frequency is coupled with some degradation elsewhere. In practical context, some harmonics are more annoying than others. Consequently, a good attenuation of these annoying harmonics will be appreciated even if it increases other harmonics as a second effect.

4 Conclusion

A key feature of the algorithm explored in this paper is its ability to realise a double purpose of regulation as well as harmonic attenuation. The examples studied are chosen purely to illustrate the functionality of the algorithm. Certainly, a further work is still to be done to examine workable implementation issues:

- The algorithm involves on-line computation making the time of execution a crucial factor in any implementation framework. In the present case, a Pentium200 platform gives the solution in about 10 *ms*. Obviously, this is not fast enough to allow real time application of the algorithm. But in the other hand, it is not too far as a preliminary result. Improvement of execution time could be achieved using neural networks to emulate the optimisation algorithm. Furthermore, it might be helpful to employ more advanced operational research techniques (more intelligent than the used combinatory one), especially if the number of vectors in U is increased.
- The algorithm has some results characterising its convergence, but it lacks a theory guaranteing this convergence. Thus, an efficient control scheme would run the algorithm under a suitable supervision layer.

Another attractive perspective of this work is to take into account robustness considerations. Actually, in this paper only the deterministic version of the algorithm is used. Whereas, the original algorithm allows a robust synthesis that can consider exogenous disturbances and parametric uncertainties. In this respect, it worths mentioning that this algorithm has already been successfully applied in its min-max version on batch processes [1] and on open-channel hydraulic systems [3].

References

1. Alamir M., Balloul I. (1999) Robust constrained control algorithm for general batch processes. Int. Jornal of Control. **72**(14), 1271–1287
2. Bornard G., Thomas J. L., Poullain S., Bethoux O. (1997) Machine asynchrone: commande à cycles limites controlés. Brevet (tous pays industrialisés). **9701346**
3. Chen M. L., D. Georges (2000) Nonlinear robust control of an open-channel hydraulic system based on an infinite-dimensional model. In: 11th IFAC Workshop Control Applications of Optimization. St. Petersburg, Russia. Accepted
4. Georges, D., de Wit C. C., Ramirez J. (1999). Nonlinear H_2 and H_∞ optimal controllers for current-fed induction motors. IEEE Trans. on Automatic Control **44**(7), 1430–1435.
5. Nicklasson, P. J., Ortega R., Espinosa-Pérez G. (1997). Passivity-based control of a classe of blondel-park transformable electric machines. IEEE Trans. on Automatic Control **42**(5).
6. Roboam, X. (1997). État de l'art de la commande à flux orienté, partie ii. In: École d'été d'automatique de Grenoble, Commande des machines à courant alternatif.
7. Roye, D. (1997). État de l'art de la commande à flux orienté, partie i. In: École d'été d'automatique de Grenoble, Commande des machines à courant alternatif.
8. von Raumer, T. (1994). Commande adaptative non linéaire de machine asynchrone. Phd thesis. Laboratoire d'Automatique de Grenoble, INPG.

Nonlinear QFT Synthesis Based on Harmonic Balance and Multiplier Theory

Alfonso Baños[1], Antonio Barreiro[2], Francisco Gordillo[3], and Javier Aracil[3]

[1] Dept. Informática y Sistemas
Universidad de Murcia
Murcia, Spain
abanos@dif.um.es

[2] Dept. Ingeniería de Sistemas y Automática
Universidad de Vigo
Vigo, Spain
abarreiro@uvigo.es

[3] Dept. Ingeniería de Sistemas y Automática
Universidad de Sevilla
Sevilla , Spain

Abstract. The problem of deriving conditions for a stabilising linear compensator in a uncertain nonlinear control system is addressed, for some types of memoryless nonlinearities like the saturation or the dead-zone. The approach is to incorporate to QFT conditions given by the application of harmonic balance and multiplier techniques, providing the designer with a very transparent tool for synthesising stabilising compensators, balancing between different possible alternatives.

1 Introduction

The existence of nonlinearities such us saturations or dead zones in control systems may give rise to the emergence of limit cycles. Well known methods are available in the literature to predict the existence of limit cycles, namely the harmonic balance and the describing function methods [11,9]. The latter has been a very popular method for decades in spite of its approximate character. Part of this success is due to the fact that is much less conservative than other rigorous methods to test stability, such as absolute stability criteria (see for example [14]). Absolute stability results provide a formal framework for the analysis of global stability, giving sufficient and usually very conservative frequency domain conditions. Recently, multiplier theory has been considered in the literature for minimising the conservativeness of previous absolute stability criteria. In a practical problem, the designer must consider the different techniques balancing between conservativeness and heuristics.

When applying absolute stability techniques, the source of conservatism arises from the confinement of certain nonlinear relation inside a cone or conic sector. The stability conditions are conservative because they hold for the whole family of functions inside this cone. Many approaches to shape the cone to the actual nonlinear relation and relax the conditions have been reported in the literature. In [12] a lot of possible multipliers (or integral quadratic constraints, IQC) are listed that can be chosen for a particular problem. These techniques have achieved a particular success in the context of antiwindup control schemes [10], under saturation or dead-zone nonlinearities. In [1,2] the problem of obtaining the less-conservative bounding cones was addressed in the context of QFT techniques for uncertain nonlinear systems.

We are mainly interested in robust nonlinear control problems, having a linear compensator to be designed, and that may be decomposed as a feedback interconnection of a linear system and a (usually memoryless) nonlinear system. An important aspect of the linear system is that it is allowed to have uncertain linear subsystems. The problem considered is to compute restrictions over the linear feedback compensator for avoiding limit cycles, and in general to obtain global stability. On the other hand, Quantitative Feedback Theory (QFT) [8] is specially well suited for dealing with uncertainty, and it has been recently shown how it can be efficiently used to adapt robust versions of classical absolute stability results such as Circle and Popov Criteria [1,2]. As a frequency domain design technique, it can also be expected that QFT can incorporate harmonic balance methods as well as stability conditions based on multiplier theory.

The goal of this work is to give a first step to provide a common QFT framework for solving the above stability problem using different frequency domain techniques, in particular harmonic balance and multiplier theory. A special class of memoryless nonlinear system will be considered here, including symmetric and asymmetric characteristics. The main idea is to substitute the nonlinear system by an equivalent frequency locus, a complex region that may depend on the frequency. In a second stage this complex locus is used jointly with the uncertain linear subsystems to compute restrictions over a stabilising compensator in the frequency domain. As a result, it will be possible to compare different techniques, and what is more important, to give the designer a transparent tool for evaluating the balance conservativeness/heuristics.

The content of the paper is as follows. In Section 2, we introduce the structure of the nonlinear feedback system that will be considered, and we address it within the robust QFT techniques, treating the nonlinearity by the describing function method. A relevant example, regarding the control of an electric motor in the presence of backlash, is developed. In Section 3, we present an extension of the harmonic balance to the case of asymmetric nonlinearities, and apply it to the example. In Section 4, we discuss the possible use and interpretation of the words 'frequency locus' for a nonlinear block. In Section 5, the technique of positive multipliers is introduced and applied to the example, and finally comparisons and conclusions are outlined.

2   Limit Cycles Analysis Based on the Harmonic Balance

Consider a SISO Lure's feedback system $L(H, N)$ where the linear part is a linear fractional transformation of a system $G(s)$, the controller to be designed, and (possibly) uncertain linear blocks $P_i(s)$, i = 1...4 (Fig. 1). The problem is to derive conditions for the existence of limit cycles, suitable for the design of $G(s)$ in order to (usually) avoid them. The problem is complicated for the existence of uncertainty in the linear subsystems $P_i(s)$, that will be considered in the framework of QFT [8], that is in the form of templates. Templates include parametric as well as unstructured uncertainty.

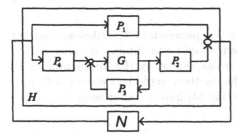

Fig. 1. The Lur'e type nonlinear system: G is the feedback compensator, $P_i, i = 1...4$ are (possibly uncertain) blocks of the plant, and N is the nonlinear block of the plant

A well-known solution for analysing the existence of limit cycles if given by the harmonic balance method [11,9], usually applied to a system $L(H, N)$ without uncertain blocks. The simplest variant of this method is the describing function method used when the nonlinear block has odd symmetry and the linear part is a low-pass filter. In this simple case, there are no limit cycles if (see for example [14])

$$1 + G(j\omega)N(a, \omega) \neq 0 \tag{1}$$

for every amplitude a and every frequency ω. Here $N(a, \omega)$ is the describing function of the nonlinear system N. A direct application to the system of Fig. 1 results in

$$1 + H(j\omega)N(a, \omega) \neq 0 \tag{2}$$

where

$$H(j\omega) = P_1(j\omega) + \frac{P_2(j\omega)G(j\omega)P_4(j\omega)}{1 + G(j\omega)P_3(j\omega)} \tag{3}$$

Eq. (2)-(3) can be integrated in QFT as a robust stability boundaries, being the describing function incorporated as an additional transfer function with uncertain parameter a. Let $L(j\omega)$ be defined as $L(j\omega) = H(j\omega)N(a, j\omega)$. Thus, condition (2) is equivalent to

$$\left| \frac{L(j\omega)}{1 + L(j\omega)} \right| < \infty \tag{4}$$

In practice, a finite bound δ have to be used, that in addition guarantees some finite distance to the bound given by (2), that is

$$\left| \frac{L(j\omega)}{1 + L(j\omega)} \right| < \delta \tag{5}$$

In this paper, δ is considered as a given parameter. Nevertheless, another approach that can be easily integrated would be to obtain the value of $\delta(\omega)$ in such a way that it includes the errors caused by the approximation of the method (see [9]). In this way, this method would be rigorous, although probably more conservative. Substituting (3) into (5), after some simple calculations we obtain

$$\left| \frac{P_1(j\omega)N(a, j\omega) + \tilde{P}(j\omega)N(a, j\omega)G(j\omega)}{1 + P_1(j\omega)N(a, j\omega) + (\tilde{P}(j\omega)N(a, j\omega) + P_3(j\omega))G(j\omega)} \right| < \delta, \tag{6}$$

where $\tilde{P}(j\omega) = P_1(j\omega)P_3(j\omega) + P_2(j\omega)P_4(j\omega)$. Thus (6) is the final condition to be satisfied by $G(j\omega)$ for every frequency ω, for every value of $P_i(j\omega), i = 1 \ldots 4$ (note that in general $P_i(j\omega)$ may be uncertain), and for every amplitude a. For any frequency, (6) defines a bound over the controller $G(j\omega)$ for avoiding limit cycles.

Example

Condition (6) can be treated in the framework of quadratic inequalities, developed in [3] and adapted in [1] to robust absolute stability problems. Although no details about computation will be given here, a realistic example is developed in the following. The nonlinear system in this example is borrowed from Example 1 in [13]. It represents an electric motor driving a load through a gear in the presence of backlash. The system is supposed to be embedded in the feedback structure of Figure 2.a, where

$$H_1(s) = \frac{1}{s(J_m s + B_m)}, H_2(s) = \frac{1}{s(J_l s + B_l)}$$

and the parameters are known to be in some intervals, $k_m \in 0.041[1, 1.2]$, $k_l \in$ 4.8[0.85, 1], $B_m \in 0.00322[1, 20]$, $B_l \in 0.00275[0.4, 1]$, $J_m = 6.39e - 6$, and $J_l =$ 0.0015. After some manipulations, the control system is transformed to a Lure's type system (Fig. 2.b), where $P_1 = k_l(H_1 + H_2)$, $P_2 = -k_l H_1$, $P_3 = k_m H_1$, and $P_4 = k_l H_1$, being the describing function $N(a)$ given as a function of the amplitude a. For the dead-zone nonlinearity $N(a)$ is a number in the interval [0,1] for any value of a, and independent of ω.

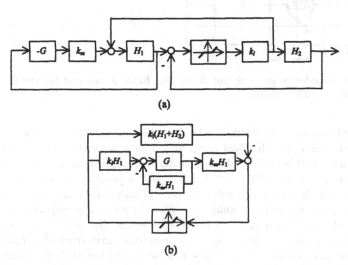

(a)

(b)

Fig. 2. (a) Modelling of the motor, (b) Transformation of the control system to a Lure's system

Equation (6) can be used to compute bounds over $G(j\omega)$ for avoiding limit cycles. Here $\delta = 10$ is used, meaning that not only (2) is satisfied for every possible combination of parameters, but also that the critical point (-1,0) is "protected" by a circle centred at -1.01 and with radius 0.1 (in general, the center is $\frac{\delta^2}{1-\delta^2}$ and the radius $\frac{\delta}{1-\delta^2}$). Results are given in Fig. 3.

Boundaries have been computed for $P_{3,0}(j\omega)G(j\omega)$, where $P_{3,0}(j\omega)$ stands for a nominal value of $P_3(j\omega)$. For each frequency, $P_{3,0}(j\omega)G(j\omega)$ must lie above the solid-line boundaries and under the dashed-line boundaries, in order to satisfy condition (6) and thus avoiding limit cycles.

3   Harmonic Balance with asymmetric nonlinearities

The above analysis works for the case in which the nonlinearity has odd symmetry, like the dead-zone nonlinearity considered in the previous Example. However, in many practical cases this requisite is not fulfilled, and thus it is

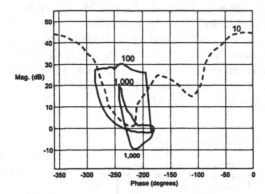

Fig. 3. Stability boundaries given by the describing function method for the frequencies 10, 100 and 1000 rad/sec.

convenient to extend the above analysis to more general situations including asymmetric nonlinearities. The dual describing function method [4] can be useful in this task. In this method the self-sustained oscillations are assumed to be of the form $y(t) = a_0 + Re(a_1 e^{j\omega t})$, where $y(t)$ represents the output of the linear part. Let the signal $y(t)$ enter in the nonlinear block, and consider its steady state periodic output signal $z(t)$. This signal can be expanded in series form as $z(t) = N_0 a_0 + Re(N_1 a_1 e^{j\omega t}) + ...$, where N_0 and N_1 are, respectively, the bias gain and the first order harmonic gain from the input of the nonlinear part to its output. They can be computed with the usual expressions of the Fourier coefficients as

$$N_0(a_0, a_1, \omega) = \frac{1}{2\pi} \int_{-\pi}^{\pi} z(t) d(\omega t)$$

$$N_1(a_0, a_1, \omega) = \frac{1}{\pi a_1} \int_{-\pi}^{\pi} z(t) e^{j\omega t} d(\omega t)$$

In order to fulfil the first-order harmonic balance the zero and first order terms of $y(t)$ must be equal to the corresponding terms of the output of the linear part to $z(t)$ (neglecting the higher harmonics)

$$1 + H(0)N_0(a_0, a_1, 0) = 0$$
$$1 + H(j\omega)N_1(a_0, a_1, \omega) = 0$$

Therefore, the condition for the no-existence of limit cycles is that one of the following inequalities is fulfilled for every value of a_0 and a_1, and for every frequency ω

$$1 + H(0)N_0(a_0, a_1, 0) \neq 0 \qquad (7)$$

$$1 + H(j\omega)N_1(a_0, a_1, \omega) \neq 0$$

Notice that the dual describing function method allows to deal with a more general problem: not only limit cycles can be detected but also equilibrium points (for $a_1 = 0$ or $\omega = 0$ and $a_0 \neq 0$). In this way, the scope of application of the method is extended since the emergence of multiple equilibria and limit cycles are the most common causes of loosing global stability [5,6]. Following a reasoning similar to the Section 2, a condition identical to (6) must be satisfied, substituting $N(a, \omega)$ by $N_1(a_0, a_1, \omega)$.

Example

Consider again the electric motor example of Section 2, where the dead-zone nonlinearity is now considered asymmetric as in Fig.4., where $b_- \leq 0$ and $b_+ \geq 0$. The difference of having an asymmetric dead-zone nonlinearity is that now we need a dual describing function, given by $N_0(a_0, a_1)$ and $N_1(a_0, a_1)$. In this case, the describing function is independent of the frequency, since the nonlinearity is memoryless. In addition, the function $N_1(a_0, a_1)$ take values in the interval $[0,1]$ for any real value of a_0 and a_1, and independently of the parameters b_- and b_+ . The easiest way to use Equation (7) to avoiding the existence of multiple equilibrium points and limit cycles is to derive conditions over $G(j\omega)$ to satisfy the second inequality, which is a sufficient condition.

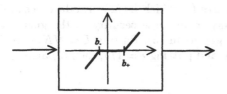

Fig. 4. Asymmetric dead-zone nonlinearity

This second condition can be embedded in the QFT framework using (6), where the function $N(a, \omega)$ is substituted by $N_1(a_0, a_1) = [0,1]$, a similar condition to the obtained in Section 2 . As a result, boundaries given in Fig. 3 are not only valid for a symmetric dead-zone, but also for any asymmetric dead-zone.

4   Frequency Locus for the Nonlinear Block

Although the main objective of this work is to obtain frequency locus for the controller $G(j\omega)$, in the form of boundaries, however, in order to analyse the degree of robustness/conservativeness achieved, it may be interesting to consider the 'frequency locus' filled by the nonlinear block N. As the nonlinear frequency response is not a so clearly defined object as the linear one, it will be an approximate object, but useful for interpretations.

In the example of Section 1, the dead-zone can be represented by an arbitrary number in $[0,1]$, so that we can identify it with $N = [0,1] \subset R$, a bounded intervalar uncertainty. The closed loop $L(H, N)$ will not have limit cycles when (2) holds, or equivalently when $H(j\omega) \neq -N^{-1}$, or:

$$H(j\omega) \cap -N^{-1} = H(j\omega) \cap (-\infty, -1] = \emptyset \tag{8}$$

Thus the linear part $H(j\omega)$ must avoid the critical locus $-N^{-1} = (-\infty, -1]$. It should be noticed that this property, derived from the describing function, is shared by many other different nonlinear blocks, including the asymmetric dead-zone, for which the first harmonic is real between 0 and 1. For example, the unit saturation also has the 'frequency locus' $N = [0, 1]$ and 'critical locus' $-N^{-1} = (-\infty, -1]$. The stability (lack of limit cycles) of the loop $L(H, N)$ is based on $H(j\omega)$ avoiding the same locus, for N being the (symmetric or asymmetric) dead-zone as well as the saturation. So there is some kind of approximation in this approach. Otherwise, the set of all stable linear $H(s)$ will be unable to discriminate dead-zone and saturation, in the sense that $H(s)$ with a dead-zone is stable if and only if $H(s)$ with a saturation is stable, which seems a rather unlikely property.

The validity of this 'frequency locus' analysis for such nonlinear blocks is strongly confirmed by the well-know validity of the harmonic balance principle for most of common practical cases (in which $H(j\omega)$ introduces a suitable low-pass effect). An additional way to increase security in the guaranteed stability is to define a robustness gap as in (5). Putting $L = HN$, omitting arguments for $H(j\omega)$ and $N(a, \omega)$, and putting $\epsilon = 1/\delta$, then (5) amounts to

$$
\begin{aligned}
|1 + (HN)^{-1}| &> \epsilon \\
(HN)^{-1} &\notin -1 + \epsilon U \\
H &\notin N^{-1}(-1 + \epsilon U)^{-1} = (-N^{-1})(-1 + \epsilon U)/(1 - \epsilon^2)
\end{aligned}
\tag{9}
$$

where U stands for the unit circle in the complex plane. Equation (9) shows the effect of introducing a robustness gap $\delta = 1/\epsilon$ in (5) on the forbidden critical locus: $-N^{-1}$ is changed by $(-N^{-1})(-1 + \epsilon U)/(1 - \epsilon^2)$. Thus, each point z in the critical locus is changed by a forbidden circle, centered at

$z/(1 - \epsilon^2)$, and with radius $|z|\epsilon/(1 - \epsilon^2)$. The union of all these circles forms the 'robust' critical locus. Fig. 5 shows this locus for the unit saturation and dead zone, for different values of $\delta = 1/\epsilon$.

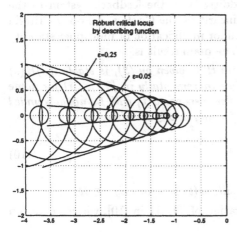

Fig. 5. Critical locus for the unit saturation and dead zone

It is worthwhile to analyse the relationship with other robust frequency bounds for $L = HN$ that have been proposed in the literature. For example, in [7] the closed loop region for $L/(1+L)$ is a rectangle, instead of a circle like in (5). In this case, it can be seen that the corresponding open loop condition is $L = HN \notin C$ where C is a clover, formed by the union of four circles centred at $-1 \pm r/2$ and $-1 \pm jr/2$ and with radius $r/2$, where $d = 1 + 1/r$ is the rectangle half-width. Then, the locus for H takes the form $H \notin (-N^{-1})C$, where C is the given clover. In fact, this locus is introduced to simplify the performance monitoring, which is not our approach, so we will considered (5) as an adequate robustness bound and frequency loci like in Fig. 5.

5   Comparison with Absolute Stability Criteria

An interesting question is about the comparison between the harmonic balance principle as developed in Sections 2 and 3, and robust absolute stability criteria. The former gives approximated solutions to the global stability problem, while the latter gives rigorous solutions at the cost of being much more conservative. Robust absolute stability in the framework of QFT has been developed in [1,2]. In this Section, we only consider memoryless nonlinearities like the dead-zone example of Section 2 or Section 3, which have the property of lying incrementally in $[0,1]$. A memoryless nonlinear system N given by a relation $y = f(x)$ is said to lie incrementally inside the sector $[k_1, k_2]$ if

$k_1 \leq \frac{f(x)-f(x')}{x-x'} \leq k_2$, for every $x, x' \in \mathbf{R}$. Our sense of I/O stability is finite gain L_2 stability. More formally, a system $H : L_{2,e} \rightarrow L_{2,e}$ is finite gain L_2 stable if for every input $x \in L_2$, the output $y = Hx$ is in L_2 and, in addition, $\|y\|_2 < k \|x\|_2$ for some constant k. Closed-loop stability is defined as I/O stability from any input entering additively to the feedback system to the rest of signals. For this type of nonlinearities, the Circle and Popov Criteria used in [1] are generalised using a minor modification of multiplier stability conditions, as found in [10] or [12]. The main result is:

Theorem: The system $L(H, \mathbf{N})$, where H is given by (9), is stable if i)$H(s)$ is stable, ii)$\mathbf{N}$ is static, odd, and lies incrementally in [0,1], iii) for some (possibly noncausal) $W(s)$, with impulse response $w(t)$ with L_1 norm bounded by $\|w(t)\|_1 < 1$, condition (6) holds for any $a, \omega \in \mathbf{R}$, where

$$N(a, \omega) = -(-1 + \frac{ja}{1 - W(j\omega)})^{-1} \tag{10}$$

and iv)if $\mathbf{N}$ is not odd then additionally $w(t) > 0$ for any $t \in \mathbf{R}$.

Proof: The proof is a minor modification of Th. 11 in [10], where condition iii was based on the inequality

$$Re((1 - W(j\omega)(H(j\omega) + 1)) > \epsilon \tag{11}$$

The limit condition $Re((1 - W(j\omega)(H(j\omega) + 1)) = 0$ defines a boundary for $H(j\omega)$. This boundary can be obtained solving the equation $1 - W(j\omega)(H(j\omega)+1) = ja$ for any $a \in \mathbf{R}$. After some simple manipulations the result is

$$1 - (-1 + \frac{ja}{1 - W(j\omega}))^{-1} H(j\omega) = 0$$

or $1 + N(a, \omega)H(j\omega) = 0$, where $N(a, \omega)$ is given by (10). Finally, from this last condition it is straightforward to prove that if (6) holds for this $N(a, \omega)$ then the original condition (11) is satisfied (note that there is an extra degree of robustness given by the parameter δ in (6)).

Following Section 4, this theorem can be interpreted in terms of a stabilising nonlinear frequency locus N given by (10), or a critical locus given by $-N^{-1}$, that is

$$-N^{-1}(a, \omega) = -1 + \frac{ja}{1 - W(j\omega)} \tag{12}$$

that in general are frequency dependent. For each frequency ω, the critical locus as given by (12) can be represented in the complex plane as straight lines passing through the -1 point, while the nonlinear frequency locus are circles.

Comparison of frequencial and critical locus

Note that for $W(s) = 0$ (no multiplier), we recover the classical (and very conservative) bound from the circle criterion, the complex half-space $Re(z) < -1$. In general, the only influence of $W(s)$ in the complex plane over the nonlinear frequency locus is to rotate the bound $Re(z) = -1$ with an angle given by the phase of $(1 - W(jw))$. By finding appropriate multipliers $W(jw)$, the conservative bound given by the Circle Criterion can be remarkable relaxed. An interesting question is how much it can be relaxed, and how close to the describing function locus $(\infty, -1]$ can be. Results in this line will clarify the limits of conservative's for both the describing function and the multiplier techniques.

The main difficulty is that the search space of possible multipliers $W(s)$ is infinite-dimensional. This was arranged in [10]in the context of antiwindup design, introducing a particular, parameterized family of multipliers in the form

$$W(s) = \frac{a_0}{s+1} + \frac{a_1}{(s+1)^2} + ... + \frac{b_0}{s-1} + \frac{b_1}{(s-1)^2} \tag{13}$$

For values of $W(j\omega)$ close to zero, N as given in (10) approaches the complex circle around $[0,1]$ (Circle Criterion). However, when $1 - W(j\omega)$ introduces a phase shift N becomes an off-axis circle, that will be referred to as $C_W(\omega)$. For stability based on multipliers, the graph of $-H^{-1}(j\omega)$ must avoid the circle $C_W(\omega)$, and for stability based on harmonic balance $-H^{-1}(j\omega)$ must avoid the real segment $[0,1] \subset C_W(\omega)$. But when $-H^{-1}(j\omega))$ lies on the lower (upper) half-plane, it is expected that the circle $C_W(\omega)$ can be moved up (down) enough so that the forbidden region on the current half-plane is very close to $[0,1]$. In this case, the excess of conservatism given by the Circle Criterion can be minimised using appropriate multipliers.

¿From (10) it is easy to obtain that the centre $c(\omega)$ and the radius $r(\omega)$ of $C_W(\omega)$ are given by

$$c(\omega) = 0.5(1 + jtan(\phi(\omega))) \tag{14}$$
$$r(\omega) = |c(\omega)|$$

where $\phi(\omega) = Angle(1 - W(j\omega))$. Note that at frequencies where $W(j\omega)$ is real there is no phase shift, that is $\phi = 0$, and we recover the circle of the Circle Criterion. Fig.6 shows the frequency nonlinear locus for a nonlinearity satisfying the conditions of the above theorem, and for $W(s) = 0.3(\frac{1}{s+1} + \frac{1}{s-1})$. It is shown the angle introduced by the multiplier as well as the boundaries for $H(j\omega)$ in the complex plane. On the other hand Fig. 7 shows circles $C_W(\omega)$.

Once the multiplier $W(j\omega)$ is fixed and N is identified with the complex locus defined by the circles $c(\omega) + r(\omega)U$ given by (14), it is straightforward to obtain boundaries for $G(j\omega)$. It suffices to apply the inequality (6) with N given by the family of circles.

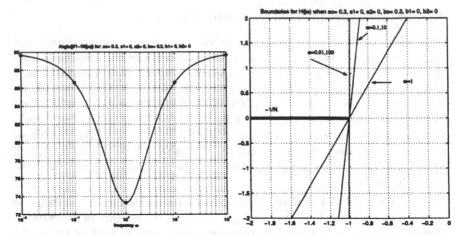

Fig. 6. Left: Phase shift for the multiplier $W(s) = 0.3(1/(s+1) + 1/(s-1))$, right: associated critical locus

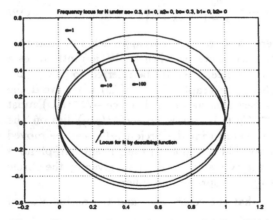

Fig. 7. Frequency locus using the multiplier $W(s) = 0.3(1/(s+1) + 1/(s-1))$

Example

For the electric motor example of Section 2, application of the stability theorem based on multipliers is valid for both the symmetric and asymmetric dead-zone nonlinearity, since they are static, and lies incrementally in [0,1]. The multiplier $W(s) = \frac{999}{s+1000}$ has been chosen, because it provides a phase shift of almost 90 between 10 and 100 rad/s. In fact, $1 - W(s) = \frac{s+1}{s+1000}$ is a lead network. Resuts are given in Fig. 8. Boundaries have been computed for $P_{3,0}(j\omega)G(j\omega)$, where $P_{3,0}(j\omega)$ stands for a nominal value of $P_3(j\omega)$. Note that for every frequency, $P_{3,0}(j\omega)G(j\omega)$ must lie above the solid-line boundaries and under the dashed-line boundaries.

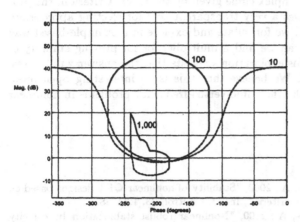

Fig. 8. Stability boundaries computed using multiplier $W(s) = 999/(s + 1000)$, for the frequencies 10,100 and 1000 rad/sec.

Boundaries of Fig. 3, given by the describing function method, can be used as a guide to measure how good is a multiplier, if the resulting boundaries are close to them. In this sense, one may conclude that the chosen multiplier give reasonable results, in the sense the corresponding boundaries, shown in Fig 8, are no overly conservative. Of course, a more rigorous study is needed in order to formalise and systematise this procedure.

Conclusions

The problem of designing stabilising linear compensators for uncertain non-linear feedback systems has been addressed. A solution for a special type of nonlinear systems has been developed in the frequency domain, using the framework of QFT. The main features of QFT are frequency discretization and an algorithmic approach for deriving boundaries under arbitrary plant uncertainty. QFT boundaries are given in the Nichols chart, thus it is easy for the designer to decide between different solutions and compare alternative formulations. As a result, it has been found a condition (Equation (6)) for the computation of restrictions over a linear compensator using standard QFT algorithms. This condition treats the nonlinearity as a region of the complex plane referred to as the nonlinear frequency locus. In this way, it has been possible to adapt and compare two different approaches to nonlinear stability: harmonic balance and multiplier techniques. As it is well-known, harmonic balance is an approximate technique giving very relaxed conditions for global stability (in a qualitative sense, avoidance of multiple equilibrium points and limit cycles), while multiplier techniques gives global I/O-stability conditions but with some degree of conservatism. Since multipliers can be interpreted as phase shifters of the complex circle given by the Circle Criterion, the proposed QFT approach gives a very transparent tool for electing appropriate multipliers. In this sense, we formulate and explore in a example-based way the 'scope question' of: can we find a multiplier for for proving stability of every stable system?. This exploration suggests that the answer will be 'yes' in many practical cases. We believe that this is an interesting open question, which has links with other time-versus-frequency problems in nonlinear stability.

References

1. Baños, A., and Barreiro, A., 2000, "Stability of nonlinear QFT designs based on robust absolute stability criteria", Int. J. Control, 73, 1,74–88.
2. Barreiro, A. and Baños, A., 2000, "Nonlinear robust stabilization by conicity and QFT techniques", Automatica, Vol. 36, No. 9.
3. Chait, Y., and Yaniv, O. 1993, "MISO computer-aided control design using the Quantitative Feedback Theory". Int. J. Robust and Nonlinear Control, 3, 47–54.
4. Cook, P.A., 1986, "Nonlinear Dynamical Systems", Prentice-Hall.
5. Cuesta, F., Gordillo, F., Aracil, J. and Ollero, A., 1999 "Global Stability Analysis of a Class of Multivariable Takagi-Sugeno Fuzzy Control Systems".IEEE Trans. Fuzzy Systems, 7, 5, 508–520.
6. Gordillo, F., Aracil J. and Ollero, A. (1999) "Robust Stability Analysis of MIMO Systems with Asymmetric Nonlinearities". Proceedings of the 14th World Congress of IFAC, Vol. E, 123–127.
7. Gustafsson, F., and S.F. Graebe, 1998, "Closed Loop Performance Monitoring in the presence of system changes and disturbances", Automatica, 34, 1311–1326.
8. Horowitz, I., 1993, Quantitative Feedback Design Theory (QFT), QFT Publications, Boulder, Colorado.
9. Khalil, H.K., 1996, Nonlinear Systems, Prentice Hall, Upper Saddle River, NJ.
10. Kothare, M.V., and M. Morari, 1999, "Multiplier Theory for Stability Analysis of Anti-Windup Control Systems", Automatica, 35, 917–928.
11. Mees, A.I.,1981, Dynamics of Feedback Systems, Wiley, New York.
12. Megretski, A. and A. Rantzer, 1997, "System Analysis via Integral Quadratic Constraints", Automatica, 42 , 819–830.
13. Oldak, S., Baril, C., and Gutman, P. O., 1994, "Quantitative design of a class of nonlinear systems with parameter uncertainty", Int. Journal of Robust and Nonlinear Control, 4, 101-117.
14. Vidyasagar, M., 1978, Nonlinear Systems Analysis, Prentice Hall, Englewoods Cliffs.

Group Invariance and Symmetries
in Nonlinear Control and Estimation

John S. Baras

Department of Electrical and Computer Engineering,
Department of Computer Science and
the Institute for Systems Research,
University of Maryland
College Park, MD 20742, USA
baras@isr.umd.edu

Abstract. We consider nonlinear filtering problems, nonlinear robust control problems and the partial differential equations that characterize their solutions. These include the Zakai equation, and in the robust control case two coupled Dynamic Programming equations. We then characterize equivalence between two such problems when we can compute the solution of one from the solution of the other using change of dependent, independent variables and solving an ordinary differential equation. We characterize the resulting transformation groups via their Lie Algebras. We illustrate the relationship of these results to symmetries and invariances in physics, Noether's theorem, and calculus of variations. We show how using these techniques one can solve nonlinear problems by reduction to linear ones.

1   Introduction

Symmetries have played an important role in mathematical physics as well as in systems and control. Symmetries in mathematical physics [1] are essential. Essentially all physics theories can be based in symmetries and symmetry properties. Some of the more celebrated results are:

(i)   Conservation laws; various physics theories.
(ii) Quantum electrodynamics, elementary particles, quarks, strings.
(iii) Quantum field theory, reductions, symmetry braking.

Symmetries have been also fundamental in systems and control. Perhaps the most well known principle has been the unifying role that equivalences of internal and external representations and associated groups of transformations play in system theory. Some of the more celebrated results are:

(a) Electrical networks and realization theory.
(b) Feedback invariants.
(c) Nonlinear Filtering and Estimation Algebra.
(d) Parameterizations of Rational Transfer Functions.
(e) Canonical forms of linear analytic systems (linear in the controls).
(f) Feedback linearization.
(g) Symmetries in multibody mechanical systems and continuum mechanics

Given the rich interplay between mathematical physics and control systems, especially variational problems and optimal control, there remain many unexplored theoretical and applied aspects of symmetries for systems and control. In this paper we describe our research in this direction with focus on stochastic estimation and nonlinear control.

Both mathematical physics and systems and control deal with differential equations (DE). Therefore, symmetry groups of differential equations and systems of differential equations provide a natural starting point for understanding the key methods and concepts. As a simple example consider a scalar ordinary differential equation (ODE):

$$F\left(x, u, \frac{du}{dx}\right) = \frac{du}{dx} - f(x, u) = 0 \ . \tag{1}$$

The left hand side of (1) can be viewed as defining a surface in $\mathbb{R}^3$ (three variables: $x, u, \frac{du}{dx}$). The middle term of (1) is the ODE and its solutions are scalar valued curves. A *Symmetry Group* of an ODE [1] is a group of transformations on (x, u) (the independent and dependent variables) which maps any solution of the DE to another solution of the DE. Similarly for systems of DEs. Thus finding symmetry groups for (1) amounts to finding transformations (diffeomorphisms)

$$\left. \begin{array}{l} H : \mathbb{R}^2 \to \mathbb{R}^2 \\ (x, u) \ \mapsto \ (\phi(x, u), \psi(x, u)) \end{array} \right\} \tag{2}$$

which permute solution curves. Finding such groups is a celebrated old problem initiated by Lie and later extended by Ovsjannikov and many others. Continuing with this simple example, if we find such a transformation H, we can extend it to derivatives using the simple observation that if a curve passes through (x, u) with slope du/dx, its image (under H) passes through (x', u') with slope du'/dx' where:

$$\begin{aligned} x' &= \phi(x, u) \\ u' &= \psi(x, u) \\ \frac{du'}{dx'} &= \frac{(\psi_x + \psi_u \ du/dx)}{(\phi_x + \phi_u \ du/dx)} \end{aligned} \tag{3}$$

The map:

$$\begin{array}{c} H' : \mathbb{R}^3 \to \mathbb{R}^3 \\ \left(x, u, \dfrac{du}{dx}\right) \ \mapsto \ \left(x', u', \dfrac{du'}{dx'}\right) \end{array} \tag{4}$$

is an extension of H.

A key theorem in the investigation of symmetry groups for ODEs is the establishment of the result that H permutes solutions of the ODE (1) iff H'

leaves the surface in $\mathbb{R}^3$, defined by the left-hand side of (1) invariant. This equivalence gives to the problem of constructing symmetry groups of ODEs a very attractive geometric foundation. The so called Lie-Ovsjannikov method [1] for constructing symmetry groups of ODEs is to find all H' that have the surface $s = f(x, u)$ as an invariant manifold.

This idea, as worked out in this simple example, can be extended directly to n^{th} order ODEs and to systems of ODEs. For an nth order ODE

$$F(x, u, u_1, u_2, \cdots, u_{n-1}) = u_n - f(x, u_1, u_2, \cdots, u_{n-1}) = 0$$
$$u_i = \frac{d^i u}{dx^i} \tag{5}$$

one extends a transformation H on (x, u), n-fold to a transformation H' on $(x, u, u_1, \cdots, u_n)$. The Lie-Ovsjannikov method (1) extends as well.

More interesting are one-parameter Lie groups which leave the solutions of (5) invariant:

$$\left.\begin{array}{l} x' = X(x, u; \epsilon) \\ u' = U(x, u; \epsilon) \end{array}\right\} \tag{6}$$

The infinitesimal generator of this group

$$X = \xi(x, u)\frac{\partial}{\partial x} + n(x, u)\frac{\partial}{\partial u} \tag{7}$$

plays a fundamental role. One extends the group n-fold to derivatives of all orders to get an nth order group and n^{th} order infinitesimal generator $X^{(n)}$ [1]. Some of the basic results of the theory are:

1. The one-parameter Lie group leaves the ODE invariant iff its nth order extension leaves the surface $F = 0$ invariant.
2. The family of all solution curves is invariant under the Lie group iff it is a symmetry group.
3. **Theorem** (Lie): The one-parameter Lie group is an invariance group of $F = 0$, iff

$$X^{(n)} = (u_n - f(x, u, u_1, \cdots, u_{n-1})) = 0, \quad \text{when}$$
$$u_n = f(x, u, u_1, \cdots, u_{n-1}) \ .$$

The consequences of these foundations were pursued by Lie who showed how to construct the Lie group (of invariance), and that the Lie algebra of infinitesimal generators determine the local Lie group.

The subject has attracted many researchers through the years. Some of the more interesting results that have been obtained are:

- The reduction of the intractable nonlinear conditions of group invariance to linear homogeneous equations, which determine the infinitesimal generators.
- Invariance of an ODE under a one-parameter Lie group of point transformations leads to reduction of the order of the ODE by one.
- Invariance of an nth order ODE under an r-parameter Lie group with solvable Lie algebra is reduced to an $(n - r)^{\text{th}}$ order ODE plus r quadratures (integrals).
- Invariance of a linear partial differential equation (PDE) under a Lie group leads to superposition of solutions in terms of transforms.

We are more interested in symmetry and invarance groups of PDEs. These transformation groups are *local* Lie groups. *Point symmetries* are point transformations on the space of independent and dependent variables. *Contact symmetries* are contact transformations acting on the space of independent, dependent variables and derivatives of dependent variables. Ovsjannikov showed that if a system of PDEs is invariant under a Lie group, we can find special solutions of the system, which are called *similarity solutions*.

A further generalization of these concepts with key significance for both systems and control and mathematical physics are the Lie-Bäcklund symmetries (or transformations) [1]. In these transformations the infinitesimal generators of the local Lie groups depend on derivatives of the dependent variables up to any finite order:

$$\left. \begin{array}{l} x' = x + \epsilon\xi(x, u, u_1, u_2, \cdots, u_p) + O(\epsilon^2) \\ u' = u + \epsilon\eta(x, u, u_1, u_2, \cdots, u_p) + O(\epsilon^2) \end{array} \right\} \tag{8}$$

It is a basic result in the theory of such transformations that the infinitesimal generators can be computed by a simple extension of Lie's aglorithmn. Another key result is that invariance of a PDE under a Lie-Bäcklund symmetry usually leads to invariance under an infinite number of symmetries (connected by recursion operators) [1].

A most celebrated results in variational problems with foundational consequences in mathematical physics is E. Noether's Theorem [1]. Euler-Lagrange equations are the governing equations of many physical systems; they are of fundamental importance in mathematical physics. Euler-Lagrange equations provide the dynamics of systems from a variational formulation (typically energy-based variational formulation). These ideas from mathematical physics have inspired many research efforts in systems and control: from stability theory, to dissipative systems, to communication network routing, to robot path planning (to mention just a few).

In this context a physical system has independent variables x in $\mathbb{R}^n$ and dependent variables u in $\mathbb{R}^m$. The system independent variables x can take values in a domain Ω of $\mathbb{R}^n$. A function is given or constructed which depends on the independent variables x, the dependent variables u, and derivatives of the dependent variables up to order k, $u_1, u_2, \cdots, u_k$. The dynamics of the system evolve so that the paths $u(x)$ correspond to extremals of the integral

$$J(u) = \int_\Omega L(x, u, u_1, u_2, \cdots, u_k) \, dx \tag{9}$$

The function L is called a *Lagrangian* and the integral $J(u)$ an *action integral*. The path $u(x) = (u^1(x), u^2(x), \cdots, u^m(x))$ describes the state evolution of the system and typically has to satisfy some boundary conditions on $\partial\Omega$. Such formulations are well known and used by control theorists and practitioners. Clearly, if $u(x)$ is an extremum of (9), any infinitesimal change

$$u(x) \mapsto u(x) + \epsilon v(x) \tag{10}$$

which also satisfies the boundary conditions, should leave $J(u)$ unchanged to order $O(\epsilon)$.

The most significant relationship of this formulation is with respect to conservation laws of a system. A *conservation law* of a system, is an equation in divergence-free form

$$div \; f = \sum_{i=1}^{n} D_i \; f_i(x, u, u_1, u_2, \cdots, u_k) = 0 \tag{11}$$

Equation (11) must hold for any extremal function $u(x)$ of (9). The vector f is called a conserved *flux* [1] since (11) implies that a net flow of f through any closed surface in the space x is zero.

Euler-Lagrange equations are the (often dynamical) equations that need to be satisfied by an extremum of (9). We refer to [1, pp. 254-257] for a concise and clear derivation. The Euler-Lagrange equations can be ODEs or PDEs dependent on the problem. As such, we may ask the question if they have symmetry groups (invariance groups). Noether's key idea and result was that in order to find conservation laws it is far more fruitful to investigate the invariance of the action integral (9). Noether considered Lie-Bäcklund transformations that leave the action integral invariant:

$$x' = x + \epsilon\xi(x, u, u_1, u_2, \cdots, u_p) + O(\epsilon^2)$$
$$u' = u + \epsilon\eta(x, u, u_1, u_2, \cdots, u_p) + O(\epsilon^2) \tag{12}$$

Noether showed that the existence of such transformations lead constructively to conservation laws of the corresponding Euler-Lagrange equations. She established the explicit relationship between the infinitesimals ξ, η and the conserved flux f. For a concise proof of this celebrated theorem we refer to [1, p. 257-260]. This celebrated theorem induced fundamental reformulations of mathematical physics, bringing a certain degree of unification. They include:

- Invariance under time translation leads to energy conservation.
- Invariance under translation or rotation in space, leads to conservation of linear or angular momentum.
- Relativity theory formulations.

The relationship with symmetry groups allows the determination of variational symmetries using Lie's algorithm. Noether's theorem resulted in many specific applications (for specific physical systems or phenomena) such as [1]:

- Conservation of the Runge-Lenz vector in Kepler's problem.
- Existence of infinity of conservation laws for the Korteweg-deVries equation and other soliton equations.

We close this brief review of the history of research on symmetry groups for ODEs, PDEs and dynamical systems by listing some more recent results and activities. Symmetry groups allow discovery of related DEs of simpler form. This for instance leads to transformations that map a given equation to a target equation. Comparing the Lie groups of symmetries admitted by each equation, actually allows the construction of the mapping. Such results have significant implications in facilitating the solution of new ODEs and PDEs using solutions of other ODEs and PDEs, known already.

For our subject, it is important to consider transformations beyond local symmetries. These are transformations where the dependence on u and the derivatives of u is global (*i.e.* not just through the instantaneous values $u(x)$). Gauge transformations in mathematical physics and quantum field theory can be such global transformations. In the theory of symmetry groups such transformations are called potentials.

Ideas, techniques and algorithms from symmetry groups have made fundamental contributions to mathematical physics. As it should be clear from this brief exposition there is great potential for similar impact and fundamental new advances by the systematic exploitation of symmetry groups in systems and control problems. Many of the advances in mathematical physics came out of application of symmetry groups in fundamental PDEs of mathematical physics. This inspires us to apply similar techniques in the fundamental PDEs of systems and control: dynamic programming, Zakai equations for nonlinear filtering, information state equations for robust control and others. In addition Noether's theorem can lead to significant advances in nonlinear optimization. The results described in the subsequent sections are a small set of what could be accomplished by such methods in systems and control.

2   Constructive Use of Symmetry Groups of PDEs: A Simple Example

An interesting, for systems and control (as we shall see) theory, application of ideas from symmetry groups is the following. Use the symmetry group of a PDE to compute easily solutions to new PDEs. This is a non-conventional use of symmetry groups developed by Rosencrans [14].

To explain the idea clearly, we use the simple example of the heat equation.

$$\frac{\partial u(t,x)}{\partial t} = \frac{\partial^2 u(t,x)}{\partial x^2}. \tag{13}$$

It is well known [7][12] that (13) is invariant under the variable transformation

$$\left. \begin{array}{l} x \longmapsto e^s x \\ t \longmapsto e^{2s} t. \end{array} \right\} \tag{14}$$

That is to say if $u(t,x)$ is a solution of (13), so is $u(e^{2s}t, e^s x)$. Clearly the initial data should be changed appropriately. So if ϕ is the initial data for u, the initial data for the transformed (under 14)) solution are $\phi(e^s x)$. This elementary invariance can be written symbolically as

$$e^{tD^2} e^{sxD} \phi = e^{sxD} e^{2^s t D^2} \phi. \tag{15}$$

Here

$$D := \frac{\partial}{\partial x} \tag{16}$$

$$D^2 := \frac{\partial^2}{\partial x^2}.$$

Often in this paper we shall give double meaning to exponentials of partial differential operators. Thus while $exp(tD^2)$ in (15) denotes the semigroups generated by D^2 [13], $exp(sxD)$ is viewed as an element of the Lie group of transformations generated by xD. It is easy to verify that

$$\phi(e^s x) = [exp(sxD)\phi](x) \ , \tag{17}$$

where we view $exp(sxD)$ as such a transformation, with parameter s. Now the association

$$(t,s) \longmapsto e^{tD^2} e^{sxD} \tag{18}$$

defines a two parameter semigroup with product rule

$$(t,s) \cdot (t_1, s_1) := (t_1 exp(-2s) + t, s + s_1) \ , \tag{19}$$

because of the invariance (15). A one parameter subgroup is

$$\left. \begin{array}{l} t = a(exp(2cr) - 1) \\ s = -cr \end{array} \right\} \tag{20}$$

where a, c are positive constants and $r > 0$ is the group parameter. To this subgroup (18) associates the one parameter semigroup of operators

$$H(r) := exp^{a(exp(2cr)-1)D^2 - crxD}$$
$$= e^{crxD} e^{a(1-exp(-2cr))D^2} . \tag{21}$$

It is straightforward to compute the infinitesimal generator of H

$$M\phi := lim_{r \to 0} \frac{H(r)\phi - \phi}{r} = 2acD^2\phi - cxD\phi . \tag{22}$$

But in view of (21) and (22) we have the operator identity

$$e^{Mt} = e^{-crtD} e^{a(1-exp(-2ct))D^2} . \tag{23}$$

To understand the meaning of (23) recall that for appropriate functions ϕ, $exp(Mt)\phi$ is the solution to the initial value problem

$$\frac{\partial}{\partial t} w(t, x) = [Mw](x) = 2ac\frac{\partial^2}{\partial x^2} w(t, x) - cx\frac{\partial}{\partial x} x(t, x) \tag{24}$$
$$w(0, x) = \phi(x)$$

Then (23) suggests the following indirect procedure for solving (24):

Step 1: Solve the simpler initial value problem

$$\frac{\partial}{\partial t} u(t, x) = \frac{\partial^2}{\partial x^2} u(t, x) \tag{25}$$
$$u(0, x) = \phi(x)$$

Step 2: Change independent variables in u to obtain w via

$$w(t, x) = u(a(1 - exp(-2ct)), exp(-ct)x) . \tag{26}$$

Here we have interpreted the exponential in (23) as a transformation of variables.

This simple example illustrates the main point of this particular application of symmetry groups: knowing that a certain partial differential equation (such as (13)) is invariant under a group of local transformations (such as (20)) can be used to solve a more difficult equation (such as (24)) by first solving the simpler equation (such as (13)) and then changing variables.

This idea has been developed by S.I. Rosencrans in [8] [14]. It is appropriate to emphasize at this point that this use of a group of invariance of a certain PDE is not traditional. The more traditional use of group invariance is discussed at length in [7] [12], and is to reduce the number of independent variables involved in the PDE. Thus the traditional use of group invariance, is just a manifestation and mathematical development of the classical similarity methods in ODE.

The point of the simple example above is to illustrate a different use of group invariance which goes roughly as follows: given a parabolic PDE

$$u_t = Lu \tag{27}$$

and a group of local transformations that leave the solution set of (27) invariant, use this group to solve a "perturbed" parabolic PDE

$$w_t = (L + P)w \tag{28}$$

by a process of variable changes and the possible solution of an ordinary (not partial) differential equation. The operator P will be referred to as the "perturbation".

One of the contributions in this paper can be viewed as an extension of the results of Rosencrans to the stochastic partial differential equations that play a fundamental role in nonlinear filtering theory.

3   The Invariance Group of a Linear Parabolic PDE.

Consider the general, linear, nondegenerate elliptic partial differential operator

$$L := \sum_{i,j=1}^{n} a_{ij}(x)\frac{\partial^2}{\partial x_i \partial x_j} + \sum_{i=1}^{n} b_i(x)\frac{\partial}{\partial x_i} + c(x)id \ . \tag{29}$$

and assume that the coefficients a_{ij}, b_i, c are smooth enough, so that $\mathcal{L}$ generates an analytic group [13], denoted by $exp(tL)$, for at least small $t \geq 0$, on some locally convex space X of initial functions ϕ and appropriate domain $Dom(L)$.

Let V be the set of solutions to

$$\frac{\partial u}{\partial t} = Lu \tag{30}$$
$$u(0, x) = \phi(x)$$

in X, as we vary ϕ. The aim is to find a local Lie transformation group G which transforms every element of V into another element of V, that is an **invariance group** of (30) or of L. Note that G induces a group $\tilde{G}$ acting on the space of functions on M with values in $\mathbb{R}^p$, denoted by $\mathcal{F}(M; \mathbb{R}^p)$. The element $\tilde{g}$ corresponding to a g in G will map the function A into A', i.e.

$$A' = \tilde{g}(A) \ . \tag{31}$$

It is easy to show [14] that G and $\tilde{G}$ are isomorphic as groups. We are interested in groups $\tilde{G}$ acting linearly. For that we need:

Definition: $\tilde{G}$ is linear if there exists a Lie group of transformations $\Sigma : M \to M$ such that for each $\tilde{g}\epsilon\tilde{G}$, there exists a $\sigma\epsilon\Sigma$, a $p \times p$ matrix "multiplier" $\nu = \nu(x,\tilde{g})$ and a solution ψ of (30) such that

$$\tilde{g}(A)(x) = \nu(x,\tilde{g})A(\sigma(x)) + \psi(x) \quad . \tag{32}$$

The meaning of (32) is rather obvious. The way $\tilde{G}$ acts on functions is basically via the "coordinate change" group Σ of M. The main result of Rosencrans [14], concerns the case of a single parabolic equation (30), $i.e..$

Theorem 3.1 [14]: Every transformation $\tilde{g}$ in the invariance group $\tilde{G}$ of a linear parabolic equation is of the form

$$u(t,x) \longmapsto \nu(p(t,x))u(p(t,x)) + \psi(x) \tag{33}$$

where p is a transformation acting on the variables (t,x), ψ a fixed solution of the parabolic equation.

Clearly for linear parabolic equations $\tilde{G}$ is always infinite dimensional since it always includes the infinite dimensional subgroup $\tilde{\mathcal{H}}$ consisting of transformations of the form

$$A \longmapsto cA + \phi \tag{34}$$

where $A\epsilon\mathcal{F}(M;\mathbb{R})$, c a scalar $\neq 0$, ϕ a fixed solution of (30). Because of (34) one says that $\tilde{G}$ acts as a **multiplier representation** of Σ upon the space of solutions of (30).

We consider now one-parameter subgroups of the invariance group G of a given partial differential equation. That is we consider subgroups of G of the form $\{X_s\}$ where s "parametrizes" the elements. According to standard Lie theory the infinitesimal generators of these one-parameter subgroups form the Lie algebra $\Lambda(G)$ of the local Lie group G [7]. We shall, using standard Lie theory notation, denote X_s by $exp(sX)$ where X is the infinitesimal generator of the one parameter group $\{X_s\}$. Thus $X\epsilon\Lambda(G)$. Clearly the elements of $\Lambda[G]$ can be considered as first order partial differential operators in $\mathbb{R}^{n+1}$

$$X = \gamma(x,u)\frac{\partial}{\partial u} - \sum_{i=1}^{n} \beta_i(x,u)\frac{\partial}{\partial x_i}. \tag{35}$$

Indeed this follows from an expansion of $exp(sX)(x,u)$ for small s. Now $\{X_s\}$ induces a one-parameter subgroup $\{\tilde{X}_s\}$ in $\tilde{G}$, acting on functions. Let $\tilde{X}$ be the infinitesimal generator of $\{\tilde{X}_s\}$. Given a function $A\epsilon\mathcal{F}(\mathbb{R}^n;\mathbb{R})$ let

$$A(s,x) := \tilde{X}_s(A)(x). \tag{36}$$

If x_i, u are transformed to x_i', u' by a specific one-parameter subgroup $exp(sX)$ of G we can expand

$$u' = A(x) + s\gamma(x, A(x) + 0(s^2)$$
$$x_i' = x_i - s\beta_i(x, A(x) + 0(s^2). \tag{37}$$

Thus

$$A(x') = A(x) - s\sum_{i=1}^{n}\beta i(x, A(x))\frac{\partial A(x)}{\partial x_i} + 0(s^2)$$

or

$$
\begin{aligned}
\tilde{X}(A)(x) &= \lim_{s\to 0}\frac{A(s, x) - A(0, x)}{s} \\
&= \lim_{s\to 0}\frac{A(s, x) - A(x)}{s} \\
&= \lim_{s\to 0}\frac{A(s, x') - A(x')}{s} \\
&= \lim_{s\to 0}\frac{u' - A(x')}{s} \\
&= \gamma(x, A(x)) + \sum_{i=1}^{n}\beta_i(x, A(x))\frac{\partial A(x)}{\partial x_i}. \tag{38}
\end{aligned}
$$

In view of (33) the condition for $\tilde{G}$ to be linear is that [14]

$$\beta_{i,u} = \gamma_{uu} = 0 \ . \tag{39}$$

The best way to characterize G (or $\tilde{G}$) is by computing its Lie algebra $\Lambda(G)$ (or $\Lambda(\tilde{G})$). A direct way of doing this is the following. By definition $\tilde{X}\epsilon\Lambda(\tilde{G})$ if

$$\mathcal{D}(A) = 0 \Longrightarrow \mathcal{D}(e^{s\tilde{X}}A) = 0 \quad \text{for small } s. \tag{40}$$

When $\mathcal{D}$ is linear this reduces to

$$\mathcal{D}(A) = 0 \Longrightarrow \mathcal{D}(\tilde{X}(A)) = 0, \tag{41}$$

since

$$\frac{d}{ds}\mathcal{D}(e^{s\tilde{X}}A) = \mathcal{D}(e^{s\tilde{X}}(A))$$

implies (40) if we set $s = 0$. It is not difficult to show that (40) leads to a system of partial differential equations for γ and β_i.

We shall consider further the determination of $\Lambda(\tilde{G})$ in the case when $\tilde{G}$ is linear, since it is the only case of importance to our interests in the present paper. Then in view of (39)

$$\begin{aligned}
\beta_i(x, u) &= \beta_i(x) \\
\gamma(x, u) &= u\delta(x) + \phi(x)
\end{aligned} \qquad (42)$$

for some β_i, δ, ϕ. Let us denote by β the vector $[\beta_1, \beta_2, \cdots, \beta_n]^T$. Then if A is a solution of (30), another solution is

$$A(s, x) = exp(s\tilde{X})A,$$

which satisfies

$$\frac{\partial}{\partial s}A(s, x) = \delta(x)A(x) + \sum_{i=1}^{n}\beta_i(x)\frac{\partial A(x)}{\partial x_i} + \phi(x)$$

$$A(0, x) = A(x) \qquad (43)$$

in view of (38) and due to the linearity assumption (42). The crucial point is that (43) is a first order hyperbolic PDE and thus it can be solved by the method of characteristics. The latter, very briefly, entails the following. Let $\epsilon(t)$ be the flow of the vector field $\sum_{i=1}^{n}\beta_i\frac{\partial}{\partial x_i}$, $i.e.$ the solution of the ODE

$$\frac{d}{dt}\epsilon(t, x) = \beta(\epsilon(t, x))$$

$$\epsilon(0, x) = x. \qquad (44)$$

Then from (43)

$$\begin{aligned}
\frac{d}{dt}A(s - t, \epsilon(t, x)) &= -\delta(\epsilon(t, x))A(s - t, \epsilon(t, x)) \\
&\quad + \phi(\epsilon(t, x))
\end{aligned}$$

and therefore

$$\begin{aligned}
A(s, x) &= exp(\int_0^s \delta(\epsilon(r, x))dr)A(\epsilon(s, x)) \\
&\quad + \int_0^s \Phi(t, x)dt
\end{aligned} \qquad (45)$$

where

$$\Phi(t, x) = exp(\int_0^t \delta(\epsilon(r, x))dr)\phi(\epsilon(t, x)). \qquad (46)$$

By comparison with (33) one can view $exp(\int_0^s \delta(\epsilon(r, x))dr)$ as the "multiplier" ν. (45) clearly displays the linearity of $\tilde{G}$ near the identity.

The most widely known example, for which $\Lambda(\tilde{G})$ has been computed explicitly is the heat equation (13). The infinitesimal generators in this case are six, as follows

$$\frac{\partial}{\partial t}, \quad 2t\frac{\partial}{\partial t} + x\frac{\partial}{\partial x}, \quad \frac{\partial}{\partial x}$$

$$1, \quad 2t\frac{\partial}{\partial x} + x, \quad 4t^2\frac{\partial}{\partial t} + 4tx\frac{\partial}{\partial x} + x^2. \tag{47}$$

Let us apply these general results to a linear parabolic equation, like (40). From Theorem 3.1, then $\tilde{G}$ is linear. The infinitesimal generators of $\tilde{G}$ are given in view of (38) (42) (note that $x_1 = t$ here) by

$$Z = \alpha(t,x)\frac{\partial}{\partial t} + \sum_{i=1}^{n}\beta_i(t,x)\frac{\partial}{\partial x_i} + \gamma(t,x)id \quad, \tag{48}$$

for some functions α, β_i, γ of t and x. If u solves (30) so does

$$v(s) = exp(sZ)u, \quad \text{for small } s. \tag{49}$$

However v is also the solution of

$$\frac{\partial}{\partial s}v = \alpha\frac{\partial}{\partial t}v + \sum_{i=1}^{n}\beta_i\frac{\partial}{\partial x_i}v + \gamma v$$

$$v(0) = u, \tag{50}$$

a first order hyperbolic PDE (solvable by the method of characteristics). Clearly since $\frac{\partial}{\partial t} - L$ is linear (40) applies and therefore

$$Zu\epsilon V \quad \text{if } u\epsilon V. \tag{51}$$

The converse is also true: if (51) holds for some first order partial differential operator, Z is a generator of $\tilde{G}$.

Now (51) indicates how to compute α, β, γ. Namely

$$(\frac{\partial}{\partial t} - L)u = 0 \tag{52}$$

implies

$$(\frac{\partial}{\partial t} - L)(\alpha u_t + \sum_{i=1}^{n}\beta_i\frac{\partial u}{\partial x_i} + \gamma u) = 0. \tag{53}$$

For $u\epsilon V$ the second reads

$$\alpha_t u_t + \sum_{i=1}^{n}\beta_{i,t}u_{x_i} + \gamma_t u + \alpha u_{tt} + \sum_{i=1}^{n}\beta_i u_{x_{i,t}} + \gamma u_t = LZu, \tag{54}$$

or

$$\frac{d}{dt}Zu = (LZ - ZL)u \ , \tag{55}$$

or

$$\frac{d}{dt}Z = [L, Z] \ \text{ on } \ V. \tag{56}$$

In (56) [,] denotes commutator and $\frac{d}{dt}Z$ is symbolic of

$$\alpha_t \frac{\partial}{\partial t} + \sum_{i=1}^{n} \beta_{i,t} \frac{\partial}{\partial x_i} + \gamma_t \ id$$

Thus the elements of $\Lambda(\tilde{G})$ in this case satisfy a Lax equation. It is immediate from (56) that Z form a Lie algebra. Furthermore it can be shown [14] that α is independent of x, *i.e.* $\alpha(t, x) = \alpha(t)$ and that every Z satisfies an ODE

$$d_l \frac{d^l Z}{dt^l} + d_{l-1} \frac{d^{l-1} Z}{dt^{l-1}} + ... + d_0 Z = 0 \tag{57}$$

where $\ell \leq dim\tilde{G}$.

4   Using the Invariance Group of a Parabolic PDE in Solving New PDEs.

In this section we use the results of the previous section, to generalize the ideas presented via the example of section 2. We follow Rosencrans [8][14].

Thus we consider a linear parabolic equation like (30) and we assume we know the infinitesimal generators Z of the nontrivial part of $\tilde{G}$. Thus if u solves (30), so does $v(s) = exp(sZ)u$ but with some new initial data, say $R(s)\phi$. That is

$$e^{sZ} e^{tL} = e^{tL} R(s) \ \text{ on } \ X. \tag{58}$$

Now $R(\cdot)$ has the following properties. First

$$\lim_{s \longrightarrow 0} R(s)\phi = \phi. \tag{59}$$

Furthermore from (58)

$$\begin{aligned} e^{tL} R(r) R(s)\phi &= e^{rZ} e^{tL} R(s)\phi = e^{rZ} e^{sZ} e^{tL} \phi \\ &= e^{(r+s)Z} e^{tL} \phi = e^{tL} R(r+s)\phi. \end{aligned} \tag{60}$$

Or

$$R(r)R(s) = R(r+s) \quad \text{for} \quad r, s \geq 0 \tag{61}$$

From (60), (61), $R(\cdot)$ is a semigroup. Let M be its generator:

$$M\phi = \lim_{s \to 0} \frac{R(s)\phi - \phi}{s}, \quad \phi \in \text{Dom}(M). \tag{62}$$

It is straightforward to compute M, given Z as in (48). Thus

$$M\phi = \alpha(0)L\phi + \sum_{i=1}^{n} \beta_i(0,x)\frac{\partial \phi}{\partial x_i} + \gamma(0,x)\phi. \tag{63}$$

Note that M is uniquely determined by the Z used in (58). The most important observation of Rosencrans [8] was that the limit as $t \to 0$ of the transformed solution $v(s) = exp(sZ)u$, call it w, solves the new initial value problem

$$\frac{\partial w}{\partial s} = Mw$$
$$w(0) = \phi. \tag{64}$$

That is

$$e^{sZ}e^{tL} = e^{tL}e^{sM} \quad \text{on} \quad X \tag{65}$$

or

$$Ze^{tL} = e^{tL}M \quad \text{on Dom}(L).$$

This leads immediately to the following generalization of discussions in section 2: To solve the initial value problem

$$\frac{\partial w}{\partial s} = Mw$$
$$w(0) = \phi \tag{66}$$

where

$$M = \alpha(0)L + \sum_{i=1}^{n} \beta_i(0,x)\frac{\partial}{\partial x_i} + \gamma(0,x)id \tag{67}$$

follow the steps given below.

Step 1: Solve $u_t = Lu, u(0) = \phi$.

Step 2: Find generator Z of $\tilde{G}$ corresponding to M and solve

$$\left.\begin{array}{l} \frac{\partial}{\partial s}v = \alpha(t)\frac{\partial}{\partial t}v + \sum_{i=1}^{n}\beta_i(t,x)\frac{\partial}{\partial x_i}v + \gamma(t,x)v \\ v(0) = u \end{array}\right\}$$

via the method of characteristics. Note this step requires the solution of ordinary differential equations only.

Step 3: Set $t = 0$ to $v(s,t,x)$.

This procedure allows easy computation of the solution to the "perturbed" problem (66)-(67) if we know the solution to the "unperturbed" problem (30). The "perturbation" which is of degree $\leq 1^{st}$, is given by the part of M:

$$P = \sum_{i=1}^{n}\beta_i(0,x)\frac{\partial}{\partial x_i} + \gamma(0,x). \quad id. \tag{68}$$

We shall denote by $\Lambda(P)$ the set of all perturbations like (68), that permit solutions of $u_t = (L + P)u$ to be computed from solutions of $u_t = Lu$, by integrating only an additional ordinary differential equation. We would like to show that $\Lambda(P)$ is a Lie algebra strongly related to the Lie algebra $\Lambda(G)$ of the invariance group of L.

Definition: The Lie algebra $\Lambda(P)$ will be called the **perturbation algebra** of the elliptic operator L.

To see the relation between $\Lambda(\tilde{G})$ and $\Lambda(P)$, observe first that each generator Z in $\Lambda(\tilde{G})$ uniquely specifies an M, via (48), (63). Conversely suppose M is given. From the Lax equation (56) we find that

$$\frac{dZ}{dt}\Big|_{t=0} = [L,Z]\,|_{t=0} = [L,M]$$

$$= \alpha_t(0)L + \sum_{i=1}^{n}\beta_{i,t}(0,x)\frac{\partial}{\partial x_i} + \gamma_t(0,x)id. \tag{69}$$

Note that the right hand side of (69) is another perturbed operator M'. Thus given an M, by repeated bracketing with L all initial derivatives of Z can be obtained. Since from (57) Z satisfies a linear ordinary differential equation, Z **can be determined from** M. So there exists a 1-1 correspondence between $\Lambda(\tilde{G})$ and the set of perturbed operators M, which we denote by $\Lambda(M)$. It is easy to see that $\Lambda(M)$ is a Lie algebra isomorphic to $\Lambda(\tilde{G})$. Indeed let Z_i correspond to M_i, $i = 1,2$. Then from (65) we have

$$e^{tL}[M_1,M_2]\phi = e^{tL}M_1M_2\phi - e^{tL}M_2M_1\phi$$
$$= Z_1e^{tL}M_2\phi - Z_2e^{tL}M_1\phi$$
$$= Z_1Z_2e^{tL}\phi - Z_2Z_1e^{tL}\phi = [Z_1,Z_2]e^{tL}\phi. \tag{70}$$

This establishes the claim. Since each perturbation P is obtained from an M by omitting the component of M that involves the unperturbed operator L, it is clear that $\Lambda(P)$ is a Lie subalgebra of $\Lambda(M)$. Moreover the dimension of $\Lambda(P)$ is one less than that of $\Lambda(M)$. In view of the isomorphism of $\Lambda(M)$ and $\Lambda(\tilde{G})$ we have established [8]:

Theorem 4.1: The perturbation algebra $\Lambda(P)$ of an elliptic operator L, is isomorphic to a Lie subalgebra of $\Lambda(\tilde{G})$ (*i.e.* of the Lie algebra of the invariance group of L). Moreover $dim\Lambda(P)) = dim(\Lambda(\tilde{G})) - 1$.

One significant question is: can we find the perturbation algebra $\Lambda(P)$ without first computing $\Lambda(\tilde{G})$, the invariance Lie algebra? The answer is affirmative and is given by the following result [8].

Theorem 4.2: Assume L has analytic coefficients. An operator P_0 of order one or less (*i.e.* of the form (69)) is in the perturbation algebra $\Lambda(P)$ of L if there exist a sequence of scalars $\lambda_1, \lambda_2, \cdots$ and a sequence of operators $P_l, P_2, \cdots$ of order less than or equal to one such that

$$[L, P_n] = \lambda_n L + P_{n+1} \ , \ n \geq 0$$

and $\sum \lambda_k t^k / k!, \sum P_k t^k / k!$ converge at least for small t.

It is an easy application of this result to compute the perturbation algebra of the heat equation in one dimension or equivalently of $L = \frac{\partial^2}{\partial x^2}$. It turns out that $\Lambda(P)$ is 5-dimensional and spanned by

$$\Lambda(P) = \mathrm{Span}(1, \ x, \ x^2, \ \frac{\partial}{\partial x}, \ x\frac{\partial}{\partial x}). \tag{71}$$

So the general perturbation for the heat equation looks like

$$P = (ax + b)\frac{\partial}{\partial x} + (cx^2 + dx + e)\mathrm{id} \tag{72}$$

where a, b, c, d, e are arbitrary constants. Note that the invariance group of the heat equation is 6-dimensional (47). It is straightforward to rework the example of section 2, along the lines suggested here.

The implications of these results are rather significant. Indeed consider the class of linear parabolic equations $u_t = Lu$, where L is of the form (29). We can define an equivalence relationship on this class by: "L_1 is equivalent to L_2 if $L_2 = L_1 + P$ where P is an element of the perturbation algebra $\Lambda^1(P)$ of L_1". Thus elliptic operators of the form (29), or equivalently linear parabolic equations are divided into equivalent classes (orbits); within each class (orbit) $\{L(k)\}$ (k indexes elements in the class) solutions to the initial value problem $u(k)_t = L(k)u(k)$ with fixed data ϕ (independent of k) can be obtained by quadrature (*i.e.* an ODE integration) from any one solution $u(k_0)$.

We close this section by a list of perturbation algebras for certain L, from [8].

Elliptic operator	Generators of perturbation
L	algebra $\Lambda(P)$
D^2	$1, x, x^2, D, xD$
xD^2	$1, x, xD$
$x^2 D^2$	$x \log xD, xD, \log x, (\log x)^2, 1$
$x^3 D^2$	$1, x^{-1}, xD$
$e^x D^2$	$1, e^{-x}, D$

Table 1. Examples of perturbation algebras.

5   Strong Equivalence of Nonlinear Filtering Problems

In this section we will apply the methods described in sections 2-4 for parabolic equations to the fundamental PDEs governing nonlinear filtering problems. As with all symmetry group methods these techniques have a strong geometric flavor.

We will only briefly discuss the focal points of our current understanding of the nonlinear filtering problem and we will refer the reader to [9] or the references [2]-[6] for details. Thus the "nonlinear filtering problem for diffusion processes" consists of a model for a "signal process" $x(t)$ via a stochastic differential equation

$$dx(t) = f(x(t))dt + g(x(t))dw(t) \tag{73}$$

which is assumed to have unique solutions in an appropriate sense (strong or weak, see [9]). In addition we are given "noisy" observations of the process $x(t)$ described by

$$dy(t) = h(x(t))dt + dv(t). \tag{74}$$

Here $w(t), v(t)$ are independent standard Wiener processes and h is such that y is a semimartingale. The problem is to compute conditional statistics of functions of the signal process $\phi(x(t))$ at time t given the data observed up to time t, *i.e.* the σ-algebra

$$\mathcal{F}_t^y = \sigma\{y(s), 0 \le s \le t\} \tag{75}$$

Clearly the maximum information about conditional statistics is obtained once we find ways to compute the conditional probability density of $x(t)$ given $\mathcal{F}_t^y$. Let us denote this conditional density by $p(t, x)$. It is more convenient to use a different function, so called unnormalized conditional density, $u(t, x)$ which produces p after normalization

$$p(t, x) = \frac{u(t, x)}{\int u(t, z)dz} \tag{76}$$

The reason for the emphasis put on u is that it satisfies a **linear** stochastic PDE driven directly by the observations. This is the so called Mortensen-Zakai stochastic PDE, which in Itô's form is

$$du(t, x) = \mathcal{L}u(t, x)dt + h^T(x)u(t, x)dy(t) \tag{77}$$

Here $\mathcal{L}$ is the adjoint of the infinitesimal generator of the diffusion process $x(\cdot)$

$$[\mathcal{L}\phi](x) = \frac{1}{2}\sum_{i,j=1}^{n}\frac{\partial^2}{\partial x_i \partial x_j}[\sigma_{ij}(x)\phi] - \sum_{i=1}^{n}\frac{\partial}{\partial x_i}[f_i(x)\phi(x)] \tag{78}$$

which is also called the Fokker-Planck operator associated with $x(\cdot)$. In (78) the matrix σ is given by

$$\sigma(x) = g(x)g(x)^T \quad, \tag{79}$$

and we shall assume that σ is positive definite, *i.e.* the elliptic operator $\mathcal{L}$ is nondegenerate. When applying geometric ideas to (77) it is more convenient to consider the Stratonovich version

$$\frac{\partial u(t, x)}{\partial t} = (\mathcal{L} - \frac{1}{2}h(x)^T h(x))u(t, x) + h^T(x)u(t, x)\frac{dy(t)}{dt}. \tag{80}$$

We shall primarily work with (80) in the present paper. Letting

$$A := \mathcal{L} - \frac{1}{2}h^T h \tag{81}$$

$$B_j := \text{Mult. by } h_j (j^{th} \text{ comp. of } h) \tag{82}$$

we can rewrite (80) as an infinite dimensional bilinear equation

$$\frac{du(t)}{dt} = (A + \sum_{j=1}^{p} B_j \dot{y}_j(t))u(t). \tag{83}$$

We shall assume that every equation of the form (80) considered has a complete existence and uniqueness theory established on a space X. Furthermore we shall assume that continuous dependence of solutions on $y(\cdot)$ has been established.

The estimation Lie algebra introduced by Brockett [2] and analyzed in [2]-[6] is the Lie algebra

$$\Lambda(E) = \text{Lie algebra generated by } A \text{ and } B_j, \ j = 1, \cdots p. \tag{84}$$

Again we shall assume that for problems considered the operators A, B_j have a common, dense invariant set of analytic vectors in X [10] and that the mathematical relationship between $\Lambda(E)$ and the existence-uniqueness theory of (80) is well understood.

We develop a methodology for recognizing mathematically "equivalent" problems. Equivalence here carries the following meaning: two nonlinear filtering problems should be equivalent when knowing the solution of one, the solution of the other can be obtained by relatively simple additional computations. Examples discovered by Beneš [11], created certain excitement for the possibility of a complete classification theory. We shall see how transparent Beneš' examples become from the point of view developed in this paper.

To make things precise consider two nonlinear filtering problems (vector)

$$dx^i(t) = f^i(x^i(t))dt + g^i(x^i(t))dw^i(t)$$
$$dy^i(t) = h^i(x^i(t))dt + dv^i(t) \; ; \; i = 1,2 \tag{85}$$

and the corresponding Mortensen-Zakai equations in Stratonovich form

$$\frac{\partial u_i(t,x)}{\partial t} = (\mathcal{L}^i - \frac{1}{2} \| h^i(x) \|^2)u_i(t,x) + h^{iT}(x)u_i(t,x)y^i(t); i = 1,2 \tag{86}$$

Definition: The two nonlinear filtering problems above are **strongly equivalent** if u_2 can be computed from u_l, and vice versa, via the following types of operations:

Type 1: $(t, x^2) = \alpha(t, x^1)$, where α is a diffeomorphism.
Type 2: $u_2(t, x) = \psi(t, x)u_1(t, x)$, where $\psi(t, x) \geq 0$ and $\psi^{-1}(t, x) \geq 0$.
Type 3: Solving a set of ordinary (finite dimensional) differential equations (i.e. quadrature).

Brockett [2], has analyzed the effects of diffeomorphisms in x-space and he and Mitter [4] the effects of so called "*gauge*" transformations (a special case of our type 2 operations) on (80). Type 3 operations are introduced here for the first time, and will be seen to be the key in linking this problem with mathematical work on group invariance methods in ODE and PDE's.

Our approach starts from the abstract version of (86)(*i.e.* (83)):

$$\frac{\partial u_i}{\partial t} = (A^i + \sum_{j=1}^{p} B_j^i \cdot y_j(t))u_i \; ; \; i = 1,2 \tag{87}$$

where A^i, B_j^i are given by (81)-(82). We are thus dealing with two parabolic equations. We will first examine whether the evolutions of the time invariant parts can be computed from one another. This is a classical problem and the methods of section 3, 4 apply. In this section we give an extension to the full equation (87) under certain conditions on B_j^i. We shall then apply this result to the examples studied by Beneš and recover the Riccati equations as a consequence of strong equivalence.

Our main result concerning equivalence (in a computational sense) of two nonlinear filtering problems is the following.

Theorem 5.1: Given two nonlinear filtering problems (see (85)), such that the corresponding Mortensen-Zakai equations (see (86)) have unique solutions, continuously dependent on $y(\cdot)$. Assume that using operations of type 1 and 2 (see definition just above) these stochastic PDEs can be transformed in bilinear form

$$\frac{\partial u_i}{\partial t} = (A^i + \sum_{j=1}^{p} B_j^i \xi_j^i(t)) u_i; \quad i = 1, 2$$

such that:

(i) $A^i, i = 1, 2$, are nondegenerate elliptic, belonging to the same equivalence class (see end of section 4)
(ii) $B_j^i, j = l, \cdots p, i = 1, 2$ belong to the perturbation algebra $\Lambda(P)$ of (i).

Then the two filtering problems are strongly equivalent.

Proof: Only a sketch will be given here. One first establishes that is enough to show computability of solutions for piecewise constant ξ, from one another, by the additional computation of solutions of an ODE. For piecewise constant ξ the solution to any one of the PDEs in bilinear form is given by

$$
\begin{aligned}
u_i &= e^{(A^i + B_{j_m}^i \xi_{j_m}^i)(t_m - t_{m-1})} \cdot e^{(A^i + B_{j_{m-1}}^i \xi_{j_{m-1}}^i)(t_{m-1} - t_{m-2})} \cdots \\
&\quad \cdots e^{(A^i + B_{j_1}^i \xi_{j_1}^i) t_1} \phi \; ; \quad i = 1, 2
\end{aligned}
\tag{88}
$$

Since A^1, A^2, belong to the same equivalence class there exist $Z^{12} \epsilon \Lambda(G)$, (where $\Lambda(G)$ is the Lie algebra of the invariance group for the class) and $P^{12} \epsilon \Lambda(P)$ (where $\Lambda(P)$ is the perturbation algebra of the class) such that (see (65)):

$$A^2 = A^1 + P^{12} \tag{89}$$

$$e^{sZ^{12}} e^{tA^1} = e^{tA^1} e^{sA^2}; \quad t, s \geq 0. \tag{90}$$

That is consider A^2 as a "perturbation" of A^1. We know by now what (89) means: to compute the semigroup generated by A^2, we first compute the semigroup generated by A^1, we then solve the ODE associated with the characteristics of the hyperbolic PDE

$$\frac{\partial v}{\partial s} = Z^{12} v \tag{91}$$

and we have

$$e^{sA_2} = [e^{sZ^{12}} e^{tA^1}]|_{t=0}. \tag{92}$$

More generally since $A^1 + B_j^1$, $A^2 + B_k^2$ belong to the same class there exist $Z_{jk}^{12} \epsilon \Lambda(G), P_{jk}^{12} \epsilon \Lambda(P)$ such that

$$A^2 + B_k^2 = A^1 + B_j^1 + P_{jk}^{12}$$
$$e^{s Z_{jk}^{12}} e^{t(A^1 + B_j^1)} = e^{t(A^1 + B_j^1)} e^{s(A^2 + B_k^2)}. \tag{93}$$

It is now apparent that if we know (88) explicitly for $i = 1$, we obtain u_2 from (93) with the only additional computations being the integration of the ODEs associated with the characteristics of the hyperbolic PDEs

$$\frac{\partial v}{\partial s} = Z_{jk}^{12} v, \ k, j = 1, ..., p. \tag{94}$$

This completes the proof.

Let us apply this result to the Beneš case. We consider the linear filtering problem (scalar x, y)

$$\left. \begin{array}{l} dx(t) = dw(t) \\ dy(t) = x(t)dt + dv(t). \end{array} \right\}$$

and the nonlinear filtering problem (scalar x, y)

$$\left. \begin{array}{l} dx(t) = f(x(t))dt + dw(t) \\ dy(t) = x(t)dt + dv(t). \end{array} \right\}$$

The corresponding Mortensen-Zakai equations in Stratonovich form are: for the linear

$$\frac{\partial u_1(t, x)}{\partial t} = \frac{1}{2} \left(\frac{\partial^2}{\partial x^2} - x^2 \right) u_1(t, x) + x\dot{y}(t) u_1(t, x); \tag{95}$$

for the nonlinear

$$\frac{\partial u_2}{\partial t} = \frac{1}{2} \left(\frac{\partial^2}{\partial x^2} - x^2 \right) u_2(t, x) - \frac{\partial}{\partial x}(f u_2) + x\dot{y}(t) u_2(t, x). \tag{96}$$

We wish to show that (95)(96) are strongly equivalent only if f (the drift) is a global solution of the Riccati equation

$$f_x + f^2 = ax^2 + bx + c. \tag{97}$$

First let us apply to (95)(96) an operation of type 2. That is let (defines v_2)

$$u_2(t, x) = v_2(t, x) exp \left(\int_0^x f(u) du \right). \tag{98}$$

The transformation (98) is global, and is an example of such more general transformations needed for systems and control problems and discussed at the end of section 1. This is like a gauge transformation from mathematical physics, or a potential transformation in symmetry group theory [1]. Then the new function v_2 satisfies

$$\frac{\partial v_2(t,x)}{\partial t} = \frac{1}{2}(\frac{\partial^2}{\partial x^2} - x^2 - V(x))v_2(t,x) + x\dot{y}(t)v_2(t,x), \qquad (99)$$

where

$$V(x) = f_x + f^2. \qquad (100)$$

Existence, uniqueness and continuous dependence on $y(\cdot)$ for (95)(96) have been established using classical p.d.e. results. We apply Theorem 5.1 to (95)(99). So

$$A^1 = \frac{1}{2}(\frac{\partial^2}{\partial x^2} - x^2)$$

$$A^2 = \frac{1}{2}(\frac{\partial^2}{\partial x^2} - x^2 - V) \qquad (101)$$

while

$$B^1 = B^2 = \text{Mult. by } x. \qquad (102)$$

From the results of section 4, the only possible equivalence class is that of the heat equation. Clearly from (72) or Table 4.1, $A^1, B^1, B^2 \epsilon \Lambda(P)$ for this class. For A^2 to belong to $\Lambda(P)$ it is necessary that V be quadratic, which is the same as f satisfying the Riccati equation (97), in view of (100).

Recall that the solution of (95) is

$$u_1(t,x) = exp(-\frac{(x-\mu(t))^2}{2\sigma(t)}) \qquad (103)$$

where

$$d\mu(t) = \sigma(t)(dy(t) - \mu(t)dt); \quad \mu(0) = \xi$$

$$d\sigma(t) = 1 - \sigma^2(t); \quad \sigma(0) = 0 \qquad (104)$$

Beneš [11], using a path integral computation showed that the solution of (99), when (97) is satisfied is given by

$$v_2(t,x) = exp(-\frac{(x-\mu(t))^2}{2\sigma(t)}) \qquad (105)$$

where

$$d\mu(t) = -(a+1)\sigma(t)\mu(t)dt - \frac{1}{2}\sigma(t)bdt + \sigma(t)dy(t)$$

$$d\sigma(t) = 1 - (a+1)\sigma^2(t). \qquad (106)$$

What we have shown here is a converse, from the point of view that strong equivalence of the linear and nonlinear filtering examples implies the Riccati equation. We also maintain that knowledge of group invariance theory makes the result immediate at the level of comparing (95) with (99).

6   Reduction of Nonlinear Output Robust Control Problems

In this section we will apply methods from symmetry groups to a problem of recent interest in nonlinear control: output robust control. As has been developed fully in [24], the nonlinear output robust control problem (in an H^∞ sense) is equivalent to a risk-sensitive partially observed stochastic control problem and to a dynamic partially observed game [20][26]. A key result in establishing these equivalences was the introduction of the information state and the nonlinear PDE that it satisfies. In this section we apply systematic methods from symmetry groups to this fundamental PDE of nonlinear robust control.

The dynamic game representation of the equivalent nonlinear robust control problem is as follows. Given the dynamical system

$$\left.\begin{array}{rcl} \dot{x}(t) & = & b(x(t), u(t)) + w(t), \quad x(0) = X_0 \\ y(t) & = & Cx(t) + v(t) \end{array}\right\} \tag{107}$$

where w, v are L^2-type disturbances. We want to find a control $u(\cdot)$, which is a non-anticipating functional of $y(\cdot)$, to minimize ($\mu > 0$)

$$J(u) = \sup_{w \in L^2} \sup_{v \in L^2} \sup_{X_0 \in L^2} \{\bar{p}(x_0) + \int_0^T [L(x(s), u(s)) \tag{108}$$

$$- \frac{1}{2}\mu(|w(s)|^2 + |v(s)|^2)]ds + \Phi(x(T))\}$$

One of the questions we want to answer, is when can we reduce this nonlinear problem to a linear one? Group invariance methods can be applied to this problem. Let us make the following structural assumptions:

$$\left\{\begin{array}{lcl} b(x, u) & = & f(x) + A(u)x + B(u) \\ f(x) & = & DF(x) \text{ for some } F \\ \frac{1}{2}|f(x)|^2 & + & f(x) \cdot (A(u)x + B(u)) \\ & = & \frac{1}{2}x^T \sum(u)x + \Lambda(u)x + \frac{1}{2}\Gamma(u) \\ \bar{p}(x) & = & -\frac{1}{2}(x - \bar{x}^T \bar{Y}^{-1}(x - \bar{x}) + \bar{\phi} + \frac{1}{\mu}F(x) \\ L(x, u) & = & \frac{1}{2}R(u) + \frac{1}{2}x^T Q(u)x \end{array}\right. \tag{109}$$

Let us next consider the information state PDE for (107)-(108):

$$\frac{\partial p}{\partial t} = -Dp.b(x, u) + \frac{\mu}{2}|Dp|^2 + L(x, u) - \frac{1}{2\mu}(y(t) - Cx)^2 \tag{110}$$

The optimal control is a memoryless function of the information state since the cost (108), can be expressed using the information state as follows [23][24]:

$$J(u) = \sup_{y \epsilon L^2}\{(P_T, \Phi) : p_0 = \bar{p}\}; \quad (p, q) = \sup_{x \epsilon \mathcal{R}^n}\{p(x) + q(x)\} \ . \tag{111}$$

That is why the information state PDE (110) is so fundamental. Under the structural assumptions (109), it is not hard to show [21][27][28] that the PDE (110) has a finitely parameterizable solution

$$p_t(x) = -\frac{1}{2\mu}(x - \hat{x}(t))^T Y(t)^{-1}(x - \hat{x}(t)) + \phi(t) + \frac{1}{\mu}F(x) \tag{112}$$

where

$$\left. \begin{aligned}
\hat{x}(t) &= (A(u(t)) + \mu Y(t)Q(u(t)) - Y(t)\textstyle\sum(u(t)))\hat{x}(t) + B(u(t)) \\
&\quad - Y(t)\Lambda(u(t)) + Y(t)C^T(y(t) - C\hat{x}(t)) \\
\hat{x}(0) &= \bar{x} \\
\dot{Y}(t) &= Y(t)A(u(t))^T + A(u(t))Y(t) \\
&\quad - Y(t)(C^T C - \mu Q(u(t)) + \textstyle\sum(u(t)))Y(t) + I \\
Y(0) &= \bar{Y} \\
\dot\phi(t) &= \tfrac{1}{2}R(u(t)) + \tfrac{1}{2}\hat{x}^T(t)Q(u(t))\hat{x}(t) + \tfrac{1}{2}\Gamma(u(t))) \\
&\quad - \tfrac{1}{\mu}(\tfrac{1}{2}\hat{x}(t)^T \textstyle\sum(u(t))\hat{x}(t) + \Lambda(u(t))\hat{x}(t) \\
&\quad - \tfrac{1}{2\mu}\mid y(t) - (C\hat{x}(t))\mid^2 \\
\phi(0) &= \bar\phi
\end{aligned} \right\} \tag{113}$$

A consequence of this is that the robust output control for the nonlinear system (107) is finite dimensional and easily implementable. The explanation for these specific results becomes clear from an invariance group perspective. Under an appropriate transformation the dynamic game (107)-(108) becomes linear, quadratic (and therefore has a well known finite dimensional controller) with state

$$\rho(t) = (\hat{x}(t), Y(t), \phi(t)) \tag{114}$$

and cost

$$J(u) = \sup_{y \epsilon L^2}\{\hat{\Phi}(\rho(T)); \rho(0) = \bar{p}\} \tag{115}$$

where

$$\left. \begin{aligned}
\hat{\Phi}(\rho) &= (p_\rho, \Phi) \\
p_\rho &= -\tfrac{1}{2\mu}(x - \hat{x})^T Y^{-1}(x - \hat{x}) + \phi + \tfrac{1}{\mu}F(x)
\end{aligned} \right\} \tag{116}$$

Similar reductions can be obtained, under similar assumptions, for the associated partially observed, risk sensitive stochastic control problem. These results can be obtained, understood and generalized by studying the group invariance of the information state PDE, and in particular using the methods of sections 3 and 4. Specifically the structural assumptions (109) are completely analogous to the Beneš structural assumptions. Requiring problem (107)-(108) to be the nonlinear equivalent to a linear quadratic problem (from the perspective of equivalence of the corresponding information state PDEs, as in sections 3, 4,) implies the structural assumptions (109). Thus it is important to study the symmetry groups (invariance properties) of the information state PDE for the linear control problem.

We shall omit the control parameter u from the notation, since it plays no part in the following calculations; in other words, instead of writing $A(u(t))$, we abbreviate to $A(t)$. The information state $p(t, x) \equiv p(t, x_1, \cdots, x_n)$ for the linear control problem satisfies the scalar PDE

$$F(t, x, p_t, \nabla p) \equiv p_t + \nabla p \cdot (Ax + b)$$
$$-|\nabla p|^2/2 + x^T Gx/2 + h \cdot x + l = 0 , \tag{117}$$

where $G \equiv G(t)$ is a symmetric matrix, $A \equiv A(t)$ is a square matrix, $b \equiv b(t)$ and $h \equiv h(t)$ are n-vectors, and $l \equiv l(t)$ is a scalar function. James and Yuliar [21] point out that there is a solution of the form

$$p(t, x) = -(x - r(t))^T W(t)(x - r(t))/2 + \phi(t), \tag{118}$$

with W symmetric. Taking the gradient of (118), substituting in (117) and equating coefficients of terms quadratic, linear, and constant in x we obtain the ODEs

$$\dot{W} = -WA - A^T W - W^2 + G ,$$
$$\dot{r} = W^{-1}(-\dot{W}r + Wb - A^T Wr - W^2 r - h) ,$$
$$\dot{\phi} = r^T(W\dot{r} + \dot{W}r/2 - Wb + W^2 r/2) - l . \tag{119}$$

The last two equations can be rewritten as

$$\dot{r} = Ar + b - W^{-1}(Gr + h) , \tag{120}$$
$$\dot{\phi} = r^T(WA - A^T W - G)r/2 - r^T h - l . \tag{121}$$

We now turn to the Lie transformation theory for the Information state PDE. We consider the invariance of (117) under an infinitesimal transformation given by a vector field of the form (note that we are not including a $\partial/\partial t$ term)

$$X \equiv \sum_{i=l}^{n} \xi^i(t, x) \frac{\partial}{\partial x_i} + \eta(t, x, p) \frac{\partial}{\partial p} . \tag{122}$$

According to Bluman and Kumei [1], Theorem 4.1.1-1, the criterion for invariance is that

$$X^{(1)} F(t, x, p_t, \nabla p) = 0 \text{ whenever}$$
$$F(t, x, p_t, \nabla p) = 0 , \qquad (123)$$

where $X^{(1)}$ is the first extended infinitesimal generator, namely

$$X^{(1)} \equiv \sum_{i=1}^{n} \xi^i \frac{\partial}{\partial x_i} + \eta \frac{\partial}{\partial p} + \eta_0^{(1)} \frac{\partial}{\partial p_t} + \sum_{i=1}^{n} \eta_i^{(1)} \frac{\partial}{\partial p_i}$$

where

$$p_i \equiv \frac{\partial}{\partial x_i} , \qquad (124)$$

$$\eta_0^{(1)}(t, x, p, p_t, \nabla p) \equiv D_t \eta - \sum_{j=1}^{n} (D_t \xi^j) p_j , \qquad (125)$$

$$\eta_i^{(1)}(t, x, p, p_t, \nabla p) \equiv D_i \eta - \sum_{j=1}^{n} (D_i \xi^j) p_j,$$
$$i = 1, \ldots, n , \qquad (126)$$

$$D_t \equiv \frac{\partial}{\partial t} + p_t \frac{\partial}{\partial p}, \quad D_i \equiv \frac{\partial}{\partial x_i} + p_i \frac{\partial}{\partial p},$$
$$i = 1, \ldots n . \qquad (127)$$

Evaluating the entries of (122) term by term,

$$\sum_{i=1}^{n} \xi^i \frac{\partial F}{\partial x_i} = (A^T (\nabla p) + Gx + h) \cdot \xi ,$$

$$\eta \frac{\partial F}{\partial p} = 0,$$

$$\eta_0^{(1)} \frac{\partial F}{\partial p_t} = \eta_t + p_t \eta_p - (\nabla p \cdot \xi_t) ,$$

$$\sum_{i=1}^{n} \eta_i^{(1)} \frac{\partial F}{\partial p_i} = \sum_{i=1}^{n} \left(D_i \eta - \sum_{j=1}^{n} (D_i \xi^j) p_j \right)$$
$$(Ax + b - \nabla p)^i$$
$$= (Ax + b - \nabla p) \cdot (\nabla \eta + \eta_p \nabla p - (\nabla p \cdot \nabla) \xi) .$$

Adding up all these terms shows that (122) gives

$$(A^T (\nabla p) + Gx + h) \cdot \xi + \eta_t + p_t \eta_p - (\nabla p \cdot \xi_t)$$
$$= -(Ax + b - \nabla p) \cdot (\nabla \eta + \eta_p \nabla p - (\nabla p \cdot \nabla) \xi) .$$

Grouping terms, we obtain:

Theorem 6.1 (Fundamental Transformation Relation): The vector fields ξ and η of the infinitesimal generator of a symmetry group of (117) must satisfy:

$$\eta_t + (Ax + b - \nabla p) \cdot \nabla \eta + (p_t + (Ax + b - \nabla p) \cdot \nabla p)\eta_p$$
$$= \nabla p \cdot \xi_t - (A^T \nabla p + Gx + h) \cdot \xi +$$
$$(Ax + b - \nabla p) \cdot (\nabla p \cdot \nabla)\xi , \tag{128}$$

where p is given by (118).

Note that only **linear** differential operators acting on ξ and η are involved, and ∇p and p_t are quadratic in x. Hence for any choice of ξ we may solve for η by the method of characteristics. That is given ξ, solving for η involves only the solution of an ODE.

We describe next how to use the fundamental transformation relation. Let

$$\varphi(\varepsilon; t, x, p)) \equiv (t, \bar{x}(\varepsilon, t, x), \bar{p}(\varepsilon, t, x, p))$$

denote the flow of the vector field

$$X \equiv \sum_{i=1}^{n} \xi^i(t, x) \frac{\partial}{\partial x_i} + \eta \frac{\partial}{\partial p} \tag{129}$$

where $\bar{x}(\varepsilon, t, x)$ are the transformed state space coordinates and $\bar{p}(\varepsilon, t, x, p) \equiv \bar{p}(t, \bar{x})$ is the information state for the transformed problem. By definition of φ,

$$\frac{d\varphi}{d\varepsilon}(\varepsilon) = X\varphi(t, \bar{x}, \bar{p}), \quad \varphi(0; (t, x, p)) = (t, x, p). \tag{130}$$

This breaks down into the system of ODEs

$$\frac{\partial \bar{x}}{\partial \varepsilon} = \xi(t, \bar{x}), \quad \bar{x}(0, t, x) = x ; \tag{131}$$

together with the scalar ODE

$$\frac{\partial \bar{p}}{\partial \varepsilon} = \eta(t, \bar{x}, \bar{p}), \quad \bar{p}(0, t, x, p) = p . \tag{132}$$

Therefore we have established the following:

Theorem 6.2: *Assume p satisfies (117). Suppose $\varepsilon \rightarrow \bar{x}\ (\varepsilon, t, x)$ is a one-parameter family of transformations of the space variable x satisfying the system of ODEs (131), for some choice of $\xi \equiv \xi(t, x)$, and that $\eta \equiv \eta(t, x, p)$ is chosen to satisfy the Fundamental Transformation Relation (128) in terms of ξ. Then the solution $\varepsilon \rightarrow \bar{p}(\varepsilon, t, x, p)$ to the ODE (132), if unique, is a one-parameter family of transformations of the information state variable p, so that (117) holds with (x, p) replaced by $(\bar{x}, \bar{p})$.*

7   A Case Explicitly Computable

The drawback of Theorem 6.2 is that it is too abstract to be of immediate practical use. Therefore we consider a more specialized situation admitting explicit computations. We shall constrain the choice of η so as to satisfy

$$\eta = -x^T W \xi . \tag{133}$$

This implies

$$\begin{aligned}
\eta_t &= -x^T \dot{W} \xi - x^T W \xi_t, \ \eta_p = 0, \\
\nabla \eta &= -(Wx \cdot \nabla)\xi - W\xi .
\end{aligned} \tag{134}$$

Now (6.24)implies

$$\begin{aligned}
&-x^T \dot{W} \xi - x^T W \xi_t - (Ax + b - \nabla p) \cdot ((Wx \cdot \nabla)\xi + W\xi) \\
&= \nabla p \cdot \xi_t - (A^T \nabla p + Gx + h) \cdot \xi \\
&\quad + (Ax + b - \nabla p) \cdot (\nabla p \cdot \nabla)\xi .
\end{aligned}$$

Rearranging terms gives

$$\begin{aligned}
&(A^T \nabla p + Gx + h - \dot{W}x - W(Ax + b - \nabla p)) \cdot \xi \\
&\qquad\qquad\qquad -(\nabla p + Wx) \cdot \xi_t
\end{aligned}$$

$$= (Ax + b - \nabla p) \cdot ((\nabla p + Wx) \cdot \nabla)\xi ,$$

$$\begin{aligned}
&(-A^T W(x - r) + Gx + h - \dot{W}x - \\
&W(Ax + b + W(x - r))) \cdot \xi - (Wr) \cdot \xi_t
\end{aligned}$$

$$= (Ax + b + W(x - r)) \cdot ((Wr) \cdot \nabla)\xi .$$

The coefficient of x in the first bracket is $-A^T W + G - \dot{W} - W(A + W) = 0$, by (119). Define the following vector functions in terms of quantities determined above:

$$\beta(t) \equiv Wr, \ \Gamma(t) \equiv A + W, \ \gamma(t) \equiv b - \beta , \tag{135}$$

$$\alpha(t) \equiv (A^T + W)Wr + h - Wb = -\beta_t , \tag{136}$$

where the last identity follows from (119) and (120)-(121), since

$$\begin{aligned}
W\dot{r} + \dot{W}r &= WAr + Wb - Gr - h + \\
&\qquad (-WA - A^T W - W^2 + G)r \\
&= -(A^T W)r - W^2 r - h + Wb .
\end{aligned}$$

Now the linear PDE which $\xi(t, x)$ must satisfy is:

$$\beta_t \cdot \xi - \beta \cdot \xi_t - (\Gamma x + \gamma) \cdot (\beta \cdot \nabla)\xi = 0 . \tag{137}$$

This can be put in an even more concise form:

Theorem 7.1: If we assume $\eta = -x^T W \xi$, then $\zeta(t, x) \equiv \beta \cdot \xi = r^T W \xi = \nabla p \cdot \xi - \eta$ must satisfy the linear first order PDE

$$\zeta_t + ((\Gamma x + \gamma) \cdot \nabla)\zeta = 0 . \tag{138}$$

Suppose ζ is a polynomial of order N in x, i.e.

$$\zeta(t, x) \equiv \sum_{k=0}^{N} \Xi^{(k)}(t)(x^{\otimes k}) , \tag{139}$$

where $x^{\otimes k} \equiv x \otimes \cdots \otimes x$ (k factors), and $\Xi^{(k)}(t)$ is a symmetric $(0, k)$-tensor. Then

$$\nabla \zeta(t, x) = \sum_{k=1}^{N} k \Xi^{(k)}(t)(\cdot \otimes x^{\otimes(k-1)}) .$$

Now (138) becomes

$$\sum_{0}^{N} \Xi_t^{(k)} x^{\otimes k} + \sum_{1}^{N} k \Xi^{(k)}((\Gamma x + \gamma) \otimes x^{\otimes(k-1)}) = 0 .$$

Equating coefficients for each power of x forces the $\{\Xi^{(k)}(t)\}$ to satisfy the following system of ODEs:

$$\Xi_t^{(N)} + N \Xi^{(N)}(\Gamma(\cdot) \otimes \cdot) = 0 ; \tag{140}$$

$$\Xi_t^{(k)} + k \Xi^{(k)}(\Gamma(\cdot) \otimes \cdot) = -(k+1)\Xi^{(k+1)}(\gamma \otimes \cdot)$$
$$\text{for } k = 1, 2, \ldots, N-1 ; \tag{141}$$

$$\Xi_t^{(0)} = \Xi^{(1)}(\gamma) . \tag{142}$$

Notice the structure of this system of ODEs. Suppose $\Xi^{(0)}(0), \cdots, \Xi^{(N)}(0)$ have been chosen. First we solve (140) for $\Xi^{(N)}(t)$; insert this solution in the right side of (141); solve (141) for $\Xi^{(N-1)}(t)$; and so on, down to $\Xi^{(0)}(t)$. Thus we have established.

Theorem 7.2: In the case when $\eta = -x^T W \xi = \nabla p \cdot \xi - \zeta$, let us assume that

$$\zeta(t,x) \equiv r^T W \xi \equiv \sum_{k=0}^{N} \Xi^{(k)}(t)(x^{\otimes k}) . \tag{143}$$

Then $\zeta(t,x)$ is completely determined by the initial conditions $\Xi^{(0)}(0), \cdots, \Xi^{(N)}(0)$ and ODEs (140)-(142). In particular, when $n = 1$ and Wr is never zero, $\xi(t,x)$ is uniquely determined by $\xi(0,x)$, assuming $\xi(t,x)$ is a polynomial of arbitrary degree in x with coefficients depending on t.

Finally we describe the procedure for computation of the transformed information state. The starting-point is the solution p given by (118) to the linear control problem. Pick an initial condition $\Xi^{(0)}(0), \cdots \Xi^{(N)}(0)$, and solve for $\zeta(t,x)$ using (140)-(142) by solving for each of the $\{\Xi^{(k)}(t)\}$. Now pick

$$\xi(t,x) \equiv \sum_{k=1}^{N} \Theta^{(k)}(t)(x^{\otimes x}) \tag{144}$$

so that $r^T W \xi = \zeta$, in other words so that

$$\Theta^{(k)}(t) = W(t)r(t) \cdot \Xi^{(k)}(t) . \tag{145}$$

Now we repeat the steps described at the end of section 6, (129)-(132), under the assumption $\eta = -x^T W \xi = \nabla p \cdot \xi - \zeta$. As before, we solve the system of ODEs

$$\frac{\partial \bar{x}}{\partial \varepsilon} = \xi(t,\bar{x}), \ \bar{x}(0,t,x) = x ; \tag{146}$$

(derived from (131)) to determine $\bar{x}(\varepsilon,t,x,p)$. Thus $\xi(t,\bar{x})$ and $\zeta(t,\bar{x})$ are now explicitly computable. Finally we determine $\bar{p}(\varepsilon,t,x,p)$ by solving the following first order PDE (derived from (131)-(132)) by the method of characteristics (see Abraham et al. [18], p. 287):

$$\frac{\partial \bar{p}}{\partial \varepsilon} = \nabla \bar{p} \cdot \xi(t,\bar{x}) - \zeta(t,\bar{x}), \ \bar{p}(0,t,x,p) = p , \tag{147}$$

which can be written out in full as

$$\frac{\partial \bar{p}}{\partial \varepsilon} = \sum_{k=0}^{N} (\nabla \bar{p} \cdot \Theta^{(k)}(t) - \Xi^{(k)}(t))(\bar{x}^{\otimes k}) . \tag{148}$$

Additional examples-cases with explicit computations can be found in [29].

168   John S. Baras

References

1. G.W. Bluman and S. Kumei (1989): *Symmetries and Differential Equations*, Springer, New York.
2. R.W. Brockett and J.M.C. Clark, "Geometry of the Conditional Density Equation", *Proc. Int. Conf. on An. and Opt. of Stoch. Syst.*, Oxford, England, 1978.
3. R.W. Brockett, "Classification and Equivalence in Estimation Theory", *Proc. 18th IEEE Conf. on Dec. and Control*, 1979, pp. 172-175.
4. S.K. Mitter, "Filtering Theory and Quantum Fields" presented in the **Conf. on Alg. and Geom. Methods in System Theory**, Bordeaux, France, 1978; in *Asterisque*, 1980.
5. S.K. Mitter, "On the Analogy Between Mathematical Problems of Non-linear Filtering and Quantum Physics", in *Richerche di Automatica*, 1980.
6. M. Hazewinkel and J. Willems (Edts), *Proc. of NATO ASI on Stoch. Syst.:The Mathematics of Filtering and Identification*, Les Arcs, France, 1980.
7. G.W. Bluman and J.C. Cole, *Similarity Methods for 'Differential Equations*, Springer-Verlag, 1974.
8. S.I. Rosencrans, "Perturbation Algebra of an Elliptic Operator", *J. of Math. Anal. and Appl.*, 56, 1976, pp. 317-329.
9. R.S. Liptser and A.N. Shiryayev, *Statistics of Random Processes I*, Springer-Verlag, 1977.
10. E. Nelson, "Analytic Vectors", *Ann. Math.*, 70, 1959, pp .572-615.
11. V. Beneš, "Exact Finite Dimensional Filters for Certain Diffusions with Nonlinear Drift", in *Stochastics*, 1980.
12. G.A. Nariboli, *Stoch. Proc. and their Appl.*, 5, 1977, pp. 157-171.
13. E. Hille and R. Phillips, **Funct. An. and Semigroups**, AMS Coll. Publ., 1957.
14. S.I. Rosencrans, *J. of Math. An. and Appl.*, 61, pp. 537-551.
15. Sophus Lie, *Arch. for Math.*, Vol VI, 1881, No. 3, Kristiana, p. 328-368.
16. L.V. Ovsjannikov, *Group Properties of Diff. Equations*, Sib. Branch of the USSR Ac. of Sci. 1962.
17. R.L. Anderson and N.H. Ibragimov. *Lie-Backlund Transformations in Applications*, SIAM, 1979.
18. R. Abraham, J.E. Marsden, and T. Ratiu (1988): *Manifolds, Tensor Analysis, and Applications*, 2nd ed., Springer, New York.
19. R.W.R. Darling (1994): *Differential Forms and Connections*, Cambridge University Press, New York.
20. M.R. James, J.S. Baras and R.J. Elliott, "Output feedback risk-sensitive control and differential games for continuous-time nonlinear systems", *32nd IEEE CDC*, San Antonio, 1993.
21. M.R. James and S. Yuliar, "A nonlinear partially observed differential game with a finite-dimensional information state", *Systems and Control Letters*, 1996.
22. M.R. James, J.S. Baras and R.J. Elliott, "Risk-sensitive control and dynamic games for partially observed discrete-time nonlinear systems", *IEEE Trans. on Automatic Control*, pp. 780-792, Vol. 39, No. 4, April 1994.
23. M.R. James and J.S. Baras, "Robust $H\infty$ output feedback control for nonlinear systems", *IEEE Trans. on Automatic Control*, pp. 1007-1017, Vol. 40, No. 6, June 1995.

24. M.R. James and J.S. Baras, "Partially observed differential games, infinite dimensional HJI Equations, and Nonlinear H_∞ Control", *SIAM Journal on Control and Optimization*, Vol. 34, No. 4, pp. 1342-1364, July 1996.

25. J.S. Baras and M.R. James, "Robust and risk-sensitive output feedback control for finite state machines and hidden markov models", *Journal of Mathematical Systems, Estimation, and Control*, Vol. 7, No. 3, pp. 371-374, 1997.

26. J.S. Baras and N. Patel, "Robust control of set-valued discrete time dynamical systems", *IEEE Transactions on Automatic Control*, Vol. 43, No. 1, pp. 61-75, January 1998.

27. J.B. Moore and J.S. Baras, "Finite-dimensional optimal controllers for nonlinear plants", *Systems and Control Letters*, 26 (1995), pp. 223-230.

28. R. Elliott and A. Bensoussan: "General Finite-Dimensional Risk-Sensitive Problems and Small Noise Limits", *IEEE Trans. on Automatic Control*, Vol. 41, pp. 210-215, January 1996.

29. J.S. Baras and R. Darling, "Finite-Dimensional Methods for Computing the Informatin State in Nonlinear Robust Control", *Proc. of 1998 IEEE Conference on Decision and Control*, pp. 343-348, Tampa, Florida, 1998.

A Globalization Procedure for Locally Stabilizing Controllers *

Jochen Behrens[1] and Fabian Wirth[2]

[1] Orthogon GmbH
   Hastedter Osterdeich 222
   28207 Bremen, Germany
   behrens@orthogon.de
[2] Centre Automatique et Systèmes
   Ecole des Mines de Paris
   77305 Fontainebleau, France
   wirth@cas.ensmp.fr**

Abstract. For a nonlinear system with a singular point that is locally asymptotically nullcontrollable we present a class of feedbacks that globally asymptotically stabilizes the system on the domain of asymptotic nullcontrollability.

The design procedure is twofold. In a neighborhood of the singular point we use linearization arguments to construct a sampled (or discrete) feedback that yields a feedback invariant neighborhood of the singular point and locally exponentially stabilizes without the need for vanishing sampling rate as the trajectory approaches the equilibrium. On the remainder of the domain of controllability we construct a piecewise constant patchy feedback that guarantees that all Carathéodory solutions of the closed loop system reach the previously constructed neighborhood.

1 Introduction

It is the aim of this paper to present a procedure to combine local stabilization procedures with global ones to obtain a globally defined feedback with desirable properties near the fixed point that are designed using inherently local arguments. It is known that asymptotic nullcontrollability is equivalent to the existence of a control Lyapunov function, see [16], and in recent years numerous papers have appeared on the question on how to construct stabilizing feedbacks from such functions, see e.g. [1,5,6,9,10,15]. A fundamental question in this area is precisely the question of the underlying solution concept for which the constructed feedback should be interpreted. Often sampling concepts are considered, but unless added structure like homogeneity

* Research partially supported by the European Nonlinear Control Network.
** This paper was written while Fabian Wirth was on leave from the Zentrum für Technomathematik, Universität Bremen, 28334 Bremen, Germany. A permanent email address is fabian@math.uni-bremen.de. The hospitality of the members of the CAS and the high quality of the coffee are gratefully acknowledged.

is used [9], the sampled feedbacks often require vanishing sampling intervals as the trajectory approaches the origin. ¿From a practical point of view, this appears to be undesirable.

For our local considerations we will rely on the work of Grüne, who has shown for linear and more generally homogeneous systems how to construct discrete feedbacks that stabilize the origin where the sampling rate can be chosen to be positive [9,10]. In fact, in [8] Grüne also shows that this procedure works locally for general nonlinear systems if the linearization is asymptotically null-controllable. Unfortunately, the results in that paper do not provide feedback invariant sets, which we will need in order to construct well defined feedbacks. Here, we will employ ideas from nonsmooth analysis already used in [5,9] to regularize a known control Lyapunov function. In a very similar vein, this also leads to local stabilization results using homogeneous approximations in [11] which implicitly provide feedback invariant sets.

The local method will be combined with ideas that are inherently global in nature and that all depend in one way or another on the construction of piece-wise constant feedbacks. This line of thought can be found in [12,3,13,1]. We follow Ancona and Bessan [1] as their approach has the advantage of guaranteeing properties of all Carathéodory solutions generated by the discontinuous feedback. A question that is unfortunately neglected in the other references on this subject, which makes some statements in [12], [3, Chapter 12], [13] somewhat imprecise. Again we will use tools from nonsmooth analysis which has the advantage that we are able to treat a more general case than the one in [1].

The theoretical interest in our procedure is that one obtains in this way a construction of a feedback that globally stabilizes the system without the need for increasingly faster sampling near the origin to ensure convergence. Also the local construction is computable as discussed in [9]. We do not discuss perturbation problems in this paper. Outer perturbations for the local and global procedures are discussed by Grüne [8], respectively Ancona and Bessan [1]. The problem of measurement noise, however, persists, see the discussion in [15].

In the following Section 2 we define the system class and make precise what we mean by a feedback in the Carathéodory and in the sampled sense. The ensuing Section 3 details a special class of Carathéodory feedbacks, that are defined by patches in a piecewise constant manner. The key in this definition is a nonsmooth "inward pointing" assumption which guarantees that there are no solutions with behavior other than the one intended in the construction of the patches. We quote some recent results showing that we can always construct a patchy feedback controlling to an open set B from its domain of attraction, for proofs which are omitted for reasons of space we refer to [4]. In Section 4 we need additional assumption on the system and on the linearization in the origin. We then construct a feedback in the sense of sampling for the nonlinear system, which renders a neighborhood of the fixed point feedback invariant and guarantees local exponential stability for the closed loop system. In the final Section 5 we show that the previous constructions can be used in a complementary fashion. In particular, the inward pointing condition is satisfied on a suitable sublevel set of a control Lyapunov function, so that we can apply the results of Sections 3 and 4 to obtain a global feedback strategy, that is "hybrid" in the sense that we employ different notions of feedback in different regions of the state space.

2   Preliminaries

We study systems of the form

$$\dot{x} = f(x, u), \tag{1}$$

where $f : \mathbb{R}^n \times U \to \mathbb{R}^n$ is continuous and locally Lipschitz continuous in x uniformly in u. Here $U \subset \mathbb{R}^m$ is a compact convex set with nonvoid interior. The unique trajectory corresponding to an initial condition $x_0 \in \mathbb{R}^n$ and $u \in \mathcal{U} := \{u : \mathbb{R} \to U \mid u \text{ measurable}\}$ is denoted by $x(\cdot; x_0, u)$.

A feedback for system (1) is a map $F : \mathbb{R}^n \to U$. If F is not continuous (as will be the case in the feedbacks we aim to construct) this immediately raises the question what solution concepts are appropriate for the solution of the discontinuous differential equation $\dot{x} = f(x, F(x))$. A number of concepts have been put forward to deal with these problems.

Definition 1 (Carathéodory closed loop system). Consider a feedback law $F : \mathbb{R}^n \to U$. For an interval $J \subset \mathbb{R}$ a *Carathéodory solution* $\gamma : J \to \mathbb{R}^n$ of

$$\dot{x} = f(x, F(x)) \tag{2}$$

is an absolutely continuous functions $x : J \to \mathbb{R}^n$ such that $\dot{x} = f(x, F(x))$ almost everywhere.

The map F is called a C-*feedback*, if we consider Carathéodory solutions of (2).

Note, that the definition of C-feedbacks does not require or guarantee that there exist any solutions to (2) or that there should be uniqueness. We simply refer the statement of necessary conditions for this to possible existence results. The second notion of solution we are interested in is the following.

Definition 2 (Sampled closed loop system). Consider a feedback law $F : \mathbb{R}^n \to U$. An infinite sequence $\pi = (t_i)_{i \geq 0}$ with $0 = t_0 < \ldots < t_i < t_{i+1}$ and $t_i \to \infty$ is called a *sampling schedule* or *partition*. The values

$$\Delta_i := t_{i+1} - t_i, \quad d(\pi) := \sup_{i \in \mathbb{N}} \Delta_i$$

are called *intersampling times* and *sampling rate*, respectively. For any sampling schedule π the corresponding *sampled* or *π-trajectory* $x_\pi(t, x_0, F)$ with initial value $x_0 \in \mathbb{R}^n$ and initial time $t_0 = 0$ is defined recursively by solving

$$\dot{x}(t) = f(x(t), F(x(t_i))), \quad t \in [t_i, t_{i+1}], \quad x(t_i) = x_\pi(t_i, x_0, F). \tag{3}$$

The map F is called a *sampled (or discrete) feedback* if we consider all sampled solutions of (3) and *h-sampled feedback* if we consider all solutions corresponding to sampling schedules π with $d(\pi) \leq h$.

Note that the definition of sampled feedbacks guarantees existence and uniqueness of π-trajectories in forward time, on the respective interval of existence. On the other hand, nothing prevents trajectories from joining at some time instant.

A specific point of this paper is that we allow to switch between different solution concepts in different regions of the state space. In order to obtain a well defined global feedback we will require, that the region where a sampled feedback is defined remains invariant under the sampled solution in the following sense.

Definition 3. A set $B \subset \mathbb{R}^n$ is called (forward) *feedback-invariant under h-sampling* for system (1) with respect to the feedback F, if for any initial condition $x_0 \in B$ and any sampling schedule π with $d(\pi) \leq h$ it holds that

$$x_\pi(t; x_0, F) \in B \quad \text{for all } t \geq 0.$$

Note that it is a peculiarity of sampled feedbacks, that the corresponding trajectories may for short times exist on regions of the state space where the sampled feedback is not defined, simply by leaving this area and returning before the next sampling instant. This is of course, somewhat undesirable and it is one of the aims of the paper to show how it can be prevented. The corresponding idea has already been used in [5], [9], [10].

3   Practical feedback stabilization with patchy feedbacks

In this section we study the system

$$\dot{x} = f(x, u), \tag{4}$$

with the properties stated in Section 2. We are interested in applying the concept of patchy feedbacks that have recently been introduced in [1]. We slightly extend the definition to allow for less regularity. In particular, we replace the definition of *inward pointing* in terms of outer normals by the appropriate concept from nonsmooth analysis.

Let B be a closed subset of $\mathbb{R}^n$, if $x \notin B$, and if a point $y \in B$ is closest to x, i.e. dist $(x, B) = \|x - y\|$, then $x - y$ is said to be a *proximal normal* to B in y. The cone generated by taking all positive multiples of these points is called *proximal normal cone* to B in y, denoted by $N_B^P(y)$, which is set to be $\{0\}$ if no proximal normal to y exists. One of the interests in this cone stems from its use in the characterization of strong invariance, see [7, Theorem 4.3.8]. This motivates our notion of patchy vector fields that slightly extends that of [1].

Definition 4. Let $\Omega \subset \mathbb{R}^n$ be an open domain with boundary $\partial\Omega$ and let $D \subset \mathbb{R}^n$ be open. A Lipschitz continuous vector field g defined on a neighborhood of cl Ω is called *inward pointing* on $(\partial\Omega) \setminus D$, if for every compact set K there exists a constant $c > 0$ such that for all $x \in K \cap \partial\Omega \setminus D$ it holds that

$$\max\{\langle \zeta, g(x) \rangle \mid \zeta \in N_\Omega^P(x)\} \leq -c\|\zeta\|.$$

The pair (Ω, g) is called a *patch* (relative to D).

Definition 5. Let $\Omega \subset \mathbb{R}^n$ be an open domain. We say that $g : \Omega \to \mathbb{R}^n$ is a *patchy vector field* if there exists a family of patches $\{(\Omega_\alpha, g_\alpha) \mid \alpha \in \mathcal{A}\}$ such that

(i) $\mathcal{A}$ is a totally ordered index set,
(ii) the open sets Ω_α form a locally finite cover of Ω,
(iii) it holds that

$$g(x) = g_\alpha(x) \quad \text{if } x \in \Omega_\alpha \setminus \bigcup_{\beta > \alpha} \Omega_\beta.$$

(iv) for every $\alpha \in \mathcal{A}$ the vector field g_α is inward pointing on $(\partial\Omega_\alpha) \setminus \bigcup_{\beta > \alpha} \Omega_\beta$.

Patchy vector fields g and solutions to the differential equation $\dot{x} = g(x)$ are discussed in detail in [1] for the case that the boundaries of Ω_α are C^1. The arguments, however, carry over to our definition. For Carathéodory solutions it can be shown that to each initial condition there exists at least one forward and at most one backward solution. Furthermore, along each Carathéodory solution x it holds that $t \mapsto \max\{\alpha \in A \mid x(t) \in \Omega_\alpha\}$ is nondecreasing and left continuous (with respect to the discrete topology on A). In an example [1, p. 457 ff.] the differences to other solution concepts are explained[1].

The concept of patchy feedbacks is now defined as follows. Assume we are given a totally order index $\mathcal{A}$, open sets $\Omega_\alpha, \alpha \in \mathcal{A}$ and functions $F_\alpha : W_\alpha \to$

---

[1] Incidentally, note that in this example not all maximal Filippov solutions are given, contrary to what is claimed. The ones that remain in $(0,0)$ for some interval $[1, r]$ and then follow the parabola to the right are missing.

$U, \alpha \in \mathcal{A}$, where W_α is an open neighborhood of Ω_α. We say that F is a patchy C-feedback for system (4), if $f(x, F(x))$ is a patchy vector field on $\Omega := \cup \Omega_\alpha$ with patches $(\Omega_\alpha, f(x, F_\alpha(x)))$.

For our discussion of practical stability we need the following definition. Given a set $Q \subset \mathbb{R}^n$ and an open set $B \subset Q$ we define the backward orbit of B relative to Q by

$$\mathcal{O}^-(B)_Q := \{y \in \mathbb{R}^n \mid \exists t \geq 0, u \in \mathcal{U} : x(t; y, u) \in B$$
$$\text{and } x(s; y, u) \in Q, s \in [0, t]\} \, .$$

Note that it is obvious by definition that $\mathcal{O}^-(B)_Q \subset Q$. Furthermore, it is an easy consequence of continuous dependence on the initial value that $\mathcal{O}_Q^-(B)$ is open, if Q is open.

Definition 6. System (4) is called *practically C-feedback controllable* if for every closed set $Q \subset \mathbb{R}^n$ and every open set $B \subset Q$ there is C-feedback $F_{B,Q} : \mathcal{O}^-(B)_{\text{int } Q} \setminus \text{cl } B \to U$ so that the system

$$\dot{x} = f(x, F_{B,Q}(x)) \, ,$$

satisfies

(i) for every $x \in \mathcal{O}^-(B)_{\text{int } Q} \setminus \text{cl } B$ there exists a Carathéodory solution γ with $\gamma(0) = x$,
(ii) for every Carathéodory solution γ starting in $x \in \mathcal{O}^-(B)_{\text{int } Q} \setminus \text{cl } B$ there is a time T such that $\gamma(T) \in \partial B$.

The following result shows that the foregoing notion of practical feedback controllability is always fulfilled. The proof of this statement follows the ideas explained in [1], with the necessary modifications for our case.

Theorem 1. *System (4) is practically C-feedback controllable using patchy C-feedbacks.*

We now state a condition with which we can avoid solutions that touch the boundary of B several times, which is essential for our goal.

Corollary 1. *Let $Q \subset \mathbb{R}^n$ be closed and $B \subset \text{int } Q$ be open and bounded. Assume that for every $x \in \partial B$ there exists a $u \in U$ such that the inward pointing condition*

$$\max\{\langle \zeta, f(y, u) \rangle \mid \zeta \in N_B^P(y)\} \leq -c\|\zeta\| \, , \tag{5}$$

is satisfied for all $y \in \partial B$ in a neighborhood of x, then there is a piecewise constant patchy C-feedback F so that for every trajectory γ starting in $x \in \mathcal{O}^-(B)_{\text{int } Q}$ of (2) there is a time T such that $\gamma(T) \in \partial B$. Furthermore, F can be chosen so that there exists no trajectory γ on an interval J such that $\gamma(t_1), \gamma(t_2) \in \partial B$ for $t_1 < t_2 \in J$ and $\gamma(t) \notin B$ else.

In [12] the author proves the above result and imposes two further conditions: that the system be affine in the controls and locally accessible. We have seen that this is indeed not necessary. However, the aim in [12] is different, as there the feedback is defined using just extremal values of the set U thereby reducing the complexity of actually designing such a patchy feedback (in particular if U is a polyhedron). A relation to this result is discussed in [4].

4 A sufficient condition for local sampled feedback stabilization with positive sampling rate

In this section we give a brief review of the result of Grüne for local stabilization at singular points. For us there remains one detail to supply, namely that the control Lyapunov functions constructed in [9] remain control Lyapunov functions for the nonlinear system, see also [11] for a similar result. Consider

$$\dot{x} = f(x, u) , \tag{6}$$

where we assume in addition to the assumptions previously stated that f : $\mathbb{R}^n \times U \to \mathbb{R}^n$ is twice differentiable in x and, furthermore, that 0 is a singular point for (6), i.e. $f(0, u) = 0$ for all $u \in U$.

We are interested in constructing a locally stabilizing feedback $F : \mathbb{R}^n \to U$ that stabilizes in the sense of sampling, see Definition 2. The local design procedure we will investigate relies on linearization techniques. We thus introduce the linearization of (6) in 0 with respect to x given by

$$\dot{z}(t) = A(u(t))z(t) , \tag{7}$$

where $A(u)$ denotes the Jacobian of $f(\cdot, u)$ in $x = 0$. We assume that $A(\cdot)$: $U \to \mathbb{R}^{n \times n}$ is Lipschitz continuous. The trajectories of (7) are denoted by $z(\cdot; z_0, u)$. It is known that asymptotic nullcontrollability of system (7), i.e. the property that for every $z \in \mathbb{R}^n$ there exists a $u \in \mathcal{U}$ such that $z(t; z, u) \to 0$, can be characterized via the use of Lyapunov exponents. We refer to [9] for details and revise just the facts essential to us. The linear system (7) can be projected to the sphere, which takes the form $\dot{s} = A(u)s - \langle s, A(u)s \rangle s$. Denoting the radial component $q(s, u) := \langle s, A(u)s \rangle$ we consider for $z \neq 0$

$$J_\delta(z, u) := \int_0^\infty e^{-\delta \tau} q(s(\tau), u(\tau)) d\tau, \text{ and } v_\delta(z) := \inf_{u \in \mathcal{U}} J_\delta(z, u) . \tag{8}$$

Then it can be shown that asymptotic nullcontrollability of (7) is equivalent to the existence of a $\delta_0 > 0$ small enough such that

$$\max_{z \in \mathbb{R}^n \setminus \{0\}} v_\delta(z) < 0, \quad \text{for all } 0 < \delta < \delta_0 . \tag{9}$$

In [9] it is shown that v_δ gives rise to a control Lyapunov function for the linearized system (7). Recall that for a continuous function $V : \mathbb{R}^n \to \mathbb{R}$ and $v \in \mathbb{R}^n$ the *lower directional derivative* of V in x in direction v is defined by

$$DV(x;v) := \liminf_{t \searrow 0, v' \to v} \frac{1}{t} \left(V(x + tv') - V(x) \right).$$

With this notation $v_\delta(y)$ is a control Lyapunov function in the following sense.

Lemma 1. *[9, Lemma 4.1] There exists a $\overline{\lambda} > 0$ such that for every $\rho \in (0, \overline{\lambda})$ there exist a $\delta_\rho > 0$ such that for every $\delta \in (0, \delta_\rho]$ the function $V_\delta : \mathbb{R}^n \to \mathbb{R}$ defined by $V_\delta(0) = 0$ and*

$$V_\delta(z) := e^{2v_\delta(z)} \|z\|^2, \qquad z \neq 0, \tag{10}$$

satisfies

$$\min_{v \in \mathrm{co}A(U)z} DV_\delta(z;v) \leq -2\rho V_\delta(z) \tag{11}$$

Note that by definition we have the homogeneity property $V_\delta(\alpha z) = \alpha^2 V_\delta(z)$, $\alpha > 0$. We follow an approach originating in [5] and employed for our special case in [9] that obtains Lipschitz continuous Lyapunov functions from V_δ. We introduce the (quadratic) *inf-convolution* of V_δ given by

$$V_\beta(x) := \inf_{y \in \mathbb{R}^n} \left[V_\delta(y) + \frac{1}{2\beta^2} \|y - x\|^2 \right]. \tag{12}$$

This function is Lipschitz for $\beta > 0$ and converges pointwise to V_δ as $\beta \to 0$. Furthermore, the homogeneity of V_δ implies that we also have $V_\beta(\alpha x) = \alpha^2 V_\beta(x)$. For $x \in \mathbb{R}^n$ we denote by $y_\beta(x)$ a minimizing vector in (12), which exists by continuity of V_δ, and introduce the vector

$$\zeta_\beta(x) := \frac{x - y_\beta(x)}{\beta^2}. \tag{13}$$

By [5, Lemma III.1] we have for all $x, v \in \mathbb{R}^n$ and all $\tau \geq 0$ that

$$V_\beta(x + \tau v) \leq V_\beta(x) + \tau \langle \zeta_\beta(x), v \rangle + \frac{\tau^2 \|v\|^2}{2\beta^2}. \tag{14}$$

which can be interpreted as a Taylor inequality for V_β in direction v. The following statement shows that we retain the property of being a control Lyapunov function if V_δ is replaced by V_β where $\delta > 0$ and $\beta > 0$ are small enough. The feedback we envisage is now given as a pointwise minimizer of $\langle \zeta_\beta(z), A(u)z \rangle$, that is $F_\beta(z)$ is chosen so that

$$\langle \zeta_\beta(z), A(F_\beta(z))z \rangle = \min_{u \in U} \langle \zeta_\beta(z), A(u)z \rangle. \tag{15}$$

This choice is of course not unique, nor can we expect that a regular choice is possible, we will however always assume that we have obtained a pointwise minimizer satisfying $F_\beta(\alpha z) = F_\beta(z)$, for $z \in \mathbb{R}^n$, $\alpha > 0$, which is easily seen to be possible.

We quote the following result by Grüne in a slightly extended manner that also states intermediate statements of the proof that we will need later on. We denote

$$A_{z_0}^t := \frac{1}{t} \int_0^t A(F_\beta(z_0)) z(\tau; z_0, F_\beta(z_0)) \, d\tau \,,$$

and

$$M := \max \left\{ 2 \max_{u \in U} \|A(u)\|, \sup_{\|x\| \le 2, u \in U} \|f(x, u)\| \right\}. \tag{16}$$

Proposition 1. *[9, Proposition 4.2] Assume that system (7) is asymptotically nullcontrollable and let $\rho \in (0, \overline{\lambda})$. Let $0 < \delta < \delta_\rho$, where δ_ρ is defined in Lemma 1. Then there exists a $\beta \in (0, 1]$, such that*

$$\min_{v \in coA(U)z} DV_\beta(z; v) \le -2\rho V_\beta(z) \tag{17}$$

for all $z \in \mathbb{R}^n$. Furthermore, there is a $\overline{t} > 0$ such that

$$V_\beta(z(t; z_0, F_\beta(z_0))) - V_\beta(z_0) \le t \langle \zeta_\beta(z_0), A_{z_0}^t \rangle + t^2 \frac{M^2}{2\beta^2} \le -\frac{3}{2} t \rho V_\beta(z_0) \,,$$

for all $z_0 \in \mathbb{R}^n$ and all $0 < t \le \overline{t}$.

The previous result contains all the arguments necessary to prove that F_β is indeed a $\overline{t}$-sampled stabilizing feedback for the linearized system (7). We now wish to carry the result over to a local statement for the nonlinear system (6).

Given that V_β is a control Lyapunov function it is no real surprise that it is also one for the nonlinear system (6) in a neighborhood of the origin as we show now. To this end we need the following technical lemma. We denote

$$\underline{\sigma} := \inf_{z \in \mathbb{R}^n \setminus \{0\}} v_\delta(z) \quad \text{and} \quad \overline{\sigma} := \sup_{z \in \mathbb{R}^n \setminus \{0\}} v_\delta(z) \,.$$

Lemma 2. *For all $z \in \mathbb{R}^n$ and all $\beta > 0$ it holds that $\|y_\beta(z) - z\| \le \sqrt{2} e^{\overline{\sigma}} \|z\| \beta$.*

Proof. By definition we have that $V_\beta(z) \le V_\delta(z)$ for all $z \in \mathbb{R}^n$. Now the assertion follows from

$$\frac{1}{2\beta^2} \|y_\beta(z) - z\|^2 = V_\beta(z) - V_\delta(y_\beta(z)) \le V_\delta(z) - V_\delta(y_\beta(z)) \le e^{2\overline{\sigma}} \|z\|^2 \,.$$

Theorem 2. *Consider the nonlinear system* (6) *and its linearization* (7). *Assume that* (7) *is asymptotically nullcontrollable. Let* $\rho \in (0, \bar{\lambda})$, $0 < \delta < \delta_\rho$ *and assume furthermore that* $\bar{t} > 0$ *and* $\beta \in (0, 1]$ *are chosen such that the assertion of Proposition 1 hold. Then there exists constants* $R > 0, \bar{t} > 0$, *such that*

$$\min_{v \in co f(x, U)} DV_\beta(x; v) \leq -\rho V_\beta(x) \tag{18}$$

for all $x \in \mathrm{cl}\, B_R(0)$ *and*

$$V_\beta(x(t; x_0, F_\beta(x_0))) - V_\beta(x_0) \leq -t\rho V_\beta(x_0), \tag{19}$$

for all $x_0 \in \mathrm{cl}\, B_R(0)$ *and all* $0 < t \leq \bar{t}$.

Proof. For the sake of abbreviation we denote $\tilde{f} := f - A$ and

$$f_{x_0}^t := \frac{1}{t} \int_0^t f(x(\tau; x_0, F_\beta(x_0)), F_\beta(x_0))\, d\tau.$$

Let M be as defined in (16). Then by decreasing $\bar{t}$ if necessary we have that $\|f_{x_0}^t\| \leq M$ for all $x_0 \in \mathrm{cl}\, B_1(0)$ and all $0 < t \leq \bar{t}$. By [3, Lemma 12.2.10 (iii)] there exists a constant $C > 0$, such that

$$\|x(t; x_0, F_\beta(x_0)) - z(t; x_0, F_\beta(x_0))\| \leq Ct\|x_0\|^2 \tag{20}$$

for all $x_0 \in \mathrm{cl}\, B_1(0)$ and all $0 < t \leq \bar{t}$. Now by Lemma 2, (20), and the definition of ζ_β we obtain for all $x_0 \in \mathrm{cl}\, B_1(0)$ and all $0 < t \leq \bar{t}$ that

$$\langle \zeta_\beta(x), f_{x_0}^t - A_{x_0}^t \rangle \leq \|\zeta_\beta(x)\| \|f_{x_0}^t - A_{x_0}^t\| = \frac{\|y_\beta(x_0) - x_0\|}{\beta^2} \|f_{x_0}^t - A_{x_0}^t\|$$

$$= \frac{\|y_\beta(x_0) - x_0\|}{t\beta^2} \left\| \int_0^t \tilde{f}(x(\tau; x_0, F_\beta(x_0)), F_\beta(x_0)) d\tau \right\|$$

$$\leq \frac{\sqrt{2}e^{\bar{\sigma}} \|x_0\|}{t\beta} \|x(t; x_0, F_\beta(x_0)) - z(t; x_0, F_\beta(x_0))\|$$

$$\leq \frac{\sqrt{2}e^{\bar{\sigma}}}{\beta} C\|x_0\|^3. \tag{21}$$

Using Proposition 1, (14), the definition of M and (21) this implies that

$$V_\beta(x(t; x_0, F_\beta(x_0))) - V_\beta(x_0) = V_\beta(x_0 + t f_{x_0}^t) - V_\beta(x_0)$$

$$\leq t\langle \zeta_\beta(x_0), f_{x_0}^t \rangle + t^2 \frac{\|f_{x_0}^t\|^2}{2\beta^2}$$

$$\leq t\langle \zeta_\beta(x_0), A_{x_0}^t \rangle + t^2 \frac{M^2}{2\beta^2} + t\langle \zeta_\beta(x_0), f_{x_0}^t - A_{x_0}^t \rangle$$

$$\leq -\frac{3}{2} t\rho V_\beta(x_0) + t\langle \zeta_\beta(x_0), f_{x_0}^t - A_{x_0}^t \rangle$$

$$\leq t\left(-\frac{3}{2}\rho V_\beta(x_0) + \frac{\sqrt{2}e^{\bar{\sigma}} C}{\beta} \|x_0\|^3 \right)$$

for all $x_0 \in \mathrm{cl}\, B_1(0)$ and all $0 < t \leq \bar{t}$. Define

$$\chi_r := \sup_{\|x\|=r} \frac{\sqrt{2}e^{\bar{\sigma}}C\|x\|^3}{\beta V_\beta(x)} = \sup_{\|x\|=1} r \frac{\sqrt{2}e^{\bar{\sigma}}C}{\beta V_\beta(x)},$$

where we have used the homogeneity of V_β. Now choose $R > 0$, such that $\chi_R \leq \frac{1}{2}\rho$ and it follows that

$$V_\beta(x(t; x_0, F_\beta(x_0))) - V_\beta(x_0) \leq t\left(-\frac{3}{2}\rho + \chi_R\right) V_\beta(x_0) \leq -t\rho V_\beta(x_0)$$

for all x_0 with $\|x_0\| \leq R$ and all $0 < t \leq \bar{t}$. This implies the assertion.

The final result of this section shows that the sublevel sets of V_β describe sets that are feedback invariant under $\bar{t}$-sampling for system (6) with respect to the $\bar{t}$-sampled feedback F_β, at least close to zero. For $r > 0$ we denote the r^2-sublevel set by $G_r^\beta := \{x \,|\, V_\beta(x) \leq r^2\}$.

Corollary 2. *Let the assumptions of Theorem 2 be satisfied. Choose $R > 0$ according to the assertions of that theorem. For any $r > 0$ such that $G_r^\beta \subset B(0, R)$ it holds that*

(i) *The set G_r^β is feedback-invariant under $\bar{t}$-sampling for system (6) with respect to the feedback F_β,*
(ii) *there exists a constant $C > 0$ such that for any $x_0 \in G_r^\beta$ and any sampling schedule π with $d(\pi) \leq \bar{t}$ it holds for the π-trajectory defined by (3) that*

$$\|x_\pi(t, x_0, F_\beta)\| \leq Ce^{-\rho t/2}\|x_0\|.$$

If we assume in addition to the assumptions of the previous Corollary 2 that $\beta < \frac{1}{\sqrt{2}}e^{-\bar{\sigma}}$, we can make possible choices for $r > 0$ and a_1, a_2 more concrete, see [4].

5   Global feedback stabilization with positive sampling rate near the origin

In this section the two ingredients of the final feedback design will be put together. We continue to consider system (6) with the additional assumptions stated at the beginning of Section 4 and the associated linearization (7) in 0.

Definition 7. System (6) is called *globally asymptotically feedback stabilizable*, if for every connected compact set $Q \subset \mathbb{R}^n$ with $0 \in \mathrm{int}\, Q$ and every open ball $B(0, r) \subset Q$ there exists a compact connected set $D \subset B(0, r)$ containing the origin, such that the following conditions are satisfied:

(i) There exists a patchy C-feedback F_1 on $\mathcal{O}^-(D)_{\mathrm{int}\,Q}\backslash D$, such that
   (a) for every $x \in \mathcal{O}^-(D)_{\mathrm{int}\,Q}\backslash D$ there exists a Carathéodory solution of

$$\dot{x} = f(x, F_1(x)) \qquad (22)$$

   with $\gamma(0) = x$ on some interval $[0, t_\gamma)$
   (b) for every solution $\gamma(\cdot)$ of (22) with $\gamma(0) = x \in \mathcal{O}^-(D)_{\mathrm{int}\,Q}\backslash D$ there exists a $T_\gamma > 0$, with

$$\gamma(T_\gamma) \in \partial D,$$

   (c) there is no solution γ of (22) with $\gamma(0) = x \in \mathcal{O}^-(D)_{\mathrm{int}\,Q}\backslash D$ such that $\gamma(t_1) \in \partial D$, $\gamma(t_2) \notin D$ for some $t_1 < t_2$.
(ii) There exists a sampling bound $\bar{t} > 0$ and a $\bar{t}$-sampled feedback F_2 on D, such that D is feedback invariant under $\bar{t}$-sampling for system (6) with respect to the feedback F_2, and such that

$$\lim_{t \to \infty} x_\pi(t; x, F_2) = 0,$$

for every $x \in D$ and every π trajectory for sampling schedules π with $d(\pi) \le \bar{t}$.

Theorem 3. *The system (6) is globally asymptotically feedback stabilizable in the sense of Definition 7 if its linearization (7) is asymptotically nullcontrollable.*

Proof. Let $Q \subset \mathbb{R}^n$ be closed with $0 \in \mathrm{int}\,Q$. Let $\bar{t}$ and $\beta \in (0, 1]$ be such that $F_2 = F_\beta$ is an $\bar{t}$-sampled exponentially stabilizing feedback on $D' = G_{r'}^\beta$. We choose an $0 < r < r'$, then the statement is obviously also true for $D = G_r^\beta$. The linear decrease statement (19) from Theorem 2 guarantees that we can satisfy the inward pointing condition (5) for ∂G_r^β. Now by Theorem 1 and Corollary 1 we have the existence of a patchy C-feedback F with ordered index $\mathcal{A}$ on $\mathcal{O}^-(G_{r'}^\beta)_Q$, stabilizing to $\partial G_{r'}^\beta$. This concludes the proof.

Remark 1. It is of course quite unrealistic from several points of view, to demand switching between the two controllers and solution concepts if the boundary of a set D is reached, especially as level sets of V_β do not lend themselves easily to computation. We can, however, relax the requirements here a bit. Take two values $0 < r_1 < r_2$ such that for a suitable β the sampled feedback F_β renders $G_{r_2}^\beta$ $\bar{t}$-sampled feedback invariant and exponentially stabilizes to 0.

Now we may just require that the switch between the Carathéodory and the sampled feedback is made somewhere in $G_{r_2}^\beta \backslash G_{r_1}^\beta$, say at a point $x \in G_r^\beta$. As for all $r_1 < r < r_2$ the set G_r^β is also $\bar{t}$-sampled feedback invariant we still have that the following π-trajectories remain in G_r^β and converge exponentially to zero. This makes the decision of switching less delicate for the price that we have concurring definitions for the feedback in a certain region of the state space, so that we need a further variable, a "switch", to remember which feedback strategy is applicable.

6 Conclusions

In this article we have presented a method to unite a local exponentially stabilizing sampled feedback with a global piecewise constant feedback interpreted in the sense of Carathéodory solutions. The key tools in this approach were methods from nonsmooth analysis in particular proximal normals and inf-convolution. In general, this approach is not restricted to the feedback types we have considered here, but can be performed for any feedback concepts that allow for the completion of the key step in our design. This consists in the construction of feedback invariant sets for the feedback F_2 that can be entered from the outside under the feedback F_1, and that have the additional property that no solutions under F_1 can move away (locally) from the feedback invariant set.

Acknowledgment: The authors would like to thank Lars Grüne for numerous helpful comments during the preparation of this manuscript.

References

1. Ancona F., Bessan A. (1999) Patchy vector fields and asymptotic stabilization. ESAIM: Control, Optim. and Calculus of Variations, **4**, 445–471
2. Aubin, J.-P., Cellina, A. (1984) Differential Inclusions: Set-Valued Maps and Viability Theory. Springer-Verlag, Berlin
3. Colonius, F., Kliemann, W. (2000) The Dynamics of Control. Birkhäuser, Boston
4. Behrens, J., Wirth, F. (2000) A globalization procedure for locally stabilizing controllers. Report 00-09. Berichte aus der Technomathematik, Universität Bremen
5. Clarke, F.H., Ledyaev, Yu.S., Sontag, E.D., Subottin, A.I. (1997) Asymptotic controllability implies feedback stabilization. IEEE Trans. Automat. Contr. **42**, 1394–1407
6. Clarke, F.H., Ledyaev, Yu.S., Rifford, L., Stern, R.J. (2000) Feedback stabilization and Lyapunov functions. SIAM J. Contr. & Opt. to appear
7. Clarke, F.H., Ledyaev, Yu.S., Stern, R.J., Wolenski, P.R. (1998) Nonsmooth Analysis and Control Theory. Springer-Verlag, New York
8. Grüne, L. (1998) Asymptotic controllability and exponential stabilization of nonlinear control systems at singular points. SIAM J. Contr. & Opt. **36**, 1585–1603
9. Grüne, L. (2000) Homogeneous state feedback stabilization of homogeneous systems. SIAM J. Contr. & Opt. **38**, 1288-1314
10. Grüne, L. (1999) Stabilization by sampled and discrete feedback with positive sampling rate. In: F. Lamnabhi-Lagarrigue, D. Aeyels, A. van der Schaft (eds.), Stability and Stabilization of Nonlinear Systems, Lecture Notes in Control and Information Sciences, Vol. 246, pp. 165-182, Springer Verlag, London
11. Grüne, L. (2000) Homogeneous control Lyapunov functions for homogeneous control systems. Proc. Math. Theory Networks Syst., MTNS 2000, Perpignan, France, CD-Rom

12. Lai R. G. (1996) Practical feedback stabilization of nonlinear control systems and applications. PhD thesis, Iowa State University, Ames, Iowa
13. Nikitin, S. (1999) Piecewise-constant stabilization. SIAM J. Contr. & Opt. **37**, 911-933
14. Ryan, E.P. (1994) On Brockett's condition for smooth stabilizability and its necessity in a context of nonsmooth feedback. SIAM J. Contr. & Opt. **32** 1597–1604
15. Sontag, E.D. (1999) Stability and stabilization: Discontinuities and the effect of disturbances. In: F.H. Clarke and R.J. Stern, eds. Nonlinear Analysis, Differential Equations, and Control, Kluwer, pp. 551-598
16. Sontag, E.D., Sussmann, H.J. (1995) Nonsmooth control-Lyapunov functions. In Proc. CDC 1995, New Orleans, USA, 2799–2805

Optimal Control and Implicit Hamiltonian Systems

Guido Blankenstein and Arjan van der Schaft

Faculty of Mathematical Sciences, Department of Systems, Signals and Control
University of Twente, P.O.Box 217
7500 AE Enschede, The Netherlands
{g.blankenstein, a.j.vanderschaft}@math.utwente.nl

Abstract. Optimal control problems naturally lead, via the Maximum Principle, to implicit Hamiltonian systems. It is shown that symmetries of an optimal control problem lead to symmetries of the corresponding implicit Hamiltonian system. Using the reduction theory described in [3,2] one can reduce the system to a lower dimensional implicit Hamiltonian system. It is shown that for symmetries coming from the optimal control problem, doing reduction and applying the Maximum Principle *commutes*.

Furthermore, it is stressed that implicit Hamiltonian systems give rise to more general symmetries than only those coming from the corresponding optimal control problem. It is shown that corresponding to these symmetries, there exist conserved quantities, i.e. functions of the phase variables (that is, q and p) which are constant along solutions of the optimal control problem. See also [19,17].

Finally, the results are extended to the class of *constrained* optimal control problems, which are shown to also give rise to implicit Hamiltonian systems.

1 The Maximum Principle

Let Q be a smooth n-dimensional manifold, with local coordinates denoted by q, and let U be a smooth m-dimensional manifold, with local coordinates denoted by u. Consider a smooth map $f : Q \times U \to TQ$, where TQ denotes the tangent bundle of Q, such that the diagram

$$Q \times U \xrightarrow{f} TQ$$
$$\rho \searrow \quad \swarrow \pi_Q$$
$$Q$$

commutes, where $\rho : Q \times U \to Q$ is the projection onto the first argument and $\pi_Q : TQ \to Q$ is the canonical projection. Then the map f defines a set of smooth vector fields $\{f(\cdot, u)\}_{u \in U}$ on Q, defined by $f(\cdot, u)(q) = f(q, u), q \in Q$. Define the nonlinear system

$$\dot{q} = f(q, u) \tag{1}$$

The next theorem gives a summary of the Maximum Principle in optimal control theory.

Theorem 1. (Maximum Principle) [15,4,8,18] *Consider the nonlinear system defined in (1), with given initial conditions $q(0) = q_0$. Consider two smooth functions $\mathcal{L} : Q \times U \to \mathbb{R}$ (called the* Lagrangian*) and $\mathcal{K} : Q \to \mathbb{R}$ and define the cost functional $J : Q^{\mathbb{R}} \times U^{\mathbb{R}} \to \mathbb{R}$*

$$J(q(\cdot), u(\cdot)) = \int_0^T \mathcal{L}(q(t), u(t))dt + \mathcal{K}(q(T)) \tag{2}$$

for some fixed and given time $T \in \mathbb{R}^+$. The free terminal point optimal control problem is given by

minimize J over $Q^{\mathbb{R}} \times U^{\mathbb{R}}$ such that $(q(\cdot), u(\cdot))$ satisfies (1), and $q(0) = q_0$.

Define the smooth Hamiltonian $H : T^*Q \times U \to \mathbb{R}$

$$H(q, p, u) = p^T f(q, u) - \mathcal{L}(q, u), \quad (q, p, u) \in T^*Q \times U \tag{3}$$

where (q, p) are local coordinates for T^*Q.

Necessary conditions for a smooth trajectory $(q(\cdot), u(\cdot)) \in Q^{\mathbb{R}} \times U^{\mathbb{R}}$ to solve the free terminal point optimal control problem are given by the existence of a smooth curve $(q(\cdot), p(\cdot)) \in (T^*Q)^{\mathbb{R}}$ satisfying the equations

$$\dot{q}(t) = \frac{\partial H}{\partial p}(q(t), p(t), u(t)) \tag{4}$$

$$\dot{p}(t) = -\frac{\partial H}{\partial q}(q(t), p(t), u(t)) \tag{5}$$

$$0 = \frac{\partial H}{\partial u}(q(t), p(t), u(t)) \tag{6}$$

for all $t \in [0, T]$, along with the initial conditions $q(0) = q_0$ and the transversality conditions

$$p(T) = -\frac{\partial \mathcal{K}}{\partial q}(q(T)) \tag{7}$$

Remark 1. Note that the Hamiltonian H in (3) can easily be defined in a coordinate free manner. Furthermore, condition (6) is usually stated as the maximization condition

$$H(q(t), p(t), u(t)) = \max_{\hat{u} \in U} H(q(t), p(t), \hat{u}) \tag{8}$$

for (almost) every time $t \in [0, T]$. However, because we assumed U to be a smooth manifold (which is quite a strong condition and leaves out for instance constrained, or bounded, control) and H a smooth function, (6) is a necessary condition for (8) to hold.

Remark 2. Notice that in the above description of the Maximum Principle we only considered so called *regular* extremals (contrary to the *abnormal* extremals). However, since, for simplicity, we only consider the *free terminal point optimal control problem*, it can be proved that every optimal solution $(q(\cdot), u(\cdot))$ corresponds to a regular extremal, [18].

In the next section, the system of equations (4)–(6) will be described in a coordinate free manner by using the notion of an implicit Hamiltonian system. Furthermore, a nice coordinate free interpretation of equation (7) will be given.

2 Implicit Hamiltonian Systems

It is well known that the cotangent bundle T^*Q has a natural interpretation as a symplectic manifold (T^*Q, ω), defined by the canonical two-form $\omega : \Gamma(T(T^*Q)) \times \Gamma(T(T^*Q)) \to C^\infty(T^*Q)$ on T^*Q (here Γ denotes the space of smooth sections, i.e. $\Gamma(T(T^*Q))$ denotes the space of smooth sections of $T(T^*Q)$, that is, the space of smooth vector fields on T^*Q). In local coordinates ω is given by $\omega = dq \wedge dp$. The map $\omega : \Gamma(T(T^*Q)) \to \Gamma(T^*(T^*Q))$ can be extended to a map $\bar\omega : \Gamma(T(T^*Q \times U)) \to \Gamma(T^*(T^*Q \times U))$ in the following way: Consider the trivial vector bundle $T^*Q \times U \overset{\rho_{T^*Q}}{\to} T^*Q$, and the trivial fiber bundle $T^*Q \times U \overset{\rho_U}{\to} U$. Define a vector field $X \in \Gamma(T(T^*Q \times U))$ on $T^*Q \times U$ to be *horizontal* if $(\rho_U)_* X = 0$, and call it *vertical* if $(\rho_{T^*Q})_* X = 0$. In local coordinates, if $X(q,p,u) = X_q(q,p,u)\frac{\partial}{\partial q} + X_p(q,p,u)\frac{\partial}{\partial p} + X_u(q,p,u)\frac{\partial}{\partial u}$, then X is horizontal if $X_u = 0$ and X is vertical if $X_q = X_p = 0$. A one-form $\alpha \in \Gamma(T^*(T^*Q \times U))$ on $T^*Q \times U$ is called horizontal if it annihilates every vertical vector field, and called vertical if it annihilates every horizontal vector field. Now, $\bar\omega$ is uniquely defined (by skew-symmetry) by requiring that $\bar\omega(X) = 0$ for every vertical vector field $X \in \Gamma(T(T^*Q \times U))$, and for every horizontal vector field $X \in \Gamma(T(T^*Q \times U))$, $\bar\omega(X)(q,p,u) = \omega(X(\cdot,\cdot,u))(q,p)$ for all $(q,p,u) \in T^*Q \times U$, where $X(\cdot,\cdot,u) \in \Gamma(T(T^*Q))$ is considered a vector field on T^*Q, for every $u \in U$. Note that $\bar\omega$ defines a presymplectic structure on $T^*Q \times U$.

Recall that a *Dirac structure* on a manifold M is defined as a linear subspace $D \subset \Gamma(TM) \times \Gamma(T^*M)$ such that $D = D^\perp$, where

$$D^\perp := \{(X_2, \alpha_2) \in \Gamma(TM) \times \Gamma(T^*M) \,|\, \langle \alpha_1, X_2 \rangle + \langle \alpha_2, X_1 \rangle = 0,$$
$$\forall (X_1, \alpha_1) \in D\}$$

where $\langle \cdot, \cdot \rangle$ denotes the natural pairing between a one-form and a vector field, see [5,7,21,20,6,3,2]. Note that D defines a smooth vector subbundle of $TM \oplus T^*M$, with $\dim D(z) = \dim M$, $\forall z \in M$.

The map $\bar{\omega}$ defines a Dirac structure D on $T^*Q \times U$ given by

$$D = \{(X, \alpha) \in \Gamma(T(T^*Q \times U)) \times \Gamma(T^*(T^*Q \times U)) \mid \alpha = \bar{\omega}(X)\} \qquad (9)$$

In local coordinates, the Dirac structure given in (9) has a very simple form

$$D = \{(X, \alpha) \in \Gamma(T(T^*Q \times U)) \times \Gamma(T^*(T^*Q \times U)) \mid$$

$$\begin{bmatrix} \alpha_q \\ \alpha_p \\ \alpha_u \end{bmatrix} = \begin{bmatrix} 0 & -I_n & 0 \\ I_n & 0 & 0 \\ 0 & 0 & 0 \end{bmatrix} \begin{bmatrix} X_q \\ X_p \\ X_u \end{bmatrix}\} \qquad (10)$$

where we denoted $X(q, p, u) = X_q(q, p, u)\frac{\partial}{\partial q} + X_p(q, p, u)\frac{\partial}{\partial p} + X_u(q, p, u)\frac{\partial}{\partial u}$
and $\alpha(q, p, u) = \alpha_q(q, p, u)dq + \alpha_p(q, p, u)dp + \alpha_u(q, p, u)du$.

Consider the Hamiltonian $H \in C^\infty(T^*Q \times U)$ defined in (3). The *implicit Hamiltonian system*, denoted by $(T^*Q \times U, D, H)$, corresponding to the Dirac structure D and the Hamiltonian H, is defined as the set of smooth solutions $(q(\cdot), p(\cdot), u(\cdot)) \in (T^*Q)^{\mathbb{R}} \times U^{\mathbb{R}}$ such that $(\dot{q}(t), \dot{p}(t), \dot{u}(t)) = X(q(t), p(t), u(t))$, $t \in \mathbb{R}$, for some smooth vector field $X \in \Gamma(T(T^*Q \times U))$, and

$$(X, dH)(q(t), p(t), u(t)) \in D(q(t), p(t), u(t)), \quad t \in \mathbb{R} \qquad (11)$$

where $dH \in \Gamma(T^*(T^*Q \times U))$ is the differential of the function H. For more information on implicit Hamiltonian systems, see [21,20,6,3,2]. Now, it is immediately clear that if the Dirac structure is given in local coordinates as in (10), then the equations (11), restricting $t \in [0, T]$, are precisely the equations (4)–(6). Thus we have shown

Proposition 1. *Every optimal control problem, applying the Maximum Principle, gives rise to a corresponding implicit Hamiltonian system.*

Remark 3. Note that equations (4)–(6) are general for all optimal control problems applying the Maximum Principle, not necessarily free terminal point problems.

Now we will give a coordinate free interpretation of the transversality conditions (7). Let $Y \in \Gamma(TQ)$ be a smooth vector field on Q. Define the smooth function $H_Y \in C^\infty(T^*Q)$ given in local coordinates (q, p) by

$$H_Y(q, p) = p^T Y(q) \qquad (12)$$

Proposition 2. *The transversality conditions (7) are equivalent to the following: for every smooth vector field $Y \in \Gamma(TQ)$ on Q, the function $H_Y \in C^\infty(T^*Q)$ at the end point $(q(T), p(T))$ of a solution of the free terminal point optimal control problem has the value*

$$H_Y(q(T), p(T)) = -(L_Y \mathcal{K})(q(T)) \qquad (13)$$

where $L_Y \mathcal{K}$ denotes the Lie derivative of the function $\mathcal{K} \in C^\infty(Q)$ with respect to the vector field Y.

Proof. Let (q, p) be local coordinates for T^*Q around $(q(T), p(T))$. If one takes $Y = \frac{\partial}{\partial q}$, then $H_Y = p$ and (13) is exactly (7). Now, let $Y(q) = Y_1(q)\frac{\partial}{\partial q_1} + \cdots + Y_n(q)\frac{\partial}{\partial q_n}$, then $H_Y(q, p) = p_1 Y_1(q) + \cdots + p_n Y_n(q)$, and it follows that

$$
\begin{aligned}
H_Y(q(T), p(T)) &= p_1(T) Y_1(q(T)) + \cdots + p_n(T) Y_n(q(T)) \\
&= -\left(\frac{\partial \mathcal{K}}{\partial q_1}(q(T)) Y_1(q(T)) + \cdots + \frac{\partial \mathcal{K}}{\partial q_n}(q(T)) Y_n(q(T))\right) \\
&= -(L_Y \mathcal{K})(q(T)) \qquad\qquad \square
\end{aligned}
$$

Proposition 2 will come in need in the next section, where we show that the momentum map, corresponding to a symmetry Lie group of the optimal control problem, will always have the value zero along solutions of the optimal control problem.

3 Symmetries of Optimal Control Problems

Consider the free terminal point optimal control problem (from now on called optimal control problem) described in theorem 1. Consider a Lie group G, with smooth action $\phi : Q \times G \to Q$ on Q, and denote the corresponding Lie algebra by $\mathcal{G}$. G is said to be a *symmetry Lie group* of the optimal control problem if [9,10,19]

$$
[f(\cdot, u), \xi_Q] = 0, \ L_{\xi_Q}\mathcal{L}(\cdot, u) = 0, \ \forall u \in U \text{ and } L_{\xi_Q}\mathcal{K} = 0, \quad \forall \xi \in \mathcal{G} \quad (14)
$$

where $\xi_Q \in \Gamma(TQ)$ denotes the vector field corresponding to $\xi \in \mathcal{G}$ and the action ϕ (i.e. the infinitesimal generator). The action of G on Q can be lifted to an action on T^*Q in the following way: let $\xi_Q \in \Gamma(TQ)$ and consider the function $H_{\xi_Q} \in C^\infty(T^*Q)$, defined in (12). Define the vector field $\dot{\xi}_Q \in \Gamma(T(T^*Q))$ by $dH_{\xi_Q} = i_{\dot{\xi}_Q}\omega$, where ω is the canonical symplectic form on T^*Q. In local coordinates, if $\xi_Q(q) = h(q)\frac{\partial}{\partial q}$ then $\dot{\xi}_Q(q, p) = h(q)\frac{\partial}{\partial q} - p^T\frac{\partial h}{\partial q}(q)\frac{\partial}{\partial p}$. Lifting all the vector fields $\xi_Q \in \Gamma(TQ), \xi \in \mathcal{G}$, to T^*Q defines an action $\dot{\phi} : T^*Q \times G \to T^*Q$ of G on T^*Q, such that the infinitesimal generators are given by $\dot{\xi}_Q, \xi \in \mathcal{G}$. From (14) it immediately follows that the lifted action $\dot{\phi}$ on T^*Q leaves the Hamiltonian (3) invariant. Indeed,

$$
\begin{aligned}
L_{\dot{\xi}_Q} H(q, p, u) &= p^T\frac{\partial f}{\partial q}(q, u)h(q) - \frac{\partial \mathcal{L}}{\partial q}(q, u)h(q) - p^T\frac{\partial h}{\partial q}(q)f(q, u) \\
&= p^T[f(\cdot, u), \xi_Q(\cdot)](q) - L_{\xi_Q}\mathcal{L}(q, u) \\
&= 0, \quad \forall(q, p) \in T^*Q, \forall u \in U \qquad\qquad (15)
\end{aligned}
$$

We will show that G is a symmetry Lie group of the implicit Hamiltonian system $(T^*Q \times U, D, H)$ defined in section 2. Therefore we need the following

Definition 1. [5,7,20,3,2] Consider a Dirac structure D on $T^*Q \times U$ (as in (9) for example). A vector field $Y \in \Gamma(T(T^*Q \times U))$ on $T^*Q \times U$ is called a symmetry of D if $(X, \alpha) \in D$ implies $([Y, X], L_Y \alpha) \in D$, for all $(X, \alpha) \in D$.

A Dirac structure is called *closed* if it satisfies some integrability conditions (generalizing the Jacobi identities for a Poisson structure), see [6] for more information. Let us just mention that the Dirac structure D given in (9) is closed (this follows immediately from proposition 4.5 in [6]). The following important result is given in [5,7].

Proposition 3. *Let D be a closed Dirac structure on $T^*Q \times U$. Let $Y \in \Gamma(T(T^*Q \times U))$ be a vector field on $T^*Q \times U$ and assume that there exists a function $F \in C^\infty(T^*Q \times U)$ such that $(Y, dF) \in D$, then Y is a symmetry of D.*

Now, consider a lifted vector field $\dot{\xi}_Q \in \Gamma(T(T^*Q))$, this can be identified with a horizontal vector field $\dot{\xi}_Q \in \Gamma(T(T^*Q \times U))$ on $T^*Q \times U$. Furthermore, the function $H_{\xi_Q} \in C^\infty(T^*Q)$ defines a function $H_{\xi_Q} \in C^\infty(T^*Q \times U)$ (by $H_{\xi_Q}(q, p, u) = H_{\xi_Q}(q, p), (q, p, u) \in T^*Q \times U$). Since $dH_{\xi_Q} = i_{\dot{\xi}_Q}\bar{\omega}$, it follows that $(\dot{\xi}_Q, dH_{\xi_Q}) \in D$, given in (9), which implies by proposition 3 that $\dot{\xi}_Q$ is a symmetry of D. It follows that every vector field $\dot{\xi}_Q, \xi \in \mathcal{G}$, is a symmetry of D, we call $\mathcal{G}$ a symmetry Lie group of D. Furthermore, since H is invariant under G, we say that G is a symmetry Lie group of the implicit Hamiltonian system $(T^*Q \times U, D, H)$. Thus we have shown that every symmetry of an optimal control problem (of the form (14)) leads to a (lifted) symmetry of the corresponding implicit Hamiltonian system.

In [3,2] we have presented a reduction theory for implicit Hamiltonian systems admitting a symmetry Lie group. This theory extends the classical reduction theory for symplectic and Poisson systems, as well as the reduction theory for constrained mechanical systems. We refer to [3,2] for further details. In this paper we want to apply this reduction theory to implicit Hamiltonian systems $(T^*Q \times U, D, H)$ of the form (3,9) subject to the (lifted) action of the symmetry Lie group G descibed above, (14). Because of the simple form of the Dirac structure D and of the (lifted) symmetries (14), the reduction will actually come down to classical symplectic reduction. We will show that after reduction of the system $(T^*Q \times U, D, H)$ to a lower dimensional implicit Hamiltonian system, one obtains the same (up to isomorphism) implicit Hamiltonian system as obtained by first reducing the optimal control problem itself (to an optimal control problem on Q/G) and then applying the Maximum Principle in theorem 1, i.e. doing reduction and applying the Maximum Principle *commutes*.

Consider the optimal control problem defined in theorem 1, and suppose it admits a symmetry Lie group G, described in (14). Assume that the quotient manifold Q/G of G-orbits on Q is well defined (e.g. the action of G on Q is free and proper), and denote the projection $\pi : Q \to Q/G$. Because of (14), every vector field $f(\cdot, u), u \in U$, on Q projects to a well defined vector field $\hat{f}(\cdot, u), u \in U$, on Q/G, defined by $\pi_* f(\cdot, u) = \hat{f}(\cdot, u), u \in U$. Furthermore, the Lagrangian $\mathcal{L}(\cdot, u) : Q \to \mathbb{R}, u \in U$, defines a well defined function $\hat{\mathcal{L}}(\cdot, u) :$ $Q/G \to \mathbb{R}, u \in U$, by $\mathcal{L}(\cdot, u) = \hat{\mathcal{L}}(\cdot, u) \circ \pi, u \in U$. Finally, the function $\mathcal{K} : Q \to \mathbb{R}$ defines a well defined function $\hat{\mathcal{K}} : Q/G \to \mathbb{R}$ by $\mathcal{K} = \hat{\mathcal{K}} \circ \pi$. Then in [9,10] it is shown that the optimal control problem on Q projects to a well defined optimal control problem on Q/G, defined by $\{\hat{f}(\cdot, u)\}_{u \in U}, \hat{\mathcal{L}}$ and $\hat{\mathcal{K}}$. The cost functional $\hat{J} : (Q/G)^{\mathbb{R}} \times U^{\mathbb{R}} \to \mathbb{R}$ becomes

$$\hat{J}(\hat{q}(\cdot), u(\cdot)) = \int_0^T \hat{\mathcal{L}}(\hat{q}(t), u(t))dt + \hat{\mathcal{K}}(\hat{q}(T))$$

where $(\hat{q}(\cdot), u(\cdot)) \in (Q/G)^{\mathbb{R}} \times U^{\mathbb{R}}$ denotes a time trajectory in $Q/G \times U$. Notice that $J(q(\cdot), u(\cdot)) = \hat{J}(\pi(q(\cdot)), u(\cdot))$ for all trajectories $(q(\cdot), u(\cdot)) \in Q^{\mathbb{R}} \times U^{\mathbb{R}}$ satisfying (1). The *reduced optimal control problem* is defined by

> minimize $\hat{J}$ over $(Q/G)^{\mathbb{R}} \times U^{\mathbb{R}}$ such that $(\hat{q}(\cdot), u(\cdot))$ satisfies
> $\dot{\hat{q}}(t) = \hat{f}(\hat{q}(t), u(t)), t \in [0, T]$, and $\hat{q}(0) = \hat{q}_0$

where $\hat{q}_0 = \pi(q_0)$. As in theorem 1 we can apply the Maximum Principle to obtain necessary conditions for a trajectory $(\hat{q}(\cdot), u(\cdot)) \in (Q/G)^{\mathbb{R}} \times U^{\mathbb{R}}$ to be a solution of the reduced optimal control problem. Define the reduced Hamiltonian

$$\hat{H}(\hat{q}, \hat{p}, u) = \hat{p}^T \hat{f}(\hat{q}, u) - \hat{\mathcal{L}}(\hat{q}, u), \quad (\hat{q}, \hat{p}, u) \in T^*(Q/G) \times U$$

where $(\hat{q}, \hat{p})$ denote local coordinates for the cotangent bundle $T^*(Q/G)$. It is easy to see that H and $\hat{H}$ are related by $\hat{H}(\cdot, u) = H(\cdot, u) \circ \pi^*, u \in U$,, where $\pi^* : T^*(Q/G) \to T^*Q$ is the pull-back defined by π. Now, applying the Maximum Principle to the reduced optimal control problem gives us the necessary conditions, for a smooth curve $(\hat{q}(\cdot), u(\cdot)) \in (Q/G)^{\mathbb{R}} \times U^{\mathbb{R}}$ to be a solution of the reduced optimal control problem, of the existence of a smooth curve $(\hat{q}(\cdot), \hat{p}(\cdot)) \in (T^*(Q/G))^{\mathbb{R}}$ satisfying the equations

$$\dot{\hat{q}}(t) = \frac{\partial \hat{H}}{\partial \hat{p}}(\hat{q}(t), \hat{p}(t), u(t)) \tag{16}$$

$$\dot{\hat{p}}(t) = -\frac{\partial \hat{H}}{\partial \hat{q}}(\hat{q}(t), \hat{p}(t), u(t)) \tag{17}$$

$$0 = \frac{\partial \hat{H}}{\partial u}(\hat{q}(t), \hat{p}(t), u(t)) \tag{18}$$

for all $t \in [0, T]$, along with the initial conditions $\hat{q}(0) = \hat{q}_0$ and the transversality conditions

$$\hat{p}(T) = -\frac{\partial \hat{\mathcal{K}}}{\partial \hat{q}}(\hat{q}(T))$$

As described in section 2, proposition 1, equations (16)–(18) give rise to an implicit Hamiltonian system $(T^*(Q/G) \times U, \hat{D}, \hat{H})$, where $\hat{D}$ is a Dirac structure on $T^*(Q/G) \times U$ defined analogously as in (9) (note that $T^*(Q/G)$, being a cotangent bundle, has a natural symplectic form $\hat{\omega}$). We will show that this implicit Hamiltonian system is isomorphic (to be defined precisely later on) to the implicit Hamiltonian system obtained by applying the reduction theory described in [3,2] to the system $(T^*Q \times U, D, H)$.

Consider the implicit Hamiltonian system $(T^*Q \times U, D, H)$ with symmetry Lie group G. The action of G on T^*Q is given by ϕ, and the infinitesimal generators are given by the lifted vector fields $\dot{\xi}_Q, \xi \in \mathcal{G}$. Corresponding to this action, there exists a Ad^*-equivariant *momentum map*, defined in local coordinates by $\Phi : T^*Q \to \mathcal{G}^*$ [1]

$$\Phi(q,p)(\xi) = p^T \xi_Q(q), \quad \forall \xi \in \mathcal{G}, \ (q,p) \in T^*Q \tag{19}$$

Define $\Phi(\xi) : T^*Q \to \mathbb{R}$ by $\Phi(\xi)(q,p) = \Phi(q,p)(\xi), (q,p) \in T^*Q, \xi \in \mathcal{G}$, then it follows that $\Phi(\xi) = H_{\xi_Q}$. Since $(\dot{\xi}_Q, dH_{\xi_Q}) \in D, \xi \in \mathcal{G}$, it follows that $(\dot{\xi}_Q, d\Phi(\xi)) \in D, \xi \in \mathcal{G}$. Now, consider a solution of the optimal control problem, i.e. $(\dot{q}(t), \dot{p}(t), \dot{u}(t)) = X(q(t), p(t), u(t))$ and

$$(X, dH)(q(t), p(t), u(t)) \in D(q(t), p(t), u(t)), \quad t \in [0, T] \tag{20}$$

see section 2. By definition, $D = D^\perp$, which implies

$$0 = (\langle X, d\Phi(\xi) \rangle + \langle \dot{\xi}_Q, dH \rangle)(q(t), p(t), u(t)) = \langle X, d\Phi(\xi) \rangle(q(t), p(t), u(t)),$$

$t \in [0, T]$, by (15), for all $\xi \in \mathcal{G}$. This means that $\Phi(\xi)$ is constant along solutions of the optimal control problem (or rather, of the implicit Hamiltonian system $(T^*Q \times U, D, H)$), see also [19,17]. Furthermore, by the transversality conditions in proposition 2 it follows that

$$\Phi(\xi)(q(T), p(T)) = -(L_{\xi_Q} K)(q(T)) = 0, \quad \xi \in \mathcal{G}$$

by (14). This means that $\Phi(\xi)$ has actually constant value zero along solutions of the optimal control problem. Since this holds for every $\xi \in \mathcal{G}$ it follows that the momentum map Φ has constant value $0 \in \mathcal{G}^*$ along solutions of the optimal control problem. Thus we have proved.

Proposition 4. *The momentum map $\Phi : T^*Q \to \mathcal{G}^*$ has constant value zero along solutions of the optimal control problem (or rather, implicit Hamiltonian system $(T^*Q \times U, D, H)$).*

Now we can use the reduction theory described in [3,2] to reduce the implicit Hamiltonian system to a lower dimensional implicit Hamiltonian system $(T^*Q \times U, D, H)$ (note that all the constant dimensionality conditions in there are satisfied). We begin by restricting the system to a levelset of the momentum map, i.e. $\Phi^{-1}(0) \times U \subset T^*Q \times U$. This gives again an implicit Hamiltonian system on $\Phi^{-1}(0) \times U$. The resticted system will have some symmetry left, corresponding to the residual subgroup $G_\mu = \{g \in G \mid \text{Ad}_g^*\mu = \mu\}, \mu \in \mathcal{G}^*$. However, since in this case $\mu = 0$, and the coadjoint action $\text{Ad}_g^* : \mathcal{G}^* \to \mathcal{G}^*$ is linear, $\forall g \in G$, it follows that $G_0 = G$. This means that the restricted system will still have G as a symmetry Lie group. Assuming that the quotient manifold $\Phi^{-1}(0)/G$ is well defined, we can project the resticted system on $\Phi^{-1}(0) \times U$ to an implicit Hamiltonian system on $\Phi^{-1}(0)/G \times U$. The resulting implicit Hamiltonian system will be denoted by $(\Phi^{-1}(0)/G \times U, \tilde{D}, \tilde{H})$. Here $\tilde{H} \in C^\infty(\Phi^{-1}(0)/G \times U)$ is defined by $H(\cdot, \cdot, u) \circ i = \tilde{H}(\cdot, u) \circ \pi_0, u \in U$, where $i : \Phi^{-1}(0) \to T^*Q$ is the inclusion map, and $\pi_0 : \Phi^{-1}(0) \to \Phi^{-1}(0)/G$ is the projection map. Since the Dirac structure D has the very special form (9), or (10), and the action of G on $T^*Q \times U$ only acts on the T^*Q-part, it can easily be seen that the reduction theory described in [3,2] in this special case comes down to symplectic reduction of the symplectic manifold (T^*Q, ω), as descibed in e.g. [14,1,12], see also example 8 in [3]. It follows that the reduced Dirac structure $\tilde{D}$ on $\Phi^{-1}(0)/G \times U$ is given by

$$\tilde{D} = \{(\tilde{X}, \tilde{\alpha}) \in \Gamma(T(\Phi^{-1}(0)/G \times U)) \times \Gamma(T^*(\Phi^{-1}(0)/G \times U)) \mid$$
$$\tilde{\alpha} = \bar{\omega}_0(\tilde{X})\}$$

where $\bar{\omega}_0 : \Gamma(T(\Phi^{-1}(0)/G \times U)) \to \Gamma(T^*(\Phi^{-1}(0)/G \times U))$ is the extension of the symplectic two form ω_0 on $\Phi^{-1}(0)/G$, defined by $\pi_0^*\omega_0 = i^*\omega$, as in section 2.

Now we will show that the two reduced implicit Hamiltonian systems $(T^*(Q/G) \times U, \hat{D}, \hat{H})$ and $(\Phi^{-1}(0)/G \times U, \tilde{D}, \tilde{H})$ are isomorphic, defined in the following way.

Definition 2. [3] Let M and N be two manifolds, and let $\tau : M \to N$ be a diffeomorphism. Let D_M be a Dirac structure on M and D_N a Dirac structure on N. Then τ is called a Dirac isomorphism if

$$(X, \alpha) \in D_M \iff (\tau_* X, (\tau^*)^{-1}\alpha) \in D_N$$

In this case we call D_M and D_N isomorphic.

Consider two implicit Hamiltonian systems (M, D_M, H_M) and (N, D_N, H_N). We call the two systems isomorphic if D_M and D_N are isomorphic, and $H_M = H_N \circ \tau$.

We begin by showing that $\tilde{D}$ and $\hat{D}$ are isomorphic (by some diffeomorphism $\tau : \Phi^{-1}(0)/G \times U \to T^*(Q/G) \times U$). First, consider the two symplectic manifolds $(\Phi^{-1}(0)/G, \omega_0)$ and $(T^*(Q/G), \hat{\omega})$ (recall that $\hat{\omega}$ denotes the canonical symplectic two-form on $T^*(Q/G)$). It is a well known result from Satzer [16], see also [11] and [13], that these two symplectic manifolds are symplectomorphic, i.e. there exists a diffeomorphism $\tau : \Phi^{-1}(0)/G \to T^*(Q/G)$ such that τ is a symplectic map, that is, $\tau^*\hat{\omega} = \omega_0$. The map τ induces a map

$(\tau, \mathrm{id}) : \Phi^{-1}(0)/G \times U \to T^*(Q/G) \times U$, which we will also denote by τ (it will be clear from the context which map is meant). Here, $\mathrm{id} : U \to U$ denotes the identity map on U. Since $\tilde{D}$ and $\hat{D}$ are completely defined by ω_0, respectively $\hat{\omega}$, it follows from the fact that $\tau^* \hat{\omega} = \omega_0$ that τ is a Dirac isomorphism between $\tilde{D}$ and $\hat{D}$.

To see that the Hamiltonians $\tilde{H}$ and $\hat{H}$ are also related, i.e. $\tilde{H} = \hat{H} \circ \tau$, we have to go a little bit deeper into the construction described in [11,13] of the map τ. In [11,13] the map τ is constructed by taking a *connection* $\nabla : TQ \to \mathcal{G}$ on the principal G-bundle $Q \xrightarrow{\pi} Q/G$. Precisely, $\nabla : (TQ, T\phi) \to (\mathcal{G}, \mathrm{Ad})$ is a homomorphism such that $\nabla(\xi_Q(q)) = \xi, \xi \in \mathcal{G}$ (where $T\phi$ represents the tangent of the map ϕ). Take arbitrary $\alpha \in \Gamma(T^*Q)$ and define $\beta \in \Gamma(T^*Q)$ by $\beta(q) = \alpha(q) - \nabla^*(\Phi(\alpha(q))), q \in Q$. Then β projects to a one-form $\hat{\alpha} \in \Gamma(T^*(Q/G))$, i.e. $\beta = \pi^* \hat{\alpha}$. Indeed,

$$\langle \beta, \xi_Q \rangle = \langle \alpha, \xi_Q \rangle - \langle \nabla^*(\Phi(\alpha)), \xi_Q \rangle = \Phi(\alpha)(\xi) - \Phi(\alpha)(\nabla(\xi_Q)) = 0, \; \xi \in \mathcal{G}$$

Define the map $\pi_\nabla : T^*Q \to T^*(Q/G)$ by $\pi_\nabla(q, \alpha_q) = (\hat{q}, \hat{\alpha}_{\hat{q}}), \alpha \in \Gamma(T^*Q)$, where $\hat{q} = \pi(q)$, and restrict this map to get the map $\pi_{\nabla_0} : \Phi^{-1}(0) \to T^*(Q/G)$. In [11] it is shown that $\Phi^{-1}(0) \xrightarrow{\pi_{\nabla_0}} T^*(Q/G)$ is a principal G-bundle. Finally, the map $\tau : \Phi^{-1}(0)/G \to T^*(Q/G)$ is defined such that the diagram

$$\begin{array}{ccc} \Phi^{-1}(0) & \xrightarrow{\pi_0} & \Phi^{-1}(0)/G \\ & \pi_{\nabla_0} \searrow \downarrow \tau & \\ & T^*(Q/G) & \end{array}$$

commutes. In [11] it is shown that $\tau : (\Phi^{-1}(0)/G, \omega_0) \to (T^*(Q/G), \hat{\omega})$ is a symplectomorphism (this does not depend on the connection ∇ choosen). Now we can prove that the Hamiltonians $\tilde{H}$ and $\hat{H}$ are related by $\tilde{H} = \hat{H} \circ \tau$. Take arbitrary $z \in \Phi^{-1}(0)/G$, then there exists a point $(q_0, p_0) \in \Phi^{-1}(0) \subset T^*Q$ such that $\pi_0(q_0, p_0) = z$. Let $\alpha \in \Gamma(T^*Q)$ be such that $(q, \alpha_q) \in \Phi^{-1}(0), q \in Q$ and $(q_0, \alpha_{q_0}) = (q_0, p_0)$. Then there exists an $\hat{\alpha} \in \Gamma(T^*(Q/G))$ such that $\alpha = \pi^* \hat{\alpha}$, and it follows that $\pi_{\nabla_0}(q_0, p_0) = \pi_{\nabla_0}(q_0, \alpha_{q_0}) = (\hat{q}_0, \hat{\alpha}_{\hat{q}_0})$. By commutation, $\tau(z) = (\hat{q}_0, \hat{\alpha}_{\hat{q}_0})$. Furthermore

$$\tilde{H}(z, u) = H(\cdot, \cdot, u) \circ i(q_0, \alpha_{q_0}) = \alpha_{q_0}^T f(q_0, u) - \mathcal{L}(q_0, u)$$

$$= \hat{\alpha}_{\hat{q}_0}^T \hat{f}(\hat{q}_0, u) - \hat{\mathcal{L}}(\hat{q}_0, u) = \hat{H}(\hat{q}_0, \hat{\alpha}_{\hat{q}_0}, u) = \hat{H} \circ \tau(z, u)$$

and it follows that $\tilde{H} = \hat{H} \circ \tau$. Thus we have shown that the two implicit Hamiltonian systems $(\Phi^{-1}(0)/G \times U, \tilde{D}, \tilde{H})$ and $(T^*(Q/G) \times U, \hat{D}, \hat{H})$ are isomorphic. In particular we have proved.

Theorem 2. *Consider an optimal control problem with symmetries described in (14). Then reducing the optimal control problem to a lower dimensional* *optimal control problem (on Q/G), called the reduced optimal control problem, and applying the Maximum Principle, is equivalent (up to isomorphism) to first applying the Maximum Principle on the original optimal control problem (on Q), and then using the reduction theory described in [3,2] to reduce the corresponding implicit Hamiltonian system.*

This result seems quite natural, and it gives a nice connection between reduction of optimal control problems on the one hand and reduction of implicit Hamiltonian systems on the other hand. To appreciate the result, notice that in general $\Phi^{-1}(\mu)/G_\mu$ and $T^*(Q/G)$ are not diffeomorphic (they do not even have the same dimension). However, because we proved that the momentum map has value zero along solutions of the optimal control problem (or, implicit Hamiltonian system), we could use the results in [16,11,13] to conclude that $(\Phi^{-1}(0)/G, \omega_0)$ and $(T^*(Q/G), \hat{\omega})$ are symplectomorphic (as symplectic manifolds), from which the final result in theorem 2 was obtained.

4   Generalized Symmetries

In the previous section we showed that symmetries of optimal control problems of the form (14) lead to symmetries of the corresponding implicit Hamiltonian system, by lifting the symmetries to the cotangent bundle. However, in general the implicit Hamiltonian system has more symmetries than only those coming from the optimal control problem. These symmetries cannot necessarily be described as the lift of a vector field on Q, so they do not correspond to a symmetry of the optimal control problem. These symmetries (of the implicit Hamiltonian system) are called *generalized symmetries* of the optimal control problem (in [19] these symmetries are called *dynamical* symmetries, contrary to lifted symmetries, called *geometrical* symmetries). In this section it is proved that to each generalized symmetry of the optimal control problem there exists (locally) a *conserved quantity*, being a function of the phase variables q and p which is constant along solutions of the optimal control problem (or, implicit Hamiltonian system). (In case of a lifted symmetry, this conserved quantity is exactly the (globaly defined) momentum map, described in section 3.) We refer to [19,17] for related results.

Consider the Dirac structure described in (9), corresponding to the implicit Hamiltonian system $(T^*Q \times U, D, H)$ obtained by applying the Maximum Principle to the optimal control problem, see section 2. Define the following (co-)distributions on T^*Q [6] (denote by $\Gamma^{\mathrm{vert}}(T(T^*Q \times U))$ the set of all vertical vector fields on $T^*Q \times U$, and by $\Gamma^{\mathrm{hor}}(T^*(T^*Q \times U))$ the set of all horizontal one-forms on $T^*Q \times U$).

$$G_0 = \{X \in \Gamma(T(T^*Q \times U)) \mid (X, 0) \in D\} = \Gamma^{\text{vert}}(T(T^*Q \times U))\}$$
$$G_1 = \{X \in \Gamma(T(T^*Q \times U)) \mid \exists \alpha \in \Gamma(T^*(T^*Q \times U)) \text{ s.t. } (X, \alpha) \in D\}$$
$$= \Gamma(T(T^*Q \times U))$$
$$P_0 = \{\alpha \in \Gamma(T^*(T^*Q \times U)) \mid (0, \alpha) \in D\} = 0$$
$$P_1 = \{\alpha \in \Gamma(T^*(T^*Q \times U)) \mid \exists X \in \Gamma(T(T^*Q \times U)) \text{ s.t. } (X, \alpha) \in D\}$$
$$= \Gamma^{\text{hor}}(T^*(T^*Q \times U))$$

Note that all the (co-)distributions are constant dimensional and that $G_0 = \ker P_1$ and $P_0 = \text{ann } G_1$, where ker denotes the kernel of a codistribution and ann the annihilator of a distribution, see [6,3]. Furthermore, the codistribution P_1 is *involutive*, defined by $G_0 = \ker P_1$ being an involutive distribution. Define the set of *admissible* functions

$$\mathcal{A}_D = \{H \in C^\infty(T^*Q \times U) \mid dH \in P_1\}$$
$$= \{H \in C^\infty(T^*Q \times U) \mid H(q, p, u) = H(q, p)\} \tag{21}$$

along with the generalized Poisson bracket on $\mathcal{A}_D$ given by $\{H_1, H_2\}_D = \langle dH_1, X_2 \rangle = -\langle dH_2, X_1 \rangle$, $H_1, H_2 \in \mathcal{A}_D$, where $(X_1, dH_1), (X_2, dH_2) \in D$ [6,3]. In [3] the following proposition is proved.

Proposition 5. *A vector field $Y \in \Gamma(T(T^*Q \times U))$ is a symmetry of D, defined in definition 1, if and only if*

- *Y is canonical with respect to $\{\cdot, \cdot\}_D$, i.e.*

$$L_Y\{H_1, H_2\}_D = \{L_Y H_1, H_2\}_D + \{H_1, L_Y H_2\}_D, \quad \forall H_1, H_2 \in \mathcal{A}_D, \tag{22}$$

- *$L_Y G_i \subset G_i$, $L_Y P_i \subset P_i$, $i = 0, 1$*

Remark 4. In [3] it is proved that the *only if*-part always holds. The *if*-part however holds under the condition that P_1 is constant dimensional and involutive, as is the case for the Dirac structure given in (9).

Now, consider the Dirac structure described in (9) and let $Y \in \Gamma(T(T^*Q \times U))$ be a symmetry of D, given in local coordinates by $Y(q, p, u) = Y_q(q, p, u)\frac{\partial}{\partial q} + Y_p(q, p, u)\frac{\partial}{\partial p} + Y_u(q, p, u)\frac{\partial}{\partial u}$. By proposition 5, $L_Y G_0 \subset G_0$ (so $[Y, \frac{\partial}{\partial u}] \in$ span $\{\frac{\partial}{\partial u}\}$), which implies that Y_q and Y_p do not depend on u, i.e. $Y_q(q, p, u) = Y_q(q, p)$ and $Y_p(q, p, u) = Y_p(q, p)$. Furthermore, $L_Y P_1 \subset P_1$, which for the given vector field Y is always satisfied as direct calculation shows. Finally, the vector field Y should satisfy (22), which together with (21) and (9) implies that $L_{\tilde{Y}}\omega = 0$, where $\tilde{Y}(q, p) = Y_q(q, p)\frac{\partial}{\partial q} + Y_p(q, p)\frac{\partial}{\partial p} \in$

$\Gamma(T(T^*Q))$ (i.e. $\tilde{Y} = (\rho_{T^*Q})_*Y$). From classical mechanics it is known that this implies that $\tilde{Y}$ is locally Hamiltonian, that is, there exists (locally) a function $\tilde{H} \in C^\infty(T^*Q)$ such that $d\tilde{H} = i_{\tilde{Y}}\omega$. It follows that any symmetry Y of D in local coordinates has the form $Y(q,p,u) = Y_q(q,p)\frac{\partial}{\partial q} + Y_p(q,p)\frac{\partial}{\partial p} + Y_u(q,p,u)\frac{\partial}{\partial u}$, where $\tilde{Y}(q,p) = (\rho_{T^*Q})_*Y \in \Gamma(T(T^*Q))$ satisfies $d\tilde{H} = i_{\tilde{Y}}\omega$ for some (locally defined) function $\tilde{H} \in C^\infty(T^*Q)$. Notice that $(Y, d\tilde{H}) \in D$ (we call Y a *Hamiltonian* symmetry). Furthermore, assume that Y is a symmetry of the implicit Hamiltonian system $(T^*Q \times U, D, H)$, that is, Y is also a symmetry of the Hamiltonian H. Then, along solutions (20) of the implicit Hamiltonian system

$$0 = (\langle d\tilde{H}, X \rangle + \langle dH, Y \rangle)(q(t), p(t), u(t)) = \langle d\tilde{H}, X \rangle(q(t), p(t), u(t)), \quad (23)$$

$t \in [0, T]$ (at least locally, that is, where $\tilde{H}$ is defined), so $\tilde{H}$ is a conserved quantity. Thus we have proved.

Proposition 6. *Consider the implicit Hamiltonian system $(T^*Q \times U, D, H)$, with D given in (9). Let $Y \in \Gamma(T(T^*Q \times U))$ be a symmetry of $(T^*Q \times U, D, H)$. Then there exists (locally) a function $\tilde{H} \in C^\infty(T^*Q)$ such that $\tilde{H}$ is constant along solutions of the implicit Hamiltonian system.*

In particular, this means that to every generalized symmetry of the optimal control problem, there exists (locally) a function of the phase variables (that is, $\tilde{H} \in C^\infty(T^*Q)$) which is constant along solutions of the optimal control problem.

In fact, we have the following converse of proposition 6 which, together with proposition 6, gives a Noether type of result between symmetries and conservation laws of optimal control problems, see also the results in [3,2].

Proposition 7. *Consider the implicit Hamiltonian system $(T^*Q \times U, D, H)$, with D given in (9). Let $\tilde{H} \in C^\infty(T^*Q)$ be a conserved quantity of $(T^*Q \times U, D, H)$, that is, $\tilde{H}$ is constant along solutions of $(T^*Q \times U, D, H)$. Then, $Y \in \Gamma^{hor}(T(T^*Q \times U))$, defined by $d\tilde{H} = i_Y\omega$, is a symmetry of $(T^*Q \times U, D, H)$.*

Proof. Since $(Y, d\tilde{H}) \in D$ it follows by proposition 3 that Y is a symmetry of D. Furthermore, since $\tilde{H}$ is a conserved quantity,

$$0 = (\langle d\tilde{H}, X \rangle + \langle dH, Y \rangle)(q(t), p(t), u(t)) = \langle dH, Y \rangle(q(t), p(t), u(t)),$$

$t \in [0, T]$, along solutions (20) of $(T^*Q \times U, D, H)$, so Y is a symmetry of H. $\square$

Remark 5. Notice that from the above proof it follows that Y is a symmetry of $(T^*Q \times U, D, H)$ only in the "weak" sense. That is, $L_Y H = 0$ only along solutions of the implicit Hamiltonian system.

Finally we remark that generalized symmetries of the optimal control problem give rise to reduction of the corresponding implicit Hamiltonian system (not of the optimal control problem itself), by factoring out these symmetries and using the conserved quantities to obtain a lower dimensional implicit Hamiltonian system, as described in the reduction theory in [3,2]. So even in the case of generalized symmetries we can do reduction to simplify the search for solutions of the optimal control problem. In general however, the reduced implicit Hamiltonian system will not like in theorem 2 correspond to a lower dimensional optimal control problem again.

Example 1. In [17] the following class of symmetries is investigated. Consider a Lie group G with action $\phi : Q \times G \to Q$ on Q. Assume that G is a symmetry Lie group of the optimal control problem in the sense that

$$\forall g \in G, \forall u \in U, \exists u_1, u_2 \in U \text{ s.t. } \quad T\phi_g \cdot f(\cdot, u) = f(\phi_g(\cdot), u_1),$$
$$T\phi_g \cdot f(\cdot, u_2) = f(\phi_g(\cdot), u), \quad \mathcal{L}(\cdot, u) = \mathcal{L}(\phi_g(\cdot), u_1) \tag{24}$$
$$\mathcal{L}(\cdot, u_2) = \mathcal{L}(\phi_g(\cdot), u), \quad \mathcal{K}(\phi_g(\cdot)) = \mathcal{K}(\cdot)$$

Notice that if $u = u_1 = u_2$ this is exactly (14). The lifted action $\dot{\phi} : T^*Q \times G \to T^*Q$ defines a symmetry of the Hamiltonian H in the sense that

$$\forall g \in G, \forall u \in U, \exists u_1, u_2 \in U \text{ s.t. } \quad H(\cdot, \cdot, u_1) \circ \dot{\phi}_g = H(\cdot, \cdot, u),$$
$$H(\cdot, \cdot, u) \circ \dot{\phi}_g = H(\cdot, \cdot, u_2)$$

or, in an infinitesimal version,

$$\forall \xi \in \mathcal{G}, \forall u \in U, \exists u_2(s) \in U^{(-\delta, \delta)} \text{ s.t. }$$
$$H(\cdot, \cdot, u) \circ \dot{\phi}(\cdot, \exp(s\xi)) = H(\cdot, \cdot, u_2(s)), \quad s \in (-\delta, \delta) \tag{25}$$

where $\exp : \mathcal{G} \to G$ denotes the exponential map of the Lie group, and where $(-\delta, \delta), \delta > 0$, denotes a small time interval. Define the momentum map $\Phi : T^*Q \to \mathcal{G}^*$ as in section 3, then in [17] it is shown that the momentum map is constant along solutions of the implicit Hamiltonian system (in their terminology called *biextremals* of the optimal control problem). In [17] this is called the *control theory version of Noether's Theorem*. In our setting, the same result follows from proposition 6. Indeed, differentiate (25) to s at $s = 0$ to get (with some abuse of notation)

$$\frac{\partial H(\cdot, \cdot, u)}{\partial(q, p)} \cdot \dot{\xi}_Q = \frac{\partial H}{\partial u}(\cdot, \cdot, u) \cdot \frac{du_2(s)}{ds}\Big|_{s=0}$$

which implies that $L_{\dot{\xi}_Q} H \in \text{span}\{\frac{\partial H}{\partial u}\}$, so there exists a vector field $Y_\xi \in \Gamma(T(T^*Q \times U))$, with $(\rho_{T^*Q})_* Y_\xi = \dot{\xi}_Q$, such that $L_{Y_\xi} H = 0$, for every $\xi \in \mathcal{G}$. Furthermore, notice that $(Y_\xi, dH_{\xi_Q}) \in D$, so Y_ξ is a generalized symmetry of the optimal control problem. It follows from proposition 6 that H_{ξ_Q} is

constant along solutions of the implicit Hamiltonian system. Since this holds
for every $\xi \in \mathcal{G}$, it follows that the momentum map Φ is constant along
solutions of the implicit Hamiltonian system, and we have obtained the same
result as in [17] (considering the smooth case). Notice that by proposition 2
the momentum map has actually the value zero along solutions of the implicit
Hamiltonian system, see also section 3.

Remark 6. Actually, (a slight modification of) theorem 2 also holds in case
the optimal control problem admits symmetries described in (24). The key
observation is that the optimal control problem can be transformed by regular
state feedback into an equivalent optimal control problem with *state-space*
symmetries described in (14) (given by the action ϕ), see [9]. Due to space
limitations we will not persue this point any further here.

5   Constrained Optimal Control Problems

In this section we generalize the results of the previous sections to constrained
optimal control problems. Consider a nonlinear system (1) subject to the
constraints $b_1(q, u) = 0, \ldots, b_k(q, u) = 0$, $b_1, \ldots, b_k \in C^\infty(Q \times U)$, or, equivalently, by defining $b(q, u) = [b_1(q, u), \ldots, b_k(q, u)]^T$,

$$b(q, u) = 0 \tag{26}$$

We will assume that the following regularity condition is satisfied ([15,4]):
the Jacobian

$$\frac{\partial b}{\partial u}(q, u) \tag{27}$$

has full row rank for all $(q, u) \in Q \times U$. Note that "holonomic" constraints $g(q) = 0$, as well as "nonholonomic" constraints $g(q, \dot{q}) = 0$, are
included in this setup, by defining $b(q, u) = \frac{\partial g}{\partial q}^T(q) f(q, u)$, respectively
$b(q, u) = g(q, f(q, u))$, assuming that the regularity condition (27) is satis-
fied (if not, one could differentiate the constraints again, and check if the
regularity condition is satisfied this time) [15,4].

Consider the cost functional given in (2). The *constrained optimal control
problem* is given by

*minimize $J(\cdot, \cdot)$ over $Q^{\mathbb{R}} \times U^{\mathbb{R}}$ such that $(q(\cdot), u(\cdot))$ satisfies (1) and (26),
and $q(0) = q_0$.*

Also in this case the Maximum Principle gives necessary conditions to be
satisfied by a solution of the constrained optimal control problem, see e.g.
[15,4]. Necessary conditions for a smooth trajectory $(q(\cdot), u(\cdot)) \in Q^{\mathbb{R}} \times U^{\mathbb{R}}$ to

solve the constrained optimal control problem are given by the existence of a smooth curve $(q(\cdot), p(\cdot)) \in (T^*Q)^{\mathbb{R}}$ satisfying the equations

$$\dot{q}(t) = \frac{\partial H}{\partial p}(q(t), p(t), u(t)) \tag{28}$$

$$\dot{p}(t) = -\frac{\partial H}{\partial q}(q(t), p(t), u(t)) + \frac{\partial b}{\partial q}(q(t), u(t))\lambda(t) \tag{29}$$

$$0 = \frac{\partial H}{\partial u}(q(t), p(t), u(t)) - \frac{\partial b}{\partial u}(q(t), u(t))\lambda(t) \tag{30}$$

along with the constraints

$$b(q(t), u(t)) = 0 \tag{31}$$

for all $t \in [0, T]$, where $H : T^*Q \times U \to \mathbb{R}$ is defined in (3), along with the initial conditions $q(0) = q_0$ and transversality conditions (7). Here, $\lambda \in \mathbb{R}^k$ are Lagrange multipliers required to keep the constraints (31) to be satisfied for all time.

Like in section 2 the equations (28–31) can be described by an implicit Hamiltonian system $(T^*Q \times U, D, H)$. This can be seen by the following. Differentiate (31) to get

$$\frac{\partial b}{\partial q}(q(t), u(t))\dot{q}(t) + \frac{\partial b}{\partial u}(q(t), u(t))\dot{u}(t) = 0 \tag{32}$$

and note that (31) is equivalent to (32), assuming that $b(q(0), u(0)) = 0$. Define the Dirac structure D in local coordinates by

$$D = \{(X, \alpha) \in \Gamma(T(T^*Q \times U)) \times \Gamma(T^*(T^*Q \times U)) \mid$$

$$\begin{bmatrix} \alpha_q \\ \alpha_p \\ \alpha_u \end{bmatrix} - \begin{bmatrix} 0 & -I_n & 0 \\ I_n & 0 & 0 \\ 0 & 0 & 0 \end{bmatrix} \begin{bmatrix} X_q \\ X_p \\ X_u \end{bmatrix} \in \text{span} \begin{bmatrix} \frac{\partial b}{\partial q}^T \\ 0 \\ \frac{\partial b}{\partial u}^T \end{bmatrix}$$

$$0 = \begin{bmatrix} \frac{\partial b}{\partial q} & 0 & \frac{\partial b}{\partial u} \end{bmatrix} \begin{bmatrix} X_q \\ X_p \\ X_u \end{bmatrix} \} \tag{33}$$

where the span is taken over $C^\infty(T^*Q \times U)$. Then the Hamiltonian equations (11) are exactly given by (28–30,32). The Dirac structure in (33) can be described in a coordinate free way by

$$D = \{(X, \alpha) \in \Gamma(T(T^*Q \times U)) \times \Gamma(T^*(T^*Q \times U)) \mid$$
$$\alpha - \bar{\omega}(X) \in \text{span}\{db\}, \ X \in \ker(db)\} \tag{34}$$

where db is the differential of b. Notice that the codistribution span $\{db\}$ is constant dimensional (by the rank condition of the Jacobian (27)), and that the distribution ker (db) is involutive (i.e. $db(X_1) = db(X_2) = 0$ then $db([X_1, X_2]) = L_{X_1}(db(X_2)) - (L_{X_1}db)(X_2) = -(d(db(X_1)))(X_2) = 0)$. Then by theorem 4.5 [6] it follows that D given in (34) is closed.

As in section 3, let G be a symmetry Lie group of the constrained optimal control problem. That is, the action of G on Q satisfies (14) as well as the condition

$$L_{\xi_Q} b(\cdot, u) = 0, \quad \text{i.e. } L_{\xi_Q} b_j(\cdot, u) = 0, \ j = 1, \ldots, k, \quad \forall \xi \in \mathcal{G} \tag{35}$$

Let $\pi : Q \to Q/G$ denote the projection map, then the function $b(\cdot, u) : Q \to \mathbb{R}^k$ projects to a well defined function $\hat{b}(\cdot, u) : Q/G \to \mathbb{R}^k, u \in U$, by $b(\cdot, u) = \hat{b}(\cdot, u) \circ \pi, u \in U$. As in section 3 the constrained optimal control problem projects to the *reduced constrained optimal control problem* defined by

> minimize $\hat{J}(\cdot, \cdot)$ over $(Q/G)^{\mathbb{R}} \times U^{\mathbb{R}}$ such that $(\hat{q}(\cdot), u(\cdot))$ satisfies
> $\dot{\hat{q}}(t) = \hat{f}(\hat{q}(t), u(t))$ and the
> constraints $\hat{b}(\hat{q}(t), u(t)) = 0, \ t \in [0, T]$, and $\hat{q}(0) = \hat{q}_0$.

Again, this reduced constrained optimal control problem gives rise to an associated implicit Hamiltonian system $(T^*(Q/G) \times U, \hat{D}, \hat{H})$, where $\hat{D}$ is a Dirac structure on $T^*(Q/G) \times U$ defined analogously as in (34) (see section 3 for the definitions of $\hat{J}$ and $\hat{H}$).

On the other hand, lifting the vector fields $\xi_Q \in \Gamma(TQ)$ to the vector fields $\dot{\xi}_Q \in \Gamma(T(T^*Q))$ turns G into a symmetry Lie group of the implicit Hamiltonian system $(T^*Q \times U, D, H)$. Indeed, $\dot{\xi}_Q \in \Gamma(T(T^*Q))$ can be identified with a horizontal vector field $\dot{\xi}_Q \in \Gamma(T(T^*Q \times U))$ on $T^*Q \times U$, and since $dH_{\xi_Q} = i_{\dot{\xi}_Q}\bar{\omega}$, where H_{ξ_Q} is defined in (12) and extended to a function on $T^*Q \times U$, it follows that $(\dot{\xi}_Q, dH_{\xi_Q}) \in D$ (notice that by (35) $\dot{\xi}_Q \in \ker (db)$), and by proposition 3 it follows that $\dot{\xi}_Q$ is a symmetry of D. Furthermore, by (15) $\dot{\xi}_Q$ is a symmetry of the Hamiltonian H.

Now we can use the reduction theory described in [3,2] to reduce the implicit Hamiltonian system $(T^*Q \times U, D, H)$ to a lower dimensional implicit Hamiltonian system, as in section 3. Since the transversality conditions are given by (7), it follows that the momentum map, defined in (19), has constant value zero along solutions of the constrained optimal control problem, see proposition 4. Following the reduction procedure analogously to the one described in example 8 [3], results in the reduced implicit Hamiltonian system $(\Phi^{-1}(0)/G \times U, \tilde{D}, \tilde{H})$, where $\tilde{H} \in C^\infty(\Phi^{-1}(0)/G \times U)$ is given as in section 3. The reduced Dirac structure $\tilde{D}$ on $\Phi^{-1}(0)/G \times U$ is given by

$$\tilde{D} = \{(\tilde{X}, \tilde{\alpha}) \in \Gamma(T(\Phi^{-1}(0)/G \times U)) \times \Gamma(T^*(\Phi^{-1}(0)/G \times U)) \mid$$
$$\tilde{\alpha} - \bar{\omega}_0(\tilde{X}) \in \text{span} \{d\tilde{b}\}, \ \tilde{X} \in \ker (d\tilde{b})\}$$

where $\tilde{b} \in C^\infty(\Phi^{-1}(0)/G \times U)$ is defined by $b(\cdot, u) \circ i = \tilde{b}(\cdot, u) \circ \pi_0, u \in U$, with $i : \Phi^{-1}(0) \to T^*Q$ the inclusion map and $\pi_0 : \Phi^{-1}(0) \to \Phi^{-1}(0)/G$ the projection map.

Consider the two implicit Hamiltonian systems $(T^*(Q/G) \times U, \hat{D}, \hat{H})$ and $(\Phi^{-1}(0)/G \times U, \tilde{D}, \tilde{H})$. As in section 3 there exists a diffeomorphism $\tau : \Phi^{-1}(0)/G \to T^*(Q/G)$ such that $\tau^*\hat{\omega} = \omega_0$ (where $\hat{\omega}$ is the canonical symplectic two-form on $T^*(Q/G)$ and ω_0 the reduced symplectic two-form on $\Phi^{-1}(0)/G$). Define the map $(\tau, \mathrm{id}) : \Phi^{-1}(0)/G \times U \to T^*(Q/G) \times U$, also denoted by τ. Then it follows that $\tilde{b} = \hat{b} \circ \tau$. Indeed, see section 3 for notation,

$$\tilde{b}(z, u) = b(\cdot, u) \circ i(q_0, \alpha_{q_0}) = b(q_0, u) = \hat{b}(\cdot, u) \circ \pi(q_0) = \hat{b} \circ \tau(z, u)$$

for all $(z, u) \in \Phi^{-1}(0)/G \times U$. This implies that τ is a Dirac isomorphism between $\tilde{D}$ and $\hat{D}$. Since also $\tilde{H} = \hat{H} \circ \tau$, see section 3, it follows that the two reduced implicit Hamiltonian systems $(\Phi^{-1}(0)/G \times U, \tilde{D}, \tilde{H})$ and $(T^*(Q/G) \times U, \hat{D}, \hat{H})$ are isomorphic. This result extends the result of theorem 2 to the case of constrained optimal control problems.

Theorem 3. *Consider a constrained optimal control problem with symmetries described in (14,35). Then reducing the constrained optimal control problem to a lower dimensional constrained optimal control problem (on Q/G), called the reduced constrained optimal control problem, and applying the Maximum Principle, is equivalent (up to isomorphism) to first applying the Maximum Principle on the original constrained optimal control problem (on Q), and then using the reduction theory described in [3,2] to reduce the corresponding implicit Hamiltonian system.*

Finally, we investigate the generalized symmetries of the implicit Hamiltonian system $(T^*Q \times U, D, H)$, corresponding to the constrained optimal control problem, and prove a (partial) analogue of proposition 6. Consider the Dirac structure D given in (34), and the corresponding (co)distributions

$$G_0 = \Gamma^{\mathrm{vert}}(T(T^*Q \times U)) \cap \ker(db)$$
$$G_1 = \ker(db)$$
$$P_0 = \mathrm{span}\{db\}$$
$$P_1 = \Gamma^{\mathrm{hor}}(T^*(T^*Q \times U)) + \mathrm{span}\{db\}$$

and note that P_1 is constant dimensional (by the full rank condition on the Jacobian (27)) and involutive (since G_0 is involutive). Proposition 5 gives necessary and sufficient conditions for a vector field $Y \in \Gamma(T(T^*Q \times U))$ to be a symmetry of D. Now, consider a vector field $Y \in \Gamma(T(T^*Q \times U))$ of the form (in local coordinates) $Y(q, p, u) = Y_q(q, p)\frac{\partial}{\partial q} + Y_p(q, p)\frac{\partial}{\partial p} + Y_u(q, p, u)\frac{\partial}{\partial u}$, i.e. Y_q and Y_p do not depend on u. This can be expressed in a coordinate free way by the condition that $[Y, \Gamma^{\mathrm{vert}}(T(T^*Q \times U))] \subset \Gamma^{\mathrm{vert}}(T(T^*Q \times U))$.

Assume that Y is a symmetry of D, then by proposition 5 it follows that Y is canonical with respect to the bracket $\{\cdot,\cdot\}_D$. Since

$$\{H \in C^\infty(T^*Q \times U) \mid H(q,p,u) = H(q,p)\} \subset$$
$$\{H \in C^\infty(T^*Q \times U) \mid dH \in \mathsf{P}_1\} = \mathcal{A}_D$$

it follows that $L_{\tilde{Y}}\omega = 0$, where $\tilde{Y} = (\rho_{T^*Q})_* Y$. This implies that $\tilde{Y}$ is locally Hamiltonian, that is, there exists (locally) a function $\tilde{H} \in C^\infty(T^*Q)$ such that $d\tilde{H} = i_{\tilde{Y}}\omega$. Now assuming that $Y \in \ker(db)$ and extending $\tilde{H}$ to a function on $T^*Q \times U$ (by $\tilde{H}(q,p,u) = \tilde{H}(q,p), \forall(q,p,u) \in T^*Q \times U$), it follows that $(Y, d\tilde{H}) \in D$. If one furthermore assumes that Y is a symmetry of the Hamiltonian H, then by (23) it follows that $\tilde{H}$ is a conserved quantity. We have the following (partial) analogue of proposition 6.

Proposition 8. *Consider the implicit Hamiltonian system $(T^*Q \times U, D, H)$, with D given in (34). Let $Y \in \Gamma(T(T^*Q \times U))$ be a symmetry of $(T^*Q \times U, D, H)$ such that $Y \in \ker(db)$ and $[Y, \Gamma^{vert}(T(T^*Q \times U))] \subset \Gamma^{vert}(T(T^*Q \times U))$. Then there exists (locally) a function $\tilde{H} \in C^\infty(T^*Q)$ such that $\tilde{H}$ is constant along solutions of the implicit Hamiltonian system.*

In particular, this means that to a generalized symmetry of the constrained optimal control problem satisfying the assumptions in proposition 8, there exists (locally) a function of the phase variables (that is, $\tilde{H} \in C^\infty(T^*Q)$) which is constant along solutions of the constrained optimal control problem. We leave it to the reader to state the corresponding converse statement, analogously to proposition 7.

Remark 7. In this paper we restricted ourselves for simplicity to the *free terminal point* (constrained) optimal control problem (i.e. $q(T) \in Q$). It is easy to prove that all the results remain true if we add the following constraint at final time: $q(T) \in Q_f \subset Q$, where Q_f is a submanifold of Q, assuming that the action of the symmetry Lie group G leaves the manifold Q_f invariant.

6   Conclusions

In this paper we showed that optimal control problems naturally lead, via the Maximum Principle, to implicit Hamiltonian systems. We showed that symmetries of the optimal control problem induce, by canonical lift to the cotangent bundle, symmetries of the corresponding implicit Hamiltonian system. The reduction theory developed for implicit Hamiltonian systems in [3,2] was used to prove that doing reduction on the optimal control problem and applying the Maximum Principle, is equivalent (up to isomorphism) to applying the reduction theory in [3,2] to the implicit Hamiltonian system corresponding to the original optimal control problem, i.e. doing reduction and applying the Maximum Principle commutes. The key observation is that the momentum map, corresponding to a symmetry Lie group of the optimal control problem, has the (constant) value zero along solutions of the optimal control problem.

Furthermore, we described the generalized symmetries of an optimal control problem, which are defined as symmetries of the corresponding implicit Hamiltonian system, and we showed that to every generalized symmetry there corresponds a conserved quantity, being a function of the phase variables which is constant along solutions of the optimal control problem. This result generalizes the results obtained in [19,17].

Finally, in the last section we generalized the results to constrained optimal control problems. The commutation of doing reduction and applying the Maximum Principle was also proved there. It was shown that under some conditions generalized symmetries of the corresponding implicit Hamiltonian system again give rise to conserved quantities.

References

1. R. Abraham and J.E. Marsden. *Foundations of Mechanics*. Benjamin / Cummings Publishing Company, second edition, 1978.
2. G. Blankenstein and A.J. van der Schaft. Reduction of implicit Hamiltonian systems with symmetry. In *Proceedings of the 5th European Control Conference, ECC'99*, Karlsruhe, 1999.
3. G. Blankenstein and A.J. van der Schaft. Symmetry and reduction in implicit generalized Hamiltonian systems. Memorandum 1489, University of Twente, Faculty of Mathematical Sciences, June 1999. Accepted for publication in Reports on Mathematical Physics.
4. A.E. Bryson and Y.-C. Ho. *Applied Optimal Control: Optimization, Estimation and Control*. John Wiley, second edition, 1975.
5. T. Courant. Dirac manifolds. *Trans. American Math. Soc.*, 319:631–661, 1990.
6. M. Dalsmo and A.J. van der Schaft. On representations and integrability of mathematical structures in energy-conserving physical systems. *SIAM J. Cont. Opt.*, 37(1):54–91, 1999.
7. I. Dorfman. *Dirac Structures and Integrability of Nonlinear Evolution Equations*. Chichester: John Wiley, 1993.
8. W.H. Fleming and R.W. Rishel. *Deterministic and Stochastic Optimal Control*. Springer-Verlag, 1975.
9. J.W. Grizzle. *The Structure and Optimization of Nonlinear Control Systems possessing Symmetries*. PhD thesis, University of Texas, 1983.
10. J.W. Grizzle and S.I. Marcus. Optimal Control of Systems Pocessing Symmetries. *IEEE Trans. Automatic Control*, 29(11):1037–1040, 1984.
11. M. Kummer. On the Constuction of the Reduced Phase Space of a Hamiltonian System with Symmetry. *Indiana Univ. Math. J.*, 30(2):281–291, 1981.
12. P. Libermann and C.-M. Marle. *Symplectic Geometry and Analytical Mechanics*. Reidel, Dordrecht, 1987.
13. J.E. Marsden. *Lectures on Mechanics*. Cambridge University Press, 1992.

14. J.E. Marsden and A. Weinstein. Reduction of symplectic manifolds with symmetry. *Rep. Math. Phys.*, 5:121–130, 1974.
15. L.S. Pontryagin, V.G. Boltyanskii, R.V. Gamkrelidze, and E.F. Mischenko. *The Mathematical Theory of Optimal Processes*. Pergamon Press, 1964.
16. W.J. Satzer. Canonical Reduction of Mechanical Systems Invariant under Abelian Group Actions with an Application to Celestial Mechanics. *Indiana Univ. Math. J.*, 26(5):951–976, 1977.
17. H.J. Sussmann. Symmetries and Integrals of Motion in Optimal Control. In A. Fryszkowski, B. Jacubczyk, W. Respondek, and T. Rzezuchowski, editors, *Geometry in Nonlinear Control and Differential Inclusions*, pages 379–393. Banach Center Publications, Math. Inst. Polish Academy of Sciences, Warsaw, 1995.
18. H.J. Sussmann. An introduction to the coordinate-free Mamimum Principle. In B. Jacubczyk and W. Respondek, editors, *Geometry of Feedback and Optimal Control*, pages 463–557. Marcel Dekker, New York, 1997.
19. A.J. van der Schaft. Symmetries in optimal control. *SIAM J. Control and Optimization*, 25(2):245–259, 1987.
20. A.J. van der Schaft. Implicit Hamiltonian Systems with Symmetry. *Rep. Math. Phys.*, 41:203–221, 1998.
21. A.J. van der Schaft and B.M Maschke. Interconnected mechanical systems, part I: geometry of interconnection and implicit Hamiltonian systems. In A. Astolfi, D.J.N Limebeer, C. Melchiorri, A. Tornambè, and R.B. Vinter, editors, *Modelling and Control of Mechanical Systems*, pages 1–15. Imperial College Press, 1997.

Robust Absolute Stability of Delay Systems

Pierre-Alexandre Bliman*

National Technical University
Department of Mathematics
Zografou Campus
157 80 Athens, Greece
Pierre-Alexandre.Bliman@inria.fr

Abstract. The present paper is devoted to the study of absolute stability of delay systems with nonlinearities subject to sector conditions, and with uncertain delays. We construct Lyapunov-Krasovskii functionals candidates for various such systems, whose decreasingness along the trajectories is expressed in terms of Linear Matrix Inequalities (LMIs). We then show that feasibility of the latter implies some frequency domain conditions, which express that circle or Popov criterion holds, whatever the nonnegative value of the delay (delay-independent criteria). Also, using a class of transformations of the system, one is able to provide LMI conditions under which circle or Popov criterion holds, whatever the value of the delay in a certain compact interval containing zero (delay-dependent criteria).

1 Introduction

As is well-known, delays in control loops may lead to bad performances or instabilities, and the study of time-delay systems has attracted large interest. One of the problems one may have to deal with, is the possible uncertainties on the value of these delays. In order to tackle this robustness issue, but also for reasons of computational simplicity, stability criteria independent of the values of the delays have been proposed, mainly for linear delay systems, see e.g. the surveys [13,17]. We are here concerned with the issue of stability of delay systems with sector-bounded nonlinearities and uncertain delays.

In the classical absolute stability theory for nonlinear, rational (that is, delay-free), systems, the usual results (circle criterion, Popov criterion) may be obtained equivalently by Lyapunov method or by frequency domain method. For delay systems, such a property is unknown. However, since at least the early

---

* On leave from I.N.R.I.A., with the support of European Commission's Training and Mobility of Researchers Programme ERB FMRXCT-970137 "Breakthrough in the control of nonlinear systems". The author is indebted to the network coordinator Françoise Lamnabhi-Lagarrigue, and the coordinator of the greek team Ioannis Tsinias, for their help and support. Permanent and corresponding address: I.N.R.I.A., Rocquencourt B.P. 105, 78153 Le Chesnay Cedex, France. Phone: (33) 1 39 63 55 68, Fax: (33) 1 39 63 57 86, Email: pierre-alexandre.bliman@inria.fr

sixties, circle [21,23] and Popov [19,15] criteria have been shown to be valid for delay systems, and Krasovskii has proposed an extension of Lyapunov method to these systems, leading in particular to some sufficient conditions for stability of linear and nonlinear delay systems independent of the value of the delay(s) [14, Section 34].

In the present paper, we attempt to establish some new sufficient conditions for absolute stability of delay systems subject to uncertainties on the delays. In a previous paper [3], considerations based on classical Lyapunov-Krasovskii functionals permitted to obtain some Linear Matrix Inequalities (LMIs) whose feasibility is sufficient for delay-independent absolute stability. It was then shown that the previous results naturally imply some frequency domain conditions, obtained from circle and Popov criteria by decoupling the terms in s and $z = e^{-hs}$ in the transfer function of the system, in a way reminiscent of some recent results on delay-independent stability of linear delay systems [9,16]. This was, to the best of our knowledge, the first attempt to introduce frequency domain interpretation of Lyapunov-Krasovskii method for nonlinear systems (the issue of the conservativity analysis of this method is examined in [4] for linear systems and in [2] for nonlinear systems).

More precisely, in [3] we provided a delay-independent version of circle criterion for systems with one delay, in other words, a version of circle criterion valid for any value of the delay $h \in [0, +\infty)$. A generalization of this result was then made for systems with two independent delays. Last, a delay-independent version of Popov criterion was finally obtained, essentially by addition of a Lur'e term to a quadratic Lyapunov-Krasovskii functional, similarly to the procedure used in the delay-free case.

In the present paper, after recalling some more details on the background in Sects. 1.1 to 1.3, we repeat the delay-independent versions of circle and Popov criteria for systems with one delay (Theorems 1 and 2 in Sect. 2), and provide shortened and simpler proofs. The new results are provided in Sect. 3, giving sufficient conditions of absolute stability for any value of the delay $h \in [0, \overline{h}]$, where $\overline{h}$ is a prescribed nonnegative upper bound. The idea is to perform a transformation of the system under study, according to some transformation whose principle has been applied to linear delay systems [18]. One reduces the initial problem to the delay-independent absolute stability of an auxiliary system with two independent delays, whose coefficients are functions of the bound $\overline{h}$ and the other data. One is then able to state circle criterion (Theorem 4) and Popov criterion (Theorem 5) on this transformed system. An effort is made all along the paper to present in a unified and methodical way the results, from the simplest to the more complex ones, and the machinery of the proofs.

Generally speaking, we do not consider here the extensively studied questions of existence and uniqueness of the solutions. In the sequel, we simply assume that there exist *global solutions* of the various delay systems under

consideration. As an example for system (9) below, this means by definition:
$\forall \phi \in \mathcal{C}([-h, 0]; \mathbb{R}^n)$, $\exists x \in \mathcal{C}([-h, +\infty)) \cap AC([0, +\infty))$, such that $x_0 = \phi$
and (9) holds a.e. on $\mathbb{R}^+ \overset{\text{def}}{=} [0, +\infty)$. In particular, the map $t \mapsto \psi(t, y(t))$ is
defined almost everywhere.

Notations Below, x_t denotes the mapping $x(\cdot + t)$, defined on $[-h, 0]$ or
$[-2h, 0]$, according to the context. The matrices I_n, 0_n, $0_{n \times p}$ are resp. the
$n \times n$ identity matrix and the $n \times n$ and $n \times p$ zero matrices, sometimes
simply abbreviated $I, 0$. The symbol $\otimes$ denotes Kronecker product, and by
$\text{diag}\{M_1, \ldots, M_p\}$ is meant the block-diagonal matrix obtained putting the
(not necessarily square) matrices M_i, $i = 1, \ldots, p$ "on the diagonal". Last,
$i = \overline{1, p}$ denotes the integers $i = 1, \ldots, p$.

1.1 Circle Criterion for Rational Systems: Equivalence Between Frequency Domain and Lyapunov Methods

Consider the negative feedback interconnection of a memoryless multivariable
time-varying nonlinearity ψ with a linear time-invariant system characterized
by its transfer function matrix $H(s)$ of size $p \times q$. Nonlinearity $\psi : \mathbb{R}^+ \times \mathbb{R}^p \to \mathbb{R}^q$ is assumed to fulfill a *sector condition*: there exists $K \in \mathbb{R}^{q \times p}$ such that

$$\forall t \in \mathbb{R}^+, \forall y \in \mathbb{R}^p, \quad \psi(t, y)^T (Ky - \psi(t, y)) \geq 0 . \tag{1}$$

The *absolute stability* problem consists in seeking conditions on H and K for
stability of the interconnection.

When H represents a strictly proper rational system (A, B, C), that is $H(s) = C(sI - A)^{-1}B$, the system may be realized as:

$$\dot{x} = Ax - B\psi(t, y), \quad y = Cx , \tag{2}$$

where $A \in \mathbb{R}^{n \times n}, B \in \mathbb{R}^{n \times q}, C \in \mathbb{R}^{p \times n}$. Searching for a Lyapunov function
of the type $V(x) = x^T Px$ for a certain positive definite matrix $P \in \mathbb{R}^{n \times n}$,
one verifies that the inequality

$$\frac{d}{dt}[V(x(t))] \leq \begin{pmatrix} x(t) \\ \psi(t, y(t)) \end{pmatrix}^T \begin{pmatrix} A^T P + PA & C^T K^T - PB \\ KC - B^T P & -2I \end{pmatrix} \begin{pmatrix} x(t) \\ \psi(t, y(t)) \end{pmatrix}$$

is fulfilled along the trajectories of (2), so the existence of a matrix $P \in \mathbb{R}^{n \times n}$
such that

$$P = P^T > 0, \quad R = \begin{pmatrix} A^T P + PA & C^T K^T - PB \\ KC - B^T P & -2I \end{pmatrix} < 0 \tag{3}$$

is sufficient for absolute stability of system (2), under sector condition (1).
Problem (3) is a Linear Matrix Inequality (LMI), for which there exist pow-
erful numerical methods of resolution [6].

Kalman-Yakubovich-Popov lemma (see e.g. [20]) permits to express *equivalently* the previous condition under a frequency domain form, namely

$$\exists \varepsilon > 0, \forall s \in \mathbb{C} \text{ with } \operatorname{Re} s \geq 0, \det(sI - A) \neq 0$$
$$\text{and } 2I + KH(s) + [KH(s)]^* > \varepsilon I, \quad (4)$$

where * denotes complex conjugation. This absolute stability criterion, the famous *circle criterion*, may also be proved directly under form (4), by use of input-output techniques [21,23].

Suppose now that ψ is time-invariant, decentralized (that is [13] $p = q$, and $\psi_i(y)$ is a function of y_i only, for $i = \overline{1,p}$), and fulfills sector condition: there exists $K = \operatorname{diag}\{K_i\} \in \mathbb{R}^{p \times p}$, $K \geq 0$, such that

$$\forall y \in \mathbb{R}^p, \quad \psi(y)^T (Ky - \psi(y)) \geq 0. \quad (5)$$

A refinement of circle criterion is then *Popov criterion*. It is obtained searching for Lyapunov function under the form

$$V(x) = x^T P x + 2 \sum_{i=1}^{p} \eta_i K_i \int_0^{(Cx)_i} \psi_i(u) \, du, \quad (6)$$

where the nonnegative diagonal matrix $\eta \overset{\text{def}}{=} \operatorname{diag}\{\eta_i\}$ is a free parameter. One then verifies that, along the trajectories of (2), the derivative of the Lyapunov function $\frac{d}{dt}[V(x(t))]$ is bounded from above by

$$\begin{pmatrix} x(t) \\ \psi(y(t)) \end{pmatrix}^T \begin{pmatrix} A^T P + PA & C^T K + A^T C^T K\eta - PB \\ KC + \eta KCA - B^T P & -2I - \eta KCB - B^T C^T K\eta \end{pmatrix} \begin{pmatrix} x(t) \\ \psi(y(t)) \end{pmatrix}$$

This leads to the LMI problem

$$\eta = \operatorname{diag}\{\eta_i\} \geq 0, \ P = P^T > 0,$$
$$R = \begin{pmatrix} A^T P + PA & C^T K + A^T C^T K\eta - PB \\ KC + \eta KCA - B^T P & -2I - \eta KCB - B^T C^T K\eta \end{pmatrix} < 0, \quad (7)$$

whose solvability is equivalent to the frequency domain condition

$$\exists \varepsilon > 0, \ \exists \eta = \operatorname{diag}\{\eta_i\} \geq 0, \ \forall s \in \mathbb{C} \text{ with } \operatorname{Re} s \geq 0, \det(sI - A) \neq 0$$
$$\text{and } 2I + (I + \eta s)KH(s) + [(I + \eta s)KH(s)]^* > \varepsilon I. \quad (8)$$

1.2   Two Absolute Stability Results for Delay Systems

Let us now consider the following delay system (where $A_0, A_1 \in \mathbb{R}^{n \times n}, B_0 \in \mathbb{R}^{n \times q}, C_0, C_1 \in \mathbb{R}^{p \times n}$):

$$\dot{x} = A_0 x + A_1 x(t - h) - B_0 \psi(t, y(t)), \ y = C_0 x + C_1 x(t - h), \quad (9)$$

which realizes the interconnection of ψ with the nonrational transfer

$$H(s) = (C_0 + e^{-sh} C_1)(sI - A_0 - e^{-sh} A_1)^{-1} B_0 . \tag{10}$$

For such a system, circle criterion [21,23] and Popov criterion [19,15] still hold: the following condition

$$\exists \varepsilon > 0, \forall s \in \mathbb{C} \text{ with } \operatorname{Re} s \geq 0, \ \det(sI - A_0 - e^{-sh} A_1) \neq 0$$
$$\text{and } 2I + KH(s) + [KH(s)]^* > \varepsilon I ,$$

instead of (4) (with H now given by (10)), or

$$\exists \varepsilon > 0, \ \exists \eta = \operatorname{diag}\{\eta_i\} \geq 0, \ \forall s \in \mathbb{C} \text{ with } \operatorname{Re} s \geq 0,$$
$$\begin{cases} \det(sI - A_0 - e^{-sh} A_1) \neq 0 \text{ and} \\ 2I + (I + \eta s)KH(s) + [(I + \eta s)KH(s)]^* > \varepsilon I . \end{cases}$$

instead of (8) in the case where the nonlinearity is time-invariant and decentralized, are sufficient for absolute stability of system (9) under hypothesis (1). Obviously, these frequency domain conditions depend explicitly upon the magnitude of the delay h.

On the other hand, it is possible to obtain absolute stability criterion by use of a quadratic Lyapunov-Krasovskii functional $V : \mathcal{C}([-h, 0]; \mathbb{R}^n) \to \mathbb{R}$ of the type [14,22]

$$V(\phi) \stackrel{\text{def}}{=} \phi^T(0) P \phi(0) + \int_{-h}^{0} \phi^T(\tau) Q \phi(\tau) \, d\tau , \tag{11}$$

where P, Q are positive definite matrices. As for the rational case, in an attempt to force the map $t \mapsto V(x_t)$ to decrease along the trajectories of (9), one is led to assume solvability of a certain LMI, whose unknowns are P and Q, see formula (12) below. The delay does not appear in this LMI: contrary to circle criterion, the obtained result is a *delay-independent* criterion.

1.3   Delay-Independent Stability of Linear Delay Systems

Important progresses have been made recently in the study of delay-independent stability conditions for linear delay systems, see [13,17]. By definition, the linear equation $\dot{x} = A_0 x + A_1 x(t - h)$ is asymptotically stable *independently of the delay* [11,12] if the roots of the equations $\det(sI - A_0 - e^{-hs} A_1) = 0$ have negative real part for any $h \geq 0$. To analyze the conditions under which this property occurs, one sees that it is necessary to disconnect the term e^{-sh} from s. Hale *et al.* [9] have shown that delay-independent (asymptotic) stability of the previous linear system is equivalent to (σ denotes the spectrum)

$$\operatorname{Re} \sigma(A_0 + A_1) < 0,$$
$$\text{and } \forall(s, z) \in j\mathbb{R} \times \mathbb{C} \text{ with } s \neq 0, |z| = 1, \ \det(sI - A_0 - zA_1) \neq 0 .$$

Niculescu *et al.* [16,17] have given numerical methods of verification of this condition, using matrix pencils. They also provide method to check a slightly stronger one, called *strong* delay-independent stability, namely

$$\mathrm{Re}\,\sigma(A_0 + A_1) < 0,$$
$$\text{and } \forall(s, z) \in j\mathbb{R} \times \mathbb{C} \text{ with } |z| = 1, \; \det(sI - A_0 - zA_1) \neq 0 \,.$$

Remark that this condition is equivalent [10, Remarks 1 and 2] to

$$\forall(s, z) \in \mathbb{C}^2 \text{ with } \mathrm{Re}\,s \geq 0, |z| \leq 1, \; \det(sI - A_0 - zA_1) \neq 0 \,.$$

2   Absolute Stability Criteria Valid for any $h \in [0, +\infty)$

2.1   Delay-Independent Circle Criterion

In [3] is given the following result.

Theorem 1 *Consider the two following statements.*

(i)   There exist $P, Q \in \mathbb{R}^{n \times n}$ such that

$$P = P^T > 0, \; Q = Q^T > 0,$$

$$R \overset{\mathrm{def}}{=} \begin{pmatrix} A_0^T P + P A_0 + Q & P A_1 & C_0^T K^T - P B_0 \\ A_1^T P & -Q & C_1^T K^T \\ K C_0 - B_0^T P & K C_1 & -2I_q \end{pmatrix} < 0 \,. \quad (12)$$

(ii)   The following property holds.

$$\exists \varepsilon > 0, \forall h \geq 0, \forall s \in \mathbb{C} \text{ with } \mathrm{Re}\,s \geq 0, \; \det(sI - A_0 - e^{-sh}A_1) \neq 0$$
$$\text{and } 2I + KH(s) + [KH(s)]^* > \varepsilon I \,, \quad (13)$$

where H is given in (10).

Suppose (i) holds. Then (ii) holds. Furthermore, the functional V defined in (11) is positive definite for any solution P, Q of (12) and, for any ψ fulfilling (1) and any $h \geq 0$,

$$\frac{d}{dt}[V(x_t)] \leq \begin{pmatrix} x(t) \\ x(t-h) \\ \psi(t, y(t)) \end{pmatrix}^T R \begin{pmatrix} x(t) \\ x(t-h) \\ \psi(t, y(t)) \end{pmatrix} \quad t - a.e. \text{ on } \mathbb{R}^+ \,, \quad (14)$$

for any global solution x of (9).

Suppose (ii) holds. Then, for any ψ fulfilling (1) and any $h \geq 0$, $\displaystyle\lim_{t \to +\infty} x(t) = 0$ for any global solution x of (9). ∎

Statement (12) (resp. (13)) must be compared with (3) (resp. (4)). Matrix R in (12) is of size $2n + q$.

Sketch of the proof. Assume that (i) holds. Define the multivariable $p \times q$ transfer function matrix H by

$$H(s, z) \overset{\text{def}}{=} (C_0 + zC_1)(sI - A_0 - zA_1)^{-1}B_0 , \tag{15}$$

to be compared with (10). We shall prove that the following property holds, from which (ii) is deduced immediatly:

$$\exists \varepsilon > 0, \ \forall (s, z) \in \mathbb{C}^2 \text{ with } \operatorname{Re} s \geq 0, |z| \leq 1, \ \det(sI - A_0 - zA_1) \neq 0$$
$$\text{and } 2I + KH(s, z) + [KH(s, z)]^* > \varepsilon I , \tag{16}$$

where H is defined in (15).
From (i) is deduced that, for any $z \in \mathbb{C}$,

$$\begin{pmatrix} I_n \\ zI_n \\ 0_{q \times n} \end{pmatrix}^* R \begin{pmatrix} I_n \\ zI_n \\ 0_{q \times n} \end{pmatrix} = (A_0^T + z^*A_1^T)P + P(A_0 + zA_1) + (1 - |z|^2)Q < 0 .$$

One deduces from this that the real part of the eigenvalues of $A_0 + zA_1$ are negative when $|z| \leq 1$. This is the first part of (16).
Denoting for sake of simplicity

$$S = S(s, z) \overset{\text{def}}{=} (sI - A_0 - zA_1)^{-1} , \tag{17}$$

the second part of (16) is deduced from the following inequalities, valid for any $(s, z) \in \mathbb{C}^2$

$$\begin{pmatrix} -SB_0 \\ -zSB_0 \\ I_q \end{pmatrix}^* \begin{pmatrix} A_0^T P + PA_0 + Q & PA_1 & -PB_0 \\ A_1^T P & -Q & 0 \\ -B_0^T P & 0 & 0 \end{pmatrix} \begin{pmatrix} -SB_0 \\ -zSB_0 \\ I_q \end{pmatrix}$$
$$= B_0^T S^*[(s + s^*)P + (1 - |z|^2)Q]SB_0 ,$$

$$\begin{pmatrix} -SB_0 \\ -zSB_0 \\ I_q \end{pmatrix}^* \begin{pmatrix} 0 & 0 & C_0^T K^T \\ 0 & 0 & C_1^T K^T \\ KC_0 & KC_1 & -2I \end{pmatrix} \begin{pmatrix} -SB_0 \\ -zSB_0 \\ I_q \end{pmatrix}$$
$$= -(2I + KH(s, z) + [KH(s, z)]^*) ,$$

and from the remark that

$$\exists \varepsilon > 0, \ R < -\varepsilon \operatorname{diag}\{0_n, 0_n, I_q\} .$$

Proposition (16) is hence proved, and this yields (ii).

To get inequality (14), differentiate V along the trajectories of (the global solutions of) (9), and add the nonnegative term $2\psi(t, y(t))^T (Ky(t) - \psi(t, y(t)))$. Suppose now that (ii) holds. To prove global asymptotic stability of (9), it is sufficient to remark that (ii) implies that the hypotheses of circle criterion are verified for any $h \geq 0$. For other details, see complete proof in [3].   □

Remark incidentally that the fact that (i) is sufficient for asymptotic stability of (9) has been here deduced directly from (ii); this furnishes an alternative to the proofs à la Krasovskii [14, e.g. Theorem 31.1].

2.2   Delay-Independent Popov Criterion

We consider here system (9) with time-invariant nonlinearity:

$$\dot{x} = A_0 x + A_1 x(t - h) - B_0 \psi(y), \quad y = C_0 x + C_1 x(t - h) . \tag{18}$$

Throughout this section is assumed that $p = q$, and that ψ is decentralized and fulfills sector condition (5) for a certain nonnegative diagonal matrix K.

In order to get a delay-independent version of Popov criterion for system (18), one may try to add to the quadratic Lyapunov-Krasovskii functional (11) the Lur'e term

$$2 \sum_{i=1}^{p} \eta_i K_i \int_0^{y_i(t)} \psi_i(u) \, du \,, \tag{19}$$

in a way similar to what is done for rational systems; matrix $\eta = \mathrm{diag}\{\eta_i\} \in \mathbb{R}^{p \times p}$, $\eta \geq 0$, is a free parameter. This permits to get sufficient stability results, in the case where the input of the nonlinearity is not delayed ($C_1 = 0$), see [8,22].

However, when the input of the nonlinearity is delayed ($y \neq C_0 x$), this addition introduces terms in $x(t - 2h)$ and $\psi(y(t - h))$, which do not permit to bound the derivative along the trajectories of (18), of the functional obtained as the sum of V as given by (11), plus the Lur'e term (19). To take this into account, one is led to introduce also more ancient state values in V. One considers finally the following Lyapunov-Krasovskii functional, defined for any $\phi \in \mathcal{C}([-2h, 0]; \mathbb{R}^n)$:

$$V(\phi) \overset{\mathrm{def}}{=} \begin{pmatrix} \phi(0) \\ \phi(-h) \end{pmatrix}^T P \begin{pmatrix} \phi(0) \\ \phi(-h) \end{pmatrix} + \int_{-h}^0 \begin{pmatrix} \phi(\tau) \\ \phi(\tau - h) \end{pmatrix}^T Q \begin{pmatrix} \phi(\tau) \\ \phi(\tau - h) \end{pmatrix} \, d\tau$$
$$+ 2 \sum_{i=1}^{p} \eta_i K_i \int_0^{(C_0\phi(0) + C_1\phi(-h))_i} \psi_i(u) \, du . \tag{20}$$

The following result is enunciated in [3].

Theorem 2 *Consider the two following statements.*

(i) There exist $\eta = \mathrm{diag}\{\eta_i\} \in \mathbb{R}^{p \times p}$, $P, Q \in \mathbb{R}^{2n \times 2n}$, such that

$$\eta \geq 0, \; P = P^T > 0, \; Q = Q^T > 0, \quad J^T R J < 0 , \tag{21}$$

where

$$J \stackrel{\text{def}}{=} \mathrm{diag} \left\{ \begin{pmatrix} 1 & 0 & 0 \\ 0 & 1 & 0 \\ 0 & 1 & 0 \\ 0 & 0 & 1 \end{pmatrix} \otimes I_n, \, I_{2p} \right\} \in \mathbb{R}^{(4n+2p) \times (3n+2p)} ,$$

$$R \stackrel{\text{def}}{=}$$
$$\begin{pmatrix} (I_2 \otimes A_0^T)P + P(I_2 \otimes A_0) + Q & P(I_2 \otimes A_1) & (I_2 \otimes C_0^T K) - P(I_2 \otimes B_0) \\ (I_2 \otimes A_1^T)P & -Q & (I_2 \otimes C_1^T K) \\ (I_2 \otimes KC_0) - (I_2 \otimes B_0^T)P & (I_2 \otimes KC_1) & -2I_{2p} \end{pmatrix}$$
$$+ \begin{pmatrix} 0_{4n \times (4n+2p)} \\ \eta K \; (C_0 A_0 \; C_1 A_0 \; C_0 A_1 \; C_1 A_1 \; -C_0 B_0 \; -C_1 B_0) \\ 0_{p \times (4n+2p)} \end{pmatrix}$$
$$+ \begin{pmatrix} 0_{4n \times (4n+2p)} \\ \eta K \; (C_0 A_0 \; C_1 A_0 \; C_0 A_1 \; C_1 A_1 \; -C_0 B_0 \; -C_1 B_0) \\ 0_{p \times (4n+2p)} \end{pmatrix}^T .$$

(ii) The following property holds

$$\exists \varepsilon > 0, \; \exists \eta = \mathrm{diag}\{\eta_i\} \in \mathbb{R}^{p \times p}, \eta \geq 0, \; \forall h \geq 0, \forall s \in \mathbb{C} \text{ with } \mathrm{Re}\, s \geq 0,$$
$$\begin{cases} \det(sI - A_0 - e^{-sh} A_1) \neq 0 \text{ and} \\ 2I + (I + \eta s)KH(s) + [(I + \eta s)KH(s)]^* > \varepsilon I , \end{cases}$$

where H is given in (10).

Suppose (i) holds. Then (ii) holds. Furthermore, the functional V defined in (20) is positive definite for any solution P, Q of (21) and, for any decentralized ψ fulfilling (5) and any $h \geq 0$,

$$\frac{d}{dt}[V(x_t)] \leq \begin{pmatrix} x(t) \\ x(t-h) \\ x(t-2h) \\ \psi(y(t)) \\ \psi(y(t-h)) \end{pmatrix}^T J^T R J \begin{pmatrix} x(t) \\ x(t-h) \\ x(t-2h) \\ \psi(y(t)) \\ \psi(y(t-h)) \end{pmatrix} \quad t-a.e. \text{ on } \mathbb{R}^+ , \tag{22}$$

for any global solution x of (18).

Suppose (ii) holds. Then, for any decentralized ψ fulfilling (5) *and any* $h \geq 0$,
$\lim_{t \to +\infty} x(t) = 0$ *for any global solution x of* (18). ∎

Matrix $J^T R J$ in (21) is of size $3n+2p$. The decomposition $J^T R J$ is introduced in order to underline the proximity with the previous results, compare with R in (7) and in (12). Writing

$$X(t) \overset{\text{def}}{=} \begin{pmatrix} x(t) \\ x(t-h) \end{pmatrix},$$

one has

$$\begin{pmatrix} X(t) \\ X(t-h) \\ \psi(t, y(t)) \\ \psi(t-h, y(t-h)) \end{pmatrix} = J \begin{pmatrix} x(t) \\ x(t-h) \\ x(t-2h) \\ \psi(t, y(t)) \\ \psi(t-h, y(t-h)) \end{pmatrix}.$$

When (12) is fulfilled for $P, Q \in \mathbb{R}^{n \times n}$, then (21) is fulfilled for $\eta = 0$ and $I_2 \otimes P, I_2 \otimes Q$: as for the delay-free case, the addition of a Lur'e term leads (for systems with decentralized, time-invariant nonlinearities) to a stronger stability result. In other words, the "delay-independent Popov criterion" in Theorem 2 is stronger than the "delay-independent circle criterion" in Theorem 1.

Sketch of the proof. Suppose (i) holds. We shall deduce (ii) from the following proposition:

$$\exists \varepsilon > 0, \ \exists \eta = \text{diag}\{\eta_i\} \in \mathbb{R}^{p \times p}, \eta \geq 0, \ \forall (s, z) \in \mathbb{C}^2 \text{ with } \operatorname{Re} s \geq 0, |z| \leq 1,$$
$$\begin{cases} \det(sI - A_0 - zA_1) \neq 0 \text{ and} \\ 2I + (I + \eta s) K H(s, z) + [(I + \eta s) K H(s, z)]^* > \varepsilon I, \end{cases} \quad (23)$$

where H is given in (15).

First, notice that, for any $z \in \mathbb{C}$,

$$0 > \begin{pmatrix} I_n \\ zI_n \\ z^2 I_n \\ 0_{2p \times n} \end{pmatrix}^* J^T RJ \begin{pmatrix} I_n \\ zI_n \\ z^2 I_n \\ 0_{2p \times n} \end{pmatrix}$$

$$= \begin{pmatrix} I_n \\ zI_n \\ zI_n \\ z^2 I_n \end{pmatrix}^* \begin{pmatrix} (I_2 \otimes A_0^T)P + P(I_2 \otimes A_0) + Q & P(I_2 \otimes A_1) \\ (I_2 \otimes A_1^T)P & -Q \end{pmatrix} \begin{pmatrix} I_n \\ zI_n \\ zI_n \\ z^2 I_n \end{pmatrix}$$

$$= \begin{pmatrix} I_n \\ zI_n \end{pmatrix}^* [(I_2 \otimes (A_0^T + z^* A_1^T))P$$

$$+ P(I_2 \otimes (A_0 + zA_1)) + (1 - |z|^2)Q] \begin{pmatrix} I_n \\ zI_n \end{pmatrix} \qquad (24)$$

$$= (A_0^T + z^* A_1^T) \begin{pmatrix} I_n \\ zI_n \end{pmatrix}^* P \begin{pmatrix} I_n \\ zI_n \end{pmatrix} + (A_0 + zA_1) \begin{pmatrix} I_n \\ zI_n \end{pmatrix}^* P \begin{pmatrix} I_n \\ zI_n \end{pmatrix}$$

$$+ (1 - |z|^2) \begin{pmatrix} I_n \\ zI_n \end{pmatrix}^* Q \begin{pmatrix} I_n \\ zI_n \end{pmatrix}$$

which permits to show, as in the proof of Theorem 1, that the eigenvalues of $A_0 + zA_1$ have negative real part whenever $|z| \leq 1$: this is the first part of (23).

On the other hand, one verifies easily that, $S(s, z)$ being defined in (17), and defining

$$w = w(s, z) \overset{\text{def}}{=} \begin{pmatrix} -S(s, z)B_0 \\ -zS(s, z)B_0 \\ -z^2 S(s, z)B_0 \\ I_p \\ zI_p \end{pmatrix}, \qquad (25)$$

one has, for any $(s, z) \in \mathbb{C}^2$,

$$w^* J^T \begin{pmatrix} (I_2 \otimes A_0^T)P + P(I_2 \otimes A_0) + Q & P(I_2 \otimes A_1) & -P(I_2 \otimes B_0) \\ (I_2 \otimes A_1^T)P & -Q & 0 \\ -(I_2 \otimes B_0^T)P & 0 & 0 \end{pmatrix} Jw$$

$$= B_0^T S(s, z)^* \begin{pmatrix} I_n \\ zI_n \end{pmatrix}^* [(s + s^*)P + (1 - |z|^2)Q] \begin{pmatrix} I_n \\ zI_n \end{pmatrix} S(s, z)B_0 ,$$

$$w^* J^T \begin{pmatrix} 0 & 0 & (I_2 \otimes C_0^T K) \\ 0 & 0 & (I_2 \otimes C_1^T K) \\ (I_2 \otimes KC_0) & (I_2 \otimes KC_1) & -2I_{2p} \end{pmatrix} Jw$$

$$= - \left\| \begin{pmatrix} 1 \\ z \end{pmatrix} \right\|^2 (2I + KH(s, z) + [KH(s, z)]^*) ,$$

and

$$w^* J^T \left[\left(\begin{matrix} 0_{4n \times (4n+2p)} \\ \eta K \left(C_0 A_0 \ C_1 A_0 \ C_0 A_1 \ C_1 A_1 \ -C_0 B_0 \ -C_1 B_0 \right) \\ 0_{p \times (4n+2p)} \end{matrix} \right) \right.$$

$$\left. + \left(\begin{matrix} 0_{4n \times (4n+2p)} \\ \eta K \left(C_0 A_0 \ C_1 A_0 \ C_0 A_1 \ C_1 A_1 \ -C_0 B_0 \ -C_1 B_0 \right) \\ 0_{p \times (4n+2p)} \end{matrix} \right)^T \right] Jw$$

$$= -\eta s K H(s,z) - [\eta s K H(s,z)]^* \ .$$

One deduces that: $\exists \varepsilon > 0, \ \exists \eta = \mathrm{diag}\{\eta_i\} \geq 0, \ \forall (s,z) \in \mathbb{C}^2$ with $\mathrm{Re}\, s \geq 0$, $|z| = 1$,

$$2I + \left(I + \frac{1}{2}\eta s \right) K H(s,z) + \left[\left(I + \frac{1}{2}\eta s \right) K H(s,z) \right]^* > \varepsilon I \ . \tag{26}$$

The previous inequality is then extended to the ball $\{z \in \mathbb{C} \ : \ |z| \leq 1\}$ by maximum modulus principle. This leads finally to the second part of (23), whose proof is now complete. One then deduces (ii).

The asymptotic stability is proved as in Theorem 1. For other details, see complete proof in [3]. $\qquad\qquad\qquad\qquad\qquad\qquad\qquad\qquad\qquad\qquad\qquad\qquad\qquad$ □

3   Absolute Stability Criteria Valid for any $h \in [0, \overline{h}]$, $\overline{h} \geq 0$ Given

3.1   Delay-Dependent Circle Criterion

We are now looking for sufficient conditions of absolute stability of, say, system (9), for any value of h in $[0, \overline{h}]$, where the upper bound $\overline{h}$ is a prescribed nonnegative quantity. For this, it is sufficient that

$$\exists \varepsilon > 0, \forall s \in \mathbb{C} \text{ with } \mathrm{Re}\, s \geq 0, \forall h \in [0, \overline{h}],$$
$$\det(sI - A_0 - A_1 e^{-sh}) \neq 0 \text{ and } 2I + K H(s) + [K H(s)]^* > \varepsilon I \ , \tag{27}$$

where H is given by (10).

Now, let D be a matrix of same size than A_0, A_1 $(D \in \mathbb{R}^{n \times n})$. From the identity

$$(sI - A_0 - A_1 e^{-sh})(I - D \frac{1 - e^{-sh}}{s})$$
$$= sI - (A_0 + D) - (A_1 - D)e^{-sh} + A_0 D \frac{1 - e^{-sh}}{s} + A_1 D e^{-sh} \frac{1 - e^{-sh}}{s} \ ,$$

and the analogous one for C_0, C_1, one deduces that, if $I - D\frac{1-e^{-sh}}{s}$ is invertible, then

$$H(s) = \left(C_0 + C_1 e^{-sh} - C_0 D\frac{1-e^{-sh}}{s} - C_1 D e^{-sh}\frac{1-e^{-sh}}{s}\right)$$

$$(sI - (A_0 + D) - (A_1 - D)e^{-sh} \tag{28}$$

$$+ A_0 D\frac{1-e^{-sh}}{s} + A_1 D e^{-sh}\frac{1-e^{-sh}}{s})^{-1} B_0$$

The previous expression leads to a "nonminimal" realization of the transfer H, using e.g. the augmented state $(x, \tilde{x})$, where $\tilde{x} \overset{\text{def}}{=} \int_{t-h}^{t} x(\tau)\, d\tau$. We now introduce the following auxiliary result.

Lemma 3 *For any $s \in \mathbb{C} \setminus \{0\}$ such that $\operatorname{Re} s \geq 0$, the following inequality holds.*

$$\left|\frac{1-e^{-s}}{s}\right| \leq 1 .$$

∎

Proof. Let $s = a + ib$ with $(a, b) \in \mathbb{R}^2 \setminus \{(0,0)\}$, $a \geq 0$. Then the desired property is verified if and only if

$$1 - 2e^{-a}\cos b + e^{-2a} - a^2 - b^2 \leq 0 .$$

The proof follows from the fact that the latter expression is identical to the sum of three nonpositive terms: $1 - 2e^{-a}\cos b + e^{-2a} - a^2 - b^2 = (2(1-\cos b) - b^2)e^{-a} + b^2(e^{-a} - 1) + ((1 - e^{-a})^2 - a^2)$. □

In consequence, define a new transfer

$$H(s, z_1, z_2) \overset{\text{def}}{=} (C_0 + z_1 C_1 - z_2\overline{h}C_0 D - z_1 z_2\overline{h}C_1 D)$$

$$(sI - (A_0 + D) - z_1(A_1 - D) + z_2\overline{h}A_0 D + z_1 z_2\overline{h}A_1 D)^{-1} B_0 , \tag{29}$$

obtained by replacing in (28) e^{-sh} by z_1 and $\frac{1-e^{-sh}}{s}$ by $\overline{h}z_2$. In order that (27) is fulfilled, it is sufficient that

$$\exists D, \exists \varepsilon > 0, \forall(s, z_1, z_2) \in \mathbb{C}^3 \text{ with } \operatorname{Re} s \geq 0, |z_1|, |z_2| \leq 1,$$

$$\begin{cases} \det(sI - (A_0 + D) - z_1(A_1 - D) + z_2\overline{h}A_0 D + z_1 z_2\overline{h}A_1 D) \neq 0 \text{ and} \\ 2I + KH(s, z_1, z_2) + [KH(s, z_1, z_2)]^* > \varepsilon I , \end{cases}$$

$$\tag{30}$$

where H is given by (29). Condition (30) may be interpreted as a condition for delay-independent absolute stability of the following auxiliary system with two independent delays

$$
\begin{cases}
\dot{x} = (A_0 + D)x + (A_1 - D)x(t - h_1) - \overline{h}A_0 Dx(t - h_2) \\
\quad - \overline{h}A_1 Dx(t - h_1 - h_2) - B_0 \psi(t, y(t)), \\
y = C_0 x + C_1 x(t - h_1) - \overline{h}C_0 Dx(t - h_2) - \overline{h}C_1 Dx(t - h_1 - h_2) .
\end{cases}
\tag{31}
$$

Remark, however, that it does not seem possible to achieve this reduction by a change of variable: the frequency domain nature of the method seems essential.

The next step towards our results consists in expressing LMI conditions sufficient for delay-independent absolute stability of (31). Such conditions for systems with two delays are given in [3], and in [5] for linear delay systems. They are obtained by "rewriting" system (31) as

$$
\begin{cases}
\dot{X} = \mathcal{A}_0 X(t) + \mathcal{A}_1 X(t - 2h_1) + \mathcal{A}_2 X(t - 2h_2) - \mathcal{B}_0 \Psi(t, Y(t)) , \\
Y = \mathcal{C}_0 X(t) + \mathcal{C}_1 X(t - 2h_1) + \mathcal{C}_2 X(t - 2h_2) ,
\end{cases}
\tag{32}
$$

where

$$
X(t) \stackrel{\text{def}}{=}
\begin{pmatrix}
x(t) \\
x(t - h_1) \\
x(t - h_2) \\
x(t - h_1 - h_2) \\
x(t - 2h_1)
\end{pmatrix} ,
\tag{33a}
$$

$$
Y(t) \stackrel{\text{def}}{=}
\begin{pmatrix}
y(t) \\
y(t - h_1) \\
y(t - h_2) \\
y(t - h_1 - h_2) \\
y(t - 2h_2)
\end{pmatrix} ,
\quad
\Psi
\left(
t,
\begin{pmatrix}
Y_1 \\
Y_2 \\
Y_3 \\
Y_4 \\
Y_5
\end{pmatrix}
\right)
\stackrel{\text{def}}{=}
\begin{pmatrix}
\psi(t, Y_1) \\
\psi(t - h_1, Y_2) \\
\psi(t - h_2, Y_3) \\
\psi(t - h_1 - h_2, Y_4) \\
\psi(t - 2h_1, Y_5)
\end{pmatrix} ,
\tag{33b}
$$

and

$$\mathcal{A}_0 \overset{\text{def}}{=} \begin{pmatrix} A_0 + D & A_1 - D & -\bar{h}A_0 D & -\bar{h}A_1 D & 0 \\ 0 & A_0 + D & 0 & -\bar{h}A_0 D & 0 \\ 0 & 0 & A_0 + D & A_1 - D & 0 \\ 0 & 0 & 0 & A_0 + D & 0 \\ 0 & 0 & 0 & 0 & A_0 + D \end{pmatrix}, \tag{34a}$$

$$\mathcal{C}_0 \overset{\text{def}}{=} \begin{pmatrix} C_0 & C_1 & -\bar{h}C_0 D & -\bar{h}C_1 D & 0 \\ 0 & C_0 & 0 & -\bar{h}C_0 D & 0 \\ 0 & 0 & C_0 & C_1 & 0 \\ 0 & 0 & 0 & C_0 & 0 \\ 0 & 0 & 0 & 0 & C_0 \end{pmatrix}, \tag{34b}$$

$$\mathcal{A}_1 \overset{\text{def}}{=} \begin{pmatrix} 0 & 0 & 0 & 0 & 0 \\ A_1 - D & 0 & -\bar{h}A_1 D & 0 & 0 \\ 0 & 0 & 0 & 0 & 0 \\ 0 & 0 & A_1 - D & 0 & 0 \\ 0 & A_1 - D & -\bar{h}A_0 D & -\bar{h}A_1 D & 0 \end{pmatrix}, \tag{34c}$$

$$\mathcal{C}_1 \overset{\text{def}}{=} \begin{pmatrix} 0 & 0 & 0 & 0 & 0 \\ C_1 & 0 & -\bar{h}C_1 D & 0 & 0 \\ 0 & 0 & 0 & 0 & 0 \\ 0 & 0 & C_1 & 0 & 0 \\ 0 & C_1 & -\bar{h}C_0 D & -\bar{h}C_1 D & 0 \end{pmatrix}, \tag{34d}$$

$$\mathcal{A}_2 \overset{\text{def}}{=} \begin{pmatrix} 0 & 0 & 0 & 0 & 0 \\ 0 & 0 & 0 & 0 & 0 \\ -\bar{h}A_0 D & -\bar{h}A_1 D & 0 & 0 & 0 \\ 0 & -\bar{h}A_0 D & 0 & 0 & -\bar{h}A_1 D \\ 0 & 0 & 0 & 0 & 0 \end{pmatrix}, \tag{34e}$$

$$\mathcal{C}_2 \overset{\text{def}}{=} \begin{pmatrix} 0 & 0 & 0 & 0 & 0 \\ 0 & 0 & 0 & 0 & 0 \\ -\bar{h}C_0 D & -\bar{h}C_1 D & 0 & 0 & 0 \\ 0 & -\bar{h}C_0 D & 0 & 0 & -\bar{h}C_1 D \\ 0 & 0 & 0 & 0 & 0 \end{pmatrix}, \tag{34f}$$

$$\mathcal{B}_0 \overset{\text{def}}{=} I_5 \otimes B_0, \quad \mathcal{K} \overset{\text{def}}{=} I_5 \otimes K. \tag{34g}$$

The size of the previous matrices is

$$\mathcal{A}_0, \mathcal{A}_1, \mathcal{A}_2 : 5n \times 5n, \quad \mathcal{B}_0 : 5n \times 5q,$$
$$\mathcal{C}_0, \mathcal{C}_1, \mathcal{C}_2 : 5p \times 5n, \quad \mathcal{K} : 5q \times 5p.$$

The trajectories of system (31) may be obtained as projections of trajectories of system (32): asymptotic stability of (32) implies asymptotic stability of (31).

System (32) having independent occurrences of the two delays, one has the benefit of a natural class of quadratic Lyapunov-Krasovskii functionals, namely

$$V(\varPhi) \stackrel{\text{def}}{=} \varPhi^T(0)\mathcal{P}\varPhi(0) + \int_{-2h_1}^{0} \varPhi^T(\tau)\mathcal{Q}_1\varPhi(\tau)\, d\tau + \int_{-2h_2}^{0} \varPhi^T(\tau)\mathcal{Q}_2\varPhi(\tau)\, d\tau ,$$

(35)

for any $\varPhi \in \mathcal{C}([-2h,0];\mathbb{R}^{5n})$, $h \stackrel{\text{def}}{=} \max\{h_1, h_2\}$. However we are only interested in those trajectories of system (32) fulfilling (33a). In order to integrate this information, we are finally led to consider for system (32) Lyapunov-Krasovskii functionals in the class

$$V(\phi) \stackrel{\text{def}}{=} \varPhi^T(0)\mathcal{P}\varPhi(0) + \int_{-2h_1}^{0} \varPhi^T(\tau)\mathcal{Q}_1\varPhi(\tau)\, d\tau + \int_{-2h_2}^{0} \varPhi^T(\tau)\mathcal{Q}_2\varPhi(\tau)\, d\tau,$$

$$\text{where } \varPhi(t) \stackrel{\text{def}}{=} \begin{pmatrix} \phi(t) \\ \phi(t-h_1) \\ \phi(t-h_2) \\ \phi(t-h_1-h_2) \\ \phi(t-2h_1) \end{pmatrix} ,$$

(36)

defined for $\phi \in \mathcal{C}([-4h,0];\mathbb{R}^n)$. Writing that the derivative of V should be negative definite along the trajectories of (32), one obtains the following result.

Theorem 4 *Consider the two following statements.*

(i)   There exist $D \in \mathbb{R}^{n\times n}$, $\mathcal{P}, \mathcal{Q}_1, \mathcal{Q}_2 \in \mathbb{R}^{5n\times5n}$ such that

$$\mathcal{P} = \mathcal{P}^T > 0, \mathcal{Q}_1 = \mathcal{Q}_1^T > 0, \mathcal{Q}_2 = \mathcal{Q}_2^T > 0,\ \mathcal{J}^T\mathcal{R}\mathcal{J} < 0 ,$$

(37)

where

$$\mathcal{J} \stackrel{\text{def}}{=} \text{diag}\left\{ \begin{pmatrix} 1&0&0&0&0 \\ 0&1&0&0&0 \\ 0&0&1&0&0 \\ 0&0&0&1&0 \\ 0&0&0&0&1 \\ 0&0&0&0&1 \end{pmatrix} \otimes I_n,\ I_{9n+5q} \right\} \in \mathbb{R}^{(15n+5q)\times(14n+5q)} ,$$

$$\mathcal{R} \stackrel{\text{def}}{=} \begin{pmatrix} \mathcal{A}_0^T\mathcal{P} + \mathcal{P}\mathcal{A}_0 + \mathcal{Q}_1 + \mathcal{Q}_2 & \mathcal{P}\mathcal{A}_1 & \mathcal{P}\mathcal{A}_2 & \mathcal{C}_0^T\mathcal{K}^T - \mathcal{P}\mathcal{B}_0 \\ \mathcal{A}_1^T\mathcal{P} & -\mathcal{Q}_1 & 0 & \mathcal{C}_1^T\mathcal{K}^T \\ \mathcal{A}_2^T\mathcal{P} & 0 & -\mathcal{Q}_2 & \mathcal{C}_2^T\mathcal{K}^T \\ \mathcal{K}\mathcal{C}_0 - \mathcal{B}_0^T\mathcal{P} & \mathcal{K}\mathcal{C}_1 & \mathcal{K}\mathcal{C}_2 & -2I_{5q} \end{pmatrix} < 0 ,$$

and $\mathcal{A}_0, \mathcal{A}_1, \mathcal{A}_2, \mathcal{B}_0, \mathcal{C}_0, \mathcal{C}_1, \mathcal{C}_2, \mathcal{K}$ are defined in formulas (34a) to (34g).

(ii) The following property holds.

$$\exists \varepsilon > 0, \forall h \in [0, \overline{h}], \forall s \in \mathbb{C} \text{ with } \operatorname{Re} s \geq 0, \ \det(sI - A_0 - A_1 e^{-sh}) \neq 0$$
$$\text{and } 2I + KH(s) + [KH(s)]^* > \varepsilon I ,$$

where H is given by (10).

Suppose (i) holds. Then (ii) holds.

Suppose (ii) holds. Then, for any ψ fulfilling (1) and any $h \in [0, \overline{h}]$, $\lim_{t \to +\infty} x(t) = 0$ for any global solution x of (9). ∎

Matrix $\mathcal{J}^T \mathcal{R} \mathcal{J}$ in (37) is of size $14n + 5q$.

Problem (37) is not convex in the unknowns $D, \mathcal{P}, \mathcal{Q}_1, \mathcal{Q}_2$, but is a LMI in $\mathcal{P}, \mathcal{Q}_1, \mathcal{Q}_2$, once D has been chosen. Remark that, contrary to what happened in Sect. 2, no Lyapunov-Krasovskii functional is available here for the initial system (9).

The matrix $\mathcal{J}$ in (37) is related to the fact that we are concerned only by the trajectories of (32) which fulfill (33a). More precisely, when (33a) is fulfilled, the 5th component of $X(t)$ and the 1st component of $X(t - 2h_1)$ are equal to $x(t - 2h_1)$, so one of the two occurrences may be omitted. This is expressed in the fact that the vector

$$\begin{pmatrix} X(t) \\ X(t - 2h_1) \\ X(t - 2h_2) \\ \Psi(t, Y(t)) \end{pmatrix}$$

is equal to $\mathcal{J}(x^T(t) \ x^T(t - h_1) \ x^T(t - h_2) \ x^T(t - h_1 - h_2) \ x^T(t - 2h_1) \ x^T(t - 3h_1) \ x^T(t - 2h_1 - h_2) \ x^T(t - 3h_1 - h_2) \ x^T(t - 4h_1) \ x^T(t - 2h_2) \ x^T(t - h_1 - 2h_2) \ x^T(t - 3h_2) \ x^T(t - h_1 - 3h_2) \ x^T(t - 2h_1 - 2h_2) \ \Psi(t, Y(t))^T)^T$, and hence contains only independent components. Use of Lyapunov-Krasovskii functionals (35) instead of (36) would lead to

$$\mathcal{P} = \mathcal{P}^T > 0, \mathcal{Q}_1 = \mathcal{Q}_1^T > 0, \mathcal{Q}_2 = \mathcal{Q}_2^T > 0, \ \mathcal{R} < 0$$

instead of (37), a condition clearly stronger.

Proof. Suppose (i) holds. In view of the consideration preceding the statement, it suffices, in order to get (ii), to prove (30).

Define (see above) $v = (1 \ z_1 \ z_2 \ z_1 z_2 \ z_1^2 \ z_1^3 \ z_1^2 z_2 \ z_1^3 z_2 \ z_1^4 \ z_2^2 \ z_1 z_2^2 \ z_2^3 \ z_1 z_2^3 \ z_1^2 z_2^2)^T \otimes I_n$. One has

$$\mathcal{J} \begin{pmatrix} v \\ 0_{5q \times n} \end{pmatrix} = \begin{pmatrix} I_{5n} \\ z_1^2 I_{5n} \\ z_2^2 I_{5n} \\ 0_{5q \times 5n} \end{pmatrix} \begin{pmatrix} I_n \\ z_1 I_n \\ z_2 I_n \\ z_1 z_2 I_n \\ z_1^2 I_n \end{pmatrix} . \tag{38}$$

Right- and left-multiplying $\mathcal{J}^T\mathcal{R}\mathcal{J}$ by $\begin{pmatrix} v \\ 0_{5q \times n} \end{pmatrix} \in \mathbb{R}^{(14n+5q) \times n}$ and its conjugate hence leads to

$$
0 > \begin{pmatrix} I_n \\ z_1 I_n \\ z_2 I_n \\ z_1 z_2 I_n \\ z_1^2 I_n \end{pmatrix}^* \begin{pmatrix} I_{5n} \\ z_1^2 I_{5n} \\ z_2^2 I_{5n} \\ 0_{5q \times 5n} \end{pmatrix}^* \mathcal{R} \begin{pmatrix} I_{5n} \\ z_1^2 I_{5n} \\ z_2^2 I_{5n} \\ 0_{5q \times 5n} \end{pmatrix} \begin{pmatrix} I_n \\ z_1 I_n \\ z_2 I_n \\ z_1 z_2 I_n \\ z_1^2 I_n \end{pmatrix}
$$

$$
= \begin{pmatrix} I_n \\ z_1 I_n \\ z_2 I_n \\ z_1 z_2 I_n \\ z_1^2 I_n \end{pmatrix}^* [(A_0^T + z_1^{2*} A_1^T + z_2^{2*} A_2^T)\mathcal{P} + \mathcal{P}(A_0 + z_1^2 A_1 + z_2^2 A_2)
$$

$$
+ (1 - |z_1|^4)\mathcal{Q}_1 + (1 - |z_2|^4)\mathcal{Q}_2] \begin{pmatrix} I_n \\ z_1 I_n \\ z_2 I_n \\ z_1 z_2 I_n \\ z_1^2 I_n \end{pmatrix}
$$

$$
\geq \begin{pmatrix} I_n \\ z_1 I_n \\ z_2 I_n \\ z_1 z_2 I_n \\ z_1^2 I_n \end{pmatrix}^* [(A_0^T + z_1^{2*} A_1^T + z_2^{2*} A_2^T)\mathcal{P}
$$

$$
+ \mathcal{P}(A_0 + z_1^2 A_1 + z_2^2 A_2)] \begin{pmatrix} I_n \\ z_1 I_n \\ z_2 I_n \\ z_1 z_2 I_n \\ z_1^2 I_n \end{pmatrix} ,
$$

when $|z_1|, |z_2| \leq 1$. Now, one may show [3], that

$$
(A_0 + z_1^2 A_1 + z_2^2 A_2) \begin{pmatrix} I_n \\ z_1 I_n \\ z_2 I_n \\ z_1 z_2 I_n \\ z_1^2 I_n \end{pmatrix}
$$

$$
= \begin{pmatrix} I_n \\ z_1 I_n \\ z_2 I_n \\ z_1 z_2 I_n \\ z_1^2 I_n \end{pmatrix} ((A_0 + D) + z_1(A_1 - D) - z_2 \overline{h} A_0 D - z_1 z_2 \overline{h} A_1 D) . \quad (39)
$$

Arguing as in the proof of Theorem 1, we then deduce the first part of (30).

To get the second part of (30), right- and left-multiply $\mathcal{J}^T \mathcal{R} \mathcal{J}$ by

$$\begin{pmatrix} -v(sI - (A_0 + D) - z_1(A_1 - D) + z_2\overline{h}A_0 D + z_1 z_2 \overline{h}A_1 D)^{-1} B_0 \\ I_q \\ z_1 I_q \\ z_2 I_q \\ z_1 z_2 I_q \\ z_1^2 I_q \end{pmatrix} \tag{40}$$

and its conjugate. Defining

$$\mathcal{S} = \mathcal{S}(s, z_1, z_2) \stackrel{\text{def}}{=} (sI - A_0 - z_1^2 A_1 - z_2^2 A_2)^{-1}, \tag{41}$$

the result of this operation writes

$$0 > \begin{pmatrix} I_q \\ z_1 I_q \\ z_2 I_q \\ z_1 z_2 I_q \\ z_1^2 I_q \end{pmatrix}^* \begin{pmatrix} \mathcal{S}B_0 \\ z_1^2 \mathcal{S}B_0 \\ z_2^2 \mathcal{S}B_0 \\ -I_{5q} \end{pmatrix}^* \mathcal{R} \begin{pmatrix} \mathcal{S}B_0 \\ z_1^2 \mathcal{S}B_0 \\ z_2^2 \mathcal{S}B_0 \\ -I_{5q} \end{pmatrix} \begin{pmatrix} I_q \\ z_1 I_q \\ z_2 I_q \\ z_1 z_2 I_q \\ z_1^2 I_q \end{pmatrix}, \tag{42}$$

where we have used the fact that

$$\begin{pmatrix} I_n \\ z_1 I_n \\ z_2 I_n \\ z_1 z_2 I_n \\ z_1^2 I_n \end{pmatrix} B_0 = B_0 \begin{pmatrix} I_q \\ z_1 I_q \\ z_2 I_q \\ z_1 z_2 I_q \\ z_1^2 I_q \end{pmatrix}. \tag{43}$$

Decomposing $\mathcal{R}$ as the sum of

$$\begin{pmatrix} A_0^T \mathcal{P} + \mathcal{P}A_0 + \mathcal{Q}_1 + \mathcal{Q}_2 & \mathcal{P}A_1 & \mathcal{P}A_2 & -\mathcal{P}B_0 \\ A_1^T \mathcal{P} & -\mathcal{Q}_1 & 0 & 0 \\ A_2^T \mathcal{P} & 0 & -\mathcal{Q}_2 & 0 \\ -B_0^T \mathcal{P} & 0 & 0 & 0 \end{pmatrix}$$

and

$$\begin{pmatrix} 0 & 0 & 0 & \mathcal{C}_0^T \mathcal{K}^T \\ 0 & 0 & 0 & \mathcal{C}_1^T \mathcal{K}^T \\ 0 & 0 & 0 & \mathcal{C}_2^T \mathcal{K}^T \\ \mathcal{K}\mathcal{C}_0 & \mathcal{K}\mathcal{C}_1 & \mathcal{K}\mathcal{C}_2 & -2I_{5q} \end{pmatrix},$$

one shows, as in Theorem 1, that, if $\operatorname{Re} s \geq 0$ and $|z_1|, |z_2| \leq 1$, then

$$
\begin{pmatrix} \mathcal{S}B_0 \\ z_1^2 \mathcal{S}B_0 \\ z_2^2 \mathcal{S}B_0 \\ -I_{5q} \end{pmatrix}^* \mathcal{R} \begin{pmatrix} \mathcal{S}B_0 \\ z_1^2 \mathcal{S}B_0 \\ z_2^2 \mathcal{S}B_0 \\ -I_{5q} \end{pmatrix}
$$

$$
\geq -(2I_{5q} + \mathcal{K}(\mathcal{C}_0 + z_1^2 \mathcal{C}_1 + z_2^2 \mathcal{C}_2)(sI - \mathcal{A}_0 - z_1^2 \mathcal{A}_1 - z_2^2 \mathcal{A}_2)^{-1} B_0
$$

$$
+ [\mathcal{K}(\mathcal{C}_0 + z_1^2 \mathcal{C}_1 + z_2^2 \mathcal{C}_2)(sI - \mathcal{A}_0 - z_1^2 \mathcal{A}_1 - z_2^2 \mathcal{A}_2)^{-1} B_0]^*) .
$$

Using (39), (43) and

$$
(\mathcal{C}_0 + z_1^2 \mathcal{C}_1 + z_2^2 \mathcal{C}_2) \begin{pmatrix} I_n \\ z_1 I_n \\ z_2 I_n \\ z_1 z_2 I_n \\ z_1^2 I_n \end{pmatrix}
$$

$$
= \begin{pmatrix} I_p \\ z_1 I_p \\ z_2 I_p \\ z_1 z_2 I_p \\ z_1^2 I_p \end{pmatrix} (C_0 + z_1 C_1 - z_2 \overline{h} C_0 D - z_1 z_2 \overline{h} C_1 D) , \tag{44}
$$

the expression in (42) is then proved to be larger or equal than

$$
- \left\| \begin{pmatrix} 1 \\ z_1 \\ z_2 \\ z_1 z_2 \\ z_1^2 \end{pmatrix} \right\|^2 (2I + KH(s, z_1, z_2) + [KH(s, z_1, z_2)]^*) ,
$$

where H is given by (29), see [3] for details. One then deduces that (30) holds, and finally (ii). The asymptotic stability is then proved as in Theorem 1. $\square$

3.2   Delay-Dependent Popov Criterion

As in Sect. 2.2, we assume throughout this section that $p = q$, and that ψ is decentralized and fulfills sector condition (5) for a certain nonnegative diagonal matrix K.

We apply now the same ideas than those exposed in Sect. 3.1, to system (18) with time-invariant decentralized nonlinearity, as in Sect. 2.2. More precisely (compare with (30)), we shall use the fact that the following condition is sufficient for absolute stability of (18), for any $h \in [0, \overline{h}]$:

$\exists D, \exists \varepsilon > 0, \exists \eta \geq 0, \forall (s, z_1, z_2) \in \mathbb{C}^3$ with $\operatorname{Re} s \geq 0, |z_1|, |z_2| \leq 1,$

$$\begin{cases} \det(sI - (A_0 + D) - z_1(A_1 - D) + z_2\overline{h}A_0 D + z_1 z_2 \overline{h} A_1 D) \neq 0 \text{ and} \\ 2I + (I + \eta s)KH(s, z_1, z_2) + [(I + \eta s)KH(s, z_1, z_2)]^* > \varepsilon I \,, \end{cases}$$

$$(45)$$

where H is given by (29).

The same considerations lead to study the time-invariant analogue of (32)

$$\begin{cases} \dot{X} = A_0 X(t) + A_1 X(t - 2h_1) + A_2 X(t - 2h_2) - B_0 \Psi(Y(t)) \,, \\ Y = C_0 X(t) + C_1 X(t - 2h_1) + C_2 X(t - 2h_2) \,, \end{cases} \qquad (46)$$

with the same notations than for the system transformation conducted in Sect. 3.1, and here

$$\Psi(Y) \stackrel{\text{def}}{=} \begin{pmatrix} \psi(Y_1) \\ \psi(Y_2) \\ \psi(Y_3) \\ \psi(Y_4) \\ \psi(Y_5) \end{pmatrix}.$$

Coherently with the choice made in Sect. 2.2 (see formula (20)), one is led to choose for system (46) Lyapunov-Krasovskii functional of the form

$$\begin{aligned}
V(\Phi) \stackrel{\text{def}}{=} &\begin{pmatrix} \Phi(0) \\ \Phi(-2h_1) \\ \Phi(-2h_2) \end{pmatrix}^T \mathcal{P} \begin{pmatrix} \Phi(0) \\ \Phi(-2h_1) \\ \Phi(-2h_2) \end{pmatrix} \\
&+ \int_{-2h_1}^{0} \begin{pmatrix} \Phi(\tau) \\ \Phi(\tau - 2h_1) \\ \Phi(\tau - 2h_2) \end{pmatrix}^T \mathcal{Q}_1 \begin{pmatrix} \Phi(\tau) \\ \Phi(\tau - 2h_1) \\ \Phi(\tau - 2h_2) \end{pmatrix} d\tau \\
&+ \int_{-2h_2}^{0} \begin{pmatrix} \Phi(\tau) \\ \Phi(\tau - 2h_1) \\ \Phi(\tau - 2h_2) \end{pmatrix}^T \mathcal{Q}_2 \begin{pmatrix} \Phi(\tau) \\ \Phi(\tau - 2h_1) \\ \Phi(\tau - 2h_2) \end{pmatrix} d\tau \\
&+ 2\sum_{i=1}^{5p} \eta_i \mathcal{K}_i \int_0^{(C_0\Phi(0) + C_1\Phi(-2h_1) + C_2\Phi(-2h_2))_i} \Psi_i(u)\, du \,. \quad (47)
\end{aligned}$$

defined for any $\Phi \in \mathcal{C}([-4h, 0]; \mathbb{R}^{5n})$, $h \stackrel{\text{def}}{=} \max\{h_1, h_2\}$. However, as in Sect. 3.1, one may use more information on Φ, as we are only interested in those trajectories of (46) constrained by

$$
X(t) = \begin{pmatrix} x(t) \\ x(t-h_1) \\ x(t-h_2) \\ x(t-h_1-h_2) \\ x(t-2h_1) \end{pmatrix}, \ \Psi(Y(t)) = \begin{pmatrix} \psi(y(t)) \\ \psi(y(t-h_1)) \\ \psi(y(t-h_2)) \\ \psi(y(t-h_1-h_2)) \\ \psi(y(t-2h_1)) \end{pmatrix}, \tag{48}
$$

for certain functions x, y. Computing the derivative of V along the trajectories of (46) for which (48) holds, yields the following result.

Theorem 5 *Consider the two following statements.*

*(i)   There exist $D \in \mathbb{R}^{n\times n}$, $\eta = \mathrm{diag}\{\eta_i\} \in \mathbb{R}^{5p\times 5p}$, $\mathcal{P}, \mathcal{Q}_1, \mathcal{Q}_2 \in \mathbb{R}^{15n\times 15n}$
such that*

$$
\eta \geq 0, \mathcal{P} = \mathcal{P}^T > 0, \mathcal{Q}_1 = \mathcal{Q}_1^T > 0, \mathcal{Q}_2 = \mathcal{Q}_2^T > 0, \tag{49}
$$
$$
and \quad \mathcal{J}^T J^T \mathcal{R} J \mathcal{J} < 0,
$$

where $\mathcal{R}$ is equal to

$$
\begin{pmatrix} (I_3 \otimes A_0^T)\mathcal{P} + \mathcal{P}(I_3 \otimes A_0) + \mathcal{Q}_1 + \mathcal{Q}_2 & \mathcal{P}(I_3 \otimes A_1) & \mathcal{P}(I_3 \otimes A_2) & -\mathcal{P}(I_3 \otimes B_0) \\ (I_3 \otimes A_1^T)\mathcal{P} & -\mathcal{Q}_1 & 0 & 0 \\ (I_3 \otimes A_1^T)\mathcal{P} & 0 & -\mathcal{Q}_2 & 0 \\ -(I_3 \otimes B_0^T)\mathcal{P} & 0 & 0 & 0 \end{pmatrix}
$$

$$
+ \begin{pmatrix} 0 & 0 & 0 & (I_3 \otimes C_0^T \mathcal{K}) \\ 0 & 0 & 0 & (I_3 \otimes C_1^T \mathcal{K}) \\ 0 & 0 & 0 & (I_3 \otimes C_2^T \mathcal{K}) \\ (I_3 \otimes \mathcal{K}C_0) & (I_3 \otimes \mathcal{K}C_1) & (I_3 \otimes \mathcal{K}C_2) & -2I_{15p} \end{pmatrix}
$$

$$
+ \left(\begin{pmatrix} 0_{(45n+15p)\times 45n} & \begin{pmatrix} A_0^T C_0^T \\ A_0^T C_1^T \\ A_0^T C_2^T \\ A_1^T C_0^T \\ A_1^T C_1^T \\ A_1^T C_2^T \\ A_2^T C_0^T \\ A_2^T C_1^T \\ A_2^T C_2^T \\ -B_0^T C_0^T \\ -B_0^T C_1^T \\ -B_0^T C_2^T \end{pmatrix} \eta \mathcal{K} \ 0_{(45n+15p)\times 10p} \end{pmatrix} + \quad \left(\quad \right)^T \right),
$$

where the void matrix stands for the previous one, where $A_0, A_1, A_2, B_0, C_0, C_1, C_2, \mathcal{K}$ are defined in formulas (34a) to (34g), and $J, \mathcal{J}$ are defined after the current statement.

(ii) *The following property holds.*

$$\exists \varepsilon > 0, \exists \eta = \mathrm{diag}\{\eta_i\} \in \mathbb{R}^{p \times p}, \eta \geq 0, \forall h \in [0, \overline{h}], \forall s \in \mathbb{C} \text{ with } \mathrm{Re}\, s \geq 0,$$

$$\begin{cases} \det(sI - A_0 - A_1 e^{-sh}) \neq 0 \text{ and} \\ 2I + (I + \eta s)KH(s) + [(I + \eta s)KH(s)]^* > \varepsilon I \,, \end{cases}$$

where H is given by (10).

Suppose (i) holds. Then (ii) holds.
Suppose (ii) holds. Then, for any decentralized ψ fulfilling (5) and any $h \in [0, \overline{h}]$, $\lim\limits_{t \to +\infty} x(t) = 0$ for any global solution x of (18).   ∎

Matrix J is introduced in (49) in order, as in (21), to obtain a nice form for $\mathcal{R}$. One has

$$J \stackrel{\mathrm{def}}{=} \mathrm{diag} \left\{ \begin{pmatrix} I_{5n} & 0 & 0 & 0 & 0 & 0 \\ 0 & I_{5n} & 0 & 0 & 0 & 0 \\ 0 & 0 & I_{5n} & 0 & 0 & 0 \\ 0 & I_{5n} & 0 & 0 & 0 & 0 \\ 0 & 0 & 0 & I_{5n} & 0 & 0 \\ 0 & 0 & 0 & 0 & I_{5n} & 0 \\ 0 & 0 & I_{5n} & 0 & 0 & 0 \\ 0 & 0 & 0 & 0 & I_{5n} & 0 \\ 0 & 0 & 0 & 0 & 0 & I_{5n} \end{pmatrix}, I_{15p} \right\} \in \mathbb{R}^{(45n+15p) \times (30n+15p)},$$

in such a way that, denoting

$$\mathcal{X}(t) \stackrel{\mathrm{def}}{=} \begin{pmatrix} X(t) \\ X(t - 2h_1) \\ X(t - 2h_2) \end{pmatrix},$$

one has

$$\begin{pmatrix} \mathcal{X}(t) \\ \mathcal{X}(t - 2h_1) \\ \mathcal{X}(t - 2h_2) \\ \Psi(Y(t)) \\ \Psi(Y(t - 2h_1)) \\ \Psi(Y(t - 2h_2)) \end{pmatrix} = J \begin{pmatrix} X(t) \\ X(t - 2h_1) \\ X(t - 2h_2) \\ X(t - 4h_1) \\ X(t - 2h_1 - 2h_2) \\ X(t - 4h_2) \\ \Psi(Y(t)) \\ \Psi(Y(t - 2h_1)) \\ \Psi(Y(t - 2h_2)) \end{pmatrix}. \tag{50}$$

Now, X and Ψ being related to the components of x and ψ by identities (48), one has

$$\left(0_n \ 0_n \ 0_n \ 0_n \ I_n \right) X(t) = \left(I_n \ 0_n \ 0_n \ 0_n \ 0_n \right) X(t - 2h_1) = x(t - 2h_1),$$

$$\left(0_n \ 0_n \ 0_n \ 0_n \ I_n \right) X(t - 2h_1) = \left(I_n \ 0_n \ 0_n \ 0_n \ 0_n \right) X(t - 4h_1) = x(t - 4h_1),$$

$$\left(0_n \ 0_n \ 0_n \ 0_n \ I_n \right) X(t - 2h_2) = \left(I_n \ 0_n \ 0_n \ 0_n \ 0_n \right) X(t - 2h_1 - 2h_2)$$
$$= x(t - 2h_1 - 2h_2),$$

$$\left(0_p \ 0_p \ 0_p \ 0_p \ I_p \right) \Psi(Y(t)) = \left(I_p \ 0_p \ 0_p \ 0_p \ 0_p \right) \Psi(Y(t - 2h_1))$$
$$= \psi(y(t - 2h_1)) .$$

One hence removes one occurrence of each of these four components, expressed twice. This permits, as in Sect. 3.1, to construct easily the (huge) matrix $\mathcal{J}$, which precisely expresses how the vector in the right-hand side of (50) is expressed as a function of $x(t)$ and $\psi(y(t))$ and their delayed values. Matrix $\mathcal{J}$ is an element of $\mathbb{R}^{(30n+15p) \times (27n+14p)}$. Finally, the size of $\mathcal{J}^T \mathcal{J}^T \mathcal{R} \mathcal{J} \mathcal{J}$ in (49) is $27n + 14p$.

Similarly to what happens with the results stated in Sect. 2, when (37) is fulfilled for $\mathcal{P}, \mathcal{Q}_1, \mathcal{Q}_2 \in \mathbb{R}^{5n \times 5n}$, then (49) is fulfilled for $\eta = 0$ and $I_3 \otimes \mathcal{P}, I_3 \otimes \mathcal{Q}_1, I_3 \otimes \mathcal{Q}_2$: the addition of a Lur'e term leads, for systems with decentralized, time-invariant nonlinearities, to a stronger stability result than the one provided by Theorem 4, i.e. the "delay-dependent Popov criterion" in Theorem 5 is stronger than the "delay-dependent circle criterion" in Theorem 4.

Proof. Suppose (i) holds. In view of the consideration preceding the statement, it suffices, in order to get (ii), to prove (45).

Analogously to the proof of Theorem 4, one defines $v \in \mathbb{R}^{27n \times n}$, expressing the vector

$$\begin{pmatrix} X(t) \\ X(t - 2h_1) \\ X(t - 2h_2) \\ X(t - 4h_1) \\ X(t - 2h_1 - 2h_2) \\ X(t - 4h_2) \end{pmatrix}$$

as a function of $x(t)$ and its delayed values, using relation (33a) and removing the repeated occurrences. By definition of v, one has (compare with (38))

$$\mathcal{J} \begin{pmatrix} v \\ 0_{14p \times n} \end{pmatrix} = \begin{pmatrix} I_{5n} \\ z_1^2 I_{5n} \\ z_2^2 I_{5n} \\ z_1^4 I_{5n} \\ z_1^2 z_2^2 I_{5n} \\ z_2^4 I_{5n} \\ 0_{15p \times 5n} \end{pmatrix} \begin{pmatrix} I_n \\ z_1 I_n \\ z_2 I_n \\ z_1 z_2 I_n \\ z_1^2 I_n \end{pmatrix} , \tag{51}$$

so

$$JJ\begin{pmatrix} v \\ 0_{14p \times n} \end{pmatrix} = \begin{pmatrix} I_{15n} \\ z_1^2 I_{15n} \\ z_2^2 I_{15n} \\ 0_{15p \times 15n} \end{pmatrix} \begin{pmatrix} I_{5n} \\ z_1^2 I_{5n} \\ z_2^2 I_{5n} \end{pmatrix} \begin{pmatrix} I_n \\ z_1 I_n \\ z_2 I_n \\ z_1 z_2 I_n \\ z_1^2 I_n \end{pmatrix}. \tag{52}$$

Right- and left-multiplying $J^T J^T \mathcal{R} J J$ by $\begin{pmatrix} v \\ 0_{14p \times n} \end{pmatrix}$ and its conjugate hence shows that the following expression is negative (compare with (24))

$$\begin{pmatrix} I_n \\ z_1 I_n \\ z_2 I_n \\ z_1 z_2 I_n \\ z_1^2 I_n \end{pmatrix}^* \begin{pmatrix} I_{5n} \\ z_1^2 I_{5n} \\ z_2^2 I_{5n} \end{pmatrix}^* [(I_3 \otimes (A_0^T + z_1^{2*} A_1^T + z_2^{2*} A_2^T)) \mathcal{P}$$

$$+ \mathcal{P}(I_3 \otimes (A_0 + z_1^2 A_1 + z_2^2 A_2))$$

$$+ (1 - |z_1|^4) \mathcal{Q}_1 + (1 - |z_2|^4) \mathcal{Q}_2] \begin{pmatrix} I_{5n} \\ z_1^2 I_{5n} \\ z_2^2 I_{5n} \end{pmatrix} \begin{pmatrix} I_n \\ z_1 I_n \\ z_2 I_n \\ z_1 z_2 I_n \\ z_1^2 I_n \end{pmatrix}.$$

Using now (39), one gets that

$$(I_3 \otimes (A_0 + z_1^2 A_1 + z_2^2 A_2)) \begin{pmatrix} I_{5n} \\ z_1^2 I_{5n} \\ z_2^2 I_{5n} \end{pmatrix} \begin{pmatrix} I_n \\ z_1 I_n \\ z_2 I_n \\ z_1 z_2 I_n \\ z_1^2 I_n \end{pmatrix}$$

$$= (I_3 \otimes (A_0 + z_1^2 A_1 + z_2^2 A_2)) \left(\begin{pmatrix} 1 \\ z_1^2 \\ z_2^2 \end{pmatrix} \otimes \begin{pmatrix} I_n \\ z_1 I_n \\ z_2 I_n \\ z_1 z_2 I_n \\ z_1^2 I_n \end{pmatrix} \right)$$

$$= \begin{pmatrix} 1 \\ z_1^2 \\ z_2^2 \end{pmatrix} \otimes \left(\begin{pmatrix} I_n \\ z_1 I_n \\ z_2 I_n \\ z_1 z_2 I_n \\ z_1^2 I_n \end{pmatrix} ((A_0 + D) + z_1(A_1 - D) - z_2 \overline{h} A_0 D - z_1 z_2 \overline{h} A_1 D) \right)$$

$$= \begin{pmatrix} I_{5n} \\ z_1^2 I_{5n} \\ z_2^2 I_{5n} \end{pmatrix} \begin{pmatrix} I_n \\ z_1 I_n \\ z_2 I_n \\ z_1 z_2 I_n \\ z_1^2 I_n \end{pmatrix} ((A_0 + D) + z_1(A_1 - D) - z_2 \overline{h} A_0 D - z_1 z_2 \overline{h} A_1 D).$$

One hence deduces, as in the proof of Theorem 2, that, if $|z_1|, |z_2| \leq 1$, then the eigenvalues of $(A_0 + D) + z_1(A_1 - D) - z_2\overline{h}A_0 D - z_1 z_2 \overline{h}A_1 D$ have negative real part: this is the first part of (45). Expressing $\begin{pmatrix} \Psi(Y(t)) \\ \Psi(Y(t - 2h_1)) \\ \Psi(Y(t - 2h_2)) \end{pmatrix}$ as a function of $\psi(y(t))$ and its delayed values, one defines now $v' \in \mathbb{R}^{14p \times n}$ in the same way than v in (51), i.e.

$$
\mathcal{J}\begin{pmatrix} 0_{27n \times p} \\ v' \end{pmatrix} = \begin{pmatrix} 0_{30n \times 5p} \\ I_{5p} \\ z_1^2 I_{5p} \\ z_2^2 I_{5p} \end{pmatrix} \begin{pmatrix} I_p \\ z_1 I_p \\ z_2 I_p \\ z_1 z_2 I_p \\ z_1^2 I_p \end{pmatrix} .
$$

This implies

$$
\mathcal{J}\mathcal{J}\begin{pmatrix} 0_{27n \times p} \\ v' \end{pmatrix} = \begin{pmatrix} 0_{45n \times 5p} \\ I_{5p} \\ z_1^2 I_{5p} \\ z_2^2 I_{5p} \end{pmatrix} \begin{pmatrix} I_p \\ z_1 I_p \\ z_2 I_p \\ z_1 z_2 I_p \\ z_1^2 I_p \end{pmatrix} = \begin{pmatrix} 0_{45n \times 15p} \\ I_{15p} \end{pmatrix} \begin{pmatrix} I_{5p} \\ z_1^2 I_{5p} \\ z_2^2 I_{5p} \end{pmatrix} \begin{pmatrix} I_p \\ z_1 I_p \\ z_2 I_p \\ z_1 z_2 I_p \\ z_1^2 I_p \end{pmatrix} .
$$

Define $w \in \mathbb{R}^{(27n+14p) \times p}$ by (compare with (25), (40))

$$
w \overset{\text{def}}{=} \begin{pmatrix} -v(sI - (A_0 + D) - z_1(A_1 - D) + z_2\overline{h}A_0 D + z_1 z_2 \overline{h}A_1 D)^{-1} B_0 \\ v' \end{pmatrix} .
$$

Using (52), (39) as in the previous page and (43) yields

$$
\mathcal{J}\mathcal{J}\begin{pmatrix} v \\ 0_{14p \times n} \end{pmatrix} (sI - (A_0 + D) - z_1(A_1 - D) + z_2\overline{h}A_0 D + z_1 z_2 \overline{h}A_1 D)^{-1} B_0
$$

$$
= \begin{pmatrix} I_{15n} \\ z_1^2 I_{15n} \\ z_2^2 I_{15n} \\ 0_{15p \times 15n} \end{pmatrix} \begin{pmatrix} I_{5n} \\ z_1^2 I_{5n} \\ z_2^2 I_{5n} \end{pmatrix} \begin{pmatrix} I_n \\ z_1 I_n \\ z_2 I_n \\ z_1 z_2 I_n \\ z_1^2 I_n \end{pmatrix}
$$

$$
(sI - (A_0 + D) - z_1(A_1 - D) + z_2\overline{h}A_0 D + z_1 z_2 \overline{h}A_1 D)^{-1} B_0
$$

$$
= \begin{pmatrix} I_{15n} \\ z_1^2 I_{15n} \\ z_2^2 I_{15n} \\ 0_{15p \times 15n} \end{pmatrix} \left[\begin{pmatrix} 1 \\ z_1^2 \\ z_2^2 \end{pmatrix} \otimes ((sI - \mathcal{A}_0 - z_1^2 \mathcal{A}_1 - z_2^2 \mathcal{A}_2)^{-1} B_0) \right] \begin{pmatrix} I_{5p} \\ z_1^2 I_{5p} \\ z_2^2 I_{5p} \end{pmatrix} \begin{pmatrix} I_p \\ z_1 I_p \\ z_2 I_p \\ z_1 z_2 I_p \\ z_1^2 I_p \end{pmatrix}
$$

$$
= \begin{pmatrix} \begin{pmatrix} I_3 \\ z_1^2 I_3 \\ z_2^2 I_3 \end{pmatrix} \otimes ((sI - \mathcal{A}_0 - z_1^2 \mathcal{A}_1 - z_2^2 \mathcal{A}_2)^{-1} B_0) \\ 0_{15p} \end{pmatrix} \begin{pmatrix} I_{5p} \\ z_1^2 I_{5p} \\ z_2^2 I_{5p} \end{pmatrix} \begin{pmatrix} I_p \\ z_1 I_p \\ z_2 I_p \\ z_1 z_2 I_p \\ z_1^2 I_p \end{pmatrix} .
$$

This shows that $J\mathcal{J}w$ is equal to

$$\left(-\begin{pmatrix} I_3 \\ z_1^2 I_3 \\ z_2^2 I_3 \end{pmatrix} \otimes ((sI - \mathcal{A}_0 - z_1^2 \mathcal{A}_1 - z_2^2 \mathcal{A}_2)^{-1} B_0) \\ I_{15p} \right) \begin{pmatrix} I_{5p} \\ z_1^2 I_{5p} \\ z_2^2 I_{5p} \end{pmatrix} \begin{pmatrix} I_p \\ z_1 I_p \\ z_2 I_p \\ z_1 z_2 I_p \\ z_1^2 I_p \end{pmatrix}.$$

Now, defining $\mathcal{S}(s, z_1, z_2)$ as in (41), right- and left-multiplication of $\mathcal{J}^T \mathcal{J} \mathcal{R} \mathcal{J} \mathcal{J}$ by w and its conjugate proves that the following expression is negative (compare with the analogous computation in the proof of Theorem 2)

$$\begin{pmatrix} I_p \\ z_1 I_p \\ z_2 I_p \\ z_1 z_2 I_p \\ z_1^2 I_p \end{pmatrix}^* \begin{pmatrix} I_{5p} \\ z_1^2 I_{5p} \\ z_2^2 I_{5p} \end{pmatrix}^* [(I_3 \otimes B_0^T \mathcal{S}^*)[(s + s^*)\mathcal{P}$$

$$+(1 - |z_1|^2)\mathcal{Q}_1 + (1 - |z_2|^2)\mathcal{Q}_2](I_3 \otimes \mathcal{S}B_0)$$

$$+I_3 \otimes (2I_{5p} + \mathcal{K}(C_0 + z_1^2 C_1 + z_2^2 C_2)\mathcal{S}B_0 + B_0^T \mathcal{S}^*(C_0^T + z_1^{2*}C_1^T + z_2^{2*}C_2^T)\mathcal{K})$$

$$+\begin{pmatrix} \eta\mathcal{K} \\ 0_{10p \times 5p} \end{pmatrix} (C_0 \ C_1 \ C_2)(I_3 \otimes s\mathcal{S}B_0)$$

$$+(I_3 \otimes s^* B_0^T \mathcal{S}^*)(C_0 \ C_1 \ C_2)^T \begin{pmatrix} \eta\mathcal{K} \\ 0_{10p \times 5p} \end{pmatrix}^T] \begin{pmatrix} I_{5p} \\ z_1^2 I_{5p} \\ z_2^2 I_{5p} \end{pmatrix} \begin{pmatrix} I_p \\ z_1 I_p \\ z_2 I_p \\ z_1 z_2 I_p \\ z_1^2 I_p \end{pmatrix}.$$

When $\operatorname{Re} s \leq 0$ and $|z_1|, |z_2| \leq 1$, this yields the negativity of

$$\begin{pmatrix} I_p \\ z_1 I_p \\ z_2 I_p \\ z_1 z_2 I_p \\ z_1^2 I_p \end{pmatrix}^* \left[\left\| \begin{pmatrix} 1 \\ z_1^2 \\ z_2^2 \end{pmatrix} \right\|^2 (2I_{5p} + \mathcal{K}(C_0 + z_1^2 C_1 + z_2^2 C_2)\mathcal{S}B_0 \right.$$

$$+B_0^T \mathcal{S}^*(C_0^T + z_1^{2*}C_1^T + z_2^{2*}C_2^T)\mathcal{K}) + s\eta\mathcal{K}(C_0 + z_1^2 C_1 + z_2^2 C_2)\mathcal{S}B_0$$

$$+ s^* B_0^T \mathcal{S}^*(C_0^T + z_1^{2*}C_1^T + z_2^{2*}C_2^T))\eta\mathcal{K} \left] \begin{pmatrix} I_p \\ z_1 I_p \\ z_2 I_p \\ z_1 z_2 I_p \\ z_1^2 I_p \end{pmatrix}. \right.$$

Using (39), (43) and (44), one then deduces that

$$
0 < \left\| \begin{pmatrix} 1 \\ z_1 \\ z_2 \\ z_1 z_2 \\ z_1^2 \end{pmatrix} \right\|^2 \left\| \begin{pmatrix} 1 \\ z_1^2 \\ z_2^2 \end{pmatrix} \right\|^2 (2I + KH(s, z_1, z_2) + [KH(s, z_1, z_2)]^*)
$$

$$
+ \left(\begin{pmatrix} I_p \\ z_1 I_p \\ z_2 I_p \\ z_1 z_2 I_p \\ z_1^2 I_p \end{pmatrix}^* \eta \begin{pmatrix} I_p \\ z_1 I_p \\ z_2 I_p \\ z_1 z_2 I_p \\ z_1^2 I_p \end{pmatrix} sKH(s, z_1, z_2) \right.
$$

$$
+ \left. \left[\begin{pmatrix} I_p \\ z_1 I_p \\ z_2 I_p \\ z_1 z_2 I_p \\ z_1^2 I_p \end{pmatrix}^* \eta \begin{pmatrix} I_p \\ z_1 I_p \\ z_2 I_p \\ z_1 z_2 I_p \\ z_1^2 I_p \end{pmatrix} sKH(s, z_1, z_2) \right]^* \right) ,
$$

where H is given by (29). Denoting $\hat{\eta} \in \mathbb{R}^{p \times p}$ the (nonnegative, diagonal) matrix

$$
\hat{\eta} \stackrel{\text{def}}{=} \begin{pmatrix} I_p \\ I_p \\ I_p \\ I_p \\ I_p \end{pmatrix}^T \eta \begin{pmatrix} I_p \\ I_p \\ I_p \\ I_p \\ I_p \end{pmatrix} ,
$$

the previous inequality implies, as in the proof of Theorem 2, the existence of $\varepsilon > 0$ such that (compare with (26)), $\forall (s, z_1, z_2) \in \mathbb{C}^3$ with $\operatorname{Re} s \leq 0$, $|z_1| = |z_2| = 1$,

$$
2I + \left(I + \frac{1}{15} \hat{\eta} s \right) KH(s, z_1, z_2) + \left[\left(I + \frac{1}{15} \hat{\eta} s \right) KH(s, z_1, z_2) \right]^* > \varepsilon I .
$$

This property is then extended to the set $\{(z_1, z_2) \in \mathbb{C}^2 : |z_1|, |z_2| \leq 1\}$ by maximum modulus principle. This provides the second part of (45).

Now, it is straightforward, from the transformation in the beginning of Sect. 3.1, to prove that (45) indeed implies (ii). This achieves the proof of the implication (i) $\Rightarrow$ (ii). The asymptotic stability is then proved as in Theorem 1.   □

3.3   Remarks on Theorems 4 and 5

Remark that the delay-free case ($h = 0$), described in Sect. 1.1, may be found taking $\overline{h} = 0$ and $D = A_1$. Also the delay-independent case ($\overline{h} = +\infty$, Theorems 1 and 2) may be found taking $D = 0$.

In order to treat the cases where the uncertainty interval on the delay is of the type $[\underline{h}, \overline{h}]$ with $\underline{h} > 0$, one may decompose the delay as $h = \underline{h} + (h - \underline{h})$. One may then approximate by a rational transfer the terms $e^{-s\underline{h}}$, and use the previous results with an uncertainty interval $[0, \overline{h} - \underline{h}]$.

One may remark that the size of the obtained LMIs, although polynomial wrt the sizes of the matrices of the problem, is already large, even for low dimension systems. However, one must not forget that the numerical verification of the properties we are interested in, and for which we have obtained here sufficient conditions, is in general NP-hard, see [7].

Of course, the techniques developed here to obtain results for systems with a unique delay, may be easily generalized to state analogue results for systems with multiple delays, at the cost of still more cumbersome formulas. As an example, to transpose the delay-dependent results obtained in Sect. 3 to systems with two independent delays h_1, h_2, the key manipulation is to adapt the transformation given in the beginning of Sect. 3.1, which permitted to consider delay-independent stability of a 2-delays systems. One writes here that

$$(sI - A_0 - A_1 e^{-sh_1} - A_2 e^{-sh_2})(I - D_1 \frac{1 - e^{-sh_1}}{s} - D_2 \frac{1 - e^{-sh_2}}{s})$$
$$= sI - (A_0 + D_1 + D_2) - (A_1 - D_1)e^{-sh_1} - (A_2 - D_2)e^{-sh_2}$$
$$+ A_0 D_1 \frac{1 - e^{-sh_1}}{s} + A_0 D_2 \frac{1 - e^{-sh_2}}{s} + A_1 D_1 e^{-sh_1} \frac{1 - e^{-sh_1}}{s}$$
$$+ A_2 D_2 e^{-sh_2} \frac{1 - e^{-sh_2}}{s} + A_1 D_2 e^{-sh_1} \frac{1 - e^{-sh_2}}{s}$$
$$+ A_2 D_1 e^{-sh_2} \frac{1 - e^{-sh_1}}{s} ,$$

and Lemma 3 permits to consider the delay-independent stability of a 4-delays auxiliary system.

At last, remark that some generalizations or refinements of the given results may be easily carried out. One may e.g. prove that the stated convergence properties are indeed uniform wrt the initial condition ϕ in bounded subset of $\mathcal{C}([-h, 0])$, wrt the nonlinearity ψ verifying sector condition (1), and wrt the initial time. Also, local stability results may be easily obtained when the sector condition holds only locally, and one may change $\psi(y)$ into $Ky - \psi(y)$, as is made in Popov theory to get nonpositive values of η [1]. Last, the case where the output of the nonlinearities is delayed too, may be considered as well.

References

1. Aizerman M.A., Gantmacher F.R. (1964) Absolute stability of regulator systems, Holden-Day Inc.

2. Bliman P.-A. (submitted, 1999) Stability of nonlinear delay systems: delay-independent small gain theorem and frequency domain interpretation of the Lyapunov-Krasovskii method
3. Bliman P.-A. (submitted, 2000) Lyapunov-Krasovskii functionals and frequency domain: delay-independent absolute stability criteria for delay systems
4. Bliman P.-A. (to appear, 2000) Lyapunov-Krasovskii method and strong delay-independent stability of linear delay systems, to appear in Proc. of IFAC Workshop on Linear Time Delay Systems, Ancona, Italy
5. Bliman P.-A., Niculescu S.-I. (submitted, 2000) A note on frequency domain interpretation of Lyapunov-Krasovskii method in control of linear delay systems
6. Boyd S., El Ghaoui L., Feron E., Balakrishnan V. (1994) Linear matrix inequalities in system and control theory, SIAM Studies in Applied Mathematics vol. 15
7. Chen J., Latchman H.A. (1995) Frequency sweeping tests for stability independent of delay, IEEE Trans. Automat. Control **40** no 9: 1640–1645
8. Gromova P.S., Pelevina A.F. (1977) Absolute stability of automatic control systems with lag, Differential Equations **13** no 8 (1978): 954-960
9. Hale J.K., Infante E.F., Tsen F.S.P. (1985) Stability in linear delay equations, J. Math. Anal. Applics **115**: 533–555
10. Hertz D., Jury E.I., Zeheb E. (1984) Stability independent and dependent of delay for delay differential systems, J. Franklin Institute **318** no 3: 143–150
11. Kamen E.W. (1982) Linear systems with commensurate time delays: stability and stabilization independent of delay, IEEE Trans. Automat. Contr. **27** no 2: 367–375
12. Kamen E.W. (1983) Correction to "Linear systems with commensurate time delays: stability and stabilization independent of delay", IEEE Trans. Automat. Contr. **28** no 2: 248–249
13. Kharitonov V.L. (1998) Robust stability analysis of time delay systems: a survey, Proc. IFAC Syst. Struct. Contr.
14. Krasovskii N.N. (1963) Stability of motion, Stanford University Press
15. Li X.-J. (1963) On the absolute stability of systems with time lags, Chinese Math. **4**: 609-626
16. Niculescu S.-I., Dion J.-M., Dugard L., Li H. (1996) Asymptotic stability sets for linear systems with commensurable delays: a matrix pencil approach, IEEE/IMACS CESA'96, Lille, France
17. Niculescu S.-I., Verriest E.I., Dugard L., Dion J.-M. (1998) Stability and robust stability of time-delay systems: a guided tour. In: Stability and control of time-delay systems, Lecture Notes in Control and Inform. Sci. 228, Springer, London: 1–71
18. Niculescu S.-I. (1999) On some frequency sweeping tests for delay-dependent stability: a model transformation case study, Proc. of 5th European Control Conference, Karlsruhe, Germany
19. Popov V.M., Halanay A. (1962) On the stability of nonlinear automatic control systems with lagging argument, Automat. Remote Control **23**: 783-786
20. Rantzer A. (1996) On the Kalman-Yakubovich-Popov lemma, Syst. Contr. Lett. **28** no 1: 7–10
21. Sandberg I.W. (1964) A frequency domain condition for stability of feedback systems containing a single time-varying nonlinear element, Bell Sys. Tech. J. Part II **43**: 1601-1608

22. Verriest E.I., Aggoune W. (1998) Stability of nonlinear differential delay systems, Mathematics and computers in Simulation **45**: 257–267
23. Zames G. (1966) On the input-output stability of nonlinear time-varying feedback systems, IEEE Trans. Automat. Contr. Part I **11** no 2: 228-238, Part II **11** no 3: 465-477

Stratification du Secteur Anormal dans la Sphère de Martinet de Petit Rayon

Bernard Bonnard[1], Emmanuel Trélat[1]

Université de Bourgogne
Dpt de Mathématiques, UFR Sciences et Techniques
BP 47870
21078 Dijon Cedex, France
{trelat, bonnard}@topolog.u-bourgogne.fr

Abstract. L'objectif de cet article est de fournir le cadre géométrique pour faire une analyse de la singularité de l'application exponentielle le long d'une direction anormale en géométrie sous-Riemannienne. Il utilise les calculs de [9], [12], et conduit dans le cas Martinet à une stratification de la singularité en secteurs Lagrangiens.

1 Introduction

Soit M une variété analytique réelle et F_1, F_2 deux champs de vecteurs analytiques. Considérons le système :

$$\frac{dq}{dt} = u_1 F_1(q) + u_2 F_2(q) \tag{1}$$

où le contrôle $u = (u_1, u_2)$ est une application L^∞ à valeurs dans un domaine C. On considère le problème du *temps minimal* avec la contrainte $u_1^2 + u_2^2 = 1$ (ou $\leqslant 1$). C'est équivalent au problème de minimiser la *longueur* d'une trajectoire de (1) pour la métrique : $l(q) = \int_0^T (u_1^2 + u_2^2)^{\frac{1}{2}} dt$ (problème *sous-Riemannien*). On considère le *système augmenté* : $\dot{q} = u_1 F_1(q) + u_2 F_2(q)$, $\dot{q}_0 = u_1^2 + u_2^2$ et on appelle *contrôle extrémal* une singularité de l'application extrémité (T fixé) associée. Elles sont paramétrées par les équations :

$$\dot{q} = \frac{\partial H_\lambda}{\partial p}, \quad \dot{p} = -\frac{\partial H_\lambda}{\partial x}, \quad \frac{\partial H_\lambda}{\partial u} = 0$$

où $H_\lambda(q, p, u) = <p, u_1 F_1(q) + u_2 F_2(q)> + \lambda(u_1^2 + u_2^2)$, $(p, \lambda) \neq (0, 0)$ et λ est une constante normalisée à 0 ou $-\frac{1}{2}$. On appelle *extrémale* une solution de ces équations et *géodésique* la projection d'une extrémale sur M. Elle est dite *stricte* si la singularité est de codimension un. On a deux types d'extrémales : les extrémales *normales* pour $\lambda = -\frac{1}{2}$, et les *anormales* pour $\lambda = 0$. Ces dernières se projettent sur les trajectoires singulières du système

(1). On note $S(0, r)$ les points à distance $r > 0$ de $q(0) = 0$ pour la métrique sous-Riemannienne, et $B(0, r)$ la boule sous-Riemannienne de rayon r.

La première étape de notre construction est de décomposer l'espace en une partie lisse et analytique et une partie non sous-analytique relativement au flot optimal.

2   La partie lisse et analytique

Les trajectoires anormales sont solutions des équations :

$$\frac{dq}{dt} = \frac{\partial H_a}{\partial p}, \quad \frac{dp}{dt} = -\frac{\partial H_a}{\partial q} \tag{2}$$

où $H_a = u_1 P_1 + u_2 P_2$, et : $u_1\{\{P_1, P_2\}, P_1\} + u_2\{\{P_1, P_2\}, P_2\} = 0$, et sont contenues dans : $P_1 = P_2 = \{P_1, P_2\} = 0$.

Notation. On identifie la condition initiale $q(0) = q_0$ à 0. Soit $t \mapsto \gamma(t)$, $t \in]-T, T[$ une trajectoire de référence. On peut supposer qu'elle est associée au contrôle $u_2 = 0$. On fait alors l'hypothèse que les conditions suivantes sont vérifiées le long de γ pour le couple (F_1, F_2) :

(H_1) $K(t) = \text{Vect } \{\text{ad}^k F_1.F_2|_\gamma \ / \ k \in \mathbb{N}\}$ (premier cône de Pontriaguine) est de codimension un pour $t \in [0, T]$ et est engendré par les $n - 1$ premiers éléments $\{F_2, \dots, \text{ad}^{n-2} F_1.F_2\}|_{\gamma(t)}$.

(H_2) $\{P_2, \{P_1, P_2\}\} \neq 0$ le long de γ.

(H_3) Si $n \geqslant 3$, pour tout $t \in [0, T]$ on suppose que :

$$F_1|_\gamma \notin \{\text{ad}^k F_1.F_2|_\gamma \ / \ k = 0 \dots n - 3\}$$

Sous ces hypothèses le vecteur adjoint $p_\gamma(0)$ est unique à un scalaire près. Identifions M localement à un voisinage U de 0 dans $\mathbb{R}^n$, et soit V un voisinage de $p_\gamma(0)$ dans l'espace projectif $P(T_0^* U)$. On peut choisir V assez petit de sorte que toutes les extrémales anormales issues de $\{0\} \times V$ vérifient les hypothèses $(H_1) - (H_2) - (H_3)$. On a :

Proposition 1. *Sous les hypothèses $(H_1) - (H_2) - (H_3)$, il existe $r > 0$ tel qu'une trajectoire anormale de longueur inférieure à r soit stricte et globalement optimale.*

On note Σ_r le secteur de U couvert par les extrémales anormales de longueur $\leqslant r$ issues de $\{0\} \times V$. La construction est visualisée sur la Fig.1.

Lemme 1 *Pour r assez petit, Σ_r forme un secteur de la boule sous-Riemannienne homéomorphe à $C \cup (-C)$, où C est un cône positif de dimension $n - 3$ pour $n \geqslant 4$ et 1 pour $n = 3$. Sa trace sur la sphère est formée de deux surfaces analytiques de dimension $n - 4$ pour $n \geqslant 4$ et réduite à deux points pour $n = 3$.*

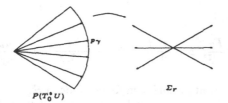

Fig. 1.

3   Le lemme de recollement

Proposition 2. *(voir [12], [28]) Soit $K(r)$ le cône de Pontriaguine évalué à l'extrémité A de la trajectoire anormale de référence, de longueur r, r assez petit. Alors $K(r)$ est un sous-espace vectoriel de codimension un et pour toute courbe lisse $\varepsilon \mapsto \alpha(\varepsilon)$, $\alpha(0) = A$, $\varepsilon \geqslant 0$ telle que :*

(i) $\alpha(\varepsilon) \subset S(0, r) \backslash \Sigma$ pour $\varepsilon \neq 0$
(ii) $\alpha(\varepsilon) \cap L(0) \neq \emptyset$ où L est le lieu de coupure

l'espace tangent à la sphère évalué en $\alpha(\varepsilon)$ tend vers $K(r)$ lorsque $\varepsilon \to 0$.

Corollaire 2 *Avec les notations de la proposition précédente, $\alpha(\varepsilon)$ n'est pas l'image d'une partie compacte du cylindre $P_1^2(0) + P_2^2(0) = \frac{1}{2}$. L'application exponentielle n'est donc pas propre le long de l'anormale.*

Une analyse plus fine utilise les estimations calculées dans le cas Martinet qui est le suivant. On considère la structure sous-Riemannienne (U, D, g) où U est un ouvert de $\mathbb{R}^3$ contenant 0, $D = \mathrm{Ker}\ \omega$, $\omega = dz - \frac{y^2}{2} dx$, $q = (x, y, z)$ et $g = a(q) dx^2 + c(q) dy^2$, a, c étant des germes de fonctions analytiques en 0 tels que $a(0) = c(0) = 1$. On note : $G_1 = \frac{\partial}{\partial x} + \frac{y^2}{2} \frac{\partial}{\partial z}$, $G_2 = \frac{\partial}{\partial y}$. Le repère orthonormé est : $F_1 = \frac{G_1}{\sqrt{a}}$, $F_2 = \frac{G_2}{\sqrt{c}}$, $F_3 = \frac{\partial}{\partial z}$, où F_1, F_2 engendrent D.

Sans nuire à la généralité de notre analyse, on peut supposer le problème *isopérimétrique*, c'est-à-dire que la métrique g ne dépend pas de z. Les géodésiques anormales se projettent dans le plan $y = 0$ et sont des droites $z = z_0$. La ligne anormale L issue de 0 et paramétrée par la longueur est $L = (\pm t, 0, 0)$. Pour $i = 1, 2, 3$ on pose : $P_i = \langle p, F_i \rangle$. Dans les coordonnées (q, P) les extrémales normales sont solutions des équations :

$$\dot{x} = \frac{P_1}{\sqrt{a}}, \ \dot{y} = \frac{P_2}{\sqrt{c}}, \ \dot{z} = \frac{y^2}{2} \frac{P_1}{\sqrt{a}}$$
$$\dot{P}_1 = \frac{P_2}{\sqrt{a}\sqrt{c}}\left(y P_3 - \frac{a_y}{2\sqrt{a}} P_1 + \frac{c_x}{2\sqrt{c}} P_2\right), \ \dot{P}_2 = -\dot{P}_1, \ \dot{P}_3 = 0$$

En paramétrant par la longueur $H_n = \frac{1}{2}(P_1^2 + P_2^2) = \frac{1}{2}$ et en introduisant les coordonnées cylindriques : $P_1 = \cos\theta$, $P_2 = \sin\theta$, $P_3 = \lambda$, on obtient les équations suivantes :

$$\dot{x} = \frac{P_1}{\sqrt{a}}, \quad \dot{y} = \frac{P_2}{\sqrt{c}}, \quad \dot{z} = \frac{y^2}{2}\frac{P_1}{\sqrt{a}}, \quad \dot{\theta} = -\frac{1}{\sqrt{a}\sqrt{c}}(\lambda y - \frac{a_y}{2\sqrt{a}}P_1 + \frac{c_x}{2\sqrt{c}}P_2)$$

Les variables x, y, z sont graduées en fonction de la règle de [7] avec les poids suivants : le poids de x, y est un et le poids de z est trois. La forme normale graduée est :

- ordre -1 : $g = dx^2 + dy^2$ (cas plat)
- ordre 0 : $g = adx^2 + cdy^2$ avec $a = (1 + \alpha y)^2$ et $c = (1 + \beta x + \gamma y)^2$

et à l'ordre 0 les équations précédentes s'écrivent :

$$\dot{x} = \frac{\cos\theta}{\sqrt{a}}, \quad \dot{y} = \frac{\sin\theta}{\sqrt{c}}, \quad \dot{z} = \frac{y^2}{2}\frac{\cos\theta}{\sqrt{a}}, \quad \dot{\theta} = \frac{1}{\sqrt{a}\sqrt{c}}(\lambda y - \alpha\cos\theta + \beta\sin\theta)$$

En introduisant le paramétrage $\sqrt{a}\sqrt{c}\frac{d}{dt} = \frac{d}{d\tau}$, les équations se projettent dans le plan (y, θ) en : $y' = \sqrt{a}\sin\theta$, $\theta' = -(\lambda y - \alpha\cos\theta + \beta\sin\theta)$, où $'$ désigne la dérivée par rapport à τ, et le système équivaut à :

$$\theta'' + \lambda\sin\theta + \alpha^2\sin\theta\cos\theta - \alpha\beta\sin^2\theta + \beta\theta'\cos\theta = 0 \qquad (3)$$

La condition initiale $q(0) = 0$ induit la contrainte $y = 0$ qui se traduit par la condition : $\theta' = \alpha\cos\theta - \beta\sin\theta$.

Pour ces calculs, le cas général est interprété comme *une déformation du cas plat*, bien que dans le cas plat la direction anormale ne soit pas stricte. En effet dans les équations précédentes, en paramétrant les solutions par $\tau\sqrt{\lambda}$, on transforme les paramètres α, β, γ de la métrique en petits paramètres. Cela revient à étudier la sphère sous-Riemannienne dans un C^0-voisinage de la trajectoire de référence.

Dans la section suivante on relève le problème sur le groupe d'Engel pour avoir une représentation uniforme.

4   Le cas Martinet relevé sur le groupe d'Engel

Si $q = (x, y, z, w)$, on considère le système de $\mathbb{R}^4$:

$$F_1 = \frac{\partial}{\partial x} + y\frac{\partial}{\partial z} + \frac{y^2}{2}\frac{\partial}{\partial w}, \quad F_2 = \frac{\partial}{\partial y}$$

On a les relations : $F_3 = [F_1, F_2] = \frac{\partial}{\partial z} + y\frac{\partial}{\partial w}$, $F_4 = [[F_1, F_2], F_1] = \frac{\partial}{\partial w}$ et $[[F_1, F_2], F_1] = 0$. Par ailleurs tous les crochets de Lie de longueur $\geqslant 4$ sont nuls. On pose :

$$L_1 = \begin{pmatrix} 0\,0\,0\,0 \\ 0\,0\,1\,0 \\ 0\,0\,0\,1 \\ 0\,0\,0\,0 \end{pmatrix}, \quad L_2 = \begin{pmatrix} 0\,1\,0\,0 \\ 0\,0\,0\,0 \\ 0\,0\,0\,0 \\ 0\,0\,0\,0 \end{pmatrix}$$

et on définit la représentation : $\rho(F_1) = L_1, \rho(F_2) = L_2$ qui permet d'identifier le système sur $\mathbb{R}^4$ au *système invariant à gauche* $\dot{R} = (u_1 L_1 + u_2 L_2)R$ sur le groupe d'Engel G_e, représenté ici par les matrices nilpotentes :

$$\begin{pmatrix} 1 & q_2 & q_3 & q_4 \\ 0 & 1 & q_1 & \frac{q_1^2}{2} \\ 0 & 0 & 1 & q_1 \\ 0 & 0 & 0 & 1 \end{pmatrix}$$

Le poids de x, y est un, le poids de z deux, et le poids de w est trois. Pour toute métrique sous-Riemannienne sur G_e, l'approximation d'ordre -1 est la métrique plate $g = dx^2 + dy^2$. Toute métrique sous-Riemannienne de Martinet s'écrit $g = adx^2 + cdy^2$ et se relève sur le groupe G_e.

Paramétrage des géodésiques dans le cas plat. Les extrémales anormales non triviales sont solutions de :

$$P_1 = P_2 = \{P_1, P_2\} = 0, \quad u_1\{\{P_1, P_2\}, P_1\} + u_2\{\{P_1, P_2\}, P_2\} = 0$$

En posant $p = (p_x, p_y, p_z, p_w)$ on obtient :

$$p_x + p_z y + p_w \frac{y^2}{2} = p_y = p_z + y p_w = p_w u_2 = 0$$

Cela implique $p_w \neq 0$ et donc $u_2 = 0$. Le flot anormal est donné par :

$$\dot{x} = u_1, \quad \dot{y} = 0, \quad \dot{z} = u_1 y, \quad \dot{w} = u_1 \frac{y^2}{2}$$

où $|u_1| = 1$ si on paramètre par la longueur.

Pour calculer les extrémales normales, on pose $P_i = \langle p, F_i(q) \rangle, i = 1, 2, 3, 4$ et $H_n = \frac{1}{2}(P_1^2 + P_2^2)$; on obtient :

$$\dot{P}_1 = P_2 P_3, \quad \dot{P}_2 = -P_1 P_3, \quad \dot{P}_3 = P_2 P_4, \quad \dot{P}_4 = 0$$

En paramétrant par la longueur $H_n = \frac{1}{2}$, on peut poser : $P_1 = \cos\theta, P_2 = \sin\theta$, et pour $\theta \neq k\pi$ il vient : $\dot{\theta} = -P_3$, $\ddot{\theta} = -P_2 P_4$. En notant $P_4 = \lambda$, cela équivaut à l'équation du pendule :

$$\ddot{\theta} + \lambda \sin\theta = 0$$

On désigne par L la ligne anormale issue de $0 : t \mapsto (\pm t, 0, 0, 0)$. *Elle n'est pas stricte* et se projette en $\theta = k\pi$.

Pour obtenir une représentation uniforme des géodésiques normales, on utilise *la fonction elliptique de Weierstrass* $\mathcal{P}$. En effet le système admet trois intégrales premières : $P_1^2 + P_2^2 = 1$, et deux Casimir : $-2P_1P_4 + P_3^2 = C$ et $P_4 = \lambda$. En utilisant $\dot{P}_1 = P_2P_3$ on obtient : $\ddot{P}_1 = -CP_1 - 3\lambda P_1^2 + \lambda$, qui équivaut pour $\dot{P}_1 \neq 0$ et $\lambda \neq 0$ à l'équation : $\dot{P}_1^2 = -\lambda(P_1^3 + \frac{C}{2\lambda}P_1^2 - P_1 - \frac{C}{2\lambda})$.
Soit $\mathcal{P}(u)$ la fonction elliptique de Weierstrass (cf [20]) solution de :

$$\mathcal{P}'(u) = -2\sqrt{(\mathcal{P}(u) - e_1)(\mathcal{P}(u) - e_2)(\mathcal{P}(u) - e_3)}$$

où les complexes e_i vérifient $e_1 + e_2 + e_3 = 0$. En posant $g_2 = -4(e_2e_3 + e_3e_1 + e_1e_2)$ et $g_3 = 4e_1e_2e_3$ on peut l'écrire : $\mathcal{P}'(u) = 4\mathcal{P}^3(u) - g_2\mathcal{P}(u) - g_3$, la fonction $\mathcal{P}(u)$ étant développable en 0 selon :

$$\mathcal{P}(u) = \frac{1}{u^2} + \frac{1}{20}q_2u^2 + \frac{1}{28}q_3u^4 + \cdots$$

La solution P_1 peut donc s'exprimer sous la forme $a\mathcal{P}(u)+b$. On peut ensuite calculer P_2 et P_3 en utilisant les intégrales premières, et x, y, z, w se calculent par quadratures. On peut retrouver les solutions oscillantes et en rotation du pendule en utilisant les fonctions elliptiques de Jacobi données par :

$$\text{cn } u = \left(\frac{\mathcal{P}(u) - e_1}{\mathcal{P}(u) - e_2}\right)^{\frac{1}{2}}, \quad \text{dn } u = \left(\frac{\mathcal{P}(u) - e_2}{\mathcal{P}(u) - e_3}\right)^{\frac{1}{2}}$$

5   Les cas Heisenberg et Martinet plat déduits du cas Engel. Eclatement en droites

On observe que les deux champs $\frac{\partial}{\partial z}$ et $\frac{\partial}{\partial w}$ commutent avec F_1 et F_2. Le cas Engel contient le cas de contact plat et le cas Martinet plat restitués par les opérations suivantes :

- En posant $p_z = 0$, on obtient les géodésiques du cas Heisenberg.
- En posant $p_w = 0$, on obtient les géodésiques du cas Martinet plat. L'interprétation est la suivante.

Lemme 3 *On obtient le cas Martinet plat (resp. Heisenberg) en minimisant la distance sous-Riemannienne par rapport à la droite (Oz) (resp. (Ow)).*

En effet la condition $p_z = 0$ (resp. $p_w = 0$) correspond alors à la *condition de transversalité*. Il est intéressant de noter que comme la distance sous-Riemannienne par rapport à une droite est plus régulière que par rapport à un point, la distance sous-Riemannienne d'Engel hérite donc au moins de toutes les singularités du cas Heisenberg et du cas Martinet plat. Une autre façon de déduire le cas Martinet plat est d'utiliser le résultat général de [7] :

Lemme 4 *Le cas Martinet plat est isométrique à $(G_{e/H}, dx^2 + dy^2)$ où H est le sous-groupe de G_e : $\{ \exp t[F_1, F_2] \ / \ t \in \mathbb{R} \}$.*

6 Stratification dans le cas Martinet

On rappelle le paramétrage explicite des géodésiques, cf [4].

Proposition 3. *Les géodésiques normales issues de 0 paramétrées par la longueur sont données par :*

- $\lambda \neq 0, u = K + t\sqrt{\lambda}$

$$x(t) = -t + \frac{2}{\sqrt{\lambda}}(E(u) - E(K)), \quad y(t) = -\frac{2k}{\sqrt{\lambda}} cn\, u$$

$$z(t) = \frac{2}{3\lambda^{\frac{3}{2}}}\left((2k^2 - 1)(E(u) - E(K)) + k'^2 t\sqrt{\lambda} + 2k^2 sn\, u\; cn\, u\; dn\, u\right)$$

où $sn\, u, cn\, u, dn\, u$ et $E(u)$ sont les fonctions elliptiques de Jacobi, et $K, E(K)$ sont les intégrales complètes.
- $\lambda = 0$

$$x(t) = t\sin\phi, \quad y(t) = t\cos\phi, \quad z(t) = \frac{t^3}{6}\sin\phi\cos^2\phi \quad où\ \phi \in]-\frac{\pi}{2}, \frac{\pi}{2}]$$

et les courbes déduites des précédentes en utilisant les symétries S_1 : $(x, y, z) \mapsto (x, -y, z)$ et $S_2 : (x, y, z) \mapsto (-x, y, -z)$. Chaque extrémale normale est minimisante jusqu'à son premier retour en $y = 0$, c'est-à-dire pour $t\sqrt{\lambda} \leqslant 2K$. Le premier temps conjugué vérifie $2K < t_{1c}\sqrt{\lambda} < 3K$, et plus précisément une simulation numérique montre que $\frac{t_{1c}\sqrt{\lambda}}{3K} \simeq 0.97$.

Conséquences. Le module k des fonctions de Jacobi est donné par : $k^2 = \frac{1-P_1(0)}{2}$. Lorsque $k \to 0$, $K(k) \to \frac{\pi}{2}$, et lorsque $k'^2 = 1 - k^2 \to 0$, $K(k) \sim \ln \frac{1}{k'}$. La trace de la sphère avec le plan de Martinet et pour $z \geqslant 0$ est représentée Fig.2 ; elle forme l'adhérence du lieu de partage. Le corollaire 2 ne s'applique pas car la géodésique anormale n'est pas stricte, néanmoins l'application exponentielle n'est pas propre. La sphère est sous-analytique en B mais pas en A. Dans ce cas dégénéré, l'application exponentielle applique tout le bord du cylindre sur la direction anormale. Lorsque $\lambda \to \infty$, les points de coupure et les points conjugués s'accumulent le long de l'anormale.

Le cas générique. Pour étudier la situation générale on utilise la forme graduée d'ordre 0 de la section 2. Elle dépend de trois paramètres α, β, γ dont le rôle géométrique est le suivant :

Lemme 5 *Pour la forme normale graduée d'ordre 0 : $g = (1 + \alpha y)^2 dx^2 + (1 + \beta x + \gamma y)^2 dy^2$, on a :*

  1. La trajectoire anormale est stricte si et seulement si $\alpha \neq 0$.

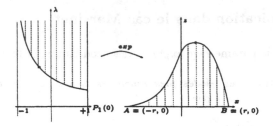

Fig. 2.

2. *Le pendule (9) est conservatif si et seulement si $\beta = 0$. Dans ce cas, le flot géodésique est intégrable par quadratures en utilisant les fonctions elliptiques de Jacobi.*

Par ailleurs les calculs de [9] conduisent à la conjecture suivante :

Conjecture. Les premiers points conjugués localisés au voisinage de l'anormale sont avant le troisième retour d'une géodésique sur le plan de Martinet.

C'est en fait un résultat qui permet de *compactifier* notre analyse car on ne considère que des trajectoires avec un nombre *uniformément borné d'oscillations* (comparer avec [1]).

Trace de la sphère avec le plan $y = 0$, au voisinage de l'anormale, si $\alpha \neq 0$ (cas strict). Elle est représentée Fig.3.

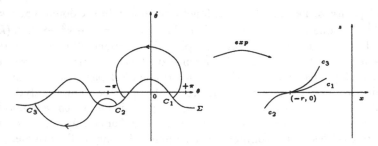

Fig. 3.

La section Σ correspond à la projection de $y = 0$ dans le plan de phase du pendule. Les trois courbes c_1, c_2, c_3 se ramifient sur la direction anormale et sont construites ainsi : la courbe c_2 est associée à des *petits déplacements* localisés au voisinage du col ; les courbes c_1, c_3 correspondent à de *grands déplacements* et il y a deux courbes car il faut considérer les trajectoires oscillantes du pendule (courbe c_1) et les trajectoires en rotation (courbe c_3). Dans le cas plat, Σ est la droite $\dot{\theta} = 0$, et seule la branche c_1 existe. Parmi les deux branches c_1 et c_3, une seule appartient à la sphère ; dans le cas de la Fig.3 c'est la courbe c_3. Les calculs de [12] montrent que le positionnement des courbes dépend de la courbure de Gauss de la métrique Riemannienne restreinte au plan (x, y), évaluée en 0. Les deux situations sont identifiées en termes de vecteurs adjoints, c'est l'équivalent de la relation $p = n\theta$ en optique, voir Fig.4.

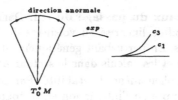

Fig. 4.

Les contacts des courbes se calculent dans la catégorie polynomiale. Les courbes c_i sont tangentes à la direction anormale car le cône de Pontriaguine $K(r)$ coïncide dans la direction anormale avec le plan (x, y). Les estimations sont les suivantes :

Proposition 4. *Posons $Z = \frac{z}{r^3}$ et $X = \frac{x+r}{2r}$, r assez petit. Alors :*

- c_1 : $Z = (\frac{1}{6} + O(r))X^3 + o(X^3)$
- c_2 : $Z = -\frac{2}{r^2\alpha^2}X^2 + o(X^2)$

On en déduit :

Proposition 5. *Dans le cas strict, la boule $B(0, r)$ de petit rayon a au voisinage de la direction anormale les propriétés suivantes :*

1. *elle est l'image par l'application exponentielle d'un secteur non compact du cylindre ($\lambda \to \infty$).*
2. *elle est homéomorphe à un secteur conique centré sur L et est en particulier simplement connexe.*
3. *elle est formée de feuilles c_2, $\bar{c}_1$ (= c_1 ou c_3) associées aux sphères $S(0, \varepsilon)$, $\varepsilon \leqslant r$ qui se recollent le long de la direction anormale, voir Fig.5.*

La transcendance du secteur. C'est un secteur non sous-analytique où les calculs nécessitent l'usage des fonctions *hyperboliques*. Dans le cas conservatif les calculs utilisent la *catégorie log-exp* et la *théorie de l'élimination* dans cette catégorie, pour les détails voir [12], [28]. Le calcul de la sphère et donc du bord du domaine est délicat car il y a un phénomène de compensation

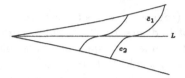

Fig. 5.

qui s'explique à l'aide des invariants micro-locaux du secteur.

Invariants micro-locaux du passage du col. Dans le cas plat, les deux valeurs propres du pendule linéarisé au voisinage de $(-\pi, 0)$ sont $(-1, +1)$ et sont donc *résonantes*. En perturbant génériquement on obtient un *spectre en bandes* $\pm 1 + O(\frac{1}{\sqrt{\lambda}})$ et les calculs dans le secteur *utilisent tout le spectre*, d'où en particulier un phénomène de stabilité. Par contre pour calculer la sphère il faut tenir compte de l'interaction entre toutes les valeurs propres du spectre pour en déduire une *moyenne* et c'est beaucoup plus complexe.

Dans nos calculs on a privilégié la coupe de $B(0, r)$ avec le plan $y = 0$ pour des raisons géométriques, mais on peut généraliser aisément.

Définition 1. On appelle 2-secteur de Martinet de la boule SR l'intersection de la boule $B(0, r)$ avec un 2-plan qui contient la direction anormale.

Les calculs montrent que les propriétés de la proposition 4 se généralisent à tout secteur de Martinet :

Proposition 6. *Tout secteur de Martinet est différentiablement représenté par la Fig. 6. Dans le cas conservatif des calculs explicites permettent d'évaluer son bord $S(0, r)$ qui est en particulier non sous-analytique.*

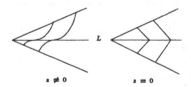

Fig. 6.

Remarque. On conjecture évidemment que la non sous-analyticité de la sphère reste vraie en général mais les calculs dans le cas non conservatif sont complexes et non standards ; en particulier il faut utiliser des cartes comme pour les calculs des cycles limites, et il y a un problème de matching.

Cut-locus. Les calculs du cas conservatif conduisent à conjecturer que la trace du lieu de coupure $L(0)$ sur la sphère, au voisinage de la direction anormale, est représentée sur la Fig.7.

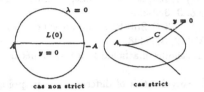

Fig. 7.

Dans le cas non strict il y a deux branches se ramifiant en A. L'extrémité C de la branche courte est un point conjugué. On peut le faire disparaître en restreignant la taille du voisinage et l'observateur voit alors deux branches de même taille. La taille de la grande branche est d'ordre r et c'est donc très différent du cas de contact.

Eclatement en droites. Les résultats descriptifs présentés ici reposent sur des calculs longs et complexes dont la difficulté principale provient de la non validité du théorème de préparation lisse au voisinage de la direction anormale. C'est une difficulté d'ordre technique. Une bonne compréhension géométrique du problème peut reposer simplement sur l'idée d'éclater en droites, en plans, etc, l'origine, comme en section 5, et d'étudier la distance sous-Riemannienne par rapport à ces objets, en utilisant la condition de transversalité. Cette idée guide d'ailleurs la représentation des variétés Lagrangiennes par des familles génératrices.

L'analyse micro-locale traduit le fait que dans la sphère sous-Riemannienne toutes les directions ne sont pas identiques et qu'il faut calculer par secteurs : secteur Riemannien, secteur de contact, secteur de Martinet... Dans chaque secteur les calculs sont différents. Par exemple dans le secteur de Martinet il faut utiliser les fonctions hyperboliques. On conçoit donc qu'en général une paramétrisation analytique de la sphère est impossible. On conjecture néanmoins l'existence d'une classe de Gevrey uniforme pour chaque problème sous-Riemannien et donc la possibilité de résoudre l'équation d'Hamilton-Jacobi-Bellman dans cette catégorie.

References

1. Agrachev A. (1999) Compactness for sub-Riemannian length minimizers and subanalyticity, Report SISSA, Trieste.

2. Agrachev A., Sarychev A. V. (1995) Strong minimality of abnormal geodesics for 2-distributions, Journal of Dynamical and Control Systems, Vol. 1, No. 2, 139-176.

3. Agrachev A., Gamkrelidze R. V. (1997) Feedback invariant control theory and differential geometry I, Regular extremals, Journal of Dynamical and Control Systems, Vol. 3, No. 3, 343-390.

4. Agrachev A. and al. (1997) Sub-Riemannian spheres in the Martinet flat case, ESAIM/COCV, Vol. 2, 377-448.

5. Arnold V. (1976) Méthodes mathématiques de la mécanique classique, Eds Mir, Moscou.

6. Arnold V. and al., Singularities of differentiable mappings, Eds Mir, Moscou.

7. Bellaïche A. (1996) Tangent space in sub-Riemannian geometry, Sub-Riemannian Geometry, Birkhäuser.

8. Bliss G. A. (1946) Lectures on the calculus of variations, U. of Chicago Press, Chicago.

9. Bonnard B., Chyba M. (1999) Méthodes géométriques et analytiques pour étudier l'application exponentielle, la sphère et le front d'onde en géométrie sous-Riemannienne de Martinet, ESAIM/COCV, Vol. 4, 245-334.

10. Bonnard B., Kupka I. (1993) Théorie des singularités de l'application entrée/sortie et optimalité des trajectoires singulières dans le problème du temps minimal, Forum Math. 5, 111-159.

11. Bonnard B., de Morant J. (1995) Towards a geometric theory in the time-minimal control of chemical batch reactors, SIAM Journal on Control and Optimization, Vol. 33, No. 5, 1279-1311.

12. Bonnard B., Trélat E. (1999) Role of abnormal minimizers in sub-Riemannian geometry, PrePrint Dijon.

13. Gromov M. (1996) Carnot-Carathéodory spaces seen from within, Sub-Riemannian Geometry, Birkhäuser.

14. Guillemin V., Sternberg S. (1984) Symplectic techniques in physics, Cambridge University Press.

15. Hrmander L. (1983) The analysis of linear partial differential operators, Springer-Verlag, New-York.

16. Jean F. (1999) Entropy and complexity of a path in sub-Riemannian geometry, rapport ENSTA.

17. Kerkovian J., Cole J. D. (1981) Perturbation methods in applied mathematics, Springer-Verlag.

18. Kupka I. (1992) Abnormal extremals, Preprint.

19. Kupka I. (1996) Géométrie sous-Riemannienne, Séminaire Bourbaki, Paris.

20. Lawden D. F. (1989) Elliptic functions and applications, Springer-Verlag, New-York.

21. Liu W. S., Sussmann H. J. (1995) Shortest paths for sub-Riemannian metrics of rank two distributions, Memoirs AMS, N564, Vol. 118.

22. Mischenko A. S. and al. (1990) Lagrangian manifolds and the Maslov operator, Springer-Verlag.

23. Moyer H. G. (1973) Sufficient conditions for a strong minimum in singular problems, SIAM Journal on Control and Optimization, 11, 620-636.

24. Naimark M. A. (1967) Linear differential operators, Frederick U. Pub. Co.

25. Nikiforov A., Ouranov V. (1982) Fontions spéciales de la physique mathématique, Eds Mir.

26. Ramis J. P., Séries divergentes et théorie asymptotique, Mémoires de la SMF.

27. Roussarie R. (1968) Bifurcations of planar vector fields and Hilbert's 16th problem, Birkhäuser, Berlin.

28. Trélat E. (2000) Some properties of the value function and its level sets for affine control systems with quadratic cost, to appear in Journal of Dynamical and Control Systems.

29. Treves F., Symplectic geometry and analytic hypo-ellipticity, Preprint.

Toward a Nonequilibrium Theory for Nonlinear Control Systems *

Christopher I. Byrnes

Department of Systems Science and Mathematics
Washington University,
St. Louis, MO 63130, USA
ChrisByrnes@seas.wustl.edu

Abstract. This paper is concerned with the development of basic concepts and constructs for a nonequilibrium theory of nonlinear control. Motivated by an example of nonstabilizability of rigid spacecraft about an equilibrium (reference attitude) but stabilizability about a revolute motion, we review recent work on the structure of those compact attractors which are Lyapunov stable. These results are illustrated and refined in a description of the asymptotic behavior of practically stabilizable systems taken form a recent work on bifurcations of the system zero dynamics. These attractors can contain periodic orbits, and necessary and sufficient condition for the existence of periodic orbits are discussed. These conditions lead to the notion of a "control one-form" and to necessary conditions for the existence of an orbitally stable periodic motion. As it turns out, even when this latter result is specialized to the equilibrium case, the criterion is new.

1 Introduction

In 1983 Brockett showed that the origin of the nonlinear nonholonomic integrator

$$\begin{cases} \dot{x}_1 = u_1 \\ \dot{x}_2 = u_2 \\ \dot{x}_3 = x_1 u_2 - x_2 u_1 \end{cases}$$

is not asymptotically stabilizable using a C^1 state or dynamic feedback law because

$$\begin{cases} u_1 = y_1 \\ u_2 = y_2 \\ x_1 u_2 - x_2 u_1 = y_3 \end{cases}$$

is not solvable for all $y = (y_1, y_2, y_3)^T$ sufficiently small. More explicitly, he proved:

Theorem 1 (Brockett [3]). *A necessary condition for the origin to be a locally asymptotically stable equilibrium of the C^1 autonomous system $\dot{x} = F(x)$ is that the map F is locally onto, i.e. the equation $F(x) = y$ is locally solvable.*

* Research Supported in Part by a Grant from AFOSR, Boeing and NSF

As a corollary he also deduced a basic result about stabilizability of $\dot{x} = f(x, u)$; viz. that $f(x.u(x)) = y$ be solvable for all sufficiently small y. We will return to this result, and several of its generalizations, at various points in this paper.

One of the interpretations of this result is that controllability, in the nonlinear sense, does not imply stabilizability of an equilibrium. Several authors (cf. [13], [12]) have pursued the philosophy that controllability, in an appropriate sense, in fact does imply asymptotic stabilizability of an equilibrium, provided more general feedback laws, e.g., time-varying or discontinuous, are allowed. In contrast, we begin this paper by considering whether stabilization, with continuously differentiable feedback, about a more general invariant set would be possible.

As an example, following [7], consider the attitude stabilization problem for a system modeled by the equations of motion of a rigid spacecraft with two actuators consisting of opposing pairs of gas jets. More specifically, consider the rigid body model of a satellite controlled by momentum exchange devices, such as momentum wheels or gas jet actuators. On the state manifold $M = SO(3) \times \mathbb{R}^3$, the evolution of an orientation, angular velocity pair (R, ω) takes the form

$$J\dot{\omega} = S(\omega)J\omega + \sum_{i=1}^{m} b_i u_i \tag{1}$$

$$\dot{R} = S(\omega)R \tag{2}$$

where J is the inertia matrix and $S(\omega)$ is the matrix representation of the cross-product, $b \to b \times \omega$; i.e.

$$S(\omega) = \begin{bmatrix} 0 & \omega_3 & -\omega_2 \\ -\omega_3 & 0 & \omega_1 \\ \omega_2 & -\omega_1 & 0 \end{bmatrix}. \tag{3}$$

Choosing principle axes (i.e. diagonalizing J), (1)-(3) can be expressed in local coordinates about a reference frame $R = [r_1, r_2, r_3]$ using Euler angles, φ, θ, ψ representing rotations about the r_1, r_2, r_3 axes, respectively. Explicitly (1)-(2) takes the form

$$\begin{bmatrix} \dot{\omega}_1 \\ \dot{\omega}_2 \\ \dot{\omega}_3 \end{bmatrix} = \begin{bmatrix} a_1 \omega_2 \omega_3 \\ a_2 \omega_1 \omega_3 \\ a_3 \omega_1 \omega_2 \end{bmatrix} + \sum_{i=1}^{m} b_i u_i, \quad b_i \in \mathbb{R}^3 \tag{4}$$

$$\begin{bmatrix} \dot{\varphi} \\ \dot{\theta} \\ \dot{\psi} \end{bmatrix} = \begin{bmatrix} \cos(\theta) & 0 & \sin(\theta) \\ \sin(\theta)\tan(\varphi) & 1 & -\cos(\theta)\tan(\varphi) \\ -\sin(\theta)\sec(\varphi) & 0 & \cos(\theta)\sec(\varphi) \end{bmatrix} \begin{bmatrix} \omega_1 \\ \omega_2 \\ \omega_3 \end{bmatrix} \tag{5}$$

Now suppose $m = 1$ or $m = 2$, so that the spacecraft is underactuated. Let $y = \begin{pmatrix} y_1 \\ y_2 \end{pmatrix}$, $y_i \in \mathbb{R}^3$ and consider solving the equation

$$\begin{bmatrix} a_1\omega_2\omega_3 \\ a_2\omega_1\omega_3 \\ a_3\omega_1\omega_2 \end{bmatrix} + \sum_{i=1}^{m} b_i u_i = y_1$$

$$\begin{bmatrix} \cos(\theta) & 0 & \sin(\theta) \\ \sin(\theta)\tan(\varphi) & 1 & -\cos(\theta)\tan(\varphi) \\ -\sin(\theta)\sec(\varphi) & 0 & \cos(\theta)\sec(\varphi) \end{bmatrix} \begin{bmatrix} \omega_1 \\ \omega_2 \\ \omega_3 \end{bmatrix} = y_2$$

with $y_2 = 0$. Since the latter matrix is invertible, $\omega_i = 0$, for $i = 1, 2, 3$. Therefore

$$\sum_{i=1}^{m} b_i u_i = y_1$$

which can only be solved for arbitrary y_1, $\|y_1\| \ll \infty$, dim span$\{b_i\} = 3$ and hence $m \geq 3$. In particular, the attitude of the underactuated satellite cannot be asymptotically stabilized.

The main results of [7] are actually stated for a broad class of nonlinear control systems, including underactuated mechanical systems. Indeed, several results on the equivalence of feedback stabilizability, feedback linearization and nonunderactuation are given which foretell more recent work on underactuated mechanical systems. In [7]. it is also shown that while stabilization about the equilibrium attitude is not possible, one can stabilize about a revolute motion. More explicitly, if two actuators are aligned with the first two principal axes, then it is shown that one can design a smooth feedback law which produces an orbitally stable motion about the third principal axis.

Motivated by this example, we ask whether, for example, the nonholonomic integrator would be stabilizable about a periodic orbit or some other compact attractor. Alternatively, are such controllable but not smoothly stabilizable systems practically stabilizable by state feedback $u_i = u_i(x, k)$? We recall that a system is said to be practically stabilizable if there exists a tunable feedback law such that for all initial conditions in any fixed compact set, the feedback gain can help to steer the trajectory to an arbitrary small neighborhood of the origin. A particularly important example of such a tunable feedback law is high gain output feedback, for a suitable choice of output function (see for example Teel and Praly [23] and Bacciotti [1]). These two questions are, of course, related by question which has remained open until the recent paper [9]:

Question 1. What are the asymptotic dynamics of a practically stabilized system inside the small ball?

In this paper, we will review recent results on the nature of Lyapunov stable attractors in $\mathbb{R}^n$, the asymptotic dynamics of practically stabilizable systems and on the geometry of periodic phenomena. It is a pleasure to thank Roger Brockett, David Gilliam, Alberto Isidori, Anders Lindquist, Clyde Martin, James Ramsey, Victor Shubov and V. Sundarapandian for their influence on my way of thinking about nonequilibrium nonlinear control theory.

2   Preliminaries on Compact Attractors

Our first result about the existence of compact attractors extends Brockett's criteria to the case of compact attractors [10].

Theorem 2 (Byrnes, Isidori, Martin). *A necessary condition for global asymptotic stabilization about a compact attractor $A \subset \mathbb{R}^n$ for a system*

$$\dot{x} = f(x, u) \qquad f \in C^1$$

by C^1 state feedback is the solvability of the equation

$$f(x, u) = y$$

for all $\|y\| \ll \infty$, for some x with $dist(x, A) \ll \infty$.

Example 1. The Brockett nonholonomic integrator is not practically stabilizable.

Taking $y = 0$ in Theorem 2 leads to a remarkable consequence.

Corollary 1. *If $A \subset \mathbb{R}^n$ is a globally asymptotically stable compact attractor for a smooth dynamical system*

$$\dot{x} = f(x), \quad x \in \mathbb{R}^n$$

then A contains an equilibrium x_0.

These results follow from

Theorem 3 (Byrnes, Isidori, Martin). *If $A \subset \mathbb{R}^n$ is a globally asymptotically stable compact attractor for a smooth dynamical system*

$$\dot{x} = f(x)$$

then

(i) A contains either an odd or an infinite number of equilibria;
(ii) If $x_0 \in A$ is an equilibrium, A is contractible to x_0.

Proof. It is well-known [25] that there exists a Lyapunov function V for $\mathcal{A}$ and $\dot{x} = f(x)$.

$$L_f V < 0 \quad \text{on} \quad \mathbb{R}^n \backslash \mathcal{A}$$

$$L_f V = 0 \quad \text{on} \quad \mathbb{R}^n \backslash \mathcal{A}$$

Set $M_c = \{x : V(x) \le c\}$. For $c > 0$, M_c is a compact n-manifold with boundary. The boundary $N_c \equiv \partial M_c$ is a compact, smooth hypersurface. For $\|y\| \ll \infty$, $F_y = -f(x) + y$ is a smooth outward normal on N_c. Define

$$\text{Ind}(F_y) = \sum_{F_y(x_0)=0} i_{x_0}(F_y)$$

where

$$i_{x_0}(F_y) = \deg\left(\frac{F_y}{\|F_y\|}\right),$$

and where

$$\frac{F_y}{\|F_y\|} : N_c \to S^{n-1},$$

is the Gauss map.

By the Poincaré-Hopf Theorem, $\text{Ind}(F_y) = \chi(M_c)$.

Claim. M_c is contractible.

Therefore $\text{Ind}(F_y) = 1$ and, in particular, F_y has either an odd number of equilibria, or an infinite number. Moreover, Wilson [25] showed that $\mathcal{A} \hookrightarrow M_c$ is a homotopy equivalence and so $\mathcal{A}$ is contractible.

Remark 1. The proof of contractibility uses Lyapunov stability and attractivity of $\mathcal{A}$ to show that any map

$$T : S^p \to M_c$$

is homotopic to a constant map (d'après the proof of Brockett's Theorem given by Byrnes-Isidori, [7])

Remark 2. The degree of the Gauss map is also known as the "curvatura integra." As in the Gauss-Bonnet Theorem, the index of the vector field can also be computed as the curvatura integra of the normalized outward normal on N_c (see Milnor [19]), leading to the more familiar formulations of the proofs of Brockett's Theorem for equilibria, based on degree theory and homotopy methods.

Remark 3. The Poincaré-Hopf Theorem also holds for continuous vector fields, so that one should certainly expect similar results in the absence of smoothness of the system, as in [26,27]. Our principal interest has been in introducing the basic ideas and results.

3   Compact Attractors and Practical Stabilizability

Consider a system

$$\begin{cases} \dot{z} = f(z) + p(z, y) \\ \dot{y} = q(z, y) + u, \end{cases}$$

in which $z \in \mathbb{R}^{n-1}$, $u \in \mathbb{R}$, $y \in \mathbb{R}$ and

$$\begin{cases} f(0) = 0 \\ p(z, 0) = 0 \\ q(0, 0) = 0. \end{cases}$$

Suppose that the *zero dynamics* $\dot{z} = f(z)$ has a globally asymptotically, but not possibly locally exponentially, stable equilibrium at $z = 0$.

Then, it is well-known that this system can be "semiglobally practically" stabilized by "high-gain" negative output feedback

$$u = -ky.$$

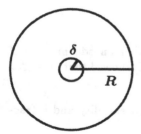

Fig. 1. The large and small balls in practical stabilizability

Remark 4. One can in fact replace the small and large balls in this construction by the sublevel sets of a Lyapunov function for the closed-loop system, so that each of the sublevel sets are positively invariant. This renders the smaller sublevel set a Lyapunov stable, compact attractor for the larger sublevel set, whose interior can be shown to be diffeomorphic to $\mathbb{R}^n$ [7]. In particular, the results of the previous section apply to the smaller sublevel set. This will be manifested, for example, in the odd number (counting multiplicities) of equilibria contained in the attractors.

Example 2. Consider $\begin{cases} \dot{z} = -z^3 + y \\ \dot{y} = -z + u \end{cases}$ This system is passive (positive real), zero-state detectable, and therefore $u = -ky$ asymptotically stabilizes the origin for any $k > 0$.

On the other hand, it has long been known that there are minimum phase systems with relative degree one for which output feedback renders the system unstable for every choice of the gain (see [6] for an example, as well as an early description of what is now known as "backstepping."). We now begin a study of what can happen in such an unstable case.

Example 3. Consider the system

$$\begin{cases} \dot{z} = -z^3 + y \\ \dot{y} = z - ky \, . \end{cases}$$

has equilibria defined via

$$z = ky, \quad -z^3 + \frac{1}{k}z.$$

$$(z, y) = (0, 0), \quad \left(k^{-1/2}, k^{-3/2}\right), \quad \left(-k^{-1/2}, -k^{-3/2}\right).$$

In fact, the origin is an unstable hyperbolic point, while the other two equilibria are hyperbolic stable points. Moreover, the set of initial conditions which generate a bounded trajectory forward and backward in time is the maximal compact invariant set, which is an attractor for bounded half-trajectories (see [15]). In this case this attractor coincides with the origin, its 1-dimensional unstable manifold and the two asymptotically stable equilibria. It follows from the general results we are about to state that this attractor is Lyapunov stable for $k \gg 0$.

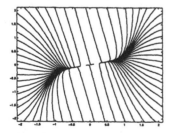

Fig. 2. The Lyapunov stable compact attractor with three equilibria

Example 4. Consider the system

$$\begin{cases} \dot{z}_1 = z_2 \\ \dot{z}_2 = -z_1 e^{z_1 z_2} + y \\ \dot{y} = z_2 - ky \, , \end{cases}$$

resulting from output feedback applied to a globally minimum phase system with unity high frequency gain. Indeed, $(z_1, z_2) = (0,0)$ is globally asymptotically stable for the zero dynamics,

$$\begin{cases} \dot{z}_1 = z_2 \\ \dot{z}_2 = -z_1 e^{z_1 z_2} \end{cases}.$$

For $k \gg 0$ the origin is unstable but there is an asymptotically stable limit cycle.

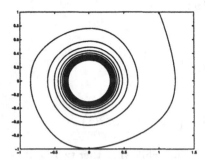

Fig. 3. Compact attractors: $k = 50$

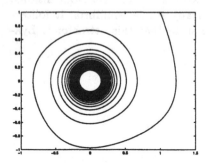

Fig. 4. Compact attractors: $k = 500$

In its linear approximation the system has two eigenvalues with positive real part for any large positive k. As a matter of fact, for any $\delta > 0$, there is $k^* > 0$ such that, if $k > k^*$, the system has an unstable equilibrium at the origin and a stable limit cycle entirely contained in B_δ

These examples illustrate the asymptotic behavior of high gain closed loop system in two cases in which the zero dynamics are *critically stable* in the mildest possible way, in a sense that we will now make precise.

Case A. The linearization of the zero dynamics at $z = 0$ has a simple zero eigenvalue and all remaining spectra is in $\mathbb{C}^-$. Moreover, the (1-dimensional) restriction of the zero dynamics to its center manifold at $z = 0$ can be expanded as

$$\dot{x} = ax^3 + \mathcal{O}(|x|^4) , \qquad a < 0 .$$

Case B. The linearization of the zero dynamics at $z = 0$ has two purely imaginary eigenvalues and all remaining spectra is in $\mathbb{C}^-$. Moreover, the (2-dimensional) restriction of the zero dynamics to its center manifold at $z = 0$ can be expanded, in polar coordinates, as (see e.g. [24, page 271])

$$\dot{r} = ar^3 + \mathcal{O}(|r|^5) , \qquad a < 0$$
$$\dot{\theta} = \omega + br^2 + \mathcal{O}(|r|^4) .$$

Case A corresponds to the situation in which the zero dynamics has a one-dimensional center manifold, corresponding to the linearization of the open loop system having a zero at 0. The key issue is, therefore, to determine whether for large k the real pole of the linearization of the closed loop system which approaches the zero at 0 of the linearization of the open loop system:

approaches this zero from the left,

approaches this zero from the right, or

cancels this zero.

Case B corresponds to the situation in which the zero dynamics has a two-dimensional center manifold, corresponding to the linearization of the open loop system having a pair of imaginary eigenvalues $\pm j\omega$. For large k, two complex-conjugate eigenvalues of the linearization of the closed loop system approach the zeros at $\pm j\omega$ of the linearization of the open loop system from the left or from the right and, again, it is key to determine which case actually occurs.

By $g(s) = n(s)/d(s)$ we denote the transfer function of the linearization of the open-loop system. In the defining equation of the root-locus, we set $k = 1/\epsilon$ obtaining $f(s, \epsilon) = \epsilon d(s) + n(s) = 0$ and consider a branch $s(\epsilon)$ for which

$$\text{Re } s(0) = 0$$

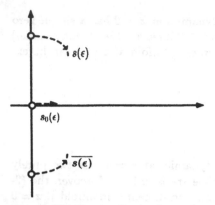

Fig. 5. The "reverse" root locus departing from imaginary open-loop zeros as k decreases from $+\infty$

If the zero is simple, implicitly differentiating $f(s, \epsilon)$ and setting ϵ equal to 0 will yield an expression for the tangent to the local branch at $\epsilon = 0$.

$$-\frac{\partial s}{\partial \epsilon}\Big|_{\epsilon=0} \cdot n'(s^*) = d(s^*)$$

Example 5. Suppose, for illustration, that $s^* = 0$ so that we are considering Case A. Substitution yields

$$d(0) \quad = \det(-J_0) = (-1)^n \det(J_0)$$

$$\text{sign}(n'(0)) = +1$$

In particular,

$$(-1)^n \det(J_0) < 0 \iff \frac{\partial s}{\partial \epsilon} > 0.$$

Theorem 4 (Byrnes, Isidori). *Consider a globally minimum phase system with positive high frequency gain $L_g h$, which has been practically stabilized by $u = -ky$. Suppose the linearization of the zero dynamics has a simple eigenvalue at 0, and all remaining spectra is in $\mathbb{C}^-$. Let J_{OL} denote the Jacobian of the open-loop system.*

1. *If $(-1)^n \det(J_{OL}) > 0$, then the origin is locally exponentially stable and there exists a k^* such that for $k > k^*$ all trajectories initialized in $B(0, R)$ tend to 0 as $t \to \infty$ (see Figure 1).*

2. *If $(-1)^n \det(J_{OL}) < 0$, there exists a k^* such that for $k > k^*$, there exists a compact, Lyapunov stable attractor $\mathcal{A}$. $\mathcal{A}$ is a 1-manifold with boundary consisting of 2 asymptotically stable equilibria. The (relative) interior of $\mathcal{A}$ consists of the origin and its one-dimensional unstable manifold.*

Remark 5. In [9] it is shown that if $\det(J) = 0$, for any $R > 0$ and $r > 0$ there exists a k^* such that for $k > k^*$ there exists a compact Lyapunov stable invariant set $\mathcal{A} \subset B_r(0)$ and all trajectories initialized in $B_R(0)$ tend to $\mathcal{A}$ as $t \to \infty$. The attractor $\mathcal{A}$ is either the origin, a critically asymptotically stable equilibrium of multiplicity three, or a 1-manifold with a boundary consisting of the origin (which is a critically unstable equilibrium of multiplicity two) and a locally exponentially stable equilibrium. It is also shown that these cases can be distinguished by the signature of a quadratic form defined on the mapping germ of the closed-loop vector field.

This analysis of the Lyapunov stability and the structure of nontrivial attractors can be motivated by an analysis of the root-locus plots of three general classes of systems. To the first class of systems, which were the objects of much research during the 1980's, belong those systems for which an appropriately defined zero dynamics exists and is exponentially stable. In particular, these systems have relative degree one and the transmission zeros of the linearization lie in the open left-half plane, as depicted below.

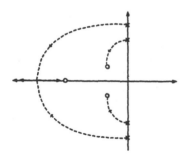

Fig. 6. The root locus plot for the linearization of an exponentially minimum phase system

These are the classes of systems for which the origin can be exponentially stabilized using high gain feedback and for which singular perturbations apply [5,17]. In the late 1980's it was clear that high gain feedback together with input-output linearizing feedback could semiglobally stabilize those systems with critically asymptotically stable zero dynamics [6,8].

By the 1990's it became widely appreciated that these classes of systems were feedback equivalent to a passive system and that passive systems themselves provided an underlying stability mechanism that was strong enough to allow a linearization with open-loop zeros and poles in the closed left-half plane (see [20,21,11] following earlier pioneering work on passivity and dissipativity in [28,16]). Of course, for passive (or positive real) systems the closed-loop poles will always tend to the imaginary axis open-loop zeros along an arc contained in the open left-half plane as depicted in Figure 7.

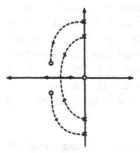

Fig. 7. The root locus plot for the linearization of a passive system

This stability property is also reflected in the inequality $(-1)^n \det(J_{OL}) > 0$. When the reverse inequality holds, one has a class of systems for which the closed-loop poles can tend to the imaginary axis open-loop zeros along an arc contained in the open right-half plane as depicted below.

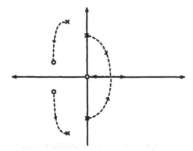

Fig. 8. The root locus plot for the linearization of a passive system undergoing a bifurcation

This class of systems is precisely the class of feedback systems undergoing a nontrivial bifurcation, which in the case discussed above is a pitch-fork bifurcation of the zero dynamics. Our next result discuss the case of a Hopf bifurcation of the nonlinear zero dynamics.

Theorem 5 (Byrnes, Isidori). *Consider case B, for the feedback law $u = -ky$, with $k > 0$ a globally minimum phase system with $L_g h > 0$. Then:*

1. *If $Re\left(d(j\omega)/n'(j\omega)\right) > 0$, then the origin is semiglobally stabilizable via $u = -ky$.*
2. *If $Re\left(d(j\omega)/n'(j\omega)\right) < 0$, then for $k > k^* > 0$ there is a compact, Lyapunov stable attractor $\mathcal{A}$. $\mathcal{A}$ is a smooth 2-manifold with boundary consisting of a periodic orbit which bounds a smooth open surface. The open surface consists of the origin and its 2-dimensional unstable manifold.*

4   The Bifurcation of the Zero Dynamics

Consider the system in normal form

$$\begin{cases} \dot{z}_1 = A_{11}z_1 + g_1(z_1, z_2) + p_1(z_1, z_2, y) \\ \dot{z}_2 = Sz_2 + g_2(z_1, z_2) + p_2(z_1, z_2, y) \\ \dot{y} = q(z_1, z_2, y) + u\,, \end{cases} \tag{6}$$

in which the matrix A_{11} is Hurwitz, the matrix S is skew symmetric, $g_i(z_1, z_2)$ vanishes together with its first partial derivatives, and $p_i(z_1, z_2, 0) = 0$, for $i = 1, 2$.

In system (6), we introduce a "high-gain" output feedback control $u = -ky$ and set $k = 1/\epsilon$, where $\epsilon > 0$ is a small number. The resulting system is a *singularly* perturbed system that can be studied via standard center-manifold methods, as in Fenichel [14].

We can add the extra equation $\dot{\epsilon} = 0$ and rescale time as $\tau = t/\epsilon$ to obtain the standard form (here " $'$ " denotes derivative with respect to τ)

$$\begin{cases} \epsilon' = 0 \\ z_1' = \epsilon A_{11}z_1 + \epsilon g_1(z_1, z_2) + \epsilon p_1(z_1, z_2, y) \\ z_2' = \epsilon Sz_2 + \epsilon g_2(z_1, z_2) + \epsilon p_2(z_1, z_2, y) \\ y' = -y + \epsilon q(z_1, z_2, y)\,. \end{cases} \tag{7}$$

The system (7) possesses an n-dimensional center manifold at $(z_1, z_2, y, \epsilon) = 0$, the graph of a mapping $y = \pi(z_1, z_2, \epsilon)$ satisfying

$$\begin{aligned} \frac{\partial \pi}{\partial z_1}&\left[\epsilon A_{11}z_1 + \epsilon g_1(z_1, z_2) + \epsilon p_1(z_1, z_2, \pi(z_1, z_2, \epsilon))\right] \\ &+ \frac{\partial \pi}{\partial z_2}\left[\epsilon Sz_2 + \epsilon g_2(z_1, z_2) + \epsilon p_2(z_1, z_2, \pi(z_1, z_2, \epsilon))\right] \\ &= -\pi(z_1, z_2, \epsilon) + \epsilon q(z_1, z_2, \pi(z_1, z_2, \epsilon)) \end{aligned} \tag{8}$$

and

$$\pi(0, 0, 0) = 0\,, \left[\frac{\partial \pi}{\partial z_1}\right]_{(0,0,0)} = 0\,, \left[\frac{\partial \pi}{\partial z_2}\right]_{(0,0,0)} = 0\,, \left[\frac{\partial \pi}{\partial \epsilon}\right]_{(0,0,0)} = 0\,. \tag{9}$$

Clearly $\pi(z_1, z_2, 0) = 0$. Moreover, since for any ϵ the point

$$(z_1, z_2, y, \epsilon) = (0, 0, 0, \epsilon)$$

is an equilibrium, we can conclude that $\pi(0, 0, \epsilon) = 0$.

After using a stable manifold argument, we can focus on the associated "reduced system", with unscaled time

$$\dot{z}_1 = A_{11} z_1 + g_1(z_1, z_2) + p_1(z_1, z_2, \pi(z_1, z_2, \epsilon))$$
$$\dot{z}_2 = S z_2 + g_2(z_1, z_2) + p_2(z_1, z_2, \pi(z_1, z_2, \epsilon)) \ .$$

We call this system the perturbed zero dynamics and note that, since $\pi(z_1, z_2, 0) = 0$ and $p_i(z_1, z_2, 0) = 0$ for $i = 1, 2$, for $\epsilon = 0$, this yields the zero dynamics of the original system. The system, augmented with $\dot{\epsilon} = 0$, is a regularly perturbed system and possesses a center manifold at $(z_1, z_2, \epsilon) = 0$, the graph of a mapping $z_1 = \gamma(z_2, \epsilon)$. Set $\sigma(z_2, \epsilon) = \pi(\gamma(z_2, \epsilon), z_2, \epsilon)$.
Moreover

$$\left[\frac{\partial \sigma^2}{\partial \epsilon \partial z_2} \right]_{(0,0)} = \left[\frac{\partial \pi^2}{\partial \epsilon \partial z_2} \right]_{(0,0,0)} \ .$$

The associated reduced system is

$$\dot{z}_2 = S z_2 + g_2(\gamma(z_2, \epsilon), z_2) + p_2(\gamma(z_2, \epsilon), z_2, \sigma(z_2, \epsilon)).$$

Cases A and B result in the two special cases $\dim(z_2) = 1$ and $S = 0$, $\dim(z_2) = 2$ and S has two nonzero purely imaginary eigenvalues. At $(z_2, \epsilon) = (0, 0)$, we then have either a pitchfork bifurcation or a Hopf bifurcation. The Lyapunov stability of both attractors follows from stable manifold theory and an extension of the reduction theorem of center manifold theory for equilibria to the case of a compact attractor [9]

5   Existence of Periodic Phenomena

In previous sections we exhibited the existence of of periodic orbits in high gain feedback systems. In this section we are interested in the existence and the geometry of periodic phenomena. For planar systems, the Poincaré-Bendixson theory gives a very complete picture. In particular, if a vector field leaves an annulus positively invariant but has no equilibria within the annulus, then the vector field possess a periodic orbit in the annulus. In [2], G. D. Birkhoff describes, without proof, a generalization of this situation. If there exists a solid n-torus M in $\mathbb{R}^n$, which is positively invariant under a vector field f, Birkhoff asserted the existence of a periodic orbit $\gamma \subset M$ provided there exists functions a_i such that

(i) $\dfrac{\partial a_i}{\partial x_j} = \dfrac{\partial a_j}{\partial x_i}, \quad i, j = 1, \cdots, n,$

(ii) $\displaystyle\sum_{i=1}^{n} a_i f_i > 0$

Indeed defining the one-form $\omega = \displaystyle\sum_{i=1}^{n} a_i dx_i$, (i) asserts that $d\omega = 0$ and (ii) asserts that the contraction $\langle f, \omega \rangle > 0$ of f with ω is positive.

In [4] a generalization of this result yields a set of necessary and sufficient conditions for the existence of a periodic orbit, a result with several corollaries. To appreciate the necessity of this kind of geometric situation, we begin with a description of the behavior of solutions of a differential equation near a periodic orbit (d'aprés Lewis, Markus, Petrowski, Sternberg). Consider a system

$$\dot{x} = f(x) \tag{10}$$

with periodic orbit γ. The stable and unstable manifolds of γ are defined as

$$W^s(\gamma) = \{x_0 : \text{dist}(x(t; x_0), \gamma) \to 0 \ (\text{exponentially as}) \ t \to \infty\}$$

and

$$W^u(\gamma) = \{x_0 : \text{dist}(x(t; x_0), \gamma) \to 0 \ (\text{exponentially as}) \ t \to -\infty\}.$$

The following result can be found in [18].

Theorem 6. W^s, W^u *are smooth invariant manifolds which intersect transversely in* γ.

W^s and W^u are disc bundles over the circle. We introduce the notation

$$\dim W^s = s, \quad \dim W^u = u,$$

and note that

$$\dim(W^s \cap W^u) = 1.$$

In particular, γ is hyperbolic if $s + u = n + 1$.

Corollary 2. γ *is exponentially orbitally stable if, and only if,* $s = n$.

These can also be constructed via the Poincaré map at each $P \in \gamma$.

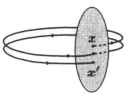
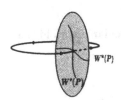

Fig. 9. $\mathcal{P} : \mathbb{D}^2 \to \mathbb{D}^2$ $\mathcal{P}(x) = x'$

Remark 6. $\dim W^s(P) = s - 1$, $\dim W^u(P) = u - 1$.

In particular

$$W^s(\gamma) = \bigcup_{P \in \gamma} W^s(P)$$

and

$$W^u(\gamma) = \bigcup_{P \in \gamma} W^u(P)$$

define codimension one foliations of $W^s(\gamma)$, $W^u(\gamma)$ respectively. We see that P hyperbolic implies $u + s = n + 1$ which implies γ is hyperbolic. If $u + s - 2 < n - 1$, then there is a center manifold $W^c(P)$ for P

$$\dim W^c(P) = n + 1 - u - s.$$

Theorem 7. *There exists a smooth invariant manifold W^c containing γ such that $W^c(\gamma) = \bigcup_{P \in \gamma} W^c(P)$ is a codimension one foliation.*

It follows that γ is orbitally stable if $u = 1$, and γ is orbitally stable on W^c. There also exists a center-stable manifold $W^{cs}(\gamma) = \bigcup_{P} W^{cs}(P)$. In the orbitally stable case, $W^{cs}(\gamma)$ is a solid torus. One can also see this in local coordinates

$$\dot{\rho} = f_1(\rho, \theta)$$
$$\dot{\theta} = f_2(\rho, \theta), \quad f_2(\rho, \theta) > 0.$$

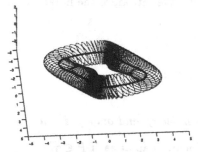

Fig. 10. A periodic orbit and its center manifold

Definition 1. A circular manifold is a manifold M, with or without boundary, such that

$$\pi_1(M) \simeq \mathbb{Z}$$

$$\pi_i(M) \simeq \{0\}, \quad i \geq 1 \tag{11}$$

Examples include circles, compact cylinders, solid tori, W^s, W^c, W^{sc}.

Remark 7. Any space satisfying (11) is an example of an Eilenberg-MacLane (classifying) space, and in this case is said to be a $K(\mathbb{Z}; 1)$.

 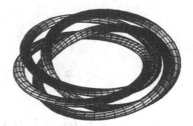

Fig. 11. Examples of compact circular manifolds

Remark 8. The second example in 11, the four-linked unknotted torus was generated in matlab using the tori4 m-file written by C. H. Edwards.

In general, then, a nonlinear system $\dot{x} = f(x)$ and a periodic solution γ give rise to

  (i) an invariant circular manifold M with f pointing inward on ∂M;
 (ii) a codimension one foliation on M defined by a closed one-form w;
(iii) w satisfying $\langle f, w \rangle > 0$.

Theorem 8 (Brockett, Byrnes). *Necessary and sufficient conditions for the existence of a periodic orbit γ are the existence of a compact, positively invariant circular submanifold M and a closed one form w on M satisfying $\langle f, w \rangle > 0$.*

Proof. Necessity: $M = \gamma$, $w = d\theta$ (although W^s, W^c, W^{sc} provide more interesting constructions)

Sufficiency: Choose a generator γ_1 for $\pi_1(M) \simeq \mathbb{Z}$ for which $\int_{\gamma_1} w = c > 0$. Without loss of generality, we may take $c = 1$.

Construct the "period map" $J : M \to S^1$ as follows. Fix $P \in M$. For any $Q \in M$ choose a path γ_2 from P to Q and consider

$$\tilde{J}_{\gamma_2}(Q) = \int_{\gamma_2} w.$$

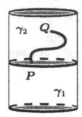

Fig. 12. The path of integration from P to Q on a circular manifold.

If γ_3 is another path

$$\gamma_3 - \gamma_2 \sim \ell\,\gamma_1$$

so

$$\tilde{J}_{\gamma_3}(Q) = \tilde{J}_{\gamma_2}(Q) + \ell.$$

Or, in $\mathbb{R}$,

$$\tilde{J}_{\gamma_3}(Q) \equiv \tilde{J}_{\gamma_2}(Q) \quad \mod \mathbb{Z}.$$

Therefore we may define the period map

$$J : M \to S^1 = \mathbb{R}/\mathbb{Z}$$

via

$$J(\theta) = \int_P^Q w.$$

From the geometry of J it follows [4] that $J^{-1}(\theta)$ is connected, is a leaf of $w = 0$, and is the domain of a Poincaré map

$$\mathcal{P}(x_0) = \Phi_T(x_0) \in J^{-1}(\theta).$$

Finally, we show that $\mathcal{P}$ has a fixed point by proving nonvanishing of its Lefschetz number, $\mathrm{Lef}(\mathcal{P}) = 1 \neq 0$, and appealing to the Lefschetz Fixed Point Theorem.

Remark 9.   1. The Lefschetz number computes the oriented intersection of the graph of $f(x)$ with the diagonal (the graph of the identity map) in $M \times M$.

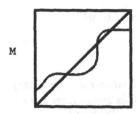

Fig. 13. The fixed point set as the intersection of the graph and the diagonal in the product manifold.

2. We have actually shown that $J^{-1}(\theta)$ is contractible and that there are either an infinite number of periodic orbits on M or, if the orbits have finite multiplicity, an odd number of orbits (counted with multiplicity).

This existence criterion for closed orbits has many corollaries and illustrations, of which we shall mention two. Consider the system

$$\dot{x} = f(x) + g(x)u \tag{12}$$

Suppose M is a circular manifold such that f, g are inward normal vector fields on ∂M. A closed, non-exact one form ω on M is a control one-form provided

$$\langle w, f(x) \rangle \leq 0 \quad \Rightarrow \quad \langle w, g(x) \rangle \neq 0.$$

Corollary 3. *If (12) has a control one-form, then there exists a state feedback law, $u = \alpha(x)$, which creates a periodic orbit for the closed-loop system.*

We next turn to the appropriate analog, for periodic orbits, of Brockett's necessary condition for asymptotically stable equilibria.

Corollary 4. *Suppose*

$$\dot{x} = f(x)$$

has an asymptotically stable periodic orbit. Then, for any jointly continuous perturbation

$$\dot{x} = f(x, \mu)$$

where for some μ_0, $f(x, \mu_0) = f(x)$, there must exist a periodic solution, for each $\mu \sim \mu_0$.

Proof. $\dot{x} = f(x, \mu)$ possesses a compact, solid torus M for which $f(x, \mu_0)$ points inward on ∂M and for which there exists a closed 1-form w satisfying

$$\langle w, f \rangle > 0.$$

These conditions are satisfied by small perturbations $f(x, \mu)$ of $f(x)$ and therefore

$$\dot{x} = f(x, \mu)$$

possesses a periodic orbit.

6   Necessary Conditions For the Existence of Asymptotically Stable Equilibria.

In the previous corollary, we presented an analog of Brockett's necessary condition, valid for periodic orbits of period $T(\mu)$. In this section, we review the recent results reported in [22] on the situation when $T(\mu) \to 0$.

Theorem 9 (Byrnes, Sundarapandian). *Suppose $\dot{x} = f(x)$ has an asymptotically stable equilibrium. Then, for any jointly continuous perturbation*

$$\dot{x} = f(x, \mu)$$

where for some μ_0, $f(x, \mu_0) = f(x)$, there must exist an equilibrium solution, for each $\mu \sim \mu_0$.

Proof. (Sketch of Proof) Let

$$M_c = V^{-1}(-\infty, c], \quad N_c = \partial M_c = V^{-1}(c),$$

$$F_\mu = -\frac{f(x, \mu}{\|f(x, \mu)\|}$$

is a unit outward normal to M_c. The Poincaré-Hopf Theorem again gives that $\mathrm{Ind}(F_\mu) = \chi(M_c) = +1$.

By defining $f(x, \mu) = F(x) - \mu$ and substituting $y = \mu$ we obtain, of course, Brockett's Theorem. Moreover, by defining $F(x, \mu) = f(x, \mu) - y$ we obtain:

Corollary 5. *Consider the nonlinear system $\dot{x} = f(x)$ where f is locally C^1 in x and μ, and $f(0, \mu_0) = 0$. A necessary condition for $x = 0$ to be a locally asymptotically stable equilibrium of the system $\dot{x} = f(x, \mu_0)$ is that for all $\mu \sim \mu_0 \in \mathbb{R}^k$, the map $f(\cdot, \mu)$ is locally onto, i.e., the equation $f(x, \mu) = y$ is locally solvable.*

Our final example shows that this new result is stronger, since the system we construct is not asymptotically stable, and passes Brockett's criterion but not the criterion we obtained by specializing from the periodic case to the case of equilibria.

Example 6. Consider the system $\dot{x} = f(x,\mu)$ for $\mu \sim \mu_0 = 1$ where

$$f(x,\mu) = \begin{bmatrix} (x_1^2 + x_2^2)^\mu \cos\left(2\mu \tan^{-1}\left(\dfrac{x_2}{|x_1|}\right)\right) \\[2mm] (x_1^2 + x_2^2)^\mu \sin\left(2\mu \tan^{-1}\left(\dfrac{x_2}{|x_1|}\right)\right) \end{bmatrix} \quad \text{for } x_2 \geq 0$$

$$f(x,\mu) = \begin{bmatrix} (x_1^2 + x_2^2)^\mu \cos\left(2\mu \tan^{-1}\left(\dfrac{-x_2}{|x_1|}\right)\right) \\[2mm] -(x_1^2 + x_2^2)^\mu \sin\left(2\mu \tan^{-1}\left(\dfrac{-x_2}{|x_1|}\right)\right) \end{bmatrix} \quad \text{for } x_2 < 0$$

We note that f is C^∞ on $\mathbb{R}^2 \backslash \{0\}$ and C^1 at 0 provided $\mu > \dfrac{1}{2}$. Now $f(x,1)$ maps the closed upper-half plane onto $\mathbb{R}^2$ and therefore satisfies the necessary condition of the corollary. We also note that $f(x,\mu)$ is not locally onto for $\mu < 1$.

These properties and claims are most easily seen using a complex variable argument. We define the complex variable

$$z = x_1 + ix_2.$$

Thus we have

$$f(z,\mu) = \begin{cases} z^{2\mu}, & \text{Im }(z) \geq 0,\ z \neq 0 \\ 0, & z = 0 \\ \bar{z}^{2\mu}, & \text{Im }(z) \leq 0,\ z \neq 0 \end{cases}$$

Therefore $f(z,1)$ is surjective, but has degree zero. Also $f(x,\mu)$ is never surjective for $\mu < 1$.

References

1. Bacciotti, A. (1992) Linear feedback: the local and potentially global stabilization of cascade systems, Proc. *2nd IFAC Symp. Nonlinear Control Systems Design*, 21–25.
2. Birkhoff, G. D. (1968) Collected Mathematical Papers, Dover, Vol. 2.
3. Brocket, R. W. (1983) Asymptotic stability and feedback stabilization, *Differential Geometric Control Theory*, R.W. Brockett, R.S. Millmann, H.J. Sussmann, eds., Birkhäuser, Boston, 181-191.
4. Brockett, R. W., Byrnes, C. I. (2000) The geometry of periodic orbits with applications to nonlinear control systems, *Preprint Department of Systems Science and mathematics, Washington University, St. Louis.*

5. Byrnes, C. I., Isidori, A. (1984) 'A Frequency Domain Philosophy for Nonlinear Systems, with Applications to Stabilization and to Adaptive Control', *Proc. of 23rd IEEE Conf. on Decision and Control*, Las Vegas.
6. Byrnes, C. I., Isidori, A. (1989) New results and examples in nonlinear feedback stabilization, *Syst. Control Lett.*, **12**, 437-442.
7. Byrnes, C. I., Isidori, A. (1991) On the attitude stabilization of rigid spacecraft, *Automatica J. IFAC* **27**, no. 1, 87-95.
8. Byrnes, C. I., Isidori, A. (1991) Asymptotic stabilization of minimum phase nonlinear systems, *IEEE Trans. Aut. Control*, **AC-36**, 1122-1137.
9. Byrnes, C. I., Isidori, A. (2000) Bifurcation analysis of the zero dynamics and the practical stabilization of nonlinear minimum-phase systems, Submitted to *Asian Journal of Control.*
10. Byrnes, C. I., Isidori, A., Martin, C.F., (2000) On the structure of globally asymptotically stable attractors, *Preprint Department of Systems Science and mathematics, Washington University, St. Louis.*
11. Byrnes, C. I., Isidori, A., Willems, J. C. (1991) Passivity, feedback equivalence, and the global stabilization of minimum phase nonlinear systems, *IEEE Trans. Autom. Contr.*, **AC-36**, 1228-1240.
12. Clarke, F. H., Ledyaev, Yu. S., Sontag, E. D., Subbotin, A. I. (1997) Asymptotic controllability implies feedback stabilization, *IEEE Trans. Aut. Control*, **42**, 1394-1407.
13. Coron, J. M., (1992) Global asymptotic stabilization for controllable systems without drift, *Math. Contr., Sig. Syst.* **5**, 295ff.
14. Fenichel, N. (1979) Geometric singular perturbation theory for ordinary differential equations, *J. Diff. Eqs.*, **31**, 53-93.
15. Hale, J. K., Magalhaes, L. T., Olvia, W. M. (1984) An introduction to infinite dimensional dynamical systems – geometric theory, *Springer-Verlag*, New York,
16. Hill, D. J., Moylan, P. J., (1976), Stability of nonlinear dissipative systems, *IEEE Trans. Aut. Contr.*, **AC-21**, 708-711.
17. Marino, R. (1985) High-gain feedback in nonlinear control systems, *Int. J. Contr.*, 42, 1369-1385.
18. Markus, L. (1960) The behavior of the solutions of a differential system near a periodic solution, *Ann. of Math.* **72**, No. 2, 245-266.
19. Milnor, J. (1997) Topology from the differentiable viewpoint, *Priceton University Press*, Princeton, NJ.
20. Ortega, R., (1989) Passivity properties for stabilization of cascaded nonlinear systems, *Automatica*, **27**, 423-424.
21. Ortega, R., Spong, M. W., (1989) Adaptive motion control of rigid robots: A tutorial, *Automatica*, **25**, 877-888.
22. Sundarpandian, V., Byrnes, C. I. (2000) Persistence of equilibria for locally asymptotically stable systems," submitted to *Int. J. of Robust and Nonlinear Control.*
23. Teel, A.R. and Praly, L. (1995) Tools for semiglobal stabilization by partial state and output feedback. *SIAM J. Control Optim.*, **33**, pp. 1443-1485.
24. Wiggins, S. (1990) Introduction to applied nonlinear dynamical systems and chaos, Springer-Verlag New York, Inc.
25. Wilson, Jr., F. W. (1967) The structure of the level surfaces of a Lyapunov function, *J. of Diff. Eqs.*, **3**, 323-329.
26. Wilson, Jr., F. W. (1969) Smoothing derivatives of functions and applications, *trans. Amer. Math. Soc.*, **139**, 413-428.

27. Wilson, Jr., F. W., Yorke, J. A. (1973) Lyaounov functions and isolating blocks, *J. Diff. Eqs.*, **13**, 106-123.
28. Willems, J.C., (1976), Dissipative dynamical systems, Part I: General theory, *Archive for Rat. Mech. and Anal.*, **45**, 321-351.

A Regularization of Zubov's Equation for Robust Domains of Attraction*

Fabio Camilli[1], Lars Grüne[2], and Fabian Wirth[3]**

[1] Dip. di Energetica, Fac. di Ingegneria
   Università de l'Aquila,
   67040 Roio Poggio (AQ), Italy
   camilli@axcasp.caspur.it
[2] Fachbereich Mathematik
   J.W. Goethe-Universität, Postfach 11 19 32,
   60054 Frankfurt a.M., Germany
   gruene@math.uni-frankfurt.de
[3] Center for Technomathematics
   University of Bremen
   28334 Bremen, Germany
   fabian@math.uni-bremen.de

Abstract. We derive a method for the computation of robust domains of attraction based on a recent generalization of Zubov's theorem on representing robust domains of attraction for perturbed systems via the viscosity solution of a suitable partial differential equation. While a direct discretization of the equation leads to numerical difficulties due to a singularity at the stable equilibrium, a suitable regularization enables us to apply a standard discretization technique for Hamilton-Jacobi-Bellman equations. We present the resulting fully discrete scheme and show a numerical example.

1 Introduction

The domain of attraction of an asymptotically stable fixed point has been one of the central objects in the study of continuous dynamical systems. The knowledge of this object is important in many applications modeled by those systems like e.g. the analysis of power systems [1] and turbulence phenomena in fluid dynamics [2,8,17]. Several papers and books discuss theoretical [19,20,5,12] as well as computational aspects [18,13,1,9] of this problem.

---

* Research supported by the TMR Networks "Nonlinear Control Network" and "Viscosity Solutions and their applications", and the DFG Priority Research Program "Ergodentheorie, Analysis und effiziente Simulation dynamischer Systeme"

** This paper was written while Fabian Wirth was a guest at the Centre Automatique et Systèmes, Ecole des Mines de Paris, Fontainebleau, France. The hospitality of all the members of the centre is gratefully acknowledged.

Taking into account that usually mathematical models of complex systems contain model errors and that exogenous perturbations are ubiquitous it is natural to consider systems with deterministic time varying perturbations and look for domains of attraction that are robust under all these perturbations. Here we consider systems of the form

$$\dot{x}(t) = f(x(t), a(t)), \quad x \in \mathbb{R}^n$$

where $a(\cdot)$ is an arbitrary measurable function with values in some compact set $A \subset \mathbb{R}^m$. Under the assumption that $x^* \in \mathbb{R}^n$ is a locally exponentially stable fixed point for all admissible perturbation functions $a(\cdot)$ we try to find the set of points which are attracted to x^* for all admissible $a(\cdot)$.

This set has been considered e.g. in [14,15,4,7]. In particular, in [14] and [7] numerical procedures based on optimal control techniques for the computation of robust domains of attraction are presented. The techniques in these papers have in common that a numerical approximation of the optimal value function of a suitable optimal control problem is computed such that the robust domain of attraction is characterized by a suitable sublevel set of this function. Whereas the method in [14] requires the numerical solution of several Hamilton-Jacobi-Bellman equations (and is thus very expensive) the method in [7] needs just one such solution, but requires some knowledge about the local behavior around x^* in order to avoid discontinuities in the optimal value functions causing numerical problems.

In this paper we use a similar optimal control technique, but start from recent results in [4] where the classical equation of Zubov [20] is generalized to perturbed systems. Under very mild conditions on the problem data this equation admits a continuous or even Lipschitz viscosity solution. The main problem in a numerical approximation is the inherent singularity of the equation at the fixed point which prevents the direct application of standard numerical schemes. Here we propose a regularization of this equation such that the classical schemes [6] and adaptive gridding techniques [11] are applicable without losing the main feature of the solution, i.e. the sublevel set characterization of the robust domain of attraction. It might be worth noting that in particular our approach is applicable to the classical Zubov equation (i.e. for unperturbed systems) and hence provides a way to compute domains of attraction also for unperturbed systems.

This paper is organized as follows: In Section 2 we give the setup and collect some facts about robust domains of attraction. In Section 3 we summarize the needed results from [4] on the generalization of Zubov's equation for perturbed system. In Section 4 we introduce the regularization technique and formulate the numerical scheme, and finally, in Section 5 we show a numerical example.

2 Robust domains of attraction

We consider systems of the following form

$$\begin{cases} \dot{x}(t) = f(x(t), a(t)), & t \in [0, \infty), \\ x(0) = x_0, \end{cases} \tag{1}$$

with solutions denoted by $x(t, x_0, a)$. Here $a(\cdot) \in \mathcal{A} = L^\infty([0, +\infty), A)$ and A is a compact subset of $\mathbb{R}^m$, f is continuous and bounded in $\mathbb{R}^n \times A$ and Lipschitz in x uniformly in $a \in A$. Furthermore, we assume that the fixed point $x = 0$ is singular, that is $f(0, a) = 0$ for any $a \in A$.

We assume that the singular point 0 is uniformly locally exponentially stable for the system (1), i.e.

(H1)   there exist constants $C, \sigma, r > 0$ such that
$\|x(t, x_0, a)\| \leq C e^{-\sigma t} \|x_0\|$ for any $x_0 \in B(0, r)$ and any $a \in A$.

The following sets describe domains of attraction for the equilibrium $x = 0$ of the system (1).

Definition 1. For the system (1) satisfying (H1) we define the *robust domain of attraction* as

$$\mathcal{D} = \{x_0 \in \mathbb{R}^n : x(t, x_0, a) \to 0 \text{ as } t \to +\infty \text{ for any } a \in \mathcal{A}\},$$

and the *uniform robust domain of attraction* by

$$\mathcal{D}_0 = \left\{ x_0 \in \mathbb{R}^n : \begin{array}{l} \text{there exists a function } \beta(t) \to 0 \text{ as } t \to \infty \\ \text{s.th. } \|x(t, x_0, a)\| \leq \beta(t) \text{ for all } t > 0, a \in \mathcal{A} \end{array} \right\}.$$

For a collection of properties of (uniform) robust domains of attraction we refer to [4, Proposition 2.4]. There it is shown in particular, that $\mathcal{D}_0$ is an open, connected and invariant set, and that the inclusion $\mathcal{D} \subset \mathrm{cl}\mathcal{D}_0$ holds.

3 Zubov's method for robust domains of attraction

In this section we discuss the following partial differential equation

$$\inf_{a \in A} \{-Dv(x)f(x, a) - (1 - v(x))g(x, a)\} = 0 \qquad x \in \mathbb{R}^n \tag{2}$$

whose solution will turn out to characterize the uniform robust domain of attraction $\mathcal{D}_0$. This equation is a straightforward generalization of Zubov's equation [20]. In this generality, however, in order to obtain a meaningful result about solutions we have to work within the framework of viscosity solutions, which we recall for the convenience of the reader (for details about this theory we refer to [3]).

Definition 2. Given an open subset Ω of $\mathbb{R}^n$ and a continuous function $H : \Omega \times \mathbb{R} \times \mathbb{R}^n \to \mathbb{R}$, we say that a lower semicontinuous (l.s.c.) function $u : \Omega \to \mathbb{R}$ (resp. an upper semicontinuous (u.s.c.) function $v : \Omega \to \mathbb{R}$) is a viscosity supersolution (resp. subsolution) of the equation

$$H(x, u, Du) = 0 \qquad x \in \Omega \tag{3}$$

if for all $\phi \in C^1(\Omega)$ and $x \in \operatorname{argmin}_\Omega(u - \phi)$ (resp., $x \in \operatorname{argmax}_\Omega(v - \phi)$) we have

$$H(x, u(x), D\phi(x)) \geq 0 \qquad (\text{resp., } H(x, v(x), D\phi(x)) \leq 0).$$

A continuous function $u : \Omega \to \mathbb{R}$ is said to be a viscosity solution of (3) if u is a viscosity supersolution and a viscosity subsolution of (3).

We now introduce the value function of a suitable optimal control problem related to (2).

Consider the functional $G : \mathbb{R}^n \times \mathcal{A} \to \mathbb{R} \cup \{+\infty\}$ and the optimal value function v given by

$$G^\infty(x, a) := \int_0^{+\infty} g(x(t), a(t))dt \quad \text{and} \quad v(x) := \sup_{a \in \mathcal{A}} 1 - e^{-G^\infty(x,a)}, \tag{4}$$

where the function $g : \mathbb{R}^n \times A \to \mathbb{R}$ is supposed to be continuous and satisfies

(H2)

    (i) For any $a \in A$, $g(0, a) = 0$ and $g(x, a) > 0$ for $x \neq 0$.

    (ii) There exists a constant $g_0 > 0$ such that
$$\inf_{x \notin B(0,r),\, a \in A} g(x, a) \geq g_0.$$

    (iii) For every $R > 0$ there exists a constant L_R such that $\|g(x, a) - g(y, a)\| \leq L_R \|x - y\|$ for all $\|x\|, \|y\| \leq R$ and all $a \in A$.

Since g is nonnegative it is immediate that $v(x) \in [0, 1]$ for all $x \in \mathbb{R}^n$. Furthermore, standard techniques from optimal control (see e.g. [3, Chapter III]) imply that v satisfies a dynamic programming principle, i.e. for each $t > 0$ we have

$$v(x) = \sup_{a \in \mathcal{A}} \{(1 - G(x, t, a)) + G(x, t, a)v(x(t, x, a))\} \tag{5}$$

with

$$G(t, x, a) := \exp\left(-\int_0^t g(x(\tau, x, a), a(\tau))d\tau\right). \tag{6}$$

Furthermore, a simple application of the chain rule shows

$$(1 - G(x, t, a)) = \int_0^t G(x, \tau, a)g(x(\tau, x, a), a(\tau))d\tau$$

implying (abbreviating $G(t) = G(x, t, a)$)

$$v(x) = \sup_{a \in \mathcal{A}} \left\{ \int_0^t G(\tau) g(x(\tau, x, a), a(\tau)) d\tau + G(t) v(x(t, x, a)) \right\} \qquad (7)$$

The next proposition shows the relation between $\mathcal{D}_0$ and v, and the continuity of v. For the proof see [4, Proposition 3.1]

Proposition 1. *Assume (H1), (H2). Then*

(i) $v(x) < 1$ if and only if $x \in \mathcal{D}_0$.
(ii) $v(0) = 0$ if and only if $x = 0$.
(iii) v is continuous on $\mathbb{R}^n$.
(iv) $v(x) \to 1$ for $x \to x_0 \in \partial \mathcal{D}_0$ and for $\|x\| \to \infty$.

We now turn to the relation between v and equation (2). Recalling that v is locally bounded on $\mathbb{R}^n$ an easy application of the dynamic programming principle (5) (cp. [3, Chapter III]) shows that and v is a viscosity solution of (2). The more difficult part is to obtain uniqueness of the solution, since equation (2) exhibits a singularity at the origin.

In order to prove the following uniqueness result we use super- and suboptimality principles, which essentially follow from Soravia [16, Theorem 3.2 (i)], see [4, Section 3] for details.

Theorem 1. *Consider the system (1) and a function $g : \mathbb{R}^n \times A \to \mathbb{R}$ such that (H1) and (H2) are satisfied. Then (2) has a unique bounded and continuous viscosity solution v on $\mathbb{R}^n$ satisfying $v(0) = 0$.*

This function coincides with v from (4). In particular the characterization $\mathcal{D}_0 = \{x \in \mathbb{R}^n \,|\, v(x) < 1\}$ holds.

We also obtain the following local version of this result.

Theorem 2. *Consider the system (1) and a function $g : \mathbb{R}^n \times A \to \mathbb{R}$. Assume (H1) and (H2). Let $\mathcal{O} \subset \mathbb{R}^n$ be an open set containing the origin, and let $v : \mathrm{cl}\,\mathcal{O} \to \mathbb{R}$ be a bounded and continuous function which is a viscosity solution of (2) on $\mathcal{O}$ and satisfies $v(0) = 0$ and $v(x) = 1$ for all $x \in \partial \mathcal{O}$.*

Then v coincides with the restriction $v|_\mathcal{O}$ of the function v from (4). In particular the characterization $\mathcal{D}_0 = \{x \in \mathbb{R}^n \,|\, v(x) < 1\}$ holds.

We end this section by stating several additional properties of v as proved in [4, Sections 4 and 5].

Theorem 3. *Assume (H1) and (H2) and consider the unique viscosity solution v of (2) with $v(0) = 0$. Then the following statements hold.*

(i) *The function v is a robust Lyapunov function for the system (1), i.e.*

$$v(x(t, x_0, a(\cdot))) - v(x_0) < 0$$

for all $x_0 \in \mathcal{D}_0 \setminus \{0\}$, all $t > 0$ and all $a(\cdot) \in \mathcal{A}$.

(ii) *If $f(\cdot, a)$ and $g(\cdot, a)$ are uniformly Lipschitz continuous in $\mathbb{R}^n$, with constants L_f, $L_g > 0$ uniformly in $a \in A$, and if there exists a neighborhood N of the origin such that for all x, $y \in N$ the inequality*

$$|g(x, a) - g(y, a)| \le K \max\{\|x\|, \|y\|\}^s \|x - y\|$$

holds for some $K > 0$ and $s > L_f / \sigma$ with $\sigma > 0$ given by (H1), then the function v is Lipschitz continuous in $\mathbb{R}^n$ for all g with $g_0 > 0$ from (H2) sufficiently large.

(iii) *If $f(x, A)$ is convex for all $x \in \mathbb{R}^n$ and $B \subset \mathcal{D}_0$ satisfies $\mathrm{dist}(B, \partial \mathcal{D}_0) > 0$, then there exists a function $g : \mathbb{R}^n \to \mathbb{R}$ satisfying (H2) such that the solution v of (2) is C^∞ on a neighborhood of B.*

4 Numerical solution

A first approach to solve equation (2) is to directly adopt the first order numerical scheme from [6] to this equation. Considering a bounded domain Ω and a simplicid grid Γ with edges x_i covering $\mathrm{cl}\Omega$ this results in solving

$$\tilde{v}(x_i) = \max_{a \in A}\{(1 - hg(x_i, a))\tilde{v}(x_i + hf(x_i, a)) + hg(x_i, a)\} \tag{8}$$

where $\tilde{v}$ is continuous and affinely linear on each simplex in the grid and satisfies $\tilde{v}(0) = 0$ (assuming, of course, that 0 is a node of the grid) and $\tilde{v}(x_i) = 1$ for all $x_i \in \partial\Omega$. Unfortunately, since also (8) has a singularity in 0 the fixed point argument used in [6] fails here and hence convergence is not guaranteed. In fact, it is easy to see that in the situation of Figure 1 (showing one trajectory and the simplices surrounding the fixed point 0 in a two-dimensional example) the piecewise linear function $\tilde{v}$ with

$$\tilde{v}(x_i) = \begin{cases} 1, & x_i \ne 0 \\ 0, & x_i = 0 \end{cases}$$

satisfies (8), since for all nodes $x_i \ne 0$ the value $x_i + hf(x_i, a)$ lies in a simplex with nodes $x_j \ne 0$, hence $\tilde{v}(x_i + hf(x_i, a)) = 1$ implying

$$(1 - hg(x_i, a))\tilde{v}(x_i + hf(x_i, a)) + hg(x_i, a) = 1 = \tilde{v}(x_i),$$

i.e. (8). As this situation may occur for arbitrarily fine grids indeed convergence is not guaranteed.

In order to ensure convergence we will therefore have to use a regularization of (2). The main idea in this is to change (2) in such a way that the "discount

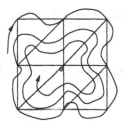

Fig. 1. A situation of non-convergence

rate" (i.e. the factor $g(x)$ in front of the zero order term $v(x)$) becomes strictly positive, and thus the singularity disappears. To this end consider some parameter $\varepsilon > 0$ and consider the function

$$g_\varepsilon(x, a) = \max\{g(x, a), \varepsilon\}.$$

Using this g_ε we approximate (2) by

$$\inf_{a \in A} \{-Dv(x)f(x, a) - g(x, a) + v(x)g_\varepsilon(x, a)\} = 0 \qquad x \in \mathbb{R}^n. \tag{9}$$

The following proposition summarizes some properties of (9). We state it in a global version on $\mathbb{R}^n$, the analogous statements hold in the situation of Theorem 2.

Proposition 2. *Let the assumptions of Theorem 1 hold and let v be the unique solution of (2) with $v(0) = 0$. Then for each $\varepsilon > 0$ equation (9) has a unique continuous viscosity solution v_ε with the following properties.*

(i) $v_\varepsilon(x) \leq v(x)$ for all $x \in \mathbb{R}^n$
(ii) $v_\varepsilon \to v$ uniformly in $\mathbb{R}^n$ as $\varepsilon \to 0$
(iii) If $\varepsilon < g_0$ from (H2)(ii) then the characterization $\mathcal{D}_0 = \{x \in \mathbb{R}^n \,|\, v_\varepsilon(x) < 1\}$ holds
(iv) If $f(\cdot, a)$ and $g(\cdot, a)$ are uniformly Lipschitz on $\mathcal{D}_0$ (uniformly in A with Lipschitz constants L_f and L_g) and g is bounded on $\mathcal{D}_0$ and satisfies the inequalities

$$|g(x, a) - g(y, a)| \leq K \max\{\|x\|, \|y\|\}^s \|x - y\| \tag{10}$$

for all $x, y \in B(0, Cr)$ and

$$|g(x, a)| \geq g_1 > L_f \tag{11}$$

for all $x \notin B(0, r/2)$ with C, σ and r from (H1), then the function v_ε is uniformly Lipschitz on $\mathbb{R}^n$.

Proof: Since the discount rate in (9) is strictly positive it follows by standard viscosity solution arguments [3, Chapter III] that there exists a unique solution v_ε which furthermore for all $t \geq 0$ satisfies the following dynamic programming principle

$$v_\varepsilon(x) = \sup_{a \in \mathcal{A}} \left\{ \int_0^t G_\varepsilon(\tau) g(x(\tau, x, a), a(\tau)) d\tau + G_\varepsilon(t) v_\varepsilon(x(t, x, a)) \right\} \quad (12)$$

with

$$G_\varepsilon(t) = G_\varepsilon(x, t, a) := \exp\left(-\int_0^t g_\varepsilon(x(\tau, x, a), a(\tau)) d\tau \right). \quad (13)$$

Since v satisfies the same principle (7) with $G(x, t, a) \geq G_\varepsilon(x, t, a)$ by (6) and $g > 0$ the stated inequality (i) follows.

In order to see (ii) observe that the continuity of g and v implies that for each $\delta > 0$ there exists $\varepsilon > 0$ such that

$$\{x \in \mathbb{R}^n \mid g_\varepsilon(x, a) \geq g(x, a) \text{ for some } a \in A\} \subset \{x \in \mathbb{R}^n \mid v(x) \leq \delta\}.$$

Now fix $\delta > 0$ and consider the corresponding $\varepsilon > 0$. Let $x \in \mathbb{R}^n$ and pick some $\gamma > 0$ and a control $a_\gamma \in \mathcal{A}$ such that

$$v(x) \leq \int_0^\infty G(x, \tau, a_\gamma) g(x(\tau, x, a_\gamma), a_\gamma(\tau)) d\tau + \gamma.$$

Now let $T \geq 0$ be the (unique) time with $v(x(T, x, a_\gamma)) = \delta$. Abbreviating $G(\tau) = G(x, \tau, a_\gamma)$ and $G_\varepsilon(\tau) = G_\varepsilon(x, \tau, a_\gamma)$ we can conclude that

$$v(x) - v_\varepsilon(x) - \gamma$$
$$\leq \int_0^\infty (G(\tau) g(x(\tau, x, a_\gamma), a_\gamma(\tau)) - G_\varepsilon(\tau) g(x(\tau, x, a_\gamma), a_\gamma(\tau))) d\tau$$
$$\leq \int_0^T \underbrace{(G(\tau) g(x(\tau, x, a_\gamma), a_\gamma(\tau)) - G_\varepsilon(\tau) g(x(\tau, x, a_\gamma), a_\gamma(\tau)))}_{=0} d\tau$$
$$+ G(T) v(x(T, x, a_\gamma)) \quad \leq \quad \delta.$$

Since $\gamma > 0$ and $x \in \mathbb{R}^n$ were arbitrary this shows (ii).

To prove (iii) let $\varepsilon < g_0$. Then for all $x \notin \mathcal{D}_0$ and all $T > 0$ there exists $a \in \mathcal{A}$ such that $G(x, t, a) = G_\varepsilon(x, t, a)$ for all $t \in [0, T]$ which immediately implies $\mathcal{D}_0 = \{x \in \mathbb{R}^n \mid v_\varepsilon(x) < 1\}$.

In order to see (iv) first note that (10) holds for g_ε for all $\varepsilon \geq 0$ (with the convention $g_0 = g$). Hence by straightforward integration using the exponential stability and (10) we can estimate

$$\left| \int_0^t g_\varepsilon(x(\tau, x, a), a(\tau)) d\tau - \int_0^t g_\varepsilon(x(\tau, y, a), a(\tau)) d\tau \right| \leq L_0 \|x - y\|$$

for all $x, y \in B(0, r)$ and some $L_0 > 0$ independent of ε and a, which also implies

$$|G_\varepsilon(x, t, a) - G_\varepsilon(y, t, a)| \leq L_0 \|x - y\|$$

for all $t \geq 0$ and consequently

$$\sup_{a \in A} \left| \int_0^\infty G_\varepsilon(x, \tau, a) g(x(\tau, x, a), a(\tau)) d\tau \right. \tag{14}$$

$$\left. - \int_0^\infty G_\varepsilon(y, \tau, a) g(x(\tau, y, a), a(\tau)) d\tau \right|$$

$$\leq \sup_{a \in A} \int_0^\infty |G_\varepsilon(x, \tau, a) - G_\varepsilon(y, \tau, a)| \underbrace{g(x(\tau, x, a), a(\tau))}_{\leq L_g C e^{-\sigma \tau} \|x\|} d\tau$$

$$+ \int_0^\infty \underbrace{G_\varepsilon(y, \tau, a)}_{\leq 1} |g(x(\tau, x, a), a(\tau)) - g(x(\tau, y, a), a(\tau))| d\tau$$

$$\leq L_1 \|x - y\| \tag{15}$$

for some suitable $L_1 > 0$ and all $x, y \in B(0, r)$, implying in particular

$$|v_\varepsilon(x) - v_\varepsilon(y)| \leq L_1 \|x - y\|.$$

For all $t > 0$ with $x(s, x, a) \notin B(0, r/2)$ and $x(s, y, a) \notin B(0, r/2)$ for all $s \in [0, t]$ we can estimate

$$|G_\varepsilon(x, t, a)| \leq e^{-tg_1}, \quad |G_\varepsilon(y, t, a)| \leq e^{-tg_1} \tag{16}$$

and using $|e^{-a} - e^{-b}| \leq \max\{e^{-a}, e^{-b}\}|a - b|$ it follows

$$|G_\varepsilon(x, t, a) - G_\varepsilon(y, t, a)|$$

$$\leq e^{-tg_1} \int_0^t |g_\varepsilon(x(\tau, x, a), a(\tau)) - g_\varepsilon(x(\tau, y, a), a(\tau))| d\tau$$

$$\leq e^{-tg_1} \int_0^t L_g e^{\tau L_f} \|x - y\| d\tau$$

$$\leq e^{-tg_1} \frac{L_g}{L_f} e^{tL_f} \|x - y\| \quad = \quad e^{t(L_f - g_1)} \frac{L_g}{L_f} \|x - y\|. \tag{17}$$

Now define $T(x, a) := \inf\{t > 0 : x(t, x, a) \in B(0, r/2)\}$. Then by continuous dependence on the initial value (recall that f is Lipschitz in x uniformly in $a \in A$) for each $x \in \mathcal{D}_0 \setminus B(0, r)$ there exists a neighborhood $\mathcal{N}(x)$ such that $x(t(x, a), y, a) \in B(0, r)$ and $x(t(y, a), x, a) \in B(0, r)$ for all $y \in \mathcal{N}(x)$ and all $a \in A$. Pick some $x \in \mathcal{D}_0 \setminus B(0, r)$ and some $y \in \mathcal{N}(x)$. Then for each $\gamma > 0$

we find $a_\gamma \in \mathcal{A}$ such that

$$
|v_\varepsilon(x) - v_\varepsilon(y)| - \gamma
$$
$$
\leq \left| \int_0^\infty G_\varepsilon(x, \tau, a_\gamma) g(x(\tau, x, a_\gamma), a_\gamma(\tau)) d\tau \right.
$$
$$
\left. - \int_0^\infty G_\varepsilon(y, \tau, a_\gamma) g(x(\tau, y, a_\gamma), a_\gamma(\tau)) d\tau \right|.
$$

Now fix some $\gamma > 0$ and let $T := \min\{T(x, a_\gamma), T(y, a_\gamma)\}$. Abbreviating $x(t) = x(t, x, a_\gamma)$ and $y(t) = x(t, y, a_\gamma)$ we can conclude that $x(T) \in B(0, r)$ and $y(T) \in B(0, r)$. Hence we can continue

$$
|v_\varepsilon(x) - v_\varepsilon(y)| - \gamma
$$
$$
\leq \left| \int_0^T G_\varepsilon(x, \tau, a_\gamma) g(x(\tau), a_\gamma(\tau)) d\tau - \int_0^T G_\varepsilon(y, \tau, a_\gamma) g(y(\tau), a_\gamma(\tau)) d\tau \right|
$$
$$
+ G_\varepsilon(x, T, a_\gamma) \left| \int_0^\infty G_\varepsilon(x(T), \tau, a_\gamma(T + \cdot)) g(x(T + \tau), a_\gamma(T + \tau)) d\tau \right.
$$
$$
\left. - \int_0^\infty G_\varepsilon(y(T), \tau, a_\gamma(T + \cdot)) g(y(T + \tau), a_\gamma(T + \tau)) d\tau \right|
$$
$$
\leq \int_0^T |G_\varepsilon(x, \tau, a_\gamma) - G_\varepsilon(y, \tau, a_\gamma)| g(x(\tau), a_\gamma(\tau)) d\tau
$$
$$
+ \int_0^T G_\varepsilon(y, \tau, a_\gamma) |g(x(\tau, x, a_\gamma), a_\gamma(\tau)) - g(y(\tau), a_\gamma(\tau))| d\tau
$$
$$
+ e^{-g_1 T} e^{L_f T} L_1 \|x - y\|
$$
$$
\leq \int_0^T |G_\varepsilon(x, \tau, a_\gamma) - G_\varepsilon(y, \tau, a_\gamma)| \underbrace{g(x(\tau), a_\gamma(\tau))}_{\leq \sup\limits_{x \in \mathcal{D}_0, a \in A} g(x, a) =: g^*} d\tau
$$
$$
+ \int_0^T e^{-\tau g_1} \underbrace{|g(x(\tau), a_\gamma(\tau)) - g(y(\tau), a_\gamma(\tau))|}_{\leq L_g e^{tL_f} \|x - y\|} d\tau
$$
$$
+ L_1 \|x - y\|
$$
$$
\leq \left(g^* \frac{L_g}{L_f(g_1 - L_f)} + \frac{L_g}{g_1 - L_f} + L_1 \right) \|x - y\|
$$

since $g_1 > L_f$. Here the first inequality follows by splitting up the integrals using the triangle inequality, the second follows by the triangle inequality for the first term and using $x(T), y(T) \in B(0, r)$, $\|x(T) - y(T)\| \leq e^{L_f T}$, and (15) for the second term, and the third and fourth inequality follow from (16) and (17).

Since $\gamma > 0$ was arbitrary the Lipschitz property follows on $\mathcal{D}_0$, thus also on $\mathrm{cl}\mathcal{D}_0$ and consequently on the whole $\mathbb{R}^n$ since $v_\varepsilon \equiv 1$ on $\mathbb{R}^n \setminus \mathcal{D}_0$. $\qquad\square$

Remark 1. Note that in general the solution v_ϵ is not a robust Lyapunov function for the origin of (1) anymore. More precisely, we can only ensure decrease of v_ϵ along trajectories $x(t, x_0, a)$ as long as $g(x(t, x_0, a), a(t)) > \epsilon$, i.e. outside the region where the regularization is effective. Hence although many properties of v are preserved in this regularization, some are nevertheless lost.

We now apply the numerical scheme from [6] to (9). Thus we end up with

$$\tilde{v}_\epsilon(x_i) = \max_{a \in A}\{(1 - hg_\epsilon(x_i, a))\tilde{v}_\epsilon(x + hf(x_i, a)) + hg(x_i, a)\} \qquad (18)$$

where again $\tilde{v}_\epsilon$ is continuous and affinely linear on each simplex in the grid and satisfies $\tilde{v}_\epsilon(0) = 0$ and $\tilde{v}_\epsilon(x_i) = 1$ for all $x_i \in \partial\Omega$.

A straightforward modification of the arguments in [3,6] yields that there exists a unique solution $\tilde{v}_\epsilon$ converging to v_ϵ as h and the size of the simplices tends to 0. Note that the adaptive gridding techniques from [11] also apply to this scheme, and that a number of different iterative solvers for (18) are available, see e.g. [6,10,11].

Remark 2. The numerical examples show good results also in the case where we cannot expect a globally Lipschitz continuous solution v_ϵ of (9). The main reason for this seems to be that in any case v_ϵ is locally Lipschitz on $\mathcal{D}_0$. In order to explain this observation in a rigorous way a thorough analysis of the numerical error is currently under investigation.

5   A numerical example

We illustrate our algorithm with a model adapted from [17]. Consider

$$\dot{x} = \begin{pmatrix} -1/25 & 1 \\ 0 & -2/25 \end{pmatrix} x + \|x\| \begin{pmatrix} 0 & -1 \\ 1 & 0 \end{pmatrix} x + \begin{pmatrix} 0 \\ ax_1 x_2 \end{pmatrix}$$

where $x = (x_1, x_2)^T \in \mathbb{R}^2$. The unperturbed equation (i.e. with $a = 0$) is introduced in order to explain the existence of turbulence in a fluid flow with Reynolds number $R = 25$ despite the stability of the linearization at the laminar solution. In [17] simulations are made in order to estimate the domain of attraction of the locally stable equilibrium at the origin. Here we compute it entirely in a neighborhood of 0, and in addition determine the effect of the perturbation term $ax_1 x_2$ for time varying perturbation with different ranges A. Figure 2 shows the corresponding results obtained with the fully discrete scheme (18), setting $g(x, a) = \|x\|^2$, $\epsilon = 10^{-10}$, $h = 1/20$. The grid was constructed adaptively using the techniques from [11] with a final number of about 20000 nodes. Note that due to numerical errors in the approximate solution it is not reasonable to take the "exact" sublevel sets $\tilde{v}_\epsilon(x) < 1$, instead some "security factor" has to be added. The domains shown in the figures are the sublevel sets $\tilde{v}_\epsilon(x) \leq 0.95$.

288   Fabio Camilli, Lars Grüne and Fabian Wirth

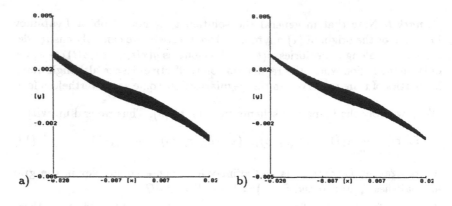

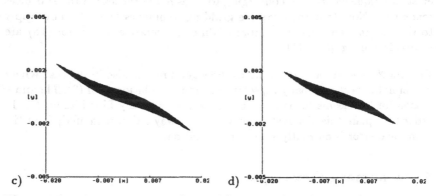

Fig. 2. Approximation of $\mathcal{D}_0$ for a) $A = \{0\}$, b) $A = [-1,1]$, c) $A = [-2,2]$, and d) $A = [-3,3]$

References

1. M. Abu Hassan and C. Storey, Numerical determination of domains of attraction for electrical power systems using the method of Zubov. *Int. J. Control* **34** (1981), 371–381.
2. J.S. Baggett and L.N. Trefethen, Low-dimensional Models of Subcritical Transition to Turbulence. *Physics of Fluids* **9** (1997), 1043–1053.
3. M. Bardi and I. Capuzzo Dolcetta, *Optimal Control and Viscosity Solutions of Hamilton-Jacobi-Bellman equations*, Birkhäuser, Boston, 1997.
4. F. Camilli, L. Grüne, and F. Wirth. A Generalization of Zubov's method to perturbed systems. Preprint 24/99, DFG-Schwerpunkt "Ergodentheorie, Analysis und effiziente Simulation dynamischer Systeme", submitted.
5. H.-D. Chiang, M. Hirsch, and F. Wu. Stability regions of nonlinear autonomous dynamical systems. *IEEE Trans. Auto. Control* **33** (1988), 16–27.
6. M. Falcone. Numerical solution of dynamic programming equations. Appendix A in: M. Bardi and I. Capuzzo Dolcetta, *Optimal Control and Viscosity Solutions of Hamilton-Jacobi-Bellman equations*, Birkhäuser, Boston, 1997.

7. M. Falcone, L. Grüne and F. Wirth. A maximum time approach to the computation of robust domains of attraction. *Proc. EQUADIFF 99, Berlin*, to appear.

8. T. Gebhardt and S. Großmann. Chaos transition despite linear stability. *Phys. Rev. E* **50** (1994), 3705–3711.

9. R. Genesio, M. Tartaglia, and A. Vicino. On the estimation of asymptotic stability regions: State of the art and new proposals. *IEEE Trans. Auto. Control* **30** (1985), 747–755.

10. R.L.V. Gonzáles and C.A. Sagastizábal, Un algorithme pour la résolution rapide d'équations discrètes de Hamilton-Jacobi-Bellman. *C. R. Acad. Sci., Paris, Sér. I* **311** (1990), 45–50.

11. L. Grüne. An adaptive grid scheme for the discrete Hamilton-Jacobi-Bellman equation. *Numer. Math.* **75** (1997), 319–337.

12. W. Hahn, *Stability of Motion*, Springer-Verlag, Berlin, 1967.

13. N.E. Kirin, R.A. Nelepin and V.N. Bajdaev, Construction of the attraction region by Zubov's method. *Differ. Equations* **17** (1982), 871–880.

14. A.D.B. Paice and F. Wirth. *Robustness analysis of domains of attraction of nonlinear systems*, Proceedings of the Mathematical Theory of Networks and Systems MTNS98, pages 353 – 356, Padova, Italy, 1998.

15. A.D.B. Paice and F. Wirth. *Robustness of nonlinear systems subject to time-varying perturbations*, In F. Colonius et al. (eds.), Advances in Mathematical Systems Theory, Birkhäuser, Boston, 2000. To appear.

16. P. Soravia, Optimality principles and representation formulas for viscosity solutions of Hamilton-Jacobi equations, I: Equations of unbounded and degenerate control problems without uniqueness, *Adv. Differ. Equ.*, **4** (1999), 275–296.

17. L.N. Trefethen, A.E. Trefethen, S.C. Reddy and T.A. Driscoll, Hydrodynamic stability without eigenvalues, *Science* **261** (1993), 578–584.

18. A. Vannelli and M. Vidyasagar. Maximal Lyapunov functions and domains of attraction for autonomous nonlinear systems. *Automatica*, **21** (1985), 69–80.

19. F.W. Wilson. The structure of the level surfaces of a Lyapunov function. *J. Differ. Equations* **3** (1967), 323–329.

20. V.I. Zubov, *Methods of A.M. Lyapunov and their Application*, P. Noordhoff, Groningen, 1964.

A Remark on Ryan's Generalization of Brockett's Condition to Discontinuous Stabilizability

Francesca Ceragioli

Dipartimento di Matematica del Politecnico di Torino
c.so Duca degli Abruzzi, 24
10129 Torino, Italy
ceragiol@calvino.polito.it

Abstract. We clarify in which sense Ryan's generalization of Brockett's condition to discontinuous stabilizability applies when solutions of the implemented system are intended in Filippov's sense. Moreover, by means of an example, we see how the interpretation of solutions of systems with discontinuous righthand side may influence a stabilizability result for a system which does not admit a continuous stabilizing feedback law.

1 Introduction

It is a common thought that Krasovskii's and Filippov's concepts of solution (see [9,10]) are not the good ones in order to deal with discontinuous stabilization problems of nonlinear systems. In order to support this idea usually two papers are mentioned: [8] and [12]. Here we are interested in the result by Ryan. We want to see in which sense it is the generalization of Brockett's necessary condition ([4]) to discontinuous stabilizability when solutions are intended either in Krasovskii's or Filippov's sense.

2 Ryan's necessary condition

Let us consider a nonlinear control system of the form

$$\dot{x} = f(x, u) \qquad (1)$$

where $x \in \mathbf{R}^n$, $u \in \mathbf{R}^m$, $f(0,0) = 0$ and f is continuous. This system is said to be (locally) asymptotically stabilizable if there exists a function $u = k(x)$, $k : \mathbf{R}^n \to \mathbf{R}^m$, called feedback law, such that the implemented system

$$\dot{x} = f(x, k(x)) \qquad (2)$$

is (locally) asymptotically stable at the origin, i.e. the origin is Lyapunov stable and attractive for all Carathéodory solutions of (2).

Note that if k is discontinuous this definition may lose sense. In the following we suppose that k is locally essentially bounded, so that the righthand side of (2) is also locally essentially bounded and its solutions can be intended in Filippov's sense.

Let us recall that if g is a locally essentially bounded vector field, by means of the Filippov operator F, we can associates to g the set-valued map defined by

$$Fg(x) = \bigcap_{\delta > 0} \bigcap_{\mu(N)=0} \overline{co}\, g(B(x,\delta) \backslash N),$$

where μ is the usual Lebesgue measure in $\mathbf{R}^n$ and $B(x,\delta)$ is the sphere centered at x with radius δ. Filippov's solutions of the differential equation $\dot{x} = g(x)$ are the solutions of the differential inclusion $\dot{x} \in Fg(x)$.

We say that system (1) is (locally) *asymptotically stabilizable in Filippov's sense* if there exists a locally essentially bounded function $k : \mathbf{R}^n \to \mathbf{R}^m$ such that the implemented system (2), with solutions intended in Filippov's sense, is (locally) asymptotically stable at the origin, i.e. the origin is Lyapunov stable and attractive for all Filippov solutions of (2).

The notion of stabilizability adpoted by Ryan in [12] is different.

First of all Ryan takes f continuous and such that

$$A \subseteq \mathbf{R}^m \text{ convex } \Rightarrow f(x,A) \subseteq \mathbf{R}^n \text{ convex}. \tag{3}$$

A condition equivalent to (3) is the following:

$$\text{for all } A \subseteq \mathbf{R}^m \quad \text{co} f(x,A) \subseteq f(x,\text{co}A). \tag{4}$$

The proof of the equivalence of these conditions easily follows from the definition of convex hull.

As admissible feedbacks, Ryan takes upper semi-continuous set-valued maps U with nonempty, compact and convex values such that $0 \in U(0)$. Note that in the case U is single valued, its upper semi-continuity implies its continuity. On the other hand if k is a single valued discontinuous essentially bounded function it is possible to associate to it an upper semi-continuous set-valued map Fk by means of Filippov's operator F. Ryan's implemented system is then the differential inclusion

$$\dot{x} \in f(x, U(x)), \tag{5}$$

which has a non-empty, upper semi-continuous, compact and convex valued righthand side.

We say that system (1) is (locally) *asymptotically stabilizable in Ryan's sense* if there exists a locally essentially bounded function $k : \mathbf{R}^n \to \mathbf{R}^m$ such that $0 \in Fk(0)$ and (5) with $U(x) = Fk(x)$ is asymptotically stable at the origin, i.e. the origin is Lyapunov stable and attractive for all solutions of (5).

In the following proposition we show a connection between stabilizability in the sense of Ryan and in the sense of Filippov .

Proposition 1 *If f satisfies (3) and system (1) is stabilizable in Ryan's sense then it is also stabilizable in Filippov's sense.*

Proof From (3) it follows that $\operatorname{co} f(x, A) \subseteq f(x, \operatorname{co} A)$ and from this, using techniques analogous to those of Paden and Sastry ([11], Theorem 1) we get that $Ff(x, k(x)) \subseteq f(x, Fk(x))$. This means that every solution of Filippov's implemented system is also a solution of Ryan's implemented system and then if the feedback law k stabilizes system (1) in Ryan's sense, it stabilizes it also in Filippov's one. □

Ryan proves the following theorem which generalizes Brockett's one ([4]).

Theorem 1 ([12]) *Let f be continuous and satisfy condition (3). If system (1) is asymptotically stabilizable (in Ryan's sense), then*
(B) *for each open neighbourhood $\mathcal{N}$ of $0 \in \mathbf{R}^n$, $f(\mathcal{N} \times \mathbf{R}^m)$ contains an open neighbourhood of 0.*

The immediate important consequence of this result is that if a control system does not satisfy Brockett's condition (B), then a discontinuous stabilizing feedback in Ryan's sense does not exist. In order to get a similar result for stabilizability in Filippov's sense we have to make a different assumption on f. We have the following proposition.

Proposition 2 *If $f(x, \operatorname{co} A) \subseteq \operatorname{co} f(x, A)$ for all $A \subseteq \mathbf{R}^m$ and system (1) is stabilizable in Filippov's sense (by means of a feedback law k such that $0 \in Fk(0)$), then it is also stabilizable in Ryan's sense.*

The proof of this proposition is analogous to that of Proposition 1.

Corollary 1 *If $f(x, \operatorname{co} A) \subseteq \operatorname{co} f(x, A)$ for all $A \subseteq \mathbf{R}^m$ and Brockett's condition (B) does not hold then system (1) is not stabilizable in Filippov's sense (by means of a feedback law such that $0 \in Fk(0)$).*

Finally we can conclude that a result analogous to Brockett's one for discontinuous stabilization in Filippov's sense holds if f satisfies the following condition

$$\text{for all } A \subseteq \mathbf{R}^m \quad \operatorname{co} f(x, A) = f(x, \operatorname{co} A). \tag{6}$$

This condition is satisfied by affine-input systems of the form:

$$\dot{x} = f(x) + G(x)u.$$

The fact that for affine-input systems there is no difference in considering stabilizability in Filippov's sense or in Ryan's sense could be also easily proven directly by means of Theorem 1 in [11]. Moreover completely analogous results could be stated for locally bounded feedback and solutions of the implemented system intended in Krasovskii's sense.

In many papers affine-input systems which do not admit continuous feedback laws are stabilized by means of discontinuous feedbacks in a sense that, though not always clear, sometimes seems to be that of of Filippov. Corollary 1 is then in contradiction with these papers. We try to clarify this apparent contradiction by means of an example.

3   An Example

Let us consider the system

$$
\begin{cases}
\dot{x} = \cos\theta u \\
\dot{y} = \sin\theta u \\
\dot{\theta} = \omega
\end{cases}
\tag{7}
$$

where $(x, y, \theta) \in \mathbf{R}^3$ is the state and $(u, \omega) \in \mathbf{R}^2$ is the control. This system can be thought to represent a unicycle positioned on the plane at the point (x, y) and with orientation θ with respect to the x-axis. This system does not satisfy Brockett's condition then a continuous stabilizing feedback does not exist. We consider the feedback law proposed by Canudas de Wit and Sørdalen in [5,6]. They introduce the following functions:

$$
\beta = \frac{y}{x} \quad x \neq 0;
$$

$$
\theta_d(x, y) = \begin{cases} 2\arctan\beta & \text{if } (x, y) \neq (0, 0) \\ 0 & \text{if } (x, y) = (0, 0), \end{cases}
$$

if C is the circle on the (x, y)-plane passing through the points (x, y) and $(0, 0)$ and with center on the y-axis, θ_d is the angle between the line tangent to C in P and the x-axis;

$$
\alpha(x, y, \theta) = \theta - \theta_d - 2\pi n(\theta - \theta_d),
$$

where $n : \mathbf{R} \to \mathbf{Z}$ is such that α belongs to $(-\pi, \pi]$ for all $(x, y, \theta) \in \mathbf{R}^3$;

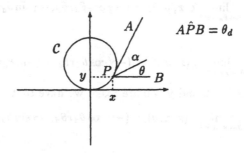

$$b_1(x, y, \theta) = \cos \theta \left(\frac{\theta_d}{\beta} - 1 \right) + \sin \theta \left(\frac{\theta_d}{2} \left(1 - \frac{1}{\beta^2} \right) + \frac{1}{\beta} \right) \quad \beta \neq 0;$$

$$b_2 = \cos \theta \frac{2\beta}{(1 + \beta^2)x} - \sin \theta \frac{2}{(1 + \beta^2)x} \quad x \neq 0;$$

$$a(x, y) = \begin{cases} x & \text{if } y = 0 \\ \frac{x^2 + y^2}{y} \arctan \frac{y}{x} & \text{if } y \neq 0, \end{cases}$$

a can be seen as the arclength of the circle C.

Canudas de Wit and Sørdalen define the feedback law

$$u(x, y, \theta) = -\gamma b_1(x, y, \theta) a(x, y)$$

$$\omega(x, y, \theta) = -b_2(x, y, \theta) u(x, y, \theta) - k\alpha(x, y, \theta),$$

where γ and k are positive constants. Note that $(0, 0) \in F(u, \omega)(0, 0, 0)$.
Let us denote the implemented system by

$$\dot{z} = g(z),$$

where $z = (x, y, \theta)^T$ is the state. We consider Filippov solutions of this system. The implemented system is discontinuous on the following surfaces:

$$\mathcal{D} = \{ z \in \mathbf{R}^3 : x = 0, y \neq 0 \}$$

$$\mathcal{E} = \{ z \in \mathbf{R}^3 : \alpha(z) = \pi \}$$

We are interested in particular in the following subsets of $\mathcal{E}$:

$$\mathcal{F}^+ = \{ z \in \mathbf{R}^3 : x > 0, \ y > 0, \ \theta = \theta_d - \pi \}$$

$$\mathcal{F}^- = \{ z \in \mathbf{R}^3 : x < 0, \ y > 0, \ \theta = \theta_d + \pi \}.$$

Let $x > 0$ and $y > 0$ be fixed. If we denote $r = \frac{x^2+y^2}{2y}$ ($|r|$ is the radius of C), we have that

$$g_+^+(x,y) = \lim_{\theta \to \theta_d - \pi^+} g(x,y,\theta) = (-\gamma \cos \theta_d r \theta_d, -\gamma \sin \theta_d r \theta_d, -\gamma \theta_d + k\pi)^T$$

and

$$g_-^+(x,y) = \lim_{\theta \to \theta_d - \pi^-} g(x,y,\theta) = (-\gamma \cos \theta_d r \theta_d, -\gamma \sin \theta_d r \theta_d, -\gamma \theta_d - k\pi)^T.$$

Analogously if $x < 0$ and $y > 0$ are fixed, we have that

$$g_+^-(x,y) = \lim_{\theta \to \theta_d + \pi^+} g(x,y,\theta) = (-\gamma \cos \theta_d r \theta_d, -\gamma \sin \theta_d r \theta_d, -\gamma \theta_d + k\pi)^T$$

and

$$g_-^-(x,y) = \lim_{\theta \to \theta_d + \pi^-} g(x,y,\theta) = (-\gamma \cos \theta_d r \theta_d, -\gamma \sin \theta_d r \theta_d, -\gamma \theta_d - k\pi)^T.$$

Let us now consider the point $(0, y, 0)$ with any fixed $y > 0$. Note that there are points both on $\mathcal{F}^+$ and $\mathcal{F}^-$ arbitrarily closed to $(0, y, 0)$. These can be obtained by letting $x \to 0^+$ and $x \to 0^-$ respectively. The Filippov set-valued map at $(0, y, 0)$ must then contain the convex hull of the following vectors:

$$g_{++}^+(y) = \lim_{x \to 0^+} g_+^+(x,y) = (\gamma \frac{\pi}{2} y, 0, -\gamma \pi + k\pi)^T,$$

$$g_{-+}^+(y) = \lim_{x \to 0^+} g_-^+(x,y) = (\gamma \frac{\pi}{2} y, 0, -\gamma \pi - k\pi)^T,$$

$$g_{+-}^-(y) = \lim_{x \to 0^-} g_+^-(x,y) = (-\gamma \frac{\pi}{2} y, 0, \gamma \pi + k\pi)^T,$$

$$g_{--}^-(y) = \lim_{x \to 0^-} g_-^-(x,y) = (-\gamma \frac{\pi}{2} y, 0, \gamma \pi - k\pi)^T.$$

The convex hull of these vectors contains the vector $(0, 0, 0)$. This means that all the points of the positive y-axis are equilibrium positions for the implemented system with solutions intended in Filippov's sense, i.e. the feedback law does not stabilize the system in Filippov's sense. Nevertheless it is possible to define conveniently the feedback law on the discontinuity surfaces in such a way that Carathéodory solutions exist and the implemented system is asymptotically stable. The same is still true if, instead of Carathéodory solutions, we consider Euler or sampling solutions (see [7,13] for definitions of these solutions).

Finally let us remark that in many other papers different strategies in order to stabilize system (7) have been proposed. Among these let us mention [2,3,1,13]. These strategies are not a priori in contradiction with Ryan's result because either the feedback is unbounded or solutions are not intended in Filippov's sense.

Acknowledgments. I want to thank professor Bacciotti for many helpful discussions.

References

1. M. Aicardi, G. Casalino, A. Bicchi and A. Balestrino, *Closed loop steering of unicycle-like vehicles via Lyapunov Techniques*, IEEE Robotics and Automation Magazine, March 1995, 27-35

2. A. Astolfi, *Exponential stabilization of a mobile robot*, Proceedings of 3rd European Control Conference, Rome, September 1995, 3092-3097

3. A. Astolfi, *Discontinuous control of nonholonomic systems*, Systems and Control Letters, 27, 1996, 37-45

4. R.W. Brockett, *Asymptotic stability and feedback stabilization*, in Differential Geometric Control Theory, R.W. Brockett, R.S. Millmann and J.H. Sussmann Eds., Birkhauser, Boston, 1983, 181-191

5. O.J. Canudas de Wit and O.J. Sørdalen, *Examples of Piecewise Smooth Stabilization of Driftless NL Systems with less Inputs than States*, Proc. Symp. on Nonlinear Control System Design, Bordeaux, France (IFAC, 1992), 57-61

6. O.J. Canudas de Wit and O.J. Sørdalen, *Exponential Stabilization of Mobile Robots with Nonholonomic Constraints*, IEEE Trans. Aut. Contr., Vol. 37, No 11, November 1992, 1791-1797

7. F.H. Clarke, Yu.S. Ledyaev, R.J. Stern and P.R. Wolenski, *Nonsoomth Analysis and Control Theory*, Springer, 1998

8. J.M. Coron and L. Rosier, *A Relation between Continuous Time-Varying and Discontinuous Feedback Stabilization*, Journal of Mathematical Systems, Estimation and Control, Vol.4, No.1, 1994, 67-84

9. A.F. Filippov, *Differential Equations with Discontinuous Righthand Sides*, Kluwer Academic Publishers, 1988

10. O.Hájek, *Discontinuous Differential Equations I,II*, Journal of Differential Equations, 32, 1979, 149-170, 171-185

11. B. Paden and S. Sastry, *A Calculus for Computing Filippov's Differential Inclusion with Application to the Variable Structure Control of Robot Manipulators*, IEEE Transaction on Circuits and Systems, Vol. Cas-34, No. 1, January 1997, 73-81

12. E.P. Ryan, *On Brockett's condition for smooth stabilizability and its necessity in a context of nonsmooth feedback*, SIAM J. Control and Optim., 32, 1994, 1597-1604

13. E.D. Sontag, *Stability and Stabilization: Discontinuities and the Effect of Disturbances*, in Nonlinear Analysis, Differential Equations and Control, F.H. Clarke and R.J. Stern editors, NATO Science Series, Series C: Mathematical and Physical Sciences, vol. 528, Kluwer Academic Publishers, 1999

Applications of Lie Algebroids in Mechanics and Control Theory

Jesús Clemente-Gallardo[1]

Faculty of Mathematical Sciences
Department of Systems, Signals and Control
University of Twente P.O. Box 217
7500 AE Enschede, The Netherlands
jesus@mailhost.math.utwente.nl

Abstract. We present some applications of the recently developed Lagrangian formalism for Lie algebroids, to mechanical systems. Examples treated are the rigid body with gravity, systems with symmetry and systems defined by an integrable Dirac structure. In this framework, we analyze the method of controlled Lagrangians for the asymptotic stabilization of mechanical systems with symmetry, which turns out to be naturally defined on the so called gauge algebroid.

1 Introduction

From the historical point of view, the original definition of Lie algebroids in the context of differential geometry is due to Pradines [13]. Roughly speaking, they are vector bundles with some particular properties that make them very similar to tangent bundles. Quite recently A Weinstein [16] proposed a generalization of the usual Lagrangian formalism (defined on tangent bundles) to include these objects. Later, E. Martínez proposed a more geometrical treatment of the problem [11], which allowed him also to construct the Hamiltonian version [12].

The aim of this paper is simply to apply their results to some simple examples which are relevant in Mechanics and in Control Theory of mechanical systems. We focus our attention in two of them, namely the heavy top and mechanical systems with symmetry, which we develop in some detail. Finally, we introduce an alternative approach to the problem of Controlled Lagrangians in mechanical systems with symmetry, which might be generalized to other systems described by means of Lie algebroids.

2 The concept of Lie algebroids

First of all we need a definition of the object we are going to work on:

Definition 1. A **Lie algebroid** on a manifold M is a vector bundle $E \to M$, in whose space of sections we define a Lie algebra structure $(\Gamma(E), [\cdot, \cdot]_E)$, and a mapping $\rho : E \to TM$ which is a homomorphism for this structure in relation with the natural Lie algebra structure of the set of vector fields $(\mathfrak{X}(M), [\cdot, \cdot]_{TM})$. We have therefore:

$$\rho([\eta, \xi]_E) = [\rho(\eta), \rho(\xi)]_{TM} \quad \forall \eta, \xi \in \Gamma(E),$$

though in the following we will omit, for simplicity, the subindex of the commutator.

Moreover, we can consider the action of the Lie bracket on the module of sections of E:

$$[\eta, f\xi] = [\eta, \xi] + (\rho(\eta)f)\,\xi \quad \forall \eta, \xi \in \Gamma(E), \quad \forall f \in C^\infty(M)$$

If we take coordinates in the base manifold $\{x^i\}$ and a basis of sections in the bundle $\{e_\alpha\}$, we can consider $\{(x^i, \lambda^\alpha)\}$ to be the coordinates for E, with respect to which we write the expression of the anchor mapping as

$$\rho(e_\alpha) = \rho^i_\alpha \frac{\partial}{\partial x^i},$$

and the expression of the Lie bracket in this base as

$$[e_\alpha, e_\beta] = C^\gamma_{\alpha\beta} e_\gamma.$$

The main idea we have to keep in mind is that a lie algebroid is a geometrical object very similar to a tangent bundle. The sections of the bundle E will play the role of vector fields (though we will need to use the anchor mapping when necessary). Can we define also analogues for other usual geometrical objects as differential forms or the exterior differential as well? The answer is affirmative:

- The analogue of one forms is easy to find, we take the sections of the dual bundle $E^* \to M$. This definition allows us to consider an action of sections of E on sections of E^* as the analogue of the inner action of vector fields on differential forms. We will denote it by i_σ for $\sigma \in \Gamma E$.
- The analogue of p-forms is also easy to define: we take simply the sections of the bundle $(E^*)^{\wedge p} \to M$.
- Finally, a basic piece of our construction will be the definition of the exterior differential d, as the operator which connects the analogue of p-forms with the analogue of $p + 1$-forms. We define it as we do for the usual case, first the action on functions, and later the action on higher order forms:
  - For functions $d : C^\infty(M) \to \Lambda^1(E)$ such that:

$$\langle df, a \rangle = \rho(a)f \quad \forall f \in C^\infty(M), a \in \Gamma(E).$$

- For higher orders forms we take the direct analogue of the usual definition $d : \Lambda^p(E) \to \Lambda^{p+1}(E)$:

$$d\theta(\sigma_1, \ldots, \sigma_{p+1}) = \sum_{i=1}^{p+1} (-1)^{i+1} \rho(\sigma_i) \theta(\sigma_1, \ldots, \hat{\sigma}_i, \ldots \sigma_{p+1})$$

$$+ \sum_{i<j} (-1)^{i+j} \theta([\sigma_i, \sigma_j], \sigma_1, \ldots, \hat{\sigma}_i, \ldots, \hat{\sigma}_j, \ldots \sigma_{p+1})$$

where by the symbol $\hat{\sigma}_i$ we mean that the corresponding section is omitted.

With these definitions, we have extended the differential calculus on manifolds to the Lie algebroid case. We can also define an analogue for the Lie derivative by using the expression for Cartan's formula

$$d_\sigma = i_\sigma \circ d + d \circ i_\sigma.$$

One of the most interesting properties which can also be extended to the Lie algebroid framework, is that, as it happens with the cotangent bundle, the dual of any Lie algebroid is a Poisson manifold:

Theorem 1. *Given a Lie algebroid E, the dual bundle E^* is a Poisson manifold.*

Proof. It can be found in [16] (direct construction), or in [11] (more sophisticated definition).

If we take a basis of sections of E^*, as the dual of the basis $\{e_\alpha\}$ of sections of E, and denote the corresponding coordinates as (x^i, μ_α), the expression of the Poisson bracket above turns out to be:

$$\{x^i, x^j\} = 0 \quad \{x^i, \mu_\alpha\} = \rho^i_\alpha \quad \{\mu_\alpha, \mu_\beta\} = c^\gamma_{\alpha\beta} \mu_\gamma$$

3   The Lagrangian formalism for Lie algebroids

In this section we will present the geometrical formalism of Lagrangian mechanics for Lie algebroids, as it has been presented in [16,11]. The main part of the construction in completely analogous to the well known framework for tangent bundles and Poisson manifolds, and hence, we will begin with a short summary of this, mainly to fix notation.

3.1   Geometrical mechanics in the usual case

Lagrangian formalism The natural framework for describing Lagrangian formalism is a tangent bundle TM. Euler-Lagrange equations are obtained from a variational principle (see [3] for a geometrical exposition of this construction and its extension to higher order Lagrangians), but they can also be formulated directly in a geometrical language by using three tools: the exterior differential d, the Liouville vector field Δ and the vertical endomorphism S (also known as almost tangent structure, see [8]).

With these elements, we can formulate Euler-Lagrange formalism as follows. First we consider a Lagrangian $L \in C^\infty(TM)$ and a set of coordinates (x^i, v^i) in TM; and define:

- The **energy function** $E_L = \Delta L - L$. The expression in coordinates reads

$$E_L = v^i \frac{\partial L}{\partial v^i} - L$$

- The one form $\theta_L = S(dL)$. In coordinates:

$$\theta_L = \frac{\partial L}{\partial v^i} dx^i$$

- The two form $\omega_L = -d\theta_L$. The corresponding coordinate expression is

$$\omega_L = -\frac{\partial^2 L}{\partial x^j \partial v^i} dx^j \wedge dx^i - \frac{\partial^2 L}{\partial v^j \partial v^i} dv^j \wedge dx^i$$

With these tools we can easily formulate geometrically Euler-Lagrange equations:

Proposition 1. *Euler-Lagrange equations are the integral curves of the second order differential equations (SODE hereafter) vector fields $\Gamma \in \mathfrak{X}(TM)$ which verify the equation:*

$$i_\Gamma \omega_L = dE_L \qquad\qquad (1)$$

3.2 Geometrical mechanics on Lie algebroids

The construction of geometrical mechanics in Lie algebroids has been recently carried out mainly by A. Weinstein [16] and E. Martínez [11,12]. Weinstein studied the problem in [16] for the first time and presented a Lagrangian framework for Lie algebroids and groupoids, while Martínez developed a more geometrical construction for Lagrangian mechanics on Lie algebroids (recovering Weinstein's results) and extended it also to the Hamiltonian formalism. We will present now briefly both approaches, though for simplicity, we will use mainly Weinstein's construction in the examples we present in the following sections.

Weinstein's construction Weinstein's construction is based in Theorem 1. The main idea is to use this property for constructing an analogous procedure to that used in the usual Lagrangian formalism.

First of all we must consider the Lagrangian of the system. It will be a function defined on the Lie algebroid E, i.e. $E \in C^\infty(E)$. If we want to use the Poisson bracket on E^* we need to transfer it to E, and the natural way of doing it is via the analogue of the Legendre transform for Lie algebroids. As in the usual case, we define $\mathcal{F}L : E \to E^*$ and it reduces to the fiber derivative defined on E. In coordinates we write:

$$\mu_\alpha = \frac{\partial L}{\partial \lambda^\alpha}.$$

With this we can define the energy function as we did in the usual case ($E_L = \mathcal{F}L - L$) and use the Poisson structure on E^* to define the dynamics:

$$\frac{dx^i}{dt} = \{x^i, E_L\}; \quad \frac{d}{dt}\left(\frac{\partial L}{\partial \lambda^\alpha}\right) = \left\{\frac{\partial L}{\partial \lambda^\alpha}, E_L\right\} \tag{2}$$

The coordinate expression of these equations, whose deduction can be found in [16], is:

$$\frac{dx^i}{dt} = \sum_\alpha \rho^i_\alpha \lambda^\alpha = \rho(\sum_\alpha \lambda^\alpha e_\alpha) \cdot x^i \tag{3}$$

$$\frac{d}{dt}\left(\frac{\partial L}{\partial \lambda^\alpha}\right) = \sum_i \rho^i_\alpha \frac{\partial L}{\partial x^i} + \sum_{\beta,\gamma} C^\gamma_{\alpha\beta} \lambda^\beta \frac{\partial L}{\partial \lambda^\gamma} \tag{4}$$

The first equation must be considered as the algebroid analogue of the condition of a SODE vector field. In this sense, Weinstein provides the concept of **admissible** vectors and curves:

Definition 2. A tangent vector $v \in T_\alpha E$ is called **admissible** if

$$T\tau(u) = \rho(v).$$

A curve is called admissible when its tangent vectors are admissible, and in this sense, a vector field in E is called a SODE if its values are all admissible vectors. Then, the solutions of Euler-Lagrange equations will be curves on the algebroid whose tangent vectors are admissible. We will treat more carefully this point in the next section.

Martínez's approach The basic element of this framework is the suitable choice of the analogue of TTM, which is the natural setting for the study of geometric objects in usual lagrangian mechanics as the dynamical vector field, or symmetries of it.

In this paper we will only introduce the objects, and we address the interested reader to the references above to find a step-by-step construction. The vector bundle we define is:

$$\mathcal{L}E = \{(a, b, v) \in E \times E \times TE | \tau(a) = \tau(b), v \in T_a E, \rho(b) = T_a \tau(v)\} \quad (5)$$

This set is a vector bundle on E (as TTM is a bundle on TM) with the vector space structure given by:

$$\lambda(a, b_1, v_1) + \mu(a, b_2, v_2) = (a, \lambda b_1 + \mu b_2, \lambda v_1 + \mu v_2).$$

Moreover, we can consider also a Lie algebra structure on the set of sections (see [11]). With this Lie algebra structure and the projection onto the third factor $\rho^1(a, b, v) = v \quad a, b \in E \, v \in TE$ we have the following result ([11]):

Theorem 2. $\mathcal{L}E$ *is a Lie algebroid.*

In order to define the analogue of the Euler-Lagrange construction we saw above for the tangent bundle, we need three elements: the exterior differential for $\mathcal{L}E$ (that we have already since we saw its definition for a general algebroid), the vertical endomorphism and the analogue of the Liouville vector field. Both elements require of the definition of vertical elements in $\mathcal{L}E$. Since E is a vector bundle, we can identify the vectors tangent to the fibers with elements of the fiber itself: given $a \in E$ we can define for any $b \in E$ the element $b_a^V \in T_a E$ as

$$b_a^V F = \frac{d}{dt} F(a + tb)|_{t=0} \forall F \in C^\infty(E).$$

It is easy to understand thus that we consider that the vertical elements in $\mathcal{L}E$ are those with vanishing second factor, since because of the condition $\rho(b) = T\tau(v) \, \forall (a, b, v) \in \mathcal{L}E$, this means that the vector $v \in T_a E$ is actually a vertical vector on the vector bundle. The set of vertical sections will define a subbundle of $\mathcal{L}E$ which will be denoted by $\text{Ver}(E)$.

Then we define the **vertical endomorphism** for Lie algebroids, as the mapping $S : \mathcal{L}E \to \text{Ver}E$ such that $S(a, b, v) = (a, 0, b_a^V)$. Analogously we will define the **Liouville section** of $\mathcal{L}E$ as

$$\Delta(a, b, v) = (a, 0, a_a^V) \quad \forall (a, b, v) \in \mathcal{L}E.$$

In this context we can reread Weinstein's definition of admissible vectors (2). The set of admissible vectors (which provided the algebroid analogue of SODE's) will be now a subbundle of $\mathcal{L}E$ defined by:

$$\text{Adm}(E) = \{(a, b, v) \in \mathcal{L}E | a = b\}.$$

We can also search for a basis of sections on $\mathcal{L}E$ which allow us to write coordinate expressions of these objects. Martínez's choice is the following:

$$\mathcal{X}_\alpha(a) = \left(a, e_\alpha(\tau(a)), \rho^i_\alpha \frac{\partial}{\partial x^i}\right) \quad \mathcal{V}_\alpha(a) = \left(a, 0, \frac{\partial}{\partial \lambda^\alpha}\right), \tag{6}$$

according to the choice of coordinates $(x^i, \lambda^\alpha, z^\alpha, v^\alpha)$ where (x^i, λ^α) identify the first and (x^i, z^α) the second element in the triplet $(a, b, v) \in \mathcal{L}E$ and the vector

$$v = \rho^i_\alpha z^\alpha \frac{\partial}{\partial x^i} + v^\alpha \frac{\partial}{\partial \lambda^\alpha}.$$

These sections play the role of $\{\frac{\partial}{\partial x^i}\}$ and $\{\frac{\partial}{\partial v^i}\}$ in the usual case at TTM (with the obvious coordinate choice).

With these assumptions, we can write the expression of SODE sections as

$$\Gamma(x, \lambda) = \lambda^\alpha \mathcal{X}_\alpha + f^\alpha(x, \lambda)\mathcal{V}_\alpha,$$

which under the anchor mapping ρ^1 goes into a vector field on TE

$$\rho^1(\Gamma)(x, \lambda) = \rho^i_\alpha \lambda^\alpha \frac{\partial}{\partial x^i} + f^\alpha(x, \lambda)\frac{\partial}{\partial \lambda^\alpha}.$$

It is simple to see that in the simplest case, where the algebroid is actually a true tangent bundle, these sections are the usual vector fields related to second order differential equations. And we know from Weinstein results, that they must be also the solutions of Euler-Lagrange equations in its algebroid version. Then we have to search for an Euler-Lagrange construction for Lie algebroid, whose solutions belong to Adm(E), such that they reduce to the usual Euler-Lagrange formalism in the case of the tangent bundle, and such that its coordinate expression coincides with Weinstein's result.

The construction of such a formalism, presented in [11], is completely analogous to the usual case. We define:

- A one form $\theta_L = S(dL)$
- The two form $\omega_L = -d\theta_L$.
- The energy function $E_L = \Delta L - L$.
- And finally, we consider Euler-Lagrange equations as the SODE sections Γ which verify:

$$i_\Gamma \omega_L = dE_L \tag{7}$$

The corresponding coordinate expression coincides with Weinstein's result (see [11]).

The main advantage of this new formalism is the fact that the construction is a close analogue of the usual geometrical description of lagrangian mechanics. This will be useful for us in the following sections, where we will need to adapt some well known geometrical constructions of the usual formalism to make them work on some interesting systems defined on Lie algebroids.

4   Relevant examples of Lie algebroids

The aim of this section is to exhibit some interesting examples of Lie algebroids, which are relevant in the description of mechanical or even control systems.

4.1   The simplest examples

- The first and simplest one is the case of the tangent bundle itself. In this case, the vector bundle $E = TM$, the base manifold is M, the Lie algebra structure is provided by the natural Lie algebra of vector fields, and the anchor mapping is the identity mapping. Of course, any mechanical system described in a tangent bundle enters trivially in the context of lie algebroids.
- The second simplest example is the case of a Lie algebra. It can be considered as a fiber bundle over a point. Obviously, the anchor mapping must be zero, while the Lie algebra structure is the natural one.
- The following example is an integrable subbundle of a tangent bundle $D \subset TM$. In this case, $E = D$, the base is M, the lie algebra structure is the restriction to D of the algebra structure of vector fields (this is why the subbundle must be integrable, in order to define a subalgebra), and the mapping ρ is the natural injection $D \hookrightarrow TM$.

  The relevance of this example for mechanical system is obvious: any constrained system, with holonomic constraints, belongs to this class.
- In a different level, we can find another interesting example when considering a Dirac structure. Following [6] one can define a Dirac structure as a subbundle of $L \subset TM \oplus T^*M$, which must be maximally isotropic for the symmetric product:

$$\langle (X, \alpha), (Y, \beta) \rangle = \langle X, \beta \rangle + \langle Y, \alpha \rangle \quad \forall X, Y \in TM \; \alpha, \beta \in T^*M$$

The definition of the Lie algebroid structure is thus as follows: we take L as our vector bundle on M, the natural projection onto TM as the anchor mapping, and the difficult part now is the definition of the Lie algebra structure. In [6], it is done as follows:

$$[(X, \alpha), (Y, \beta)] = ([X, Y], \mathcal{L}_X \beta - \mathcal{L}_Y \alpha - d\alpha(Y))$$

Applications of Dirac structure are many, but we will mention particularly those related with control theory as are the implicit Hamiltonian systems [15].

4.2   More detailed examples of mechanical systems

We will present now a couple of more developed examples in which Lie algebroids help to describe in a suitable way mechanical systems whose usual description in terms of tangent or cotangent bundles turns out to be more involved or less elegant (though this must be considered as a subjective point of view, of course) than their Lie algebroid formulation.

The heavy top The first case that we will consider is the heavy rigid body. It is well known [10,1] that the more direct formulation of the top dynamics is done in $TSO(3)$ or $T^*SO(3)$ depending on the formalism we use (Lagrangian or Hamiltonian respectively). In those formulations a fixed reference frame, external to the rigid body is used, and the dynamics of the top has six degrees of freedom. The case without gravity is well known to be more suitably solved by a change in the reference frame which passes to a set of axes centered in the body: the equations we obtain then are the well known Euler equations for the rigid body. The problem with gravity is more involved since we must take into account the potential term in the lagrangian. Our experience with the case without gravity suggests us to consider a set of axes which move with the body. The expression for the Lagrangian for such a case will be:

$$L = \omega I \omega - Mg\gamma \cdot r^0_{CM} \quad \omega \in \mathfrak{so}(3) \ \gamma \in S^2 \tag{8}$$

where r^0_{CM} is the position of the center of mass in the body axes.

This Lagrangian is not a function defined on a tangent bundle, but on the manifold $S^2 \times \mathfrak{so}(3)$. This manifold can be considered as an example of the Lie algebroid known as **action algebroid**, taking the following choices (see [16]):

- The bundle E is taken to be the trivial bundle $S^2 \times \mathfrak{so}(3)$, as a bundle on the two dimensional sphere.
- The Lie algebra structure on the sections is naturally taken from $\mathfrak{so}(3)$. A basic fact to be taken into account is the existence of a natural action $\Phi : SO(3) \times S^2 \to S^2$ which can be used also for the elements of the Lie algebra via the exponential mapping $\exp : \mathfrak{so}(3) \to SO(3)$.
- Finally the anchor mapping is obtained from the representation of the Lie algebra on the set of vector fields defined by the action above. We define the fundamental vector fields $\rho : S^2 \times \mathfrak{so}(3) \to TS^2$

$$\rho(s,\omega) = X^\omega_s = \frac{d}{dt}\Phi(\exp t\omega, s)|_{t=0} \in T_s S^2 \quad \forall s \in S^2 \ \omega \in \mathfrak{so}(3)$$

The well known representation of the algebra on the two dimensional sphere gives us the desired result.

The dynamics can be thus easily described on the Lie algebroid framework. If we use Weinstein construction the corresponding Euler-Lagrange equations define the five dimensional system:

$$\dot{\theta} = \sin\phi\,\omega_1 + \cos\phi\,\omega_2$$
$$\dot{\phi} = \cos\phi\cot\theta\,\omega_1 + \sin\phi\cot\theta\,\omega_2 + \omega_3$$
$$I_1\dot{\omega}_1 = Mgr_0\sin\theta\sin\phi + (I_2 - I_3)\omega_2\omega_3$$
$$I_2\dot{\omega}_2 = -Mgr_0\sin\theta\cos\phi + (I_3 - I_1)\omega_1\omega_3$$
$$I_3\dot{\omega}_3 = (I_1 - I_2)\omega_2\omega_1$$

for (θ, ϕ) spherical coordinates on S^2. It can be easily seen that this dynamics coincide with other approaches to the problem as are [10,1], but it is the simplest Lagrangian treatment known to us. In [11] an equivalent result is obtained in the geometric approach.

Systems with symmetry Let us consider now a system whose configuration space is a principal fiber bundle $P \to Q$ with structural group G. Let us suppose a system whose dynamics is described with a Lagrangian which is cyclic in the coordinate corresponding to the fiber of P. Such a system can be described also as a system on TP/G instead of TP, but if we want to use the usual Lagrangian formalism, we must choose the first setting in order to have a pure tangent bundle. The Lagrangian formalism on lie algebroids allows us to work with the system defined on the reduced tangent bundle, directly, without using objects defined on the whole bundle TP that we must quotient in order to work with them. Of course, one may argue that it is not necessary to take the Lie algebroid version, and that is true, but we consider a more natural and formally elegant framework for the description of this kind of systems this **gauge algebroid**. A precise definition of the geometrical structure hidden behind the gauge algebroid can be found in [5]. In this short paper we will present only the direct definition, enumerating the elements which define the Lie algebroid structure:

- The vector bundle is taken to be the aforementioned TP/G, a vector bundle over Q, whose sections are the G–invariant vector fields. Local coordinates are given by those on Q, the corresponding velocities, and the elements of the Lie algebra $\mathfrak{g}$ (see [9]).
- The Lie algebra structure on the sections is trivially defined by using the natural Lie algebra structure of sections TP restricted to G–invariant vector fields.
- And finally the anchor mapping is taken to be the differential of the projection which defines the original principal bundle which constitutes the configuration space. It is evident that it goes properly to the quotient by G and it can be easily proved that it is an homomorphism for the Lie algebra structure we chose $\rho = (T\pi)|_{\text{quotient}} : TP/G \to TQ$.

We consider now a particular example which will be useful later: the well known example of the inverted pendulum on a cart. It is very well known the usual Lagrangian description:

$$L(\theta, s, \dot{\theta}, \dot{s}) = \frac{1}{2}(M + m)\dot{s}^2 + \frac{1}{2}ml^2\dot{\theta}^2 + ml\cos\theta\dot{s}\dot{\theta} - mgl\cos\theta \, , \qquad (9)$$

where this Lagrangian is supposed to be defined on the full bundle $T(S^1 \times \mathbb{R})$ which is the usual setting for the Lagrangian description of this system. But after the discussion above, it is evident that we can also consider this cyclic Lagrangian as a function defined on the gauge algebroid $T(S^1 \times \mathbb{R})/\mathbb{R}$ which is the most natural framework for the analysis of this system. It is a particularly simple example of Lie algebroid, since the corresponding Lie algebra on the set of sections is commutative and the anchor mapping is a projection.

The expression for L will then be exactly the same as (9), but now it is considered as a function of the Lie algebroid, where the unnecessary dependence in the group variable (s for the cart) has been removed:

$$L(\theta, \dot{\theta}, \dot{s}) = \frac{1}{2}(M + m)\dot{s}^2 + \frac{1}{2}ml^2\dot{\theta}^2 + ml\cos\theta\dot{s}\dot{\theta} - mgl\cos\theta \, . \qquad (10)$$

By using (2) we can write the dynamics associated to (10) as

$$\frac{d\theta}{dt} = \dot{\theta}$$

$$\frac{d}{dt}(ml^2\dot{\theta} + ml\cos\theta\dot{s}) + ml\sin\theta\dot{s}\dot{\theta} = mgl\sin\theta$$

$$\frac{d}{dt}\left((M + m)\dot{s} + ml\cos\theta\dot{\theta}\right) = 0$$

5   The controlled Lagrangian method

This method for the stabilization of systems with symmetry was recently proposed [2]. Given a system with symmetry, where the Lagrangian is cyclic in the group variable (as in the case we explained above), we suppose we control the group variables dynamics by fixing externally a generalized force u. For example, in the case of the inverted pendulum and the cart we have:

$$\frac{d}{dt}(ml^2\dot{\theta} + ml\cos\theta\dot{s}) + ml\sin\theta\dot{s}\dot{\theta} = mgl\sin\theta$$

$$\frac{d}{dt}\left((M + m)\dot{s} + ml\cos\theta\dot{\theta}\right) = u \qquad (11)$$

that can be also written in the form $\dot{x}^i = \Phi^i(x, \dot{x}, u)$ using the inverse of the mass matrix to transform the system above.

The main idea of the method is to find a feed-back law such that the dynamics above matches the dynamics given by a different Lagrangian (denoted by L_ρ^σ

in [2]) without explicit controls. In the reference above, a detailed algorithm is provided for the process in general systems with symmetry, in order to define, with this new Lagrangian, a system where some interesting fixed points become (asymptotically) stable. We can divide it in two parts: the choice of the suitable Lagrangian to reproduce the dynamics (the so called **matching theorem**) and the choice of the suitable control law for the stabilization. In the rest of the paper we will focus our attention in the first part (the second part will be studied in a future paper).

Our aim is to present a different approach to the matching problem, based in the Inverse Problem of Lagrangian dynamics. The Inverse Problem can be formulated as follows: given a system of differential equations on some manifold M (we will consider now only the second order case), does it exist a Lagrangian description of this dynamics? The set of conditions that must verify the system of differential equations are usually known as Helmholtz conditions (see [14] for a detailed exposition).

There are two interesting geometrical interpretations of these equations [7,4]. We follow now [7] and we consider that the dynamical vector field Γ defines a nonlinear connection (in general) on the bundle TTM. Crampin obtains Helmholtz equations as the expressions of a two form ω of maximal rank, which is preserved by the proposed dynamics and such that the vertical subspaces of $T(TM)$ are Lagrangian for ω and for $i_{X_h}\, d\omega$ where X_h any horizontal vector for the connection defined by the dynamical vector field. The construction of the Lagrangian is based, roughly speaking, on Poincaré Lemma (see [7,14]).

How can we generalize these approaches to the case of Lie algebroids? The definition of the connection given by the dynamics, will take place now on the bundle $\mathcal{L}E$. We have seen above how we can define vertical sections on this bundle, and in [11] it can be also found the definition of complete lifts of sections. This is all we need to adapt Crampin's formalism to this setting, and then we can define, for a section $\sigma \in \Gamma E$, its horizontal lift $\sigma^h \in \Gamma(\mathcal{L}E)$ as:

$$\sigma^h = \frac{1}{2}\left(\sigma^c + [\sigma^v, \Gamma]\right)$$

where Γ is the dynamical section in $\mathrm{Adm}(E) \subset \mathcal{L}E$, $\sigma^c \in \Gamma(\mathcal{L}E)$ is the complete lift of σ and $\sigma^v \in \Gamma(\mathcal{L}E)$ is its vertical lift.

It is not difficult to check that this definition makes sense, and that it does define a connection on the algebroid. Now we can extend the conditions for the existence of a lagrangian to the algebroid framework by borrowing them from the case above. We will consider thus the corresponding forms of maximal rank, preserved by the action of the dynamical section and such that the vertical subspaces of $\mathcal{L}E$ are lagrangian for the form and for $i_{\sigma_h}\, d\omega$ (where σ_h is any horizontal section). The problem arises now in the definition of the Lagrangian, since Poincaré Lemma will depend on the cohomological properties of the exterior differential of the algebroid, which are not known in general. Then, the analysis must be carried out case by case.

As an application, we will consider the equations (11), since in this case, Poincaré Lemma is valid. If we consider a feed-back law for the controls, they can be considered as a system of differential equations defined on $E = T(S^1 \times \mathbb{R})/\mathbb{R}$. If we apply the methods above, i.e. we consider the connection associated to this dynamics, search for a two form defined on $\mathcal{L}E$ such that it is preserved by it, and impose the conditions on the horizontal subspaces, we obtain a kind of Helmholtz equations suitably quotiented.

The conditions on the two form and its properties, if we consider $\omega = g_{\alpha\beta}\mathcal{X}^\alpha \wedge \mathcal{Y}^\beta$ (where $\{\mathcal{X}^\alpha, \mathcal{Y}^\alpha\}$ is the dual basis to the corresponding one on $\mathcal{L}E$ defined by horizontal and vertical sections), are that it is a symmetric non degenerate tensor, such that:

$$\frac{\partial g_{\alpha\beta}}{\partial \lambda^\gamma} = \frac{\partial g_{\alpha\gamma}}{\partial \lambda^\beta}$$

$$d_\Gamma g_{\alpha\beta} + \frac{1}{2}\frac{\partial \Phi^\gamma}{\partial \lambda^\beta}g_{\alpha\gamma} + \frac{1}{2}\frac{\partial \Phi^\gamma}{\partial \lambda^\alpha}g_{\beta\gamma} = 0$$

$$g_{\alpha\gamma}\left[d_\Gamma\left(\frac{\partial \Phi^\gamma}{\partial \lambda^\beta}\right) - 2\frac{\partial \Phi^\gamma}{\partial x^i}\rho_i^\beta - \frac{1}{2}\frac{\partial \Phi^\delta}{\partial \lambda^\beta}\frac{\partial \Phi^\gamma}{\partial \lambda^\delta}\right] =$$

$$g_{\beta\gamma}\left[d_\Gamma\left(\frac{\partial \Phi^\gamma}{\partial \lambda^\alpha}\right) - 2\frac{\partial \Phi^\gamma}{\partial x^i}\rho_i^\alpha - \frac{1}{2}\frac{\partial \Phi^\delta}{\partial \lambda^\alpha}\frac{\partial \Phi^\gamma}{\partial \lambda^\delta}\right]$$

where the Greek index runs on the set $\{\dot{\theta}, \dot{s}\}$ and $i = \theta$.

These equations must be also considered as equations for the control law, which is hidden in the section components and in Φ. If we consider the matrix g as the mass matrix of the new Lagrangian, we can simplify the system by searching for solutions whose mass matrix coincides with the mass matrix proposed in [2], where only a connection one form $k = k(\theta)d\theta$ on the algebroid in unknown (the connection they choose is actually a connection on the gauge algebroid). Moreover, we will have no fiber dependence in the mass matrix and hence the system above becomes much simpler. The resulting system turns out to be a first order differential equation for the control law and the connection of the mass matrix, from which we obtain two free constants that we have chosen trivially in this case, but might be useful in the stabilization part of the process.

If we want an equivalent expression for the mass matrix (the same formal dependence with respect to the unknown connection) the solution is as follows (see [2] for the notation):

$$k\ (\theta) = \frac{l^2 m \sqrt{m + 2M - m\cos(2\theta)}}{\sqrt{2M}(M + m)\sigma}$$

$$u\ (\theta) = \operatorname{arctanh}\left(\frac{\sqrt{2M}lm\cos\theta}{\sigma\sqrt{m + 2M - m\cos(2\theta)}}\right) \times$$

$$\exp\left(\frac{\sqrt{m + 2M - m\cos(2\theta)}}{l^2 m^2 M - (m + 2M)\sigma^2 + m(l^2 mM + \sigma^2)\cos(2\theta)}\right) \times$$

$$\exp\left(\frac{\sqrt{l^2 m^2 M - (m + 2M)\sigma^2 + m(l^2 mM + \sigma^2)\cos(2\theta)}}{l^2 m^2 M - (m + 2M)\sigma^2 + m(l^2 mM + \sigma^2)\cos(2\theta)}\right) \times$$

$$\exp\left(-\frac{\sqrt{2}lm(m + M)g\sigma\sqrt{m + 2M - m\cos(2\theta)}}{l^2 m^2 M - (m + 2M)\sigma^2 + m(l^2 mM + \sigma^2)\cos(2\theta)}\right) \times$$

$$\exp\left(-\frac{\sqrt{1 + \frac{2l^2 m^2 M\cos^2\theta}{\sigma^2(-m - 2M + m\cos(2\theta))}}\sin\theta}{l^2 m^2 M - (m + 2M)\sigma^2 + m(l^2 mM + \sigma^2)\cos(2\theta)}\right)$$

It is simple to verify that the corresponding mass matrix is non degenerate. As an advantage of working on the Lie algebroid directly, the Lagrangian we obtain is again a function of the gauge algebroid, and hence cyclic in the group variable (s in our example). Evidently, this is a control law much more complicated than that obtained in [2], but, it must be considered just as a preliminary result.

The objective of this small example was to show how the procedure works, and that the use of this method on Lie algebroids may be useful. A better solution for this example, as well as the extension to other interesting cases is our following objective.

Acknowledgements:

The author wishes to thank Laboratoire des Signaux et Systèmes where this work was begun. He is also deeply indebted to F. Lamnabhi-Lagarrigue, A.J. van der Schaft, E. Martínez and J. F. Cariñena for their support and many illuminating discussions.

References

1. M. Audin. *Spinning tops*. Cambridge University Press, 1996.
2. A.M. Bloch, N.E. Leonard, and J.E. Marsden. Controlled lagrangians and the stabilization of mechanical systems I. *IEEE Trans. Auto. Control*, To appear.
3. Jose F. Carinena and C. López. Geometric study of hamilton's variational principle. *Rev. in Math. Phys.*, 3(4):379–401, 1991.

4. J. F. Cariñena and E. Martínez Generalized Jacobi equation and Inverse Problem in Classical Mechanics In V.V. Dodonov and V. Man'ko, editors, *Integral systems, Solid State physics and theory of Phase transitions, Part II: Symmetries and Algebraic structures in Physics*, p. 59-65, Nova Science, 1992

5. A. Cannas da Silva, K. Hartshorn, and A. Weinstein. *Lectures on Geometric Models for Noncommutative algebras*. U.C. Berkeley, 1998.

6. T.J. Courant. Dirac manifolds. *Trans. of the AMS*, 319(2):631–661, June 1990.

7. M. Crampin. On the inverse problem of Lagrangian dynamics. *J. Phys. A:Math. and Gen.*,14 2567-2575, 1981.

8. J. Grifone. Structure presque tangente et connexions I. *Ann. Inst. Fourier*, XXII(1):287–334, 1972.

9. K. Mackenzie. *Lie groupoids and Lie algebroids in differential geometry*. London Math. Soc. Lecture Notes Series **124**, Cambridge University Press, 1987.

10. J.E. Marsden and T. Ratiu. *Introduction to Mechanics and Symmetry*. AMS, second edition, 1994.

11. E. Martínez. Lagrangian mechanics on lie algebroids. Preprint, Universidad de Zaragoza, 1999.

12. E. Martínez. Hamiltonian mechanics on lie algebroids. Preprint, Universidad de Zaragoza, 1999.

13. J. Pradines. Théorie de Lie pour les groupoïdes différentiables. Relations entre propriétés locales et globales. *C. R. Acad. Sci. Paris Sér. I Math.*, 263:907–910, 1966.

14. R.M. Santilli. *Foundations of Theoretical Mechanics I*. Springer-Verlag, 1978.

15. M. Dalsmo and A.J. van der Schaft . On representations and integrability of mathematical structures in energy-conserving physical systems. *SIAM J. Control and Optimization*, 37(1), 54-91, 1999.

16. A. Weinstein. Lagrangian mechanics and groupoids. *Fields Institut Com.*, pages 207–231, 1996.

Observer Design for Locally Observable Analytic Systems: Convergence and Separation Property

Fabio Conticelli[1] and Antonio Bicchi[2]

[1] Scuola Superiore Sant'Anna
56126 Pisa, Italy
bicchi@ing.unipi.it
[2] Robotics Interdepartmental Research Center "E. Piaggio"
University of Pisa
Via Diotisalvi, 2
56126 Pisa, Italy
contice@sssup.it

Abstract. This paper presents a novel nonlinear observer, which exhibits a local separation property. In fact, if there exists a stabilizing static state feedback, the designed observer permits to achieve local practical stability of the closed-loop system, if the real state has been substituted with the current estimated one. The observer requires only that the nonlinear system must be locally observable for the considered real analytic input function.

1 Introduction

1.1 Main Results

In this paper, we consider general nonlinear systems of the form:

$$\dot{\mathbf{x}} = \mathbf{f}(\mathbf{x}, \mathbf{u})$$
$$\mathbf{y}(\mathbf{x}) = \mathbf{h}(\mathbf{x}) \quad , \tag{1}$$

where $\mathbf{x} \in \mathbf{X}$, an open subset of $\Re^n$ containing the origin $\mathbf{x} = 0$, is the state vector, $\mathbf{u} \in \Re^m$ is the control input vector, $\mathbf{y} \in \Re^p$ is the output vector. The vector field $\mathbf{f}(\mathbf{x}, \mathbf{u})$ and the output map $\mathbf{h}(\mathbf{x})$ are assumed real analytic in the following, and $\mathbf{f}(0, 0) = 0$.

The main contribution of this paper is twofold:

- a novel local nonlinear observer is presented, which ensures practical stability of the trivial equilibrium in the observation error dynamics;
- a local separation property is achieved for the considered class of nonlinear analytic system, i.e. locally observable systems in the sense of [27,30].

The main idea in the observer design is the use of higher-order output time derivatives (henceforth called "observables") taken from the observability space associated to (1) in the assumption that the rank condition on the observability matrix is satisfied for the considered real analytic input function. Connections with strong observability under piecewise constant inputs are also highlight. Then, the observables of order higher than one are estimated in the observer design by using high-pass filters. Practical stability is guaranteed since the introduced persistent perturbations can be made arbitrarily small.

Moreover, a local separation property for the considered class of nonlinear system is proven. In particular if there exists a stabilizing state feedback, the equilibrium with the estimated state feedback remains locally practically stable. Finally an example is reported. The considered system is drift-less, and not uniformly locally observable. We stress that the techniques proposed in [22,1] can not be applied to the presented example. Our approach ensures local practical output stabilization of the trivial equilibrium. Even if the proposed analysis is local, the simulation results show that the region of attraction has a noticeable extension.

The paper is organized as follows. In Sect. 2, local observability of analytic nonlinear systems is briefly reminded. Sect. 3 presents a local separation property, which derives naturally from the application of the proposed nonlinear observer. In Sect. 4 the proposed framework has been applied to a simple, but meaningful, example. In Sect. 5 the major contribution of the paper is summarized and future investigations are outlined.

1.2 Related Work

A first approach to design an observer is to transform the original nonlinear system into another one for which the design is known. Transformations, which have been proposed in the literature, are the system immersion [8] which permits to obtain a bilinear system if the observation space is finite dimensional, and the linearization by means output injection [16,17,19] assuming that particular differential–geometric conditions on the system vector fields are verified. Rank conditions under which the dynamics of the observation error is linear, i.e. the original system can be transformed into the observer canonical form, are also investigated in [27]. Results on bilinear observers are presented in [5,10]. Extension of the Luenberger filter in a nonlinear setting, by using the time derivatives of the input, has been proposed in [30].

Early results on the observer design of bilinear systems without bad inputs are reported in [29]. Gauthier et al. [9,3] generalized the results in the case of input-affine nonlinear systems without bad inputs and applied the approach to biological reactors. The first step is to write the input affine nonlinear system in a so-called normal observation form. However, this form requires that the trivial input is an universal input [2] for the system, and also that a diffeomorphism can be constructed using the Lie derivatives of the output along the drift nonlinear term. Results on the normal observation form have been provided also by Tsinias [25,26]. In [6] the authors consider single input - single output input-affine nonlinear systems, and in the case of relative degree equal to the dimension of the state space n, full-rankness of the observability matrix, and global Hödel conditions on appropriate functions, it is shown the global asymptotic convergence of the estimated state; in case of relative degree less than n, stronger conditions on the admissible inputs have to be assumed.

High-gain techniques have been applied in the field of nonlinear observers. Early results are due to Tornambè [24], which proposed an approach based on high-gain approximate cancellation of the nonlinearity. A high gain observer which estimates the output derivatives combined with a globally bounded state feedback control law permits to obtain semiglobal stabilization by output feedback [7,14,20,22,23] in case of uniformly observable input-affine nonlinear systems [22]. In the recent reference [1] the authors employ a separation principle of a certain class of nonlinear systems, showing that with the estimated state feedback it is possible the performance recovery of the real state feedback, i.e. the asymptotic stability of the equilibrium, the region of attraction, and trajectories. Even if the results obtained concern stability analysis in the large, the class of considered systems is quite restricted. As we will show in the following, the approach presented in [1] cannot be applied to the presented example.

2 The Class of Systems

Let us associate to the nonlinear system (1) the following extended output map $\Phi : \Re^n \times \Re^{l_p\,m} \to \Re^{(l_p+1)\,p}$, defined as

$$\Phi(\mathbf{x}, \mathbf{v}) = \begin{pmatrix} \phi_0(\mathbf{x}) \\ \phi_1(\mathbf{x}, \mathbf{v}_0) \\ \cdots \\ \phi_{l_p}(\mathbf{x}, \mathbf{v}_0, \ldots, \mathbf{v}_{l_p-1}) \end{pmatrix} \tag{2}$$

where $\mathbf{v} = [\mathbf{v}_0^T, \ldots, \mathbf{v}_{l_p-1}^T]^T = [\mathbf{u}^T, \ldots, \mathbf{u}^{(l_p}-1)^T]^T \in \Re^{l_p\,m}$ is the extended input vector (i.e. the input vector and its time derivatives up to order $l_p -$ 1), and l_p is an integer such that $(l_p + 1)\,p \geq n$. The observable of order $i = 0, \ldots, l_p$ is $\phi_i(\mathbf{x}, \mathbf{v}_0, \ldots, \mathbf{v}_{i-1})$, the $i - th$ output time derivative. These functions are defined recursively as

$$\phi_0(\mathbf{x}) = \mathbf{h}(\mathbf{x}) \tag{3}$$

$$\phi_1(\mathbf{x}, \mathbf{v}_0) = \frac{\partial \phi_0(\mathbf{x})}{\partial \mathbf{x}} \mathbf{f}(\mathbf{x}, \mathbf{v}_0)$$

$$\cdots$$

$$\phi_{l_p}(\mathbf{x}, \mathbf{v}_0, \ldots, \mathbf{v}_{l_p-1}) = \frac{\partial \phi_{l_p-1}}{\partial \mathbf{x}} \mathbf{f}(\mathbf{x}, \mathbf{v}_0)$$

$$+ \sum_{j=0}^{l_p-2} \frac{\partial \phi_{l_p-1}}{\partial \mathbf{v}_j} \mathbf{v}_{j+1} \ .$$

Assumption 1 *There exists an integer l_p such that $(l_p + 1)p \geq n$ and the map $\Phi(\mathbf{x}, \mathbf{v})$ satisfies the rank condition:*

$$rank\left(\frac{\partial \Phi(\mathbf{x}, \mathbf{v})}{\partial \mathbf{x}}\right) = n \ , \tag{4}$$

$\forall(\mathbf{x}, \mathbf{v})$ *in an open neighborhood* $\mathbf{X}^0 \times \mathbf{V}^0$ *of* $\mathbf{X} \times \Re^{l_p m}$.

Defined $\mathbf{J}(\mathbf{x}, \mathbf{v}) = \left(\frac{\partial \Phi(\mathbf{x}, \mathbf{v})}{\partial \mathbf{x}}\right)$, this matrix is referred to as the Extended Output Jacobian (EOJ) associated to the nonlinear system (1) in the following. We now give the following motivation of the introduced assumption, based on the Implicit Function Theorem. Let us introduce the map $\mathbf{F} : \Re^{[(l_p+1)p+l_p m]} \times \Re^n \to \Re^{(l_p+1)p}$, defined as

$$\mathbf{F}(\overline{\mathbf{z}}, \overline{\mathbf{x}}) = \begin{pmatrix} \overline{\mathbf{z}}_0 - \phi_0(\overline{\mathbf{x}}) \\ \overline{\mathbf{z}}_1 - \phi_1(\overline{\mathbf{x}}, \overline{\mathbf{z}}_{l_p+1}) \\ \cdots \\ \overline{\mathbf{z}}_{l_p} - \phi_{l_p}(\overline{\mathbf{x}}, \overline{\mathbf{z}}_{l_p+1}, \ldots, \overline{\mathbf{z}}_{2l_p}) \end{pmatrix} \ , \tag{5}$$

where $\overline{\mathbf{z}}_k \in \Re^p$, $k = 0, \ldots, l_p$, $\overline{\mathbf{z}}_k \in \Re^m$, $k = l_p + 1, \ldots, 2l_p$, and $\overline{\mathbf{x}} \in \Re^n$. Consider the extended nonlinear system associated to (1):

$$\dot{\mathbf{x}} = \mathbf{f}(\mathbf{x}, \mathbf{v}_0) \tag{6}$$
$$\dot{\mathbf{v}}_0 = \mathbf{v}_1$$
$$\cdots$$
$$\dot{\mathbf{v}}_{l_p-1} = \nu$$
$$\xi = \Phi(\mathbf{x}, \mathbf{v}) \ ,$$

where $(\mathbf{x}, \mathbf{v}) \in \mathbf{X} \times \Re^{l_p m}$ is the extended state vector, $\nu \in \Re^m$ is the control input vector, and $\xi \in \Re^{(l_p+1)p}$ is the extended output vector. At any time instant $t \geq t^0 \geq 0$ the flow of the above nonlinear system satisfies the equation

$$\mathbf{F}(\mathbf{z}, \mathbf{x}) = 0 \ , \tag{7}$$

where $\mathbf{z} = [\xi^T \ \mathbf{v}^T]^T \in \Re^{l_p(l_p+1)p m}$, $\xi = [\xi_0^T, \ldots, \xi_{l_p}^T] \in \Re^{(l_p+1)p}$, and $\mathbf{x} \in \mathbf{X}$. If the assumption 1 is satisfied, given a point $(\mathbf{z}^0, \mathbf{x}^0)$, $\mathbf{z}^0 = [\xi^{0T} \ \mathbf{v}^{0T}]^T$, such that $(\mathbf{x}^0, \mathbf{v}^0) \in \mathbf{X}^0 \times \mathbf{V}^0$, there exist n observables taken from the $(l_p + 1)p$

ones in $\Phi(\mathbf{x},\mathbf{v})$ which are linear independent in $\mathbf{X}^0 \times \mathbf{V}^0$. By the Implicit Function Theorem, there exist two open neighborhoods, namely $\mathbf{A}^0 = \mathbf{I}^0 \times \mathbf{V}^0$ of $\mathbf{z}^0$, $\mathbf{X}^0$ of $\mathbf{x}^0$, being $\mathbf{I}^0$ an open neighborhood of ξ^0, an unique map $\mathbf{g} : \mathbf{A}^0 \to \mathbf{X}^0$, with $\mathbf{g}(\mathbf{z}) = \mathbf{x}$, such that $\mathbf{F}(\mathbf{z},\mathbf{g}(\mathbf{z})) = 0$.

Remark 1. In the case of an uniformly observable SISO (Single Input Single Output) nonlinear system [22,18], given the map $[y \; \dot{y} \; \dots y^{(n-1)}]^T = \Phi(\mathbf{x},\mathbf{v}) \in \Re^n$, being $\mathbf{x} \in \Re^n$ the state vector, and $\mathbf{v} = [u \; \dot{u} \; \dots u^{(n-2)}]^T \in \Re^{n-1}$ the extended input vector, the Assumption 1 is satisfied for each $(\mathbf{x},\mathbf{v}) \in \Re^n \times \Re^{n-1}$. Thus, uniform observability is a sufficient condition for our observer design.

Assume that the input function $\mathbf{u}(t)$, $t \geq t_0 \geq 0$ of the considered nonlinear system (1) is real analytic. The nonlinear system can be viewed as a time-varying analytic system without input

$$\dot{\mathbf{x}} = \mathbf{f}(\mathbf{x},\mathbf{u}(t)) = \bar{\mathbf{f}}(\mathbf{x},t) \qquad (8)$$
$$\mathbf{y}(\mathbf{x}) = \mathbf{h}(\mathbf{x}) \; . \qquad (9)$$

We re-call the following definition [13,30,27].

Definition 1. The system (8) is locally observable at $\mathbf{x}^0$ in the interval $[t_0, \; T]$, $\mathbf{x}^0 \in \mathbf{X}$, $T > t_0 \geq 0$, if given the output function $\mathbf{y}(t)$, $t \in [t_0, \; T]$, then $\mathbf{x}^0$ can be uniquely distinguished in a small neighborhood.

The class of nonlinear systems which are locally observable at $\mathbf{x}^0$ in the interval $[t_0, \; T]$ is determined by the following proposition [27,30].

Theorem 1. *The system (8) is locally observable at $\mathbf{x}^0$ in the interval $[t_0, \; T]$ if and only if there exist a neighborhood $\overline{\mathbf{X}}^0$ of $\mathbf{x}^0$, and an p-tuple of integers $(k_1, \dots, k_p)$, called the observability indices, such that:*

- $k_1 \geq k_2 \geq k_p > 0$ *and* $\sum_{i=1}^{p} k_i = n$;
- *defined the differential operator:* $\mathcal{N}^0 \, \mathbf{w} = \mathbf{w}$, *and*

$$\mathcal{N}\mathbf{w} = \bar{\mathbf{f}}^T \frac{\partial \mathbf{w}^T}{\partial \mathbf{x}} + \mathbf{w} \frac{\partial \bar{\mathbf{f}}}{\partial \mathbf{x}} + \frac{\partial \mathbf{w}}{\partial t}$$

where $\mathbf{w}(\mathbf{x},t) = (w_1(\mathbf{x},t), \dots, w_p(\mathbf{x},t))$, $w_i(\mathbf{x},t)$ *are real analytic time-varying functions, the observability matrix* $\mathbf{Q}(\mathbf{x},t) \in \Re^{n \times n}$ *defined as:*

$$\mathbf{Q}(\mathbf{x},t) = \begin{pmatrix} dh_1(\mathbf{x}) \\ \mathcal{N}dh_1(\mathbf{x}) \\ \dots \\ \mathcal{N}^{(k_1-1)}dh_1(\mathbf{x}) \\ \dots \\ dh_p(\mathbf{x}) \\ \dots \\ \mathcal{N}^{(k_p-1)}dh_p(\mathbf{x}) \end{pmatrix}, \qquad (10)$$

is nonsingular $\forall \mathbf{x} \in \overline{\mathbf{X}}^0$, *and* $t \in [t_0, T]$, *being* $dh_i = \frac{\partial h_i}{\partial \mathbf{x}}, i = 1, \ldots, p$ *the exact differentials associated to the output.*

The proof of the above proposition uses results of linear time-varying systems [21], and of perturbed differential equations [12], see also [13,27].

We are now in the position to determine the class of nonlinear analytic systems which satisfies Assumption 1.

Theorem 2. *Assume that the input function* $\mathbf{u}(t)$, $t \geq t_0 \geq 0$ *of the analytic nonlinear system (1) is real analytic. Assumption 1 holds if and only if the system (8) is locally observable at every* $\mathbf{x}^0 \in \mathbf{X}^0$ *in the interval* $[t_0, T]$, *for some* $T > t_0 \geq 0$.

Proof. If Assumption 1 holds, by using the Implicit Function Theorem, for every real analytic input function, given the output $\mathbf{y}(t)$, $t \geq t_0 \geq 0$, there exists an unique function $\mathbf{x}(t)$, $t \geq t_0 \geq 0$ which undergoes the equation $\mathbf{F}(\mathbf{z}, \mathbf{x}) = 0$. This is in fact the unique implicit function $\mathbf{g}(\mathbf{z}) = \mathbf{x}$, such that $\mathbf{F}(\mathbf{z}, \mathbf{g}(\mathbf{z})) = 0$. Hence, the system (8) is locally observable at every $\mathbf{x}^0 \in \mathbf{X}^0$ in the interval $[t_0, T]$, for some interval $T - t_0 > 0$ sufficiently small. Notice that since the rows of the observability matrix $\mathbf{Q}(\mathbf{x}, t)$ are differentials which appear as rows in the EOJ matrix $\mathbf{J}(\mathbf{x}, \mathbf{v})$, after a possible reordering of the indices of the output variables, there exist the observability indices which satisfy the conditions of Theorem 1 in $[t_0, T]$.

Vice versa, if the system (8) is locally observable at every $\mathbf{x}^0 \in \mathbf{X}^0$ in the interval $[t_0, T]$, for some interval $T - t_0 > 0$ sufficiently small, then by choosing $l_p = k_1 - 1$, where k_1 is the higher observability index, then Assumption 1 holds by construction.

The concept of local observability at $\mathbf{x}^0$ in the time interval $[t_0, T]$, $T > t_0$ with the assumption of real analytic input can be related to the observability of nonlinear systems under piecewise constant input. We remind that [28] the nonlinear system (1) is strongly observable at $\mathbf{x}^0$, if the autonomous system

$$\dot{\mathbf{x}} = \mathbf{f}^{\overline{\mathbf{u}}}(\mathbf{x}) = \mathbf{f}(\mathbf{x}, \overline{\mathbf{u}}) \tag{11}$$

$$\mathbf{y}(\mathbf{x}) = \mathbf{h}(\mathbf{x}) \ . \tag{12}$$

with $\overline{\mathbf{u}}$ constant, is locally weakly observable [11], for all $\overline{\mathbf{u}}$ of interest. In the analytic case, if the nonlinear system is weakly controllable, then the system is local weakly observable at $\mathbf{x}^0 \in \mathbf{X}^0$ if and only if the observability rank condition is satisfied [11]. If this is the case, strong observability implies that $\dim (d\mathcal{O})(\mathbf{x}^0) = n$, $\forall \overline{\mathbf{u}}$ of interest. If a system is strongly observable at $\mathbf{x}^0 \in \mathbf{X}^0$, then every constant input function distinguishes between nearby states, which imply that every C^∞ function (and in particular analytic) also distinguishes (see [28]). Hence if the nonlinear system(1) is strongly observable at every $\mathbf{x}^0 \in \mathbf{X}^0$ then there exists $T > t_0$ such that it is locally observable at $\mathbf{x}^0$ in the interval $[t_0, T]$ under real analytic input. We can state the following.

Theorem 3. *Assume that the system (1) is weakly controllable and analytic. If the system is strongly observable at every $\mathbf{x}^0 \in \mathbf{X}^0$ (under piecewise constant inputs), i.e. $\dim(d\mathcal{O})(\mathbf{x}^0) = n$, $\forall \bar{\mathbf{u}}$ of interest, then the system (1), under real analytic input functions, satisfies Assumption 1.*

Proof. This is a straightforward consequence of the above discussion and Theorem 2.

3 A Local Separation Property

Assume that there exists a static state feedback which ensures the local asymptotic stability of the trivial equilibrium.

Assumption 2 *Consider the nonlinear system (1), there exist two functions $\alpha : \mathbf{X} \to \Re^m$, and $V : \mathbf{X} \to \Re$ both, at least, of class C^1, such that $V(\mathbf{x})$ is positive definite, and $\frac{\partial V}{\partial \mathbf{x}} \mathbf{f}(\mathbf{x}, \alpha(\mathbf{x})) < 0$ in an open neighborhood of the origin $\mathbf{x} = 0$.*

We remind the following result, for the proof, see for example [4,18,15].

Theorem 4. *Consider the extended nonlinear system in Eq. (6), if Assumption 2 holds, then there exists a function $\bar{\alpha} : \mathbf{X} \times \Re^{l_p m} \to \Re^m$ of class, at least, C^1, such that the equilibrium $(\mathbf{x}, \mathbf{v}) = (0, 0)$ of the closed-loop system*

$$\dot{\mathbf{x}} = \mathbf{f}(\mathbf{x}, \mathbf{v}_0) \tag{13}$$
$$\dot{\mathbf{v}}_0 = \mathbf{v}_1$$
$$\cdots$$
$$\dot{\mathbf{v}}_{l_p - 1} = \bar{\alpha}(\mathbf{x}, \mathbf{v}) \quad ,$$

is asymptotically stable.

Denote with s the Laplace variable, $\mathbf{Y}(s)$ the output Laplace transform. $\hat{\xi} = [\mathbf{y}^T, \psi_1^T, \ldots, \psi_{l_p}^T]^T$ is the estimated extended output vector, being $\psi_i \in \Re^p$, $i = 1, \ldots, l_p$ the estimation of the $i - th$ output derivative, and T is a small positive constant. Indicate with $\mathbf{Q}$ a positive definite matrix, and, in virtue of Assumption 1, with $\mathbf{J}^+ = (\mathbf{J}^T \mathbf{J})^{-1} \mathbf{J}^T$ the left pseudo-inverse of the EOJ. Consider the observer

$$\dot{\hat{\mathbf{x}}} = \mathbf{f}(\hat{\mathbf{x}}, \mathbf{v}_0) + \mathbf{P}(\hat{\mathbf{x}}, \mathbf{v})(\hat{\xi} - \Phi(\hat{\mathbf{x}}, \mathbf{v})) \tag{14}$$

$$\mathbf{P}(\hat{\mathbf{x}}, \mathbf{v}) = \left(\mathbf{Q} + \left(\frac{\partial \mathbf{f}}{\partial \mathbf{x}}(\hat{\mathbf{x}}, \mathbf{v}_0) \right) \right) \mathbf{J}^+(\hat{\mathbf{x}}, \mathbf{v})$$

$$\psi_i(s) = \frac{s^i}{(1 + T s)^i} \mathbf{Y}(s), \quad \psi_i(0) = 0, \quad i = 1, \ldots, l_p$$

The above equation can be rewritten as

$$\dot{\mathbf{x}} = \mathbf{f}(\hat{\mathbf{x}}, \mathbf{u}) + \mathbf{P}(\hat{\mathbf{x}}, \mathbf{v})\,(\xi - \varPhi(\hat{\mathbf{x}}, \mathbf{v})) - \mathbf{P}(\hat{\mathbf{x}}, \mathbf{v})\epsilon^{*} \quad ,$$

where $\epsilon^{*} = [0^{T}, \epsilon_{1}^{T}, \ldots, \epsilon_{l_{p}}^{T}]^{T} \in \Re^{(l_{p}+1)\,p}$, $\epsilon_{i} = \mathbf{y}^{(i)} - \psi_{i}$, $i = 1, \ldots, l_{p}$ is the introduced persistent perturbation due to the estimated observables. The main result of this paper is the following.

Theorem 5. *If Assumptions 1, and 2 hold, and T is chosen sufficiently small, the equilibrium $(\mathbf{x}, \hat{\mathbf{x}} - \mathbf{x}) = (0,0)$ of the closed loop system*

$$\dot{\mathbf{x}} = \mathbf{f}(\mathbf{x}, \mathbf{v}_{0}) \tag{15}$$
$$\dot{\mathbf{v}}_{0} = \mathbf{v}_{1}$$
$$\cdots$$
$$\dot{\mathbf{v}}_{l_{p}-1} = \overline{\alpha}(\hat{\mathbf{x}}, \mathbf{v})$$
$$\mathbf{y}(\mathbf{x}) = \mathbf{h}(\mathbf{x})$$
$$\dot{\hat{\mathbf{x}}} = \mathbf{f}(\hat{\mathbf{x}}, \mathbf{v}_{0}) + \mathbf{P}(\hat{\mathbf{x}}, \mathbf{v})\,(\hat{\xi} - \varPhi(\hat{\mathbf{x}}, \mathbf{v}))$$
$$\mathbf{P}(\hat{\mathbf{x}}, \mathbf{v}) = (\mathbf{Q} + \left(\frac{\partial \mathbf{f}}{\partial \mathbf{x}}(\hat{\mathbf{x}}, \mathbf{v}_{0})\right)) \mathbf{J}^{+}(\hat{\mathbf{x}}, \mathbf{v})$$
$$\psi_{i}(s) = \frac{s^{i}}{(1 + T\,s)^{i}}\,\mathbf{Y}(s), \ \psi_{i}(0) = 0, \ i = 1, \ldots, l_{p}$$

is locally practically stable, i.e. $\forall \epsilon > 0$ there exist $\delta_{1} > 0$, $K > 0$, and T which depends on K, such that if $\|[\mathbf{x}^{T}(0), \hat{\mathbf{x}}^{T}(0) - \mathbf{x}^{T}(0)]^{T}\| < \delta_{1}$, then $\|\epsilon^{*}(t)\| < K$, $t > 0$, and the solutions of (15) satisfy the condition

$$\|[\mathbf{x}^{T}(t, 0, \epsilon^{*}), \ \hat{\mathbf{x}}^{T}(t, 0, \epsilon^{*}) - \mathbf{x}^{T}(t, 0, \epsilon^{*})]^{T}\| < \epsilon, \ t > 0 \quad .$$

We begin with the following Lemmas.

Lemma 1. *Consider the nonlinear system $\dot{\mathbf{x}} = \mathbf{f}(\mathbf{x}) + \mathbf{g}(\mathbf{x})\mathbf{u}(t)$, where $\mathbf{x} \in \Re^{n}$, $\mathbf{u}(t) \in \Re^{m}$, and $\mathbf{f}(\mathbf{x})$, $\mathbf{g}(\mathbf{x})$ are smooth vector fields. Assume that the origin of $\dot{\mathbf{x}} = \mathbf{f}(\mathbf{x})$ is a locally asymptotically stable equilibrium. Then $\forall \epsilon > 0$, there exist $\delta_{1} > 0$ and $K > 0$ such that if $\|\mathbf{x}(0)\| < \delta_{1}$ and $\|\mathbf{u}(t)\| < K$, $t \geq t^{0} \geq 0$, the solution $\mathbf{x}(t, t^{0}, \mathbf{u})$ of $\dot{\mathbf{x}} = \mathbf{f}(\mathbf{x}) + \mathbf{g}(\mathbf{x})\mathbf{u}(t)$ satisfies the condition: $\|\mathbf{x}(t, t^{0}, \mathbf{u})\| < \epsilon$, $t \geq t^{0} \geq 0$.*

For the proof see, for example, [18].

Lemma 2. *Consider the nonlinear system (1), if Assumption 1 holds, and T is chosen sufficiently small, the equilibrium $\mathbf{e} = \hat{\mathbf{x}} - \mathbf{x} = 0$ of the observation error dynamics deriving from the observer (14) is locally practically stable, i.e. $\forall \epsilon_{\mathbf{e}} > 0$ there exist $\delta_{1} > 0$, $K > 0$, and T which depends on K, such that if $\|\mathbf{e}(0)\| < \delta_{1}$, then $\|\epsilon^{*}(t)\| < K$, $t > 0$, and the observation error satisfies the condition $\|\mathbf{e}(t, 0, \epsilon^{*})\| < \epsilon_{\mathbf{e}}$, $t > 0$.*

Proof. By using Eq. (14), the dynamics of the observation error $e = \hat{x} - x$ results

$$
\begin{aligned}
\dot{e} &= \left(\frac{\partial f}{\partial x}(\hat{x}, v_0)\right) - P(\hat{x}, v)\left(\frac{\partial \Phi}{\partial x}(\hat{x}, v)\right) \\
&\quad + \left(p_f(v_0, e) + P(\hat{x}, v)\, p_\Phi(v, e)\right) \\
&= -Q e - P(\hat{x}, v)\epsilon^* + P(\hat{x}, v)\, p_\Phi(v, e) - p_f(v_0, e)
\end{aligned}
\tag{16}
$$

where ϵ^* is the introduced perturbation due to the observables estimation, and the functions $p_f(v_0, e)$ and $p_\Phi(v, e)$ vanish at $e = 0$ with their first order partial derivatives, i.e. satisfy the conditions

$$
\lim_{e \to 0} \frac{p_f(v_0, e)}{\|e\|} = 0, \quad \lim_{e \to 0} \frac{p_\Phi(v, e)}{\|e\|} = 0 \; .
\tag{17}
$$

We now prove by induction on the order of the output derivatives that the error ϵ^* can be reduced to an arbitrarily small perturbation if the constant T is sufficiently small. In fact, the Laplace transform of ϵ_1 results: $\epsilon_1(s) = \dot{Y}(s) - \frac{s}{1+Ts} Y(s) = T \frac{s}{1+Ts} \dot{Y}(s) - \frac{y(0^+)}{1+Ts}$, where $Y(s)$ denotes the output Laplace transform, and (with an abuse of notation) $\dot{Y}(s) = s\, Y(s) - y(0^+)$ indicates the output derivative Laplace transform. $\forall t \geq 0$, $\epsilon_1(t) = T\dot{y}_f(t) - \frac{1}{T} y(0^+) \exp(\frac{-t}{T})$, where $\dot{y}_f(t)$ is the output derivative filtered by $\frac{s}{1+Ts}$. Since the high-pass filter $\frac{s}{1+Ts}$ is Bounded Input Bounded Output (BIBO), there exists $M_{\dot{y}} > 0$ such that $\|\dot{y}_f(t)\| < M_{\dot{y}}, \forall t \geq 0$. Let us consider the term $\frac{1}{T} y(0^+) \exp(\frac{-t}{T})$ in the output derivative error $\epsilon_1(t)$, it holds: $\lim_{T\to 0} \frac{1}{T} \|y(0^+)\| \exp(\frac{-t}{T}) = 0, \forall t > 0$, which means that $\forall \epsilon^* > 0$, there exists $\delta > 0$, such that $T < \delta$, implies $\frac{1}{T} \|y(0^+)\| \exp(\frac{-t}{T}) < \epsilon^*$. Hence $\forall \bar{\epsilon} > 0$, fixed $\epsilon^* < \bar{\epsilon}$, choose $T < \min\{\delta, \frac{\bar{\epsilon}-\epsilon^*}{M_{\dot{y}}}\}$, then $\|\epsilon_1(t)\| < T M_{\dot{y}} + \epsilon^* < \bar{\epsilon}, \forall t > 0$. Hence $\lim_{T\to 0} \|\epsilon_1(t)\| = 0, \forall t > 0$. Assume that $\lim_{T\to 0} \|\epsilon_k(t)\| = 0, \forall t > 0$, where k is chosen in the open indices set $(1, \ldots, l_p)$. Since $\psi_{k+1}(s) = \frac{s^{k+1}}{(1+Ts)^{k+1}} Y(s) = \frac{s}{(1+Ts)} \psi_k(s)$, and $s\, \psi_k(s) = s\, Y^{(k)}(s) - s\, \epsilon_k(s) = Y^{(k+1)}(s) + y^{(k)}(0^+) - s\, \epsilon_k(s)$, where $Y^{(k+1)}(s)$ denotes the Laplace transform of the $(k+1)-th$ output derivative, it follows that:

$$
\begin{aligned}
\psi_{k+1}(s) &= \frac{s}{(1+Ts)} \psi_k(s) = \frac{1}{(1+Ts)}\left(Y^{(k+1)}(s)\right. \\
&\quad \left. + y^{(k)}(0^+) - s\, \epsilon_k(s)\right) ,
\end{aligned}
\tag{18}
$$

and

$$
\begin{aligned}
\epsilon_{k+1}(s) &= Y^{(k+1)}(s) - \psi_{k+1}(s) \\
&= T\frac{s}{(1+Ts)} Y^{(k+1)}(s) - \frac{y^{(k)}(0^+)}{(1+Ts)} + \frac{s}{(1+Ts)}\epsilon_k(s)
\end{aligned}
\tag{19}
$$

the above expression, if $y^{(k+1)}(t)$ is bounded and by using the induction assumption, immediately yields $\lim_{T \to 0} \|\epsilon_{k+1}(t)\| = 0$, $\forall t > 0$.

Let us consider the nonlinear system in Eq. (16). It easy to show that, if $\epsilon^*(t) = 0$, $t \geq 0$, the origin $e = 0$ (of the unperturbed system) is locally asymptotically stable. In fact, consider the quadratic Lyapunov function candidate $V = \frac{1}{2} e^T e$, its time derivative results:

$$\dot{V} = e^T \dot{e} \tag{20}$$

$$= e^T \left[\left(\frac{\partial f}{\partial x}(\hat{x}, v_0) \right) - P(\hat{x}, v) \left(\frac{\partial \Phi}{\partial x}(\hat{x}, v) \right) \right] e$$

$$+ e^T \left(p_f(v_0, e) + P(\hat{x}, v) \, p_\phi(v, e) \right)$$

$$= -e^T Q e + e^T \left(p_f(v_0, e) + P(\hat{x}, v) \, p_\phi(v, e) \right) \quad .$$

Then, due to Eq. (17), there exists an open neighborhood of the trivial equilibrium in which $\dot{V}$ is negative definite. By applying the Lyapunov's direct method, the claim follows. Lemma 1 indicates that $\forall \epsilon_e > 0$ there exist $\delta_1 > 0$ and $K > 0$ such that if the initial observation error is sufficiently small, i.e $\|e(0)\| < \delta_1$ and choosing T sufficiently small such that $\|\epsilon_k(t)\| < \frac{K}{l_p}$, $k = 1, \ldots, l_p$, $t > 0$, it follows $\|\epsilon^*\| < \sum_{k=1}^{l_p} \|\epsilon_k(t)\| < K$, and the observation error satisfies the condition $\|e(t, 0, \epsilon^*)\| < \epsilon_e$, $t > 0$.

Proof of Theorem 5.

Proof. In the coordinates $(x, e) = (x, \hat{x} - x)$ the system (15) results

$$\dot{x} = f(x, v_0)$$

$$\dot{v}_0 = v_1$$

$$\ldots$$

$$\dot{v}_{l_p-1} = \overline{\alpha}(e + x, v)$$

$$y(x) = h(x)$$

$$\dot{e} = -Qe - P(\hat{x}, v)\epsilon^* + P(\hat{x}, v) \, p_\phi(v, e)$$

$$- p_f(v_0, e) \quad . \tag{21}$$

As first step , we prove that if $\epsilon^* = 0$ the equilibrium of the unperturbed system (21) is asymptotically stable.

Let us denote, for convenience, $w = [x^T \ v^T]^T \in X \times \Re^{l_p m}$, $\hat{w} = [\hat{x}^T \ v^T]^T$, $\overline{\alpha}(w) = \overline{\alpha}(x, v)$ and

$$\tilde{f}(w, \overline{\alpha}(w)) = \begin{pmatrix} f(x, v_0) \\ \dot{v}_0 \\ \ldots \\ \overline{\alpha}(w) \end{pmatrix} \quad , \tag{22}$$

notice also that $\hat{\mathbf{w}} - \mathbf{w} = [\mathbf{e}^T,\ \mathbf{0}^T]^T$ is only a function of the observation error $\mathbf{e}$. Since

$$\tilde{\mathbf{f}}(\mathbf{w}, \overline{\alpha}(\hat{\mathbf{w}})) - \tilde{\mathbf{f}}(\mathbf{w}, \overline{\alpha}(\mathbf{w})) = \left(\frac{\partial \tilde{\mathbf{f}}}{\partial \overline{\alpha}}(\overline{\alpha}(\mathbf{w})) \right)$$

$$(\overline{\alpha}(\hat{\mathbf{w}}) - \overline{\alpha}(\mathbf{w})) + \mathbf{p}_{\tilde{\mathbf{f}}}(\mathbf{w}, \overline{\alpha}(\hat{\mathbf{w}}) - \overline{\alpha}(\mathbf{w})) \qquad (23)$$

and

$$\overline{\alpha}(\hat{\mathbf{w}}) - \overline{\alpha}(\mathbf{w}) = \left(\frac{\partial \overline{\alpha}}{\partial \mathbf{w}}(\mathbf{w}) \right) (\hat{\mathbf{w}} - \mathbf{w}) + \mathbf{p}_{\overline{\alpha}}(\mathbf{w}, \mathbf{e}) \qquad (24)$$

where the functions $p_{\tilde{\mathbf{f}}}(\mathbf{w}, \overline{\alpha}(\hat{\mathbf{w}}) - \overline{\alpha}(\mathbf{w}))$ and $p_{\overline{\alpha}}(\mathbf{w}, \mathbf{e})$ vanish at $\mathbf{e} = 0$ with their first order partial derivatives, i.e. satisfy the conditions:

$$\lim_{\mathbf{e} \to 0} \frac{p_{\tilde{\mathbf{f}}}(\mathbf{w}, \overline{\alpha}(\hat{\mathbf{w}}) - \overline{\alpha}(\mathbf{w}))}{\|\overline{\alpha}(\hat{\mathbf{w}}) - \overline{\alpha}(\mathbf{w})\|} = 0, \quad \lim_{\mathbf{e} \to 0} \frac{p_{\overline{\alpha}}(\mathbf{w}, \mathbf{e})}{\|\mathbf{e}\|} = 0 \ .$$

Using Eqs. (23), (24) the closed-loop system (21) is written as

$$\dot{\mathbf{w}} = \tilde{\mathbf{f}}(\mathbf{w}, \overline{\alpha}(\mathbf{w})) + \tilde{\mathbf{p}}(\mathbf{w}, \mathbf{e}) \qquad (25)$$

$$\dot{\mathbf{e}} = -\mathbf{Q}\mathbf{e} + \mathbf{P}(\hat{\mathbf{x}}, \mathbf{v})\, \mathbf{p}_\phi(\mathbf{v}, \mathbf{e}) - \mathbf{p}_f(\mathbf{v}_0, \mathbf{e}) \quad , \qquad (26)$$

where

$$\tilde{\mathbf{p}}(\mathbf{w}, \mathbf{e}) = \left(\frac{\partial \tilde{\mathbf{f}}}{\partial \overline{\alpha}}(\overline{\alpha}(\mathbf{w})) \right) \left[\left(\frac{\partial \overline{\alpha}}{\partial \mathbf{w}}(\mathbf{w}) \right) (\hat{\mathbf{w}} - \mathbf{w}) + \mathbf{p}_{\overline{\alpha}}(\mathbf{w}, \mathbf{e}) \right] \qquad (27)$$

$$+ \ \mathbf{p}_{\tilde{\mathbf{f}}}(\mathbf{w}, \overline{\alpha}(\hat{\mathbf{w}}) - \overline{\alpha}(\mathbf{w}))$$

is such that $\tilde{\mathbf{p}}(\mathbf{w}, 0) = 0$, $\forall \mathbf{w} \in \mathbf{X} \times \Re^{l_p\,m}$.

From Theorem 4, the system $\dot{\mathbf{w}} = \tilde{\mathbf{f}}(\mathbf{w}, \overline{\alpha}(\mathbf{w})) + \tilde{\mathbf{p}}(\mathbf{w}, 0)$ has an asymptotically stable equilibrium at $\mathbf{w} = 0$, since also $\dot{\mathbf{e}} = -\mathbf{Q}\mathbf{e} + \mathbf{P}(\hat{\mathbf{x}}, \mathbf{v})\, \mathbf{p}_\phi(\mathbf{v}, \mathbf{e}) - \mathbf{p}_f(\mathbf{v}_0, \mathbf{e})$ has an asymptotically stable equilibrium at $\mathbf{e} = 0$, then, as a consequence of a known property deriving from the center manifold theory [18], the cascade system (25), i.e. the unperturbed system ($\epsilon^* = 0$) in Eq. (15), has an asymptotically stable equilibrium at $(\mathbf{w}, \mathbf{e}) = (0, 0)$.

Consider the perturbed system

$$\dot{\mathbf{w}} = \tilde{\mathbf{f}}(\mathbf{w}, \overline{\alpha}(\mathbf{w})) + \tilde{\mathbf{p}}(\mathbf{w}, \mathbf{e})$$

$$\dot{\mathbf{e}} = -\mathbf{Q}\mathbf{e} - \mathbf{P}(\hat{\mathbf{x}}, \mathbf{v})\epsilon^* + \mathbf{P}(\hat{\mathbf{x}}, \mathbf{v})\, \mathbf{p}_\phi(\mathbf{v}, \mathbf{e}) - \mathbf{p}_f(\mathbf{v}_0, \mathbf{e}) \qquad (28)$$

$\forall \epsilon > 0$ there exist $\delta_1 > 0$ and $K > 0$ such that if the initial state is sufficiently small, i.e $\|[\mathbf{x}(0)^T, \mathbf{e}(0)^T]^T\| < \delta_1$, then since from Lemma 2 there exists $T > 0$ sufficiently small such that $\|\epsilon^*(0)\| < K$, $t > 0$, from Lemma 1, the solutions of (28) satisfy the condition $\|[\mathbf{x}^T(t, 0, \epsilon^*),\ \mathbf{e}^T(t, 0, \epsilon^*)]^T\| < \epsilon$, $t > 0$.

4   Example

Consider the model of a holonomic vehicle, which is able only to measure the distance from the origin of a priori fixed reference frame:

$$\dot{\mathbf{x}} = \mathbf{u}$$
$$y(\mathbf{x}) = \frac{1}{2}\left(x_1^2 + x_2^2\right) \quad , \tag{29}$$

where $\mathbf{x} = [x_1, \ x_2]^T \in \mathbf{X}$, an open subset of $\Re^2$ containing the origin, is the state vector, $\mathbf{u} = [u_1, \ u_2]^T \in \Re^2$ is the control input vector, $y \in \Re$ is the output vector. We consider the problem of finding an output feedback control law which locally stabilizes the origin $\mathbf{x} = 0$. First notice that the first approximation of the above model around the trivial equilibrium is non–observable, but the observability rank condition is satisfied $\forall \mathbf{x} \in \mathbf{X}$. It follows that, since the model is analytic and weakly controllable, it is also weakly locally observable $\forall \mathbf{x} \in \mathbf{X}$. Fixed $l_p = 1$, the EOJ matrix results:

$$\mathbf{J}(\mathbf{x}, \mathbf{v}_0) = \begin{bmatrix} x_1 & x_2 \\ u_1 & u_2 \end{bmatrix} \quad , \tag{30}$$

the singularity manifold (i.e. $\det\left(\mathbf{J}(\mathbf{x}, \mathbf{v}_0)\right) = 0$) is given by the set $S = \{(\mathbf{x}, \mathbf{u}) \in \mathbf{X} \times \Re^2 : \ x_1 u_2 - x_2 u_1 = 0\}$. This analysis indicates that the simple holonomic vehicle (29) is not uniformly observable [22,18], in fact an input $u_b(t)$, $t \geq t_0 \geq 0$, such that, at a certain time instant $\bar{t} \geq t_0$, satisfies the condition $x_1(\bar{t})\, u_2(\bar{t}) - x_2(\bar{t})\, u_1(\bar{t}) = 0$, i.e. the point $(\mathbf{x}, \mathbf{u}_b)$ lies on the manifold S at time $\bar{t}$, causes a loss of rank in the EOJ matrix, the set of inputs $u_b(t)$ which satisfy the above property are the *bad inputs*. The assumption 1 is verified only in the region $(\mathbf{X} \times \Re^2) \setminus S$. Moreover, the system can not globally transformed in an element of the class of nonlinear systems described in the recent reference [1]. In fact, the map

$$\bar{x}_1(\mathbf{x}, \mathbf{u}) = y(\mathbf{x}) = \frac{1}{2}\left(x_1^2 + x_2^2\right)$$
$$\bar{x}_2(\mathbf{x}, \mathbf{u}) = \dot{y}(\mathbf{x}) = x_1 u_1 + x_2 u_2$$
$$z_1(\mathbf{x}, \mathbf{u}) = u_1$$
$$z_2(\mathbf{x}, \mathbf{u}) = u_2 \quad , \tag{31}$$

does not define a global diffeomorphism in the extended state space $\mathbf{X} \times \Re^2$.

We apply the observer design and the local separation property presented in Theorem 5, in the case $(l_p + 1)\, p = n = 2$, i.e. $l_p = 1$. This is is referred to as the *full observer* design. A stabilizing state feedback is simply $\mathbf{u} = \alpha_0(\mathbf{x}) = [-k\, x_1\ -k\, x_2]^T$, $k > 0$. By using Theorem 4 with $V(\mathbf{x}) = \frac{1}{2}\mathbf{x}^T\mathbf{x}$, a stabilizing state feedback of the extended system:

$$\dot{\mathbf{x}} = \mathbf{v}_0$$
$$\dot{\mathbf{v}}_0 = \nu \quad , \tag{32}$$

is $\nu = \overline{\alpha}(\mathbf{x}, \mathbf{v}_0) = \dot{\alpha}_0(\mathbf{x}, \mathbf{v}_0) - \mathbf{x} - \lambda\,(v_0 - \alpha_0(\mathbf{x})), \lambda > 0$, with $\dot{\alpha}_0(\mathbf{x}, \mathbf{v}_0) = -k\,\mathbf{v}_0$. ¿From Theorem 5, the output feedback given by the equations

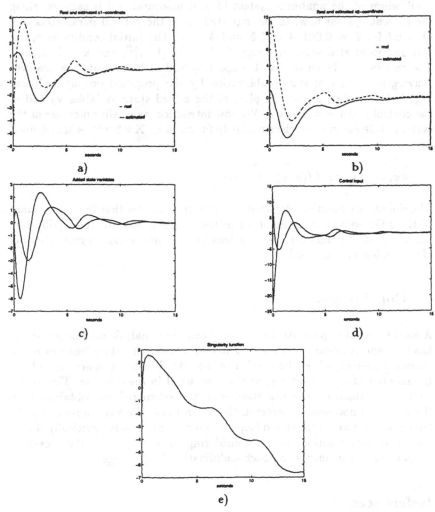

Fig. 1. Example: output feedback stabilization. a) real and estimated x_1 component; b) real and estimated x_2 component; c) added state variables $\mathbf{v}_0$; d) control input $\overline{\alpha}(\mathbf{x}, \mathbf{v}_0)$; e) singularity index of the EOJ matrix: $\log s(\mathbf{x}, \mathbf{v}_0)$.

$$\dot{\hat{\mathbf{x}}} = \mathbf{v}_0 + \mathbf{P}(\hat{\mathbf{x}}, \mathbf{v}_0)\,(\hat{\xi} - \Phi(\hat{\mathbf{x}}, \mathbf{v}_0))$$
$$\mathbf{P}(\hat{\mathbf{x}}, \mathbf{v}_0) = \mathbf{Q}\,\mathbf{J}^+(\hat{\mathbf{x}}, \mathbf{v}_0)$$
$$\dot{\psi}_1(s) = \frac{s}{(1 + T\,s)}\,\mathbf{Y}(s), \quad \psi_1(0) = 0$$

$$\nu = \overline{\alpha}(\hat{\mathbf{x}}, \mathbf{v}_0) \tag{33}$$

where $\mathbf{J}(\hat{\mathbf{x}}, \mathbf{v}_0)$ is reported in (30), ensures practical stability of the trivial equilibrium of the nonlinear system (32), if assumption 1 is satisfied along the current system flow. In the reported trial, the control parameters are: $\mathbf{Q} = 0.5\,\mathbf{I}_2$, $T = 0.001$, $k = 1.5$, and $\lambda = 1$. The initial conditions of the real and estimated state are respectively $\mathbf{x} = [1,\ 1]^T$, and $\hat{\mathbf{x}} = [-2,\ 3]^T$. The results are shown in Fig. 1. Figs. 1.a) and 1.b) indicate the observer convergence and real state stabilization by the proposed output feedback. Fig. 1.c) and 1.d) report the plots of the added state variables $\mathbf{v}_0$ and of the control input $\nu = \overline{\alpha}(\hat{\mathbf{x}}, \mathbf{v}_0)$. We now introduce, as quality measure of the current state estimation, the singularity function $s : \mathbf{X} \times \Re^{l_p\, m} \to \Re_+$, defined as

$$s(\mathbf{x}, \mathbf{v}) = \sqrt{\det\left(J(\mathbf{x}, \mathbf{v})^T J(\mathbf{x}, \mathbf{v})\right)} \ .$$

The singularity function, depicted in Fig. 1.e), indicates that the determinant of the EOJ matrix becomes small in the neighborhood of the equilibrium $(\mathbf{x},\ \mathbf{v}_0) = (0,\ 0)$. Extensive simulations have shown a meaningful extension of the region of attraction.

5   Conclusions

A local separation property for locally observable analytic nonlinear systems has been shown. First, it it necessary to find a stabilizing state feedback of the extended system, where the number of the added state variables depends on the number of the output derivatives considered in the observer. The second step is to substitute the real state with the estimated one obtained from the proposed nonlinear observer in the control law. We have proven that the trivial equilibrium of the closed loop system remains locally practically stable. As future investigation, we are considering an extension of the presented framework to the output feedback stabilization in the large.

References

1. A.N. Atassi and H.K. Khalil: "A separation principle for the stabilization of a class of nonlinear systems", IEEE Trans. Automatic Control, vol. 44, no. 9, pp. 1672-1687, 1999.
2. G. Bonard, F. Celle-Couenne and G. Gilles: "Observability and observers", Nonlinear Systems, Vol.1, Edited by A.J. Fossard and D. Normand-Cyrot, Chapman & Hall, 1995.
3. G. Bonard and H. Hammouri: "A high gain observer for a class of uniformly observable systems", 30th IEEE Conference on Decision and Control, 1991.

4. C.I. Byrnes, A. Isidori: "New results and examples in nonlinear feedback stabilization", Systems & Control Letters, n. 12, pp. 437-442, 1989.
5. F. Celle, J.P. Gauthier, and D. Kazakos: "Orthogonal representations of nonlinear systems and input-output maps", Systems & Control Letters, n. 7, pp. 365-372, 1986.
6. G. Ciccarella, M. Dalla Mora, and A. Germani: "A Luenberger-like observer for nonlinear systems", International Journal of Control, vol. 57, no. 3, pp. 537-556, 1993.
7. F. Esfandiari and H.K. Khalil: "Output feedback stabilization of fully linearizable systems", Int. Journal of Control, vol. 56, pp. 1007-1037, 1992.
8. M. Fliess and I. Kupka: "A finiteness criterion for nonlinear input-output differential systems", SIAM Journal Control and Optimization, n. 21, pp. 721-729, 1983.
9. J.P. Gauthier, H. Hammouri and S. Othman: "A simple observer for nonlinear systems. Application to bioreactors", IEEE Trans. Automatic Control, vol. 36, n. 6, 1992.
10. H. Hammouri, and J.P. Gauthier: "Bilinearization up to output injection", Systems & Control Letters, n. 11, pp. 139-149, 1988.
11. R. Hermann and A.J. Krener: "Nonlinear controllability and observability ", IEEE Trans. Automatic Control, vol. 22, n. 5, pp. 728-740, 1977.
12. M.R. Hestenes and T. Guinn: "An Embedding Theorem for Differential Equations", Journal of Optimization Theory and Applications, vol. 2, no. 2, 1968.
13. M. Hwang and J.H. Seinfeld: "Observability of Nonlinear Systems", Journal of Optimization Theory and Applications, vol. 10, no. 2, pp. 67-77, 1972.
14. H.K. Khalil and F. Esfandiari: "Semiglobal stabilization of a class of nonlinear systems using output feedback", IEEE Trans. Automatic Control, vol. 38, pp. 1412-1415, 1993.
15. M. Krstić, I. Kanellakopoulos and P. Kokotović: "Nonlinear and Adaptive Control Design", John Wiley & Sons, Inc., 1995.
16. A.J. Krener and A. Isidori: "Linearization by output injection and nonlinear observers", Systems & Control Letters, n. 3, pp. 47-52, 1983.
17. A.J. Krener and W. Respondek: "Nonlinear observers with linear error dynamics", SIAM Journal of Control and Optimization, n. 23, pp. 197-216, 1985.
18. A. Isidori: "Nonlinear Control Systems", 3rd ed., Springer Verlag, 1995.
19. J. Levine, R. Marino: "Nonlinear system immersion, observers and finite dimensional filters", Systems & Control Letters, n. 7, pp. 137-142, 1986.
20. Z. Lin and A. Saberi: "Robust semi-global stabilization of minimum-phase input-output linearizable systems via partial state and output feedback", IEEE Trans. Automatic Control, vol. 40, pp. 1029-1041, 1995.
21. L.M. Silverman and H.E. Meadows: "Controllability and Observability in Time-Variable Linear Systems", SIAM Journal on Control, vol. 5, no. 1, 1967.
22. A.R. Teel and L. Praly: "Global stabilizability and observability imply semi-global stabilizability by output feedback", Systems & Control Letters, n. 22, pp. 313-325, 1994.
23. A.R. Teel and L. Praly: "Tools for semi-global stabilization by partial state and output feedback", SIAM Journal Control and Optimization, vol. 33, 1995.
24. A. Tornambè: "Use of asymptotic observers having high-gains in the state and parameter estimation", 28th IEEE Conference on Decision and Control, Tampa, Florida, pp. 1791-1794, 1989.

25. J. Tsinias: "Further results on the observer design problem", Systems & Control Letters, n. 14, pp. 411-418, 1990.
26. J. Tsinias: "A theorem on global stabilization of nonlinear systems by linear feedback", Systems & Control Letters, n. 17, pp. 357-362, 1990.
27. Xia, Xiao-Hua and Gao, Wei-Bin: "Nonlinear observer design by observer error linearization", SIAM Journal on Control and Optimization, n. 27, pp. 199-216, 1989.
28. A.J. van der Shaft: "Observability and Controllability for Smooth Nonlinear Systems", SIAM Journal of Control and Optimization, n. 20, n0. 3, pp. 338-354, 1982.
29. D. Williamson: "Observability of bilinear systems, with applications to biological control", Automatica, vol. 13, pp. 243-254, 1977.
30. M. Zeitz: "The extended Luenberger observer for nonlinear systems", Systems and Control letters, vol. 9, pp. 149-156, 1987.

An H_∞-suboptimal Fault Detection Filter for Bilinear Systems*

Claudio De Persis[1] and Alberto Isidori[1][2]

[1] Department of Systems Science and Mathematics
Washington University
St. Louis, MO 63130, USA
depersis@zach.wustl.edu

[2] Dipartimento di Informatica e Sistemistica
Università di Roma "La Sapienza"
00184 Rome, Italy
isidori@giannutri.caspur.it

Abstract. We address the problem of fault detection and isolation in presence of noisy observations for bilinear systems. Our solution rests on results derived in the geometric theory of fault detection.

1 Introduction

We consider in this paper the problem of fault detection and isolation for systems modeled by equations of the form

$$\dot{x} = Ax + \sum_{i=1}^{m} u_i N_i x + \ell(x)m + \sum_{i=1}^{d} p_i(x) w_i \tag{1}$$
$$y = Cx,$$

in which $x \in \mathbb{R}^n$, $y \in \mathbb{R}^p$, $m \in \mathbb{R}$, and

$$\mathcal{P} = \text{span}\{p_1, \ldots, p_d\}$$

is "independent of x". Note that we say that a distribution Δ (codistribution Ω) on $\mathbb{R}^n$ is independent of x if there is a subspace $\mathcal{D}$ of $\mathbb{R}^n$ (a subspace $\mathcal{W}$ of the dual space of $\mathbb{R}^n$) such that, in the natural basis $\{\partial/\partial x_1, \ldots, \partial/\partial x_n\}$ of the tangent space to $\mathbb{R}^n$ at x, $\Delta(x) = \mathcal{D}$ (in the natural basis $\{dx_1, \ldots, dx_n\}$ of the cotangent space to $\mathbb{R}^n$ at x, $\Omega(x) = \mathcal{W}$). In these cases, it turns out quite useful to identify Δ with a matrix D whose columns span $\mathcal{D}$ (and Ω with a matrix W whose rows span $\mathcal{W}$).

---

* Research supported in part by ONR under grant N00014-99-1-0697, by AFOSR under grant F49620-95-1-0232, by DARPA, AFRL, AFMC, under grant F30602-99-2-0551, and by MURST.

Solving the problem of fault detection and isolation for (1) means designing a filter (a *residual generator*) of the form

$$\dot{\hat{x}} = \hat{f}(\hat{x}, y) + \hat{g}(\hat{x}, y)u$$
$$r = \hat{h}(\hat{x}, y)$$
(2)

such that, in the cascaded system (1)-(2), the *residual* $r(\cdot)$ depends "non-trivially" on (i.e is *affected by*) the input $m(\cdot)$, depends "trivially" on (i.e. is *decoupled from*) the inputs u and w and asymptotically converges to zero whenever $m(\cdot)$ is identically zero (see [4]).

Following Theorem 3 of [5], in which the study of the problem of fault detection and isolation for systems of the form

$$\dot{x} = A(u)x + \psi(u, y) + e_1(x)\nu_1 + e_2(x)\nu_2$$
$$y = Cx$$
(3)

($\nu_i \in \mathbb{R}$, $i = 1, 2$, and $A(u)$ is a matrix of analytic functions) was initiated, it is known that the problem of fault detection and isolation is solvable provided that a constant matrix L and a matrix $D(u)$ of analytic functions exist such that the subspace $\mathcal{V}$, defined as the largest subspace of $\text{Ker}\{LC\}$ which is invariant under $A(u) + D(u)C$ for all $u \in U$, satisfies: (a) $e_1(x) \notin \mathcal{V}$ and (b) $e_2(x) \in \mathcal{V}$ for all x.

In [2], we introduced the notion of observability codistribution for systems of the form (3) and we proved that the existence of the matrices $L, D(u)$ above can easily be characterized in terms of this notion. Here, we examine some details of this approach for systems of the form (1) and we use the results to address the problem of fault detection in the case in which the measurements are affected by noise v, i.e. $y = Cx + v$, proposing a filter in which the influence of the noise on the residual is attenuated. Related work on the problem of noise attenuation in fault detection, even though in a different setting, can be found in [1].

2   Observability codistributions

The concept of observability codistributions for an input-affine nonlinear system and the related construction algorithms have been introduced in [3] (cf. formulas (8), (12) and Proposition 2). In the case of state-affine systems, the corresponding concept and algorithms are those presented in [2] (cf. formulas (4) and (5)). Bilinear systems (1) can be indifferently seen as a special subclass of the systems dealt with in either [2] or [3]. For these systems, the algorithms in question assume the special form described in what follows. First of all note that, for a system of the form (1), conditioned invariant distributions and observability codistributions are independent of x. Also, for notational convenience, set $N_0 = A$.

Given a fixed subspace $\mathcal{P}$ of $I\!\!R^n$, consider the non-decreasing sequence of subspaces (of $I\!\!R^n$)

$$\mathcal{S}_0 = \mathcal{P}$$
$$\mathcal{S}_{i+1} = \mathcal{S}_i + \sum_{j=0}^{m} N_j(\mathcal{S}_i \cap \mathrm{Ker}\{C\}) \tag{4}$$

with $i = 0, \ldots, n-1$, and set $\mathcal{S}_*^{\mathcal{P}} = \mathcal{S}_{n-1}$.

Given a fixed subspace $\mathcal{Q}$ of the dual space of $I\!\!R^n$, consider the non-decreasing sequence of subspaces (of the dual space of $I\!\!R^n$)

$$\mathcal{Q}_0 = \mathcal{Q} \cap \mathrm{span}\{C\}$$
$$\mathcal{Q}_{i+1} = \mathcal{Q} \cap (\sum_{j=0}^{m} \mathcal{Q}_i N_j + \mathrm{span}\{C\}), \tag{5}$$

with $i = 0, \ldots, n-1$, and set o.s.a.$(\mathcal{Q}) = \mathcal{Q}_{n-1}$, where the acronym "o.s.a." stands for "observability subspace algorithm".

Finally, set

$$\mathcal{Q}_{\mathcal{P}}^* = \text{o.s.a.}((\mathcal{S}_*^{\mathcal{P}})^\perp) \ .$$

The following results describe the main properties of the subspaces thus defined.

Lemma 1. *Consider the system (1). The subspace $\mathcal{S}_*^{\mathcal{P}}$ is the minimal element (with respect to subspace inclusion) of the family of all subspaces of $I\!\!R^n$ which satisfy*

$$\mathcal{S} \supset \mathcal{P}$$
$$N_i(\mathcal{S} \cap \mathrm{Ker} C) \subset \mathcal{S}, \qquad i = 0, \ldots, m \ . \tag{6}$$

Remark 1. In the terminology of [3], the distribution $\Delta : x \mapsto \mathcal{S}_*^{\mathcal{P}}$ is the minimal *conditioned invariant distribution* containing the distribution $P :$ $x \mapsto \mathcal{P}$.

Lemma 2. *The subspace $\mathcal{Q}_{\mathcal{P}}^*$ is the maximal element (with respect to subspace inclusion) of the family of all subspaces of the dual space of $I\!\!R^n$ which satisfy*

$$\mathcal{Q} \subset \mathcal{P}^\perp$$
$$\mathcal{Q} N_i \subset \mathcal{Q} + \mathrm{span}\{C\}, \qquad i = 0, 1, \ldots, m \tag{7}$$
$$\mathcal{Q} = \text{o.s.a.}(\mathcal{Q}).$$

Remark 2. In the terminology of [2], [3], a codistribution $Q : x \mapsto \mathcal{Q}$ fulfilling the last two properties of (7) is an *observability codistribution* for (1), and hence the codistribution $Q^* : x \mapsto \mathcal{Q}_{\mathcal{P}}^*$ is the maximal observability codistribution contained in the codistribution $P^\perp : x \mapsto \mathcal{P}^\perp$. ◁

3   A natural candidate for residual generation

Specializing the results of [2] or [3] to system (1), we first observe that the following holds.

Proposition 1. *Consider system (1), set $A(u) := A + \sum_{i=1}^{m} N_i u_i$ and $\mathcal{P} := \text{span}\{p_1, \ldots, p_d\}$. The following conditions are equivalent:*

  (i) there exist a constant matrix L and a matrix $D(u)$ of analytic functions such that the largest subspace $\mathcal{V}$ contained in $\text{Ker}\{LC\}$ and invariant under $A(u) + D(u)C$, for all $u \in \mathbb{R}^m$, satisfies: (a) $\ell(x) \notin \mathcal{V}$ and (b) $\mathcal{P} \subset \mathcal{V}$.

  (ii) $\ell(x) \notin (\mathcal{Q}_{\mathcal{P}}^*)^{\perp}$;

  (iii) there exists a change of coordinates $\tilde{x} = \text{col}(x_1, x_2, x_3) = Tx$ on the state space and a change of coordinates $\tilde{y} = \text{col}(y_1, y_2) = Hy$ on the output space which transform (1) into a system of the form:

$$\dot{x}_1 = A_{11}x_1 + A_{12}x_2 + \sum_{i=1}^{m}(N_{11}^i x_1 + N_{12}^i x_2)u_i + \ell_1(\tilde{x})m$$

$$\dot{x}_2 = \sum_{k=1}^{3} A_{2k}x_k + \sum_{i=1}^{m}\left(\sum_{k=1}^{3} N_{2k}x_k\right)u_i + \ell_2(\tilde{x})m + p_2(\tilde{x})w$$

$$\dot{x}_3 = \sum_{k=1}^{3} A_{3k}x_k + \sum_{i=1}^{m}\left(\sum_{k=1}^{3} N_{3k}x_k\right)u_i + \ell_3(\tilde{x})m + p_3(\tilde{x})w \qquad (8)$$

$$y_1 = C_1 x_1$$

$$y_2 = x_2 \,,$$

in which ℓ_1 is nonzero, and the subsystem

$$\dot{x}_1 = A_{11}x_1 + \sum_{i=1}^{m} u_i N_{11}^i x_1$$

$$y_1 = C_1 x_1 \,, \qquad (9)$$

is observable.

Proof. The proof of the various implications can be obtained by specializing the proofs of similar statements given in [2] and [3]. It is worth stressing that, in the implication (ii)$\Rightarrow$(iii), the form (8) can be obtained in the following way: define $x_1 = T_1 x$ with T_1 a matrix whose rows span $\mathcal{Q}_{\mathcal{P}}^*$ and define $y_1 = H_1 y$ with H_1 a matrix such that the rows of the matrix $H_1 C$ span $\mathcal{Q}_{\mathcal{P}}^* \cap \text{span}\{C\}$. Then, the form (8) derives from the properties

$$\mathcal{Q}_{\mathcal{P}}^* N_i \subset \mathcal{Q}_{\mathcal{P}}^* + \text{span}\{C\}, \ i = 0, 1, \ldots, m \,,$$

$\mathcal{Q}_{\mathcal{P}}^* \subset \mathcal{P}^{\perp}$, $\ell(x) \notin (\mathcal{Q}_{\mathcal{P}}^*)^{\perp}$, and from the fact that the chosen isomorphisms preserve the structure of the system. Note in particular that, since $\mathcal{Q}_{\mathcal{P}}^*$ is spanned by the rows of T_1, necessarily $\ell_1(x) \neq 0$, because otherwise $\ell(x) \in (Q_P^*)^\perp$. The observability of the subsystem (9) derives from the fact that Q_P^* is the maximal element of the family of all subspaces of the dual space of $\mathbb{R}^n$ which satisfy (7).

The implication (iii)$\Rightarrow$(i) is easily obtained by setting $D(u) := D_0 + \sum_{i=1}^m D_i u_i$, with

$$D_i = T^{-1} \begin{pmatrix} 0 & -N_{12}^i \\ 0 & 0 \\ 0 & 0 \end{pmatrix} H, \quad i = 0, 1, \ldots, m, \tag{10}$$

and

$$L = H_1. \tag{11}$$

In fact, by Proposition 2 in [2], the subspace $\mathcal{V} := \text{Ker}\{T_1\}$ is the largest subspace of $\text{Ker}\{LC\}$ which is invariant under $A(u) + D(u)C$ for all $u \in \mathbb{R}^m$. Since the disturbance term does not appear in the x_1-subsystem, the inclusion $P \subset \mathcal{V}$ must hold. On the other hand, since $\ell_1 \neq 0$, the vector field $\ell(x)$ cannot belong to $\mathcal{V}$ for all x. ◁

The condition (i) is precisely the condition assumed in [5], Theorem 3, to establish the existence of a solution of the fundamental problem of residual generation for system (1). In this respect, the equivalences established in the Proposition above provide: a *conclusive* test – condition (ii) – to determine whether or not the subspace $\mathcal{V}$ and the pair of matrices $D(u), L$ in (i) *ever exist*, and a straightforward *procedure* – based on the change of coordinates described in (iii) – to actually identify this subspace and to construct such matrices.

Special form (8), and in particular subsystem (9), leads itself to an immediate interpretation of the construction proposed in [5]. As a matter of fact, it is easy to check that, *if $D(u), L$ are chosen as indicated in the proof of the Proposition above*, the residual generator of [5] reduces to a system of the form

$$\dot{\hat{x}} = A_{11}\hat{x} + \sum_{i=1}^m u_i N_{11}^i \hat{x} + A_{12}y_2 + \sum_{i=1}^m u_i N_{12}^i y_2 - GC_1^T(C_1\hat{x} - y_1)$$
$$r = y_1 - C_1\hat{x}, \tag{12}$$

in which G is a suitable matrix of functions of time. Indeed, the latter is nothing else than an observer for x_1, yielding an "observation error" $e = x_1 - \hat{x}$ obeying

$$\dot{e} = (A_{11} + \sum_{i=1}^m u_i N_{11}^i - GC_1^T C_1)e + \ell_1(\tilde{x})m$$
$$r = C_1 e. \tag{13}$$

If G is chosen, as suggested in [5], in such a way that the latter is exponentially stable for $m = 0$, then (12) solves the problem of residual generation.

4   Fault detection with noisy observations

In the approach to residual generation summarized in the previous section the observed output y that feeds the residual generator is supposed to be noise-free. If this is not the case, as in any realistic setup, then the important problem of attenuating the effect that this noise may have on the residual r must be addressed. Following the fundamental results of [1], the problem of noise attenuation in the design of residual generators can be cast as a problem of state estimation in a game-theoretic, or H_∞, setting. To this end, the special features of form (8) prove to be particularly helpful.

Consider the case in which the observed output of system (1) is corrupted by a measurement noise v, i.e.

$$\dot{x} = Ax + \sum_{i=1}^{m} u_i N_i x + \ell(x)m + \sum_{i=1}^{d} p_i(x)w_i \qquad (14)$$
$$y = Cx + v ,$$

and let $v(\cdot)$ be a function in $\mathcal{L}_2[0, t_1)$, where $[0, t_1)$ is a fixed time interval (with $t_1 \leq \infty$). Proposition 1 shows, among the other things, that in the special new coordinates used to obtain (8) we can write

$$\dot{x}_1 = A_{11}x_1 + A_{12}x_2 + \sum_{i=1}^{m}(N_{11}^i x_1 + N_{21}^i x_2)u_i + \ell_1(\tilde{x})m \qquad (15)$$
$$y_1 = C_1 x_1 + v_1$$
$$y_2 = x_2 + v_2 .$$

Ideally, in the absence of the fault signal (the input m to (15)), the output of the residual residual generator must be identically zero, in spite of a possible nonzero disturbance w (whose role in (15) is taken by the input x_2) and of the observation noise. In practice, in view of the way the noise affects the observations, both x_2 and v are going to influence the residual when $m = 0$. As a matter of fact, if the second component y_2 of the observed output is corrupted by noise, it is no longer possible to compensate – via output injection – the effect of x_2 on the dynamics of the observation error, as it was done in the residual generator (12). Thus, in this case, the problem becomes a problem of attenuating the effect of both x_2 and v on the residual signal, which for instance – as in [1] – can be cast in a game-theoretic setting.

Motivated by the approach of [1], we address in what follows the problem of finding a filter with state variable $\hat{x}$, driven by the observed outputs (y_1, y_2), such that, for some fixed positive number γ and a given choice of weighting positive definite matrices Q, M, V, P_0,

$$\frac{\int_0^{t_1} \|C_1(x_1 - \hat{x})\|_Q^2 dt}{\int_0^{t_1} [\|v_2\|_{M^{-1}}^2 + \|v_1\|_{V^{-1}}^2]dt + \|x_1(0) - \hat{x}_0\|_{P_0}^2} \leq \gamma^2, \tag{16}$$

for all the signals v_1 and v_2 and for all the initial conditions $x_1(0)$, subject to the constraints

$$\dot{x}_1 = A_1 x_1 + A_2 x_2 \tag{17}$$
$$y_1 = C_1 x_1 + v_1 \tag{18}$$
$$y_2 = x_2 + v_2, \tag{19}$$

in which, for convenience, we have set $A_1 := A_{11} + \sum_{i=1}^{m} N_{11}^i u_i$ and $A_2 := A_{12} + \sum_{i=1}^{m} N_{12}^i u_i$.

To solve this problem, we consider the cost function

$$J = \int_0^{t_1} [\|C_1(x_1 - \hat{x})\|_Q^2 - \gamma^2 (\|y_2 - x_2\|_{M^{-1}}^2 + \|y_1 - C_1 x_1\|_{V^{-1}}^2)]dt \\ - \|x_1(0) - \hat{x}_0\|_{\Pi_0}^2 \tag{20}$$

(where $\Pi_0 := \gamma^{-2} P_0$) and the differential game

$$\min_{\hat{x}} \max_{(y_1, y_2)} \max_{x_2} \max_{x_1(0)} J \leq 0$$

subject to the differential constraint (17).

Standard methods show that, if the differential Riccati equation

$$\dot{\Pi} + A_1^T \Pi + \Pi A_1 + \gamma^{-2} \Pi A_2 M A_2^T \Pi + C_1^T (Q - \gamma^2 V^{-1}) C_1 = 0, \tag{21}$$

with initial condition $\Pi(0) = \Pi_0$, has a solution $\Pi(t)$ defined and nonsingular for all $t \in [0, t_1]$, then the attenuation requirement (16) is fulfilled by an estimate $\hat{x}$ of x_1 provided by the estimator

$$\dot{\hat{x}} = A_1 \hat{x} + A_2 y_2 + \gamma^2 \Pi^{-1} C_1^T V^{-1} (y_1 - C_1 \hat{x}), \qquad \hat{x}(0) = \hat{x}_0. \tag{22}$$

Comparing the dynamics of the estimator (22) with that of the residual generator (12) and bearing in mind the expressions of A_1 and A_2, we see that (22) corresponds to choosing in (12) a "gain matrix" G of the form

$$G = \gamma^2 \Pi^{-1} C_1^T V^{-1},$$

where Π is the solution of the differential Riccati equation (21). Indeed, this particular choice guarantees the attenuation properties expressed by (16).

It is interesting to examine the special case in which $t_1 = \infty$. Note, to this end, that the estimation error $e = x_1 - \hat{x}$ obeys

$$\dot{e} = (A_1 - \gamma^2 \Pi^{-1} C_1^T V^{-1} C_1)e + \ell_1(\tilde{x})m - A_2 v_2 - \gamma^2 \Pi^{-1} C_1^T V^{-1} v_1 . \quad (23)$$

In order to (16) make sense for $t_1 = \infty$, we require $e(\cdot)$ to be in $\mathcal{L}_2[0, \infty)$. Assume that $u(t)$ and $x(t)$ are bounded on $[0, \infty)$ and $m(\cdot)$ is in $\mathcal{L}_2[0, \infty)$. If also $\Pi^{-1}(t)$ is bounded on $[0, \infty)$, the "input"

$$\ell_1(\tilde{x})m - A_2 v_2 - \gamma^2 \Pi^{-1} C_1^T V^{-1} v_1$$

to (23) is in $\mathcal{L}_2[0, \infty)$ and hence $e(\cdot)$ is in $\mathcal{L}_2[0, \infty)$ if the homogeneous linear system

$$\dot{e} = (A_1 - \gamma^2 \Pi^{-1} C_1^T V^{-1} C_1)e \quad (24)$$

is *exponentially stable*.

Conditions implying the exponential stability of the latter can be derived from the works [6] and [7]. In fact, recall that if $\Pi(t)$ is defined and nonsingular for all $t \in [0, \infty)$ the matrix $P(t) := \gamma^2 \Pi^{-1}(t)$ satisfies the differential Riccati equation

$$\dot{P} = AP + PA^T + BB^T + P[\gamma^{-2} L^T L - C^T C]P \quad (25)$$

in which

$$A := A_1, \qquad B := A_2 M^{1/2}, \qquad C := V^{-1/2} C_1, \qquad L := Q^{1/2} C_1 .$$

The following result of [7] provides the requested condition.

Lemma 3. *Let $P(t)$ be a symmetric solution of (25) satisfying*

$$0 < \beta_1 I \leq P(t) \leq \beta_2 I$$

for all $t \in [0, \infty)$, for some β_1, β_2, and such that the system

$$\dot{p} = (A + P[\gamma^{-2} L^T L - C^T C])p$$

is exponentially stable. Then, the system

$$\dot{e} = (A - PC^T C)e \quad (26)$$

is exponentially stable.

In fact, by definition of A, C, P, system (26) coincides with the homogeneous linear system (24) and, hence, the hypotheses of the previous lemma guarantee that the estimator (22) renders (16) fulfilled over the infinite horizon.

5 Conclusion

We have proposed, within a geometric framework, a solution to the problem of fault detection and isolation for systems of the form (1) in the presence of noisy observations.

References

1. Chung W.C., Speyer J. L. (1998) A game theoretic fault detection filter. IEEE Transactions on Automatic Control **43**, 143-161
2. De Persis C., Isidori A. (2000) An addendum to the discussion on the paper "Fault Detection and Isolation for State Affine Systems". European Journal of Control
3. De Persis C., Isidori A. (2000) On the observability codistributions of a nonlinear system. Systems & Control Letters
4. De Persis C., Isidori A. (1999) A differential-geometric approach to nonlinear fault detection and isolation. Submitted
5. Hammouri H., Kinnaert M., El Yaagoubi E.H. (1998) Fault detection and isolation for state affine systems. European Journal of Control **4**, 2-16
6. Nagpal K.M., Khargonekar P.P. (1991) Filtering and Smoothing in an H^∞ Setting. IEEE Transactions on Automatic Control **36**, 152-166
7. Ravi R., Pascoal A.M., Khargonekar P.P. (1992) Normalized coprime factorizations for linear time-varying systems. Systems & Control Letters **18**, 455-465

Adaptive Control of Feedback Linearizable Systems by Orthogonal Approximation Functions

Domitilla Del Vecchio, Riccardo Marino, and Patrizio Tomei

Università di Roma "Tor Vergata"
Dipartimento di Ingegneria Elettronica
Via di Tor vergata 110
00133 Roma, Italia
ddomitilla@hotmail.com, Marino@ing.uniroma2.it

Abstract. The problem addressed in this paper is the control of a class of SISO feedback linearizable systems containing unknown nonlinearities with known bound on a given compact set. The unknown nonlinearities are locally approximated by a finite sum of Orthogonal basis functions so that by virtue of Bessel inequality a bound on the norm of the optimal weights derives directly from the bound of the nonlinearity. Projection algorithms can then be used to assure that the parameters estimates remain inside the set in which the optimal weights lie. The proposed control algorithm is repeatable since the set of initial conditions for which given performances are guaranteed is explicitly determined; no *a priori* knowledge on the bound of the norm of the optimal weights is required.

1 Introduction

In this paper a repeatable control design methodology for a class of feedback linearizable plants with unknown nonlinearities satisfying the matching condition is proposed. By assuming that an upper bound on the modulus of the nonlinear function on a given compact set is known, orthogonal functions are used to approximate the unknown nonlinearities on the given compact set. By virtue of Bessel inequality, a bound on the norm of the optimal weights is obtained which allows us to use projection algorithms in the parameter updating law. We determine the set of the initial conditions and the values of the control parameters that assure the state vector to remain inside the compact set in which the approximation holds with bounded control signals. Since the control algorithm guarantees that the state vector remains in the compact set where the approximation holds, its internal adaptation algorithm can reconstruct the approximation of the unknown nonlinearities when there is sufficiently persistency of excitation so that smaller tracking errors are obtained without increasing the control gains. With our assumptions none of the previously proposed controllers works for the class of systems considered. In fact [11] requires the knowledge of an upper bound for the nonlinear function outside a compact set and of an upper bound for the approximation error; this latter is also required in [5], while in [10] it is not required but it is needed to determine the set in which the state vector will converge; [1] requires the measure of the state derivative and a known bound on the norm of the optimal weights. This bound is not required in [8] but it necessary to determine the size of the neighbourhood around zero in which the tracking error converges. Since the bound on the approximation error holds only locally, with the exception of [8] where a bounding function for this error is assumed to be global, it is a key point to assure the state vector to remain in the desired bounded set. In this paper we guarantee this through a constructive control algorithm without any dependence on the initialization of the approximation network and without introducing any forgetting factor in the adaptation dynamics, as in [5] and in [8], that obstructs the convergence of the parameter estimates to their true valu es.

2   Problem Statement

Consider single-input single-output uncertain feedback linearizable systems in the matching structure

$$\dot{x}_i = x_{i+1} \qquad i = 1, ..., n-1 \, ,$$
$$\dot{x}_n = f(x_1, x_2 .... x_n) + \frac{\psi(x)}{\nu} u + d(t) \tag{1}$$
$$y = x_1$$

in which $x = (x_1 ....x_n)^T$ belonging to R^n is the state vector, u is the scalar control input, y is the scalar output to be controlled, $f(x) : R^n \to R$ is an unknown nonlinear function belonging to $C(A)$, $d(t)$ is a scalar bounded exogenous disturbance, $\psi(x)$ is a known function, ν is an unknown constant parameter. We make the following assumptions.

A1. The bound for the disturbance is known: $|d(t)| \le D, \ \forall t \in R$.

A2. A bound for the modulus of $f(x)$ is known when x belongs to a known compact set $A = I_{n[-r_a, r_a]} \subset R^n$, i.e.: $|f(x)| \le B_f, \forall x \in A \subset R^n$.

A3. ν is a positive constant parameter with known bounds:
$0 < \nu_m \le \nu \le \nu_M$.

A4. We consider smooth bounded reference trajectories $y_r(t)$ with bounded time derivatives $y_r^{(1)}, ..., y_r^{(n)}$ such that $|y_r^{(i)}(t)| \le \varrho_{ri}$ for $0 \le i \le n$ with $\varrho_{ri} \le r_a$.

Moreover we define the vector $y_r = (y_r, ..., y_r^{(n-1)})^T \in R^n$.

If $f(x)$ and ν are known and $d(t)$ is zero then the feedback linearizing control input is:

$$u_{id} = \frac{\nu}{\psi(x)}(-f(x) + y_r^{(n)} - K^T(x - y_r)) \tag{2}$$

so that the closed loop system becomes linear and by a proper choice of K it becomes asymptotically stable. Since both $f(x)$ and ν are not known (2) cannot be used. Since $f(x)$ is not linearly parameterized, i.e. $f(x) \neq \theta^T \Phi(x)$ with $\theta \in R^p$ and $\Phi(x)$ a known regressor vector, nonlinear adaptive techniques ([7], [14]) cannot be used. We will use orthogonal functions to generate functional approximations for the unknown $f(x)$.

2.1 Approximation network structure

Let us recall some preliminary definitions from [4].
Definition 1. A set of bounded continuous functions $\{\varphi_k(x)\} := \phi$ for x belonging to the set A is said to be a set of orthogonal functions if

$$\int_A \varphi_i(x)\varphi_j(x)dx = \begin{cases} mis(A) & \text{for } i = j \\ 0 & \text{for } i \neq j \end{cases} \tag{3}$$

with $mis(A)$ is the measure of the set A .
Definition 2. Let

$$\theta_k := \frac{(f, \varphi_k)}{mis(A)} \tag{4}$$

be the Fourier coefficient, where the operator $(\ ,\ )$ is the scalar product between elements of the linear euclidean space of the functions belonging to $C(A)$ and it is defined as $(f, \varphi_k) := \int_A f(x)\varphi_k(x)dx$. Now let us consider a general approximation with N terms $S_N(x) = \sum_{k=0}^{N-1} \alpha_k \varphi_k(x)$, it can be proved ([4]) that the minimum distance between S_N and $f(x)$ is reached when $\alpha_k = \theta_k$ and the following equality holds:

$$\|f - S_N\|^2 = \|f\|^2 - \sum_{k=0}^{N-1} \theta_k^2 \cdot mis(A) \tag{5}$$

from which the Bessel inequality follows

$$\sum_{k=0}^{N-1} \theta_k^2 \leq \frac{\|f\|^2}{mis(A)} \tag{6}$$

where $\|f\|^2 = (f, f) = \int_A f^2(x)dx$.
Definition 3. The set ϕ of orthogonal functions is said to be closed inside the space of the bounded functions f in $C(A)$ if the Parseval equality holds:

$$\sum_{k=0}^{\infty} \theta_k^2 = \frac{\|f\|^2}{mis(A)}. \tag{7}$$

Since in the space of orthogonal functions the properties of closure and completeness are equivalent, we recall from [4] the following result.

The set of orthogonal functions ϕ, $x \in A \subset R^n$, is complete in the space $C(A)$ if and only if there is no element in $C(A)$ different from zero that is orthogonal to all the functions of the set ϕ. In other words $(\varphi_k, g) = 0$ for every k is equivalent to $g(x) \equiv 0$ in A, which is equivalent to $\int_A \varphi_k(x)g(x)dx = 0$ for every k.

An example of a complete set of orthogonal functions belonging to $C(A)$ is obtained by generalizing the concept of Fourier series ([4]):

$$f(x_1, x_2, \ldots, x_n) =$$

$$\sum_{m_1,m_2\ldots m_n=-\infty}^{\infty} c_{m_1 m_2 \ldots m_n} exp(j(m_1 x_1 + \ldots + m_n x_n)2\pi/T) \qquad (8)$$

$$c_{m_1 m_2 \ldots m_n} = \frac{1}{T^n} \int_{-T/2}^{T/2} \ldots$$

$$\ldots \int_{-T/2}^{T/2} f(x_1, x_2 \ldots x_n) exp(-j(m_1 x_1 + \ldots + m_n x_n)2\pi/T)dx_1 \ldots dx_n. \qquad (9)$$

where $A = I_{n[-T/2, T/2]}$. It can be easily shown that the set $\phi = \{exp(j(m_1 x_1 + \ldots + m_n x_n)2\pi/T)\}$ is complete and the functions are orthogonal. The number of approximating functions N when $x \in R^n$ is determined as follows. Be m the number of harmonics, including the zero frequency, chosen to approximate the function $f(x)$, then $N = 2m^n - 1$. The resulting approximation with N terms is given by

$$S_N = \theta_0 + \sum_{i=1}^{m^n-1} \theta_{ic}\sqrt{2}cos[(M_i^T x)2\pi/T] + \theta_{is}\sqrt{2}sin[(M_i^T x)2\pi/T]$$

where $M_i \in R^n$ is the harmonics vector whose coordinates are natural numbers and each coordinate can assume values belonging to $[0, m-1]$ so that $M_i \neq M_j$ for $i \neq j$. By defining

$$\varphi_0 = 1, \quad \varphi_{2i}(x) = \sqrt{2}cos[\frac{(M_i^T x)2\pi}{T}], \quad \varphi_{2i-1}(x) = \sqrt{2}sin[\frac{(M_i^T x)2\pi}{T}] (10)$$

that satisfy (3), S_N can be also rewritten as $S_N(x) = \sum_{k=0}^{N-1} \theta_k \varphi_k(x)$, where θ_k is defined in (4). If by hypothesis A is the hypercube of side r_a, in the above relations $T = 2r_a$.

So by considering (7) and (5) and the fact that $(f-S, f-S) = 0$ only if $f-S$ is zero, where we call $S = lim_{N \to \infty} S_N$, we obtain that $f(x) = \sum_{k=0}^{\infty} \theta_k \varphi_k(x)$, $\forall x \in A$ that can be written in the following property.

Property 1. If we denote by ϕ the complete set of orthogonal functions, considering the set of functions f in $C(A)$ as specified in A2 and taking (4) into account, then for every $\varepsilon \geq 0 \ \exists \ N(\varepsilon) > 0$ such that

$$\text{for } N \geq N(\varepsilon) \ \ |f(x) - \sum_{k=0}^{N-1} \theta_k \varphi_k(x)| \leq \varepsilon , \qquad \forall x \in A . \tag{11}$$

This result shows that networks with one layer and functions φ_i in ϕ, complete set of orthogonal functions, can approximate functions f in $C(A)$ with increased precision by sufficiently increasing the number of nodes. Therefore in our control problem we will consider a network that assures the approximation of $f(x)$ in the given compact set A with approximation error $\varepsilon(x)$, such that $|\varepsilon(x)| \leq \varepsilon$ for every x in A , i.e.:

$$f(x) = \sum_{k=0}^{N-1} \theta_k \varphi_k(x) + \varepsilon(x) , \qquad \forall x \in A \tag{12}$$

in which since $f(x)$ is unknown, the constant parameters θ_i, as defined in (4), are also unknown.

From the bound assumed for $|f|$ when x is in A , we can establish a bound on the norm of the parameter vector $\theta = (\theta_0, ..., \theta_N - 1)^T$, in fact by (6) and A2 it follows that

$$\sum_{k=0}^{N-1} \theta_k^2 \leq \frac{1}{mis(A)} \int_A f^2(x) dx \leq \frac{B_f^2}{mis(A)} \int_A dx = B_f^2$$

so that

$$\sum_{k=0}^{N-1} \theta_k^2 \equiv \|\theta\|^2 \leq B_f^2. \tag{13}$$

Notice that the bound does not depend on N.

2.2   The tracking problem

From (12) we consider an estimation of $f(x)$ given by

$$\hat{f}(x) = \sum_{k=0}^{N-1} \hat{\theta}_k \varphi_k(x) := \hat{\theta}^T \Phi(x) , \qquad \forall x \in A \tag{14}$$

where $\hat{\theta} = (\hat{\theta}_0, ..., \hat{\theta}_{N-1})^T$ is the estimate of $\theta = (\theta_0, ..., \theta_{N-1})^T$ and θ_k, $0 \leq k \leq N-1$, are given in (4) and $\Phi = (\varphi_0, ..., \varphi_{N-1})$. At this point (14) is the estimate of the functional approximation $\theta^T \Phi(x)$ of $f(x)$ in A with the parameter vector bound given by (13); the approximation error $\varepsilon(\boldsymbol{x})$ may be viewed as a disturbance and the robust adaptive techniques proposed in [6] are applicable provided that we can guarantee that $\boldsymbol{x}$ remains in A for all t . The tracking problem is formulated as in [6], that is to design a state feedback control and to determine the set of initial conditions so that: (i) $\boldsymbol{x}(t) \in A \ \forall t \geq 0$; (ii) all the signal of the closed-loop system remain bounded; (iii) the output y is forced to enter asymptotically into an arbitrarily small neighbourhood of the smooth bounded reference signal satisfying A4; (iv) arbitrarily improvable transient performances of the tracking error are given in terms of both the approximation error and the parameter estimation error; (v) when $\varepsilon(\boldsymbol{x})$ and $d(t)$ are zero the asymptotic tracking error is zero.

3   Main Result

Let A1-A4 hold. By virtue of (12) and by virtue of A3 we rewrite system (1) as

$$
\begin{aligned}
\dot{x}_i &= x_{i+1} \qquad i = 1, ..., n-1 \\
\dot{x}_n &= \sum_{i=0}^{N-1} \theta_i \varphi_i(\boldsymbol{x}) + R(\boldsymbol{x}, t) + \frac{\psi(\boldsymbol{x})}{\nu_0 + b} u \\
y &= x_1
\end{aligned}
\tag{15}
$$

in which $R(\boldsymbol{x}, t) = d(t) + \varepsilon(\boldsymbol{x})$ with $|\varepsilon(\boldsymbol{x})| \leq \varepsilon \ \forall \boldsymbol{x} \in A$ and $\nu_0 := (\nu_M + \nu_m)/2$ so that $\nu = \nu_0 + b$ with ν_0 known and b the unknown parameter satisfying

$$
|b| \leq \frac{\nu_M - \nu_m}{2} := b_M .
\tag{16}
$$

Theorem 3.1 Given system (15) under assumptions A1-A4 and (16) for any arbitrary real $\varepsilon > 0$, there exists a state feedback dynamic control

$$
u\left(x_1, ..., x_n, k, \lambda_1, y_r, ..., y_r^{(n)}, vec\hat{\theta}, \hat{b}\right)
$$
$$
\dot{\hat{\theta}} = \omega_1(x_1, ..., x_n, y_r, ..., y_r^{(n-1)}, \hat{\theta}), \quad \dot{\hat{b}} = \omega_2(x_1, ..., x_n, y_r, ..., y_r^{(n-1)}, \hat{b})
$$

parameterized by the positive real k such that for every $k \geq \bar{k}(B_f, D, \nu_M, \nu_m)$, for any initial condition $\boldsymbol{x}(0) \in B(r) \subset A$, with $0 < r < r_a$, $\hat{\theta}(0) \in B(B_f) \subset R^N$, $\hat{b}(0) \in B(b_M) \subset R$ and for every reference signal with $y_r \in B(r)$ the following properties hold:

(i) $\boldsymbol{x}(t) \in A$, $\|\hat{\theta}(t)\| \leq B_f + \alpha$, $|\hat{b}(t)| \leq b_M + \alpha \ \forall t \geq 0$;
(ii) for all $t \geq 0$ the L_∞ inequality is verified

$$
|y(t) - y_r(t)| \leq \alpha_1(0) exp(-\lambda_1 t) +
$$
$$
+ \frac{1}{\sqrt{k}}(\varepsilon + D + \sqrt{N} sup_{\tau \in [0,t]} \|\bar{\theta}(\tau)\| + sup_{\tau \in [0,t]} |\bar{b}(\tau)|) ;
$$

(iii) for all $t \geq 0$ the L_2 inequality is verified

$$\int_0^t (y(\tau) - y_r(\tau))^2 d\tau \leq \alpha_2(0) + \frac{1}{k}\int_0^t (\varepsilon^2 + D^2 + N\|\bar{\theta}(\tau)\|^2 + \bar{b}^2)d\tau ;$$

in which $\alpha_1(0)$ and $\alpha_2(0)$ are nonnegative constants depending on the initial condition $x(0)$, α is an arbitrarily small positive constant, λ_1 is a positive control parameter, $\bar{\theta} = \theta - \hat{\theta}$ with $\theta = (\theta_0, ..., \theta_{N-1})^T$ and θ_i given in (4), $\bar{b} = b - \hat{b}$;

(iv) if $d(t) = 0$ and $\varepsilon(x) = 0$ then $lim_{t \to \infty}(y(t) - y_r(t)) = 0$

We will provide a constructive proof for Theorem 3.1 in which both the dynamic control and the radius r of the ball $B(r)$ are explicitly given. First we proceed by assuming $x(t) \in A$ for every time so that A2 and (12) hold. We then show that the above assumption is verified for a specific choice of the ball $B(r)$ and of the control parameter k. Before proving Theorem 3.1 we give the following results.

Lemma 3.1 Given system (15), there exists a state feedback control

$$u = \frac{1}{\psi(x)}((\nu_0 + \hat{b})v + \bar{v}) , \tag{17}$$

a global filtered transformation

$$\begin{aligned} e_1 &= x_1 - y_r \\ e_i &= x_i - x_i^* - y_r^{(i-1)} \qquad i = 2, ..., n , \end{aligned} \tag{18}$$

an adaptation dynamics

$$\dot{\hat{\theta}} = \omega_1(x_1, ..., x_n, y_r, ...y_r^{(n-1)}, \hat{\theta}) , \quad \dot{\hat{b}} = \omega_2(x_1, ..., x_n, y_r, ...y_r^{(n-1)}, \hat{b}) ,$$

with $\omega_1(\cdot)$ and $\omega_2(\cdot)$ Lipschits continuous functions, and $n + 1$ smooth functions

$$x_2^*(x_1, y_r, v_1), ..., x_n^*(x_1, ..., x_{n-1}, y_r, ..., y_r^{(n-2)}, v_{n-1}),$$

$$v(x_1, ..., x_n, y_r, ..., y_r^{(n)}, \hat{\theta}, v_n), \quad \bar{v}(x_1, ..., x_n, y_r, ..., y_r^{(n-1)})$$

transforming system (15) into:

$$\begin{aligned} \dot{e}_i &= -\lambda_i e_i + e_{i+1} + v_i \qquad i = 1, ..., n - 1 \\ \dot{e}_n &= -\lambda_n e_n + \bar{\theta}^T \boldsymbol{\Phi}(x) + R(x, t) + v_n + \frac{1}{\nu}(-\bar{b}v + \bar{v}) , \end{aligned} \tag{19}$$

where $v_i(x_1, ..., x_{i-1}, y_r, ...y_r^{(i-2)})$ for $1 \leq i \leq n - 1$ and $v_n(x1, ..., x_n, y_r, ..., y_r^{(n-1)})$ are n smooth functions, and λ_i are positive control parameters.

Proof (sketch). Consider the time derivatives of (18), put

$$x_{i+1}^* = -\lambda_i e_i + \dot{x}_i^* + v_i \qquad i = 1, ..., n - 1 \tag{20}$$

and

$$v = \dot{x}_n^* - \hat{\theta}^T \Phi(x) - \lambda_n e_n + v_n + y_r^{(n)} \tag{21}$$

so that we obtain (19). For the complete proof see [6].

Lemma 3.2 Given the transformations (18) and the relations (20), if $v_i = -e_{i-1}$ for $2 \leq i \leq n-1$ then there exists an inferior triangular matrix A such that

$$e = A(x - y_r). \tag{22}$$

where $x = (x_1, ..., x_n)^T$, $y_r = (y_r, ..., y_r^{(n-1)})^T$, $e = (e_1, ..., e_n)^T$.
Proof (sketch). Defining $x^* = (x_1^*, ..., x_n^*)^T$ we can write

$$e = x - y_r - x^* \tag{23}$$

now by considering (20) and (19), we derive that:

$$x^* = \bar{A}(x - y_r) \tag{24}$$

where

$$\bar{A} = \begin{pmatrix} 0 & 0 & \cdots & & \cdots & 0 \\ a_{21} & 0 & \cdots & & \cdots & 0 \\ a_{31} & a_{32} & 0 & & \cdots & 0 \\ \vdots & & \cdots\cdots & & \ddots & 0 \\ a_{n1} & & \cdots\cdots & & a_{n(n-1)} & 0 \end{pmatrix}$$

and its coefficients a_{ij} are implicitly defined by the λ_i that are arbitrarily chosen. By considering (24) and (23) we conclude that $e = x - y_r - \bar{A}(x - y_r) = (I - \bar{A})(x - y_r)$ and defining $A := (I - \bar{A})$ we obtain (22).

Proof of Theorem 3.1 (sketch). By applying Lemma 3.1 we determine a global filtered transformation (18) and a state feedback control u (17), where v is given by (21), transforming system (15) into system (19). In particular we choose:

$$\dot{\hat{\theta}} = \omega_1 = \gamma_1 Proj(\Phi e_n, \hat{\theta}) , \qquad \dot{\hat{b}} = \omega_2 = \gamma_2 Proj(v e_n, \hat{b}) \tag{25}$$

where $Proj(y, \hat{\Theta})$ is the Lipschitz continuous projection defined as

$$Proj(y, \hat{\Theta}) = \begin{cases} y & \text{if } p(\hat{\Theta}) \leq 0 \\ y & \text{if } p(\hat{\Theta}) \geq 0 \text{ and } (gradp(\hat{\Theta}), y) \leq 0 \\ y_p & \text{if } p(\hat{\Theta}) > 0 \text{ and } (gradp(\hat{\Theta}), y) > 0 \end{cases} \qquad (26)$$

where

$$p(\hat{\Theta}) := \frac{\|\hat{\Theta}\|^2 - r_\Omega^2}{\alpha^2 + 2\alpha r_\Omega}, \quad y_p = y[I - \frac{p(\hat{\Theta})gradp(\hat{\Theta})gradp(\hat{\Theta})^T}{\|gradp(\hat{\Theta})\|^2}]$$

with α an arbitrary positive constant and r_Ω is the radius of the region Ω in which Θ is supposed to be, i.e. $\Omega = B(r_\Omega)$. When $\Theta = \theta$ $r_\Omega = r_{\Omega_1}$ and by (13)

$$r_{\Omega_1} = B_f, \qquad (27)$$

while when $\Theta = b$ $r_\Omega = r_{\Omega_2}$ and by (16)

$$r_{\Omega_2} = b_M \qquad (28)$$

In this way if $\hat{\Theta}(0) \in \Omega$ then the following properties hold:
1) $\|\hat{\Theta}(t)\| \leq r_\Omega + \alpha$, $\forall t \geq 0$;
2) $Proj(y, \hat{\Theta})$ is Lipschitz continuous;
3) $\|Proj(y, \hat{\Theta})\| \leq \|y\|$;
4) $\Theta Proj(y, \Theta) \geq \Theta^T y$.
(The proof of the afore mentioned properties are in [9]).
The signals v_i for $i \leq 1 \leq n$ are chosen as

$$\begin{array}{ll} v_1 = 0 & v_n = -\frac{3}{4}ke_n - e_{n-1} \\ v_i = -e_{i-1} \quad i = 2, ..., n-1 & \bar{v} = -\frac{1}{4}ke_n(\frac{v_M^2}{v_m}) \end{array} \qquad (29)$$

in which $v_M := max_{x \in A}(v)$ and it is computable by considering from (24) that $\dot{x}_n^* = a_{n1}(x_2 - y_r^{(1)}) + ... + a_{n(n-1)}(x_n - y_r^{(n)})$ and that A4 holds. The resulting control input is linear with respect to the error e.

In order to prove property (i), since by virtue of property 1) $\hat{\theta}$ and $\hat{b}$ remain bounded respectively in Ω_1 and Ω_2 for every time, we proceed by considering a Lyapunov function that does not include the parameter estimates:

$$W = \frac{1}{2} \sum_{i=1}^{n} e_i^2 \qquad (30)$$

and computing the time derivatives by taking (19) into account, considering (21) and (29) we finally obtain

$$\dot{W} = -\sum_{i=1}^{n} \lambda_i e_i^2 + \bar{\theta}^T \Phi e_n + e_n R - \frac{3}{4}ke_n^2 + \frac{1}{\nu}(-\tilde{b}ve_n - \frac{1}{4}ke_n(\frac{v_M^2}{v_m})) \ . \ (31)$$

Now we are going to consider the worst case in which $|f(x)|$ reaches the maximum value B_f, and the network causes an additional error instead of attenuating the effect of $f(x)$ on the error dynamics. Therefore if the control can constrain x inside A for every time when the endogenous and exogenous disturbances are maximum then x will surely remain in A in every other circumstance. To do this we complete the squares in (31) taking property 1) of the *Proj* operator into account:

$$\dot{W} \leq -\lambda_1 \|e\|^2 + \frac{B_f^2 + N(B_f + \alpha)^2 + 4(b_M + \alpha)^2 + D^2}{k} \qquad (32)$$

so that

$$W(t) \leq W(0)exp(-2\lambda_1 t) + [B_f^2 + N(B_f + \alpha)^2 + 4(b_M + \alpha)^2 + D^2]/(2\lambda_1 k),$$

where we have chosen $\lambda_1 \leq \lambda_i$ for every $2 \leq i \leq n$.

Recalling that $2W(t) = \sum_{i=1}^{n} e_i^2(t) = \|e(t)\|^2$ for every $t \geq 0$ we obtain

$$\|e(t)\|^2 \leq max\{2W(0), \frac{B_f^2 + N(B_f + \alpha)^2 + 4(b_M + \alpha)^2 + D^2}{\lambda_1 k}\} \qquad (33)$$
$$:= \rho_e^2 ,$$

for all $t \geq 0$. Result (33) gives a bound for e, but our interest is concerned with the bound for x since the functional approximation that led us to the latter result holds only when x is inside A. Therefore we are going to prove that for a proper choice of the reference signal and of the initial conditions of the system there exists a value for k that constrains $x(t)$ to remain inside A for every time. To do this we shall combine results (33) and (22):

$$e^T e = (x - y_r)^T A^T A(x - y_r) \geq \lambda_m \|x - y_r\|^2$$

where $\lambda_m = \lambda_{min}(A^T A)$ is the minimum eigenvalue of $A^T A$, so that

$$\|e\| \geq \sqrt{\lambda_m}\|x - y_r\| \geq \sqrt{\lambda_m}(\|x\| - \|y_r\|) , \qquad (34)$$

supposing without loss of generality $\lambda_1 \geq 1$ we have from (33)

$$\|e\| \leq \|e(0)\| + \frac{\sqrt{B_f^2 + N(B_f + \alpha)^2 + 4(b_M + \alpha)^2 + D^2}}{\sqrt{k}}$$

and since by (22) $\|e(0)\| \leq \sqrt{\lambda_M}(\|x(0)\| + \|y_r(0)\|)$ where $\lambda_M = \lambda_{Max}(A^T A)$, by using (34) it follows that

$$\|x\| \leq \frac{\sqrt{B_f^2 + N(B_f + \alpha)^2 + 4(b_M + \alpha)^2 + D^2}}{\sqrt{\lambda_m k}}$$
$$+ \|y_r\| + \|x(0)\|\sqrt{\frac{\lambda_M}{\lambda_m}} + \|y_r(0)\|\sqrt{\frac{\lambda_M}{\lambda_m}}.$$

Since by hypothesis both y_r and $x(0)$ are contained in $B(r)$, the ball centered in the origin of radius r, we derive that

$$\|x(t)\| \leq \delta + r(1 + 2\sqrt{\frac{\lambda_M}{\lambda_m}}) , \qquad \forall t \geq 0 \qquad (35)$$

where

$$\delta := \frac{\sqrt{B_f^2 + N(B_f + \alpha)^2 + 4(b_M + \alpha)^2 + D^2}}{\sqrt{k\lambda_m}} .$$

Since we assumed that $A = B(r_a)$, in order to assure $\|x(t)\| \leq r_a$ for every $t \geq 0$, we use (35) by requiring that

$$\delta + r(1 + 2\sqrt{\frac{\lambda_M}{\lambda_m}}) \leq r_a$$

from which it is sufficient that

$$r \leq \frac{r_a - \delta}{1 + 2\sqrt{\frac{\lambda_M}{\lambda_m}}} < r_a. \qquad (36)$$

From the latter we conclude that once δ is fixed, by taking $k \geq [B_f^2 + N(B_f + \alpha)^2 + 4(b_M + \alpha)^2 + D^2]/(\delta^2\lambda_m) := \bar{k}$, if both y_r and $x(0)$ are inside $B(r)$, then property (i) is assured.

To show properties (ii) and (iii) we complete the squares in (31) in order to account separately for the effects of $\tilde{\theta}$, $\tilde{b}$ and ε, so that we find

$$\dot{W} \leq -2\lambda_1 W + \frac{\varepsilon^2 + D^2 + N\|\bar{\theta}\|^2 + \tilde{b}^2}{k} \qquad (37)$$

from which by integrating up to time t and since $\tilde{y}^2 \leq 2W(t)$ for every $t \geq 0$, we derive that

$$\tilde{y}^2 \leq 2W(0)exp(-2\lambda_1 t) + \frac{\varepsilon^2 + D^2 + N sup_{\tau\epsilon[0,t]}\|\bar{\theta}\|^2 + sup_{\tau\epsilon[0,t]}\|\tilde{b}\|^2}{2k}$$

that is (ii) with $\alpha_1(0) = \sqrt{2W(0)}$. By integrating with respect to time (37) in which we have considered again that $\tilde{y}^2 \leq 2W(t)$ we find

$$\int_0^t \tilde{y}(\tau)d\tau \leq W(0) + \int_0^t \frac{\varepsilon^2 + D^2 + N\|\bar{\theta}\|^2 + \tilde{b}^2}{k}d\tau$$

that is (iii) with $\alpha_2(0) = W(0)$.

To show property (iv) consider the time derivative of

$$V = W + \frac{1}{2}(\frac{1}{\gamma_1}\tilde{\theta}^T\tilde{\theta} + \frac{1}{\gamma_2\nu}\tilde{b}^2)$$

that is $\dot{V} \leq -\sum_{i=1}^{n} \lambda_i e_i^2 + \tilde{\theta}^T \Phi e_n + e_n R - (1/\nu)\tilde{b}ve_n - (1/\gamma_1)\tilde{\theta}^T\dot{\hat{\theta}} - (1/\gamma_2\nu)\dot{\tilde{b}}$
, and by (25) and property 4), by assuming $\varepsilon(x) = 0$ and $d(t) = 0$ we find

$$\dot{V} \leq -\sum_{i=1}^{n} \lambda_i e_i^2. \tag{38}$$

Integrating (38) with respect to time, we obtain

$$lim_{t\to\infty} \int_0^t \sum_{i=1}^{n} \lambda_i e_i(\tau)^2 d\tau \leq V(0) - V(\infty) \leq \infty.$$

Since from (25) and property 1) $\dot{e}_i$, $1 \leq i \leq n$, are bounded, we can apply Barbalat's Lemma [7] [12] [13] from which it follows that $lim_{t\to\infty} e_i(t) = 0$, $1 \leq i \leq n$, that implies in particular that $lim_{t\to\infty}(y(t) - y_r(t)) = 0$.

4   Conclusions

For a class of feedback linearizable systems (1) with A1-A4, a robust adaptive state feedback control (17), (21), (25), (29) is proposed to solve a tracking problem. The control makes use of orthogonal approximating functions (10) and of the on-line updating algorithm for the estimates of the corresponding weights (4). The innovative features of the control algorithm are: 1) the number of approximating functions is a free parameter and no requirement is imposed on the optimal value of the corresponding weights and on the approximation error; 2) it is repeatable since the initial conditions and the class of reference signals such that properties (i), (ii) and (iii) of Theorem 3.1 hold are explicitly given; 3) it guarantees that the state remains within the approximation region in which the bound on the unknown nonlinearity is given; 4)it guarantees properties (ii) and (iii) for the tracking error which account for the potential improvement due to a smaller approximation error ε and a small parameter estimation error $\tilde{\theta}$.

References

1. G. Arslan and T. Basar, "Robust Output Tracking for Strict-Feedback Systems Using Neural-Net Based Approximators for Nonlinearities", *Proc. IEEE Conf. Decision and Control*, pp. 2987-2992 Phoenix, 1999
2. A. R. Barron, "Universal Approximation Bounds for Superpositions of a Sigmoidal Function", IEEE *Trans. Information Theory*, vol. 39, no. 3, pp. 930-945,1993
3. G. Cybenko, "Approximation by Superpositions of a Sigmoidal Function", *Mathematics of Control, Signals and Systems*, pp. 303-314, Springer-Verlag New York Inc, 1989

4. A. N. Kolmogorov, S. V. Fomin, *Elementy teorii funktsij i funktsional'nogo analiza*. Mosca, 1980
5. A. Yesildirek and F. L. Lewis, "Feedback Linearization Using Neural Networks", *Automatica*, vol. 31, no. 11, pp. 1659-1664, 1995
6. R. Marino and P. Tomei, "Robust Adaptive State-Feedback Tracking for Nonlinear Systems", IEEE *Trans. Automatic Control*, vol. 43, no. 1, pp. 84-89, 1998
7. R. Marino and P. Tomei, *Nonlinear Control Design-Geometric, Adaptive and Robust*. London: Prentice Hall, 1995
8. M. M. Polycarpou and M. J. Mears, "Stable adaptive tracking of uncertain systems using nonlinearly parametrized on-line approximators", *Int.J.Control*, vol. 70, no. 3, pp. 363-384, 1998
9. J. Pomet and L. Praly, "Adaptive nonlinear regulation: Estimation from the Lyapunov equation", IEEE *Trans. Automatic Control*, vol. 37, pp. 729-740, 1992
10. G. A. Rovithakis, "Robustifying Nonlinear Systems Using High-Order Neural Network Controllers", IEEE *Trans. Automatic Control*, vol. 44, no. 1, pp. 102-108, 1999
11. R. M. Sanner and J. J. E. Slotine, "Gaussian Networks for Direct Adaptive Control", IEEE *Trans. Neural Networks*, vol. 3, no. 6, pp. 837-863, 1992
12. V. Popov, *Hyperstability of Control Systems*. Berlin: Springer-Verlag, 1973
13. K. S. Narendra and A. M. Annaswamy, *Stable Adaptive Systems*. Englewood Cliffs, NJ: Prentice Hall, 1989
14. Kristic, M., I. Kanellakopoulos, and P. V. Kokotovic (1995), *Nonlinear and Adaptive Control design*. J. Wiley, New York
15. L. K. Jones, "A simple lemma on greedy approximation in Hilbert space and convergence rates for projection pursuit regression and neural network training", *Ann. Statist.*, vol. 20, pp. 608-613, Mar. 1992

Sampled-data Low-gain Integral Control of Linear Systems with Actuator and Sensor Nonlinearities

Thomas Fliegner, Hartmut Logemann, and Eugene P. Ryan

Department of Mathematical Sciences
University of Bath
Claverton Down, Bath BA2 7AY, United Kingdom
T.Fliegner@maths.bath.ac.uk

Abstract. Non-adaptive (but possibly time-varying) and adaptive sampled-data low-gain integral control strategies are derived for asymptotic tracking of constant reference signals for exponentially stable, finite-dimensional, single-input, single-output, linear systems subject to a globally Lipschitz, nondecreasing actuator nonlinearity and a locally Lipschitz, nondecreasing, affinely sector-bounded sensor nonlinearity (the conditions on the sensor nonlinearities may be relaxed if the actuator nonlinearity is bounded). In particular, it is shown that applying error feedback using a sampled-data low-gain controller ensures asymptotic tracking of constant reference signals, provided that (a) the steady-state gain of the linear part of the continuous-time plant is positive, (b) the positive controller gain sequence is ultimately sufficiently small and not of class l^1 and (c) the reference value is feasible in a natural sense.

1 Introduction

The paper complements a sequence [5,7–9] of recent results pertaining to the problem of tracking constant reference signals, by low-gain integral control, for linear (uncertain) systems subject to actuator and/or sensor nonlinearities. All these results are related to the well-known principle that closing the loop around an exponentially stable, finite-dimensional, continuous-time, single-input, single-output linear plant Σ_c, with transfer function G_c, compensated by an integral controller (see Fig. 1), will lead to a stable closed-loop system which achieves asymptotic tracking of arbitrary constant reference signals, provided that the modulus $|k|$ of the integrator gain k is sufficiently small and $kG_c(0) > 0$ (see [4,13,16]). Therefore, if a plant is known to be exponentially stable and if the sign of $G_c(0)$ is known (this information can be obtained from plant step response data), then the problem of tracking by low-gain integral control reduces to that of tuning the gain parameter k. Such a controller design approach ('tuning regulator theory' [4]) has been successfully applied in process control, see, for example, [3,12]. Furthermore, the problem of tuning the integrator gain adaptively has been addressed in

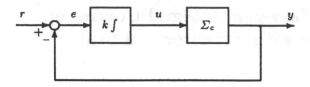

Fig. 1. Low-gain control system

a number of papers for finite-dimensional [2,5,14,15] and infinite-dimensional systems [7,8,10], with actuator constraints treated in [7,8,14] and actuator and sensor nonlinearities in [5].

The purpose of this paper is to derive sampled-data low-gain integral control strategies for asymptotic tracking of constant reference signals for exponentially stable, single-input, single-output, linear systems with transfer function G_c satisfying $G_c(0) > 0$ and with monotone nonlinearities, φ and ψ, in the input and output channel. In particular, our aim is to establish the efficacy of the control structure in Fig. 2, wherein S and H denote standard sampling and hold operations, respectively, and LG represents the (discrete-time) low-gain integral control law

$$u_{n+1} = u_n + k_n e_n , \qquad u_0 \in \mathbb{R} , \tag{1}$$

where the gain sequence $(k_n) \subset \mathbb{R}$ is either prescribed or updated adaptively. The classes of nonlinearities under consideration contain standard nonlinear-

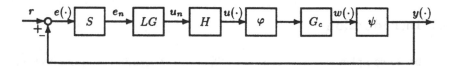

Fig. 2. Sampled-data low-gain control with actuator and sensor nonlinearities

ities important in control engineering such as saturation and deadzone.

The main result of the paper in the non-adaptive case shows that tracking of reference values r, which are feasible in an entirely natural sense, is achieved by (1) provided that the positive gain sequence (k_n) is ultimately sufficiently small and is not of class l^1. In the adaptive case it is shown that a simple adaptation law produces a gain sequence (k_n) such that (1) ensures tracking of feasible reference values.

The main concern of the paper is sampled-data control of continuous-time systems. The essence of our approach is to invoke results on discrete-time low-gain control which will be summarized without proofs in Sect. 2; proofs will appear elsewhere (see [6]). These discrete-time results will subsequently be applied to demonstrate the usefulness of the control structure in Fig. 2 for achieving the tracking objective for the considered class of nonlinear continuous-time systems. The performance of the sampled-data controller is illustrated by means of an example.

2   Low-Gain Control for Linear Discrete-Time Systems with Actuator and Sensor Nonlinearities

In [5], the problem of tracking constant reference values, by low-gain integral control, has been solved for a class of continuous-time linear systems subject to actuator and sensor nonlinearities. These results have discrete-time counterparts (see [6]) which are crucial in solving the sampled-data tracking problem and which will, therefore, be recorded in this section for later use.

Consider a discrete-time, single-input, single-output system of the following form:

$$x_{n+1} = Ax_n + B\varphi(u_n) , \quad x_0 \in \mathbb{R}^N , \tag{2a}$$

$$w_n = Cx_n + D\varphi(u_n) , \tag{2b}$$

$$y_n = \psi(w_n) = \psi(Cx_n + D\varphi(u_n)) . \tag{2c}$$

2.1   The Class S of Discrete–Time Linear Systems

In (2a), A is assumed to be power-stable, i.e., each eigenvalue of A has modulus strictly less than one. Furthermore, the transfer function G, given by

$$G(z) = C(zI - A)^{-1}B + D ,$$

is assumed to satisfy $G(1) > 0$. Thus, the underlying class of linear discrete-time systems $\Sigma = (A, B, C, D)$ is

$$S := \{\Sigma = (A, B, C, D) \mid A \text{ power-stable}, \ G(1) = C(I - A)^{-1}B + D > 0\} .$$

We denote the l^∞-gain of $\Sigma = (A, B, C, D) \in S$ by Γ_Σ, and so

$$0 < G(1) \leq \Gamma_\Sigma = \sum_{j=0}^{\infty} |CA^jB| + |D| < \infty .$$

If G is the transfer function of a system $\Sigma = (A, B, C, D) \in S$, then it is readily shown that

$$1 + \kappa \operatorname{Re} \frac{G(z)}{z-1} \geq 0 \quad \forall z \in \mathbb{C} \quad \text{with } |z| > 1 \tag{3}$$

for all sufficiently small $\kappa > 0$, see [11, Theorem 2.5]. Define

$$\kappa^* := \sup\{\kappa > 0 \mid (3) \text{ holds}\} . \tag{4}$$

2.2 The Class $\mathcal{N}$ of Actuator/Sensor Nonlinearities

The following sets of monotone, non-decreasing nonlinearities are first introduced:

$$\mathcal{M} := \{f : \mathbb{R} \to \mathbb{R} \mid f \text{ locally Lipschitz and non-decreasing}\} ,$$

$$\mathcal{M}(\lambda) := \{f \in \mathcal{M} \mid 0 \le (f(\xi) - f(0))\xi \le \lambda\xi^2 \ \forall \xi \in \mathbb{R}\} ,$$

$$\mathcal{M}_L(\lambda) := \{f \in \mathcal{M} \mid f \text{ is globally Lipschitz with Lipschitz constant } \lambda\} .$$

Clearly, $\mathcal{M}_L(\lambda) \subset \mathcal{M}(\lambda) \subset \mathcal{M}$.

Actuator/Sensor Nonlinearities. We are now in a position to make precise the class $\mathcal{N}$ of actuator/sensor nonlinearities: a pair (φ, ψ) is in $\mathcal{N}$ if $\varphi \in \mathcal{M}_L(\lambda_1)$ for some $\lambda_1 > 0$, $\psi \in \mathcal{M}$ and at least one of the following holds:

$$\text{(i) } \varphi \text{ is bounded , or (ii) } \psi \in \mathcal{M}(\lambda_2) \text{ for some } \lambda_2 > 0 . \tag{5}$$

Critical Points and Critical Values. Let $f : \mathbb{R} \to \mathbb{R}$ be locally Lipschitz. For $\xi, \nu \in \mathbb{R}$, the Clarke [1] directional derivative $f^\circ(\xi; \nu)$ of f at ξ in direction ν is given by

$$f^\circ(\xi; \nu) := \limsup_{\substack{\theta \to \xi \\ h \downarrow 0}} \frac{f(\theta + h\nu) - f(\theta)}{h} .$$

Define $f^-(\cdot) := -f^\circ(\cdot; -1)$ (if f is continuously differentiable with derivative f', then $f^- \equiv f'$). By upper semicontinuity of f°, f^- is lower semicontinuous. Let $f \in \mathcal{M}$: a point $\xi \in \mathbb{R}$ is said to be a *critical point* (and $f(\xi)$ is said to be a *critical value*) of f if $f^-(\xi) = 0$.[1] We denote, by $\mathcal{C}(f)$, the set of critical values of f.

2.3   The Tracking Objective and Feasibility

Given $\Sigma = (A, B, C, D) \in \mathcal{S}$ and $(\varphi, \psi) \in \mathcal{N}$, the tracking objective is to determine, by feedback, an input sequence $(u_n) \subset \mathbb{R}$ such that, for given $r \in \mathbb{R}$, the output y_n of (2) has the property $y_n \to r$ as $n \to \infty$. Clearly, if this objective is achievable, then r is necessarily in the closure of $\operatorname{im}\psi$. We will impose a stronger condition, namely,

$$\Psi^r \cap \overline{\Phi} \ne \emptyset, \text{ where } \Psi^r := \{v \in \mathbb{R} \mid \psi(G(1)v) = r\}, \ \Phi := \operatorname{im}\varphi , \ \overline{\Phi} := \operatorname{clos}(\Phi)$$

---

[1] If f is merely locally Lipschitz, but not in $\mathcal{M}$, then it would be natural to deem $\xi \in \mathbb{R}$ a critical point of f if 0 belongs to the (Clarke [1]) generalized gradient $\partial f(\xi)$ of f at ξ. We remark that, in the case of $f \in \mathcal{M}$, $0 \le \min\{\partial f(\xi)\} = f^-(\xi)$. Therefore, for functions $f \in \mathcal{M}$, the latter concept of a critical point coincides with that given above.

and refer to the set

$$\mathcal{R} := \{r \in \mathbb{R} \mid \Psi^r \cap \overline{\Phi} \neq \emptyset\}$$

as the set of *feasible reference values*. It may be shown that $r \in \mathcal{R}$ is close to being a necessary condition for tracking insofar as, if tracking of r is achievable whilst maintaining boundedness of $\varphi(u_n)$, then $r \in \mathcal{R}$.

2.4   Discrete-Time Low-Gain Control

Let $\Sigma = (A, B, C, D) \in \mathcal{S}$ and $(\varphi, \psi) \in \mathcal{N}$. To achieve the objective of tracking feasible reference values $r \in \mathcal{R}$, we will investigate integral control action of the form

$$u_{n+1} = u_n + k_n(r - \psi(Cx_n + D\varphi(u_n))) = u_n + k_n(r - y_n), \qquad (6)$$

where $(k_n) \subset \mathbb{R}$ is a sequence of gain parameters (possibly constant) which is either prescribed or determined adaptively.

Prescribed Gain. Henceforth, we assume that the gain sequence (k_n) satisfies

$$(k_n) \in \mathcal{G} := \{g \mid g : \mathbb{N}_0 \to (0, \infty), \ (g_n) \text{ is bounded}\},$$

where $\mathbb{N}_0 := \mathbb{N} \cup \{0\}$. An application of the control law (6) leads to the following system of nonlinear difference equations

$$x_{n+1} = Ax_n + B\varphi(u_n), \quad x_0 \in \mathbb{R}^N, \qquad (7a)$$

$$u_{n+1} = u_n + k_n(r - \psi(Cx_n + D\varphi(u_n))), \quad u_0 \in \mathbb{R}. \qquad (7b)$$

Before presenting the main results, we describe a convenient family of projection operators. Specifically, with each $p \in [0, \infty]$, we associate an operator $\Pi_p : \mathcal{M} \to \mathcal{M}$, with the property $\Pi_p \circ \Pi_p = \Pi_p$ (whence the terminology projection operator), defined as follows:

$$\text{if} \quad p < \infty, \text{ then} \quad \Pi_p f : \xi \mapsto \begin{cases} f(-p), & \xi < -p \\ f(\xi), & |\xi| \leq p \ ; \\ f(p), & \xi > p \end{cases}$$

$$\text{if} \quad p = \infty, \text{ then} \quad \Pi_p f = \Pi_\infty f := f.$$

We are now in a position to state the main discrete-time tracking result in the non-adaptive situation.

Theorem 1. *Let $\Sigma = (A, B, C, D) \in \mathcal{S}$, $(\varphi, \psi) \in \mathcal{N}$ and $r \in \mathcal{R}$. Define*

$$p^* := \Gamma_\Sigma \sup_{\xi \in \Phi} |\xi| \in (0, \infty].$$

There exists $k^* > 0$, independent of (x_0, u_0) and $(k_n) \in \mathcal{G}$, such that for all $(k_n) \in \mathcal{G}$ with the properties $\limsup_{n \to \infty} k_n < k^*$ and $K_n := \sum_{j=0}^n k_j \to \infty$ as $n \to \infty$, and for every $(x_0, u_0) \in \mathbb{R}^N \times \mathbb{R}$, the following hold for the solution $n \mapsto (x_n, u_n)$ of the initial-value problem (7):

(i) $\lim_{n \to \infty} \varphi(u_n) =: \varphi^r \in \Psi^r \cap \overline{\Phi}$,

(ii) $\lim_{n \to \infty} x_n = (I - A)^{-1} B \varphi^r$,

(iii) $\lim_{n \to \infty} y_n = r$, where $y_n = \psi(Cx_n + D\varphi(u_n))$,

(iv) if $\Psi^r \cap \overline{\Phi} = \Psi^r \cap \Phi$, then $\lim_{n \to \infty} \text{dist}(u_n, \varphi^{-1}(\varphi^r)) = 0$,

(v) if $\Psi^r \cap \overline{\Phi} = \Psi^r \cap \text{int}(\Phi)$, then (u_n) is bounded ,

(vi) if $\Psi^r \cap \overline{\Phi} = \Psi^r \cap \Phi$, $\Psi^r \cap \mathcal{C}(\varphi) = \emptyset$, $r \notin \mathcal{C}(\psi)$ and (k_n) is ultimately non-increasing, then the convergence in (i) to (iii) is of order ρ^{-K_n} for some $\rho > 1$.

Let $\lambda_1 > 0$ be a Lipschitz constant for φ. If $\Pi_{(p^* + \delta)} \psi \in \mathcal{M}_L(\lambda_2)$ for some $\delta > 0$ and some $\lambda_2 > 0$, then k^* may be chosen as $k^* = \kappa^* / (\lambda_1 \lambda_2)$, where κ^* is given by (4).

Remark 1. It is easy to construct examples for which $(k_n) \in \mathcal{G}$ satisfies $\limsup_{n \to \infty} k_n < k^*$ with (K_n) bounded and $\varphi(u_n)$ converges to $\varphi^* \notin \Psi^r \cap \overline{\Phi}$ (and so the tracking objective is not achieved). The hypothesis that (K_n) is unbounded ensures convergence of $\varphi(u_n)$ to an element of $\Psi^r \cap \overline{\Phi}$ (and hence the tracking objective is achieved).

Remark 2. Statement (iv) implies that (u_n) converges as $n \to \infty$ if the set $\varphi^{-1}(\varphi^r)$ is a singleton, which, in turn, will be true if $\Psi^r \cap \mathcal{C}(\varphi) = \emptyset$.

Adaptive Gain. Whilst Theorem 1 identifies conditions under which the tracking objective is achieved through the use of a prescribed gain sequence, the resulting control strategy is somewhat unsatisfactory insofar as the gain sequence is selected *a priori*: no use is made of the output information from the plant to update the gain. We now consider the possibility of exploiting this output information to generate, by feedback, an appropriate gain sequence. Let $\mathcal{L}$ denote the class of locally Lipschitz functions $\mathbb{R}_+ \to \mathbb{R}_+$ with value zero only at zero and which satisfy a particular growth condition near zero, specifically:

$$\mathcal{L} := \{ f \mid f : \mathbb{R}_+ \to \mathbb{R}_+ ,$$
$$f \text{ locally Lipschitz}, \ f^{-1}(0) = \{0\}, \ \liminf_{\xi \downarrow 0} \xi^{-1} f(\xi) > 0 \} . \quad (8)$$

Let $\chi \in \mathcal{L}$ and let the gain sequence (k_n) be generated by the following adaptation law:

$$k_n = \frac{1}{l_n}, \quad l_{n+1} = l_n + \chi(|r - y_n|), \quad l_0 > 0 . \quad (9)$$

This leads to the feedback system

$$x_{n+1} = Ax_n + B\varphi(u_n), \quad x_0 \in \mathbb{R}^N , \quad (10a)$$

$$u_{n+1} = u_n + l_n^{-1}[r - \psi(Cx_n + D\varphi(u_n))], \quad u_0 \in \mathbb{R} , \quad (10b)$$

$$l_{n+1} = l_n + \chi(|r - \psi(Cx_n + D\varphi(u_n))|), \quad l_0 \in (0, \infty) . \quad (10c)$$

We now arrive at the main adaptive discrete-time tracking result.

Theorem 2. *Let $\Sigma = (A, B, C, D) \in \mathcal{S}$, $(\varphi, \psi) \in \mathcal{N}$, $\chi \in \mathcal{L}$ and $r \in \mathcal{R}$. For each $(x_0, u_0, l_0) \in \mathbb{R}^N \times \mathbb{R} \times (0, \infty)$, the solution $n \mapsto (x_n, u_n, l_n)$ of the initial-value problem (10) is such that statements (i) to (v) of Theorem 1 hold. Moreover, if $\Psi^r \cap \overline{\Phi} = \Psi^r \cap \Phi$, $\Psi^r \cap \mathcal{C}(\varphi) = \emptyset$ and $r \notin \mathcal{C}(\psi)$, then the monotone gain $(k_n) = (l_n^{-1})$ converges to a positive value.*

3   Sampled-Data Control of Linear Systems with Actuator and Sensor Nonlinearities

In this section we shall apply the results of Sect. 2 to solve the continuous-time low-gain tracking problem, by sampled-data integral control, for the class of systems introduced below.

3.1   The Class of Continuous-Time Systems

Tracking results will be derived for a class of finite-dimensional (state space $\mathbb{R}^N$) single-input ($u(t) \in \mathbb{R}$), single-output ($y(t) \in \mathbb{R}$), continuous-time (time domain $\mathbb{R}_+ := [0, \infty)$), real linear systems $\Sigma_c = (A_c, B_c, C_c, D_c)$ having a nonlinearity in the input and output channel:

$$\dot{x} = A_c x + B_c \varphi(u) , \quad x_0 := x(0) \in \mathbb{R}^N , \tag{11a}$$

$$w = C_c x + D_c \varphi(u) , \tag{11b}$$

$$y = \psi(w) = \psi(C_c x + D_c \varphi(u)) . \tag{11c}$$

In (11a), A_c is assumed to be Hurwitz, i.e., each eigenvalue of A_c has negative real part. Furthermore, the transfer function G_c, given by

$$G_c(s) = C_c(sI - A_c)^{-1} B_c + D_c ,$$

is assumed to satisfy $G_c(0) > 0$. The underlying class of real linear systems $\Sigma_c = (A_c, B_c, C_c, D_c)$ is denoted

$$\mathcal{S}_c := \{ \Sigma_c = (A_c, B_c, C_c, D_c) \mid A_c \text{ Hurwitz}, \ G_c(0) = D_c - C_c A_c^{-1} B_c > 0 \} .$$

Given $\Sigma_c = (A_c, B_c, C_c, D_c) \in \mathcal{S}_c$ and $(\varphi, \psi) \in \mathcal{N}$, the tracking objective is to determine, by sampled-data control, a sequence $(u_n) \subset \mathbb{R}$ such that, for given $r \in \mathbb{R}$, the output $y(\cdot)$ of (11), resulting from applying the input $u(\cdot)$ given by

$$u(t) = u_n \quad \text{for } t \in [n\tau, (n+1)\tau) , \ \tau > 0 , \ n \in \mathbb{N}_0 , \tag{12}$$

has the property $y(\cdot) \to r$ as $t \to \infty$. As in Sect. 2, it will not be possible to track arbitrary $r \in \mathbb{R}$. Instead, we require $r \in \mathcal{R}_c$ where

$$\mathcal{R}_c := \{ r \in \mathbb{R} \mid \Psi_c^r \cap \overline{\Phi} \neq \emptyset \}$$

and

$$\Psi_c^r := \{v \in \mathbb{R} \mid \psi(G_c(0)v) = r\} \,, \quad \Phi := \operatorname{im} \varphi \,, \quad \overline{\Phi} := \operatorname{clos}(\Phi) \,.$$

Again, it may be shown that $r \in \mathcal{R}_c$ is close to being necessary for tracking insofar as, if tracking of r is achievable whilst maintaining boundedness of $\varphi(u)$, then $r \in \mathcal{R}_c$.

3.2 Discretization of the Continuous-Time Plant

Given a sequence $(u_n) \subset \mathbb{R}$, we define a continuous-time signal $u(\cdot)$ by the standard hold operation (12) where $\tau > 0$ denotes the sampling period. If this signal is applied to the continuous-time system (11), then, defining

$$x_n := x(n\tau) \,, \quad w_n := w(n\tau) \quad \text{and} \quad y_n := y(n\tau) \qquad \forall n \in \mathbb{N}_0 \,, \tag{13}$$

the discretization of (11) is given by

$$x_{n+1} = Ax_n + B\varphi(u_n) \,, \quad x_0 := x(0) \in \mathbb{R}^N \,, \tag{14a}$$
$$w_n = Cx_n + D\varphi(u_n) \,, \tag{14b}$$
$$y_n = \psi(w_n) = \psi(Cx_n + D\varphi(u_n)) \,, \tag{14c}$$

where

$$A = e^{A_c \tau} \,, \quad B = (e^{A_c \tau} - I)A_c^{-1}B_c \,, \quad C = C_c \quad \text{and} \quad D = D_c \,. \tag{15}$$

Since A_c is Hurwitz, the eigenvalues of A have, for arbitrary $\tau > 0$, magnitude strictly less than one, i.e., the matrix A is power-stable. Furthermore, the transfer function G of $\Sigma = (A, B, C, D)$ satisfies

$$G(1) = C_c(I - e^{A_c \tau})^{-1}(e^{A_c \tau} - I)A_c^{-1}B_c + D_c = G_c(0) \,. \tag{16}$$

It follows from (16) that

$$\Psi^r = \Psi_c^r \quad \text{and} \quad \mathcal{R} = \mathcal{R}_c \,,$$

where $\Psi^r = \{v \in \mathbb{R} \mid \psi(G(1))v = r\}$ and $\mathcal{R}$ denotes the set of feasible reference values for the discretization $\Sigma = (A, B, C, D)$ of $\Sigma_c = (A_c, B_c, C_c, D_c)$.

3.3 Prescribed Gain

We first treat the case of a prescribed gain sequence $(k_n) \in \mathcal{G}$. For this purpose, consider the following sampled-data low-gain controller for (11)

$$u(t) = u_n \,, \qquad \text{for } t \in [n\tau, (n+1)\tau) \,, \ n \in \mathbb{N}_0 \,, \tag{17a}$$
$$y_n = y(n\tau) \,, \qquad n \in \mathbb{N}_0 \,, \tag{17b}$$
$$u_{n+1} = u_n + k_n(r - y_n) \,, \qquad u_0 \in \mathbb{R} \,. \tag{17c}$$

Theorem 3. *Let* $\Sigma_c = (A_c, B_c, C_c, D_c) \in \mathcal{S}_c$, $(\varphi, \psi) \in \mathcal{N}$, $\tau > 0$ *and* $r \in \mathcal{R}_c$. *Define*

$$p^* := \Gamma_\Sigma \sup_{\xi \in \Phi} |\xi| \in (0, \infty],$$

where $\Sigma = (A, B, C, D)$ *denotes the discretization of* $\Sigma_c = (A_c, B_c, C_c, D_c) \in \mathcal{S}_c$ *with sampling period* τ.
There exists $k^* > 0$, *independent of* (x_0, u_0) *and* $(k_n) \in \mathcal{G}$, *such that for all* $(k_n) \in \mathcal{G}$ *with the properties* $\limsup_{n \to \infty} k_n < k^*$ *and* $K_n := \sum_{j=0}^n k_j \to \infty$ *as* $n \to \infty$, *and for every* $(x_0, u_0) \in \mathbb{R}^N \times \mathbb{R}$ *the following hold for the solution of the closed loop system given by* (11) *and* (17):

- (i) $\lim_{t \to \infty} \varphi(u(t)) =: \varphi^r \in \Psi_c^r \cap \overline{\Phi}$,
- (ii) $\lim_{t \to \infty} x(t) = -A_c^{-1} B_c \varphi^r$,
- (iii) $\lim_{t \to \infty} y(t) = r$, *where* $y(t) = \psi(C_c x(t) + D_c \varphi(u(t)))$,
- (iv) *if* $\Psi_c^r \cap \overline{\Phi} = \Psi_c^r \cap \Phi$, *then* $\lim_{t \to \infty} \text{dist}(u(t), \varphi^{-1}(\varphi^r)) = 0$,
- (v) *if* $\Psi_c^r \cap \overline{\Phi} = \Psi_c^r \cap \text{int}(\Phi)$, *then* $u(\cdot)$ *is bounded*,
- (vi) *if* (k_n) *is ultimately non-increasing,* $\Psi_c^r \cap \overline{\Phi} = \Psi_c^r \cap \Phi$, $\Psi_c^r \cap \mathcal{C}(\varphi) = \emptyset$ *and* $r \notin \mathcal{C}(\psi)$, *then the convergence in* (i) *to* (iii) *is of order* $\exp(-\alpha K(t))$ *for some* $\alpha > 0$, *where* $K : \mathbb{R}_+ \to \mathbb{R}_+$ *is defined by* $K(t) = K_n$ *for* $t \in [n\tau, (n+1)\tau)$.

Let $\lambda_1 > 0$ *be a Lipschitz constant for* φ. *If* $\Pi_{(p^*+\delta)}\psi \in \mathcal{M}_L(\lambda_2)$ *for some* $\delta > 0$ *and some* $\lambda_2 > 0$, *then* k^* *may be chosen as* $k^* = \kappa^*/(\lambda_1 \lambda_2)$, *where* κ^* *is given by* (4).

Remark 3.
(i) Statement (iv) implies that $u(t)$ converges as $t \to \infty$ if $\Psi_c^r \cap \mathcal{C}(\varphi) = \emptyset$.
(ii) An immediate consequence of Theorem 3 is the following: if $(k_n) \in \mathcal{G}$ is chosen such that, as $n \to \infty$, k_n tends to zero sufficiently slowly in the sense that $(k_n) \notin l^1(\mathbb{R}_+)$, then the tracking objective is achieved.
(iii) If $\Pi_{(p^*+\delta)}\psi \in \mathcal{M}_L(\lambda_2)$ for some $\delta > 0$ and some $\lambda_2 > 0$, we may infer that the tracking objective is achievable by a *constant* gain sequence with $k_n = k \in (0, \kappa^*/(\lambda_1 \lambda_2))$ for all $n \in \mathbb{N}_0$, where $\lambda_1 > 0$ is a Lipschitz constant of φ.

Proof of Theorem 3. It follows from the previous subsection that x_n, w_n, y_n (given by (13)) and u_n satisfy (14) with (A, B, C, D) given by (15). Moreover, A is power-stable and $G(1) = G_c(0) > 0$. Hence $\Sigma = (A, B, C, D) \in \mathcal{S}$. Thus, by part (i) of Theorem 1,

$$\lim_{n \to \infty} \varphi(u_n) = \varphi^r \in \Psi^r \cap \overline{\Phi}.$$

Since the continuous-time signal u is given by (12), assertions (i), (iv) and (v) follow. Assertion (ii) is a consequence of (i) and the Hurwitz property of A. Assertion (iii) follows from (i), (ii) and continuity of ψ.

To show (vi), note that by Theorem 1(vi), there exist $M > 0$ and $\rho > 1$ such that

$$|\varphi(u_n) - \varphi^r| \le M\rho^{-K_n} \quad \text{and} \quad \|x_n - x^r\| \le M\rho^{-K_n}$$

for all $n \in \mathbb{N}_0$, where $x^r = -A_c^{-1} B_c \varphi^r$. For all $t \in [n\tau, (n+1)\tau)$, $\varphi(u(t)) = \varphi(u_n)$ and so $|\varphi(u(t)) - \varphi^r| \le M\rho^{-K_n} = M \exp(-\alpha K_n)$, where $\alpha = \ln \rho > 0$. Therefore, the convergence in (i) is of order $\exp(-\alpha K(t))$. Define $\zeta(t) := x(t) - x^r$. Then $\zeta(\cdot)$ satisfies

$$\dot\zeta(t) = A_c \zeta(t) + B_c(\varphi(u(t)) - \varphi^r)$$

and thus

$$\zeta(t) = e^{A_c(t - n\tau)} \zeta(n\tau) + \int_{n\tau}^{t} e^{A_c(t-s)} B_c(\varphi(u_n) - \varphi^r) ds, \quad t \in [n\tau, (n+1)\tau).$$

We may now conclude that

$$\|\zeta(t)\| \le$$
$$\max_{0 \le s \le \tau} \|e^{A_c s}\| M e^{-\alpha K_n} + \tau \max_{0 \le s \le \tau} \|e^{A_c s}\| \|B_c\| M e^{-\alpha K_n}, \quad t \in [n\tau, (n+1)\tau)$$

which proves that convergence in (ii) is of order $\exp(-\alpha K(t))$. From the convergence of order $\exp(-\alpha K(t))$ in (i) and (ii), it is straightforward to show, invoking the hypothesis $r \notin \mathcal{C}(\psi)$ and the local Lipschitz property of ψ, that the convergence in (iii) is also of order $\exp(-\alpha K(t))$. $\square$

3.4   Adaptive Gain

In this subsection, our goal is to establish the efficacy of the adaptation law (9) to generate an appropriate gain sequence in the sampled-data setting. Hence, consider the following adaptive sampled-data low-gain controller for (11)

$$u(t) = u_n, \quad \text{for } t \in [n\tau, (n+1)\tau), \ n \in \mathbb{N}_0, \tag{18a}$$

$$y_n = y(n\tau), \quad n \in \mathbb{N}_0, \tag{18b}$$

$$u_{n+1} = u_n + l_n^{-1}(r - y_n), \quad u_0 \in \mathbb{R}, \tag{18c}$$

$$l_{n+1} = l_n + \chi(|r - y_n|), \quad l_0 \in (0, \infty). \tag{18d}$$

Theorem 4. *Let $\Sigma_c = (A_c, B_c, C_c, D_c) \in \mathcal{S}_c$, $(\varphi, \psi) \in \mathcal{N}$, $\tau > 0$ and $r \in \mathcal{R}_c$. For all $(x_0, u_0, l_0) \in \mathbb{R}^N \times \mathbb{R} \times (0, \infty)$, the solution $(x(\cdot), u(\cdot), (l_n))$ of the closed-loop system given by (11) and (18) is such that statements (i) to (v) of Theorem 3 hold. If moreover $\Psi_c^r \cap \overline{\Phi} = \Psi_c^r \cap \Phi$, $\Psi_c^r \cap \mathcal{C}(\varphi) = \emptyset$ and $r \notin \mathcal{C}(\psi)$, the monotone gain sequence $(k_n) = (l_n^{-1})$ converges to a positive value.*

Proof. The discretization $\Sigma = (A, B, C, D)$ of $\Sigma_c = (A_c, B_c, C_c, D_c) \in \mathcal{S}_c$ is of class $\mathcal{S}$. Thus, the theorem is an immediate consequence of Theorem 2. $\square$

3.5 Example

Consider the second-order system

$$\dot{x}_1 = x_2 , \quad \dot{x}_2 = -ax_2 - bx_1 + \varphi(u) , \quad w = x_1 , \quad y = \psi(w) ,$$

where $a, b > 0$, the sensor nonlinearity $\psi \in \mathcal{M}$ is the cubic $\psi : w \mapsto w^3$, and the actuator nonlinearity $\varphi \in \mathcal{M}_L(1)$ is of saturation type, defined as follows

$$u \mapsto \varphi(u) := \begin{cases} -1 , & u < -1 \\ u , & u \in [-1, 1] \\ 1 , & u > 1 , \end{cases}$$

and so $\overline{\Phi} = [-1, 1]$. The transfer function G_c of the associated linear system Σ_c is given by

$$G_c(s) = \frac{1}{s^2 + as + b} , \quad \text{with } G_c(0) = \frac{1}{b} > 0 .$$

Since $\Psi_c^r = \{v \in \mathbb{R} \mid \psi(G_c(0)v) = r\} = \{br^{1/3}\}$, we have $\Psi_c^r \cap \overline{\Phi} \neq \emptyset$ if and only if $-b^{-3} \leq r \leq b^{-3}$. Thus, the set $\mathcal{R}_c$ of feasible reference values is given by $\mathcal{R}_c = [-b^{-3}, b^{-3}]$.

Let $\chi \in \mathcal{L}$. By Theorem 4, it follows that the adaptive sampled-data controller (18) achieves the tracking objective for each feasible reference value $r \in \mathcal{R}_c = [-b^{-3}, b^{-3}]$. Moreover, if r is both non-zero (so that $r \notin \mathcal{C}(\psi) = \{0\}$) and in the interior of $\mathcal{R}_c$ (so that $\Psi_c^r \cap \mathcal{C}(\varphi) = \emptyset$), then the adapting gain (k_n) converges to a positive value. For purposes of illustration, let $a = 2$, $b = 1$ (in which case $\mathcal{R}_c = [-1, 1]$), $\tau = 1$ and let $\chi = \varphi|_{\mathbb{R}_+}$ (the restriction of the saturation function to $\mathbb{R}_+$). For initial data $(x_1(0), x_2(0), u_0, l_0) = (0, 0, 0, 1)$

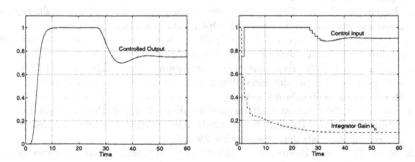

Fig. 3. Performance under adaptive sampled-data control

and the feasible reference value $r = 0.75$, Fig. 3 (generated using SIMULINK Simulation Software under MATLAB) depicts the system performance under adaptive sampled-data control. The convergence of the gain to a positive limiting value is evident.

Acknowledgement. This work was supported by the UK Engineering & Physical Sciences Research Council (Grant Ref: GR/L78086).

References

1. Clarke F. H. (1983) Optimization and Nonsmooth Analysis. Wiley, New York
2. Cook P. A. (1992) Controllers with universal tracking properties. Proc. of Int. IMA Conf. on Control: Modelling, Computation, Information. Manchester
3. Coppus G. W. M., Sha S. L., and Wood R. K. (1983) Robust multivariable control of a binary distillation column. IEE Proceedings, Pt. D **130**, 201–208
4. Davison E. J. (1976) Multivariable tuning regulators: the feedforward and robust control of a general servomechanism problem. IEEE Trans. Auto. Control **21**, 35–47
5. Fliegner T., Logemann H., and Ryan E. P. (1999) Low-gain integral control of continuous-time linear systems subject to input and output nonlinearities. Mathematics Preprint 99/24[1], University of Bath, (submitted for publication)
6. Fliegner T., Logemann H., and Ryan E. P. (1999) Discrete-time and sampled-data integral control of linear systems with input and output nonlinearities. Mathematics Preprint 99/26[1], University of Bath, (submitted for publication)
7. Logemann H. and Ryan E. P. (2000) Time-varying and adaptive integral control of infinite-dimensional regular systems with input nonlinearities. SIAM J. Control & Optim. **38**, 1120–1144
8. Logemann H. and Ryan E. P. (1998) Time-varying and adaptive discrete-time low–gain control of infinite-dimensional linear systems with input nonlinearities. Mathematics Preprint 98/20[1], University of Bath, (to appear in Mathematics of Control, Signals, and Systems)
9. Logemann H., Ryan E. P., and Townley S. (1998) Integral control of infinite-dimensional linear systems subject to input saturation. SIAM J. Control & Optim. **36**, 1940–1961
10. Logemann H. and Townley S. (1997) Low-gain control of uncertain regular linear systems. SIAM J. Control & Optim. **35**, 78–116
11. Logemann H. and Townley S. (1997) Discrete-time low-gain control of uncertain infinite-dimensional systems. IEEE Trans. Auto. Control **42**, 22–37
12. Lunze J. (1987) Experimentelle Erprobung einer Einstellregel für PI-Mehrgrößenregler bei der Herstellung von Ammoniumnitrat-Harnstoff-Lösung. Messen Steuern Regeln **30**, 2–6
13. Lunze J. (1988) Robust Multivariable Feedback Control. Prentice Hall, London
14. Miller D. E. and Davison E. J. (1993) An adaptive tracking problem with a control input constraint. Automatica **29**, 877–887
15. Miller D. E. and Davison E. J. (1989) The self-tuning robust servomechanism problem. IEEE Trans. Auto. Control **34**, 511–523
16. Morari M. (1985) Robust stability of systems with integral control. IEEE Trans. Auto. Control **30**, 574–577

---

[1] Preprints in this series are available on the World Wide Web at the URL
    http://www.maths.bath.ac.uk/MATHEMATICS/preprints.html
and by anonymous ftp from ftp.maths.bath.ac.uk/pub/preprints.

State Feedbacks Without Asymptotic Observers and Generalized PID Regulators*

Michel Fliess[1], Richard Marquez[2],**, and Emmanuel Delaleau[2]

[1] Centre de Mathématiques et Leurs Applications
École Normale Supérieure de Cachan
61, avenue du Président Wilson
94235 Cachan, France
fliess@cmla.ens-cachan.fr
[2] Laboratoire des Signaux et Systèmes
CNRS–Supélec–Université Paris-Sud
Plateau de Moulon
3, rue Joliot-Curie
91192 Gif-sur-Yvette, France
{marquez,delaleau}@lss.supelec.fr

Abstract. For constant linear systems we introduce the class of *exact integral observers* which yield *generalized PID regulators* with good robustness properties. When utilized in conjunction with static state feedbacks they permit to bypass classical asymptotic observers. Three illustrative examples are examined:

1. a classical PID controller where we replace the derivative term by appropriate integral ones;
2. a generalized PID for a second order system with a non-trivial zero dynamics;
3. a <u>real</u> DC motor without mechanical sensors.

Our approach, which is mainly of algebraic flavour, is based on the module-theoretic framework for linear systems and on operational calculus in Mikusiński's setting. We conclude by discussing possible extensions, especially nonlinear ones.

1 Introduction

In spite of many fundamental works, nonlinear observers certainly remain one of the most challenging control problems. We are here launching a new programme for trying to elucidate this topic. We will of course start with continuous-time constant linear systems not only for obvious simplicity's

* Work partially supported by the European Commission's Training and Mobility of Researchers (TMR) under contract ERBFMRXT-CT970137. One author (RM) was also partially supported by the *Consejo Nacional de Investigaciones Científicas y Tecnológicas (CONICIT)*, Venezuela.
** On temporary leave from Departamento de Sistemas de Control, Facultad de Ingeniería, Universidad de Los Andes, Mérida 5101, Venezuela.

sake but also for explaining our viewpoint where asymptotic observers are replaced by exact integral ones, i.e., by expressions where we only deal with measured quantities and their iterated integrals. This endeavour, which rests on a novel exploitation of observability, is completing almost ten years of flatness-based control synthesis where the core was a revisited notion of controllability (see [14] and the references therein).

Consider the constant linear input-output system

$$sx = Ax + Bu \qquad (1a)$$
$$y = Cx \qquad (1b)$$

where s stands for $\frac{d}{dt}$. Rewrite (1a) as

$$x = A\frac{x}{s} + B\frac{u}{s}$$

By induction we obtain for any $\mu \geq 1$

$$x = A^{\mu}s^{-\mu}x + \sum_{i=1}^{\mu} A^{i-1}Bs^{-i}u \qquad (2)$$

Assume that (1) is observable. From the constructibility property, which is equivalent to observability (cf. [23]), we know that any component of the state $x = (x_1, \ldots, x_n)$ may be expressed as a finite linear combination of the input and output variables, $u = (u_1, \ldots, u_m)$ and $y = (y_1, \ldots, y_p)$, and of their derivatives. Thus (2) yields, for μ large enough,

$$\begin{pmatrix} x_1 \\ \vdots \\ x_n \end{pmatrix} = P \begin{pmatrix} y_1 \\ \vdots \\ y_p \end{pmatrix} + Q \begin{pmatrix} u_1 \\ \vdots \\ u_m \end{pmatrix} \qquad (3)$$

where P and Q are respectively $n \times p$ and $n \times m$ matrices whose entries are polynomials in the variable s^{-1}. Formula (3) means that the components of the state may be recovered *via* a finite linear combination of iterated integrals of the input and output variables. Those integrals will be given by equating $\frac{f}{s}$ to $\int_{-\infty}^{t} f(\tau)d\tau$, where f is some time function. In this interpretation, which is based on the operational calculus *via* the two-sided Laplace transform (see, e.g., [30]), a non-zero constant function on $(-\infty, +\infty)$ is not admissible. This leads to the following convention[1] which is one of the roots of our approach:

Any constant linear system is at rest for $t \leq t_0$, $t_0 \in (-\infty, +\infty)$.

Then, (3), which we call an *exact integral observer*, may be regarded as a feasible alternative to classical asymptotic observers.

---

[1] The very same convention was also employed by the main father of operational calculus, Oliver Heaviside (see, e.g., [30]).

Consider the static state feedback

$$
\begin{pmatrix} u_1 \\ \vdots \\ u_m \end{pmatrix} = -\mathcal{K} \begin{pmatrix} x_1 \\ \vdots \\ x_n \end{pmatrix} \tag{4}
$$

where the gain $\mathcal{K}$ is an $m \times n$ matrix. Replacing x by its expression in (3) yields, for (4), what we call a *generalized PID regulator*[2]. Its robustness[3] with respect to perturbations is obtained by adding a finite linear combination of iterated integrals of the output variables. In particular robustness is satisfied when, for some reasons, the lower bounds of integration in the iterated integrals is not $-\infty$ but some finite t_1. This remark applies to the operational calculus based on the one-sided Laplace transform where one should replace (1a) by

$$
sx = Ax + Bu + x(0) \tag{5}
$$

Note that in this case (3) is no more an exact observer.

Our first mathematical tool is the module-theoretic setting for finite-dimensional linear systems, as developed since [11], with one more ingredient, namely *localization*, which is classic in commutative algebra (see, e.g., [22]) and was already extensively employed for delay systems [18]. The second tool is operational calculus in Mikusiński's setting [28,29] (see also [10,35]). It not only permits to deal with module theory in a simple and natural manner, without analytical difficulties, but it also provides the features we need of operational calculus *via* the two-sided Laplace transform. Operational calculus leads us to the second convention:

The module corresponding to a control system is free[4].

Consider as a matter of fact a torsion element τ. It satisfies $\psi\tau = 0$, $\psi \in \mathbb{R}[s, s^{-1}]$, $\psi \neq 0$. The only solution of this equation is 0 and τ would be pointless. It implies that any system variable is influenced by, at least, a perturbation variable. Stability analysis becomes in this setting a consequence of robustness with respect to some peculiar perturbations[5].

This communication is organized as follows. In the next two sections we review the module-theoretic setting. We define the abstract notion of regulator,

---

[2] Those generalized PID regulators are defined for multivariable systems (compare with [1,33]).

[3] There exists a vast literature on robust asymptotic observers. Consider, for example, disturbance estimation [21], and sliding mode [9,32] approaches.

[4] For more general systems like delay ones, where the ground ring R is no more a principal ideal domain (cf. [18,19]), freeness should be replaced by torsion freeness.

[5] Besides approaching stability and robustness under the very same perspective, observer and controller designs are now included in a *unified* controller synthesis.

and the state-variable realization. Generalized PID regulators are introduced
in Sect. 4. After a short overview of Mikusiński's operational calculus in
Sect. 5, the next one is devoted to stability and robustness. Sects. 7.1, 7.2,
and 8 are examining three illustrative examples:

1. In the classical PID regulator [1,33] we are replacing the derivative term
   sy by a suitable linear combination of integral ones of the form $s^{-1}y$ and
   $s^{-1}u$. We thus avoid taking the derivative of the measured quantity like
   in the PI regulator, although we are retaining the advantages of PIDs
   with respect to the transient response.
2. We are introducing a generalized PID controller for a second order system
   given by $\frac{b_0+b_1s}{a_0+a_1s+s^2}$, i.e., with a non-trivial zero dynamics.
3. A <u>real</u> DC motor without mechanical sensors.

In the conclusion we sketch some possible extensions of our work to delay
systems and especially to nonlinear ones.

2   Systems over Principal Ideal Domains

What follows is, with the exception of the regulator, borrowed from
[4,11,13,16,18,19] and adapted if necessary to the case of modules over principal ideal domains.

2.1   Generalities

Let R be a principal ideal domain[6]. An R–*linear system* Λ, or an R–*system*
for short, is a finitely generated R–module. An R–*linear dynamics* $\mathcal{D}$ is an
R–system, which is equipped with an *input* $u = (u_1,\ldots,u_m) \subset \mathcal{D}$ such
that the quotient module $\mathcal{D}/\mathrm{span}_R(u)$ is torsion[7]. The input u is said to be
independent if, and only if, the R–module $\mathrm{span}_R(u)$ is free, of rank m. This
condition is assumed to be satisfied in the sequel. An *input-output* R–*system*
S is an R–dynamics equipped with an *output* $y = (y_1,\ldots,y_p) \subset S$.

---

[6] The usual mathematical abbreviation for a principal ideal domain is PID. We
will of course not use it in order to avoid any confusion with PIDs in engineering!

[7] An element ξ of an R-module M, where R is a commutative domain, is said to
be *torsion* (see, e.g., [22]) if, and only if, there exists $\epsilon \in R$, $\epsilon \neq 0$, such that
$\epsilon\xi = 0$. The set of all torsion elements is a submodule M_{tor}, called the *torsion
submodule*. The module M is said to be torsion if, and only if, $M = M_{\mathrm{tor}}$. It
is *torsion free* if, and only if, $M_{\mathrm{tor}} = \{0\}$. If R is a principal ideal domain and
if M is finitely generated, freeness and torsion freeness are equivalent (see, for
example, [22]).

2.2 Controllability and Flat Outputs

The R–system Λ is said to be *controllable* if, and only if, the R–module Λ is free. Any basis of the free module Λ is called a *flat output*.

2.3 Observability

The input-output R–system S is said to be *observable* if, and only if, the R–modules S and $\mathrm{span}_R(u, y)$ coincide.

2.4 Regulator

An R–*linear regulator* of the R–system Λ is a short exact sequence (cf. [22]) of R–modules

$$0 \to N \to \Lambda \to T \to 0$$

where

- T is torsion,
- N is free.

The freeness of N is equivalent to the following fact: the restriction of the mapping $\Lambda \to T$ to Λ_{tor} is one-to-one. This is equivalent saying that a torsion element is not altered by a regulator.

2.5 Localization

Let S be a multiplicative subset of R. Consider the localized ring $S^{-1}R$. The functor $S^{-1}R \otimes_R$ from the category of finitely generated R–modules to that of finitely generated $S^{-1}R$–modules is called a *localization* functor (see, e.g., [22]). This functor is not faithful as it maps to $\{0\}$ any torsion R–module T such that there exists $\sigma \in S$ verifying, for any $\tau \in T$, $\sigma\tau = 0$.

For any R–system Λ, the $S^{-1}R$–system $_S\Lambda = S^{-1}R \otimes_R \Lambda$ is the S–*localized* system. For any $\lambda \in \Lambda$, $_S\lambda = 1 \otimes \lambda \in {}_S\Lambda$ is the S–*localized* element of λ. Take an input-output R–system S. The S–localized system $_SS$ is also input-output: its input (resp. output) is the S–localized input (resp. output) of S. The next result is clear.

Proposition 1. *The S–localized system of a controllable R–system is also controllable. The S–localized system of an observable input-output R–system is also observable. The converse does not hold in general.*

Example 1. If the multiplicative set S above is $R\backslash\{0\}$, $S^{-1}R$ is the quotient field K of R. The functor $K\otimes_R$, from the category of finitely generated R–modules to that of finite-dimensional K–vector spaces, is called the *Laplace functor* [13] and corresponds to the simplest type of localization (see, e.g., [22]). This functor is not faithful, i.e., it maps any torsion R–module to $\{0\}$. The K–vector space $K\otimes_R \Lambda = \hat{\Lambda}$ is called the *transfer vector space* of the R–system Λ. The mapping $\Lambda \to \hat{\Lambda}$, $\lambda \mapsto 1\otimes\lambda = \hat{\lambda}$ is the *(formal) Laplace transform*, and $\hat{\lambda}$ is the (formal) Laplace transform of λ [13]. A matrix $H \in K^{p\times m}$ such that

$$\begin{pmatrix} \hat{y}_1 \\ \vdots \\ \hat{y}_p \end{pmatrix} = H \begin{pmatrix} \hat{u}_1 \\ \vdots \\ \hat{u}_m \end{pmatrix}$$

is called a *transfer matrix* of S. If u is independent, $\hat{u}_1, \ldots, \hat{u}_m$ is a basis of $\hat{\Lambda}$, and H is uniquely defined. Then, the input-output R–system S is said to be *left invertible* (resp. *right invertible*) if, and only if, H is so.

2.6 Perturbed Systems

An R–system Λ^{pert} is said to be *perturbed* if, and only if, a *perturbation* $\boldsymbol{\pi} = (\pi_1, \ldots, \pi_q)$, $\pi_1, \ldots, \pi_q \in \Lambda$, has been distinguished. The *unperturbed* R–system corresponds to the quotient module $\Lambda = \Lambda^{\text{pert}}/\text{span}_R(\boldsymbol{\pi})$. For a perturbed R–dynamics $\mathcal{D}^{\text{pert}}$, the input is the union of the control variables $u = (u_1, \ldots, u_m)$ and of the perturbation variables $\boldsymbol{\pi}$ such that

- the control and perturbation variables are assumed not to interact, i.e., $\text{span}_R(u) \cap \text{span}_R(\boldsymbol{\pi}) = \{0\}$,
- the quotient module $\mathcal{D}/\text{span}_R(\varphi(u))$ is torsion, where φ denotes the canonical epimorphism $\mathcal{D}^{\text{pert}} \to \mathcal{D}$.

From $\text{span}_R(u) \cap \text{span}_R(\boldsymbol{\pi}) = \{0\}$ it is immediate that the restriction of φ to $\text{span}_R(u)$ defines an isomorphism $\text{span}_R(u) \to \text{span}_R(\varphi(u))$. By a slight abuse of notation, we will therefore not distinguish in the sequel between $\varphi(u)$ and u.

A regulator (2.4) is said to be *not affecting the perturbation* if, and only if, the intersection of the image of N in Λ with $\text{span}_R(\boldsymbol{\pi})$ is trivial, i.e., $\{0\}$.

3 Classic State-Variable Representation

3.1 Unperturbed State-Variable Representation

Here R is the principal ideal domain $\mathbb{R}[s]$ of polynomials over the field $\mathbb{R}$ of real numbers in the indeterminate s (s will of course in the sequel correspond to the time derivation d/dt). An input-output $\mathbb{R}[s]$–system $\mathcal{S}$ may be defined by the *Kalman* state variable representation [11]

$$s \begin{pmatrix} x_1 \\ \vdots \\ x_n \end{pmatrix} = A \begin{pmatrix} x_1 \\ \vdots \\ x_n \end{pmatrix} + B \begin{pmatrix} u_1 \\ \vdots \\ u_m \end{pmatrix} \tag{6a}$$

$$\begin{pmatrix} y_1 \\ \vdots \\ y_p \end{pmatrix} = C \begin{pmatrix} x_1 \\ \vdots \\ x_n \end{pmatrix} + \sum_{\alpha=0}^{\nu} D_\alpha s^\alpha \begin{pmatrix} u_1 \\ \vdots \\ u_m \end{pmatrix} \tag{6b}$$

where $A \in \mathbb{R}^{n \times n}$, $B \in \mathbb{R}^{n \times m}$, $C \in \mathbb{R}^{p \times n}$, $D_\alpha \in \mathbb{R}^{p \times p}$. We know that the dimension n of the state vector is nothing else than the dimension of the torsion module $\mathcal{S}/\mathrm{span}_{\mathbb{R}[s]}(u)$, when viewed as a $\mathbb{R}$–vector space.

3.2   The Static State Feedback as a Regulator

The next result is a reformulation, in our language, of a well known fact in control engineering.

Proposition 2. *The static state feedback (4) is a regulator.*

Proof. The torsion $\mathbb{R}[s]$–module T in (2.4) is defined by

$$s \begin{pmatrix} \underline{x}_1 \\ \vdots \\ \underline{x}_n \end{pmatrix} = (A + BK) \begin{pmatrix} \underline{x}_1 \\ \vdots \\ \underline{x}_n \end{pmatrix} \tag{7}$$

The module $N \subset \mathcal{S}$ is the kernel of the mapping $\mathcal{S} \to T$ defined by (4) and (6a), i.e., by $x_\alpha \mapsto \underline{x}_\alpha$, and $u_\beta \mapsto -\sum_\gamma k_{\beta\gamma} \underline{x}_\gamma$, where $K = (k_{\beta\gamma})$. It is free, since the uncontrollable Kalman subspace, i.e., the torsion submodule, is not affected by (4).

3.3   Perturbed State-Variable Representation

Pulling back (6) from $\mathcal{S}$ to $\mathcal{S}^{\mathrm{pert}}$ yields [17]

$$s \begin{pmatrix} x_1^{\mathrm{pert}} \\ \vdots \\ x_n^{\mathrm{pert}} \end{pmatrix} = A \begin{pmatrix} x_1^{\mathrm{pert}} \\ \vdots \\ x_n^{\mathrm{pert}} \end{pmatrix} + B \begin{pmatrix} u_1 \\ \vdots \\ u_m \end{pmatrix} + \begin{pmatrix} \varpi_1 \\ \vdots \\ \varpi_n \end{pmatrix} \tag{8a}$$

$$\begin{pmatrix} y_1^{\mathrm{pert}} \\ \vdots \\ y_p^{\mathrm{pert}} \end{pmatrix} = C \begin{pmatrix} x_1^{\mathrm{pert}} \\ \vdots \\ x_n^{\mathrm{pert}} \end{pmatrix} + \sum_{\alpha=0}^{\nu} D_\alpha s^\alpha \begin{pmatrix} u_1 \\ \vdots \\ u_m \end{pmatrix} + \begin{pmatrix} \omega_1 \\ \vdots \\ \omega_p \end{pmatrix} \tag{8b}$$

where $\varpi_1, \ldots, \varpi_n, \omega_1, \ldots, \omega_n \in \mathrm{span}_{\mathbb{R}[s]}(\pi)$.

The output y of S is said to be *good* if, and only if, the matrix C in (8b) is of rank p. A more intrinsic definition reads as follows: The $\mathbb{R}$–vector space spanned by the residues of the components of y in $S/\mathrm{span}_{\mathbb{R}[s]}(u)$ is p–dimensional. The next result will be useful for the robustness analysis of Theorem 2.

Proposition 3. *Consider a perturbed input-output $\mathbb{R}[s]$–system S^{pert} such that the output of the corresponding unperturbed system S is good. Then there exists a state variable representation (8) such that the ω_is in (8b) are identically 0.*

Proof. From $\mathrm{rk}(C) = p$, we may choose a state $\tilde{x} = (\tilde{x}_1, \ldots, \tilde{x}_n)$ such that the corresponding output matrix $\tilde{C} = (\tilde{c}_{ij})$ verifies

$$\tilde{c}_{ij} = \begin{cases} 1 & i = j, \ 1 \le i \le p \\ 0 & i \ne j \end{cases}$$

Then an obvious perturbation dependent state transformation yields the answer.

Remark 1. Perturbation dependent state transformations, which are quite natural in our module-theoretic setting, were already utilized in [17] for obtaining the *perturbed Brunovský canonical form*[8].

4   Generalized PID Regulators

4.1 s–Localization

The principal ideal domain of Laurent polynomials $\mathbb{R}[s, s^{-1}]$, which are finite sums of the form $\sum a_\nu s^\nu$, $\nu \in \mathbb{Z}$, $a_\nu \in \mathbb{R}$, is the localized ring of $\mathbb{R}[s]$ with respect to the multiplicative subset $\{s^\nu \mid \nu \ge 0\}$. The *order* of $\epsilon \in \mathbb{R}[s, s^{-1}]$, $\epsilon \ne 0$, is $\alpha \in \mathbb{Z}$ such that $s^{-\alpha}\epsilon$ belongs to $\mathbb{R}[s]$ and possesses a non-zero constant term, i.e., $s^{-\alpha}\epsilon \in \mathbb{R}[s]\backslash s\mathbb{R}[s]$. The order of 0 is, by definition, $+\infty$.

Call s–localization the localization with respect to $\{s^\nu \mid \nu \ge 0\}$. Write $_s\Lambda$ the s–localized system of the $\mathbb{R}[s]$–system Λ and $_s\lambda \in {_s\Lambda}$ the s–localized element of $\lambda \in \Lambda$.

From now on and for the sake of simplicity we will consider a controllable and observable $\mathbb{R}[s]$–system S with input $u = (u_1, \ldots, u_m)$ and output $y = (y_1, \ldots, y_p)$. We assume u to be independent. The canonical $\mathbb{R}[s]$–linear morphism $_s\iota : S \to {_sS}$, $\lambda \mapsto {_s\lambda}$, is injective. With a slight abuse of notation, we will therefore consider S as a subset of $_sS$.

A s–*generalized PID regulator* of S is a regulator of $_s\Lambda$.

---

[8] This perturbed Brunovský form was the key ingredient for understanding the robustness of our flatness-based predictive control strategy (see [17] and the references therein).

4.2   Static State Feedbacks as PID Regulators

For any $\mu \geq 1$, equation (6a) yields (2). From the observability assumption we know that $x_1, \ldots, x_n$ are $\mathbb{R}[s]$–linear combination of the components of u and y. Thus, from (2), we obtain the following crucial theorem.

Theorem 1. *The Kalman state in (6a) may be expressed by (3) where $P \in \mathbb{R}[s, s^{-1}]^{n \times p}$, $Q \in \mathbb{R}[s, s^{-1}]^{n \times m}$. The order[9] of P may be chosen to be any integer ≥ 0. The order of Q may be chosen ≥ 1.*

The next corollary follows from Proposition 2.

Corollary 1. *The static state feedback (4) may be written as an s-generalized PID regulator*

$$\begin{pmatrix} u_1 \\ \vdots \\ u_m \end{pmatrix} = -\mathcal{K} \left(P \begin{pmatrix} y_1 \\ \vdots \\ y_p \end{pmatrix} + Q \begin{pmatrix} u_1 \\ \vdots \\ u_m \end{pmatrix} \right)$$

Remark 2. Corollary 1 may be of course extended to other types of state feedbacks.

5   Some Algebraic Analysis

5.1   Mikusiński's Operational Calculus

Consider the commutative ring $\mathcal{C}$ of continuous functions $[0, +\infty) \to \mathbb{C}$ with respect to the addition $+$

$$(f + g)(t) = f(t) + g(t)$$

and the convolution product $\star$

$$(f \star g)(t) = (g \star f)(t) = \int_0^t f(\tau)g(t - \tau)d\tau = \int_0^t g(\tau)f(t - \tau)d\tau$$

It follows from a famous result due to Titchmarsh (cf. [28,29,35]) that there are no zero divisors in $\mathcal{C}$. Call *Mikusiński's field* the quotient field $\mathcal{M}$ of $\mathcal{C}$. Any element of $\mathcal{M}$ is an *operator*.

Example 2. The neutral element 1 of $\mathcal{M}$ is the analogue of the Dirac measure in L. Schwartz's distribution theory.

---

[9] The *order* of a matrix with entries in $\mathbb{R}[s, s^{-1}]$ is the smallest order of its entries.

Example 3. The inverse in $\mathcal{M}$ of the Heaviside function is the operator s which satisfies the usual rules of operational calculus. Take a C^1–function $f : \mathbb{R} \to \mathbb{C}$ with left bounded support. Then sf is the derivative $\frac{df}{dt}$. The subfield $\mathbb{C}(s) \subset \mathcal{M}$ of rational functions in the indeterminate s with complex coefficients possesses the usual meaning. Take for instance a locally Lebesgue integrable function $g : \mathbb{R} \to \mathbb{C}$, with left bounded support. Then $\frac{g}{s} = \int_{-\infty}^{t} g(\sigma)d\sigma$ has also a left bounded support.

Example 4. The exponential $e^{-\vartheta s}$, $\vartheta \geq 0$, represents the usual delay operator. Its inverse $e^{\vartheta s}$ is the forward shift operator.

We refer to [29] for the notion of *regular* operators and their *supports*. Write $\mathcal{M}_{\mathrm{reg}}^{\mathrm{left}} \subset \mathcal{M}$ the set of regular operators with left bounded supports; $\mathcal{M}_{\mathrm{reg}}^{\mathrm{left}}$ is a ring with respect to the addition and the convolution. Note that a locally Lebesgue integrable function with left bounded support belongs to $\mathcal{M}_{\mathrm{reg}}^{\mathrm{left}}$.

5.2   Trajectories of Mikusiński's Systems

A *Mikusiński's R–system* is an R–system Λ where R is a subring[10] of $\mathcal{M}$. This is the case for $\mathbb{R}[s]$ or $\mathbb{R}[s, s^{-1}]$. Following [12] call $\mathcal{M}$*-trajectory*[11] of Λ any R–module morphism $\Lambda \to \mathcal{M}$. If $\mathcal{M}_{\mathrm{reg}}^{\mathrm{left}}$ is also an R–module (this is verified with $R = \mathbb{R}[s]$ or $R = \mathbb{R}[s, s^{-1}]$), a *trajectory with left bounded support*, or a $\mathcal{M}_{\mathrm{reg}}^{\mathrm{left}}$*–trajectory*, of the R–system Λ is an R–module morphism $\Lambda \to \mathcal{M}_{\mathrm{reg}}^{\mathrm{left}}$.

6   Robustness and Stability

The next result is an extension of familiar properties of PIDs.

Theorem 2. *Take a perturbed input-output $\mathbb{R}[s]$–system S^{pert} such that the corresponding unperturbed system S verifies the following two properties:*

- *it is controllable and observable,*
- *the output is good.*

Take a mapping of the perturbation variables in $\mathcal{M}$ such that

- *their images are finite sums of operators of the form $\frac{a(s)}{b(s)}e^{\vartheta s}$, $\vartheta \in \mathbb{R}$, $a, b \in \mathbb{R}[s]$, $b \neq 0$, $(a, b) = 1$,*
- *the real parts of the roots of b are ≤ 0.*

---

[10] The R-module Λ is, according to the introduction, assumed to be free.
[11] See [20] for connections with controllability.

Then there exists an s-generalized PID regulator, which is not affecting the perturbation, such that the canonical images of the components of the output are equal to finite sums of operators of the form $\frac{c(s)}{d(s)}e^{\vartheta s}$, $c,d \in \mathbb{R}[s]$, $d \neq 0$, $(c,d) = 1$, where the real parts of the roots of d are < 0. If the real parts of the roots of b are < 0, then there exists an s-generalized PID regulator, which is not affecting the perturbation, such that the canonical images of the components of the input and the output are as above.

Proof. Applying Corollary 1 to $\mathcal{S}^{\text{pert}}$ yields an error due to the perturbations. In order to counteract it, apply Proposition 3 and extend the state variables with a finite number of terms of the form $s^{-\alpha}y_j$, $j = 1,\ldots,p$, $1 \leq \alpha \leq \Delta$. The corresponding equation reads $s(s^{-\alpha}y_j) = s^{1-\alpha}y_j$ for $\alpha \geq 2$. For $\alpha = 1$ we obtain $s(s^{-1}y_j) = y_j$ where we replace y_j by its expression with respect to the state, control and perturbation variables. Note that if derivatives of the control variables appear they might be eliminated by the standard procedure of [11]. Place the poles of the extended system by a static state feedback where the state variables x are dealt with *via* Corollary 1. The integer Δ is chosen such that the resulting expression for y_j does not exhibit any pole at $s = 0$.

Remark 3. Assume that the measurements start at $t > 0$. Then one is left with (5), where $x(0)$ should be regarded as a perturbation which, according to the previous result, may easily be counteracted. As already stated in the introduction, stability analysis is being included as a particular case in the study of perturbation attenuation.

By utilizing high gain techniques we obtain the following corollary.

Corollary 2. $\mathcal{S}^{\text{pert}}$ *is as above. Take a mapping of the perturbation variables in $\mathcal{M}$ such that their images are finite sums as above and of a C^{∞}-function with left bounded support and which is bounded as well as all its derivatives. Then, for any $\delta > 0$, there exists an s-generalized PID regulator, which is not affecting the perturbation, such that the canonical images of the components of y may be written as finite sums as above and of a C^{∞}-function $f(t)$ such that $\lim_{t\to+\infty}|f(t)| < \delta$.*

7 Implementation of a PID Without Derivator

Figure 1 is representing a PID controller acting on an input-output system subject to a constant perturbation[12] $\pi = \varpi e^{-\vartheta s}/s$, $\varpi \in \mathbb{R}$, $\vartheta \geq 0$. Following the flatness-based philosophy of predictive control [17], we want to transfer an initial equilibrium point to another one by following a desired trajectory.

---

[12] This perturbation is in fact a step function. It is equal to ϖ for $t > \vartheta$ and, according to our fundamental assumption, to 0 for $t \leq \vartheta$.

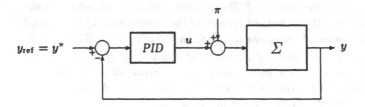

Fig. 1. The block diagram of a PID regulator

7.1 A Simple Linear System Without Zero Dynamics

Consider the second order model

$$s^2 y + a_1 s y + a_0 y = u + \varpi \frac{e^{-\vartheta s}}{s}$$

Note that y is a flat output of the unperturbed system. Define the errors $y_e = y - y^*$, and $u_e = u - u^*$. The quantities (u^*, y^*) satisfy the nominal system, where the disturbance is taken to be 0. We obtain

$$s^2 y_e + a_1 s y_e + a_0 y_e = u_e + \varpi \frac{e^{-\vartheta s}}{s} \tag{9}$$

Utilize a classical PID controller [1] for pole placement

$$u_e = -k_\mathrm{D} s y_e - k_\mathrm{P} y_e - \frac{k_\mathrm{I}}{s} y_e \tag{10}$$

where k_D, k_P and k_I are the coefficients related to the derivative, proportional and integral action. In the implementation of the derivative, several approximation methods can be used (see [1]). Nevertheless, sensitivity to high frequency perturbations is their major drawback; thus, a generalized PID controller is to be designed. From (9), with $\varpi = 0$, we have

$$s y_e = -a_1 y_e - \frac{a_0}{s} y_e + \frac{1}{s} u_e \tag{11}$$

Substituting this expression in (10) yields

$$
\begin{aligned}
u_e &= -k_\mathrm{D} \left(-a_1 y_e - \frac{a_0}{s} y_e + \frac{1}{s} u_e \right) - k_\mathrm{P} y_e - \frac{k_\mathrm{I}}{s} y_e \\
&= (a_1 k_\mathrm{D} - k_\mathrm{P}) y_e + \frac{a_0 k_\mathrm{D} - k_\mathrm{I}}{s} y_e - \frac{k_\mathrm{D}}{s} u_e
\end{aligned}
$$

which does not include any derivator.

To guarantee zero steady-state error, a double integral term in the PID controller has to be added. A PID controller of the form

$$u_e = -k_\mathrm{D} s y_e - k_\mathrm{P} y_e - \frac{k_{\mathrm{I}_1}}{s} y_e - \frac{k_{\mathrm{I}_2}}{s^2} y_e \tag{12}$$

yields, from (9), (12) and (11), a closed-loop dynamics given by

$$\left(s^2 + (a_1 + k_D)s + (a_0 + k_P) + \frac{k_{I_1}}{s} + \frac{k_{I_2}}{s^2}\right) y_e = \left(1 + \frac{k_D}{s}\right) \varpi \frac{e^{-\vartheta s}}{s}$$

which implies for a suitable choice of the coefficients $\lim_{t \to +\infty} y_e(t) = 0$.

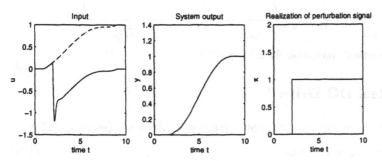

Fig. 2. Numerical simulation for Example 7.1

7.2   A Linear System With Zero Dynamics

Consider a second order model with a non-trivial zero dynamics

$$\left(s^2 + a_1 s + a_0\right) y = (b_1 s + b_0) u + \varpi \frac{e^{-\vartheta s}}{s} \quad , \quad b_1 \neq 0$$

where y is no more a flat output of the unperturbed system. With y_e and u_e given as before, apply a controller given by (12). The unperturbed estimation of the derivative term sy_e

$$sy_e = -a_1 y_e - \frac{a_0}{s} y_e + b_1 u_e + \frac{b_0}{s} u_e$$

yields the generalized PID controller

$$u_e = (a_1 k_D - k_P)y_e + \frac{(a_0 k_D - k_{I_1})}{s} y_e - \frac{k_{I_2}}{s^2} y_e - \frac{b_1 k_D}{s} u_e - \frac{b_0 k_D}{s^2} u_e$$

Thus, the closed-loop system dynamics results

$$\left((1 + b_1 k_D)s^2 + (a_1 + b_0 k_D + b_1 k_P)s + (a_o + b_0 k_P + b_1 k_{I_1})\right.$$

$$\left. + \frac{(b_0 k_{I_1} + b_1 k_{I_2})}{s} + \frac{b_0 k_{I_2}}{s^2}\right) y_e = \left(1 + b_1 k_D + \frac{b_0 k_D}{s}\right) \varpi \frac{e^{-\vartheta s}}{s}$$

The steady state error is 0 for a suitable choice of the coefficients.

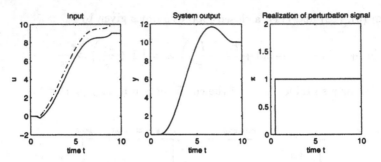

Fig. 3. Numerical simulation for Example 7.2

8 A Real DC Drive[13]

The equations of the real DC motor depicted in fig. 4 are (see, e.g., [25])

$$J\,s\omega = -f\omega + k_c I - C_r \tag{13a}$$
$$L\,sI = -k_e\omega - RI + U \tag{13b}$$

where U is the command signal, I the armature current, ω the motor shaft angular velocity, J the motor inertia, k_c the motor torque constant, k_e the back-emf constant, f the viscous friction coefficient, R and L the armature resistance and inductance, C_r the load torque. We want to control ω without measuring it[14].

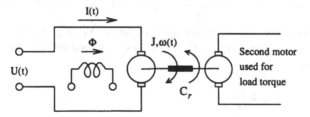

Fig. 4. DC motor diagram

The control law

$$U = U^\star - k_I\left(\frac{\omega}{s} - \frac{\omega^\star}{s}\right) \tag{14}$$

where $U^\star = \left(JL\,s^2 + (Lf + RJ)\,s + (k_e k_c + Rf)\right)\omega^\star)/k_c$, stabilizes the motor speed ω around the desired speed $\omega^\star$ if the gain k_I is chosen such that $JLs^3 + (Lf + RJ)s^2 + (k_e k_c + Rf)s + k_c k_I$ is Hurwitz.

---

[13] See [26] for more details.
[14] This is a typical problem of sensorless control (see, e.g., [5]).

Following the same lines as before, ω/s can be obtained from the non-perturbed equation (13b)

$$\frac{\omega}{s} = \frac{1}{k_e}\left(-LI - R\frac{I}{s} + \frac{U}{s}\right) \tag{15}$$

Thus, the control law (14)–(15) results

$$U = \frac{k_I}{k_I + k_e s}\left(sLI + RI + k_e\omega^\star + \frac{k_e}{k_I}sU^\star\right) \tag{16}$$

where the low pass filter $k_I/(k_I + k_e s)$ attenuates high frequency disturbances[15].

The experiment reported in fig. 5 shows the performance of the generalized PID under a load torque C_r, at starting and braking phases, on a real DC motor. The nominal behaviour is represented by dashed lines $(--)$. A speed control with a good transient behaviour is thus obtained in the absence of any mechanical sensor.

9   Conclusion

The conjunction between flatness-based linear predictive control [17] and these new PID controllers[16] should be most useful in many practical instances (see, also, [15]) and therefore play a role in the future of PID control (see [2]). In order to facilitate the practical implementation of our PIDs, computer-aided methods (see [1,3,33]) have to be designed. From the case-studies examined here it is clear that computer algebra will be an important ingredient.

The generalization to other systems is now being investigated. For a class of delay systems, which is often arising in practice, this extension is quite immediate[17] thanks to the methods from [18].

Preliminary results on nonlinear systems will be presented soon. They will of course rely on an interpretation of nonlinear observability which is analogous to the one used here. This interpretation [7,8] (see, also, [6]) was given in the context of differential algebra and reads as follows: an input-output system is observable if, and only if, any system variable may be expressed as a function

---

[15] In steady-state, (16) corresponds, for $s = 0$, to the *armature feedback*, or *IR-compensation*, $U = \tilde{R}I + k_e\omega^\star$ often used for speed control of DC motors without a tachometer (see, for example, the manual *CN0055 & CN0055B: DC to DC Negative Resistance speed control. Centent Company, Santa Ana, CA, 1997*, available at http://www.centent.com).

[16] Classical PID regulators were already employed in [17] (see, also, [27]).

[17] Many problems related to PIDs with delays and Smith predictors (see, e.g., [34]) should then become tractable (see [15] for a preliminary study on *flatness-based Smith predictors*).

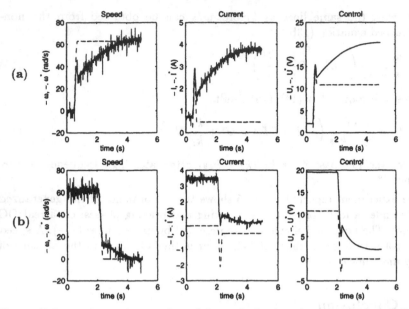

Fig. 5. DC motor speed control: real (*straight line*) and nominal (*dashed line*) behaviour at (a) start-up and (b) braking phases

of the input and output variables and of their derivatives up to some finite order. Ritt's *characteristic sets* (see [31] and [24]) will play an important role in the calculations.

References

1. Aström K.J., Hägglund T. (1988) PID Controllers: Theory, Design, and Tuning. Instrument Society of America, Research Triangle Park, NC
2. Åström K.J., Hägglund T. (2000) The Future of PID Control. In: Digital Control: Past, Present and Future of PID Control, IFAC Workshop at Terrasa. 19–30
3. Aström K.J., Wittenmark, B. (1984) Computer Controlled Systems. Prentice Hall, Englewood Cliffs, NJ
4. Bourlès H., Fliess M. (1997) Finite Poles and Zeros of Linear Systems: an Intrinsic Approach. Internat J Control 68:897–922
5. Buja G.S., Menis R., Valla M.I. (1995) Disturbance Torque Estimation in a Sensorless DC Drive. IEEE Trans Ind Electron 42:351–357
6. Diop S. (1993) Closedness of Morphisms of Differential Algebraic Sets: Applications to System Theory. Forum Math 5:33–47
7. Diop S., Fliess M. (1991) On Nonlinear Observability. In: Proc. 1st European Control Conference at Grenoble. Hermès, Paris, 152–157

8. Diop S., Fliess M. (1991) Nonlinear Observability, Identifiability, and Persistent Trajectories. In: IEEE Conference on Decision and Control at Brighton. 714–719

9. Edwards C., Spurgeon S.K. (1998) Sliding Mode Control: Theory and Applications. Taylor & Francis, London

10. Erdélyi A. (1962) Operational Calculus and Generalized Functions. Holt Rinehard Winston, New York

11. Fliess M. (1990) Some Basic Structural Properties of Generalized Linear Systems. Systems Control Lett 15:391–396

12. Fliess M. (1992) A Remark on Willems' Trajectory Characterization of Linear Controllability. Systems Control Lett 19:43–45

13. Fliess M. (1994) Une Interprétation Algébrique de la Transformation de Laplace et des Matrices de Transfert. Linear Alg Appl 203–204:429–442

14. Fliess M. (2000) Variations sur la Notion de Contrôlabilité. In: Quelques Aspects de la Théorie du Contrôle. Journée Annuelle Soc Math France, Paris, 47–86

15. Fliess M. (2000) Sur des Pensers Nouveaux Faisons des Vers Anciens. In: Actes Conférence Internationale Francophone d'Automatique at Lille

16. Fliess M., Lévine J., Martin P., Rouchon P. (1995) Flatness and Defect of Nonlinear Systems: Introductory Theory and Examples. Internat J Control 61:1327–1361

17. Fliess M., Marquez R. (2000) Continuous-Time Linear Predictive Control and Flatness: A Module-Theoretic Setting with Examples. Internat J Control 73:606–623

18. Fliess M., Mounier H. (1998) Controllability and Observability of Linear Delay Systems: an Algebraic Approach. ESAIM COCV 3:301–314

19. Fliess M., Mounier H. (1999) Tracking Control and π–Freeness of Infinite Dimensional Linear Systems. In: Picci G., Gilliam D.S. (Eds.) Dynamical Systems, Control, Coding, Computer Vision. Birkhäuser, Basel, 45–68

20. Fliess M., Mounier H. (to appear) A Trajectorian Interpretation of Torsion-Free Controllability

21. Franklin G.F., Powell J.D., Naeini A. E. (1994) Feedback Control of Dynamic Systems, 3rd edn. Addison Wesley, Reading, MA

22. Jacobson N. (1974 & 1980) Basic Algebra I & II. Freeman, San Francisco

23. Kailath T. (1980) Linear Systems. Prentice-Hall, Englewood Cliffs, NJ

24. Kolchin, E.R. (1973) Differential Algebra. Academic Press, New York

25. Louis J.-P., Multon B., Lavabre M. (1988) Commande de Machines à Courant Continu à Vitesse Variable. Techniques de l'Ingénieur, D 3610:1–14, D 3611:1–24, D 3612:1–13

26. Marquez R., Delaleau E., Fliess, M. (2000) Commande par PID Généralisé d'un Moteur Électrique sans Capteur Mécanique. In: Conférence Internationale Francophone d'Automatique at Lille

27. Marquez R., Fliess M. (2000) From PID to Model Predictive Control: A Flatness Based Approach. In: Digital Control: Past, Present and Future of PID Control, IFAC Workshop at Terrasa. 534–539

28. Mikusiński J. (1983) Operational Calculus, 2nd edn, Vol. 1. PWN, Warsaw, & Oxford University Press, Oxford

29. Mikusiński J., Boehme T.K. (1987) Operational Calculus, 2nd edn, Vol. 2. PWN, Warsaw, & Oxford University Press, Oxford

30. van der Pol B., Bremmer H. (1955) Operational Calculus Based on the Two-Sided Laplace Integral, 2nd edn. Cambridge University Press, Cambridge
31. Ritt J.F. (1950) Differential Algebra. American Mathematical Society, New York
32. Slotine J.J.E., Li W. (1991) Applied Nonlinear Control. Prentice-Hall, Englewood Cliffs, NJ
33. Tan K.K., Wang Q.-G., Lee T.H., Hägglund T.J. (1999) Advances in PID Control. Springer, London
34. Wang Q.-G., Lee T.H., Tan K.K.(1999) Finite Spectrum Assigment for Time-Delay Systems. Springer, London
35. Yosida K. (1984) Operational Calculus. Springer, New York

Eigenstructure of Nonlinear Hankel Operators

Kenji Fujimoto[1] and Jacquelien M.A. Scherpen[2]

[1] Kyoto University
   Uji, Kyoto 611-0011, Japan
[2] Delft University of Technology
   P.O. Box 5031
   2600 GA Delft, The Netherlands
   j.m.a.scherpen@its.tudelft.nl

Abstract. This paper investigates the eigenstructure of Hankel operators for nonlinear systems. It is proved that the variational system and Hamiltonian extension can be interpreted as the Gâteaux differentiation of dynamical input-output systems and their adjoints respectively. We utilize this differentiation in order to clarify the eigenstructure of the Hankel operator, which is closely related to the Hankel norm of the original system. The results in the paper thus provide new insights to the realization and balancing theory for nonlinear systems.

1 Introduction

The Hankel operator is one of the most important tools in linear control theory. Indeed they are extremely useful for balancing and model reduction of linear systems. The nonlinear extension of balancing theory were firstly introduced in [8], and a lot of Hankel operator results followed, see e.g. [5,6,10,11]. In particular, singular value functions which are a nonlinear extension of the Hankel singular values in the linear case play an important role in the nonlinear version of Hankel theory. Recently it was shown that singular value functions are closely related to Hankel operators [5,10]. The purpose of this paper is to clarify the eigenstructure of the Hankel operator, and to give a new characterization of Hankel singular value functions for nonlinear systems.

Nonlinear adjoint operators can be found in the mathematics literature, e.g. [1], and they are expected to play a similar role in the nonlinear systems theory. So called nonlinear Hilbert adjoints were introduced in [5,10] as a special class of nonlinear adjoints. The existence of such operators in input-output sense was shown in [6] and their state-space realizations are preliminary available in [4] where the main interest is the Hilbert adjoint extension with an emphasis on the use of port-controlled Hamiltonian systems.

Here, firstly we consider these adjoint operators from a variational point of view, and provide a formal justification for the use of Hamiltonian extensions via Gâteaux differentiation. Secondly we investigate both the eigenstructure of the self adjoint of Hankel operators and the eigenstructure of the Gâteaux derivative of the square norm of Hankel operators. It is shown that the eigen-structure of these operators is closely related. Furthermore this eigenstructure derives an alternative definition of the singular value functions, which has a relationship with the Hankel norm for nonlinear systems, other than the singular value functions given in [8] have.

2 Variational systems as Gâteaux derivative of dynamical systems

This section is devoted to the state-space characterization of Gâteaux differentiation of dynamical systems and their adjoints via variational systems and Hamiltonian extensions. Some of these results are reviewed from [4,9]. Let us consider an input-output system $\Sigma : L_2^m[t^0, t^1] \to L_2^r[t^0, t^1]$ defined on a (possibly infinite) time interval $[t^0, t^1]$ which has a state-space realization

$$y = \Sigma(u) : \begin{cases} \dot{x} = f(x, u) & x(t^0) = x^0 \\ y = h(x, u) \end{cases} \tag{1}$$

with $x(t) \in \mathbb{R}^n$, $u(t) \in \mathbb{R}^m$ and $y(t) \in \mathbb{R}^r$. Here we assume the origin is an equilibrium, i.e. $f(0, 0) = 0$ and $h(0, 0) = 0$ hold.

Before giving the Hamiltonian extension of Σ, we introduce the variational system of Σ. It is given by

$$y_v = \Sigma_v(u, u_v) : \begin{cases} \dot{x} = f(x, u) & x(t^0) = x^0 \\ \dot{x}_v = \frac{\partial f}{\partial x} x_v + \frac{\partial f}{\partial u} u_v & x_v(t^1) = x_v^1 \\ y_v = \frac{\partial h}{\partial x} x_v + \frac{\partial h}{\partial u} u_v \end{cases} \tag{2}$$

The input-state-output set (u_v, x_v, y_v) are so called variational input, state and output respectively and they represent variation along the trajectory (u, x, y) of the original system Σ.

The Hamiltonian extension Σ_a of Σ is now given by a Hamiltonian control system [2] which has an adjoint form of the variational system. It is given by

$$y_a = \Sigma_a(u, u_a) : \begin{cases} \dot{x} = \frac{\partial H}{\partial p}^T = f(x, u) & x(t^0) = x^0 \\ \dot{p} = -\frac{\partial H}{\partial x}^T = -\left(\frac{\partial f}{\partial x}^T p + \frac{\partial h}{\partial x}^T u_a\right) & p(t^1) = p^1 \\ y_a = \frac{\partial H}{\partial u}^T = \frac{\partial f}{\partial u}^T p + \frac{\partial h}{\partial u}^T u_a \end{cases} \tag{3}$$

with the Hamiltonian $H(x, p, u, u_a) := p^T f(x, u) + u_a^T h(x, u)$.

Gâteaux derivative is important for understanding the meaning of the variational systems and the Hamiltonian extensions. Also, Gâteaux differentiation of Hankel operators plays an important role in the analysis of the properties of Hankel operators, which is the topic of Section 4. To this end, we state the definition of Gâteaux differentiation.

Definition 1. Suppose X and Y are Banach spaces, $U \subseteq X$ is open, and $T : U \to Y$. Then T has a Gâteaux derivative at $x \in X$ if, for all $\xi \in U$ the following limit exits:

$$dT(x)(\xi) = \lim_{\varepsilon \to 0} \frac{T(x + \varepsilon \xi) - T(x)}{\varepsilon} = \frac{d}{d\varepsilon} T(x + \varepsilon \xi)|_{\varepsilon=0} . \tag{4}$$

We write $dT(x)(\xi)$ for the Gâteaux derivative of T at x in the "direction" ξ.

Perhaps more well-known than the Gâteaux derivative is the Fréchet derivative, which is especially useful for analysis of nonlinear static functions. Fréchet differentiation is a special class of Gâteaux differentiation. In the sequel, we concentrate on Gâteaux differentiation, since that is the most suitable for our framework. It is also noted that Fréchet derivative is linear from its definition and it does not depend on the "direction" to be differentiated, whereas Gâteaux derivative is not necessarily linear.

Theorem 1. *Suppose that $\Sigma : u \mapsto y$ in (1) is input-affine and has no direct feed-through, i.e. $f(x,u) \equiv g_0(x) + g(x)u$ and $h(x,u) \equiv h(x)$ for some analytic functions g_0, g and h, and that Σ is Gâteaux differentiable, namely there exists a neighborhood $\mathcal{U}_v \subseteq L_2^m[t^0, t^1]$ of 0 such that*

$$u \in L_2^m[t^0, t^1], \ u_v \in \mathcal{U}_v \subseteq L_2^m[t^0, t^1] \ \Rightarrow \ y_v \in L_2^r[t^0, t^1]. \tag{5}$$

Then there holds

$$\Sigma_v(u, u_v) = d\Sigma(u)(u_v) \tag{6}$$

with the variational system Σ_v given in (2).

For the reason of space we omit the proof. See [9] for the details. The Hamiltonian extension Σ_a also has a relation with Gâteaux derivative, and provides a justification for the fact that it is called the adjoint form of the variational system in [2]. Let Σ^* represent the nonlinear Hilbert adjoint of Σ, see [10,4], then we can give the differential version of Proposition 2 in [4],

Theorem 2. *Suppose that the assumptions in Theorem 1 hold and that*

$$u \in L_2^m[t^0, t^1], u_a \in L_2^r[t^0, t^1] \ \Rightarrow \ |x(t^1)| < \infty, |p(t^0)| < \infty. \tag{7}$$

Let $\hat{\Sigma}_a$ denote the mapping $((x(t^0), u), (p(t^1), u_a)) \mapsto (p(t^0), y_a)$ of the system Σ_a and let $\hat{\Sigma}_v$ denote the mapping $((x(t^0), u), (x_v(t^0), u_v)) \mapsto (x_v(t^1), y_v)$ of the system Σ_v. Then there holds

$$\hat{\Sigma}_a((x^0, u), (p^1, u_a)) = (\hat{\Sigma}_v(x^0, u))^*(p^1, u_a). \tag{8}$$

Proof. The proof follows similar arguments as Proposition 2 in [4], i.e., it uses the port-controlled Hamiltonian systems structure. Let the Hamiltonian function be given by $H_v = p^T x_v$, and denote $x^0 := x(t^0)$, $x^1 := x(t^1)$, $p^0 := p(t^0)$ and $p^1 := p(t^1)$ for simplicity. Then we have

$$
\frac{dH_v}{dt} = p^T \dot{x}_v + x_v^T \dot{p} = p^T \left(\frac{\partial f}{\partial x} x_v + g(x) u_v \right) - x_v^T \left(\frac{\partial f}{\partial x}^T p + \frac{\partial h}{\partial x}^T u_a \right)
$$

$$
= p^T g(x) u_v - x_v^T \frac{\partial h}{\partial x}^T u_a = y_a^T u_v - y_v^T u_a.
$$

This reduces to

$$
\langle y_a, u_v \rangle_{L_2^m} - \langle y_v, u_a \rangle_{L_2^r} = \int_{t^0}^{t^1} \left(y_a^T u_v - y_v^T u_a \right) dt = \int_{t^0}^{t^1} \frac{dH_v}{dt} dt
$$

$$
= H_v|_{t=t^1} - H_v|_{t=t^0} = \langle x_v^1, p^1 \rangle_{\mathbb{R}^n} - \langle x_v^0, p^0 \rangle_{\mathbb{R}^n}.
$$

Therefore

$$
\langle (x_v^1, y_v), (p^1, u_a) \rangle_{\mathbb{R}^n \times L_2^r} = \langle (x_v^0, u_v), (p^0, y_a) \rangle_{\mathbb{R}^n \times L_2^m} \tag{9}
$$

holds with the inner product on $\mathbb{R}^n \times L_2$. Substituting $(x_v^1, y_v) = \hat{\Sigma}_v(x^0, u)$ (x_v^0, u_v) and $(p^0, y_a) = \hat{\Sigma}_a(x^0, u)(p^1, u_a)$ yields

$$
\langle \hat{\Sigma}_v(x^0, u)(x_v^0, u_v), (p^1, u_a) \rangle_{\mathbb{R}^n \times L_2^r} = \langle (x_v^0, u_v), \hat{\Sigma}_a(x^0, u)(p^1, u_a) \rangle_{\mathbb{R}^n \times L_2^m}.
$$

This implies (8) and completes the proof.   □

This adjoint system has the same structure as port-controlled adjoint systems [4]. However its state-space realization is unique and coordinate independent, in contrast with the port-controlled adjoint systems. Setting $x_v(t^0) = 0$ and $p(t^1) = 0$ in Theorem 2, we obtain the following corollary.

Corollary 1. *Suppose that the assumptions in Theorem 1 hold and that*

$$
u \in L_2^m[t^0, t^1], \; u_a \in L_2^r[t^0, t^1] \quad \Rightarrow \quad |x(t^1)| < \infty, \; |p(t^0)| < \infty. \tag{10}
$$

Then there holds

$$
\Sigma_a(u, u_a) = (d\Sigma(u))^*(u_a) \tag{11}
$$

with the Hamiltonian extension Σ_a given in (3).

3   The Hankel operator and its derivative

This section gives the state-space realizations of Gâteaux derivative of some operators. We only consider time invariant input-affine nonlinear systems without direct feed-through of the form

$$
\Sigma : \begin{cases} \dot{x} = f(x) + g(x)u \\ y = h(x) \end{cases} \tag{12}
$$

defined on the time interval $(-\infty, \infty)$. Here Σ is L_2-stable in the sense that $u \in L_2^m(-\infty, 0]$ implies that $\Sigma(u)$ restricted to $[0, \infty)$ is in $L_2^r[0, \infty)$.

Although the original definitions of the observability and controllability operators [5] are given by Chen-Fliess functional expansion [3], one can employ state-space systems to describe them which are operators of $\mathbb{R}^n \to L_2^r[0, \infty)$ and $L_2^m[0, \infty) \to \mathbb{R}^n$, specifically:

$$y = \mathcal{O}_\Sigma(x^0) : \begin{cases} \dot{x} = f(x) \quad x(0) = x^0 \\ y = h(x) \end{cases} \tag{13}$$

$$x^1 = \mathcal{C}_\Sigma(u) : \begin{cases} \dot{x} = f(x) + g(x)\mathcal{F}_-(u) \quad x(-\infty) = 0 \\ x^1 = x(0) \end{cases} \tag{14}$$

$\mathcal{F}_- : L_2(-\infty, \infty) \to L_2(-\infty, 0]$ and $\mathcal{F}_+ : L_2(-\infty, \infty) \to L_2[0, \infty)$ denote the flipping operators defined by

$$\mathcal{F}_-(u)(t) := \begin{cases} u(-t) & t \in (-\infty, 0] \\ 0 & t \in [0, \infty) \end{cases}, \quad \mathcal{F}_+(u)(t) := \begin{cases} 0 & t \in (-\infty, 0] \\ u(-t) & t \in [0, \infty) \end{cases}.$$

Furthermore the Hankel operator $\mathcal{H}_\Sigma : L_2^m[0, \infty) \to L_2^r[0, \infty)$ of Σ is given by $\mathcal{H}_\Sigma := \Sigma \circ \mathcal{F}_-$ and $\mathcal{H}_\Sigma = \mathcal{O}_\Sigma \circ \mathcal{C}_\Sigma$ holds. This has been proven in [5,6], along with a deeper and more detailed analysis of the Hankel operator. The adjoints of their derivatives can be obtained using Theorem 2.

Theorem 3. *Consider the operator Σ in (12). Suppose that the assumptions in Theorems 1 and 2 hold. Then state-space realizations of $(d\mathcal{O}_\Sigma(x^0))^*$: $L_2^r[0, \infty) \, (\times \mathbb{R}^n) \to \mathbb{R}^n$, $(d\mathcal{C}_\Sigma(u))^*$: $\mathbb{R}^n \, (\times L_2^m[0, \infty)) \to L_2^m[0, \infty)$ and $(d\mathcal{H}_\Sigma(u))^* : L_2^r[0, \infty)(\times L_2^m[0, \infty)) \to L_2^m[0, \infty)$ are given by*

$$p^0 = (d\mathcal{O}_\Sigma(x^0))^*(u_a) : \begin{cases} \dot{x} = f(x) & x(0) = x^0 \\ \dot{p} = -\frac{\partial f}{\partial x}^T(x)\, p - \frac{\partial h}{\partial x}^T(x)\, u_a & p(\infty) = 0 \\ p^0 = p(0) \end{cases}$$

$$y_a = (d\mathcal{C}_\Sigma(u))^*(p^1) : \begin{cases} \dot{x} = f(x) + g(x)\mathcal{F}_-(u) & x(-\infty) = 0 \\ \dot{p} = -\frac{\partial(f+g\mathcal{F}_-(u))}{\partial x}^T(x)\, p & p(0) = p^1 \\ y_a = \mathcal{F}_+(g^T(x)\, p) \end{cases}$$

$$y_a = (d\mathcal{H}_\Sigma(u))^*(u_a) : \begin{cases} \dot{x} = f(x) + g(x)\, \mathcal{F}_-(u) & x(-\infty) = 0 \\ \dot{p} = -\frac{\partial(f+g\mathcal{F}_-(u))}{\partial x}^T p - \frac{\partial h}{\partial x}^T u_a & p(\infty) = 0 \\ y_a = \mathcal{F}_+(g^T(x)\, p) \end{cases}.$$

This theorem is differential version of Proposition 3 in [4]. The proof of this theorem is obtained by applying the adjoint Hamiltonian extensions of Section 2, and using techniques from [4]. It is readily checked that for linear systems, the above characterizations yield the well-known state space characterizations of these operators.

4   Eigenstructure of Hankel operators

This section clarifies the eigenstructure related to the Hankel operator utilizing the Gâteaux derivative. Further we will give a new definition of singular value functions which are closely related to the eigenstructure and the Hankel norm of nonlinear operators. In order to proceed, we need to define the following energy functions:

Definition 2. The observability function $L_o(x)$ and the controllability function $L_c(x)$ of Σ in (12) are defined by

$$L_o(x^0) := \frac{1}{2} \int_0^\infty \|y(t)\|^2 dt, \; x(0) = x^0, \; u(t) \equiv 0 \tag{15}$$

$$L_c(x^1) := \min_{\substack{u \in L_2^m[0,\infty) \\ x(-\infty)=0, x(0)=x^1}} \frac{1}{2} \int_{-\infty}^0 \|u(t)\|^2 dt. \tag{16}$$

In the sequel, we concentrate on the system Σ in (12) and, suppose that the assumptions of Theorems 1 and 2 hold and that there exist well-defined observability and controllability functions L_o and L_c. These functions are closely related to observability and controllability operators and Gramians in the linear case. In [8] these functions have been used for the definition of balanced realizations and singular value functions of nonlinear systems. Also they fulfill corresponding Hamilton-Jacobi equations, in a similar way as the observability Gramian and the inverse of the controllability Gramian are solutions of a Lyapunov/Riccati equation. We first review what we mean by input-normal/output-diagonal form, see [8]:

Theorem 4. [8] Consider a system (f, g, h) that fulfills certain technical conditions. Then there exists on a neighborhood $U \subset V$ of 0, a coordinate transformation $x = \psi(z)$, $\psi(0) = 0$, which converts the system into an input-normal/output-diagonal form, where

$$\tilde{L}_c(z) := L_c(\psi(z)) = \frac{1}{2} z^T z,$$

$$\tilde{L}_o(z) := L_o(\psi(z)) = \frac{1}{2} z^T \mathrm{diag}(\tau_1(z), \ldots, \tau_n(z)) z$$

with $\tau_1(z) \geq \ldots \geq \tau_n(z)$ being the so called smooth singular value functions on $W := \psi^{-1}(U)$.

Now, we can state the result from [5,6] that relates the singular value functions to the Hankel operator:

Theorem 5. *[5] Let (f, g, h) be an analytic n dimensional input-normal/ output-diagonal realization of a causal L_2-stable input-output mapping Σ on a neighborhood W of 0. Define on W the collection of component vectors $\tilde{z}_j = (0, \ldots, 0, z_j, 0, \ldots, 0)$ for $j = 1, 2, \ldots, n$, and the functions $\hat{\sigma}^2(z_j) = \tau(\tilde{z}_j)$. Let v_j be the minimum energy input which drives the state from $z(-\infty) = 0$ to $z(0) = \tilde{z}_j$ and define $\hat{v}_j = \mathcal{F}_+(v_j)$. Then the functions $\{\hat{\sigma}_j\}_{j=1}^n$ are singular value functions of the Hankel operator $\mathcal{H}_\Sigma$ in the following sense:*

$$\langle \hat{v}_j, \mathcal{H}_\Sigma^*(\mathcal{H}_\Sigma(\hat{v}_j), \hat{v}_j)\rangle_{L_2} = \hat{\sigma}_j^2(z_j)\langle \hat{v}_j, \hat{v}_j\rangle_{L_2}, \quad j = 1, 2, \ldots n. \tag{17}$$

The above result is quite limited in the sense that it is dependent on the coordinate frame in which the system is in input-normal/output-diagonal form. Moreover, in the linear case,

$$\mathcal{H}_\Sigma^* \circ \mathcal{H}_\Sigma(v_j) = \sigma_i^2 v_j \tag{18}$$

holds for each eigenvector v_j. In the nonlinear case, on the other hand, such a relation does not hold. Now we consider an eigenstructure of the operator $u \mapsto (d\mathcal{H}_\Sigma(u))^* \circ \mathcal{H}_\Sigma(u)$, instead of (18), characterized by

$$(d\mathcal{H}(v))^* \circ \mathcal{H}(v) = \lambda\, v \tag{19}$$

where $\lambda \in \mathbb{R}$ is an eigenvalue and $v \in L_2^m[0, \infty)$ the corresponding eigenvector. This eigenstructure has a close relationship with the Hankel norm of Σ defined by

$$\|\Sigma\|_H := \max_{\substack{u \in L_2^m[0,\infty) \\ u \neq 0}} \frac{\|\mathcal{H}_\Sigma(u)\|_2}{\|u\|_2}. \tag{20}$$

Theorem 6. *Let $v \in L_2^m[0, \infty)$ denote the input which achieves the maximization in the definition of Hankel norm in (20), namely $\|\mathcal{H}_\Sigma(v)\|_2/\|v\|_2 = \|\Sigma\|_H$ holds. Then v satisfies (19) with the eigenvalue $\lambda = \|\Sigma\|_H^2$.*

Proof. The derivative of $\|\mathcal{H}_\Sigma(u)\|_2/\|u\|_2$ (in the direction δu) satisfies

$$\begin{aligned}
d\left(\frac{\|\mathcal{H}_\Sigma(u)\|_2}{\|u\|_2}\right)(\delta u) &= \frac{\|u\|_2 d(\|\mathcal{H}_\Sigma(u)\|_2)(\delta u) - \|\mathcal{H}_\Sigma(u)\|_2 d(\|u\|_2)(\delta u)}{\|u\|_2^2} \\
&= \frac{\|u\|_2/\|\mathcal{H}_\Sigma(u)\|_2 \langle (d\mathcal{H}_\Sigma(u))^* \circ \mathcal{H}_\Sigma(u), \delta u\rangle - \|\mathcal{H}_\Sigma(u)\|_2/\|u\|_2\langle u, \delta u\rangle}{\|u\|_2^2} \\
&= \frac{\langle (d\mathcal{H}_\Sigma(u))^* \circ \mathcal{H}_\Sigma(u) - (\|\mathcal{H}_\Sigma(u)\|_2/\|u\|_2)^2 u,\ \delta u\rangle}{\|u\|_2\, \|\mathcal{H}_\Sigma(u)\|_2} \equiv 0
\end{aligned} \tag{21}$$

for all variations δu at $u = v$ because it is a critical point. This reduces to

$$(d\mathcal{H}(v))^* \circ \mathcal{H}(v) \equiv (\|\mathcal{H}_\Sigma(v)\|_2/\|v\|_2)^2 v = \|\Sigma\|_H^2 v \tag{22}$$

and thus proves the theorem. □

Theorem 6 clarifies that the eigenstructure (19) characterizes a necessary condition for the input v to achieve the maximization in the definition of the Hankel norm (20). In the remainder of this section, precise properties of the eigenstructure (19) are investigated. Firstly the following lemma gives the complete characterization of the eigenvectors.

Lemma 1. *A pair $\lambda \in \mathbb{R}$ and a nonzero $v \in L_2^m[0,\infty)$ is a pair of an eigenvalue and an eigenvector of the mapping $u \mapsto (d\mathcal{H}_\Sigma(u))^* \circ \mathcal{H}_\Sigma(u)$ if and only if there exists $x^0 \in \mathbb{R}^n$ such that*

$$
\begin{cases}
\dot{x} = -f(x) - g(x)g^T(x)p & x(0) = x^0 \\
\dot{p} = \left. \frac{\partial(f+gu)}{\partial x}\right|_{u=g^Tp}^T p & p(0) = \frac{1}{\lambda}\frac{\partial L_a}{\partial x}^T(x^0) . \\
v = g^T(x)p
\end{cases}
\tag{23}
$$

Proof. The sufficiency is straightforwardly obtained, and therefore, we only prove the necessity here. We can observe

$$
y_a = (d\mathcal{H}_\Sigma(u))^* \circ \mathcal{H}_\Sigma(u) = (d\mathcal{C}_\Sigma(u))^* \circ (d\mathcal{O}_\Sigma(\mathcal{C}_\Sigma(u)))^* \circ \mathcal{O}_\Sigma \circ \mathcal{C}_\Sigma(u). \tag{24}
$$

Let $x^0 := \mathcal{C}_\Sigma(u)$. Then (24) reduces to

$$
y_a = (d\mathcal{C}_\Sigma(u))^* \circ (d\mathcal{O}_\Sigma(x^0))^* \circ \mathcal{O}_\Sigma(x^0). \tag{25}
$$

Next we consider the Gâteaux derivative of $L_o(x^0)$ in the direction ξ

$$
\frac{\partial L_o(x^0)}{\partial x^0} \xi = dL_o(x^0)(\xi) = \frac{1}{2}d(\|\mathcal{O}_\Sigma(x^0)\|_2^2)(\xi)
$$
$$
= \langle \mathcal{O}_\Sigma(x^0), d\mathcal{O}_\Sigma(x^0)(\xi)\rangle_{L_2^r} = \langle (d\mathcal{O}_\Sigma(x^0))^* \circ \mathcal{O}_\Sigma(x^0), \xi\rangle_{\mathbb{R}^n}.
$$

This means

$$
d\mathcal{O}_\Sigma(x^0))^* \circ \mathcal{O}_\Sigma(x^0) = \frac{\partial L_o(x^0)}{\partial x^0}^T . \tag{26}
$$

We obtain from (25) that $y_a = (d\mathcal{C}_\Sigma(u))^*(\partial L_o(x^0)/\partial x^0)$. It follows from Theorem 3 that the the state space realization of this operator is given by

$$
\begin{cases}
\dot{x} = f(x) + g(x)\,\mathcal{F}_-(u) & x(-\infty) = 0 \\
\dot{p} = -\frac{\partial(f+g\mathcal{F}_-(u))}{\partial x}^T p & p(0) = \frac{\partial L_a}{\partial x}^T(x^0) . \\
y_a = \mathcal{F}_+(g^T(x)\,p)
\end{cases}
\tag{27}
$$

Suppose $u = v$ and $y_a = \lambda v$, i.e. the pair λ and v is a pair of an eigenvalue and an eigenvector. Then we have

$$
v = \frac{1}{\lambda}y_a = \frac{1}{\lambda}g^T(x)p. \tag{28}
$$

Let $\bar{p} := (1/\lambda)p$, then the reverse-time expression of (27) reduces to

$$
\begin{cases}
\dot{x} = -f(x) - g(x)g^T(x)\bar{p} & x(0) = x^0 \quad (x(\infty) = 0) \\
\dot{\bar{p}} = \frac{1}{\lambda}\dot{p} = \left.\frac{\partial(f+gu)}{\partial x}\right|^T_{u=g^T\bar{p}}\bar{p} & \bar{p}(0) = \frac{1}{\lambda}\frac{\partial L_o}{\partial x}^T(x^0) \\
v = g^T(x)\bar{p}
\end{cases}
\tag{29}
$$

This is identical to (23) and this completes the proof.   □

Lemma 1 gives the characterization of all pairs of eigenvalues and eigenvectors. Next we concentrate on a special class of all pairs which are closely related to the energy functions $L_o(x)$ and $L_c(x)$.

Lemma 2. *Suppose there exist $\lambda \in \mathbb{R}$ and a nonzero $x^0 \in \mathbb{R}^n$ such that*

$$
\frac{\partial L_o}{\partial x}(x^0) = \lambda \, \frac{\partial L_c}{\partial x}(x^0).
\tag{30}
$$

Then λ is the eigenvalue of the mapping $u \mapsto (d\mathcal{H}_\Sigma(u))^* \circ \mathcal{H}_\Sigma(u)$ corresponding to the eigenvector

$$
v = \mathcal{C}^\dagger_\Sigma(x^0)
\tag{31}
$$

with $\mathcal{C}^\dagger_\Sigma : \mathbb{R}^n \to L^m_2[0,\infty)$ the pseudo-inverse of $\mathcal{C}_\Sigma$ defined by

$$
\mathcal{C}^\dagger_\Sigma(x^1) := \arg \min_{\mathcal{C}_\Sigma(u)=x^1} \|u\|_2.
\tag{32}
$$

Proof. The state-space realization of $\mathcal{C}^\dagger_\Sigma$ is given in [4] as

$$
y_a = \mathcal{C}^\dagger_\Sigma(x^0) : \begin{cases}
\dot{x} = -f(x) - g(x)g^T(x)\frac{\partial L_c}{\partial x}^T x(0) = x^0 \\
y_a = g^T(x)\frac{\partial L_c}{\partial x}^T
\end{cases}
\tag{33}
$$

If the condition (30) holds, then it follows from Lemma 1 that $p(t) \equiv (\partial L_c/\partial x)^T(x(t))$ holds along the trajectory of $(d\mathcal{H}_\Sigma(u))^* \circ \mathcal{H}_\Sigma(u)$ and this proves the lemma.   □

Though Lemma 2 gives a sufficient condition of the pairs of eigenvalues and eigenvectors which are closely related to the energy functions, it does not deal with the existence of the solutions. We will now continue to investigate the necessity. In order to proceed, we define two scalar functions

$$
\rho_{\max}(c) := \max_{\substack{u \in \text{Im }\mathcal{C}^\dagger_\Sigma(\mathbb{R}^n) \\ \|u\|_2 = c}} \frac{\|\mathcal{H}_\Sigma(u)\|_2}{\|u\|_2} = \max_{\substack{u \in L^m_2[0,\infty) \\ \|u\|_2 = c}} \frac{\|\mathcal{H}_\Sigma(u)\|_2}{\|u\|_2}
\tag{34}
$$

$$
\rho_{\min}(c) := \min_{\substack{u \in \text{Im }\mathcal{C}^\dagger_\Sigma(\mathbb{R}^n) \\ \|u\|_2 = c}} \frac{\|\mathcal{H}_\Sigma(u)\|_2}{\|u\|_2}.
\tag{35}
$$

$\rho_{\max}(c)$ is closely related to the Hankel norm because $\|\Sigma\|_H = \max_{c>0} \rho_{\max}(c)$ holds. Namely $\rho_{\max}(c)$ represents the Hankel norm under the fixed input magnitude $\|u\|_2 = c$. Furthermore it should be noticed that $\rho_{\max}$ and $\rho_{\min}$ exactly coincide with the maximum and minimum Hankel singular values in the linear case. Therefore these functions are an alternative nonlinear extension of Hankel singular values other than $\tau_i(x)$'s in Theorem 4.

Theorem 7. *Let $u_{\max}(c)$ and $u_{\min}(c)$ denote the inputs which achieve the maximization and minimization in the definition of $\rho_{\max}(c)$ and $\rho_{\min}(c)$ in (34) and (35) respectively, namely they satisfy $\|u_i\|_2 = c$ and $\|\mathcal{H}_\Sigma(u_i)\|_2/\|u_i\|_2 = \rho_i(c)$ ($i = \{\max, \min\}$). Suppose the energy functions $L_o(x)$ and $L_c(x)$, and the related functions $\rho_{\max}(c)$ and $\rho_{\min}(c)$ are sufficiently smooth. Then $u_{\max}(c)$ and $u_{\min}(c)$ are the eigenvectors of $u \mapsto (d\mathcal{H}_\Sigma(u))^* \circ \mathcal{H}_\Sigma(u)$ with respect to the following eigenvalues $\lambda_{\max}(c)$ and $\lambda_{\min}(c)$ respectively.*

$$\lambda_i(c) = \rho_i^2(c) + \frac{c}{2}\frac{d\rho_i^2(c)}{dc} \quad i = \{\max, \min\} \tag{36}$$

Further both the pair $\lambda_{\max}(c)$ and $C_\Sigma(u_{\max}(c))$, and the pair $\lambda_{\min}(c)$ and $C_\Sigma(u_{\min}(c))$ satisfy the condition (30).

Proof. Let i denote the index such that $i \in \{\max, \min\}$. Firstly we define $x_i(c) := C_\Sigma(u_i(c))$ and show the existence of $\lambda_i(c)$ such that

$$\frac{\partial L_o}{\partial x}(x_i(c)) = \lambda_i(c)\frac{\partial L_c}{\partial x}(x_i(c)). \tag{37}$$

To this effect, let the level set of $L_c(x)$ be given by $\mathcal{X}_{L_c=k} := \{ x \mid L_c(x) = k \}$. Then $x_i(c) \in \mathcal{X}_{L_c=\frac{c^2}{2}}$ follows from the fact that $u_i(c)$ is the input which minimizes the input energy under the constraint $\|u\|_2 = c$. Take a curve $\eta(s) \in \mathcal{X}_{L_c=\frac{c^2}{2}}$ parametrized by a scalar variable s such that $\eta(0) = x_i(c)$ holds. Since $\eta(s)$ is contained in the level set $\mathcal{X}_{L_c=\frac{c^2}{2}}$,

$$\frac{dL_c(\eta(s))}{ds} = \frac{\partial L_c(\eta)}{\partial \eta}\frac{d\eta(s)}{ds} = 0 \tag{38}$$

holds along $\eta(s)$. Next we can observe the following relations

$$\max_{\substack{u \in \mathcal{L}_2^\dagger(\mathbb{R}^n) \\ \|u\|_2=c}}\frac{\|\mathcal{H}_\Sigma(u)\|_2}{\|u\|_2} = \max_{x \in \mathcal{X}_{L_c=\frac{c^2}{2}}}\sqrt{\frac{L_o(x)}{L_c(x)}}, \quad \min_{\substack{u \in \mathcal{L}_2^\dagger(\mathbb{R}^n) \\ \|u\|_2=c}}\frac{\|\mathcal{H}_\Sigma(u)\|_2}{\|u\|_2} = \min_{x \in \mathcal{X}_{L_c=\frac{c^2}{2}}}\sqrt{\frac{L_o(x)}{L_c(x)}}.$$

This implies $\eta(0) = x_i(c)$ maximizes (minimizes) the value (L_o/L_c) in the level set $\mathcal{X}_{L_c=\frac{c^2}{2}}$. Therefore we obtain

$$\frac{d\frac{L_o(\eta(s))}{L_c(\eta(s))}}{ds}\bigg|_{s=0} = \frac{2}{c^2}\frac{dL_o(\eta(s))}{ds}\bigg|_{s=0} = \frac{2}{c^2}\frac{\partial L_o(\eta)}{\partial \eta}\frac{d\eta(s)}{ds}\bigg|_{s=0} = 0. \tag{39}$$

The equations (38) and (39) have to hold for all curves $\eta(s) \in \mathcal{X}_{L_c = \frac{c^2}{2}}$. Namely both $(\partial L_o / \partial x)$ and $(\partial L_c / \partial x)$ are orthogonal to the tangent space of $\mathcal{X}_{L_c = \frac{c^2}{2}}$ at $x = x_i(c)$. Because the tangent space is $(n-1)$-dimensional, we can conclude that $(\partial L_o / \partial x)$ and $(\partial L_c / \partial x)$ are linearly dependent at $x = x_i(c)$. Therefore there exists a scalar constant $\lambda_i(c)$ such that (37) holds. Remember that $u_i(c)$ can be described by $u_i(c) = C_{\Sigma}^{\dagger}(x_i(c))$. It follows directly from Lemma 2 that $u_i(c) = C_{\Sigma}^{\dagger}(x_i(c))$ is the eigenvector of $u \mapsto (d\mathcal{H}_{\Sigma}(u))^* \circ \mathcal{H}_{\Sigma}(u)$ with respect to $\lambda_i(c)$.

Secondly we prove the equation (36). From the above discussion, for any vector $\xi \in \mathbb{R}^n$ which is not orthogonal to the tangent space of $\mathcal{X}_{L_c = \frac{c^2}{2}}$ at $x = x_i(c)$, $\lambda_i(c)$ can be expressed as

$$\lambda_i(c) = \frac{\frac{\partial L_o}{\partial x} \xi}{\frac{\partial L_c}{\partial x} \xi}. \tag{40}$$

Let ξ be the directional derivative $(d\zeta(s)/ds)$ of another curve $\zeta(s) = x_i(s)$ passing through the maximizing (minimizing) state. Namely it goes across the level set $\mathcal{X}_{L_c = \frac{c^2}{2}}$ through $x = x_i(c) (= \zeta(c))$. Then we can obtain

$$\lambda_i(c) = \left. \frac{\frac{\partial L_o(\zeta)}{\partial \zeta} \frac{d\zeta(s)}{ds}}{\frac{\partial L_c(\zeta)}{\partial \zeta} \frac{d\zeta(s)}{ds}} \right|_{s=c} = \left. \frac{\frac{dL_o(\zeta(s))}{ds}}{\frac{dL_c(\zeta(s))}{ds}} \right|_{s=c} = \frac{\frac{d(\rho_i^2(c) \, c^2/2)}{dc}}{\frac{d(c^2/2)}{dc}} = \rho_i^2(c) + \frac{c}{2} \frac{d\rho_i^2(c)}{dc}.$$

This completes the proof.   □

Further we can extend this result for general singular value functions.

Theorem 8. *Suppose the energy functions $L_o(x)$ and $L_c(x)$ are sufficiently smooth and the Jacobian linearization of the system Σ has n distinct Hankel singular values. Then there exists locally $2n$ smooth singular value functions $\rho_i^j(c)$'s, $i \in \{1, 2, \ldots, n\}$, $j \in \{+, -\}$ such that $\min\{\rho_i^+(c), \rho_i^-(c)\} > \max\{\rho_{i+1}^+(c), \rho_{i+1}^-(c)\}$, ($\max\{\rho_1^+(c), \rho_1^-(c)\} = \rho_{\max}(c)$, $\min\{\rho_n^+(c), \rho_n^-(c)\} = \rho_{\min}(c)$) and that there exists $x_i^j(c)$'s satisfying*

$$L_c(x_i^j(c)) = \frac{c^2}{2}, \quad L_o(x_i^j(c)) = \frac{c^2 \rho_i^{j^2}(c)}{2} \tag{41}$$

$$\frac{\partial L_o}{\partial x}(x_i^j(c)) = \lambda_i^j(c) \frac{\partial L_c}{\partial x}(x_i^j(c)) \tag{42}$$

with $\lambda_i^j(c) := \rho_i^{j^2}(c) + (d\rho_i^{j^2}(c)/dc)c/2$.

Proof. Suppose the state-space realization is in input-normal form. The assumption that the Jacobian linearization of the system has n distinct Hankel singular values implies that the hyper plane $L_o(x) = k > 0$ has the form of a hyper ellipsoid whose radiuses are distinct when k is small. This means that there exist $2n$ critical points such that (38) and (39) hold for any curve $\eta(s) \in \mathcal{X}_{L_c=k}$. The number of the critical points (which include local maximums, minimums and saddle points) are same at lease in a neighborhood of the origin. This proves the existence of $\lambda_i^j(c)$'s and $x_i^j(c)$'s satisfying (42). The property (41) can be proved in a similar way as in the proof of Theorem 7. Further the order $\min\{\rho_i^+(c), \rho_i^-(c)\} > \max\{\rho_{i+1}^+(c), \rho_{i+1}^-(c)\}$ follows from the fact that $\rho_i^+(0) = \rho_i^-(0) = \sigma_i$ holds with the Hankel singular value σ_i of the Jacobian linearization of the system. This completes the proof. □

The following fact similar to Theorem 5 follows immediately.

Corollary 2. *Suppose the assumptions in Theorem 8 hold. Then $\rho_i^j(c)$'s and $u_i^j(c) := \mathcal{C}_\Sigma^\dagger(x_i^j(c))$'s satisfy*

$$\langle u_i^j(c), \mathcal{H}_\Sigma^*(\mathcal{H}_\Sigma(u_i^j(c)), u_i^j(c)))\rangle_{L_2} = \rho_i^{j\,2}(c) \ \langle u_i^j(c), u_i^j(c)\rangle_{L_2}. \tag{43}$$

Unfortunately, the pair $\rho_i^{j\,2}(c)$ and $u_i^j(c)$ is not the pair of an eigenvalue and an eigenvector of the operator $u \mapsto \mathcal{H}_\Sigma^*(\mathcal{H}_\Sigma(u), u)$ itself, which is due to several non-uniqueness issues [7,4]. There are other pairs ρ's and u's such that the relation (43) holds, and even the pairs given in Theorem 5 are not unique [7]. However, the eigenstructure given in Theorems 6, 7 and 8 is particularly important because it is closely related to the critical points of the Hankel operator. Furthermore $x_i^j(c)$'s play the role of the coordinate of the balanced realization. This fact may be especially important for applications in model reduction for nonlinear dynamical systems.

In the linear case $\lambda_i^j(c)$'s coincide with the square of singular value functions $\rho_i^{j\,2}(c)$'s respectively whereas $\rho_i^j(c)$'s themselves are identical to the Hankel singular values. Indeed the equation (30) reduces to

$$x^T Q P = \lambda \, x^T \tag{44}$$

with observability and controllability Gramians Q and P respectively. This equation implies the linear case result given in Theorem 8.1 in [12].

5   Conclusions

We studied the eigenstructure of Hankel operators for nonlinear systems via Hamiltonian extensions. It was proved that the variational system and Hamiltonian extension can be interpreted as the Gâteaux differentiation of dynamical input-output systems and their adjoints respectively. We utilized this differentiation in order to clarify the eigenstructure of the self adjoint of the Hankel operator, which is closely related to the Hankel norm of the original system. In future research the application to model reduction of nonlinear systems of this new concept will be investigated.

References

1. Batt, J. (1970). Nonlinear compact mappings and their adjoints. *Math. Ann.*, 189:5–25.
2. Crouch, P. E. and van der Schaft, A. J. (1987). *Variational and Hamiltonian Control Systems*, volume 101 of *Lecture Notes on Control and Information Science*. Springer-Verlag, Berlin.
3. Fliess, M. (1974). Matrices de Hankel. *J. Math. Pures. Appl.*, pages 197–222.
4. Fujimoto, K., Scherpen, J. M. A., and Gray, W. S. (2000). Hamiltonian realizations of nonlinear adjoint operators. *Proc. IFAC Workshop on Lagrangian and Hamiltonian Methods for Nonlinear Control*, pages 39–44.
5. Gray, W. S. and Scherpen, J. M. A. (1998). Hankel operators and Gramians for nonlinear systems. *Proc. 37th IEEE Conf. on Decision and Control*, pages 3349–3353.
6. Gray, W. S. and Scherpen, J. M. A. (1999a). Hankel operators, singular value functions and Gramian generalizations for nonlinear systems. Submitted.
7. Gray, W. S. and Scherpen, J. M. A. (1999b). On the nonuniqueness of balanced nonlinear realizations. *Proc. 38th IEEE Conf. on Decision and Control*.
8. Scherpen, J. M. A. (1993). Balancing for nonlinear systems. *Systems & Control Letters*, 21:143–153.
9. Scherpen, J. M. A., Fujimoto, K., and Gray, W. S. (2000). On adjoints and singular value functions for nonlinear systems. *Proc. 2000 Conf. Information Sciences and Systems*, FP8, pages 7–12.
10. Scherpen, J. M. A. and Gray, W. S. (1999). On singular value functions and Hankel operators for nonlinear systems. *Proc. ACC*, pages 2360–2364.
11. Scherpen, J. M. A. and Gray, W. S. (2000). Minimality and similarity invariants of a nonlinear state space realization. Accepted for *IEEE Trans. Autom. Contr.*
12. Zhou, K., Doyle, J. C., and Glover, K. (1996). *Robust and Optimal Control*. Prentice-Hall, Inc., Upper Saddle River, N.J.

Distributed Architecture for Teleoperation over the Internet

Denis Gillet, Christophe Salzmann, and Pierre Huguenin

Swiss Federal Institute of Technology
CH – 1015 Lausanne, Switzerland
Denis.Gillet@epfl.ch

Abstract. This paper discusses challenges in enabling teleoperation over the Internet for the specific case of real systems which exhibit fast dynamics. A distributed client-server architecture is proposed to provide the necessary level of interactivity for supervision and tuning, without compromising the essential control tasks. Enabling features include real-time simulation and augmented-reality visualization to enhance the user perception of the distant ongoing operations. The proposed approach is illustrated through the remote control of an inverted pendulum.

1 Introduction

To operate critical facilities efficiently, the process and manufacturing industries increasingly request global monitoring, as well as for sustainable support and immediate service from their suppliers. Among these services, telemaintenance is the most requested. As a consequence, new industrial network implementations are increasingly based on the Internet protocol to ensure a global and continuous access. However, these new capabilities are not yet exploited extensively due to the lack of predictability of the transmission delay inherent to the Internet and the lack of standardized deployment methodologies.

To contribute to the fulfilment of the market expectations related with emerging real-time services, an architecture is proposed to enable the remote supervision and remote tuning of distributed controlled systems, including their embedded or external control devices. These applications are among the most critical ones in industrial environments, since they handle the functions related with the dynamic behavior of real equipment.

The industrial equipment previously mentioned is just one example of physical systems that can be teleoperated. Didactic setups available in instructional laboratories can also be teleoperated in a flexible learning context to provide students with the necessary resources for practical experimentation, as described in [1], [2], and [3].

The architecture implemented for the teleoperation of both industrial equipment and didactic setups relies on a computer-based client-server scheme with Internet as a communication channel.

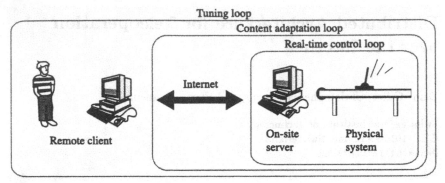

Fig. 1. Client-server Control Loops

The remote operators are provided with client software to observe and to pilot the physical system via the Internet. The on-site server is the computer that stands for the communication and operation interface with the real system and its instrumentation. To overcome the drawback of the operators not being present at the equipment location, additional devices have to be integrated, such as cameras to broadcast video views of the system and of its environment.

With such an architecture for teleoperation over the Internet, there are three imbricated control loops to implement (Fig. 1). The first one, dedicated to the real-time control of the real system, has to be implemented at the server side. A common and dangerous mistake is to close this loop across the Internet. Even if such an approach is challenging from a robust control point-of-view, it is unacceptable from an industrial point-of-view. Just think about the effect of the loss of a single or numerous measurement samples for stability and safety. The second one is the control loop designed to adapt the transmitted content to the available bandwidth in order to guarantee a sufficient quality of service. This is merely to ensure the reception of the information essential to carry out the desired operations according to their respective priority. The last loop is closed by the user for setting the conditions of operation or for tuning the distant real-time control loop. This is the loop the most sensitive to the transmission delay, because wrong actions can be performed by operators if the reactions of the system are not noticed fast enough for a valid interpretation.

In Sect. 2, the inverted pendulum and its local control are introduced. Various ways to handle the transmission delay that occurs when switching from local to remote control are described in Sect. 3. Then, details about the client-server architecture and the necessary features needed at the operator side to provide a sufficient quality of service are given in Sects. 4 and 5, respectively. Finally, concluding remarks and perspectives are expressed in Sect. 6.

2 The Inverted Pendulum and its Local Control

The pendulum described in this section is an inverted pendulum with two degrees of freedom. It emulates, in two dimensions, a juggler trying to keep a broomstick in equilibrium on his fingertip. The control objective is to simultaneously keep the pendulum stick in the upright position and the supporting cart at the center of the rail, using a single actuator.

The inverted pendulum shown in Fig. 2 is made up of a 4-meter horizontal rail mounted on top of a case that contains both the electronics and the power supply. A cart, moved by an electric motor by means of a metallic belt, can slide along this rail. The rotational axis of the pendulum stick is on this cart. No actuator exists for this axle. The only way to change the stick position is to accelerate the cart. Hardware security switches that turn the power off can be found at each end of the rail.

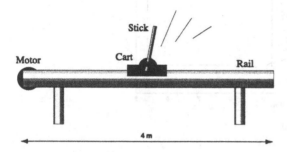

Fig. 2. The inverted pendulum

This laboratory-scale system can be controlled by any type of computer equipped with a National Instruments PCI-1200 family DAQ board. This portability feature turns the setup into an ideal support for educational demonstrations or trade shows.

The underlying state-space multivariable control principle aims at minimizing the energy spent by the controller to keep the stick raised and the cart at the center of the rail. The implementation of the corresponding linear quadratic regulator (LQR) relies on both angular and longitudinal position measurements, as well as the respective velocity estimates.

A representative physical model of the system is required for the controller design, as well as a good knowledge of its physical parameters. It is worth developing a good model because it leads to a much more accurate control than empirical methods. Moreover, the model can also be used for real-time simulation as needed for advanced teleoperation.

The swing-up algorithm implemented to move the stick from the lower stable vertical position to the upper unstable one is a very simple bang bang procedure: a constant voltage value is chosen to be applied to the motor and its sign is worked out adequately depending on the stick position in order to induce and amplify the stick oscillations.

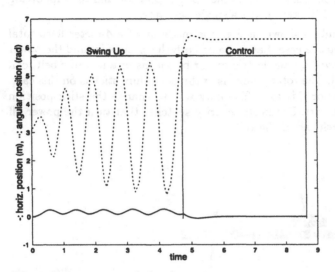

Fig. 3. Swing-up control signal

The swing-up procedure chosen is quasi-optimal from a time point-of-view. Its sequence and its intuitive justification are given as follows. First, the cart accelerates and the stick swings up due to its inertia. Then, the cart reaches its maximal velocity and the stick swings down by gravity. Once the stick crosses the lower point of its trajectory, it has accumulated the maximal possible amount of kinetic energy. Thus, it is the optimal time to inverse the voltage polarity (for a more accurate solution, the dynamics of the electrical drive has to be taken into account). This sequence has to be repeated until the stick reaches a pre-defined interval around the vertical position (Fig. 3). In this restricted control sector, the state-space controller can be activated to stabilize the pendulum.

The swing-up tuning parameters are the constant voltage value and the control sector size. The voltage level is a trade-off between the horizontal range of the cart displacement (the boundaries of the rail have not to be reached) and the angular velocity of the stick. The control sector size depends on the damping the controller can introduce and its robustness to the system nonlinearities, such as the ones due to the angular position of the stick.

The controller parameters that can be modified by local operators, in addition to the swing-up tuning ones, are the sampling period and the four state feedback gains. The same settings have to be available when remote operations are considered. Enabling fast remote prototyping of the controller requires that the effect of any change in these parameter values can be noticed as soon as possible by the operators, even in the case of a significant transmission delay.

3   Transmission Delay Handling

The varying transmission delay (approximately half the Round Trip Time) in Internet communication is the most important problem to handle when implementing teleoperation solutions. Depending on how the information packets transmitted are routed and how many routers are crossed, the delay can vary from a few milliseconds to hundreds (Fig. 4).

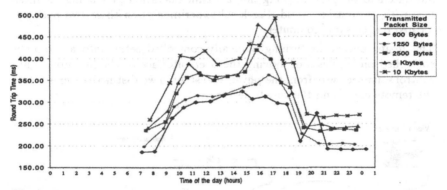

Fig. 4. Example of the transmission delay variation in Internet communication

The main difficulty for the operator occurs if the time constants of the remote system are of the same order of magnitude or faster than the time delay. Unfortunately, this condition appears to be true for most of the mechatronics systems.

The solution to partially alleviating this difficulty relies on an intelligent characterization and management of the data streams transmitted, such as the measurement stream, the video stream, the tuning stream and the coordination stream [4]. Such schemes aim to use optimally the available bandwidth. Since the transmission delay is incompressible, the additional solutions only aim to reduce its annoying effect on the sensorimotor behavior of the operators.

There are two classes of additional solutions, the client-side and the server-side ones. The client-side solutions mainly rely on predicting the remote ongoing operations to help the operator in anticipating the effect of its actions. The server-side solutions rely on intelligent context analysis carried out automatically by the server to feed back synthetic and composite results that can reduce the reaction time of the operators, once such a high level of information becomes available to them.

In automatic control, where the dynamic models are usually known and where the context of operation is always predefined, client-side solutions are more interesting. In mobile robotics, where the context of operation may evolve significantly and where more intelligence is embedded in the on-site system, server-side solutions can be more convenient. The teleoperation of an inverted pendulum clearly belongs to the first class.

4   Client–Server Architecture

The teleoperation server software is made of two distinctive parts, the user interface and the part handling the real-time operations, including the interface with the outside world. For local experimentation these two parts are located on the same computer.

The main concept in turning a locally-controlled setup into a remotely-controlled one consists of moving the user interface of the monitoring and control software away from the physical system. Two distinctive parts result: the remote client and the on-site server.

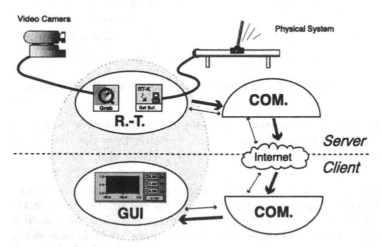

Fig. 5. Software components at the client and server sides

The remote client is a computer equipped with the functionalities necessary to observe and to act on the physical system. The client application is used to generate excitation signals and observe corresponding responses. The main objective of such an interface is to provide the user with a general view of the equipment, and to allow full control of the operations.

The on-site server is the computer located near the real system and equipped with the hardware interface to the sensors and actuators. The video camera and microphone can be seen as sensors. The server application receives the client commands and transmits them to the equipment. It also returns its physical and operational states to the client, including an image.

Three components (Fig. 5) are necessary to build the client and the server applications [4]. The client and the server application can be designed by adding a communication component to the two components used in local implementation: the real-time and the GUI ones. The client application is made up of the GUI and the communication components. The server application is made up of the real-time and the communication components. The server may require a basic user interface for supervision of the ongoing operations. The communication component allows the client and server applications to exchange information with other computers distributed in different geographical locations. This module also takes care of security issues regarding network management. For example, it prevents unauthorized access and schedules login to avoid conflicts.

By isolating carefully these modules in the development process, it is easy to port a local solution to a remote one, or port the remote solution to different physical systems.

5   Operator–Side Features

The user interface at the client side typically features an area showing a scope that displays all the relevant signals (see label 1 in Fig. 6), and also includes areas with dialog boxes that permit setting the controller parameters and specifying the access rights of the users (labels 2 and 3 in Fig. 6). In more advanced implementations the control algorithm may also be presented as a user option [5].

To enhance the user perception, a virtual representation of key parts of the real system is superimposed on the video image (see the dotted lines in area 4 in Fig. 6). Such a composite view is called augmented reality. Useful information can be added, such as the reference signal for the controller (triangle drawn on the target track position), the force applied on the cart (length of the horizontal line drawn at the base of the cart), or even a virtual hand used to apply a disturbance on the pendulum (see button 5 and the hand displayed in area 4 which appears when the button is pressed). Different views of the distant equipment can be selected. When the bandwidth available is not large enough to ensure a sufficiently high video throughput, the display is limited to the virtual representation which in turn can be animated using real data

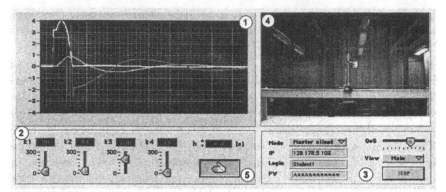

Fig. 6. User Interface of the remote client software

or, in the worst case, using samples generated by a real-time simulator. In the pendulum example, the animation of the virtual image can be achieved from knowledge of only two variables, namely, the angular and longitudinal positions of the pendulum.

The implementation of a real-time simulator in the client software permits the user to carry out off-line simulations for pre-validation purposes. In addition, the simulator can be used to provide synthetic data that can be posted on the user interface when packet losses occur during transmission. This gives continuity to the display, and provides the user with a sense of real-time behavior. In these cases, it is important to post an indicator that announces that the data shown is provided by a model due to the absence of reliable network data.

6   Concluding Remarks and Perspectives

The teleoperation of controlled systems that exhibit fast dynamic behavior is a challenging application when it is carried out over the Internet. It constitutes an innovative real-time service that brings new opportunities for both industry and academia. In industry it enables the monitoring and the maintenance of critical facilities. In academia it allows access to and sharing of laboratory resources for collaborative research and flexible education.

The client-server architecture implemented that features three hierarchical control loops poses the basis for efficient operations that require a high degree of interaction. It constitutes a framework for the development of adaptive schemes that cope with varying bandwidth and transmission delay, as well as quality of service constraints.

The inverted pendulum is a convenient introductory example that permits underlining the principal requirements related with the teleoperation of controlled mecatronic systems. The most important ones are: the necessity to ensure a cadenced stream of information to reproduce the dynamic behavior of the on-site system, the need for hybrid representation to enhance the operator perception of the ongoing operations, and the possibility to remotely perturb the systems for validation purposes.

References

1. Gillet D., Salzmann Ch., Longchamp R., Bonvin D. (1997) Telepresence: An Opportunity to Develop Practical Experimentation in Automatic Control Education. In: European Control Conference, Brussels, Belgium
2. Bhandari A., Shor M.H. (1998) Access to an Instructional Control Laboratory Experiment through the World Wide Web. In: American Control Conference, Philadelphia, USA
3. Overstreet J.W., Tzes A. (1999) Internet-Based Client/Server Virtual Instruments Designs for Real-Time Remote-Access Control Engineering Laboratory. In: American Control Conference, San Diego, USA, 1472–1476
4. Salzmann Ch., Latchman H.A., Gillet D.; Crisalle O.D. (1998) Requirements for Real-Time Experimentation over the Internet. In: International Conference on Engineering Education, Rio de Janeiro, Brazil
5. Piguet Y.,Gillet D. (1999) Java-Based Remote Experimentation for Control Algorithms Prototyping. In: American Control Conference, San Diego, USA, 1465–1469

Singular L-Q Problems and the Dirac-Bergmann Theory of Constraints

Manuel Guerra*

Universidad Tecnica de Lisboa - ISEG
R. do Quelhas 6
1200 Lisboa, Portugal
mguerra@iseg.utl.pt

Abstract. We apply Dirac's theory of constraints to singular linear-quadratic optimal control problems and compare the results with our previous characterization of the "generalized optimal solutions" for this class of problems. The generalized optimal control(s) for a singular L-Q problem is the sum of a real-analytic function with a distribution of order r concentrated at the extremes of the time interval and the corresponding trajectory is the sum of an analytic function with a distribution of order $(r - 1)$. The Dirac's approach provides an alternative method to find the continuous term of the generalized optimal solution(s) by solving a set of equations involving Poisson brackets but omits the distributional term . We show how the Dirac-Bergmann approach can be modified to allow for the distributional terms.

1 Introduction

The Dirac-Bergmann theory of constraints provides a method for finding singular extremals for a problem of variations in Lagrangian or Hamiltonian mechanics. In it's original form, it was first developed in the early 1950's by P.A.M. Dirac [2], [3]. Later, it was recast in the language of symplectic geometry by Gotay, Nester and Hinds [4], who generalized the theory to cases where the phase space has no natural symplectic structure. Volckaert and Aeyels [7] pointed out that in singular optimal control problems, the Pontryagin's Maximum Principle generates constraints affecting the pair state-costate in a way similar to that devised by Dirac. Hence it should be possible to apply a Dirac-Bergmann-type theory to this kind of problems.

In [7] the authors raise some questions concerning the meaning of *gauge freedom* in the context of optimal control problems and speculate if this could be related to abnormal extremals. Also, it is known that some singular optimal control problems have discontinuous "generalized optimal trajectories", and it was suggested that this could be related to gauge freedom.

In order to address these issues, we turned to the special case of singular linear-quadratic (L-Q) problems. This is an useful example because the appli-

---

* This work is part of the author PhD project at the University of Aveiro, under supervision of A. Sarychev

cation of the Dirac-Bergmann approach to these problems is straightforward and we can give an exhaustive characterization of the generalized optimal solutions using different methods [5].

It turns out that gauge freedom is related to none of the phenomena indicated above. It is the consequence of the existence of an infinite-dimensional subspace of the space of controls in which the cost functional is constant.

In [5] we showed that the generalized optimal trajectories of a singular L-Q problem are sums of an analytic function with a distribution concentrated at the extremes of the time interval, and showed how the generalized optimal solutions can be approximated by ordinary (suboptimal) solutions. The Dirac-Bergmann approach yields the analytical arc in the optimal generalized trajectory(ies), but gives no information about the distributional terms when they are present. We show how generalized solutions can be introduced in the Dirac-Bergmann framework. This gives some further insight about the structure of the generalized optimal solutions.

This paper is organized as follows. In the section 2 we give a brief account of our method to solve singular L-Q problems and present its main results. In the section 3 we apply Dirac's method to the same type of problem, we explore the relationship between both methods and show how generalized optimal solutions can be introduced in Dirac's method.

2   Generalized Solutions for Singular L-Q Problems

Consider an optimal control problem of the type

$$
\begin{aligned}
&J(x, u) \to \min \\
&\dot{x} = Ax + Bu, \qquad x(0) = \overline{x}, \qquad x(T) = \overline{\overline{x}}, \\
&u \in L_2^k[0, T], \qquad x \in AC^n[0, T],
\end{aligned}
\tag{1}
$$

where $J(x, u) = \int_0^T x(\tau)'Px(\tau) + 2u(\tau)'Qx(\tau) + u(\tau)'Ru(\tau) \, d\tau$, $A \in \mathbb{R}^{n \times n}$, $B \in \mathbb{R}^{n \times k}$ and $Q \in \mathbb{R}^{k \times n}$ are arbitrary matrices, $P \in \mathbb{R}^{n \times n}$, $R \in \mathbb{R}^{k \times k}$ are symmetric matrices, $\overline{x} \in \mathbb{R}^n$, $\overline{\overline{x}} \in \mathbb{R}^n$ are fixed points and $T \in]0, +\infty[$ is fixed. For each $u \in L_2^k[0, T]$, let $x_{u, \overline{x}}$ denote the corresponding trajectory of the system $\dot{x} = Ax + Bu$, $x(0) = \overline{x}$. Also, let $J_{\overline{x}}(u)$ denote the functional $u \mapsto J(x_{u, \overline{x}}, u)$, with domain in $L_2^k[0, T]$. Utilizing this notation, one may represent the problem (1) in the form

$$
J_{\overline{x}}(u) \to \min, \qquad u \in L_2^k[0, T], \qquad x_{u, \overline{x}}(T) = \overline{\overline{x}}.
$$

¿From the functional theoretical point of view, the problem is the one of finding a minimizer of a functional in the affine subspace $\mathcal{U}_{\overline{x}, \overline{\overline{x}}} = \{u \in L_2^k[0, T] : x_{u, \overline{x}}(T) = \overline{\overline{x}}\}$. Below, we provide $L_2^k[0, T]$ with a weaker topology and extend both maps $u \mapsto J_{\overline{x}}$ and $u \mapsto x_{u, \overline{x}}$ onto the topological completion of $L_2^k[0, T]$

relative to this topology. This space of generalized controls is a subspace of a Sobolev space, $H_{-r}^k [0,T]$, with $r \leq n$. The problem (1) has finite infimum if and only if the extended functional has a minimum in the closure of $\mathcal{U}_{\overline{x},\overline{\overline{x}}}$ with respect to the new topology.

2.1   Generalized Controls of Class $H_{-i}^k [0,T]$

Consider a continuous linear operator $f : L_2^k [0,T] \mapsto \mathcal{H}$, where $\mathcal{H}$ is an Hilbert space with norm $\|\cdot\|_{\mathcal{H}}$. If f is injective, then the functional $u \mapsto \|u\|_f = \|fu\|_{\mathcal{H}}$ is a norm in $L_2^k [0,T]$ and defines a topology that is weaker then the usual topology of $L_2^k [0,T]$. The topological completion of $L_2^k [0,T]$ with respect to the topology of $\|\cdot\|_f$ is denoted by H_f. In particular, let $\phi : L_2^k [0,T] \mapsto L_2^k [0,T]$ denote the primitivation operator, i.e.

$$\phi u(t) = \int_0^t u(\tau)\, d\tau, \qquad u \in L_2^k [0,T], \; t \in [0,T].$$

This generates a sequence of linear operators $\phi^i : L_2^k [0,T] \mapsto L_2^k [0,T], i \in \mathbb{N}$, defined by $\phi^0 = Id$, $\phi^i = \phi_\circ \phi^{i-1}$, $i = 1,2,3,....$ For these operators we keep the usual notation, i.e., we denote H_{ϕ^i} by $H_{-i}^k [0,T]$, $i \in \mathbb{N}$.

The input-trajectory map can be written as

$$x_{u,\overline{x}}(t) = e^{tA}\overline{x} + \int_0^t e^{(t-\tau)A} Bu(\tau)\, d\tau, \qquad u \in L_2^k [0,T], \; t \in [0,T].$$

The singularity of the problem means that $\ker(R) \neq \{0\}$. Decomposing the control space into $\ker(R)$ and $\ker(R)^\perp$, let Π^+ and Π^0 denote the orthogonal projections from $\mathbb{R}^k$ onto $\ker(R)^\perp$ and $\ker(R)$, respectively. By integrating $\Pi^0 u$ and leaving $\Pi^+ u$ unchanged, one obtains:

$$x_{u,\overline{x}}(t) = e^{tA}\overline{x} + \int_0^t e^{(t-\tau)A}(B\Pi^+ u(\tau) + AB\Pi^0 \phi u(\tau))\, d\tau + B\Pi^0 \phi u(t).$$

If we consider the operator $\varphi : L_2^k [0,T] \mapsto L_2^k [0,T]$ defined by $\varphi u = \Pi^+ u + \Pi^0 \phi u$, we have:

$$x_{u,\overline{x}} = z_{\varphi u, \overline{x}} + B\Pi^0 \varphi u, \tag{2}$$

where $z_{v,\overline{x}}$ denotes the trajectory of the system $\dot{z} = Az + (B\Pi^+ + AB\Pi^0)v$, $z(0) = \overline{x}$. Let

$$B_1 = B\Pi^+ + AB\Pi^0; \qquad Q_1 = \Pi^+ Q + \Pi^0 (B'P - QA);$$

$$\begin{aligned} R_1 = {} & R + \Pi^+(QB - B'Q')\Pi^0 + \Pi^0(B'Q' - QB)\Pi^+ + \\ & + \Pi^0(B'PB - QAB - B'A'Q')\Pi^0. \end{aligned}$$

By substituting (2) in the definition of $J_{\overline{x}}$ and performing all possible integrations by parts aiming to eliminate the terms that contain $\Pi^0 u$, one obtains

$$J_{\overline{x}}(u) = \int_0^T z'_{\varphi u,\overline{x}} P z_{\varphi u,\overline{x}} + 2\varphi u' Q_1 z_{\varphi u,\overline{x}} + \varphi u' R_1 \varphi u + \tag{3}$$
$$+ (\Pi^0 u)'(QB - B'Q')\Pi^0 \phi u \, d\tau +$$
$$+ 2(\Pi^0 \varphi u)(T)' Q z_{\varphi u,\overline{x}}(T) + (\Pi^0 \varphi u)(T)' QB(\Pi^0 \varphi u)(T).$$

In order to proceed we need the following Proposition [1, Appendix 2].

Proposition 1. *In order to exist some linear subspace, $\mathcal{V}$, of finite codimension in $L_2^k[0,T]$, satisfying $\inf\limits_{u \in \mathcal{V}} J_0(u) > -\infty$, the following conditions must hold*

$$\Pi^0 (QB - B'Q')\Pi^0 = 0 \tag{4}$$

$$R_1 \geq 0. \; \Diamond \tag{5}$$

These conditions are known as *Goh* and *generalized Legendre-Clebsch* conditions, respectively.

The transformations (2) and (3) must have one of the following outcomes:

(i)   one of the conditions (4) or (5) fails;
(ii) the condition (4) holds and (5) is strictly positive;
(iii) both the conditions (4) and (5) hold, but (5) is only semidefinite.

In the case (i), it follows that $\inf\limits_{u \in \mathcal{U}_{\overline{x},\overline{\overline{x}}}} J_{\overline{x}}(u) = -\infty$ and hence there exists no optimal solution in whatever sense. In the cases (ii) and (iii), the right-hand side of (2) and (3) give the unique continuous extensions of the maps $u \mapsto x_{u,\overline{x}}$ and $u \mapsto J_{\overline{x}}(u)$ onto the space $H_{\widehat{\varphi}}$, where $\widehat{\varphi} : L_2^k[0,T] \mapsto L_2^k[0,T] \times \ker(R)$ is the operator defined by $\widehat{\varphi}u = (\varphi u, (\Pi^0 \varphi u)(T))$. It is clear that $H_{\widehat{\varphi}} \subset H_{-1}^k[0,T]$ and the closure of $\mathcal{U}_{\overline{x},\overline{\overline{x}}}$ with respect to the topology of $H_{\widehat{\varphi}}$ is $\overline{\mathcal{U}}_{\overline{x},\overline{\overline{x}}} = \left\{ u \in H_{\widehat{\varphi}} : z_{\varphi u}(T) + B(\Pi^0 \varphi u)(T) = \overline{\overline{x}} \right\}$. Let $B^{\#}$ denote the left inverse of $B\Pi^0$, i.e., $B^{\#} B \Pi^0 v = v$, for all $v \in \ker(B\Pi^0)^\perp$. Then, for all $u \in \overline{\mathcal{U}}$, the equality (3) reduces to

$$J_{\overline{x}}(u) = \int_0^T z'_{\varphi u,\overline{x}} P z_{\varphi u,\overline{x}} + 2(\varphi u)' Q_1 z_{\varphi u,\overline{x}} + (\varphi u)' R_1 (\varphi u) \, d\tau +$$
$$+ (\overline{\overline{x}} - z_{\varphi u,\overline{x}}(T))' B^{\#'} Q(\overline{\overline{x}} + z_{\varphi u,\overline{x}}(T)).$$

Thus, we may consider φu as a new control and this reduces the problem (1) into a new L-Q problem

$$J_1(z,v) + \left(\overline{\overline{x}} - z(T)\right)' B^{\#'} Q\left(\overline{\overline{x}} + z(T)\right) \to \min$$
$$\dot{z} = Az + B_1 v, \qquad z(0) = \overline{x}, \qquad z(T) \in \overline{\overline{x}} + B\Pi^0\left(\mathbb{R}^k\right),$$
$$v \in L_2^k[0,T], \qquad z \in AC^n[0,T],$$

with $J_1(z, v) = \int_0^T z(\tau)'Pz(\tau) + 2v(\tau)'Q_1 z(\tau) + v(\tau)'R_1 v(\tau) \, d\tau$.

If (ii) holds, then this is a regular L-Q problem. If (iii) holds, this is a new singular L-Q problem and we may repeat the procedure, obtaining a larger space of generalized controls, $H_{\widehat{\varphi_2}} \subset H_{-2}^k [0, T]$, etc.

2.2   The General Reduction Procedure

There exist problems which can not be reduced to a regular problem by any finite sequence of transformations of the kind outlined above. However, all problems of the type (1) can be reduced by a finite sequence of transformations of the more general type

$$u \mapsto \phi L(u - Fx_u) + (Id - L)(u - Fx_u),$$

where L is an idempotent matrix mapping $\mathbb{R}^k$ into $\ker(R)$, and $F \in \mathbb{R}^{k \times n}$ (see [5] for details). Thus, starting with $(A_0, B_0, P_0, Q_0, R_0) = (A, B, P, Q, R)$, we consider a sequence of transformations, $\{\varphi_i, \ i = 1, 2, ...\}$, corresponding to a sequence of L-Q problems with matrices $(A_i, B_i, P_i, Q_i, R_i) \ i = 1, 2, ...$ For any such sequence, the following Proposition holds.

Proposition 2. *For J_0 to have finite infimum in some linear subspace of finite codimension in $L_2^k [0, T]$, it is necessary that all the following conditions hold for all $i \geq 0$.*

$$\Pi_i^0 (Q_i B_i - B_i' Q_i') \Pi_i^0 = 0; \tag{6}$$

$$R_i \geq 0; \tag{7}$$

$$v'Q_i \left(B_i \ A_i B_i \ \cdots \ A_i^{n-1} B_i \right) = 0, \qquad \forall v \in \ker(B_i) \cap \ker(R_i). \tag{8}$$

For a given $\overline{x} \in \mathbb{R}^n$, the existence of some $T > 0$ and some $\overline{\overline{x}} \in \mathbb{R}^n$ such that $\inf\limits_{u \in \mathcal{U}_{\overline{x}, \overline{\overline{x}}}} J_{\overline{x}}(u) > -\infty$ also implies

$$v'Q_i \left(\overline{x} \ A_i \overline{x} \ \cdots \ A_i^{n-1} \overline{x} \right) = 0, \qquad \forall v \in \ker(B_i) \cap \ker(R_i). \ \Diamond \tag{9}$$

In [5] we showed how to chose a sequence of transformations such that each φ_i can be partitioned $\varphi_i = \left(\varphi_i^B, \varphi_i^N \right)$, and the corresponding matrices have the structure $B_i = \left(\widehat{B}_i \ 0 \right)$, $Q_i = \begin{pmatrix} \widehat{Q}_i \\ 0 \end{pmatrix}$, $R_i = \begin{pmatrix} R_i^B & 0 \\ 0 & R_i^N \end{pmatrix}$, with $\widehat{B}_i$ having full column rank. For this particular sequence we have the following stronger version of the Proposition 2 :

Proposition 3. *For J_0 to have finite infimum in some linear subspace of finite codimension in $L_2^k [0, T]$, it is necessary and sufficient the existence of an integer $r \leq n$ such that the conditions (6) to (8) hold for all $i \leq r$ and either $R_r^B > 0$ or $B_r = 0$. $\Diamond$*

We call the number r the *"order of singularity"* of the problem.

If the conditions of the Proposition 3 hold and the conditions $(8,9)$ hold for all $i \leq r$, then the extension of the functional onto $H_{\widehat{\varphi_r}}$ satisfies

$$J_{\overline{x}}(u) = J_{\overline{x}}^r \left(\varphi_r^B u \right) + \langle \varphi_r^N u, R_r^N \varphi_r^N u \rangle + C_r (\overline{x}, \overline{\overline{x}}), \qquad \forall u \in \overline{\mathcal{U}}_{\overline{x},\overline{\overline{x}}},$$

where $J_{\overline{x}}^r(v) = \int_0^T x_{v,\overline{x}}^{r\prime} P_r x_{v,\overline{x}}^r + 2v' \widehat{Q}_r x_{v,\overline{x}}^r + v' R_r^B v \, d\tau$, $x_{v,\overline{x}}^r$ denotes the trajectory of the system $\dot{x} = A_r x + \widehat{B}_r v$, $x(0) = \overline{x}$, and $C_r(\cdot) : \mathbb{R}^{2n} \mapsto \mathbb{R}$ is a quadratic form. This reduces the problem (1) to the new problem:

$$\begin{aligned}
&\int_0^T z' P_r z + 2v' \widehat{Q}_r z + v' R_r^B v \, d\tau \to \min \\
&\dot{z} = A_r z + \widehat{B}_r v, \qquad z(0) = \overline{x}, \qquad z(T) = \Upsilon \overline{\overline{x}} + (Id - \Upsilon) x_{0,\overline{x}}^r(T), \qquad (10) \\
&v \in L_2^h[0,T], \qquad z \in AC^n[0,T],
\end{aligned}$$

where Υ is a projection of $\mathbb{R}^{2n}$ into the controllable space of (A_r, B_r) (this is always noncontrollable). Thus, the generalized optimal controls for the problem (1) are exactly the generalized controls that can be represented as $\widehat{u} = \varphi_r^{-1} ((\widehat{v}, 0) + \omega)$, where $\widehat{v} \in L_2^h[0,T]$ is optimal for the transformed problem (10) and ω is any square-integrable function such that $\omega(t) \in \ker(R_r)$ a.e. in $[0,T]$. Since the transformed problem (10) is either regular or trivial, this shows that the problem (1) has some generalized optimal solution if and only if $\inf_{u \in \mathcal{U}_{\overline{x},\overline{\overline{x}}}} J_{\overline{x}}(u) > -\infty$.

2.3   Properties of the Generalized Optimal Solutions

Using the reduction procedure outlined above, it is possible to characterize the problems of the type (1) whose infimum is finite. The generalized optimal solution(s) for those problems can be computed, either from the Maximum Principle or from the solution of an adequate Riccati differential equation. Approximations of the generalized optimal control by smooth functions can also be computed, as well as the correspondent trajectories [5]. Below we present just a few results that are most important for the discussion of Dirac's approach presented in the next section.

Theorem 1. *If* $\inf_{u \in \mathcal{U}_{\overline{x},\overline{\overline{x}}}} J_{\overline{x}}(u) > -\infty$, *then there exists at least one generalized optimal control that is the sum of a real-analytic function in $[0,T]$ and a distribution of order $j \leq r$, concentrated at $t = 0$ and $t = T$. The corresponding trajectory is the sum of a real-analytic function in $[0,T]$ and a distribution of order $(j-1)$, concentrated at $t = 0$ and $t = T$.* $\diamond$

The presence of distributional terms in the generalized optimal trajectory is important because it implies that any minimizing sequence of continuous trajectories must be unbounded in a very precise sense [5].

Any discontinuities and/or distributional terms that may be present in any generalized trajectory must lie in $span\left(B\Pi^0, B_1\Pi_1^0, ..., B_{r-1}\Pi_{r-1}^0\right)$. This coincides with the space of *jump directions* described by Jurdjevic [6]. Let $\overline{X}$, $\overline{\overline{X}}$ denote two translations of $span\left(B\Pi^0, B_1\Pi_1^0, ..., B_{r-1}\Pi_{r-1}^0\right)$.

In the next two results we are using the sequence of transformations required to obtain the Proposition 3.

Proposition 4. *Assume that the problem* (1) *satisfies* $\inf\limits_{u\in\mathcal{U}_{\overline{x},\overline{\overline{x}}}} J_{\overline{x}}(u) > -\infty$.

for all $\overline{x}$, $\overline{\overline{x}} \in \mathbb{R}^n$ *such that* $\overline{\overline{x}}$ *can be reached from* $\overline{x}$, *and that there exists some point in* $\overline{\overline{X}}$ *that can be reached from some point in* $\overline{X}$. *Then, there exist unique* $\overline{a} \in \overline{X}$, $\overline{\overline{a}} \in \overline{\overline{X}}$, *and a unique real-analytic control, w, for which the following conditions hold*

(i) *w is optimal with respect to* $(\overline{a}, \overline{\overline{a}})$;

(ii) $\varphi_r^N w = 0$, *and* $\varphi_r w(t) \in \ker\left(R_r\right)^\perp$ *a.e. in* $[0, T]$.

For each $\overline{x} \in \overline{X}$, $\overline{\overline{x}} \in \overline{\overline{X}}$, *an optimal generalized control is* $\hat{u} = w + u$ *where u is the unique distribution that satisfies all the following conditions*

(iii) $x_{w+u,\overline{x}}(T) = \overline{\overline{x}}$;

(iv) *u is a distribution of order* $j \le r$ *concentrated in* $\{0\} \cup \{T\}$;

(v) $\varphi_r^N(w + u) = 0$, *and* $\varphi_r(w + u)(t) \in \ker\left(R_r\right)^\perp$ *a.e. in* $[0, T]$.

This control also satisfies

(vi) $x_{w+u,\overline{x}} = x_{w,\overline{a}} + \Delta$,

where Δ is a distribution of order $(j-1)$ *concentrated in* $\{0\}\cup\{T\}$. *The corresponding cost is* $J_{\overline{x}}(w + u) = J_{\overline{a}}(w) + (\overline{a} - \overline{x})' M(\overline{a} + \overline{x}) + (\overline{\overline{x}} - \overline{\overline{a}})' M(\overline{\overline{x}} + \overline{\overline{a}})$, *where M is a real matrix depending on* (A, B, P, Q, R). ◊

The conditions (ii) and (v) above mean that, if the generalized optimal controls are not unique, then w and $(u + w)$ are the smallest generalized optimal controls with respect to the norm $\|\cdot\|_{\widehat{\varphi_r}}$.

Proposition 5. *In order to* $\inf\limits_{u\in\mathcal{U}_{\overline{x},\overline{\overline{x}}}} J_{\overline{x}}(u) > -\infty$, *for all* $\overline{x}$, $\overline{\overline{x}}$ *such that* $\overline{\overline{x}}$ *can be reached from* $\overline{x}$, *it is necessary and sufficient that the conditions of the Proposition 3 hold,* $v'Q_i = 0$ *for all* $v \in \ker\left(B_i\right) \cap \ker\left(R_i\right)$ *and all* $i \le r$, *and either* $B_r = 0$ *or the transformed problem* (10) *is regular and has no conjugate point in the interval* $[0, T]$. ◊

3   The Dirac-Bergmann Approach

3.1   The Dirac's Constraints

For any pair of smooth functions $f(x, \xi, u) : \mathbb{R}^{2n+k} \mapsto \mathbb{R}^h$, $g(x, \xi, u) : \mathbb{R}^{2n+k} \mapsto \mathbb{R}^l$, we define the Poisson bracket $[f, g]$ to be the matrix whose entries are the functions of (x, ξ, u) defined by $[f, g]_{i,j} = [f_i, g_j] = \frac{\partial f_i}{\partial x} \left(\frac{\partial g_j}{\partial \xi} \right)' - \frac{\partial f_i}{\partial \xi} \left(\frac{\partial g_j}{\partial x} \right)'$, $1 \leq i \leq h$, $1 \leq j \leq l$. For simplicity, we assume that the coordinates of the control space $(\mathbb{R}^k)$ are such that

$$R = \begin{pmatrix} R^+ & 0 \\ 0 & 0 \end{pmatrix}, \qquad B = (B^+ \ B^0), \qquad Q = \begin{pmatrix} Q^+ \\ Q^0 \end{pmatrix}, \qquad u = \begin{pmatrix} u^+ \\ u^0 \end{pmatrix},$$

with $R^+ > 0$.

$u \in \mathbb{R}^k$ maximizes the Hamiltonian function at the point $(x, \xi) \in \mathbb{R}^{2n}$ if and only if $u = \frac{1}{2} (R^+)^{-1} (B^{+\prime} \xi - 2Q^+ x)$, and $B^{0\prime} \xi - 2Q^0 x = 0$. Hence, the Maximum Principle states that any absolutely continuous optimal trajectory for the problem (1) must be the projection into the state-space of some trajectory of the linear control system

$$\left(\dot{x}, \dot{\xi} \right) = \left[Id, H^+ \right] + \left[Id, H^0 \right] u^0 \tag{11}$$

satisfying

$$H^0 (x, \xi) = 0, \qquad \forall t \in [0, T], \tag{12}$$

where $H^+(x, \xi) = x' \left(Q^{+\prime} (R^+)^{-1} Q^+ - P \right) x + \xi' \left(A - B^+ (R^+)^{-1} Q^+ \right) x + \frac{1}{4} \xi' B^+ (R^+)^{-1} B^{+\prime} \xi$, $H^0(x, \xi) = B^{0\prime} \xi - 2Q^0 x$. The condition (12) implies $\frac{d}{dt} H^0 (x, \xi) = 0$ a.e. in $[0, T]$. Using the Hamiltonian equation (11), this reduces to $\left[H^0, H^+ \right] (x, \xi) + \left[H^0, H^0 \right] (x, \xi) u^0 = 0$, a.e. in $[0, T]$. Since the higher order derivatives of H^0 must also be identically zero, this generates a sequence of equations involving higher order Poisson brackets. Dirac called the condition (12) the *primary constraint*, and the conditions that arise by differentiating H^0 i times he called the *secondary constraints of order i*. When we proceed through the sequence of constraints, one of the following must obtain:

(i) the secondary constraints up to order j solve to give u^0 as a function of (x, ξ);

(ii) the secondary constraints up to order j leave at least some coordinates of u^0 unsolved but the restriction of order $(j + 1)$ can be expressed as a linear combination of all previous restrictions (including the primary restriction).

Hence the procedure must terminate in no more then $2n$ steps. Since all constraints can either be solved to u^0 or are linear homogeneous in (x, ξ), L-Q problems never generate incompatible sets of constraints. If (i) obtains, then we just have to substitute u^0 in the Hamiltonian equation (11) to obtain the extremal trajectories. If (ii) holds, then for each pair $(\bar{x}, \bar{\bar{x}})$, we have an infinite set of extremal trajectories, parameterized by the set of square integrable functions that map $[0, T]$ into the remaining free coordinates of u^0. In this case, the problem (1) is said to have gauge freedom.

3.2 Relationship Between Dirac's Approach and the Generalized Control Approach

In order to see the relationship between Dirac's approach and the previous one, consider a sequence of transformations as described in the Section 2, starting with

$$\varphi_1 u = \varphi_1\left(u^+, u^0\right) = \left(u^+ + \left(R^+\right)^{-1}\left(Q^+ B^0 - B^{+\prime} Q^{0\prime}\right)\phi u^0, \phi u^0\right).$$

This yields $R_1 = \begin{pmatrix} R^+ & 0 & 0 \\ 0 & R_1^+ & 0 \\ 0 & 0 & 0 \end{pmatrix}$, with R_1^+ nonsingular. Proceeding in the same way, we obtain a sequence of transformations such that the matrices of the corresponding transformed problems take the form

$$R_i = \begin{pmatrix} R^+ & \cdots & 0 & 0 \\ \vdots & \ddots & \vdots & \vdots \\ 0 & \cdots & R_i^+ & 0 \\ 0 & \cdots & 0 & 0 \end{pmatrix}, \qquad B_i = \left(B^+ \cdots B_i^+ \ B_i^0\right), \qquad Q_i = \begin{pmatrix} Q^+ \\ \vdots \\ Q_i^+ \\ Q_i^0 \end{pmatrix},$$

with all the R_j^+ nonsingular. The corresponding transformed control is $\varphi_i u = \left(v^+, v_1^+, ..., v_i^+, v_i^0\right)$. For each one of these transformed problems, the "maximized Hamiltonian" is

$$H_i\left(x, \xi, v_i^0\right) = H_i^+\left(x, \xi\right) + v_i^{0\prime} H_i^0\left(x, \xi\right),$$

where $H_i^0\left(x, \xi\right) = B_i^{0\prime}\xi - 2Q_i^0 x$, $H_i^+\left(x, \xi\right) = x'(\sum_{j=0}^{i} Q_j^{+\prime}\left(R_j^+\right)^{-1} Q_j^+ - P)x +$

$\xi'(A - \sum_{j=0}^{i} B_j^+\left(R_j^+\right)^{-1} Q_j^+)x + \frac{1}{4}\xi'(\sum_{j=0}^{i} B_j^+\left(R_j^+\right)^{-1} B_j^{+\prime})\xi$. We also consider the functions $G_i(x, \xi) = B_i^{+\prime}\xi - 2Q_i^+ x$, $i = 1, 2, ...$ Utilizing this notation, one obtains

$$\begin{array}{ll} [H_i^0, H_i^0] = 2\left(B_i^{0\prime} Q_i^{0\prime} - Q_i^0 B_i^0\right), & [H_i^0, H_i^+] = \begin{pmatrix} -G_{i+1} \\ -H_{i+1}^0 \end{pmatrix}, \\ \left[[H_i^0, H_i^+], H_i^0\right] = \begin{pmatrix} 2R_{i+1}^+ & 0 \\ 0 & 0 \end{pmatrix}, & i = 0, 1, 2, ... \end{array} \qquad (13)$$

Hence we can give the following formulation for the Goh $(4, 6)$ and generalized Legendre-Clebsch $(5, 7)$ conditions.

Proposition 6. *For J_0 to have finite infimum in some linear subspace of finite codimension in $L_2^k[0, T]$, it is necessary that the following conditions hold for all $i \geq 0$.*

$$[H_i^0, H_i^0] = 0, \qquad [[H_i^0, H_i^+] H_i^0] \geq 0. \Diamond$$

Assuming that the conditions of the Proposition 6 hold and using the equalities (13), the $(2i - 1)^{th}$ secondary constraint reduces to

$$\left[\cdots \left[\left[H_{i-1}^0, H_{i-1}^+\right], H_{i-2}^+\right], \cdots, H^+\right] = 0, \tag{14}$$

while the $(2i)^{th}$ secondary constraint reduces to

$$u_i^+ = \frac{1}{2}\left(R_i^+\right)^{-1}\left[\cdots \left[\left[G_i, H_{i-1}^+\right], H_{i-2}^+\right], \cdots, H^+\right]; \tag{15}$$

$$\left[\cdots \left[\left[H_i^0, H_{i-1}^+\right], H_{i-2}^+\right], \cdots, H^+\right] = 0. \tag{16}$$

Thus, both approaches proceed by computing essentially the same sequence of matrices.

It is clear that the analytic trajectory described in the Proposition 4 must be the projection into the state space of some solution of the Hamiltonian system (11) that satisfies all the Dirac's constraints. If $R_r > 0$, then there exists some $T > 0$ such that the generalized optimal solution exists and is unique for each $(\bar{x}, \bar{\bar{x}})$ such that $\bar{\bar{x}}$ can be reached from $\bar{x}$. In this case the solution of the Hamiltonian system (11) that satisfies all the Dirac's constraints and $x(0) = \bar{a}$, $x(T) = \bar{\bar{a}}$ must be unique. This implies that the Dirac's set of constraints completely solves u^0 as a function of (x, ξ). Conversely, if the Dirac's constraints completely solve u^0 as a function of (x, ξ), and $\inf_{u \in \mathcal{U}_{\bar{x}, \bar{\bar{x}}}} J_{\bar{x}}(u) > -\infty$, then the generalized optimal solution must be unique. It follows that the sequence of transformations required by the Proposition 3 yields $R_r > 0$ if and only if the sequence of transformations implicit in the Dirac-Bergmann approach yields $R_r > 0$. Thus, the problems that have an infinite dimensional set of optimal solutions are exactly the problems that have finite infimum and have gauge freedom. The case when the optimal solution is not unique but the set of optimal solutions has finite dimension can only arise when T is the first conjugate point of the transformed problem (10).

Contrary to the generalized control approach, the Dirac-Bergmann approach does not give a general characterization of the problems which satisfy $\inf_{u \in \mathcal{U}_{\bar{x}, \bar{\bar{x}}}} J_{\bar{x}}(u) > -\infty$.

3.3 Generalized Trajectories in the Dirac-Bergmann Framework

In the rest of this section we assume that (A, B) is controllable and the generalized optimal solution exists and is unique for each $\overline{x}$, $\overline{\overline{x}} \in I\!\!R^n$. This is not essential but avoids the lengthy analysis of the degeneracies that arise when the generalized optimal solution is not unique and/or exists only for some pairs $(\overline{x}, \overline{\overline{x}})$.

Let (x, ξ) denote some trajectory of the Hamiltonian system (11) with $(x(0), \xi(0)) = (\overline{x}, \overline{\xi})$, and let $u = (u^+, u^0)$ denote the corresponding control. Using the same argument we used to extend the map $u \mapsto x_{u,\overline{x}}$, one may show that

$$(x, \xi) = (x^1, \xi^1) + [Id, H^0] \phi u^0,$$

where (x^1, ξ^1) denotes the trajectory of the system $(\dot{x}, \dot{\xi}) = [Id, H^+] - [Id, [H^0, H^+]] \phi u^0$, $(x(0), \xi(0)) = (\overline{x}, \overline{\xi})$. Using the Proposition 6, it follows that (x, ξ) satisfies the primary constraint (12) if and only if (x^1, ξ^1) does. By differentiating $H^0 (x^1, \xi^1)$, it follows that (x^1, ξ^1) must satisfy

$$[H^0, H^+] (x^1, \xi^1) + [[H^0, H^+], H^0] (x^1, \xi^1) \phi u^0 = 0,$$

that is,

$$\phi u_1^+ = \frac{1}{2} (R_1^+)^{-1} G_1 (x^1, \xi^1); \tag{17}$$

$$H_1^0 (x^1, \xi^1) = 0. \tag{18}$$

Using (17), one obtains

$$(\dot{x}^1, \dot{\xi}^1) = [Id, H_1^+] (x^1, \xi^1) + [Id, H_1^0] \phi u_1^0.$$

Repeating the same procedure, one obtains

$$(x, \xi) = (x^r, \xi^r) + \sum_{i=0}^{r-1} [Id, H_i^0] \phi^{i+1} u_i^0, \tag{19}$$

where (x^r, ξ^r) satisfies $(\dot{x}^r, \dot{\xi}^r) = [Id, H_r^+] (x^r, \xi^r)$, $(x^r(0), \xi^r(0)) = (\overline{x}, \overline{\xi})$. The Dirac's set of constraints reduces to

$$H_{i-1}^0 (x^r, \xi^r) = 0, \qquad 1 \leq i \leq r. \tag{20}$$

$$\phi^i u_i^+ = \frac{1}{2} (R_i^+)^{-1} \left(G_i (x^r, \xi^r) + \sum_{j=i}^{r-1} [G_i, H_j^0] \phi^{j+1} u_j^0 \right), \quad 1 \leq i \leq r. \tag{21}$$

The Hamiltonian system (11) is linear. Hence, the argument we used to obtain the extension of the map $u \mapsto x_{u,\overline{x}}$ can also be used to obtain the unique extension onto $H_{\widehat{\varphi}_r}$ of any input-to-phase-trajectory map, $u \mapsto \left(x_{u,\overline{x},\overline{\xi}}, \xi_{u,\overline{x},\xi} \right)$. For any $u \in H_{\widehat{\varphi}_r}$, all the terms in the equations $(19, 20, 21)$ are well defined. Thus we define a *generalized extremal* to be an image of some generalized control, $\widehat{u} \in H_{\widehat{\varphi}_r}$, by some map $u \mapsto \left(x_{u,\overline{x},\overline{\xi}}, \xi_{u,\overline{x},\overline{\xi}} \right)$, such that $\left(\widehat{u}, x_{\widehat{u},\overline{x},\overline{\xi}}, \xi_{\widehat{u},\overline{x},\overline{\xi}} \right)$ satisfies the equations $(19, 20, 21)$.

Let $\mathcal{H} = \{(a, \alpha) \in \mathbb{R}^{2n} : H_{i-1}^0 (a, \alpha) = 0, \ i = 1, 2, ..., r\}$, $\mathcal{D} = \{(a, \alpha) \in \mathbb{R}^{2n} : H_{i-1}^0 (a, \alpha) = 0, \ G_i (a, \alpha) = 0, \ i = 1, 2, ..., r\}$, $\mathcal{J} = span \left(\left[Id, H^0 \right], \left[Id, H_1^0 \right], ..., \left[Id, H_{r-1}^0 \right] \right)$. It is possible to prove that $\mathcal{J} \subset \mathcal{H}$, the projection of $\mathcal{H}$ into the state space is the whole state space and $\mathcal{D}$ is the set of all $(a, \alpha) \in \mathbb{R}^{2n}$ that satisfy all the Dirac's constraints $(14, 16)$.

A generalized extremal has the following structure: it may start at any point $(x(0), \xi(0)) \in \mathcal{H}$, but "jumps" at time $t = 0$ to a point $(x(0^+), \xi(0^+)) \in ((x(0), \xi(0)) + \mathcal{J}) \cap \mathcal{D}$. In the interval $]0, T[$ the generalized extremal coincides with the unique absolutely continuous trajectory of the Hamiltonian equation (11), starting at $(x(0^+), \xi(0^+))$, and lying in $\mathcal{D}$. This trajectory reaches a point $(x(T^-), \xi(T^-))$, from which it is possible to "jump" to any point $\left(\overline{\overline{x}}, \overline{\overline{\xi}} \right) \in (x(T^-), \xi(T^-)) + \mathcal{J}$.

Not surprisingly, we have the following correspondence between generalized extremals and generalized optimal trajectories.

Proposition 7. *For each $\overline{x}, \overline{\overline{x}} \in \mathbb{R}^n$ there exists one unique generalized extremal that satisfies $x(0) = \overline{x}$, $x(T) = \overline{\overline{x}}$. The projection into the state space of this generalized extremal is the generalized optimal trajectory for the pair $\left(\overline{x}, \overline{\overline{x}} \right)$.* ◊

For each $\overline{x} \in \mathbb{R}^n$, let $\mathcal{H}_{\overline{x}} = \{(x, \xi) \in \mathcal{H} : x = \overline{x}\}$. For each $\left(\overline{x}, \overline{\xi} \right) \in \mathcal{H}$, let $\left(x_{\overline{x},\overline{\xi}}, \xi_{\overline{x},\overline{\xi}} \right)$ denote the generalized extremal such that $(x(0), \xi(0)) = \left(\overline{x}, \overline{\xi} \right)$ (this is uniquely defined in the interval $[0, T[$). The following Proposition gives a geometric interpretation to the Proposition 4.

Proposition 8. *For each fixed $\overline{x} \in \mathbb{R}^n$, the map $\overline{\xi} \mapsto \left(x_{\overline{x},\overline{\xi}}(0^+), \xi_{\overline{x},\overline{\xi}}(0^+) \right)$ is one-to-one from $\mathcal{H}_{\overline{x}}$ onto $(\mathcal{H}_{\overline{x}} + \mathcal{J}) \cap \mathcal{D}$. The map $(\overline{a}, \overline{\alpha}) \mapsto x_{\overline{a},\overline{\alpha}}(T^-)$ is one-to-one from $(\mathcal{H}_{\overline{x}} + \mathcal{J}) \cap \mathcal{D}$ onto the quotient space $\mathbb{R}^n / span \left(B^0, B_1^0, ..., B_{r-1}^0 \right)$.* ◊

The equations $(19, 20, 21)$ can be used to compute approximations of the generalized optimal solution by (suboptimal) ordinary solutions in a way similar to [5].

References

1. Agrachev, A.A.; Sarychev, A.V.(1996) Abnormal Sub-Riemanian Geodesics: Morse Index and Rigidity. Ann. Inst. Poincaré - Analyse non linéaire, Vol.13 N°6, 635–690.

2. Dirac, P.A.M.: Generalized Hamiltonian Dynamics (1950) Can. J. Math. 2, pp. 129–148.

3. Dirac, P.A.M. (1964) Lectures on Quantum Mechanics. Belfer Graduate School of Science.

4. Gotay, M.J.; Nester, J.M. Hinds, G.(1978) Presymplectic manifolds and the Dirac-Bergmann theory of constraints. J. Math Phys. 19(11), 2388–2399.

5. Guerra, M (2000) Highly Singular L-Q Problems: Solutions in Distribution Spaces. J. Dynamical Control Systems, Vol.6, N°2, 265–309.

6. Jurdjevic, V. (1997) Geometric Control Theory. Cambridge University Press.

7. Volckaert, K; Aeyels, D. (1999) The Gotay-Nester algorithm in Singular Optimal Control. Proc 38th Conf. on Decision & Control, Phoenix, Arizona USA, 873–874.

Robust Tracking of Multi-variable Linear Systems under Parametric Uncertainty *

Veit Hagenmeyer

Laboratoire des Signaux et Systèmes
CNRS-Supélec
Plateau de Moulon
3, rue Joliot-Curie
91192 Gif-sur-Yvette Cedex, France
hagenmey@lss.supelec.fr

Abstract. In this article the robustness of tracking controllers acting on multi-variable linear systems under parametric uncertainty is investigated. After a system transformation we design PID-like tracking controllers for the resulting subsystems taking nominal parameters into account. Robustness is studied under parametric uncertainty: it splits into the robust stability of an autonomous linear system and into the robust stability of the same autonomous system being perturbed by so-called "quasi-exogenous" signals. The article is concluded by a DC drive example.

1 Introduction

Tracking control of controllable linear systems has been studied profoundly in the last thirty years (refer to [6], [10], [11] and the references therein). These studies were accompanied by the question of robustness of the developed controls: a series of papers by Schmitendorf and coworkers (see [19], [20], [21] and [12]) provide a good overview of the attained results. In these papers constructive tools for choosing the right controls with respect to given bounded uncertainties are given.

The work presented in this article differs from the aforementioned results with respect to two points: first, we follow a different philosophy in the sense, that we design a relatively simple tracking control for the nominal system and provide thereafter an analysis tool for determining its robustness with respect to the a priori given bounded parametric uncertainty. Second, the parametric uncertainty can enter the linear state equation nonlinearly in the respective system and input matrices.

We extend in this article the results already obtained for the SISO case [8] to the MIMO case: by transforming the system into its controller form we develop control laws for the nominal value of the uncertain parameters. These

---
* This work was financially supported by the German Academic Exchange Service (DAAD)

control laws decouple the respective subsystems, inject a desired trajectory behaviour and stabilize the system around the desired trajectory by using PID-like control techniques. We investigate tracking stability in analyzing robustness of the tracking error equation under the uncertainty of the parameters. In contrast to the SISO case, the decoupled nominal subsystems of the MIMO linear systems recouple when the uncertainty intervals are taken into account.

Nevertheless, the robustness problem splits into two parts: first as the stability of a linear autonomous system under endogenous (parametric) uncertainty, second the stability of the same system being perturbed by so called "quasi-exogenous" signals at the input. For both parts known stability criteria are stated which lead to the main result of the article.

The paper is organized as follows: after formulating the robust tracking control problem in section 2, we establish the tracking control laws for the nominal system in section 3. In the following section 4 we investigate the according tracking error equation and analyze thereafter its stability in section 5. The main result is stated in section 6, followed by important remarks on controller design and tracking performance in section 7. After treating a DC drive example in section 8, we conclude the paper with a discussion in section 9.

2   Problem Formulation

Given the MIMO linear system

$$\dot{x}(t) = A(p)x(t) + B(p)u(t), \quad x(0) = x_0 \tag{1}$$

where the time $t \in \mathbb{R}_0^+$, the state $x(t) \in \mathbb{R}^n$, the input $u(t) \in \mathbb{R}^r$ and the system parameters $p \in \mathbb{R}^q$, which are constant in time, but not known by their exact values:

$$p = p_o + \tilde{p}, \quad \tilde{p}_i \in [\underline{p}_i, \overline{p}_i], \quad i = 1, \ldots, q \tag{2}$$

where p_o is the nominal value of the parameters. The system matrix is $A(p) : \mathbb{R}^q \mapsto \mathbb{R}^{n \times n}$ and the input matrix $B(p) : \mathbb{R}^q \mapsto \mathbb{R}^{n \times r}$. The state vector is assumed to be known via full state measurements. The fundamental hypothesis we impose on the system is that of system controllability for all possible p in (2). In addition, it is assumed, without loss of generality, that the r columns of $B(p)$ are linearly independent. We establish the following definition:

Definition 1 *The tracking of the desired trajectory* $t \mapsto x^d(t)$ *is robust in the presence of parametric uncertainty with respect to the interval parameters in (2), if for all parameters p in (2) the system (1) under the tracking control law the state remains bounded.*

The control task we study in this paper is to track a given sufficiently smooth desired trajectory $t \mapsto x^d(t)$ and to guarantee robustness for this tracking.

Remark 1 *If a linear system (1) is subject to a tracking control law, its tracking error dynamics represent a non-autonomous system under parametric uncertainty: the calculations in this paper show, that the tracking error system is perturbed by a so-called "quasi-exogenous" perturbation at the input; the quasi-exogenous perturbation term follows necessarily from the desired trajectory injection under parametric uncertainty.*

3 Tracking Control

3.1 System Transformation

The theorem noted below holds for interval parameter systems (see [16], [22] and [18] for nominal linear control systems):

Theorem 1 *Suppose the system (1) is controllable, with fixed controllability indices* ρ_i, $i = 1, 2, \ldots, r$, *for all p in (2). Then there is a nonsingular transformation*

$$z(t) = S(p)x(t) \tag{3}$$

of the state vector $x(t)$ *and a nonsingular transformation of the input vector* $u(t)$ *which reduce the system to a coupled set of r single-input subsystems. Each subsystem is in the single-input controller form. Furthermore, all additional coupling enters each subsystem at its input.*

Thus, the system (1) can be transformed into

$$\dot{z}(t) = \tilde{A}(p)z(t) + \tilde{B}(p)u(t) \tag{4}$$

The transformed system matrix $\tilde{A}(p)$ reads as

$$\tilde{A}(p) = \begin{bmatrix} A_{1,1}(p) & A_{1,2}(p) & \cdots & A_{1,r}(p) \\ A_{2,1}(p) & A_{2,2}(p) & \cdots & A_{2,r}(p) \\ \vdots & \vdots & \ddots & \vdots \\ A_{r,1}(p) & A_{r,2}(p) & \cdots & A_{r,r}(p) \end{bmatrix} \tag{5}$$

where each

$$A_{i,i}(p) = \begin{bmatrix} 0 & 1 & 0 & \cdots & 0 \\ 0 & 0 & 1 & \cdots & 0 \\ \vdots & & & \ddots & \vdots \\ 0 & 0 & 0 & \cdots & 1 \\ \alpha_{i,i,1}(p) & \alpha_{i,i,2}(p) & \alpha_{i,i,3}(p) & \cdots & \alpha_{i,i,\rho_i}(p) \end{bmatrix}, \quad i = 1, \cdots, r \tag{6}$$

is an $\mathbb{R}^{\rho_i \times \rho_i}$ matrix and each

$$A_{i,j}(p) = \begin{bmatrix} 0 & 0 & \cdots & 0 \\ \vdots & & \ddots & \vdots \\ 0 & 0 & \cdots & 0 \\ \alpha_{i,j,1}(p) & \alpha_{i,j,2}(p) & \cdots & \alpha_{i,j,\rho_j}(p) \end{bmatrix}, \quad i,j = 1, \cdots, r, \ j \neq i \tag{7}$$

is an $\mathbb{R}^{\rho_i \times \rho_j}$ matrix. Furthermore it follows from [16] and [18] that $\tilde{B}(p)$ can be represented as

$$\tilde{B}(p) = \hat{B}\Gamma(p) \tag{8}$$

Thereby is $\Gamma(p) \in \mathbb{R}^{r \times r}$ an upper-triangular matrix with elements denoted by $\gamma_{i,j}$, $i,j = 1, \ldots, r$. The matrix $\hat{B}$ in (8) takes the following form (defining $\beta_i = [0 \ 0 \cdots 0 \ 1]^T \in \mathbb{R}^{\rho_i}$, $i = 1, \ldots, r$)

$$\hat{B} = \text{diag}(\beta_1, \ldots, \beta_r) \tag{9}$$

Hence, each subsystem can be represented as

$$\dot{z}_{i,j} = z_{i,j+1}, \quad j \in \{1, 2, \ldots, \rho_i - 1\}$$
$$\dot{z}_{i,\rho_i} = \sum_{k=1}^{\rho_i} \alpha_{i,i,k}(p) z_{i,k} + \sum_{j \neq i} \sum_{k=1}^{\rho_j} \alpha_{i,j,k}(p) z_{j,k} + \sum_{k=i}^{r} \gamma_{i,k}(p) u_k \tag{10}$$

for $i = 1, \ldots, r$. The $z_{i,1}$, $i = 1, \ldots, r$ are often the to-be-controlled variables in real applications (see [3],[4]). For these the given controller form is natural and easy to implement (see for instance the DC drive example in section 8).

3.2 Control Law Design

The tracking feedback control law to be developed is based on the nominal parameters p_o given in (2). First we execute the nominal system transformation

$$\bar{z}(t) = S(p_o)x(t) \tag{11}$$

as in (3), to find the representation of the nominal system $\dot{\bar{z}}(t) = \tilde{A}(p_o)\bar{z}(t) + \tilde{B}(p_o)u(t)$ as in (4) with according nominal matrices $\tilde{A}(p_o)$ in (5) and $\tilde{B}(p_o) = \hat{B}\Gamma(p_o)$ in (8). Since $\Gamma(p_o)$ is upper-triangular and invertible for all p under the assumption of controllability, a new but equivalent set of system inputs can be defined as $v = \Gamma(p_o)u$. The nominal subsystems read therefore (compare with (10))

$$\dot{\bar{z}}_{i,j} = \bar{z}_{i,j+1}, \ \ j \in \{1, 2, \ldots, \rho_i - 1\}$$
$$\dot{\bar{z}}_{i,\rho_i} = \sum_{k=1}^{\rho_i} \alpha_{i,i,k}(p_o)\bar{z}_{i,k} + \sum_{j \neq i}\sum_{k=1}^{\rho_j} \alpha_{i,j,k}(p_o)\bar{z}_{j,k} + v_i \tag{12}$$

for $i = 1, \ldots, r$. The tracking feedback control law is constructed via the *Brunovský form* [22] for this nominal parameter set, that is, we decouple the nominal subsystems by canceling the coupling terms at the input and second inject the desired nominal behaviour at the inputs of the decoupled chains of integrators. The desired trajectory $t \mapsto \bar{z}^d$ and the desired $t \mapsto \dot{\bar{z}}_{i,\rho_i}(t)$ to be injected are calculated from the desired behaviour $x^d(t)$ using the nominal system transformation (11) to get the expressions for the outputs of the respective chain of integrators $\bar{z}_{i,1}^d(t)$, $i = 1, \ldots, r$. Thereafter these are derived ρ_i- times with respect to time. We stabilize the system around the desired trajectory using a PID-like controller in placing the poles. Defining the nominal tracking error by

$$\bar{e}(t) = \bar{z}^d(t) - \bar{z}(t) \tag{13}$$

the control law with respect to the new input v can then be represented as

$$v_i = -(\sum_{k=1}^{\rho_i} \alpha_{i,i,k}(p_o)\bar{z}_{i,k} + \underbrace{\sum_{j \neq i} \sum_{k=1}^{\rho_j} \alpha_{i,j,k}(p_o)\bar{z}_{j,k}}_{\text{(i)}}) + \underbrace{\overset{.}{\bar{z}}{}^d_{i,\rho_i}}_{\text{(ii)}}$$

$$+ \underbrace{\lambda_{i,0} \int_0^t \bar{e}_{i,1} d\tau + \sum_{k=1}^{\rho_i} \lambda_{i,k} \bar{e}_{i,k}}_{\text{(iii)}}, \quad i = 1, \ldots, r \qquad (14)$$

(i) : cancellation of nominal terms

(ii) : desired behaviour injection

(iii) : stabilization around desired trajectory

The coefficients $\lambda_{i,j}$, $i = 1, \ldots, r$, $j = 0, \ldots, \rho_i$ are chosen, such that the corresponding characteristic polynomials $s^{\rho_i+1} + \lambda_{i,\rho_i} s^{\rho_i} + \cdots + \lambda_{i,0}$, $i = 1, \ldots, r$ are Hurwitz. The problem to be solved in this article can now be stated as: does the control law (14), which was developed for the nominal case $\alpha_{i,j,k}(p_o)$ and $\gamma_{i,k}(p_o, p_o)$ in (10), stabilize the linear feedback system containing the unknown but bounded parameters $\tilde{p}$ around the desired trajectory $\bar{z}^d$?

4   The Tracking Error Equation

In this section, we are going to establish the tracking error equation resulting from the problem posed beforehand. The necessary calculations are straightforward, but messy and very tedious. Therefore we go on in describing the algorithm of calculation and present the structure of the result thereafter, but we refrain from giving the exact formulæ for every single coefficient[1] within the resulting equation for the general case (see section 8 for a specific example).

The algorithm to find the tracking error equation is as follows:

1. Plug (14) and $u = \Gamma(p_o)^{-1} v$ in (4) to get a linear differential equation system in z which contains expressions of $\bar{z}^d$, $\bar{z}$ (remember $\bar{e} = \bar{z}^d(t) - \bar{z}$) and r-integrals. We denote this linear integro-differential equation system in z as $\mathcal{Z}$ for the ongoing.

2. Use (3) and (11) to find

$$\bar{z} = \Sigma(p, p_o)z \qquad (15)$$

with $\Sigma(p, p_o) = S(p_o)S(p)^{-1}$.

---

[1] We propose them to be calculated by symbolic computation for higher order applications.

3. Plug (15) in $\mathcal{Z}$ to get $\bar{\mathcal{Z}}$, which contains only $\bar{z}^d$ as exogenous variables.
4. Derive $\bar{\mathcal{Z}}$ one time with respect to time to get $\hat{\mathcal{Z}}$, an augmented system of dimension $n + r$.
5. Define the real tracking error as

$$e(t) = \bar{z}^d(t) - z(t) \tag{16}$$

6. Use $\hat{\mathcal{Z}}$ and (16) to find the searched for tracking error system. It is of the form

$$\dot{e}(t) = \Pi(p)e(t) + \hat{B}\Psi(\dot{\bar{z}}^d(t), p) \tag{17}$$

The tracking error system matrix $\Pi(p) \in \mathbb{R}^{(n+r)\times(n+r)}$ is of the structure of $\tilde{A}(p)$ (5), $\hat{B}$ is as defined as in (9) with $\beta_i = [0\ 0 \cdots 0\ 1]^T \in \mathbb{R}^{\rho_i+1}$, $i = 1, \ldots, r$, and $\Psi(\dot{\bar{z}}^d(t), p) : \mathbb{R} \times \mathbb{R}^q \mapsto \mathbb{R}^{r \times r}$, where $\Psi(\dot{\bar{z}}^d(t), p)$ is diagonal and every diagonal element is a linear function of $\dot{\bar{z}}^d(t)$.

We remark that the respective subsystems of the tracking error equation are not only recoupled under parametric uncertainty, moreover the controllers for each subsystem are intertwined in their effects on the different subsystems. Hence, the robustness problem has to be divided into two questions:

1. Is the autonomous tracking error system $(\Psi(\dot{\bar{z}}^d(t), p) = 0$ in (17)) perturbed by endogenous parametric uncertainties stable?
2. Does this linear system stay stable under the quasi-exogenous perturbations $\Psi(\dot{\bar{z}}^d(t), p)$?

Remark 2 *We call this perturbation as being "quasi-exogenous" to make the difference with respect to real exogenous perturbations which will be included in this theory in a forthcoming publication.*

5   Stability Analysis

5.1   Stability of the Autonomous Part of the Tracking Error System

In this section we study the stability properties of the internal part of the tracking error system. For this purpose, we set $\Psi(\dot{\bar{z}}^d(t), p) = 0$ in (17) to get

$$\dot{e}(t) = \Pi(p)e(t) \tag{18}$$

This resulting equation represents a linear time-invariant system of dimension $\mathbb{R}^{n+r}$ containing interval parameters. Their stability property has been intensively studied in the literature (see [1] and [2] for instance) and depends on the magnitude of both the additive and multiplicative perturbation of the desired coefficients $\lambda_{i,j}$. To determine the exponential stability of the resulting characteristic polynomial of the $\mathbb{R}^{n+r}$-system with coefficients containing the interval expressions $\tilde{p}$, we use one of the following methods:

Kharitonov's Theorem If the coefficients of the characteristic polynomial are algebraically independent with respect to the interval parameters (any two coefficients are never functions of the same interval parameter), the stability can be deduced analytically in a necessary and sufficient way by the famous theorem by Kharitonov [13]. If the coefficients are algebraically dependent with respect to the interval parameters, Kharitonov's theorem gives a sufficient result when calculating each coefficient explicitly via interval analysis [17] and treating the interval coefficients thereafter as being algebraically independent (see for instance the DC drive example in section 8). The theorem can be found in a nice version in [1].

The Theorem of Frazer and Duncan In the case, in which the coefficients of the characteristic polynomial are algebraically dependent with respect to the interval parameters, we are able to determine the stability algebraically by the theorem of Frazer and Duncan [7]. We use the version presented in [1]: starting from the knowledge of one stable nominal linear time-invariant system, the theorem consists basically in determining the regularity of the $n + r$-th Hurwitz-matrix for all p in (2). Sometimes the determinant of this matrix may be difficult to be calculated or the coefficients may depend in a non-continuous way of the uncertain parameters, then we propose to apply the following algorithm of Walter and Jaulin [23].

The Algorithm of Walter and Jaulin A numerical algorithm was developed by Walter and Jaulin [23], which is based on interval analysis and projects the coefficients being dependent of interval parameters in the pole domain via set inversion. It determines in a necessary and sufficient way the stability of the interval characteristic polynomial resulting from the calculations above.

5.2 Stability of the Tracking Error System being Subject to the Quasi-exogenous Perturbation

After having assured the stability of the endogenously perturbed system in the previous subsection 5.1, we are going to study the stability of the linear system (17) including the quasi-exogenous signals $\Psi(\dot{\bar{z}}^d(t), p)$ (see Remark 2 for the nomenclature). We see that the quasi-exogenous perturbations $\Psi(\dot{\bar{z}}^d(t), p)$ do not depend on $\bar{z}^d(t)$, but on $\dot{\bar{z}}^d(t)$. We remark, that this quasi-exogenous perturbation is bounded as a sum of bounded terms, since we do not consider in this work neither unbounded desired trajectories nor unbounded desired trajectory derivatives. We remember the following theorem:

Theorem 2 ([18]) *Suppose a time-invariant linear state equation is controllable. Then the state equation is uniformly bounded-input bounded-output stable if and only if it is exponentially stable.*

Controllability follows from the basic assumptions, exponential stability was already established in the previous subsection 5.1. Thus, the linear tracking error system being stable with respect to the endogenous part of the perturbation is always stable in the bounded-input bounded-output sense with respect to the quasi-exogenous part of the perturbation.

6   Main Result

In this section, we state the main theorem, which solves the problem formulated above and follows from the content of the preceding sections:

Theorem 3 *Given the linear system (1) under the assumption of controllability for all parameters p in (2). The tracking imposed after system transformation by the presented PID-like control technique (14) is robust with respect to the given uncertainty intervals of p in (2), if and only if the autonomous linear tracking error system (18) is exponentially stable.*

Remark 3 *The exponential stability of (18) can for instance be analyzed by applying one of the different methods presented in subsection 5.1.*

7   Remarks on Controller Design and Tracking Performance

Remark 4 *The choice of the nominal values p_o in (2) for the given uncertainty intervals influences the magnitude of both the additive and multiplicative perturbation of the desired coefficients $\lambda_{i,j}$. This effect can be taken into account whilst the controller design phase: eventually different nominal values for the uncertainty intervals can be chosen to assure stability without changing the desired coefficients $\lambda_{i,j}$.*

Remark 5 *If we design the desired trajectories such that $\forall t \geq t^\star : \dot{\bar{z}}^d = 0$, we are able to stabilize the system around these desired trajectories from $t^\star$ onwards without persistent tracking error: the equation (17) reduces in this case to the linear autonomous system (18) in subsection 5.1 as $\forall t \geq t^\star$: $\Psi(\dot{\bar{z}}^d(t), p) = 0$.*

Remark 6 *Scaling the desired trajectories with respect to time, that is decelerating or accelerating them, results in higher or lesser magnitudes of $\dot{\bar{z}}^d$. Hence, in slowing down the desired trajectories, we are able to reduce the magnitude of the quasi-exogenous perturbation $\Psi_i(\dot{\bar{z}}^d(t), p)$ in (17) and therefore the resulting tracking error without changing the nominal poles of the closed loop system. If rapid desired trajectories are to be designed, the quasi-exogenous perturbation stemming from parametric uncertainty will be of higher magnitude, therefore we have to use higher controller gains to keep the tracking error in a reasonable range.*

Thus, in real applications in which saturating control inputs and measurement noise have to be considered, we are able to trade off the magnitude of the controller gains and the velocity of the desired trajectories to fulfill the desired control performance and the given physical constraints at the same time (see [5] for a profound discussion of this subject).

8   Example: a Separately Excited DC Drive

In this section, we apply the theory presented in the previous sections to a separately excited DC drive model. We first linearize its nonlinear model (see for instance [15]) around the nominal values of the drive and neglect thereafter the dynamics of the electrical part (it being much faster than the mechanical part [14]). The linearized and singularly perturbed form of the DC drive reads as follows

$$
\dot{x} = \begin{bmatrix} -\dfrac{c^2 \Phi_{s_o}^2 + B R_r}{J R_r} & \dfrac{c I_{r_o} R_r - c^2 \Phi_{s_o} \omega_o}{J R_r} \\ 0 & -\dfrac{R_s}{L_s} \end{bmatrix} x + \begin{bmatrix} \dfrac{c \Phi_{s_o}}{J R_r} & 0 \\ 0 & 1 \end{bmatrix} u \tag{19}
$$

Thereby $x_1 = \omega - \omega_o$, $x_2 = \Phi_s - \Phi_{s_o}$, $u_1 = U_r - U_{r_o}$ and $u_2 = U_s - U_{s_o}$ and ω, I_r Φ_s represent the angular velocity, the rotor current and the stator flux respectively. The rotor tension U_r and the stator tension U_s are the two controls of the system. The values of the nominal velocity ω_o, of the nominal rotor current I_{r_o}, of the nominal flux Φ_{s_o} at ω_0 (which can be determined by minimization of energy losses [9]) and the values of the nominal inputs U_{r_o} and U_{s_o} are represented in Table 1.

Nominal states/inputs	Value	Unit
ω_o	104.7198	rad/s
I_{r_o}	0.4	A
Φ_{s_o}	12	Wb
U_{r_o}	26.6654	V
U_{s_o}	32	V

Table 1. Nominal values of the states and the inputs of the DC motor

The inertia of the drive is denoted by J, the viscous friction coefficient by B, the motor constant by c. The coefficients L_r and L_s represent the rotor and stator inductance respectively; the rotor resistance and the stator resistance are denoted as R_r and R_s.

Since the viscous friction coefficient B is not very well known, it can be represented by $B = B_o + \tilde{B}$. Due to termic effects, the resistances can double its value whilst in operation, hence we denote $R_r = R_{r_o} + \tilde{R}_r$, where R_{r_o} is the value identified in cold state. All parameter values can be found in Table 2.

Parameter	Value	Unit
L_r	9.61	mH
L_s	45	H
R_{r_o}	0.69	Ω
$\tilde{R}_r$	[0, 0.69]	Ω
R_s	120	Ω
J	11.4×10^{-3}	kg m^2
c	0.021	Wb/rad
B_o	0.0084	Nm/rad
$\tilde{B}$	[-0.0024, 0.0016]	Nm/rad

Table 2. Parameters of the DC motor

The derived model is valid for desired trajectories in the neighborhood of the nominal set point. We choose the transformation matrix $S(p)$ in (3) as

$$S(B, R) = \begin{bmatrix} 1 & 0 \\ 0 & 1 \end{bmatrix} \tag{20}$$

which is in this case independent of any parametric uncertainty. Therefore we have $\bar{z} = z$ in view of (3), (11) and (15), which simplifies the ongoing. The transformed system can thus be represented both for the controller design and the robustness analysis in z-coordinates by

$$\dot{z} = \begin{bmatrix} -\dfrac{c^2\Phi_{s_o}^2 + BR_r}{JR_r} & \dfrac{cI_{r_o}R_r - c^2\Phi_{s_o}\omega_o}{JR_r} \\ 0 & -\dfrac{R_s}{L_s} \end{bmatrix} z + \begin{bmatrix} \dfrac{c\Phi_{s_o}}{JR_r} & 0 \\ 0 & 1 \end{bmatrix} u \tag{21}$$

For the controller design we use the nominal model (see for the notation (10))

$$\dot{z}_{1,1} = -\frac{c^2\Phi_{s_o}^2 + B_oR_{r_o}}{JR_{r_o}} z_{1,1} + \frac{cI_{r_o}R_{r_o} - c^2\Phi_{s_o}\omega_o}{JR_{r_o}} z_{2,1} + \frac{c\Phi_{s_o}}{JR_{r_o}} u_1$$

$$\dot{z}_{2,1} = -\frac{R_{s_o}}{L_{s_o}} z_{2,1} + u_2 \tag{22}$$

The new input v can be defined by

$$\frac{c\Phi_{s_o}}{JR_{r_o}} u_1 := v_1 \tag{23a}$$

$$u_2 := v_2 \tag{23b}$$

which yields for u

$$u_1 = \frac{JR_{r_o}}{c\Phi_{s_o}} v_1 \tag{24a}$$

$$u_2 = v_2 \tag{24b}$$

We design the tracking feedback control as in (14), that is

$$v_1 = \frac{c^2\Phi_{s_o}^2 + B_oR_{r_o}}{JR_{r_o}} z_{1,1} - \frac{cI_{r_o}R_{r_o} - c^2\Phi_{s_o}\omega_o}{JR_{r_o}} z_{2,1} + \dot{z}_{1,1}^d +$$

$$\lambda_{1,0} \int_0^t e_{1,1} d\tau + \lambda_{1,1} e_{1,1} \tag{25a}$$

$$v_2 = \frac{R_s}{L_s} z_{2,1} + \dot{z}_{2,1}^d + \lambda_{2,0} \int_0^t e_{2,1} d\tau + \lambda_{2,1} e_{2,1} \tag{25b}$$

where the $\lambda_{i,j}$ are chosen such that the corresponding characteristic polynomials

$$s^2 + \lambda_{1,1}s + \lambda_{1,0} \tag{26a}$$

$$s^2 + \lambda_{2,1}s + \lambda_{2,0} \tag{26b}$$

are Hurwitz. See Table 3 for our choice, which places the poles for the two uncoupled nominal closed loop subsystems at a double multiplicity of $-10\frac{1}{s}$ and a double multiplicity of $-5\frac{1}{s}$ respectively (the eigenvalues of the nominal open loop system are at $-8.81\frac{1}{s}$ and $-2.67\frac{1}{s}$).

Coefficient	Value	Unit
$\lambda_{1,0}$	- 100	$\frac{1}{s^2}$
$\lambda_{1,1}$	- 20	$\frac{1}{s}$
$\lambda_{2,0}$	- 25	$\frac{1}{s^2}$
$\lambda_{2,1}$	- 10	$\frac{1}{s}$

Table 3. Desired coefficients

Using (25) and (22)(remember again $z = z^d - e$ (16)) , we get for the tracking error system

$$
\dot{e}_{1,1} = \frac{R_r\tilde{B} + \tilde{R}_r B_o + \tilde{R}_r\tilde{B}}{J(R_{r_o} + \tilde{R}_r)}(z_{1,1}^d - e_{1,1}) - \frac{cI_{r_o}\tilde{R}_r}{J(R_{r_o} + \tilde{R}_r)}(z_{2,1}^d - e_{2,1})
$$

$$
-(1 - \frac{R_{r_o}}{(R_{r_o} + \tilde{R}_r)})\dot{z}_{1,1}^d - \lambda_{1,0}\int_0^t e_{1,1}d\tau - \lambda_{1,1}e_{1,1} \qquad (27)
$$

$$
\dot{e}_{2,1} = -\lambda_{2,0}\int_0^t e_{2,1}d\tau - \lambda_{2,1}e_{2,1} \qquad (28)
$$

To study the stability of this system, we derive one time with respect to time to get

$$
\dot{e}_{1,1} = e_{1,2}
$$

$$
\dot{e}_{1,2} = \frac{R_r\tilde{B} + \tilde{R}_r B_o + \tilde{R}_r\tilde{B}}{J(R_{r_o} + \tilde{R}_r)}(\dot{z}_{1,1}^d - e_{1,2}) - \frac{cI_{r_o}\tilde{R}_r}{J(R_{r_o} + \tilde{R}_r)}(\dot{z}_{2,1}^d - e_{2,2})
$$

$$
-(1 - \frac{R_{r_o}}{(R_{r_o} + \tilde{R}_r)})\ddot{z}_{1,1}^d - \lambda_{1,0}e_{1,1} - \lambda_{1,1}e_{1,2} \qquad (29a)
$$

$$
\dot{e}_{2,1} = e_{2,2}
$$

$$
\dot{e}_{2,2} = -\lambda_{2,0}e_{2,1} - \lambda_{2,1}e_{2,2} \qquad (29b)
$$

We go on in investigating the stability of the endogenously perturbed system, that is we set $\dot{z}_{i,1}^d = 0$, $i = 1, 2$ and $\ddot{z}_{1,1}^d = 0$ in (29) to get the autonomous error system

$$
\dot{e} =
\begin{bmatrix}
0 & 1 & 0 & 0 \\
-\lambda_{1,0} & -\dfrac{R_r\tilde{B}+\tilde{R}_r B_o+\tilde{R}_r\tilde{B}}{J(R_{r_o}+\tilde{R}_r)} & -\lambda_{1,1} & 0 & \dfrac{cI_{r_o}\tilde{R}_r}{J(R_{r_o}+\tilde{R}_r)} \\
0 & 0 & 0 & 1 \\
0 & 0 & -\lambda_{2,0} & -\lambda_{2,1}
\end{bmatrix} e
\tag{30}
$$

$$
=
\begin{bmatrix}
0 & 1 & 0 & 0 \\
\kappa_1 & \kappa_2 & 0 & \kappa_4 \\
0 & 0 & 0 & 1 \\
0 & 0 & \kappa_5 & \kappa_6
\end{bmatrix} e
\tag{31}
$$

The characteristical polynomial of this system, which contains the interval parameters $\tilde{B}$ and $\tilde{R}_r$, can be calculated as

$$
s^4 + (-\kappa_6 - \kappa_2)s^3 + (-\kappa_5 - \kappa_1 + \kappa_2\kappa_6)s^2 + (\kappa_2\kappa_5 + \kappa_1\kappa_6)s + \kappa_1\kappa_6
\tag{32}
$$

To answer the question whether this characteristical polynomial is Hurwitz, we apply the algorithm of Walter and Jaulin presented in section 5.1. The answer is positive, the characteristical polynomial is robustly Hurwitz. After having established the stability of (30) with respect to the endogenous perturbation, we assured equally the stability of the quasi-exogenously perturbed system (29) following the results of subsection 5.2.

9   Conclusions

In this article we investigate the robustness of tracking controllers working on controllable linear MIMO systems containing parametric uncertainty which is a priori known within given intervals. After having transformed the linear systems into their respective controller form we establish a nominal tracking control law via PID-like control techniques.

We analyze the stability of the tracking error equation and see, that the problem splits into the analysis of a linear autonomous system under endogenous parameter perturbation and the analysis of the same system being affected by so called quasi-exogenous bounded signals at the input.

For the robustness of the endogenously perturbed system we recall first Kharitonov's theorem, second a theorem by Frazer and Duncan and third a numeric algorithm established by Walter and Jaulin. The first theorem yields only a sufficient result in our general case, the second theorem is necessary and sufficient, but sometimes its analytic calculation may be very difficult, the algorithm finally gives necessary and sufficient results in a numerically guaranteed way.

After the robustness of the linear autonomous systems being affected by parametric uncertainty has been proven, the robustness of the quasi-exogenously perturbed system follows directly from BIBO - stability of linear systems. Thus the main theorem which solves the problem of robust tracking of linear systems under parametric uncertainty is stated. Important remarks are thereafter deduced from the obtained results. They point at the relation between robust stability and controller design and discuss the tracking performance under parametric uncertainty.

In the example of DC drive we apply our theoretical results and show the tracking robustness of its linearized and singularly perturbed model. In forthcoming publications we will study extensions of the work presented here to linear systems being perturbed by real exogenous perturbations and to linear systems being controlled by partial state feedback.

Acknowledgments

The author is very grateful to Dr. Emmanuel Delaleau for many intensive discussions and helpful suggestions. He is, moreover, thankful to Richard Marquez for important advises with respect to the presentation of the material, to Dr. Luc Jaulin for having him introduced to interval analysis and to Dr. Michel Kieffer for the software implementing the algorithm of Walter and Jaulin.

The author wants to thank Prof. H. Nijmeijer for comments helping to improve the readability of this article.

References

1. Ackermann, J.(1993) Robust Control:Systems with Uncertain Physical Parameters. Springer-Verlag, London
2. Barmish, B. R.(1994) New Tools for Robustness of Linear Systems. Macmillan, New York
3. Bitauld, L., Fliess, M. and Lévine, J. (1997) Flatness based control synthesis of linear systems: an application to windshield wipers. Proceedings of ECC, Brussels
4. Delaleau, E. (1997) Suivi de trajectoires pour les systèmes linéaires. Actes Coll. Cetsis - Eea, Orsay, 149–154
5. Fliess, M. and Marquez, R. (2000) Continuous-time linear predictive control and flatness: a module-theoretic setting with examples. Internat. J. Control 73, 606–623
6. Francis, B. A. (1977) The linear multivariable regulator problem. SIAM Contr. Opt. 15, 486–505
7. Frazer, R. and Duncan, W. (1929) On the criteria for the stability of small motions. In Proceedings of the Royal Society A 124, 642–654

8. Hagenmeyer, V. (2000) Robust tracking of linear systems under parametric uncertainty. Proceedings of MTNS 2000, Perpignan

9. Hagenmeyer, V., Kohlrausch, P. and Delaleau, E. (2000) Flatness based control of the separately excited DC drive. In "Nonlinear Control in the Year 2000" (this very book), eds. Isidori, A., Lamnabhi-Lagarrigue, F. and Respondek, W., Springer-Verlag, London

10. Hunt, L. R., Meyer, G. and Su, R.(1996) Noncausal inverses for linear systems. IEEE Trans. Automat. Contr. 41, 608–611

11. Hunt, L. R., Meyer, G. and Su, R.(1997) Driven dynamics of time-varying linear systems. IEEE Trans. Automat. Contr. 42, 1313–1317

12. Hopp, T. H., and Schmitendorf, W. E.(1990) Design of a linear controller for robust tracking and model following. Trans. ASME, J. Dynamic Syst., Measurement, and Contr. 112, 552–558

13. Kharitonov, V. L. (1978) Asymptotic stability of an equilibrium position of a family of systems of linear differential equations. Differentsial'nye Uravneniya 14, 2086–2088

14. Kokotović, P., Khalil, H. K. and O'Reilly, J. (1986) Singular Perturbation Methods in Control: Analysis and Design. Academic Press, London

15. Leonhard, W. (1996) Control of Electrical Drives. Springer, Berlin

16. Luenberger, D. G.(1967) Canonical Forms for Linear Multivariable Systems. IEEE Trans. Automat. Contr., 290–293

17. Moore, R. E.(1979) Methods and Applications of Interval Analysis. SIAM, Philadelphia

18. Rugh, W. J. (1996) Linear System Theory (2nd edition). Prentice-Hall, Upper Saddle River

19. Schmitendorf, W. E. and Barmish, B. R. (1986) Robust asymptotic tracking for linear systems with unknown parameters. Automatica 22, 335–360

20. Schmitendorf, W. E. and Barmish, B. R. (1987) Guaranteed output stability for systems with constant disturbances. Trans. ASME, J. Dynamic Syst., Measurement, and Contr. 109, 186–189

21. Schmitendorf, W. E. (1987) Methods for obtaining robust tracking control laws. Automatica 23, 675–677

22. Sontag, E. D. (1998) Mathematical Control Theory (2nd edition). Springer-Verlag, New York

23. Walter, E. and Jaulin, L. (1994) Guaranteed Characterization of Stability Domains Via Set Inversion. IEEE Trans. Automat. Contr. 39, 886–889

Flatness-based Control of the Separately Excited DC Drive*

Veit Hagenmeyer, Philipp Kohlrausch**, and Emmanuel Delaleau

Laboratoire des Signaux et Systèmes
C.N.R.S.–Supélec–Université Paris-sud
Plateau de Moulon, 3 rue Joliot-Curie
91 192 Gif-sur-Yvette cedex, France
{hagenmey, delaleau}@lss.supelec.fr

Abstract. Due to the flatness of the separately excited DC drive a novel control scheme that achieves copper loss minimization can be designed. It makes use of an on-line replanification of desired trajectories by using the information of a fast converging load torque observer. Simulation results show the performance of the proposed control scheme.

1 Introduction

The work presented here focuses on the separately excitated DC drive and its differential flatness property. As the separately excitated DC drive configuration has two inputs (the stator and rotor voltage respectively) available, it offers a second degree of freedom with respect to the serially and parallely excited ones. Nowadays power electronics being far less expensive than some years ago, we study the advantages of this second degree of freedom combined with a nonlinear, flatness based control strategy. Thus we are able to use the nonlinear model without any simplifications and to design suitable desired trajectories for the rotor speed and the stator flux for first accelerating, second maintaining a constant angular velocity and third braking the system under the aspect of energy loss minimization with respect to copper losses due to the Joule's effect.

Up to now energy loss minimization of the separately excited DC drive has only used linearized variants of the model and studied either the acceleration motion or the constant angular velocity motion. For the first time, we give a control scheme in which first the full nonlinearity of the model is respected and second in which the different motions of acceleration, constant angular speed and electrical braking are all treated by the same method. Furthermore,

---

* The work of V.H. was financially supported by the Nonlinear Control Network (NCN) and by the German Academic Exchange Service (DAAD). The work of P.K. was financially supported by the Nonlinear Control Network (NCN).

** Student at the Institut für Regelungs- und Steuerungstheorie Technische Universität Dresden, Germany.

we can start the drive from rest without any discontinuous jerk and without premagnetizing the machine. Finally we are able to reinject the kinematic energy won whilst braking the drive electrically in the power supply system.

The paper is organized as follows: After having briefly exposed the mathematical model of the motor in Sec. 2, we establish its differential flatness property in Sec. 3. We present the flatness based control of the DC drive which achieves the minimization of the copper losses in Sec. 4. We give some simulation results in Sec. 5 and conclude the article with a final discussion in Sec. 6.

2   Model of the Motor

The physical properties which underly the dynamic behavior of DC motors have been studied thoroughly and can be found in [11] for example. Fig. 1 depicts the equivalent circuit of this machine.

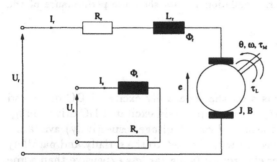

Fig. 1. Equivalent circuit of a DC machine

The following system of differential and algebraic equations reigns their dynamical domain:

$$\frac{d\theta}{dt} = \omega \tag{1a}$$

$$J\frac{d\omega}{dt} = \tau_m - B\omega - \tau_l \tag{1b}$$

$$U_s = R_s I_s + \frac{d\Phi_s}{dt} \tag{1c}$$

$$U_r = R_r I_r + L_r \frac{dI_r}{dt} + e \tag{1d}$$

$$\tau_m = c\Phi_s I_r \tag{1e}$$

$$\Phi_s = f(I_s) \tag{1f}$$

$$e = c\Phi_s \omega \tag{1g}$$

where θ and ω respectively denote the angular position and angular velocity of the shaft; τ_m is the electromagnetic torque produced by the motor and τ_l the load torque. The rotor and stator voltages are expressed by U_r and U_s, the corresponding resistances by R_r, R_s; the rotor inductance is denoted by L_r; Φ_s is the stator flux, I_r and I_s are the currents of the rotor circuit and the stator circuit respectively; e represents the back-electromagnetic force induced in the armature winding. The parameter J expresses the moment of inertia of the rotor, B the viscous friction coefficient, c represents a constant dependent of the spatial architecture of the drive.

The relation between the flux Φ_s and the stator current I_s is nonlinear as there is a saturation effect: $\Phi_s = f(I_s)$. The function $f(I_s)$ is shown in Fig. 2. Remark that this function is strictly increasing, thus bijective with an inverse f^{-1} itself strictly monotonic increasing. For small stator currents, the flux is itself small and it is possible to use the common approximation $\Phi_s \simeq L_s I_s$ with L_s as the constant auto-inductance coefficient. We do not take into account the hysteresis effects which are quite negligible.

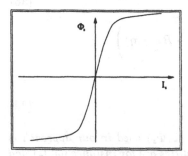

Fig. 2. Typical magnetization curve of a DC motor

Using (1b–1g), the state variable model for the separately excited DC motor for speed control[1] is:

$$J\frac{d\omega}{dt} = c\Phi_s I_r - B\omega - \tau_l \tag{2a}$$

$$L_r\frac{dI_r}{dt} = U_r - R_r I_r - c\Phi_s\omega \tag{2b}$$

$$\frac{d\Phi_s}{dt} = U_s - R_s f^{-1}(\Phi_s) \tag{2c}$$

---

[1] The ongoing studies are quite similar in the case of an angular position control. They are not carried out for the sake of simplicity.

3   Differential Flatness

Recall that a nonlinear system $\dot{x} = f(x, u)$, $u(t) \in \mathbb{R}^m$, is said to be *(differentially) flat* [7,8] if there exists $y = (y_1, \ldots, y_m)$ such that:

1. $y = h(x, u, \ldots, u^{(\alpha)})$;
2. $x = \phi(y, \dot{y}, \ldots, y^{(\beta)})$ and $u = \psi(y, \dot{y}, \ldots, y^{(\beta+1)})$;
3. the components of y are differentially independent.

The separately excited DC motor (2) is flat [7] with the flat output[2] $y = (\omega, \Phi_s)$. To prove this fact one simply assumes the quite reasonable hypothesis that η is an (unknown) function of time which can possibly depend on the mechanical variables (ω or θ) but not on any electrical variable. Thus, every variable of (2) can be expressed as a function of the flat output and a finite number of its derivatives:

$$I_s = f^{-1}(\Phi_s) \tag{3a}$$

$$I_r = \frac{J\dot{\omega} + B\omega + \eta}{c\Phi_s} \tag{3b}$$

$$U_r = L_r \left(\frac{1}{c\Phi_s}(J\ddot{\omega} + B\dot{\omega} + \dot{\eta}) - \frac{\dot{\Phi}_s}{c\Phi_s^2}(J\dot{\omega} + B\omega + \eta) \right)$$

$$+ c\omega\Phi_s + R_r \frac{1}{c\Phi_s}(J\dot{\omega} + B\omega + \eta) \tag{3c}$$

$$U_s = \dot{\Phi}_s + R_s f^{-1}(\Phi_s) \tag{3d}$$

Remark 1 *Notice, that the flat output $y = (\omega, \Phi_s)$ used in our approach is not unique. Other physical meaningful possibilities are for instance (ω, I_r) and (ω, e). We choose (ω, Φ_s) in view of Φ_s appearing in both nonlinear products in (2a) and (2b) and therefore being the physical link between the electrical and the mechanical part of the system.*

Remark 2 *Remark also that (3c) is singular at $\Phi_s = 0$. This stems from the intrinsic physical singularity in which the motor cannot operate. See 4.4 for the avoidance of this singularity.*

4   Minimization of the Copper Losses

We go on in studying the design of an optimal performance trajectory for both components of the flat output of the separately excited DC motor as the second degree of freedom at the input is nowadays usable power electronics being far less expensive than some years ago. We will denote the reference

---

[2] It would be $y = (\theta, \Phi_s)$ in case of a position control.

trajectories of the components of the flat output ω and Φ_s as ω_d and Φ_{sd} respectively. Both ω_d and Φ_{sd} are functions of time.

Energy loss minimization in DC and AC drives has been intensively studied in a founding article by Kusko and Galler [10]. They generally showed that DC motor losses can be reduced considerably by independently controlling the stator and field current. Several publications following this article can be found, which all use linear models of the respective DC drive for energy loss minimization. They can be divided into two different groups: one studies the acceleration motion via optimal control techniques [5], the other establishes control methods for minimizing the losses whilst maintaining a constant angular velocity applying optimal regulator theory (see [2], [3], [4]). A profound survey can be found in [6].

For the first time we propose an energy loss minimization control scheme in which all nonlinearities of the DC drive model are taken into account. The most important nonlinearity for this task seems us to be the saturation curve for the relation between stator flux and stator current (see Fig. 2). Furthermore we present for the first time an energy loss minimization control method which treats the different operating motions (as acceleration, constant angular velocity and electrical braking) at the same time. This is only possible in view of the flatness property.

4.1 Nominal Control

From the flatness property, (3c) and (3d) one obtains the nominal (open-loop) control:

$$U_{rd} = L_r \left(\frac{1}{c\Phi_{sd}}(J\ddot{\omega}_d + B\dot{\omega}_d) - \frac{\dot{\Phi}_{sd}}{c\Phi_{sd}^2}(J\dot{\omega}_d + B\omega_d + \hat{\eta}) \right)$$
$$+ c\omega_d\Phi_{sd} + R_r\frac{1}{c\Phi_{sd}}(J\dot{\omega}_d + B\omega_d + \hat{\eta}) \qquad (4a)$$

$$U_{sd} = \dot{\Phi}_{sd} + R_s f^{-1}(\Phi_{sd}) \qquad (4b)$$

where ω_d and Φ_{sd} are the references trajectories of ω and Φ_s that will be design below (§ 4.3) in order to avoid singularities, i.e. that U_{rd} and U_{sd} remain everywhere defined and bounded. The variable $\hat{\eta}$ denotes an estimation of the value of the load torque. We will present in the next section the construction of a simple observer for η under the assumption that $\dot{\eta} = 0$. Consequently, we find (4a) and (4b) from (3c) and (3d) by simply replacing the components of the flat output by their respective reference values and by setting $\eta = \hat{\eta}$ and $\dot{\eta} = 0$.

Remark 3 *Notice that the injection of the estimation of the disturbance $\hat{\eta}$ in (4a) facilitates the work of the closed loop controller. With this choice, the feedback has only to cope with the unknown transients of the disturbance as its instantaneous mean value $\hat{\eta}$ is estimated on-line by an observer.*

4.2 Observer for the Load Torque

In the preceding section we supposed throughout the calculations the load torque perturbation to be known. Hence we go on in constructing a rapidly converging observer for the load torque η using the measured state as it is done in [1]:

$$\frac{d\hat{\omega}}{dt} = -\frac{B}{J}\hat{\omega} - \hat{\eta} - l_1(\omega - \hat{\omega}) + \frac{c}{J}\Phi_s I_r \tag{5a}$$

$$\frac{d\hat{\eta}}{dt} = l_2(\omega - \hat{\omega}) \tag{5b}$$

Note that the product $\Phi_s I_r$ in (5a) is available by measuring both currents I_r and I_s. Thereafter, Φ_s is determined using the magnetization curve (see Fig. 2) and the measurement of I_s.

Subtracting (5a) and (5b) from (2a) and $\dot{\eta} = 0$ results in the linear error dynamics

$$\frac{d}{dt}\begin{pmatrix} \varepsilon_1 \\ \varepsilon_2 \end{pmatrix} = \begin{pmatrix} (l_1 - B/J) & -1/J \\ l_2 & 0 \end{pmatrix}\begin{pmatrix} \varepsilon_1 \\ \varepsilon_2 \end{pmatrix} \tag{6}$$

with $\varepsilon_1 = \omega - \hat{\omega}$ and $\varepsilon_2 = \eta - \hat{\eta}$.

The observer gains l_1 and l_2 are chosen such that the linear error dynamics converge faster to zero than the dynamics of the observed drive. Using the knowledge of $\hat{\eta}$, we feedback on-line this information to the desired trajectory generator, and therefore to the nominal open loop control.

4.3 Reference Trajectory Generation

In view of the nominal control using the flatness approach (§ 4.1), we have to design the reference trajectories $t \mapsto \omega_d$ and $t \mapsto \Phi_{sd}$ for both components of the flat output.

For the angular speed, this choice is quite obvious since we are designing a speed control. We can for example divide the working time interval $[t_{ini}, t_{fin}]$ in nonempty subintervals $[t_i, t_{i+1}]$, $i = 0, \ldots, N-1$, with $t_0 = t_{ini}$ and $t_N = t_{fin}$. At each point of linking one can impose a given speed $\omega_d(t_i) = \omega_{di}$. In each subinterval, the function ω_d can simply be chosen as a polynomial of time: $\omega_d(t) = \sum_{j=0}^{l_i} w_{ij}(t - t_i)^i/(i!)$, $i = 0, \ldots, N-1$, where $w_{ij} \in \mathbb{R}$. In order to obtain a continuous nominal stator voltage (4b), one has to choose in each interval a polynomial of degree at least 5 satisfying the 6 following connection conditions[3]: $\omega_d^{(k)}(t_{j-1}^-) = \omega_d^{(k)}(t_j^+)$, $j = i, i+1$, $k = 0, 1, 2$. These conditions

---

[3] We use the standard notations: $f(x_o^+) = \lim\limits_{\substack{x \to x_o \\ x > x_o}} f(x)$ and $f(x_o^-) = \lim\limits_{\substack{x \to x_o \\ x < x_o}} f(x)$.

lead to linear systems of equations which permit to obtain the coefficients w_{ij} of the polynomials defining the reference speed trajectory. This choice of the speed trajectory implies that the jerk $\ddot{\omega}$ is everywhere continuous.

For the trajectory of the stator flux, $\Phi_{s,d}$ the choice is not directly related to the speed control. We will show that it is possible to choose its value as a function of the reference speed ω_d and the estimated load torque $\hat{\eta}$ in order to achieve copper loss minimization. We denote by $P_J = R_r I_r^2 + R_s I_s^2$ the total instantaneous power dissipated by the Joule's effect that are the copper losses in the rotor and stator circuit of the motor. As P_J is a function of the variables of the system (2) which is a flat system, it is possible to express P_J in terms of the flat output, using the expressions of the currents I_r and I_s (from (3b) and (3a)). So, $P_J = R_r \left(\frac{J\dot{\omega} + B\omega + \eta}{c\Phi_s} \right)^2 + R_r \left(f^{-1}(\Phi_s) \right)^2$. Remembering that $\tau_m = J\dot{\omega} + B\omega + \eta$, we finally obtain:

$$P_J = R_r \left(\frac{\tau_m}{c\Phi_s} \right)^2 + R_s \left(f^{-1}(\Phi_s) \right)^2 \tag{7}$$

Remark 4 *Remembering that the torque produced by the motor is $\tau_m = c\Phi_s I_r$, one sees that a given torque can be produced by various choices of Φ_s and I_r. All these different values of Φ_s and I_s lead to different values of the copper losses P_J. We will show below that one choice of Φ_s and I_s results in a minimum of P_J.*

From (2a) one deduces that the mechanical torque produced by the motor $t \mapsto \tau_{m*} = J\dot{\omega}_d + B\omega_d + \eta_*$ is fixed for a given desired speed trajectory $t \mapsto \omega_d$ under a given load torque behavior $t \mapsto \eta_*$. consequently, the power dissipated by the Joule's effect (7) can be thought as a function of the stator flux only.

If $\tau_m \neq 0$, P_J is the sum of two squares, the first of which is a strictly decreasing continuous function of the absolute value of the flux $|\Phi_s|$ whilst the second one is a strictly increasing continuous function of $|\Phi_s|$ (cf. Fig. 2). As a consequence, for a given τ_{m*}, P_J admits a unique minimum for $\Phi_s > 0$ and a unique minimum for $\Phi_s < 0$ which is the solution of the following equation:

$$\frac{1}{2} \frac{dP_J}{d\Phi_s} = -R_r \frac{\tau_{m*}^2}{c^2\Phi_s^3} + R_s f^{-1}(\Phi_s) \frac{df^{-1}\Phi_s}{d\Phi_s} = 0 \tag{8}$$

If $\tau_m = 0$, P_J is a strictly increasing of $|\Phi_s|$ and P_J is minimum for $\Phi_s = 0$. However, if one has to generate a speed trajectory which leads to $\tau_m(t_x) = 0$ in only one point of time t_x, as it is the case for electromechanical braking, it is not a good idea to set at this point $\Phi_{s,d}(t_x) = 0$ in view of the time demanding operation of demagnetizing and remagnetizing the drive. In order to change the sign of the motor torque τ_m it is better to change the sign of the rotor current only. This is the option that we selected; consequently, we always choose the positive solution of (8).

Solving numerically (8) leads to the reference stator flux Φ_{sd} at any point of time. As we also need the first derivatives of the flux in (4b), one can think of differentiating (8) with respect to time and then solving numerically the obtained expression. However, this leads to tedious calculations. As a consequence, we have preferred to find a good approximation of the solution of (8) of the form $\Phi_{sd} = h(\tau_{m*}) \simeq \alpha\sqrt{\tanh \beta \tau_{m*}}$. The coefficients α and β have been well fitted from numerical solutions of (8) inside the range of operation of the motor. Therefore it is possible to obtain $\dot{\Phi}_{sd}$ by the analytic expression $\dot{\Phi}_{sd} = h'(\tau_{m*})\dot{\tau}_{m*}$.

Nevertheless, (8) is not defined for $\Phi_s = 0$ and we have to study the corresponding singularities.

4.4   Study of Singularities

In order to study the singularities at small flux it is convenient to linearize the flux characteristics $\Phi_s = f(I_s) \simeq L_s I_s$. In this case it is even possible to give analytical expressions for Φ_{sd} and $\dot{\Phi}_{sd}$:

$$\Phi_{sd} = \sqrt[4]{\frac{R_r}{R_s} \frac{L_s^2 \tau_{m*}^2}{c^2}} \tag{9a}$$

$$\dot{\Phi}_{sd} = \frac{1}{2}\sqrt[4]{\frac{R_r L_s^2}{R_s c^2}} \frac{\dot{\tau}_{m*}}{\sqrt{|\tau_{m*}|}} \tag{9b}$$

Notice that (9a) is similar to the one given in [10]. Our approach is, however, more general as we also consider transients by calculating $\dot{\Phi}_{sd}$ and by using the flatness property.

When starting the system from rest, both the motor torque τ_m and its first time derivative are zero. In this case the expression (9b) must not become singular, hence we show that $\lim_{t\to 0+}\left(\frac{\dot{\tau}_{m*}(t)}{\sqrt{\tau_{m*}(t)}}\right) = 0, \forall \tau_{m*}(t) \geq 0$. Using a polynomial of order 5 to design ω_d as described above:

$$\frac{\dot{\tau}_{m*}(t)}{\sqrt{\tau_{m*}(t)}} = \frac{\sum_{i=0}^{4} b_i t^i}{\sqrt{\sum_{j=0}^{5} a_j t^j}} = \frac{t^2(\sum_{i=2}^{4} b_i t^{i-2})}{t^{3/2}\sqrt{\sum_{j=3}^{5} a_j t^{j-3}}}$$

whereas the coefficients $a_j = 0$, $j = 0, 1, 2$ (and therefore $b_i = 0$, $i = 0, 1$) from considerations of initial conditions for ω_d being equal to zero. In the neighborhood of $t \to 0+$, $\frac{\dot{\tau}_{m*}(t)}{\sqrt{\tau_{m*}(t)}} \simeq \frac{b_2}{\sqrt{a_3}} t^{1/2} \xrightarrow[t\to 0+]{} 0$.

Therefore there does not occur any singularity for the design of the desired trajectories when accelerating the DC drive from the resting position.

As for the braking of the drive to stillstand, the same analysis with respect to the singularity can be led.

4.5 Tracking Control

Using the flatness property of the separately excited DC motor, we are able to exactly linearize its model (2) by endogenous feedback [7]. Notice that in the present case the linearizing feedback is a static one as the model is static state linearizable [9]. Introducing the new input v, we establish the following:

$$U_r = \frac{L_r}{c\Phi_s}\left(Jv_2 - cI_r v_1 + \frac{B}{J}(c\Phi_s I_r - B\omega - \hat{n})\right) + R_r I_r + c\Phi_s \omega \qquad (10a)$$

$$U_s = R_s f^{-1}(\Phi_s) + v_1 \qquad (10b)$$

to get the following linear system to be controlled:

$$\dot{\Phi}_s = v_1 \qquad (11a)$$

$$\ddot{\omega} = v_2 \qquad (11b)$$

Using (11a) and the designed desired trajectories for ω_d and Φ_{sd}, we develop the controller for the new input v as follows:

$$v_1 = \dot{\Phi}_{sd} + \lambda_{11}(\Phi_s - \Phi_{sd}) + \lambda_{12}\int_0^t (\Phi_s - \Phi_{sd})d\tau \qquad (12a)$$

$$v_2 = \ddot{\omega}_d + \lambda_{21}(\dot{\omega} - \dot{\omega}_d) + \lambda_{22}(\omega - \omega_d) + \lambda_{23}\int_0^t (\omega - \omega_d)d\tau \qquad (12b)$$

whereas λ_{ij} to be chosen appropriately. The integral term is set in order to achieve a good tracking under disturbance and parameter uncertainties. The complete control scheme is depicted in Fig. 3.

5 Simulation Results

In this section we present the results of the preceding sections for a specific separately excited DC motor. We first give the parameters of this DC drive, second show the desired trajectories which were discussed in § 4.3, third present the results of the observer and its effect on the on-line trajectory generation of § 4.2 and conclude finally with the simulation results of the controller developed in § 4.5.

The parameters used for the simulations correspond to those identified on a real DC motor: $L_r = 9.61\,\mathrm{mH}$, $L_s = 45\,\mathrm{H}$, $R_r = 0.69\,\Omega$, $R_s = 120\,\Omega$, $J = 11.4 \cdot 10^{-3}\,\mathrm{kg\,m^2}$, $c = 0.021\,\mathrm{s.I.}$, $B = 0.0084\,\mathrm{Nm/rad}$. The saturation curve has been approximated by $f(I_s) = L_s I_o \tanh(I_s/I_o)$ with $L_s I_o = 19\,\mathrm{Wb}$.

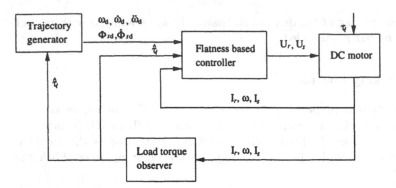

Fig. 3. Novel scheme linking a trajectory generator, a flatness based controller and a load torque observer

First we want to accelerate the DC drive without load torque ($\eta = 0$) from rest to a velocity of 100 rad/s in one second and brake it to still-stand in the same time. The resulting desired trajectories are presented in Fig.4. We see, that the field weakening normally used in a serially excited DC drive by actuating a switch is automatically imposed on the separately excited configuration following our optimal desired trajectory design using the differential flatness property: when accelerating the drive the flux is led to its maximum (see (2a)) where the desired motor torque τ_{m*} is also maximal. For maintaining a certain angular velocity the flux is thereafter reduced to the necessary level. During the braking interval there exists a point for which $\tau_m = 0$. Therefore, we lead at the midpoint of the braking interval the flux to its necessary maximum (see again (2a)) at the point where τ_{m*} is extreme.

To show the effect of braking the system under the condition of $\tau_m(t) > 0$ for $t \in [4\,\mathrm{s}, 5\,\mathrm{s}]$, we apply a constant load torque at $t = 2.5\,\mathrm{s}$ (see Fig. 5). In this case we apply the optimization method already used for the desired acceleration, we see that the desired flux is reduced for the braking maneuver, which is due to the braking effect of the load torque. The latter is estimated by the observer (5a) established in section 4.2, which shows an excellent convergence behavior. The chosen observer coefficients are $l_1 = -1000\,\mathrm{s}^{-1}$ and $l_2 = -3125\,\mathrm{Nm/rad}$, which places both observer poles at $-500\,\mathrm{s}^{-1}$.

The estimated load torque $\hat{\eta}$ is used in a novel way both in the desired trajectory generator and in the flatness based controller (see Fig. 3).

The flatness based controller was established in section 4.5, its coefficients are $\lambda_{11} = 10\,\mathrm{s}^{-1}$, $\lambda_{12} = 20\,\mathrm{s}^{-2}$, $\lambda_{21} = 30\,\mathrm{s}^{-1}$, $\lambda_{22} = 300\,\mathrm{s}^{-2}$ and $\lambda_{23} = 1000\,\mathrm{s}^{-3}$, which places the poles of (12a) at $-2\,\mathrm{s}^{-1}$ and $-5\,\mathrm{s}^{-1}$ and all three poles of (12b) at $-10\,\mathrm{s}^{-1}$ respectively.

We see in Fig. 5, that the controller corrects quickly the errors stemming from the unmodelled dynamics of η. To show the performance of the developed flatness based control technique minimizing copper losses, we depict in Fig. 6

the response of the closed loop system to several load torque perturbations. We remark, that the novel control scheme finds for every point of operation consisting of the pair of desired velocity and given load torque an optimal balance between the rotor current and stator flux. All states and the inputs stay within their given physical bounds.

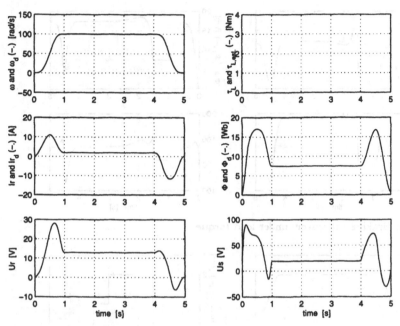

Fig. 4. Desired trajectories

6   Conclusions

In this article the relation of the separately excitated DC drive and its differential flatness property is investigated. Having two inputs available, we propose a nonlinear flatness based control scheme under the aspect of energy loss minimization with respect to copper losses. Hereby we solve a longstanding problem for this subject in using the full nonlinear model of the drive and presenting a control strategy which is applicable to all different motions (accelerating, maintaining a constant angular speed and electrical braking) of the drive. Our method yields moreover the advantage of starting the drive from rest without any discontinuous jerk and without the necessity of pre-magnetizing the machine and the possibility to reinject the kinematic energy won whilst braking the drive electrically into the power supply system.

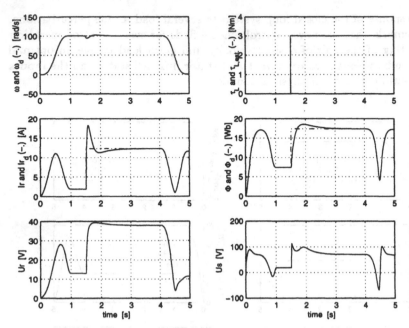

Fig. 5. Braking the system under load torque

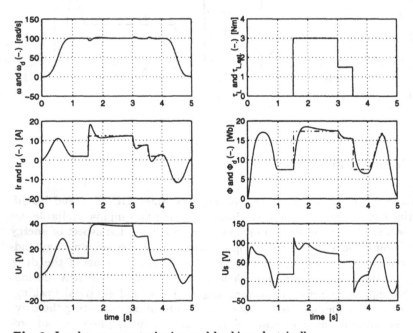

Fig. 6. Load torque perturbation and braking electrically

After having designed the desired trajectories for copper loss minimization, we establish an observer for the load torque and present thereafter a nonlinear flatness based controller. Thereby the estimate of the load torque is used in a novel way for both an on-line redesign of the desired trajectories and for the enhancement of the flatness based controller.

We apply the proposed control method to a specific example: we first show results of the desired trajectory generation, second present the effect of the on-line redesign of the desired trajectories under load torque perturbations and finally depict the performance of the nonlinear flatness based controller.

An implementation of the described control strategy to a real separately excited DC drive is currently undertaken and will be presented in a forthcoming publication. Applications of the work specified here concerns, for instance, traction drives like in subway trains for which the DC machine is still widely used.

Acknowledgement

The authors want to thank Dr. J. Rudolph for helpful comments with respect to the organization of the paper.

References

1. Chiasson J. (1994) Nonlinear differential-geometric techniques for control of a series DC motor. IEEE Trans Control Systems Technology 2:35–42
2. Egami T., Tsuchiya T. (1986) Efficiency-optimized speed-control system based on improved optimal regulator theory. IEEE Trans Ind Electron 33:114–125
3. Egami T., Tsuchiya T. (1987) Efficiency-optimized speed control system with feed-forward compensation. IEEE Trans Ind Electron 34:216–226
4. Egami T., Wang J., Tsuchiya T. (1985) Efficiency-optimized speed control system synthesis method based on improved optimal regulator theory — application to separately exited DC motor system. IEEE Trans Ind Electron 32:372–380
5. Famouri P., Cooley W. (1994) Design of DC traction motor drives for high efficiency under accelerating conditions. IEEE Trans Ind Appl 30:1134–1138
6. Famouri P., Wang J. (1997) Loss minimization control of a DC motor drive. Electric Machines and Power Systems 25:525–537
7. Fliess M., Lévine J., Martin P., Rouchon P. (1995) Flatness and defect of nonlinear systems: introductory theory and examples. Internat J Control 61:1327–1361
8. Fliess M., Lévine J., Martin P., Rouchon P. (1999) A Lie-Bäcklund approach to equivalence and flatness of nonlinear systems. IEEE Trans Automat Control 44:922–937
9. Jakubczyk B., Respondek W. (1980) On linearization of control systems. Bull Acad Pol Sci Set Sci Math 28:517–522
10. Kusko A., Galler D. (1983) Control means for minimization of losses in AC and DC motor drives. IEEE Trans Ind Appl 19:561–570

452   Veit Hagenmeyer et al.

11. Leonhard W. (1996) Control of Electrical Drives, 2nd edn. Springer, Braunschweig
12. Müller G. Elektrische Maschinen: Theorie rotierender elektrischer Maschinen, VEB Verlag Technik, Berlin

State Detection and Stability for Uncertain Dynamical Systems

Mohamed Ali Hammami

Faculty of Sciences of Sfax
Department of Mathematics
Route Soukra BP 802
Sfax 3018 Tunisia
Mohamed.Hammami@fss.rnu.tn

Abstract. In this paper, we study the problems of observer configuration and feedback stability for a certain class of uncertain dynamical systems whose nominal part is linear. Under some conditions on the nonlinear part, we show that the system is detectable and stabilizable by a continuous controller. A nonlinear observer can be designed provided that the system is detectable, that guarantees the observation error is globally exponentially stable. This observer design incorporates only the bound of the nonlinearities (uncertainties), and does not require exact knowledge concerning the structure of the plant nonlinearities. Furthemore, a continuous feedback control is proposed to exponentially stabilizes nonlinear dynamical systems using the Lyapunov approach, based on the stabilizability of the nominal system.

1 Introduction

Consider the nonlinear dynamical system

$$\begin{cases} \dot{x} = F(x, u) \\ y = h(x) \end{cases} \tag{1}$$

where t is the time, $x(t)$ is the state, u is the control input and y is the output of the system. The observation of nonlinear systems has first been addressed in the literature by Thau [1]. Latter, Kou et al [2] and Xia and Gao [3] have introduced and studied the notion of exponential observer. A state observer for the system (1) is a dynamic system

$$\dot{\hat{x}} = G(\hat{x}, y, u) \tag{2}$$

which is expected to produce the estimation $\hat{x}(t)$ of the state $x(t)$ of the system. As a necessary condition for the existence of an exponential observer, the linearized counterpart of the nonlinear system must be detectable. A variety of methods has been developed for constructing nonlinear observers. The Lyapunov-like method ([4], [5]) and observer canonical form method ([6], [7], [8], [9]), where necessary and sufficient conditions for the existence of coordinate transformation have been established. Even, if these conditions are satisfied, the construction of the observer still remains a difficult task due to the need to solve a set of simultaneous partial differential equations to obtain the actual transformation function. Besides, using Lyapunov techniques, this problem is also solved by ([10], [11], [12], [13]) and for uncertain systems by ([14], [15]). For our case, we want to supply an upper bound of the nonlinearity which guarantees the stability of the reconstruction of the system state. For the stability problem, stabilization of dynamical systems with uncertainties (nonlinearities) has been widely studied ([16], [17], [18]). A common approach is to treat the uncertainties in the system deterministically. Then, when the uncertainties are characterized by known bounds, one can use the Lyapunov direct method to design a state feedback control.

2   Problem formulation

In this section, we recall some definitions and results concerning the observer design (state detection) and stability for nonlinear dynamical systems.

The system (2) is called a (local) exponential observer for the system (1), if for all input signal u,

$$\hat{x}(0) = x(0) \text{ implies } \hat{x}(t) = x(t), \text{ for } t \geq 0$$

and

$$\forall \varepsilon > 0, \ \exists \delta > 0 \text{ such that } \|\hat{x}(0) - x(0)\| < \delta$$

implies

$$\|\hat{x}(t) - x(t)\| < \varepsilon \exp(-\lambda t), \ \lambda > 0, \ \forall t > 0.$$

If in addition, $\|\hat{x}(t) - x(t)\|$ tends to zero as t tends to infinity for any initial condition $\|\hat{x}(0) - x(0)\|$, then (2) is called a global exponential observer. The above conditions are known as the two basic requirements of state observers. Therefore, the function G should have the following form

$$G(\hat{x}, y, u) = F(\hat{x}, u) + k(\hat{x}, y, u)$$

where $F(., .)$ is the function describing the plant and $k(., ., .)$ is a nonlinear function that satisfies, on the state space and for all input signal u, the equality, $k(x, h(x), u) = 0$. This means that the observer and the plant have the same dynamics under the condition that the output function $\hat{y} = h(\hat{x})$ copies the output function $h(x)$. It should be remarked that, the literature on nonlinear observers is nearly exclusively devoted to exponential observers, that is, observers whose state converges exponentially to the state of the plant.

Furthemore, the approximating system of (1) at the equilibrium point is given by

$$\begin{cases} \dot{x} = Ax + Bu \\ y = Cx \end{cases} \tag{3}$$

where $A = \frac{\partial F}{\partial x}(0,0)$, $B = \frac{\partial F}{\partial u}(0,0)$, and $C = \frac{\partial h}{\partial x}(0)$. The system (3) is said detectable if there exists a matrix K such that the matrix $(A - KC)$ is exponentially stable. A sufficient condition for (3) to be detectable is that it is observable. A linear system is observable if its observability matrix has full rank, i.e, $rank(C, CA, ..., CA^{n-1}) = n$.

Let the state observation error e be defined by $e = \hat{x} - x$. The dynamics of e are given by the following error system :

$$\dot{e} = F(x + e, u) - F(x, u) + k(x + e, y, u)$$

Note that, the origin $e = 0$ is an equilibrium point of these dynamics, and this system is required to be (locally) exponentially stable for every input signal u. In particular, for $u = 0$, the origin $x = 0$ is an equilibrium point of the system to be observed and the error equation becomes : $\dot{e} = (A - LC)e + o(e^2)$. These dynamics are exponentially stable if and only if the first order matrix of the above equation is Hurwitz. This yields the following property : there exists a local exponential observer of the form (2) for (1), if and only if the linear approximation at the origin (3) is detectable.

In this paper, we consider nonlinear dynamical systems described by the following state equations :

$$\begin{cases} \dot{x} = F(t, x, u) = Ax + Bu + f(t, x, u) \\ y = h(x) = Cx \end{cases} \tag{4}$$

where $x \in \Re^n$, $u \in \Re^q$, $y \in \Re^p$, A, B and C are respectively $(n \times n)$, $(n \times q)$, $(p \times n)$ known constant matrices and f is Lipschitz continuous represents the uncertainties in the plant. In ([19], [20]), Vidyasagar introduced the local asymptotic observer called "weak detector" specifying properties of the function $g(., ., ., .)$ defining the observer dynamic.

Definition 1 The system (4) is said to be weakly detectable at the origin if one can find a function $g : \Re^+ \times \Re^n \times \Re^p \times \Re^q \to \Re^n$ and a function $\Phi : \Re^+ \times \Re^n \times \Re^n \to \Re^+$, such that the following properties hold :

(i) $g(.,.,.,.) \in C^{\infty}$ (i.e, g is smooth) and $g(t,0,0,0) = 0$.

(ii) There exist three functions ψ_1, ψ_2, $\psi_3 : \Re^+ \to \Re^+$ continuous and strictly increasing such that $\psi_1(0) = \psi_2(0) = \psi_3(0) = 0$ and $\exists \varepsilon > 0$, such that $\forall t \geq 0$, $\forall x$, $\hat{x}$, u, $\|x\| < \varepsilon$, $\|\hat{x}\| < \varepsilon$, $\|u\| < \varepsilon$,

$$\psi_1(\|\hat{x} - x\|) \leq \Phi(t, x, \hat{x}) \leq \psi_2(\|\hat{x} - x\|)$$

and

$$\frac{\partial \Phi}{\partial t}(t, x, \hat{x})F(t, x, u) + \frac{\partial \Phi}{\partial x}(t, x, \hat{x})F(t, x, u) + \frac{\partial \Phi}{\partial \hat{x}}g(t, \hat{x}, h(x), u) \leq -\psi_3(\|\hat{x} - x\|).$$

As an example, consider a detectable linear system (3), where the pair (A, C) is detectable. According to this fact, this system is weakly detectable. Indeed, the function $g(.,.,.)$ defined as,

$$g(\hat{x}, y, u) = Az + Bu - L(C\hat{x} - y)$$

it defines a Luenberger observer. Since, (A, C) is detectable, there exists L such that $(A - LC)$ is Hurwitz, i.e, there exists a matrix K and symmetric and positive definite matrices P and Q such that,

$$(A - LC)^T P + P(A - LC) = -Q, \ Q > 0.$$

Defining the functions $\Phi(.,.)$ and ψ_i by,

$$\Phi(\hat{x} - x) = \psi_1(\|\hat{x} - x\|) = \psi_2(\|\hat{x} - x\|) = (\hat{x} - x)^T P(\hat{x} - x)$$

$$\psi_3(\|\hat{x} - x\|) = (\hat{x} - x)^T Q(\hat{x} - x).$$

One can check that, the requirements of weak detectability are satisfied. Vidyasagar, shows that for weakly detectable systems (4), the following "weak detector" can be set up,

$$\dot{\hat{x}} = g(\hat{x}, y, u) = g(\hat{x}, h(x), u) \tag{5}$$

where $g(.,.,.)$ satisfy the conditions of definition 1. When the solution trajectories $x(t)$ of (4), the solution trajectories $\hat{x}(t)$ of (5) and the input u are bounded by a sufficiently small $\varepsilon > 0$, it is proved that $(\hat{x}(t) - x(t))$ tends to zero when t tends to infinity. It should be remarked that, in definition 1 the "weak detectability" does not require the linear approximation of the system (1) to be detectable. Here, we will examine the detectability of a class of dynamical system in the global sense, where we suppose that the pair (A, C) of the nominal system is detectable.

For the stability purpose, we shall say that the uncertain system (4) is feedback stable (in the global exponential stability sense), if there exists a continuous controller $u = u(t)$ such that the system (4) with $u(t)$ is globally exponentially stable. In fact, an uncertain system has in general no equilibrium point. By convection, stability of uncertain system is then studied with respect to the equilibrium point of the nominal system (see [18]). Provided that all states are available, the state feedback controller can be represented by a nonlinear function

$$u = u(t, x).$$

Our goal, is to find a such feedback controller, that can guarantee the exponential stability of nonlinear dynamic system in the presence of uncertainties $f(t, x, u)$.

In order to study the detectability and stability of the system (4), we will need the following lemma [21] which gives an exponential rate of convergence.

Lemma 1 *Given a continuous system*

$$\dot{x}(t) = F(t, x(t)) \tag{6}$$

where $x(t)$ is an $n \times 1$ vector, let $V(t, x)$ be the associated Lyapunov function with the following properties :

$$\lambda_1 \|x(t)\|^2 \leq V(t, x) \leq \lambda_2 \|x(t)\|^2, \ \forall (t, x) \in \Re \times \Re^n \tag{7}$$

$$\dot{V}(t, x) \leq -\lambda_3 \|x(t)\|^2 + \varepsilon e^{-\beta t}, \ \forall (t, x) \in \Re \times \Re^n, \tag{8}$$

where λ_1, λ_2, λ_3, ε and β are positive scalar constants. If the Lyapunov function satisfies (7) and (8), the state $x(t)$ of (6) is globally exponentially stable [22], in the sense that

$$\|x(t)\| \leq ((\frac{\lambda_2}{\lambda_1} \|x(0)\|^2 + \frac{\varepsilon}{\lambda_1} t)e^{-\lambda t})^{\frac{1}{2}}, \ if \ \beta = \lambda,$$

$$\|x(t)\| \leq ((\frac{\lambda_2}{\lambda_1} \|x(0)\|^2 e^{-\lambda t} + \frac{\varepsilon}{\lambda_1 (\lambda - \beta)})(e^{-\beta t} - e^{-\lambda t}))^{\frac{1}{2}}, \ if \ \beta \neq \lambda$$

where $\lambda = \frac{\lambda_3}{\lambda_2}$.

3   Detectability

Tsinias in [23], generalize the well known theorem of Vidyasagar [19] on the local stabilizability of nonlinear systems using state detection. For our case, in the presence of uncertainties, we give a definition of detectability, as in definition 1, where we treat here the global case.

Definition 2 Consider the system (1) with the output $y = h(x)$, $y \in \Re^p$ where $p < n$ and is a continuous function. We say that (4) is detectable, if there exists a continuous mapping

$$G : \Re^+ \times \Re^n \times \Re^p \times \Re^q \to \Re^n \text{ with } G(t,0,0,0) = 0,$$

a continuously differentiable function $V : \Re^+ \times \Re^n \times \Re^n \to \Re^+$ and real functions $\beta_i : \Re^+ \to \Re^+$, $\beta_i(0) = 0$, $\{i = 1,2,3\}$ of class $\mathcal{K}_\infty$ (namely, β_i is continuous, strictly increasing, with $\beta_i(0) = 0$ and $\beta_i(s) \to +\infty$ as $s \to +\infty$), such that

$$F(t,x,u) = G(t,x,h(x),u), \quad \forall t \geq 0, \forall x \in \Re^n, \ \forall u \in \Re^q$$

$$\beta_1(\|\hat{x} - x\|) \leq V(t,x,\hat{x}) \leq \beta_2(\|\hat{x} - x\|) \tag{9}$$

and

$$\frac{\partial V}{\partial t}F(t,x,u) + \frac{\partial V}{\partial x}F(t,x,u) + \frac{\partial V}{\partial \hat{x}}G(t,\hat{x},h(x),u) \leq -\beta_3(\|\hat{x} - x\|) \tag{10}$$

for every $t \geq 0$, $u \in \Re^q$ and $(x, \hat{x}) \in \Re^n \times \Re^n$.

It turns out that, if (4) is detectable, then the system

$$\dot{\hat{x}} = G(t,\hat{x},h(x),u)$$

is an observer for (4), i.e, the origin of the error equation

$$\dot{e} = G(t,e+x,h(x),u) - F(t,x,u), \ e = \hat{x} - x.$$

is globally asymptotically (exponentially) stable.

With the model given in (4), the problem is to show the detectability of the system, so that the conception of a continuous observer is possible, with input $y(t)$ such that the estimates denoted by $\hat{x}(t)$ converge to $x(t)$ asymptotically fast. If $\beta_i \in \mathcal{K}_\infty$, we obtain an asymptotic observer. In order to obtain an exponential one, we shall find an exponential estimation (i.e, $\|\hat{x}(t) - x(t)\| \leq \lambda_1 exp - \lambda_2 t$, $\lambda_1 > 0$, $\lambda_2 > 0$). Therefore, we shall assume the following assumptions.

$(\mathcal{A}_1)$ The pair (A, C) is detectable, then there exists a matrix L such that $Re\lambda(A_L) < 0$, where $A_L = A - LC$, a Lyapunov function for the nominal part can be chosen as $x^T P x$, where $P = P^T > 0$ is such that,

$$PA_L + A_L^T P = -Q < 0 \qquad (11)$$

$(\mathcal{A}_2)$ There exists a symetric positive definite matrix $Q(n \times n)$ and a function ϕ (see [14], [15]) where $\phi(.,.,.) : \Re^+ \times \Re^n \times \Re^q \to \Re^p$, such that

$$f(t, x, u) = P^{-1}C^T \phi(t, x, u),$$

where P is the unique positive definite solution to the Lyapunov equation (11).

$(\mathcal{A}_3)$ There exits a positive scalar function $\rho(t, u)$ ([14], [15]) such that, $\|\phi(t, x, u)\| \leq \rho(t, u)$, where $\|.\|$ denotes the Euclidean norm on $\Re^n$.

Then, one can state the following theorem.

Theorem 1. *Suppose that $(\mathcal{A}_1)$, $(\mathcal{A}_2)$ and $(\mathcal{A}_3)$ hold. Then, the system (4) is detectable with*

$$G(t, \hat{x}, y, u) = A\hat{x} + Bu + \varphi(t, \hat{x}, y, u) - L(C\hat{x} - y)$$

where

$$\varphi(t, \hat{x}, y, u) = -\frac{P^{-1}C^T(C\hat{x} - y)\rho(t, u)^2}{\|C\hat{x} - y)\|\rho(t, u) + \alpha e^{-\beta t}} \quad with \quad \alpha > 0, \ \beta > 0 \qquad (12)$$

and

$$V(x, \hat{x}) = (\hat{x} - x)^T P(\hat{x} - x).$$

Proof First, remark that V satisfies the condition (9). Next, as the argument used in proving the detectability of linear systems, in the section one, we will use the function V to show that (10) is satisfied for the system (4). Let,

$$\partial V(x, \hat{x}) = (\dot{\hat{x}} - \dot{x})^T P(\hat{x} - x) + (\hat{x} - x)^T P(\dot{\hat{x}} - \dot{x})$$

Then,

$$\partial V(x,\hat{x}) = \frac{\partial V}{\partial x}\Big(Ax + Bu + f(t,x,u)\Big) + \frac{\partial V}{\partial \hat{x}}\Big(G(t,\hat{x},y,u)\Big)$$

Thus,

$$\partial V(x,\hat{x}) = \Big(A\hat{x} + \varphi(t,\hat{x},y,u) - L(C\hat{x} - Cx) - Ax - f(t,x,u)\Big)^T P(\hat{x} - x)$$
$$+ (\hat{x} - x)^T P\Big(A\hat{x} + \varphi(t,\hat{x},y,u) - L(C\hat{x} - Cx) - Ax - f(t,x,u)\Big)$$

which implies that,

$$\partial V(x,\hat{x}) = (\hat{x} - x)^T (A - LC)^T P(\hat{x} - x) + (\hat{x} - x)^T P(A - LC)(\hat{x} - x)$$
$$+ 2(\hat{x} - x)^T P\Big(\varphi(t,\hat{x},y,u) - f(t,x,u)\Big)$$

Since $(\mathcal{A}_1)$ holds, then by (11) one gets,

$$\partial V(x,\hat{x}) = -(\hat{x} - x)^T Q(\hat{x} - x) + 2(\hat{x} - x)^T P\Big(\varphi(t,\hat{x},y,u) - f(t,x,u)\Big)$$

Tacking into account, $(\mathcal{A}_2)$, $(\mathcal{A}_3)$ and (12), we obtain

$$\partial V(x,\hat{x}) \leq -(\hat{x} - x)^T Q(\hat{x} - x) + 2\|C(\hat{x} - x)\|\rho(t,u)$$
$$+ 2(\hat{x} - x)^T P\frac{-P^{-1}C^T(C\hat{x} - y)\rho(t,u)^2}{\|C\hat{x} - y\|\rho(t,u) + \alpha e^{-\beta t}}$$

$$\partial V(x,\hat{x}) \leq -(\hat{x} - x)^T Q(\hat{x} - x) + 2\|C(\hat{x} - x)\|\rho(t,u)$$
$$- 2\frac{\|C(\hat{x} - x)\|^2 \rho(t,u)^2}{\|C\hat{x} - y\|\rho(t,u) + \alpha e^{-\beta t}}$$

$$\partial V(x,\hat{x}) \leq -(\hat{x} - x)^T Q(\hat{x} - x) + 2\|C\hat{x} - y\|\rho(t,u)$$
$$- 2\frac{\|C\hat{x} - y\|^2 \rho(t,u)^2}{\|C\hat{x} - y\|\rho(t,u) + \alpha e^{-\beta t}}$$

It follows that,

$$\partial V(x,\hat{x}) \leq -(\hat{x} - x)^T Q(\hat{x} - x) + 2\frac{\|C\hat{x} - y\|\rho(t,u).\alpha e^{-\beta t}}{\|C\hat{x} - y\|\rho(t,u) + \alpha e^{-\beta t}}$$

Now using the fact that, $\forall a > 0$, $\forall b > 0$, $\frac{ab}{a+b} < b$, it follows that

$$\partial V(x, \hat{x}) \leq -\lambda_{min}(Q)\|\hat{x} - x\|^2 + 2\alpha e^{-\beta t},$$

for all $(x, \hat{x}) \in \Re^n \times \Re^n$ and $u \in \Re^q$. Hence, the condition (10), which means that $(\hat{x}(t) - x(t))$ tends to zero globally and asymptotically, follows by using the fact that $e^{-\beta t}$ is bounded and the latter inequality holds for any $\alpha > 0$ (which can be choosen arbitrary small) so it implies (10). Hence, the system (10) is detectable. Therefore, from lemma 1, one can obtain an exponential rate of convergence and the system $\dot{\hat{x}} = G(t, \hat{x}, y, u)$, with $\varphi(t, \hat{x}, y, u)$ given in (12), becomes an exponential observer for the system (4). Note that, the theorem 1 includes the linear case which treated in the section two.

4   Stability

Suppose that the system (4) satisfies the following assumption.

(A_4) The pair (A, B) is stabilizable, then there exists a matrix K such that $Re\lambda(A_K) < 0$, where $A_K = A + BK$, a Lyapunov function for the nominal part can be chosen as $x^T P x$, where $P = P^T > 0$ is such that

$$PA_K + A_K^T P = -Q < 0 \tag{13}$$

Under the above assumption, by considering the derivative of $V(x) = x^T P x$ along the trajectories of the system (4), one can reach conclusions about the sign definiteness of $\dot{V}$ by imposing certain restrictions on the uncertainties $f(t, x, u)$. As an example, if the nonlinearities is norm bounded by a known positive function $\rho(t, x)$ which satisfies $\rho(t, x) \leq k(t)\|x\|$, with $k(t) < \frac{1}{2}\frac{\lambda_{min}(Q)}{\lambda_{max}(P)}$, where $\lambda_{min}(.)$ and $\lambda_{max}(.)$ are respectively the minimum and the maximum eigenvalue of the matrix $(.)$, then the system is globally exponentially stable. Therefore, using this idea, we give a sufficient condition on the uncertainties to ensure the stability of system (4).

Assume that,

(A_5) There exist some positive constant k_1, k_2, such that

$$\|f(t, x, u)\| \leq k_1\|x\| + k_2\|u\|, \quad \forall x \in \Re^n, \forall u \in \Re^q \tag{14}$$

Let $k_0 = k_1 + k_2\|K\| > 0$. Then, under the latter two conditions, we have the following theorem.

Theorem 2. *Suppose that the assumptions* (A_4), (A_5) *hold and* k_0 *satisfies* $k_0 < \frac{1}{2}\frac{\lambda_{min}(Q)}{\lambda_{max}(P)}$. *Then, the system* (4) *with the controller* $u(t) = Kx(t)$, *is globally exponentially stable.*

Proof Consider the Lyapunov function $V = x^T P x$ which satisfies (13). The derivative of V along the trajectories of (4), satisfies

$$\dot{V} = 2x^T P(Ax + Bu + f(t, x, u)) = 2x^T PAx + 2x^T PBu + 2x^T Pf(t, x, u)$$

Now, using the controller $u(t)$ and the Lyapunov equation (13), we obtain

$$\dot{V} \le -x^T Qx + 2\|x\|\|P\|\|f(t, x, u(t))\|$$

Thus, from (14)

$$\dot{V} \le -x^T Qx + 2\|x\|\|P\|\Big(k_1\|x\| + k_2\|u(t)\|\Big)$$

Using the form of $u(t)$, it follows that

$$\dot{V} \le -x^T Qx + 2\|P\|\Big(k_1\|x\| + k_2\|Kx\|\Big)$$

$$\dot{V} \le -\lambda_{min}(Q)\|x\|^2 + 2\|P\|(k_1 + k_2\|K\|)\|x\|^2$$

$$\dot{V} \le \Big(-\lambda_{min}(Q) + 2\|P\|(k_1 + k_2\|K\|)\Big)\|x\|^2$$

$$\dot{V} \le \Big(-\lambda_{min}(Q) + 2\lambda_{max}(P)k_0\Big)\|x\|^2$$

Since, $k_0 < \frac{1}{2}\frac{\lambda_{min}(Q)}{\lambda_{max}(P)}$, it follows that the system (4) is globally exponentially stable.

Note that, the ratio $\frac{\lambda_{min}(Q)}{\lambda_{max}(P)}$ is maximized with the choice $Q = I$ (see Vidyasagar [20]).

Finally, consider the system (4) and suppose that $p = q$, the assumptions $(\mathcal{A}_2)$, $(\mathcal{A}_4)$ hold and the following condition,

$(\mathcal{A}_6)$ There exits a positive scalar function $\rho(t, x)$ ([16]), such that

$$\|\phi(t, x, u)\| \leq \rho(t, x), \ \forall t, \ \forall x, \forall u.$$

The design of the controller to stabilize the system represented by equations (4) subject to the bound given in assumption $(\mathcal{A}_6)$, is described by

$$u(t) = Kx(t) + \psi(t, x) \tag{15}$$

where ψ is defined as follows :

$$\psi(t, x) = -\frac{1}{2} \frac{B^T Px \rho(t, x)^2}{\|B^T Px\| \rho(t, x) + \varepsilon_0 e^{-\beta t}}, \ \varepsilon_0 > 0, \ \beta > 0. \tag{16}$$

Then, one can proof that,

Theorem 3. *Suppose that the assumptions $(\mathcal{A}_2)$, $(\mathcal{A}_4)$, $(\mathcal{A}_6)$ hold and that*

$$B^T P = C \tag{17}$$

then, the system (4), with the controller (15), is globally exponentially stable.

The controller given in (15) consists of $Kx(t)$ which is a linear state feedback which stabilizes the nominal system, and $\psi(t, x)$ is continuous nonlinear state feedback which used to produce an exponential stability of the whole system in the presence of nonlinearity (uncertainties $f(t, x, u)$). In [24], the authors characterize the class of linear systems specified by $(\mathcal{A}_4)$ and the condition (17), they prove that, for linear system, there exists a matrix K satisfying $(\mathcal{A}_4)$ and (17) $\Longleftrightarrow$ the system is stabilizable, stable invertible and moreover, its leading Markov parameter CB is symmetric positive definite. Also, in [25] and [26] the authors used this condition to show the stability by output feedback for a class of nonlinear system. The proof is nearly identical to the proof of the theorem presented in [15].

Proof As the proof of theorem 2, let considers the Lyapunov function $V = x^T Px$ which satisfies (13). The derivative of V along the trajectories of (4) , satisfies $\dot{V} = 2x^T PAx + 2x^T PBu + 2x^T Pf(t, x, u)$. Thus, using the controller $u = Kx(t) + \psi(t, x(t))$, we obtain

$$\dot{V} = 2x^T PAx + 2x^T PB(Kx + \psi(t, x)) + 2x^T Pf(t, x, Kx + \psi(t, x))$$

One can verify, by (A_2), (A_4), (A_6) and (16), that

$$\dot{V} \leq -x^T Q x + 2\|Cx\|\rho(t,x) - 2x^T PB \frac{B^T Px\rho(t,x)^2}{\|B^T Px\|\rho(t,x) + \varepsilon_0 e^{-\beta t}}$$

$$\dot{V} \leq -x^T Q x + 2\|B^T Px\|\rho(t,x) - \frac{2\|B^T Px\|^2\rho(t,x)^2}{\|B^T Px\|\rho(t,x) + \varepsilon_0 e^{-\beta t}}$$

$$\dot{V} \leq -x^T Q x + \frac{2\|B^T Px\|\rho(t,x).\varepsilon_0 e^{-\beta t}}{\|B^T Px\|\rho(t,x) + \varepsilon_0 e^{-\beta t}}$$

Now using the fact that, $\forall a > 0$, $\forall b > 0$, $\frac{ab}{a+b} < b$, it follows that

$$\dot{V} \leq -x^T Q x + 2\varepsilon_0 e^{-\beta t} \tag{18}$$

Applying lemma 1 to (18), with $\lambda_3 = \lambda_{min}(Q)$, $\lambda_1 = \lambda_{min}(P)$, $\lambda_2 = \lambda_{max}(P)$ and $\varepsilon = 2\varepsilon_0$, yields theorem 3.

In the above theorem, it is assumed that both $u(t)$ and $y(t) \in \Re^p$ and $B^T P = C$. This result is known as the Kalman-Yakubovich lemma [22]. Since the Lyapunov equation (13) can be satisfied for any arbitrary symmetric positive definite matrix Q, the Kalman-Yakubovich states that given any open-loop strictly stable linear system, one can construct an infinity of dissipative input-output maps, simply by using compatible choices of inputs and outputs. In particular, given the system's phisical inputs and the associated matrix B, one can choose an infinity of outputs from which the linear system will look dissipative. A simple choice is that the output controller is linear. Suppose that, the triple (A, B, C) are output controllable, there exists a matrix D such that $Re\lambda(A_D) < 0$, where $A_D = A - BDC$, and there exits a positive scalar function $\rho(t, y)$ [15], such that $\|\phi(t, x, u)\| \leq \rho(t, y)$, $\forall t$, $\forall x$, $\forall u$. Then, the output controller $u(t) = Dy + \psi(t, y)$, where ψ is defined as, $\psi(t, y) = -\frac{1}{2}\frac{y\rho(t,y)^2}{\|y\|\rho(t,y)+\varepsilon_0 e^{-\beta t}}$, $\varepsilon_0 > 0$, $\beta > 0$, stabilizes the system (4). For more general systems, using this lemma, the authors in [27] give an output controller which stabilizes globally and exponentially a class of uncertain nonlinear dynamic systems.

References

1. F.Thau, Observing the states of nonlinear dynamic systems, Int.Journal.Control, 17 (1993) 471-479.
2. S.Kou, D.Elliott and T.Tarn, Exponential observers for nonlinear dynamic systems, Inform and Control, 29 (3) (1976) 204-216.

3. Xia-hua Xia and Wei-bing Gao, On exponential observers for non linear systems, Syst.Cont.Lett, 11 (1988) 319-325.
4. D.G.Luenberger, An introduction to observers, IEEE Trans.Autom.Control, 16 (1971) 596-602.
5. R.E.Kalman and R.S.Bycy, New results in linear filtring and prediction theory, Tr.ASME ser D.J.Basic Eng, 83 (1961) 95-108.
6. D.Bestle and M.Zeitz, Canonical form observer design for nonlinear time variable systems, Int.J.Cont, 38 (1983) 419-431.
7. A.J.Krener and A.Isidori, Linearization by output injection and nonlinear observers, Syst.Cont.Lett, 3 (1983) 47-52.
8. M.Zeitz, Observability canonical (phase-variable) form for nonlinear time-variable systems, Int.J.Syst.Sci, vol 15 n 9 (1984) 949-958.
9. A.J.Krener and W.Respondek, Nonlinear observers with linearizable error dynamics, SIAM.J.Cont.Optim, 23 (1985) 197-216.
10. J.P.Gauthier and Y.Kupka, A separation principle for bilinear systems with dissipative drift, IEEE.Trans.Aut.Cont, AC vol37 (1992) 12 pp 1970-1974.
11. R.Chabour and H.Hammouri, Stabilization of planar bilinear systems using an observer configuration, App.Math.Lett, 6 (1993) pp 7-10.
12. A.Iggidr and G.Sallet, Exponential stabilization of nonlinear systems by an estimated state feedback, ECC'93 (1993) 2015-2018.
13. M.A.Hammami, Stabilization of a class of nonlinear systems using an observer design, Proc of 32nd IEEE Conf.Dec.Cont, San Antonio Texas (1993) 1954-1959.
14. B.Walcott and S.Zak, State observation of nonlinear uncertain dynamical systems, IEEE.Trans.Aut.Cont, 32 (2) (1987) 166-170.
15. D.M.Dawson, Z.Qu and J.C.Carroll, On the state observation and output feedback problems for nonlinear uncertain dynamic systems, Syst.Cont.Lett, 18 (1992) 217-222.
16. H.K.Khalil, Nonlinear Systems, Mac Millan (1992).
17. Z.Qu, Global stabilization of nonlinear systems with a class of unmatched uncertainties. Syst.Cont.Lett 18 (1992) 301-307.
18. H.Wu and K.Mizukami, Exponential stability of a class of nonlinear dynamic systems with uncertainties. Syst.Cont.Lett 21 (1993) 307-313.
19. M.Vidyasagar, On the stabilization of nonlinear systems using state detection, IEEE Trans.Auto.Cont, AC-25 (1980) 504-509.
20. M.Vidyasagar, Nonlinear systems analysis, Practice Hall, 2nd edition (1993).
21. Z.Qu and D.Dawson, Continuous feedback control guaranteeing exponential stability for uncertain dynamic systems, IEEE Conf.Dec.Cont, (1991).
22. J.Slotine and W.Li, Applied nonlinear control, Prentice Hall, Engl-Cliffs, (1991).
23. J.Tsinias, A generalization of Vidyasagar's theorem on stabilizability using state detection, Syst.Cont.Lett, 17 (1991) 37-42.
24. A.Saberi, P.V.Kokotovic and J.Sussmann, Global stabilization of partially linear composite systems, SIAM.J.Cont.Optim, vol 28, N6 (1990) 1491-1503.
25. B.L.Walcott, M.J.Corless and H.Zak, Comparative study of non linear state observation techniques, Int.J.C, vol 45 N6 (1987) 2109-2132.
26. A.Steinberg and M.Corless, Output feedback stabilization of uncertain dynamical systems, IEEE Trans.Aut.Cont, vol AC30 N10 (1985) 1025-1027.
27. A.Benabdallah and M.A.Hammami, On the output feedback stability for nonlinear uncertain control systems, submitted.

Controllability Properties of Numerical Eigenvalue Algorithms

Uwe Helmke[1] and Fabian Wirth[2]*

[1] Mathematisches Institut
   Universität Würzburg
   97074 Würzburg, Germany
   helmke@mathematik.uni-wuerzburg.de
[2] Zentrum für Technomathematik
   Universität Bremen
   28334 Bremen, Germany
   fabian@math.uni-bremen.de

Abstract. We analyze controllability properties of the inverse iteration and the QR-algorithm equipped with a shifting parameter as a control input. In the case of the inverse iteration with real shifts the theory of universally regular controls may be used to obtain necessary and sufficient conditions for complete controllability in terms of the solvability of a matrix equation. Partial results on conditions for the solvability of this matrix equation are given. We discuss an interpretation of the system in terms of control systems on rational functions. Finally, first results on the extension to inverse Rayleigh iteration on Grassmann manifolds using complex shifts is discussed.

For many numerical matrix eigenvalue methods such as the QR algorithm or inverse iterations shift strategies have been introduced in order to design algorithms that have faster (local) convergence. The shifted inverse iteration is studied in [3,4,15] and in [17,18], where the latter references concentrate on complex shifts. For an algorithm using multidimensional shifts for the QR-algorithm see the paper of Absil, Mahony, Sepulchre and van Dooren in this book.

In this paper we interpret the shifts as control inputs to the algorithm. With this point of view standard shift strategies as the well known Rayleigh iteration can be interpreted as feedbacks for the control system. It is known (for instance in the case of the inverse iteration or its multidimensional analogue, the QR-algorithm) that the behavior of the Rayleigh shifted algorithm can be very complicated, in particular if it is applied to non-Hermitian matrices A [4]. It is therefore of interest to obtain a better understanding of the underlying control system, which up to now has been hardly studied.

---

* This paper was written while Fabian Wirth was a guest at the Centre Automatique et Systèmes, Ecole des Mines de Paris, Fontainebleau, France. The hospitality of all the members of the centre is gratefully acknowledged.

Here we focus on controllability properties of the corresponding systems on projective space for the case of inverse iteration, respectively the Grassmannian manifold for the QR-algorithm. As it turns out the results depend heavily on the question whether one uses *real* or *complex* shifts. The controllability of the inverse iteration with complex shifts has been studied in [13], while the real case is treated in [14].

In Section 1 we introduce the shifted inverse power iteration with real shifts and the associated system on projective space and discuss its forward accessibility properties. In particular, there is an easy characterization of the set of universally regular control sequences, that is those sequences with the property, that they steer every point into the interior of its forward orbit. This will be used in Section 2 to give a characterization of complete controllability of the system on projective space in terms of solvability of a matrix equation. In Section 3 we investigate the obtained characterization and interpret it in terms of the characteristic polynomial of A. Some concrete cases in which it is possible to decide based on spectral information whether a matrix leads to complete controllable shifted inverse iteration are presented in Section 4. An interpretation of these results in terms of control systems on rational functions is given in Section 5. In Section 6 we turn to the analysis of the shifted QR algorithm. We show that the corresponding control system on the Grassmannian is never controllable except for few cases. The reachable sets are characterized in terms of Grassmann simplices. We conclude in Section 7.

1   The shifted inverse iteration on projective space

We begin by reviewing recent results on the shifted inverse iteration which will motivate the ideas employed in the case of the shifted QR algorithm. Let A denote a real $n \times n$-matrix with spectrum $\sigma(A) \subset \mathbb{C}$. The *shifted inverse iteration* in its controlled form is given by

$$x(t+1) = \frac{(A - u_t I)^{-1} x(t)}{\|(A - u_t I)^{-1} x(t)\|}, \quad t \in \mathbb{N}, \tag{1}$$

where $u_t \notin \sigma(A)$. This describes a nonlinear control system on the $(n-1)$-sphere. The trajectory corresponding to a normalized initial condition x_0 and a control sequence $u = (u_0, u_1, \dots)$ is denoted by $\phi(t; x_0, u)$. Via the choice $u_t = x^*(t) A x(t)$ we obtain from (1) the Rayleigh quotient iteration studied in [3], [4].

If the initial condition x_0 for system (1) lies in an invariant subspace of A then the same holds true for the entire trajectory $\phi(t; x_0, u)$, regardless of the control sequence u. In order to understand the controllability properties from x_0 it would then suffice to study the system in the corresponding invariant subspace. Therefore we may restrict our attention to those points not lying in a nontrivial invariant subspace of A, i.e. those $x \in \mathbb{R}^n$ such that

$\{x, Ax, \ldots, A^{n-1}x\}$ is a basis of $\mathbb{R}^n$. Vectors with this property are called *cyclic* and a matrix A is called cyclic if it has a cyclic vector, which we will always assume in the following. To keep notation short let us introduce the union of A-invariant subspaces

$$\mathcal{V}(A) := \bigcup_{AV \subset V, 0 < dim V < n} V.$$

Using the fact that the interesting dynamics of (1) are on the unit sphere and identifying opposite points (which give no further information) we then define our state space of interest to be

$$M := \mathbb{RP}^{n-1} \setminus \mathcal{V}(A), \tag{2}$$

where $\mathbb{RP}^{n-1}$ denotes the real projective space of dimension $n-1$. The natural projection from $\mathbb{R}^n \setminus \{0\}$ to $\mathbb{RP}^{n-1}$ will be denoted by $\mathbb{P}$. Thus M consists of the 1-dimensional linear subspaces of $\mathbb{R}^n$, defined by the cyclic vectors of A. Since a cyclic matrix has only a finite number of invariant subspaces, $\mathcal{V}(A)$ is a closed algebraic subset of $\mathbb{R}^n$. Moreover, M is an open and dense subset of $\mathbb{RP}^{n-1}$. The system on M is now given by

$$\xi(t+1) = (A - u_t I)^{-1} \xi(t), \quad t \in \mathbb{N} \tag{3}$$
$$\xi(0) = \xi_0 \in M,$$

where $u_t \in U := \mathbb{R} \setminus \sigma(A)$ (the set of admissible control values). We denote the space of finite and infinite admissible control sequences by U^t and $U^{\mathbb{N}}$, respectively. The solution of (3) corresponding to the initial value ξ_0 and a control sequence $u \in U^{\mathbb{N}}$ is denoted by $\varphi(t; \xi_0, u)$. The forward orbit of a point $\xi \in M$ is then given by

$$\mathcal{O}^+(\xi) := \{\eta \in M \mid \exists t \in \mathbb{N}, u \in U^t \text{ such that } \eta = \varphi(t; \xi, u)\}.$$

Similarly, the set of points reachable exactly in time t is denoted by $\mathcal{O}_t^+(\xi)$. System (3) is called *forward accessible* [2], if the forward orbit $\mathcal{O}^+(\xi)$ of every point $\xi \in M$ has nonempty interior and *uniformly forward accessible (in time t)* if there is a $t \in \mathbb{N}$ such that $\text{int}\, \mathcal{O}_t^+(\xi) \neq \emptyset$ for all $\xi \in M$. Note that $\text{int}\, \mathcal{O}^+(\xi) \neq \emptyset$ holds iff there is a $t \in \mathbb{N}$ such that $\text{int}\, \mathcal{O}_t^+(\xi) \neq \emptyset$. Sard's theorem implies then the existence of a control $u \in U^t$ such that

$$\text{rk}\, \frac{\partial \varphi(t; \xi, u)}{\partial u} = n - 1.$$

A pair $(\xi, u) \in M \times U^t$ is called *regular* if this rank condition holds. The control sequence $u \in U^t$ is called *universally regular* if (ξ, u) is a regular pair for every $\xi \in M$. By [16, Corollaries 3.2 & 3.3] forward accessibility is equivalent to the fact that the set of universally regular control sequences

U_{reg}^t is open and dense in U^t for all t large enough. (For a precise statement we refer to [16].)

The following result shows forward accessibility for (3) and gives an easy characterization of universally regular controls.

Lemma 1. *System* (3) *is uniformly forward accessible in time* $n - 1$. *A control sequence* $u \in U^t$ *is universally regular if and only if there are* $n - 1$ *pairwise different values in the sequence* $u_0, \ldots, u_{t-1}$.

2 Controllability of the projected system

By the results of the previous section we know that every point in M has a forward orbit with interior points and it is reasonable to wonder about controllability properties of system (3). As usual, we will call a point $\xi \in M$ controllable to $\eta \in M$ if $\eta \in \mathcal{O}^+(\xi)$. System (3) is said to be completely controllable on a subset $N \subset M$ if for all $\xi \in N$ we have $N \subset \mathcal{O}^+(\xi)$.

In order to analyze the controllability properties of (3) we introduce the following definition of what can be thought of as regions of approximate controllability in $M \subset \mathbb{RP}^{n-1}$. A *control set* of system (3) is a set $D \subset M$ satisfying

(i) $D \subset \mathrm{cl}\,\mathcal{O}^+(\xi)$ for all $\xi \in D$.
(ii) For every $\xi \in D$ there exists a $u \in U$ such that $\varphi(1; x, u) \in D$.
(iii) D is a maximal set (with respect to inclusion) satisfying (i).

An important subset of a control set D is its *core* defined by

$$\mathrm{core}(D) := \{\xi \in D \mid \mathrm{int}\,\hat{\mathcal{O}}^-(\xi) \cap D \neq \emptyset \text{ and } \mathrm{int}\,\hat{\mathcal{O}}^+(\xi) \cap D \neq \emptyset\}.$$

Here $\hat{\mathcal{O}}^-(\xi)$ denotes the points $\eta \in \mathbb{RP}^{n-1}$ such that there exist $t \in \mathbb{N}$, $u_0 \in \mathrm{int}\,U^t$ such that $\varphi(t; \eta, u_0) = \xi$ and (η, u_0) is a regular pair. By this assumption it is evident that on the core of a control set the system is completely controllable.

We are now in a position to state a result characterizing controllability of (3), see [14].

Theorem 1. *Let* $A \in \mathbb{R}^{n \times n}$ *be cyclic. Consider the system* (3) *on* M. *The following statements are equivalent:*

(i) *There exists a* $\xi \in M$ *such that* $\mathcal{O}^+(\xi)$ *is dense in* M.
(ii) *There exists a control set* $D \subset M$ *with* $\mathrm{int}\,D \neq \emptyset$.
(iii) M *is a control set of system* (3).
(iv) *System* (3) *is completely controllable on* M.

(v) There exists a universally regular control sequence $u \in U^t$ such that

$$\prod_{s=0}^{t-1}(A - u_s I)^{-1} \in \mathbb{R}^* I.$$ (4)

The unusual fact about the system we are studying is thus that by the universally regular representation of one element of the system's semigroup we can immediately conclude that the system is completely controllable. Furthermore, already the fact that there is a control set of the system implies complete controllability on the whole state space M. On the other hand it is worth pointing out, that if the conditions of the above theorem are not met, then no forward orbit of (3) is dense in M.

For brevity we will call a cyclic matrix A *II-controllable* (for inverse iteration controllable), if A satisfies any of the equivalent conditions of Theorem 1.

We have another simple characterization of II-controllability in terms of the existence of a universally regular periodic orbit through a cyclic vector v of A. This may come as a surprise.

Corollary 1. *Let $A \in \mathbb{R}^{n \times n}$ be cyclic with cyclic vector v and characteristic polynomial q_A. Consider the system (3) on M. The following statements are equivalent:*

(i) the matrix A is II-controllable.
(ii) There exist $t \in \mathbb{N}$, $u \in U_{reg}^t$ such that $\mathbb{P}v$ is a periodic point for system (3) under the control sequence u.

3 Polynomial characterizations of II-controllability

As has already become evident in the last result of the previous section the question of II-controllability is closely linked to properties of real polynomials. We will now further investigate this relationship. Here we follow the ideas for the complex case in [13] and discuss comparable results for the real case, see [14].

In the following theorem we use the notation $p \wedge q = 1$ to denote the fact that the two polynomials $p, q \in \mathbb{R}[z]$ are coprime.

Theorem 2. *Let $A \in \mathbb{R}^{n \times n}$ be cyclic with characteristic polynomial q. Consider the system (3) on M. The following statements are equivalent:*

(i) the matrix A is II-controllable.
(ii) For every $B \in \Gamma_A := \{p(A) \mid p \in \mathbb{R}[z], p \wedge q = 1\}$ there exist $t \in \mathbb{N}$, $u \in U_{reg}^t$, $\alpha \in \mathbb{R}^*$ such that

$$B = \alpha \prod_{s=0}^{t-1}(A - u_s I),$$

i.e. $\Gamma_A = \Gamma_A^{\mathbb{R}} := \{p(A) \mid p(z) = \alpha \prod_{s=0}^{t-1}(z - u_s), u_s \in \mathbb{R}, p \wedge q = 1, \alpha \in \mathbb{R}^*\}$.

(iii) *For every* $p \in \mathbb{R}[z], p \wedge q = 1$ *there exist* $t \in \mathbb{N}$, $u \in U_{reg}^t$, $\alpha \in \mathbb{R}^*$ *such that*

$$p(z) = \alpha \prod_{s=0}^{t-1}(z - u_s) \mod q(z).$$

(iv) *There exists a monic polynomial* f *with only real roots and at least* $n - 1$ *pairwise different roots,* $\alpha \in \mathbb{R}^*$ *and* $r(z) \in \mathbb{R}[z]$ *such that*

$$f(z) = \alpha + r(z)q(z).\tag{5}$$

Remark 1. ¿From (5) it is easy to deduce the following statement: If for a cyclic matrix A with characteristic polynomial q there exists a monic polynomial f with only real roots that are *all* pairwise distinct such that (5) is satisfied, then there is a neighborhood of A consisting of II-controllable matrices. The reason for this is that, keeping α and $r(z)$ fixed, small changes in the coefficients of q will only lead to small changes in the coefficients of f, and the assumption guarantees that all polynomials in a neighborhood of f have simple real roots.

As an immediate consequence of Theorem 1 we obtain a complete characterization of the reachable sets of the inverse power iteration given by

$$\xi(t+1) = (A - u_t I)^{-1}\xi(t), \quad t \in \mathbb{N}, \qquad \xi(0) = \xi_0 \in \mathbb{RP}^{n-1},\tag{6}$$

for II-controllable matrices $A \in \mathbb{R}^{n \times n}$. This extends a result in [13] for real matrices.

Corollary 2. *Let* A *be II-controllable with characteristic polynomial* q_A, *then*
(i) *for each* $\xi = \mathbb{P}x \in \mathbb{RP}^{n-1}$ *we have*

$$\mathcal{O}^+(\xi) = \mathbb{P} \bigcap_{x \in V, AV \subset V} V \setminus \bigcup_{x \notin V, AV \subset V} V,$$

$$\mathrm{cl}\,\mathcal{O}^+(\xi) = \mathbb{P} \bigcap_{x \in V, AV \subset V} V = \mathbb{P}\,\mathrm{span}\{x, Ax, A^2x, \ldots, A^{n-1}x\}.$$

(ii) *There is a one-to-one correspondence between*

a) *The forward orbits of system (6).*
b) *The closures of the forward orbits of system (6).*
c) *The* A-*invariant subspaces of* $\mathbb{R}^n$.
d) *The factors of* $q_A(z)$ *over the polynomial ring* $\mathbb{R}[z]$.

4 Conditions for II controllability

The result of the previous section raises the question which cyclic matrices A admit a representation of the form (4) or equivalently when (5) is possible. With respect to this question we have the following preliminary results.

Proposition 1. *Let $A \in \mathbb{R}^{n \times n}$ be cyclic with characteristic polynomial q_A.*

(i) *A is not II-controllable, if it satisfies one of the following conditions*
   (a) *A has a nonreal eigenvalue of multiplicity $\mu > 1$.*
   (b) *A has a real eigenvalue of multiplicity $\mu > 2$.*
(ii) *A is II-controllable, if $\sigma(A) \subset \mathbb{R}$ and no eigenvalue has multiplicity $\mu > 2$.*

In general, a complete characterization of the set of cyclic matrices that is not II-controllable is not known. Several examples, showing obstructions to this property in terms of the location of the eigenvalues are discussed in detail in [14]. These are obtained via the following result.

Proposition 2. *Let $A \in \mathbb{R}^{n \times n}$ be cyclic.*

(i) *If for two eigenvalues $\lambda_1, \lambda_2 \in \sigma(A)$ we have $\operatorname{Re} \lambda_1 = \operatorname{Re} \lambda_2, |\lambda_1| \neq |\lambda_2|$ then A is not II-controllable.*
(ii) *If the spectrum $\sigma(A)$ is symmetric with respect to rotation by a root of unity, i.e. $\sigma(A) = \exp(2\pi i/m)\sigma(A)$ (taking into account multiplicities) and two eigenvalues of A^m satisfy the condition of (i) then A is not II-controllable.*
   If, furthermore, m is even, then it is sufficient that for two eigenvalues of A^m we have

$$|\lambda_1 - u| < |\lambda_2 - u|,$$

   for all $u > 0$ in order that A is not II-controllable.

Using this corollary it is easy to construct examples of matrices that are not II-controllable. Such are e.g. the companion matrices of the polynomial $p(z) = z(z^2 + 1)$ and the 7-th degree polynomial whose roots are 0, the three cubic roots of i and their respective complex conjugates. Using the last statement, one sees that the matrix corresponding to $p(z) = (z^2 - 1)(z^2 + 1)$ is not II-controllable. Many more examples like this can be constructed, some more examples are discussed in [14].

For the case $n \leq 3$ the following complete result can be given.

Proposition 3. *Let $A \in \mathbb{R}^{n \times n}$ be cyclic.*

(i) *If $n = 1, 2$ then A is II-controllable.*
(ii) *If $n = 3$ then A is II-controllable if and only if the eigenvalues $\lambda_1, \lambda_2, \lambda_3$ of A do not have a common real part, i.e. do not satisfy $\operatorname{Re} \lambda_1 = \operatorname{Re} \lambda_2 = \operatorname{Re} \lambda_3$.*

5   Control system on rational functions

There is an interesting reformulation of the inverse Rayleigh iteration as an equivalent control system on rational function spaces. This connects up with the work by Brockett and Krishnaprasad [9] on scaling actions on rational functions, as well as with divided difference schemes in interpolation theory. Let $(c, A) \in \mathbb{R}^{1 \times n} \times \mathbb{R}^{n \times n}$ be an observable pair and let $q(z) := \det(zI - A)$ denote the characteristic polynomial. Let

$$\mathrm{Rat}(q) := \left\{ \frac{p(z)}{q(z)} \in \mathbb{R}(z) \,\middle|\, \deg p < \deg q \right\}$$

denote the real vectorspace of all strictly proper real rational functions with fixed denomination polynomial $q(z)$. The map

$$\varphi : \mathbb{R}^n \to \mathrm{Rat}(q)$$
$$x \mapsto c(zI - A)^{-1}x$$

defines a bijective isomorphism between $\mathbb{R}^n$ and $\mathrm{Rat}(q)$. We use it to transport the inverse Rayleigh iteration onto $\mathrm{Rat}(q)$. Let $\mathbb{P}(\mathrm{Rat}(q))$ denote the associated projective space; i.e. two rational functions $g_1(z), g_2(z) \in \mathrm{Rat}(q)$ define the same element $\mathbb{P}g_1 = \mathbb{P}g_2$ in $\mathbb{P}(\mathrm{Rat}(q))$ if and only if g_1 and g_2 differ by a nonzero constant factor. Then φ induces a homeomorphism

$$\phi : \mathbb{RP}^{n-1} \to \mathbb{P}(\mathrm{Rat}(q))$$
$$\phi(\mathbb{P}x) := \mathbb{P}c(zI - A)^{-1}x.$$

Let $(\mathbb{P}x_t)_{t \in \mathbb{N}}$ denote the sequence in $\mathbb{RP}^{n-1}$ generated by the inverse power iteration (3). Then, for

$$g_t(z) := c(zI - A)^{-1}x_t$$

we obtain the divided difference scheme

$$\mathbb{P}g_{t+1}(z) = \mathbb{P}c(zI - A)^{-1}(A - u_t I)^{-1}x_t$$
$$= \mathbb{P}\left(\frac{g_t(z) - g_t(u_t)}{z - u_t} \right).$$

Conversely, if $g_0(z) = c(zI - A)^{-1}x_0 \in \mathrm{Rat}(q)$ and $(\mathbb{P}g_t)_{t \in \mathbb{N}}$ is recursively defined by

$$g_{t+1}(z) = \frac{g_t(t) - g_t(u_t)}{z - u_t}, t \in \mathbb{N}_0,$$

then $g_t \in \mathrm{Rat}(q)$ for all $t \in \mathbb{N}$ and

$$g_t(z) = c(zI - A)^{-1}x_t$$

for a sequence $(\mathbb{P}x_t)_{t\in\mathbb{N}_0}$ generated by the inverse Rayleigh iteration. Thus the inverse Rayleigh iteration (3) on $\mathbb{R}\mathbb{P}^{n-1}$ is equivalent to the divided difference control system

$$\mathbb{P}g_{t+1}(z) = \mathbb{P}\left(\frac{g_t(z) - g_t(u_t)}{z - u_t}\right) \tag{7}$$

on $\mathbb{P}(\mathrm{Rat}(q))$.

Equivalently, we can reformulate this algorithm as a control system on polynomials of degree $< n$. To this end let

$$\mathbb{R}_n[z] := \{p \in \mathbb{R}[z] \mid \deg p < n\}$$

denote the vectorspace of polynomials of degree $< n$ and let $\mathbb{P}(\mathbb{R}_n[z])$ denote the associated projective space. Note that for any polynomial $p \in \mathbb{R}_n[z]$ and $u \in \mathbb{R}$

$$\hat{P}(z) := \frac{p(z)q(u) - p(u)q(z)}{z - u}$$

is again a polynomial of degree $< n$. Thus

$$\mathbb{P}(p_{t+1}(z)) = \mathbb{P}\left(\frac{p_t(z)q(u_t) - p_t(u_t)q(z)}{z - u_t}\right) \tag{8}$$

defines a control system on $\mathbb{P}(\mathbb{R}_n[z])$.

Since

$$\mathbb{P}\left(\frac{p_{t+1}}{q}(z)\right) = \mathbb{P}\left(\frac{\frac{p_t(z)}{q(z)} - \frac{p_t(u_t)}{q(u_t)}}{z - u_t}\right)$$

we see (for $q(u_t) \neq 0$) that the control systems (7) and (8) are equivalent.

6 Inverse iteration on flag manifolds

A well known extension of the inverse Rayleigh iteration (3) is the QR-algorithm. To include such algorithms in our approach we have to extend the analysis to inverse iterations on partial flag manifolds. A full analysis is beyond the scope of this paper and will be presented elsewhere.

For simplicity we focus on the complex case. Recall, that a partial *flag* in $\mathbb{C}^n$ is an increasing sequence $\{0\} \neq V_1 \subsetneqq \ldots \subsetneqq V_k \subset \mathbb{C}^n$ of $\mathbb{C}$-linear subspaces of $\mathbb{C}^n$. The *type* of the flag $(V_1, \ldots, V_k)$ is specified by the k-tuple $a = (a_1, \ldots, a_k)$ of dimensions

$$a_i = \dim_{\mathbb{C}} V_i, \quad i = 1, \ldots, k.$$

Thus

$$1 \leq a_1 < \ldots < a_k \leq n.$$

For any such sequence of integers $a = (a_1, \ldots, a_k)$, $1 \leq a_1 < \ldots < a_k \leq n$, let $\text{Flag}(a, \mathbb{C}^n)$ denote the set of all flags $(V_1, \ldots, V_k)$ of type a. The set $\text{Flag}(a, \mathbb{C}^n)$ is called a (partial) *flag manifold*. It is indeed a compact complex manifold. For $k = 1$, $a := a_1$, we obtain the *Grassmann manifold* $G_a(\mathbb{C}^n)$ as a special case while for $k = n$ and $a = (1, 2, \ldots, n)$ we obtain the (full) flag manifold

$$\text{Flag}(\mathbb{C}^n) := \text{Flag}((1, \ldots, n), \mathbb{C}^n).$$

For any linear map $A : \mathbb{C}^n \to \mathbb{C}^n$ and any sequence (u_t), $u_t \notin \sigma(A)$, of complex numbers we obtain the *inverse iteration on flag manifolds*

$$(A - u_t I)^{-1} : \text{Flag}(a, \mathbb{C}^n) \to \text{Flag}(a, \mathbb{C}^n).$$

This defines a nonlinear control system on the flag manifold. The reachable sets are again easily seen to be equal to

$$\mathcal{R}_A(\mathcal{V}) := \{ \prod_{i=0}^{t-1} (A - u_i I)^{-1} \mathcal{V} \mid t \in \mathbb{N}, u_i \notin \sigma(A) \}$$

$$= \Gamma_A^{\mathbb{C}} \cdot \mathcal{V} \qquad\qquad , \mathcal{V} \in \text{Flag}(a, \mathbb{C}^n)$$

where the semigroup

$$\Gamma_A^{\mathbb{C}} = \{ p(A) \mid p(z) = \alpha \prod_{i=0}^{t-1} (z - u_i), \alpha \in \mathbb{C}^*, q(u_i) \neq 0 \}$$

is defined as in Section 3. Since $\dim \Gamma_A^{\mathbb{C}} \leq n$ and since $\Gamma_A^{\mathbb{C}}$ acts with a stabilizer of dimension ≥ 1 on $\text{Flag}(a, \mathbb{C})$ we obtain

$$\dim \mathcal{R}_A(\mathcal{V}) \leq n - 1, \quad \forall \mathcal{V} \in \text{Flag}(a, \mathbb{C}^n).$$

Now $\dim \text{Flag}(a, \mathbb{C}^n) > n - 1$, except for the cases $k = 1$, $k = n - 1$ or $n = 2$. Thus we conclude

Proposition 4. *Except for $k = 1$, $k = n - 1$ or $n = 2$, the reachable sets of the inverse iteration on the flag manifold $\text{Flag}(a, \mathbb{C}^n)$ have empty interior.*

Moreover, since the QR-algorithm with shifts is equivalent to the inverse iteration

$$(A - u_t I)^{-1} : \text{Flag}(\mathbb{C}^n) \to \text{Flag}(\mathbb{C}^n)$$

we obtain

Corollary 3. *The QR-algorithm with origin shifts is not locally accessible nor controllable, if $n \geq 3$.*

We now describe in more detail the structure of the reachable sets. For simplicity we assume that $A \in \mathbb{C}^{n \times n}$ is a diagonal matrix with has distinct eigenvalues and we focus on the inverse iteration on Grassmann manifolds

$$(A - u_t I)^{-1} : G_k(\mathbb{C}^n) \to G_k(\mathbb{C}^n).$$

For any full rank matrices $X \in \mathbb{C}^{n \times k}$ let

$$[X] := \text{Im} X \in G_k(\mathbb{C}^n)$$

denote the k–dimensional subspace spanned by the columns of X.

For any increasing sequence α of integers $1 \leq \alpha_1 < \ldots < \alpha_r \leq n$ let X_α denote the $r \times k$ submatrix formed by the rows $\alpha_1, \ldots, \alpha_r$ of X.

Definition 1. (a) Two complex linear subspaces $[X], [Y]$ in $\mathbb{C}^n$ of dimension k are called *rank equivalent* if

$$rk X_\alpha = rk Y_\alpha$$

for all $1 \leq \alpha_1 < \ldots < \alpha_r \leq n$ and $r = 1, \ldots, n$.

(b) Rank equivalence defines an equivalence relation on $G_k(\mathbb{C}^n)$. The equivalence classes are called *Grassmann simplices* of $G_k(\mathbb{C}^n)$.

For example, the following two matrices span rank equivalent subspaces.

$$X = \text{span} \begin{pmatrix} 1 & 0 \\ 3 & 4 \\ 0 & 1 \\ 0 & 2 \end{pmatrix}, \qquad Y = \text{span} \begin{pmatrix} 1 & 0 \\ 1 & 1 \\ 0 & 1 \\ 0 & 1 \end{pmatrix}.$$

The stabilization of Grassmann manifolds into Grassmann simplices has been introduced by Gelfand et.al. [7], [8]. For us they are of interest because of the following fact. (Remember that A is diagonal!)

Lemma 2. *Every reachable set $\mathcal{R}_A([X])$ in $G_k(\mathbb{C}^n)$ is contained in a Grassmann simplex.*

To obtain a more precise description of reachable sets and Grassmann simplices we consider a projection of the Grassmannian on a polytope.

For any subset $\alpha = \{\alpha_1, \ldots, \alpha_r\} \subset \bar{n} := \{1, \ldots, n\}$, $1 \leq \alpha_1 < \ldots < \alpha_r \leq n$, let

$$e_\alpha := e_{\alpha_1} + \cdots + e_{\alpha_r}$$

where e_i, $1 \leq i \leq n$, denotes the i-th standard basis vector of $\mathbb{C}^n$. For any full rank matrix $X \in \mathbb{C}^{n \times k}$ define

$$\mu(X) := \frac{\displaystyle\sum_{1 \leq \alpha_1 < \ldots < \alpha_k \leq n} |\det X_\alpha|^2 e_\alpha}{\displaystyle\sum_{\substack{1 \leq \beta_1 \leq \ldots \leq \beta_r \leq n \\ 1 \leq r \leq n}} |\det(X_\beta X_\beta^T)|}.$$

Then $\mu(X) = \mu(XS^{-1})$ for any invertible matrix $S \in \mathbb{C}^{k \times k}$ and thus $\mu(X)$ defines a smooth map

$$\mu : G_k(\mathbb{C}^n) \to \mathbb{R}^n, \ \mu([X]) := \mu(X)$$

on the Grassmann manifold. We refer to it as the *moment map* on $G_k(\mathbb{C}^n)$. It is easily seen that the image of μ in $\mathbb{R}^n$ is a convex polytope. More precisely we have

$$\mu(G_k(\mathbb{C}^n)) = \Delta_{k,n}$$

where $\Delta_{k,n}$ denotes the hypersimplex

$$\Delta_{k,k} := \{(t_1, \ldots, t_n) \in \mathbb{R}_+^n \mid t_1 + \ldots + t_n = k\}.$$

The following result by Gelfand et.al. [7] describes the geometry of Grassmann simplices in terms of the moment map.

Theorem 3. *(a) Every reachable set $\mathcal{R}_A([X])$, $[X] \in G_k(\mathbb{C}^n)$, is contained in a Grassmann simplex. More precisely, two subspaces $[X], [Y] \in G_k(\mathbb{C}^n)$ are rank equivalent if and only if*

$$\mu(\mathcal{R}_A([X])) = \mu(\mathcal{R}_A([Y])).$$

(b) $\mu(\overline{\mathcal{R}_A([X])})$ is a compact polytope in $\mathbb{R}^n$ with vertices $\{e_\alpha \mid \det X_\alpha \neq 0\}$. It is a closed subface of $\Delta_{k,n}$.

(c) There is a bijective correspondence between
  (i) p–dimensional reachable sets in $\overline{\mathcal{R}_A([X])}$.
  (ii) Open p–dimensional faces of $\mu(\overline{\mathcal{R}_A([X])})$.

7 Conclusions

Controllability properties of inverse iteration schemes provide fundamental limitation for any numerical algorithm defined by them in terms of suitable feedback strategies. In the complex case, reachable sets for the inverse iteration on projective space $\mathbb{CP}^{n-1}$ correspond bijectively to invariant subspaces of A. Moreover, complete controllability holds if and only if A is cyclic, see [13]. The real case is considerably harder and only partial results for complete controllability in terms of necessary or sufficient conditions are given.

Differences also occur for inverse iteration on Grassmannians or flag manifolds. The algorithms are never controllable, in particular the QR-algorithm is seen to be not controllable. Reachable sets are contained in Grassmann simplices and their adherence relation is described by the combinations of faces of a hypersimplex.

References

1. Albertini, F. (1993) Controllability of discrete-time nonlinear systems and some related topics, PhD thesis, Grad. School New Brunswick Rutgers, New Brunswick, New Jersey.
2. Albertini, F., Sontag, E. (1993) Discrete-time transitivity and accessibility: analytic systems, SIAM J. Contr. & Opt. **31**, 1599–1622
3. Batterson, S. (1995) Dynamical analysis of numerical systems. Numer. Linear Algebra Appl. **2**, 297-310
4. Batterson, S., Smillie, J. (1990) Rayleigh quotient iteration for nonsymmetric matrices. Math. Comput. **55**, 169-178
5. Batterson, S., Smillie, J. (1989) Rayleigh quotient iteration fails for nonsymmetric matrices. Appl. Math. Lett. **2**, 19-20
6. Batterson, S., Smillie, J. (1989) The dynamics of Rayleigh quotient iteration. SIAM J. Numer. Anal. **26**, 624-636
7. Gelfand, I.M., Serganova, V.V. (1987) Combinatorial geometries and torus strata on homogeneous compact manifolds. Uspekhi Math. Nauk. **42**, 107–134
8. Gelfand, I.M., Goresky, R.M., MacPherson, R.D., Serganova, V.V. (1987) Combinatorial geometries, convex polyhedra and Schubert cells. Adv. in Math. **63**, 301–316
9. Brockett, R.W., Krishnaprasad, P.S. (1980) A scaling theory for linear systems, IEEE Trans. Autom. Control **AC-25**, 197-207
10. Colonius, F., Kliemann, W. (1993) Linear control semigroups acting on projective space. J. Dynamics Diff. Equations **5**, 495–528
11. Colonius, F., Kliemann, W. (1996) The Lyapunov spectrum of families of time varying matrices. Trans. Amer. Math. Soc. **348**, 4389–4408
12. Fuhrmann, P. A. (1996) A Polynomial Approach to Linear Algebra, Springer Publ., New York
13. Fuhrmann, P. A., Helmke, U. (2000) On controllability of matrix eigenvalue algorithms: The inverse power method, Syst. Cont. Lett., to appear.
14. Helmke, U., Wirth, F. (2000) On controllability of the real shifted inverse power iteration, Syst. Cont. Lett., to appear.
15. Ipsen, I. C. F. (1997) Computing an eigenvector with inverse iteration. SIAM Rev. **39**, 254-291
16. Sontag, E. D., Wirth F. R (1998) Remarks on universal nonsingular controls for discrete-time systems, Syst. Cont. Lett. **33**, 81-88
17. Suzuki, T (1992) Inverse iteration method with a complex parameter. Proc. Japan Acad., Ser. A **68**, 68-73

18. Suzuki, T (1994) Inverse iteration method with a complex parameter. II. in Yajima, K. (ed.), Spectral scattering theory and applications. Proceedings of a conference on spectral and scattering theory held at Tokyo Institute of Technology, Japan, June 30-July 3, 1992 in honour of Shige Toshi Kuroda on the occasion of his 60th birthday. Tokyo: Kinokuniya Company Ltd.. Adv. Stud. Pure Math. **23**, 307-310

19. Wirth, F. (1998) Dynamics of time-varying discrete-time linear systems: Spectral theory and the projected system. SIAM J. Contr. & Opt. **36**, 447-487

20. Wirth, F. (1998) Dynamics and controllability of nonlinear discrete-time control systems. In 4th IFAC Nonlinear Control Systems Design Symposium (NOLCOS'98) Enschede, The Netherlands, 269 - 275

On the Discretization of Sliding-mode-like Controllers

Guido Herrmann*, Sarah K. Spurgeon, and Christopher Edwards

Control Systems Research Group
University of Leicester
Leicester LE1 7RH, U.K.
gh17@sun.engg.le.ac.uk

Abstract. Stability of a discretized continuous sliding-mode based state feedback control is proved using an L_2-gain analysis result for linear continuous-time systems with sampled-data output. It has been shown before that a strictly proper linear continuous-time system with sampled-data output has finite L_2-gain. This gain converges to the L_2-gain associated with the continuous-time output when the sampling period approaches $+0$. This result is incorporated in the analysis of the discretized sliding-mode based control applying techniques from non-linear L_2-gain theory. The result is then compared to a Lyapunov function analysis based approach. In contrast to the Lyapunov function technique, the sampling-time constraint vanishes for a stable plant if no control is used. Numerical results are demonstrated for a particular example, the control of the non-linear inverted pendulum.

1 Introduction

Analysis of sampled-data implementations of continuous-time control systems has been of interest for a long period. This is because controllers are easily developed in continuous time but often need to be applied via sampled-data technology. Non-linear systems often rely on Lyapunov function based analysis techniques [1–5]. Such an approach of [3,4] has been useful for determining sampling-frequencies which are sufficient for stable, robust closed loop control using a discretized non-linear control. The technique has been applied successfully to a sliding-mode based control. In contrast, linear time-invariant (LTI) systems are particularly useful for the application of $\mathcal{H}_\infty$ techniques [6]. An extensive frame-work of theory has been developed for linear, sampled-data control of linear systems [6–8]. An important result for the discretization of LTI systems is that any stable linear system (A, B, C) followed by a sample-and-hold process has finite L_p-induced norm for constant sampling time $\tau > 0$. Furthermore, the L_p-norm of the discretized output signal approaches the L_p-norm of the continuous output signal of (A, B, C) as $\tau \to +0$. This result has also been shown for linear time-varying systems [7].

---
* G. Herrmann is supported by a grant of the European Commission (TMR-grant, project number: FMBICT983463).

L_2-gain analysis techniques have proved to be useful for non-linear systems as the basis for non-linear $\mathcal{H}_\infty$-theory. Furthermore, a Lyapunov function can be derived from the Hamilton-Jacobi inequality, which is a fundamental tool of L_2-gain theory. In particular, exponentially stabilizing non-linear controls imply a finite L_2-gain. Thus, a non-linear continuous-time sliding-mode based control derived from [9] will be investigated with respect to its sampled-data implementation. This continuous-time state feedback control exhibits exponential stability properties and is robust with respect to matched parametric uncertainty. The states ultimately reach a cone-shaped boundary layer around a sliding-mode plane. L_2-gain techniques are applied to show that the discretized control is stable and robust to the same class of uncertainty if the sampling time τ is chosen small enough. The L_p-gain result for LTI systems of [6] is incorporated, implying that a small gain relationship needs to be satisfied for stability of the non-linear control.

A practically valid example, the control of an inverted pendulum, is used to compare the results with those using a Lyapunov function based approach [3,9]. It will be seen that both approaches have their advantages and disadvantages from a theoretical and practical point of view. In the Lyapunov function based approach, an upper bound for the norm difference between the time-sampled state $x(t_i)$ and the value of the continuous signal $x(t)$ is found for each particular finite sampling interval $[t_i, t_{i+1}]$:

$$\sup_{t_\alpha \in [t_i, t_{i+1}]} (\|x(t_\alpha) - x(t_i)\|) \leq \tau \mathcal{K} \|x(t)\|, \ \mathcal{K} \geq 0, \ t \in [t_i, t_{i+1}]. \tag{1}$$

The upper bound decreases with decreasing sampling time τ. A problem of this approach [3,2,4] is that the relation (1) can only be derived if an upper bound $\bar{\tau}$ on the sampling-time $\tau < \bar{\tau}$ is assumed. For (1) being valid, this upper bound $\bar{\tau}$ remains finite for any control even if the non-linear system is stable and the applied control is zero. Thus, a minimum sampling frequency is always indirectly imposed. This minimum sampling frequency is not necessary in the L_2-induced norm approach.

This paper is structured as follows. The continuous sliding-mode-like control is introduced in Sect. 2. Section 3 deals with the proof of stability of the discretized control while the example of the control of the inverted pendulum is provided in Sect. 4. The results of [6] for sampled data LTI systems are given in the Appendix.

2   A Sliding-Mode-Like Continuous Control Law

Linear, uncertain systems shall be considered

$$\dot{x} = Ax + Bu + \mathbf{F}(t, x) \tag{2}$$

where $x \in \mathbb{R}^n$, $u \in \mathbb{R}^m$ and the known matrix pair (A, B) defines the nominal linear system, which is assumed to be controllable with B of full rank. The unknown function $\mathbf{F}(.,.) : \mathbb{R} \times \mathbb{R}^n \rightarrow \mathbb{R}^n$ models parametric uncertainties and non-linearities of the system lying in the range space of B, whilst unmatched and actuator uncertainty has been neglected for simplicity in contrast to [9]. Under certain assumptions on the class of functions, $\mathcal{F}$, to which the uncertainty $\mathbf{F}$ belongs, it is possible to define a linear transformation $\tilde{T}$ such that the system (2) becomes:

$$\dot{z}_1 = \Sigma z_1 + A_{12}\phi, \tag{3a}$$
$$\dot{\phi} = (\Theta + \Delta_\Theta) z_1 + (\Omega + \Delta_\Omega) \phi + B_2 u, \tag{3b}$$

where

$$\tilde{z} = \tilde{T} x = \left[z_1^T \ \phi^T \right]^T, \tag{4}$$

and Σ is a stable design matrix. The sub-system (3a) defines *the null-space dynamics* and the sub-system (3b) represents the *range-space dynamics*. The matrices Δ_Θ and Δ_Ω are bounded so that

$$\Delta_\Theta(t, \tilde{z}) : \mathbb{R} \times \mathbb{R}^n \rightarrow \mathbb{R}^{m \times (n-m)}, \ \|\Delta_\Theta\| \leq K_\Theta,$$
$$\Delta_\Omega(t, \tilde{z}) : \mathbb{R} \times \mathbb{R}^n \rightarrow \mathbb{R}^{m \times m}, \ \|\Delta_\Omega\| \leq K_\Omega.$$

The above represent the matched uncertainty and disturbances. The uncertain matrices stem from the transformation of $\mathbf{F}(.,.)$ and the associated non-negative bounds K_Θ and K_Ω are assumed known and finite. It is assumed that $\mathbf{F}(.,.)$ is Caratheodory to ensure existence of solution (see [3,10]). In the absence of any uncertainty in the null-space dynamics and if $\phi(t) = 0$, the null-space dynamics reduce to $\dot{z}_1(t) = \Sigma z_1(t)$, which corresponds to the system attaining an ideal sliding mode. It can be shown similar to [9], that the sliding-mode-like control law outlined below forces the system states $\tilde{z}$ to satisfy $\phi(t) \approx 0$ and the control is robust to matched parametric uncertainty. This control has two parts:

$$u(t) = u_L^C(z_1(t), \phi(t)) + u_{NL}^C(z_1(t), \phi(t)) \tag{5}$$

where $u_L^C(\cdot)$ and $u_{NL}^C(\cdot)$ are the linear and the non-linear control components. The linear controller part

$$u_L^C(z_1(t), \phi(t)) \stackrel{def}{=} -B_2^{-1} \left(\Theta z_1(t) + (\Omega - \Omega^*)\phi(t) \right) \tag{6}$$

mainly enhances reachability of a cone shaped layer around the sliding mode, where Ω^* is a stable design matrix and P_2 satisfies $P_2 \Omega^* + \Omega^{*T} P_2 = -I_m$, where I_m is the $(m \times m)$ identity matrix. Thus, a Lyapunov function $V_2(t)$ for the analysis of the range-space dynamics is

$$V_2(t) \stackrel{def}{=} \frac{1}{2}\phi^T(t) P_2 \phi(t), \tag{7}$$

and

$$V_1(t) \overset{def}{=} \frac{1}{2} z_1{}^T(t) P_1 z_1(t) \tag{8}$$

with $P_1 \Sigma + \Sigma^T P_1 = -I_{(n-m)}$ is appropriate for the null-space dynamics. The non-linear control component

$$\boldsymbol{u}_{NL}^C(\boldsymbol{z}_1(t), \boldsymbol{\phi}(t)) \overset{def}{=} \begin{cases} \frac{-\varrho(\boldsymbol{z}_1(t), \boldsymbol{\phi}(t)) B_2^{-1} P_2 \boldsymbol{\phi}(t)}{\|P_2 \boldsymbol{\phi}(t)\| + \delta(\|\boldsymbol{z}_1(t)\| + \|\boldsymbol{\phi}(t)\|)} & for \ \left[\boldsymbol{z}_1^T \ \boldsymbol{\phi}^T \right]^T \neq 0 \\ 0 & for \ \left[\boldsymbol{z}_1^T \ \boldsymbol{\phi}^T \right]^T = 0 \end{cases}, \tag{9}$$

achieves robustness by counteracting the matched uncertainties and ensuring reachability of a cone shaped layer around the sliding mode. The constant $\delta \in \mathbb{R}^+$ is a design parameter which is implicitly bounded by the maximum size of the cone shaped boundary layer. The expression

$$\delta(\|\boldsymbol{z}_1(t)\| + \|\boldsymbol{\phi}(t)\|)$$

results in the cone shaped layer around $\phi = 0$, which is defined by the relation $V_2 = \omega^2 V_1$, where $\omega \in \mathbb{R}^+$ is a small, positive design dependent constant. Hence, if $V_2 \geq \omega^2 V_1$ holds, then the states are outside the cone shaped layer. The constant ω

$$\omega \overset{def}{=} \frac{\lambda_{max}(P_2^{\frac{1}{2}}) \delta / \lambda_{min}(P_1^{\frac{1}{2}})}{\left(\lambda_{max}(P_2^{\frac{1}{2}}) \lambda_{min}(P_2^{\frac{1}{2}})(\gamma_1 - 1) - \delta \right)}$$

has to be positive, which can be ensured by imposing the constraint

$$\lambda_{min}(P_2^{\frac{1}{2}}) \lambda_{max}(P_2^{\frac{1}{2}}) > \frac{\delta}{\gamma_1 - 1},$$

which guarantees that ω remains positive. The function $\varrho(\boldsymbol{z}_1(t), \boldsymbol{\phi}(t))$ is defined as:

$$\varrho(\boldsymbol{z}_1(t), \boldsymbol{\phi}(t)) \overset{def}{=} \gamma_1(\eta_1 \|P_2 \boldsymbol{\phi}(t)\| + \eta_2 \|\boldsymbol{z}_1(t)\|). \tag{10}$$

where $\gamma_1 > 1$. The gains η_1 and η_2 in (10) have been defined so that they ensure robustness with respect to the matched disturbances and reachability of the cone shaped sliding mode layer

$$\eta_1 \overset{def}{=} \max\left(\sup_{\Delta_\Omega} \left(\frac{1}{2} \lambda_{max} \left\{ P_2^{-1} \Upsilon^T + \Upsilon P_2^{-1} \right\} \right), 0 \right), \tag{11}$$

$$\eta_2 \overset{def}{=} \sup_{\Delta_\Theta} (\|\Delta_\Theta\|) + \omega^2 \|P_1 A_{12} P_2^{-1}\| + \frac{\omega}{2} \|P_1^{-\frac{1}{2}}\| \|P_2^{-\frac{1}{2}}\|, \tag{12}$$

$$\Upsilon \overset{def}{=} ((1 - \gamma_2)\Omega^* + \Delta_\Omega), \ 1 \geq \gamma_2 > 0. \tag{13}$$

However, it is also possible to adjust the parameter $1 \geq \gamma_2 > 0$ for a compromise of using linear control for reachability ($\gamma_2 = 1$) or robustness ($\gamma_2 \to 0$).

A proof of stability [9] makes it necessary to show that the null space Lyapunov function $V_1(t)$ in (8) is decreasing as soon as the states have entered the cone shaped layer, i.e. by imposing a quadratic stability constraint which will limit the choice of ω with an upper bound. An implicit bound on ω will be given in Sect. 3. Since the control presented here is a modification of [9], it is easily possible to use a theorem of [9] for proving stability of the continuous time control. Further, it can be shown that $V_2 - \omega^2 V_1$ will become ultimately smaller than an arbitrary $\varepsilon > 0$ (sliding-mode-like performance).

3   Discretizing the Sliding-Mode-Like Control Using the L_2-Gain Approach

Instead of the continuous-time control $u_L^C(z_1, \phi) + u_{NL}^C(z_1, \phi)$ (6),(9), a time-discretized control may be applied to the linear, uncertain system (3a-3b):

$$
\begin{aligned}
\forall t \in (t_i, t_{i+1}] : u(t) &= u_L^\Delta(t, z_1, \phi) + u_{NL}^\Delta(t, z_1, \phi) \\
&= u_L^C(z_1(t_i), \phi(t_i)) + u_{NL}^C(z_1(t_i), \phi(t_i)), \\
u(t_0) &= 0,
\end{aligned}
\tag{14}
$$

where the sampling instants $t_i = i\tau$, $i = 0, 1, 2, 3, ...$ are defined by a constant sampling time τ. To symbolize the change in methodology from continuous to discrete-time, the superscript Δ(=discrete) has been introduced. Observe that the sliding-mode-like control $u_L^C(z_1, \phi) + u_{NL}^C(z_1, \phi)$ of (5-6) and (9) is globally Lipschitz [3] with Lipschitz constant $\mathcal{K}_u$. Due to this characteristic, it is possible to apply techniques from L_2-gain analysis to prove that the discrete controlled system is asymptotically stable for a sufficiently small sampling time:

Theorem 1 *There exists a $\hat{\tau}$ small enough such that for any $\tau < \hat{\tau}$, the discretized control (14) stabilizes the system (3a-3b) asymptotically. The L_2-norm of $\begin{bmatrix} z_1^T & \phi^T \end{bmatrix}^T$ for the discrete controlled system is finite. The constraints on the sampling time τ vanishes, $\hat{\tau} \to +\infty$, for a control with Lipschitz constant $\mathcal{K}_u \to +0$ and an exponentially stable system (3a-3b).*

Note the contrast to [3] where relations similar to (1) need to hold and therefore the upper bound on the sampling time τ remains finite for any control even if the control is 0 and the system is open-loop stable. The proof of Theorem 1 is sketched below.

Proof 1
Step I: The existence of solution of (2) has been proved by [3] for the time

interval $[0, \infty)$ implying that any signal occurring in the system is Lebesgue integrable and has finite L_2-norm on the finite horizon $T < \infty$.

Step II: A global Lyapunov function can be provided by non-smooth theory using a max-Lyapunov function as described in [3,4]:

$$V(z_1(t), \phi(t)) = \max\left(V_1(z_1(t)) + kV_2(\phi(t)), \ (k\omega^2 + 1)V_1(z_1(t))\right)$$
$$= \begin{cases} V_1(z_1(t)) + kV_2(\phi(t)) & for \ V_2 \geq \omega^2 V_1 \\ (k\omega^2 + 1)V_1(z_1(t)) & for \ V_2 \leq \omega^2 V_1 \end{cases}$$

(15)

where $k > 0$ has to be determined [3]. It has been shown in [3] that this function is differentiable for almost all t since the max-Lyapunov function is globally Lipschitz in z_1 and ϕ and the solution $z_1(t)$ and $\phi(t)$ is absolutely continuous in t [11]. Thus, the equality

$$\int^t \dot{V}(z_1(s), \phi(s))ds = (V(z_1(t), \phi(t)) + const.)$$

applies. This integrability feature of $\dot{V}(z_1(t), \phi(t))$ allows the non-smooth max-Lyapunov function to be used in a similar fashion to the usual Lyapunov function. Further, it can be shown similar to [3,11] that the non-smooth max-Lyapunov function satisfies for the continuous-time controlled system the relation

$$\dot{V} \leq -\tilde{\vartheta}V, \ \tilde{\vartheta} > 0$$

for almost all t. This implies exponentially fast decay of $V(z_1(t), \phi(t))$ and exponential stability of the continuous-time controlled system.

Step III: The differential system

$$\dot{\hat{z}}_1 = \Sigma \hat{z}_1 + A_{12}\dot{\phi},$$

(16)

$$\dot{\hat{\phi}} = \left[(\Theta + \Delta_\Theta)\hat{z}_1 + (\Omega + \Delta_\Omega)\hat{\phi} + B_2\left(u_L^c + u_{NL}^c\right)\right] + B_2\hat{u},$$

(17)

with input perturbation $\hat{u} \in \mathbb{R}^m$ is now investigated. The system is derived from (3a-3b) using the continuous-time control (6),(9). Observe that the system (16-17) has a Caratheodory-type solution if the input $\hat{u}$ is of Caratheodory type. Further, the auxiliary output for (16-17) may be introduced:

$$\tilde{y} = \underbrace{\left(\begin{bmatrix} \Sigma & A_{12} \\ 0 & \Omega^* \end{bmatrix} + \begin{bmatrix} 0 & 0 \\ \Delta_\Theta & \Delta_\Omega \end{bmatrix} + \begin{bmatrix} 0 & 0 \\ 0 & \psi \end{bmatrix} - F\right)}_{\tilde{C}} \begin{bmatrix} z_1 \\ \phi \end{bmatrix},$$

(18)

where

$$\psi = \begin{cases} \dfrac{-\varrho(z_1(t), \phi(t))P_2}{\|P_2\phi(t)\| + \delta(\|z_1(t)\| + \|\phi(t)\|)} & for \ \left[z_1^T \ \phi^T\right]^T \neq 0 \\ 0 & for \ \left[z_1^T \ \phi^T\right]^T = 0 \end{cases}$$

Note that the scalar ψ derived from (9) and (10) is bounded

$$0 \leq |\psi| \leq \tilde{\varrho}_{max}.$$

Thus, $\|\tilde{C}\|$ is bounded and it is permissible to define the stable matrix

$$F = \begin{bmatrix} \Sigma & A_{12} \\ 0 & \Omega^* - \frac{\tilde{\varrho}_{max} P_2}{2} \end{bmatrix}.$$

The output matrix $\tilde{C}$ is partitioned as follows:

$$\tilde{C} = \begin{bmatrix} \tilde{C}_1 & \tilde{C}_2 \end{bmatrix}, \quad \tilde{C}_1 \in \mathbb{R}^{n \times (n-m)}, \quad \tilde{C}_2 \in \mathbb{R}^{n \times m}.$$

In the following, the max-Lyapunov function $V(\hat{z}_1(t), \hat{\phi}(t))$ from (15) and the respective sub-functions V_1 in (7) and $V_1 + kV_2$ in (7-8) shall be investigated with respect to the newly introduced states $\hat{z}_1$ and $\hat{\phi}$. As a result, a constant γ shall be derived determining the input-output characteristics from $\hat{u}$ to $\tilde{y}$. Thus, it will be seen that γ is used similar to an L_2-gain.

In similarity to [9], it can be shown for $V_2 \geq \omega^2 V_1$

$$\hat{\phi}^T P_2 \left(B_2 u_{NL}^C(\hat{z}_1(t), \hat{\phi}(t)) + \frac{\varrho(\hat{z}_1(t), \hat{\phi}(t))}{\gamma_1} \right) \leq 0.$$

Since $\varrho(\hat{z}_1(t), \hat{\phi}(t))$ is an upper bound of the norm of the uncertainty, the uncertainty is compensated for $V_2 \geq \omega^2 V_1$ [9]. In this case, the derivative $\dot{V}_1(\hat{z}_1(t)) + k\dot{V}_2(\hat{\phi}(t))$ satisfies for almost all t considering a perturbation $\hat{u}$ and $\Sigma^T P_1 + P_1 \Sigma = -I$ as well as $(\Omega^*)^T P_2 + P_2 \Omega^* = -I$:

$$
\begin{aligned}
\dot{V}_1(\hat{z}_1(t)) + k\dot{V}_2(\hat{\phi}(t)) &\leq \begin{bmatrix} \hat{z}_1(t) \\ \hat{\phi}(t) \end{bmatrix}^T \begin{bmatrix} \frac{-I}{2} & \frac{P_1 A_{12}}{2} \\ \frac{A_{12}^T P_1}{2} & -k\frac{\gamma_2 I}{2} \end{bmatrix} \begin{bmatrix} \hat{z}_1(t) \\ \hat{\phi}(t) \end{bmatrix} + k\hat{\phi}^T(t) P_2 B_2 \hat{u} \\
&= \begin{bmatrix} \hat{z}_1(t) \\ \hat{\phi}(t) \\ \frac{1}{\epsilon}\hat{u} \end{bmatrix}^T \underbrace{\begin{bmatrix} \frac{-I+\hat{\vartheta}P_1}{2} + \frac{\tilde{C}_1^T \tilde{C}_1}{\epsilon^2} & \frac{P_1 A_{12}}{2} + \frac{\tilde{C}_1^T \tilde{C}_2}{\epsilon^2} & 0 \\ \frac{A_{12}^T P_1}{2} + \frac{\tilde{C}_2^T \tilde{C}_1}{\epsilon^2} & -k\frac{\gamma_2 I - \hat{\vartheta}P_2}{2} + \frac{\tilde{C}_2^T \tilde{C}_2}{\epsilon^2} & \frac{kP_2 B_2}{2}\epsilon \\ 0 & \frac{kB_2^T P_2}{2}\epsilon & -\gamma^2 I_m \end{bmatrix}}_{\Psi_1} \begin{bmatrix} \hat{z}_1(t) \\ \hat{\phi}(t) \\ \frac{1}{\epsilon}\hat{u} \end{bmatrix} \\
&\quad - \hat{\vartheta}(V_1(\hat{z}_1(t) + kV_2(\hat{\phi}(t))) + \frac{\gamma^2}{\epsilon^2}\|\hat{u}\|^2 - \frac{1}{\epsilon^2}\|\tilde{y}\|^2.
\end{aligned}
\tag{19}
$$

The scalars $\epsilon > 0$, $\hat{\vartheta} > 0$ and $\gamma \geq 0$ have been introduced to ensure that the symmetric matrix Ψ_1 is negative semi-definite $\Psi_1 \leq 0$. For that reason, $\frac{1}{\epsilon}$ and $\hat{\vartheta}$ should not be chosen too large to ensure that the principal minors of Ψ_1 are negative semi-definite for any value of $\tilde{C}$. Similar arguments need to be made for $k > 0$. Further, the value of γ should be chosen as small as possible so that $\Psi_1 \leq 0$ for any value of $\tilde{C}$. As mentioned ealier, the scalar γ will be used similar to an L_2-gain.

For $V_2 \leq \omega^2 V_1$, the derivative $\dot{V}_1$ is investigated; it holds for almost all t:

$$
\dot{V}_1 = \frac{1}{2} \begin{bmatrix} \hat{z}_1(t) \\ \hat{\phi}(t) \end{bmatrix}^T \underbrace{\begin{bmatrix} -I + \bar{\vartheta} P_1 + \frac{2\tilde{C}_1^T \tilde{C}_1}{(k\omega^2+1)\epsilon^2} & P_1 A_{12} + \frac{2\tilde{C}_1^T \tilde{C}_2}{(k\omega^2+1)\epsilon^2} \\ A_{12}^T P_1 + \frac{2\tilde{C}_2^T \tilde{C}_1}{(k\omega^2+1)\epsilon^2} & \frac{2\tilde{C}_2^T \tilde{C}_2}{(k\omega^2+1)\epsilon^2} \end{bmatrix}}_{\tilde{\Psi}} \begin{bmatrix} \hat{z}_1(t) \\ \hat{\phi}(t) \end{bmatrix}
$$

$$
- \bar{\vartheta} V_1 - \frac{1}{(k\omega^2+1)\epsilon^2} \|\tilde{y}\|^2 \tag{20}
$$

Now it is of interest to determine for which $\bar{\vartheta} > 0$, $\omega > 0$ and $\epsilon > 0$ the inequality $\tilde{\Psi} < 0$ holds when $V_2(\hat{\phi}(t)) - \omega^2 V_1(\hat{z}_1(t)) \leq 0$ is satisfied. According to the $\mathcal{S}$-procedure in [12, Sect. 2.6.3], the latter is satisfied if and only if there exists a $\xi \geq 0$ so that

$$
\tilde{\Psi} - \xi \left(\frac{V_2(\hat{\phi})}{\omega^2} - V_1(\hat{z}_1) \right) \leq 0
$$

for all $\hat{\phi} \in \mathbb{R}^m$ and $\hat{z}_1 \in \mathbb{R}^{(n-m)}$. This is equivalent to finding a $\xi \geq 0$ so that:

$$
\begin{bmatrix} \hat{z}_1(t) \\ \hat{\phi}(t) \end{bmatrix}^T \underbrace{\begin{bmatrix} -I + \bar{\vartheta} P_1 + \frac{2\tilde{C}_1^T \tilde{C}_1}{(k\omega^2+1)\epsilon^2} + \xi P_1 & P_1 A_{12} + \frac{2\tilde{C}_1^T \tilde{C}_2}{(k\omega^2+1)\epsilon^2} \\ A_{12}^T P_1 + \frac{2\tilde{C}_2^T \tilde{C}_1}{(k\omega^2+1)\epsilon^2} & -\frac{\xi P_2}{\omega^2} + \frac{2\tilde{C}_2^T \tilde{C}_2}{(k\omega^2+1)\epsilon^2} \end{bmatrix}}_{\Psi_2} \begin{bmatrix} \hat{z}_1(t) \\ \hat{\phi}(t) \end{bmatrix} \leq 0.
$$

If $\bar{\vartheta} > 0$, $\omega > 0$ and $\frac{2}{(k\omega^2+1)\epsilon^2}$ are chosen small enough then there always exists a $\xi \geq 0$ so that $\Psi_2 \leq 0$. Assuming $\Psi_2 \leq 0$, $V_2(\hat{\phi}(t)) - \omega^2 V_1(\hat{z}_1(t)) \leq 0$, it follows for almost all t:

$$
\dot{V}_1 = \frac{1}{2} \begin{bmatrix} \hat{z}_1(t) \\ \hat{\phi}(t) \end{bmatrix}^T \Psi_2 \begin{bmatrix} \hat{z}_1(t) \\ \hat{\phi}^T(t) \end{bmatrix} + \xi \left(-V_1 + \frac{1}{\omega^2} V_2 \right) - \bar{\vartheta} V_1 - \frac{\|\tilde{y}\|^2}{(k\omega^2+1)\epsilon^2}
$$

$$
\leq - \bar{\vartheta} V_1 - \frac{1}{(k\omega^2+1)\epsilon^2} \|\tilde{y}\|^2 \tag{21}
$$

Thus, the sub-functions V_1 and $V_1 + kV_2$ of the max-Lyapunov functions $V(\hat{z}_1(t), \hat{\phi}(t))$ (15) have been investigated with respect to their time derivative (19),(21) at those points where they define the max-Lyapunov function (15). Suppose $\bar{\vartheta} > 0$, $\hat{\vartheta} > 0$, k, ω, γ and ϵ have been determined so that the matrices Ψ_1 and Ψ_2 are negative semi-definite, then it is possible using non-smooth analysis to show that for almost all t the derivative $\dot{V}(\hat{z}_1(t), \hat{\phi}(t))$

satisfies:

$$\dot{V}(\hat{z}_1(t), \hat{\phi}(t)) \leq -\min(\hat{\vartheta}, (k\omega^2 + 1)\bar{\vartheta})V(\hat{z}_1(t), \hat{\phi}(t)) + \frac{\gamma^2}{\epsilon^2}\|\hat{u}\|^2 - \frac{1}{\epsilon^2}\|\tilde{y}\|^2$$

(22)

As discussed in Step I with respect to non-smooth theory and Lebesgue integration, an inequality can be implied from (22) for any T:

$$V(\hat{z}_1(t=T), \hat{\phi}(t=T)) - V(\hat{z}_1(t_0), \hat{\phi}(t_0)) \leq \frac{1}{\epsilon^2}\gamma^2 \int_{t_0}^{T} \|\hat{u}(s)\|^2 \, ds$$

$$-\frac{1}{\epsilon^2}\int_{t_0}^{T} \|\tilde{y}(s)\|^2 \, ds - \min(\hat{\vartheta}, (k\omega^2 + 1)\bar{\vartheta}) \int_{t_0}^{T} V(\hat{z}_1(s), \hat{\phi}(s)) ds$$

(23)

Step IV: The system dynamics (16-17) may be rewritten in terms of the signal $\tilde{y}$ in (18) and an auxiliary output $\hat{u}$:

$$\begin{bmatrix} \dot{\hat{z}}_1 \\ \dot{\hat{\phi}} \end{bmatrix} = F \begin{bmatrix} \hat{z}_1 \\ \hat{\phi} \end{bmatrix} + \left(\tilde{y} + \begin{bmatrix} 0 \\ B_2 \end{bmatrix} \hat{u} \right),$$

$$\hat{u} = u_L^\Delta(t, \hat{z}_1, \hat{\phi}) + u_{NL}^\Delta(t, \hat{z}_1, \hat{\phi}) - u_L^C(\hat{z}_1(t), \hat{\phi}(t)) - u_{NL}^C(\hat{z}_1(t), \hat{\phi}(t)).$$

(24)

The signal $u_L^\Delta(t, \hat{z}_1, \hat{\phi}) + u_{NL}^\Delta(t, \hat{z}_1, \hat{\phi})$ results as in (14) from a sample and hold process operating on the non-linear control $u_L^C(\hat{z}_1, \hat{\phi}) + u_{NL}^C(\hat{z}_1, \hat{\phi})$, which is now a function of $[\hat{z}_1^T(t) \; \hat{\phi}^T(t)]^T$. Considering

$$\tilde{y} + \begin{bmatrix} 0 \\ B_2 \end{bmatrix} \hat{u}$$

as the input signal, the dynamics (24) have for any finite horizon $T < \infty$, as discussed in the Appendix, a finite L_2-gain applying the linear system (F, I, I) to (34). Thus, it holds using the Lipschitz constant K_u for the continuous control and the relation $\sum_{k=1}^{4}|a_k| \leq \sqrt{4\sum_{k=1}^{4}|a_k|^2}$:

$$\sqrt{\int_{t_0}^{T} \|\hat{u}(s)\|^2 \, ds}$$

$$\leq K_u \gamma_{\Delta-I}(F, I, I) \left(\sqrt{\int_{t_0}^{T} \|\tilde{y}\|^2 \, ds} + \|B_2\| \sqrt{\int_{t_0}^{T} \|\hat{u}\|^2 \, ds} \right) + K_u \beta$$

$$\leq 2\sqrt{K_u^2(\gamma_{\Delta-I}(F, I, I))^2 \left(\int_{t_0}^{T} \|\tilde{y}\|^2 \, ds + \|B_2\|^2 \int_{t_0}^{T} \|\hat{u}\|^2 \, ds \right) + K_u^2 \beta^2}.$$

(25)

Note that for $\gamma_{\Delta-I}(F,I,I)$ and $\beta = \beta(F,I,I,[\hat{z}_1^T(t_0)\ \hat{\phi}^T(t_0)]^T,T)$ holds:

$$\lim_{\tau\to 0}(\gamma_{\Delta-I}(F,I,I)) = 0,\quad \lim_{\tau\to 0}(\beta(F,I,I,[\hat{z}_1^T(t_0)\ \hat{\phi}^T(t_0)]^T,T)) = 0,\quad (26)$$

as discussed in the Appendix with respect to the theory of [6].

Step V: Consider now the inequality (25). Both sides of the inequality may be squared and multiplied by the constant:

$$\frac{\gamma^2}{\epsilon^2(1 - 4\mathcal{K}_u^2(\gamma_{\Delta-I}(F,I,I))^2\,\|B_2\|^2)},$$

which has to be positive. By (26), this is satisfied if the sampling time τ has been chosen small enough so that $2\mathcal{K}_u^2(\gamma_{\Delta-I}(F,I,I))^2\,\|B_2\|^2 < 1$. The resulting inequality can now be added to (23) and equivalently rewritten so that for any T:

$$V(\hat{z}_1(t=T),\hat{\phi}(t=T)) - V(\hat{z}_1(t_0),\hat{\phi}(t_0))$$

$$\leq -\min(\hat{\vartheta},(k\omega^2+1)\bar{\vartheta})\int_{t_0}^T V(\hat{z}_1(s),\hat{\phi}(s))ds$$

$$+\frac{1}{\epsilon^2}\frac{\gamma^2\left(\int_{t_0}^T\|\hat{u}(s)\|^2\,ds - \int_{t_0}^T\|\tilde{u}(s)\|^2\,ds\right)}{1 - 4\mathcal{K}_u^2(\gamma_{\Delta-I}(F,I,I))^2\,\|B_2\|^2}$$

$$+\frac{1}{\epsilon^2}\left(\frac{4\gamma^2\mathcal{K}_u^2(\gamma_{\Delta-I}(F,I,I))^2}{1 - 4\mathcal{K}_u^2(\gamma_{\Delta-I}(F,I,I))^2\,\|B_2\|^2} - 1\right)\int_{t_0}^T\|\tilde{y}(s)\|^2\,ds$$

$$+\frac{1}{\epsilon^2}\frac{\gamma^2 4\mathcal{K}_u^2\beta^2}{1 - 4\mathcal{K}_u^2(\gamma_{\Delta-I}(F,I,I))^2\,\|B_2\|^2}\qquad (27)$$

The perturbation $\hat{u}$ may now be chosen to be $\hat{u} = \tilde{u}$, which makes the system (16-17) and (24) equivalent to the discrete controlled system (3a-3b). Hence by (26), if the sampling frequency has been chosen small enough so that

$$4\mathcal{K}_u^2(\gamma_{\Delta-I}(F,I,I))^2\|B_2\|^2 < 1,$$

$$\gamma\frac{2\mathcal{K}_u\gamma_{\Delta-I}(F,I,I)}{\sqrt{1 - 4\mathcal{K}_u^2(\gamma_{\Delta-I}(F,I,I))^2\|B_2\|^2}} \leq 1\qquad (28)$$

then it follows from (27):

$$V(\hat{z}_1(t=T),\hat{\phi}(t=T)) - V(\hat{z}_1(t_0),\hat{\phi}(t_0))$$

$$\leq -\min(\hat{\vartheta},(k\omega^2+1)\bar{\vartheta})\int_{t_0}^T V(\hat{z}_1(s),\hat{\phi}(s))ds$$

$$+\frac{1}{\epsilon^2}\frac{\gamma^2 4\mathcal{K}_u^2\beta^2}{1 - 4\mathcal{K}_u^2(\gamma_{\Delta-I}(F,I,I))^2\,\|B_2\|^2}.\qquad (29)$$

Observe that the right hand side of (25) has the value 0 and the constraint on τ in (28) vanishes if the Lipschitz constant $\mathcal{K}_u$ is zero. This would be satisfied if there is no control necessary to stabilize the system. Subsequently, it follows

$$
\min(\hat{\vartheta}, (k\omega^2 + 1)\bar{\vartheta}) \lim_{T \to \infty} \int_{t_0}^{T} V(\hat{z}_1(s), \hat{\phi}(s))ds
$$
$$
\leq V(\hat{z}_1(t_0), \hat{\phi}(t_0)) + \frac{1}{\epsilon^2} \frac{\gamma^2 4 \mathcal{K}_u^2 \beta^2}{1 - 4\mathcal{K}_u^2 (\gamma_{\Delta - I}(F, I, I))^2 \|B_2\|^2}. \tag{30}
$$

It can be easily derived from (15) that there exists a constant c_V for the non-smooth max-Lyapunov function, so that:

$$
c_V \left\| \begin{bmatrix} \hat{z}_1^T & \hat{\phi}^T \end{bmatrix} \right\|^2 \leq V(\hat{z}_1, \hat{\phi})
$$

This implies that the state vector $\begin{bmatrix} \hat{z}_1^T & \hat{\phi}^T \end{bmatrix}^T$ has finite L_2-norm. Thus, asymptotic stability of the closed loop system follows. Note that $\lim_{\tau \to 0+} \beta = 0$ and $\lim_{\tau \to 0+} \gamma_{\Delta - I} = 0$. $\blacksquare$

4 Examples of Discretized Sliding-Mode Based Controllers

The L_2-gain technique is now demonstrated and compared to the Lyapunov function based discretization methodology of [4] using both the inverted pendulum system and a stable linear plant. First, consider a simple model of the inverted pendulum, which is formed by a light rod and a heavy mass attached to one end of the rod with the pivot of rotation at the other end:

$$
\ddot{\theta}(t) = 0.5 \cdot \sin(\theta(t)) - b\dot{\theta}(t) + u(t), \quad 0.09 \leq b \leq 0.11.
$$

where θ is the angle of rotation, b the damping coefficient and $u(t)$ the control torque. The system may be expressed in the form of (2) with

$$
A = \begin{bmatrix} 0 & 1 \\ 0.5 & -0.1 \end{bmatrix}, \quad B = \begin{bmatrix} 0 \\ 1 \end{bmatrix}, \quad [\Delta_\Theta \; \Delta_\Omega] = [0.5(\text{sinc}(\theta(t)) - 1) \quad -b + 0.1]
$$
$$
\tag{31}
$$

where

$$
\text{sinc}(t) = \begin{cases} \frac{\sin(t)}{t} & \text{if } t \neq 0 \\ 1 & \text{if } t = 0 \end{cases}.
$$

It is also of interest to investigate the discretization approaches on a stable model in the nominal case when the uncertainty is set to zero and $\eta_2 \neq 0$, the non-linear control gain, is only used for reachability of the sliding-mode. Thus, the system satisfies in this case:

$$A = \begin{bmatrix} -0.5 & 1 \\ 0 & -0.5 \end{bmatrix}, \quad B = \begin{bmatrix} 0 \\ 1 \end{bmatrix}, \quad [\Delta_\Theta \ \ \Delta_\Omega] = 0. \tag{32}$$

The non-linear controllers for both systems are chosen to satisfy $\gamma_1 = 1.6$, $\eta_1 = 0$. By demanding $\eta_1 = 0$ (11), the linear controller is used to compensate for the parametric uncertainty $(-b+0.1)$ of Ω. Therefore assuming $\eta_1 = 0$, the value of γ_2 $(0 < \gamma_2 \leq 1)$ must be chosen as large as possible, so that the linear control is also used for reachability. Provided the matrix Σ (4) determining the sliding-mode performance, the linear control matrix Ω^* (6) and the respective Lyapunov matrices P_1 (8) and P_2 (7) have been set, then the features of the sliding-mode cone and the parameter ω can be determined using nominal system and controller parameters only:

$$\underbrace{-\lambda_{max} \left(P_1 \Sigma + \Sigma^T P_1 \right)}_{=1} - 2 |P_1 A_{12} P_2^{-\frac{1}{2}}| |\omega| |P_1^{\frac{1}{2}}| = 0.85$$

For the considered single input, second order systems, this constraint is necessary and sufficient so that for any state satisfying $V_2 \leq \omega^2 V_1$ the derivative $\dot{V}_1$ holds for almost all t: $\dot{V}_1 \leq -0.85 |z_1|^2$. This ensures consistency with former results [4] and comparability of the control of the nominal and the uncertain systems and implies stable performance for both examples (20). Further, it is necessary to ensure that $\gamma \geq 0$ in the analysis of Sect. 3 is as small as possible since it is used in a similar manner to an L_2-gain. Thus, for the L_2-based analysis, the value of $\hat{\vartheta}$ (19) is chosen to satisfy $0.5 \geq P_1 \hat{\vartheta}$ and $0.5\gamma_2 \geq P_2 \hat{\vartheta}$, and $\bar{\vartheta}$ (20) holds $0.85/2 = P_1 \bar{\vartheta}$. Hence, this choice determines a trade-off between low sampling frequency and controller performance. For the Lyapunov function based analysis technique a similar trade-off needs to be made [3,4]. There a constant κ, $0 < \kappa < 1$ is used. This constant has been set to 0.5. The constant k from the max-Lyapunov function (15) has been adjusted via numerical methods so that the sampling frequency bound $1/\hat{\tau}$ for each controller is minimal.

Note that the results (Fig. 1) for the L_2-norm based technique are more conservative than those calculated for the Lyapunov function analysis based approach for most choices of $\Sigma = \Omega^*$. The sampling frequency which is sufficient to stabilize robustly the unstable inverted pendulum (31) is generally for the L_2-norm based methodology more than ten times higher than for the Lyapunov function based technique. A similar result applies to the stable system with no uncertainties (32) where the non-linear control is only used for reachability of the sliding-mode region. The reason for the more conservative results is connected to the conservative estimate of the L_2-gain γ with respect to the output $\tilde{y}$ (16-17), (18). Using the Matlab$^{\copyright}$-LMI toolbox, the

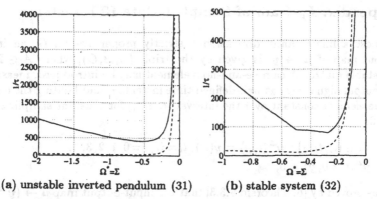

(a) unstable inverted pendulum (31)   (b) stable system (32)

Fig. 1. Values of sufficient sampling frequencies against changing controller poles $\Sigma = \Omega^*$ using the L_2-norm based approach (*line*) and the Lyapunov function analysis based approach (*dashed*)

matrix $\tilde{C}^T\tilde{C}$ was approximated by a polytopic set. Furthermore in the Lyapunov function based approach, the very large value of the Lipschitz constant of the non-linear control u_{NL}^C (9) could be compensated by the upper estimate of the sampling error $\sup_{t_\alpha \in [t_i, t_{i+1}]} (\|x(t_\alpha) - x(t_i)\|)$ (1) [3], which is not straightforward for the L_2-norm technique.

5   Conclusion

This work has applied tools known from non-linear L_2-gain analysis and linear optimal sampled-data control to a non-linear sliding-mode-like control. It has been seen that these tools can be used to show asymptotic stability of the discretized non-linear control. A non-smooth Lyapunov function has been used to derive a differential dissipation inequality and subsequently L_2-stability of the discretized control. An advantage of this analysis technique in comparison to a Lyapunov function analysis based technique is that the sampling-time constraint vanishes for a stable plant if no control is used. However, an example of a robustly controlled inverted pendulum has shown that the presented methodology might have disadvantages when compared to the Lyapunov function analysis based technique: the results may be too conservative.

6   Acknowledgements

The authors would like to acknowledge the support for G. Herrmann from the European Commission (TMR-grant, project number: FMBICT983463).

Appendix: L_p-Gain of Sampled-Data LTI Systems

Suppose a linear, finite dimensional, strictly proper system G, an input-output map of $u \to y$, is given by the triple (A, B, C), where $A \in \mathbb{R}^{n \times n}$ is a stable matrix. A sample-and-hold element may be introduced, measuring the output signal $y(t)$ at well-defined time instants t_i and holding the output of the element constant over the interval $(t_i, t_{i+1}]$ for constant sampling time $\tau > 0$:

$$\forall t \in (t_i, t_{i+1}] : y^{\Delta}(t, y) = y(t_i),\ t_i = i\tau,\ i = 0, 1, 2, 3, \dots$$
$$y^{\Delta}(t = 0, y) = 0,$$

It was proved by [6, Theorem 9.3.3] that the input-output map $u \to (y^{\Delta} - y)$ has finite L_p-gain, $(p \in [1, \infty])$. The L_p-gain of the input-output map $u \to (y^{\Delta} - y)$ is bounded from above by:

$$\gamma_{\Delta-I}(A, B, C) \overset{def}{=} \int_0^{\infty} f_H(t)dt + \tau\phi(0), \qquad (33)$$

where the function f_H and the term $\phi(0)$ satisfy:

$$f_H(t) \overset{def}{=} \sup_{a \in (0, \tau)} \left(\left\| Ce^{At}B - Ce^{A(t-a)}B \right\| \right),$$

$$\phi(0) \overset{def}{=} \sup_{t \in [0, \tau)} \left(\left\| Ce^{At}B \right\| \right)$$

and $\| \cdot \|$ is the induced Euclidean norm. This implies for any Lebesgue-integrable input signal $u(t) \in L_p[0, T]$ for the in/finite horizon $(T \leq \infty)$:

$$\left(\int_0^T \left\| y^{\Delta}(t) - y(t) \right\|^p dt \right)^{\frac{1}{p}} \leq \gamma_{\Delta-I}(A, B, C) \left(\int_0^T \| u(t) \|^p \right)^{\frac{1}{p}}$$
$$+ \beta(A, B, C, x(0), T), \qquad (34)$$

The bias term $\beta(A, B, C, x(0), T)$ is bounded above by:

$$\beta(A, B, C, x(0), T) \leq \left(\int_0^T \sup_{a \in (0, \tau)} \left(\left\| Ce^{At}(I - e^{-Aa})x(0) \right\|^p \right) dt \right)^{\frac{1}{p}}.$$

Note the dependence of β on the intial state $x(0)$ of the linear system. It can be seen that the upper limit of the L_p-gain $(\int_0^{\infty} f_H(t)dt + \tau\phi(0))$ converges to 0 for $\tau \to 0$ using the following upper estimates of f_H and the term $\phi(0)$

$$f_H(t) \leq \|B\| \|C\| \left\| e^{tA} \right\| \left| e^{\tau\|A\|} - 1 \right|, \quad \phi(0) \leq \|B\| \|C\| e^{\tau\|A\|}.$$

Note that $\int_0^{\infty} \left\| e^{tA} \right\| dt$ is finite for stable A. A similar argument also shows that $\lim_{\tau \to 0+} \beta = 0$.

References

1. Itkis U. (1976) Control systems of variable structure. John Wiley and Sons, New York
2. Djenoune S., El-Moudni A., Zerhouni N. (1998) Stabilization and regulation of a class of nonlinear singularly perturbed system with discretized composite feedback. International Journal of Systems Science, 29(4):419–434
3. Herrmann G., Spurgeon S. K., Edwards C. (1999) A new approach to discretization applied to a continuous non-linear, sliding-mode-like control using non-smooth analysis. Accepted in IMA Journal of Mathematical Control and Information
4. Herrmann G., Spurgeon S. K., Edwards C. (1999) Discretisation of sliding mode based control schemes. In Proceedings of the 38th Conference on Decision and Control, Phoenix, 4257–4262
5. Herrmann G., Spurgeon S. K., Edwards C. (2000) Discretization of a non-linear, continuous-time control law with small control delays - a theoretical treatment. In Proceedings of the Fourteenth International Symposium on Mathematical Theory of Networks and Systems, France
6. Chen T., Francis B. (1995) Optimal Sampled-Data Control Systems. Communications and Control Engineering Series (CCES). Springer-Verlag, London
7. Iglesias P. (1995) Input-output stability of sampled-data linear time-varying systems. IEEE Transactions on Automatic Control, 40(9):1647–1650
8. Keller J., Anderson B. D. O. (1992) A new approach to the discretization of continuous-time controllers. IEEE Transactions on Automatic Control, 37(214–223):1241–1243
9. Herrmann G., Spurgeon S. K., Edwards C. (1998) A new non-linear, continuous-time control law using sliding-mode control approaches. In Proceedings of the Fifth International Workshop on Variable Structure Systems, Long Boat Key, Florida, 50–56
10. Ryan, E. P.,Corless, M. (1984) Ultimate boundedness and asymptotic stability of a class of uncertain dynamical systems via continuous and discontinuous feedback control. IMA Journal of Mathematical Control and Information, 1:223–242
11. Herrmann G. (1999) Contributions to discrete-time non-linear control using sliding mode control approaches. Internal Second Year PhD-Research Report, Leicester University, Department of Engineering
12. Boyd S., El-Ghaoui L., Feron E., Balakrishnan V. (1994) Linear Matrix Inequalities in System and Control Theory. Society for Industrial and Applied Mathematics, Philadelphia
13. Khalil H. K. (1992) Nonlinear Systems, 2nd edn. Macmillan Publishing Company, New York

Nonlinear Adaptive State Space Control for a Class of Nonlinear Systems with Unknown Parameters

Christian Hintz, Martin Rau, and Dierk Schröder

Technical University Munich
Institute for Electrical Drive Systems
Arcisstr. 21
80333 Muenchen, Germany
Christian.Hintz@ei.tum.de - Martin.Rau@ei.tum.de

Abstract. In this paper, we present an identification method for mechatronic systems consisting of a linear part with unknown parameters and an unknown nonlinearity (systems with an isolated nonlinearity) . Based on this identification method we introduce an adaptive state space controller in order to generate an overall linear system behavior. A structured recurrent neural network is used to identify the unknown parameters of the known signal flow chart. The control concept starts from a nonlinear canonical form and takes advantage of the online identified parameters of the plant.

The novelty of this approach is the simultaneous identification of the parameters of the linear part and the nonlinearity, the use of prior structural and parameter knowledge and the ability to completely compensate the nonlinearity.

1 Introduction

Many technical plants in the field of motion control and electrical drive systems are modeled by linear differential equations. Nevertheless, most of these systems contain nonlinearities, which are unknown or partially unknown, e.g. friction, spring characteristics, backlash or excentricities. Some parameters of the linear part of the system may also be unknown. Any plant containing these unknown linear parameters and nonlinearities cannot be controlled optimally by a linear controller. Hence, it is desirable to identify the nonlinearity and the unknown linear parameters for means of improving the controller design. This should be done online to automate initial identification as well as to cope with parameter drift in the plant.

2 Identification with structured recurrent networks

Starting from the signal flow chart of a mechatronic system, containing only elementary operators (gains, sums, integrators) and the unknown nonlinearity, the system is mapped into a structured recurrent neural network. In this mapping, sums are represented by perceptrons with linear activation, integrators by time delays, gains by network weights and the nonlinearity by an additional neural network, e.g. a Radial Basis Function Network (RBF), a General Regression Neural Network (GRNN) or a Multi Layer Perceptron Network (MLP). The basic signal flow chart elements and their corresponding network elements are depicted in figure 1. The unknown trainable parameters

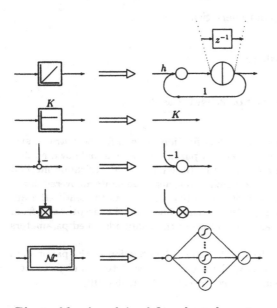

Fig. 1. Mapping of signal flow chart elements

are the gains of the signal flow chart and the weights of the additional neural network for the nonlinearity. These parameters are adapted by a gradient based learning algorithm which uses only the system's output error.
The system under consideration is of the following form:[1]

$$\dot{x}_c = A_c \cdot x_c + b_c \cdot u + k_c \cdot \mathcal{NL}(x_c) \qquad y_c = c_c^T \cdot x_c \tag{1}$$

It is called *system with an isolated nonlinearity* [5] because there is only one nonlinearity with a fixed input coupling into the plant. The system in (1) is discretized for the identification purpose by *Euler–Forward* and results in:

$$x[k+1] = A \cdot x[k] + b \cdot u[k] + k \cdot \mathcal{NL}(x[k]) \qquad y[k] = c^T \cdot x[k] \tag{2}$$

[1] A_c, A system matrix, b_c, b input coupling vector, c_c, c output coupling vector, x_c continuous time system state vector, $x[k]$ discrete system state vector, k_c, k coupling vector for static nonlinearity $\mathcal{NL}$, u input signal, y_c, y output signal

The structured recurrent network is implemented in analogy to the plant's structure in (2):

$$\hat{x}\,[k+1] = \hat{A} \cdot \hat{x}\,[k] + \hat{b} \cdot u\,[k] + k \cdot \widehat{\mathcal{NL}}\,(\hat{x}\,[k]) \qquad \hat{y}\,[k] = \hat{c}^T \cdot \hat{x}\,[k] \qquad (3)$$

The discretization of the continuous system simplifies the development of the learning laws of the proposed structured recurrent neural network. In representation (3) the estimated nonlinearity $\widehat{\mathcal{NL}}$ is identified e.g. by a RBF–Network or a MLP–Network with the weights $\hat{\Theta}$.

To find a gradient based learning rule for the unknown parameters $\hat{A}, \hat{b}, \hat{c}$ and $\hat{\Theta}$, it is convenient to introduce a weight vector w.

$$w^T = (\, \hat{a}^T \ \hat{b}^T \ \hat{c}^T \ \hat{\Theta}^T \,) \qquad \hat{a} = \text{vec}\left(\hat{A}\right) \qquad (4)$$

The operator "vec" combines all columns of a matrix into one vector. The adaptation error is defined as:

$$e\,[k] = y\,[k] - \hat{y}\,[k] \qquad (5)$$

The signal $y\,[k]$ is equal to the signal y_c at time instant $k \cdot h$, where h is the sample time. In order to minimize the square error cost function

$$E\,[k] = \frac{1}{2}e^2\,[k] \qquad (6)$$

it is necessary to evaluate the gradient $\frac{dE}{dw}$. With this gradient it is possible to use the following learning rule:

$$w\,[k+1] = w\,[k] - \eta \cdot \frac{dE\,[k]}{dw\,[k]} = w\,[k] + \eta \cdot \frac{d\hat{y}\,[k]}{dw\,[k]} \cdot e\,[k] \qquad \eta > 0 \qquad (7)$$

The main task is to calculate $d\hat{y}\,[k]\,/dw\,[k]$. This is done in analogy to real time recurrent learning [1,10,11]. The detailed steps are shown in [3] and in section 4 for a mechatronic drive train.

In order to get correct identification results, the following facts have to be observed:

- Due to the fact that only the output error is used for identification, it is only possible to identify the real physical parameters if the system is unique. A system is unique in the sense of the given signal flow chart if, and only if, one combination of the parameters w results in $e\,[k] = 0$ for $k \to \infty$. If the system is not unique a variety of parameter combinations can result in the same input–output behavior for the signal flow chart under consideration. Most drive systems are unique, and therefore the identified parameters are identical to the real physical parameters.

- If the system shows globally integrating behavior, all previous input signals contribute to the current output signal. Even if the identification algorithm finds the correct parameters, it is not possible to reduce the output error to zero, because at the beginning of the identification procedure the wrong parameters lead the system's states towards a wrong region, and due to the integrating behavior the error of the states is not able to tend to zero. Since the output error will never be zero, the algorithm is not able to find the correct parameters. To overcome this problem the algorithm has to reset the estimated states such, that the output error becomes zero, even with the wrong parameters. Since the parameters tend towards their true values, the state error tends to zero, and eventually the output error stays zero. The reset values of the estimated states are calculated in analogy to the observability matrix for linear systems [2]. This technique leads to a faster identification result even if the system does not show globally integrating behavior.

- In most applications, the parameters are not completely unknown, e.g the physical parameters cannot be outside a certain range. It is useful to interrupt the identification process for a specific parameter if this parameter tends to leave this range.

3   Linearizing Neuro–Control (LNC)

For linear systems, there are many canonical forms with different properties for special applications. Here, we want to extend the linear controllable canonical form to nonlinear systems. This is done in analogy to the input–state linearization of nonlinear systems [4,8,9]. The linear form for SISO systems is described by equation (8), where the parameters α_i and c_i are the coefficients of the corresponding transfer function.

$$\dot{\bar{x}} = \begin{bmatrix} 0 & 1 & 0 & 0 & \dots & 0 \\ 0 & 0 & 1 & 0 & \dots & 0 \\ \vdots & & & \ddots & & \vdots \\ 0 & \dots & & & 1 & 0 \\ 0 & \dots & & & 0 & 1 \\ -\alpha_0 & -\alpha_1 & \dots & & & -\alpha_{n-1} \end{bmatrix} \cdot \bar{x} + \begin{bmatrix} 0 \\ 0 \\ \vdots \\ 0 \\ 1 \end{bmatrix} \cdot u \qquad (8)$$

$$y = \begin{bmatrix} c_1 & c_2 & \dots & c_{n-1} \end{bmatrix} \cdot x$$

To transform a system from an arbitrary state space description to the controllable canonical form, a linear transformation $\bar{x} = T \cdot x$ is necessary. The transformation is possible, if the controllability matrix has full rank. The advantage of this special state space description is that linear state space controllers can be designed easily, e.g. by pole placement.

An extension to nonlinear systems is the nonlinear controllable canonical form (NCCF). Its equations for a system of order n are

$$\dot{\bar{x}} = \bar{f}(\bar{x}) + \bar{g}(\bar{x}) \cdot u \qquad (9)$$
$$y = \bar{h}(\bar{x})$$

with

$$\bar{f}(\bar{x}) = \begin{bmatrix} \bar{x}_2 \\ \bar{x}_3 \\ \vdots \\ \bar{x}_n \\ \bar{\mathcal{F}}(\bar{x}) \end{bmatrix} \qquad \bar{g}(\bar{x}) = \begin{bmatrix} 0 \\ 0 \\ \vdots \\ 0 \\ \bar{\mathcal{G}}(\bar{x}) \end{bmatrix} \qquad (10)$$

The transformation from an arbitrary nonlinear state space equation with state vector x can be achieved by a nonlinear state transformation which has to be a **diffeomorphism in the whole range of operation** of the system in order to be able to invert the transformation ($\bar{x} = v(x_c)$ and $x_c = v^{-1}(\bar{x})$). The calculation of $v(x)$ will not be discussed in this paper, for further studies see [4,8,9]. The nonlinear state space transformation $v(x_c)$ is only possible if the *controllability condition* for nonlinear systems as described in [8] holds. It is assumed, that the system under consideration is already given in the nonlinear controllable canonical form. We will apply the following control law with the reference value w to the system.

$$u = -\frac{1}{\bar{\mathcal{G}}(\bar{x})} \left(\bar{\mathcal{F}}(\bar{x}) + \alpha_0 \bar{x}_1 + \alpha_1 \bar{x}_2 + \ldots + \alpha_{n-1}\bar{x}_n - \gamma w \right) \qquad (11)$$
$$= r(\bar{x}) + m(\bar{x}) \cdot w$$

This control law compensates all nonlinear effects ($\bar{\mathcal{F}}$ and $\bar{\mathcal{G}}$) and results in a linear differential equation with α_i as coefficients.

$$\dot{\bar{x}}_n + \alpha_{n-1}\bar{x}_n + \alpha_{n-2}\bar{x}_{n-1} + \ldots + \alpha_0 \bar{x}_1 = \gamma \cdot w \qquad (12)$$

The dynamics of the controlled system is linear and can be adjusted by an appropriate choice of the coefficients α_i (pole placement or other linear techniques). The parameter γ can be chosen such, that stationary accuracy is reached for $\dot{\bar{x}} = 0$.

If a complete system model is available, it is possible to apply the proposed control concept in (11) in order to compensate the nonlinearity and to gain an overall linear system behavior. To show the benefit of the proposed control method, a nonlinear state space controller as defined in (11) is applied to the system in section 4. Figure 2 shows the step responses for three different types of controllers. $N2_{lin}$ is the desired step response of the system in section 4, where $\mathcal{NL}(x_3)$ is set to zero. The linear state space controller was adjusted according to the damping optimum criterion [7]. $N2_{nl}$ was achieved by the

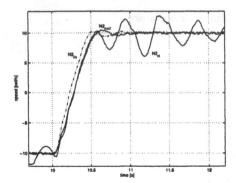

Fig. 2. Comparision of step responses

same linear state space controller, but the nonlinearity was set to $3 \cdot \arctan(2 \cdot N_2)$. The nonlinearity was not taken into account in the controller design and therefore, the control result is not optimal. $N2_{ncct}$ is the control result with the proposed controller. It is nearly identical to the desired step response $N2_{lin}$.

The control concept in equation (11) requires

- information about the nonlinearity, which is provided by the proposed identification procedure
- information about the parameters of the linear part of the system, which are also provided by the proposed identification procedure
- all state variables of the system which can be provided by an adaptive observer [6]
- a symbolically calculated control law with specified linear dynamics (α_i)

The overall structure of the presented linearizing neuro–control concept is depicted in figure 3.

The adaptive component of the controller results from the adaptation of the identificator to the unknown nonlinearity and the parameters of the linear part. Both, the identified nonlinearity and the identified linear parameters are provided to the control law and the adaptive observer. The additional adaptive observer [6] is necessary to avoid the utilization of the estimated states of the identificator, which can be reset. These steps of the estimated states cause disturbances in the controller.

4   Simulation Example

Many subsystems in electro–mechanical drives can be modeled as multi–body systems. Our sample system is a two–body system coupled by a damped elastic spring. The first body is the rotor of a motor, the second body includes

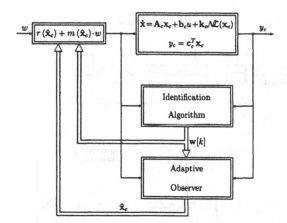

Fig. 3. Linearizing neuro–control

all rotating parts of the load. The objective of the controller design is the speed control of the second body. As a nonlinearity, we use a speed dependent torque at the second body, e.g. a friction torque. The system's state space description in continuous time is

$$
\dot{x} = \underbrace{\begin{bmatrix} -\frac{d}{J_1} & -\frac{c}{J_1} & \frac{d}{J_1} \\ 1 & 0 & -1 \\ \frac{d}{J_2} & \frac{c}{J_2} & -\frac{d}{J_2} \end{bmatrix}}_{A} \cdot x + \underbrace{\begin{bmatrix} \frac{1}{J_1} \\ 0 \\ 0 \end{bmatrix}}_{b} \cdot u + \underbrace{\begin{bmatrix} 0 \\ 0 \\ -\frac{1}{J_2} \end{bmatrix}}_{k} \cdot \underbrace{3 \cdot \arctan(2 \cdot N_2)}_{\mathcal{N\!L}(x_3)}
$$

$$
y = \underbrace{\begin{bmatrix} 0 & 0 & 1 \end{bmatrix}}_{c^T} \cdot x \tag{13}
$$

with the state vector

$$
x^T = \begin{bmatrix} N_1 & \Delta\varphi & N_2 \end{bmatrix} \tag{14}
$$

where N_1 is the speed of the motor, N_2 the speed of the load and $\Delta\varphi$ the torsion angle between the two bodies. The input signal u is the active motor torque (torque at the first body). The signal flow description of this system is depicted in the upper part of figure 4. The lower part of figure 4 shows the mapping of the continuous–time signal flow chart to a discrete–time structured neural network, i.e. all prior knowledge about the system's structure is used by setting the appropriate neurons to their known fixed values. Two steps have to be performed to create the network.

1. Discretization of the continuous–time signal flow chart with the sample time h, e.g. by *Euler–Forward*
2. Mapping of the signal flow chart to a recurrent neural network: gains become network weights and sums become perceptrons.

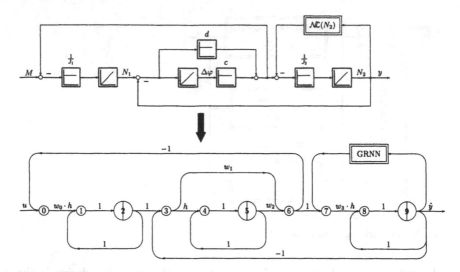

Fig. 4. Mapping of the signal flow chart of a two–body system to a structured neural network

In the mapping in figure 4, perceptrons #0, #1, #3, #4, #6, #7 and #8 contain linear activation functions, perceptrons #2, #5 and #9 are one step time delays. The output of the perceptrons with linear activation functions is calculated according to equation (15).

$$o_j = \sum_i in_i \qquad (15)$$

in_i are the input signals of neuron j; o_j is the output signal of neuron j. The detailed structure of neurons #2, #5 and #9 is depicted in figure 5. It consists of a neuron with a linear activation function and a one step time delay.

Fig. 5. Detailed structure of the time delay neurons

As it can be seen in figure 4, the elements a_{21}, a_{22} and a_{23} of the system matrix $\mathbf{A}$ are fixed network weights, since they are not physical parameters, but given by the structure of the system. The unknown parameter vector $\boldsymbol{w}$ for the system under consideration is

$$\boldsymbol{w} = \begin{bmatrix} w_0 & w_1 & w_2 & w_3 & \Theta \end{bmatrix}^T = \begin{bmatrix} 1/J_1 & c & d & 1/J_2 & \mathcal{NL} \end{bmatrix}^T \qquad (16)$$

As an example, the calculation of the adaptation law for the unknown parameter w_0 is shown. The output signal $\hat{y}[k]$ of the identificator is equal to $o_9[k-1]$. First, the outputs o_j of each neuron have to be calculated (see equation (17)).

$$o_9[k] = o_8[k] \qquad\qquad o_4[k] = h \cdot o_3[k] + o_5[k-1]$$
$$o_8[k] = w_3 \cdot h \cdot o_7[k] + o_9[k-1] \qquad o_3[k] = o_2[k-1] - o_9[k-1]$$
$$o_7[k] = o_6[k] + \mathcal{NL}(y[k]) \qquad\qquad o_2[k] = o_1[k] \qquad\qquad\qquad (17)$$
$$o_6[k] = w_2 \cdot o_5[k-1] + w_1 \cdot o_3[k] \qquad o_1[k] = w_0 \cdot h \cdot o_0[k] + o_2[k-1]$$
$$o_5[k] = o_4[k] \qquad\qquad\qquad o_0[k] = u[k] - o_6[k]$$

The network weight w_0 is adapted with the term $\partial(o_9[k-1])/\partial w_0$. This gradient has to be calculated from the neuron's outputs o_j and gradients from one time instance before ($[k-2]$). Equation (18) shows the necessary steps.

$$\frac{\partial o_9[k-1]}{\partial w_0} = w_3 \cdot h \cdot \frac{\partial o_7[k-1]}{\partial w_0} + \frac{\partial o_9[k-2]}{\partial w_0}$$
$$\frac{\partial o_7[k-1]}{\partial w_0} = \frac{\partial o_6[k-1]}{\partial w_0}$$
$$\frac{\partial o_6[k-1]}{\partial w_0} = w_2 \cdot \frac{\partial o_5[k-2]}{\partial w_0} + w_1 \cdot \frac{\partial o_3[k-1]}{\partial w_0}$$
$$\frac{\partial o_5[k-1]}{\partial w_0} = h \cdot \frac{\partial o_3[k-1]}{\partial w_0} + \frac{\partial o_5[k-2]}{\partial w_0} \qquad\qquad (18)$$
$$\frac{\partial o_3[k-1]}{\partial w_0} = \frac{\partial o_2[k-2]}{\partial w_0} - \frac{\partial o_9[k-2]}{\partial w_0}$$
$$\frac{\partial o_2[k-1]}{\partial w_0} = h \cdot o_0[k-1] + w_0 \cdot h \cdot \frac{\partial o_0[k-1]}{\partial w_0} + \frac{\partial o_2[k-2]}{\partial w_0}$$
$$\frac{\partial o_0[k-1]}{\partial w_0} = -w_2 \cdot \frac{\partial o_5[k-2]}{\partial w_0} - w_1 \cdot \frac{\partial o_2[k-2]}{\partial w_0} + w_1 \cdot \frac{\partial o_9[k-2]}{\partial w_0}$$

At the beginning of the identification process, all gradients are set to zero. At the next timesteps, the gradients of the timesteps before are known and the algorithm can start. The gradients at the current timestep are calculated recursively by the gradients of the past.

The adaptation law for the parameter w_0 can therefore be written as:

$$w_0[k+1] = w_0[k] + \eta \cdot e[k] \cdot \frac{\partial o_9[k-1]}{\partial w_0} \qquad\qquad (19)$$

The recursive calculation of $\partial o_9[k-1]/\partial w_0$ is achieved by evaluating equation (18). In this example, the nonlinearity is identified by a Multi Layer Perceptron Network (MLP). The gradients for the weights of the MLP–Network are calculated in the same way as those for the linear parameters.

4.1   Simulation Results

Figures 6 to 10 show the identification results (linear parameters and nonlinearity) for the two–body system in equation (13).

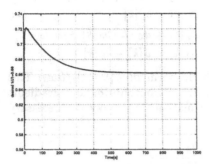

Fig. 6. Identification results for the linear parameter w_0

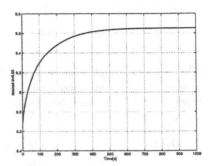

Fig. 7. Identification results for the linear parameter w_1

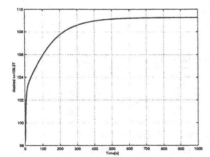

Fig. 8. Identification results for the linear parameter w_2

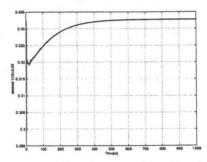

Fig. 9. Identification results for the linear parameter w_3

Figure 11 shows the difference (control error) between a linear reference system (the two–body system without isolated nonlinearity) and the nonlinear system controlled by the proposed control concept (LNC). As the estimated parameters converge towards their true values, the control error tends to zero. This shoes, that the LNC–controller is able to compensate the nonlinearity even without exact prior parameter knowledge.

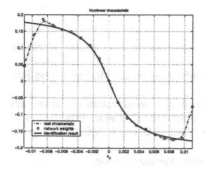

Fig. 10. Identification results for the nonlinear characteristic

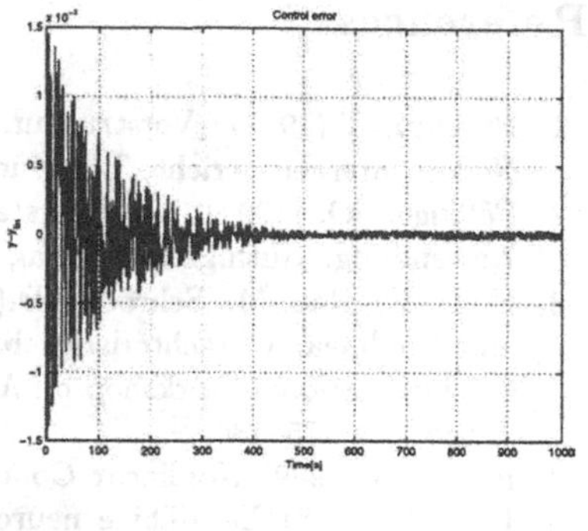

Fig. 11. Control error

5 Conclusion

The main benefit of the presented identification approach is the ability to include the whole prior knowledge about the system into the structured neural network. This includes the possibility to specify parameter regions as well as fixed values to accelerate the identification procedure. In the field of motion control, one can usually use fixed values when describing the relation between speed and position. A main property of the method are interpretable identification results, especially the gains of the signal flow chart and an approximation of the nonlinearity. This is an advantage compared to other nonlinear input–output (black box) identification methods. If the input–output behavior can only be achieved by a single combination of the unknown parameters, the identification result is unique and therefore this method provides an estimate of the non-measurable system states. In many problems in mechatronic systems, the result is unique and the identified parameters are identical to the physical parameters, e.g. spring constants, momenta of inertia or friction characteristics. Simultaneous identification of the parameters of the linear part and the nonlinearity is another feature of the presented method. Therefore it can be used in online applications to identify time–variant parameters.

The control concept results in an adaptive nonlinear state space controller, which takes advantage of the identification result. Both, identification and control are online methods and can operate simultaneously.

To conclude, we summarize the novelty of this approach: Simultaneous identification of the parameters of the linear part and the nonlinearity, interpretable identification results and prior structural and parameter knowledge can be included. The controller considers the identified parameters and the nonlinearity. The controlled system shows globally linear behavior.

References

1. Brychcy, T.(1997) Vorstrukturierte Verallgemeinerte Rekurrente Neuronale Netze. Interner Bericht, TU-München
2. Föllinger, O. (1992) Regelungstechnik, Einführung in die Methoden und ihre Anwendung. Hüthigbuch Verlag
3. Hintz, C., Rau, M., Schröder, D.(2000) Combined Identification of Parameters and Nonlinear Chrachteristics based on Input–Output Data. Proceedings of 6th International Workshop on Advanced Motion Control (AMC2000) Nagoya, Japan, pp. 175–180
4. Isidori, A.(1989) Nonlinear Control Systems. Springer–Verlag
5. Lenz, U.(1998) Lernfähige neuronale Beobachter für eine Klasse nichtlinearer dynamischer Systeme und ihre Anwendung zur intelligenten Regelung von Verbrennungsmotoren. Dissertation, Lehrstuhl für Elektrische Antriebssysteme, TU München
6. Schröder, D.(1999) Intelligent Observer and Control Design for Nonlinear Systems. Springer–Verlag
7. Schröder, D.(1995) Elektrische Antriebe 2. Springer–Verlag
8. Slotine, J., Weiping, L.(1991) Applied Nonlinear Control. Prentice–Hall
9. Sommer, R.(1979) Entwurf nichtlinearer, zeitvarianter Systeme durch Polvorgabe. Regelungstechnik, Oldenbourg Verlag, Vol. 12, pp. 393–399
10. Williams, R.J., Zipser, D.(1990) Gradient–Based Learning Algorithms for Recurrent Connectionist Networks. Tech. Rep. NU-CSS90-9, College of Computer Science, Northeastern University, Boston, MA
11. Zell, A.(1994) Simulation Neuronaler Netze. Addison-Wesley

An Observer View on Synchronization

Henri J.C. Huijberts[1] and Henk Nijmeijer[2]

[1] Department of Engineering
Queen Mary and Westfield College
Mile End Road
London E1 4NS, United Kingdom
h.j.c.huijberts@qmw.ac.uk
[2] Department of Mechanical Engineering
Eindhoven University of Technology
P.O. Box 513
5600 MB Eindhoven, The Netherlands
h.nijmeijer@tue.nl

Abstract. Synchronization of complex/chaotic systems is reviewed from a dynamical control perspective. It is shown that the notion of an observer is essential in the problem of how to achieve synchronization between two systems on the basis of partial state measurements of one of the systems. An overview of recent results on the design of synchronizing systems using observers is given. Examples are given to demonstrate the main results.

Keywords: synchronization, dynamics, observers, chaotic systems

1 Introduction

Probably one of the earliest detailed accounts on synchronized motion was made by Christiaan Huygens, who around 1650 describes in his notebook [19] an experiment where two identical pendulum clocks are attached to the same (flexible) bar. In a short while, these clocks then exhibit synchronized motion, even when initialized at different phases. The explanation by Huijgens is remarkably accurate since by that time the differential calculus needed to describe the clocks' motion was still to be developed. Many other examples of synchronized motion have been described after the 17th century. For instance, Rayleigh describes in his famous treatise "The theory of sound" [29] in 1877 that two organ tubes may produce a synchronized sound provided the outlets are close to each other. Early this century the Dutch scientist B. van der Pol, studied synchronization of certain (electro-) mechanical systems, see [36]. Actually, rotating bodies, or more generally rotating mechanical structures form a very important and special class of systems that, with or without the interaction through some coupling, exhibit synchronized motion. In fact, synchronization of oscillating physical systems is by today an important subject in some of the major physics journals. An illuminating survey on synchronization of a wide variety of mostly (electro-) mechanical systems is given in [3]. Also [26] contains a rich class of motivating and illustrative examples of synchronizing systems. The growing interest in synchronization – and the above mentioned surveys are illustrative for this – was probably caused by the paper by Pecora and Carroll [27] where, among others, secure communication as a potential application has been indicated. Although sofar it is still questionable whether this application can be fully realized, the Pecora and Carroll paper has formed an impulse for much research along these lines.

On the other hand, for mechanical systems synchronization is of utmost importance as soon as two machines have to cooperate. Typically robot coordination, cf. [5] and cooperation of manipulators, see [24] form important illustrations of the same goal, where it is desired that two or more mechanical systems, either identical or different, are asked to work in synchrony.

The purpose of this paper is to address the synchronization problem from a control theory perspective. More specifically, the paper addresses the synchronization problem from the perspective of (nonlinear) observer design. Hopefully, the paper initiates further interest in dynamical control methods in the study of synchronization problems.

The paper is organized as follows. In the following section, the synchronization problem will be introduced and it will be shown that under certain circumstances the synchronization problem can be viewed as an observer problem. Section 3 explores some of the methods to design nonlinear observers. In Section 4, static and dynamic methods to design synchronizing systems for nonlinear discrete-time systems are discussed. Section 5 contains some conclusions.

2   Synchronization as an Observer Problem

In this paper, we consider a nonlinear discrete-time or continuous-time system of the form

$$\begin{cases} \sigma x = f(x) \\ y = h(x) \end{cases} \tag{1}$$

where $x \in \mathbf{R}^n$, $y \in \mathbf{R}$ and f and h are smooth. For discrete-time systems, σ denotes the time-shift, i.e., $\sigma x(k) = x(k+1)$, while for continuous-time systems σ denotes differentiation, i.e., $\sigma x(t) = \dot{x}(t)$.

Besides (4), we consider a second system that is driven by (4):

$$\sigma z = g(y, z) \tag{2}$$

where $z \in \mathbf{R}^n$ and g is smooth. System (1) is the so-called *transmitter* (or *master*), and system (2) is the *receiver* (or *slave*). Synchronization of (1) and (2) occurs if, no matter how (1) and (2) are initialized, we have that asymptotically their states will match, i.e.,

$$\lim_{t \to +\infty} \|x(t) - z(t)\| = 0 \qquad (3)$$

Typically, the receiver (2) depends on (1) via the drive signal $y = h(x)$, which explains the transmitter/receiver terminology.

We illustrate this definition by means of an example.

Example 1: Following [27], we consider the Lorenz system

$$\begin{cases} \dot{x}_1 = \sigma(x_2 - x_1) \\ \dot{x}_2 = rx_1 - x_2 - x_1 x_3 \\ \dot{x}_3 = -bx_3 + x_1 x_2 \end{cases} \qquad (4)$$

The system (4) is known to exhibit complex or chaotic motions for certain parameters $\sigma, r, b > 0$. With the system (4) viewed as the *transmitter*, or *master system*, we introduce the drive signal

$$y = x_1 \qquad (5)$$

which can be used at the *receiver*, or slave system, to achieve asymptotic synchronization. This means, as in [27], that we take as receiver dynamics

$$\begin{cases} \dot{z}_1 = \sigma(z_2 - z_1) \\ \dot{z}_2 = ry - z_2 - yz_3 \\ \dot{z}_3 = -bz_3 + yz_2 \end{cases} \qquad (6)$$

Notice that (6) consists of a copy of (4), with state (z_1, z_2, z_3) and where in the (z_2, z_3)-dynamics the known signal $y = x_1$, see (5), is substituted for z_1. Introducing the error variables $e_1 = x_1 - z_1$, $e_2 = x_2 - z_2$, $e_3 = x_3 - z_3$ we obtain the error dynamics

$$\begin{cases} \dot{e}_1 = \sigma(e_2 - e_1) \\ \dot{e}_2 = -e_2 - ye_3 \\ \dot{e}_3 = -be_3 + ye_2 \end{cases} \qquad (7)$$

which is a linear time-varying system. The stability of $(e_1, e_2, e_3) = (0, 0, 0)$ is straightforwardly checked using the Lyapunov-function

$$V(e_1, e_2, e_3) = \frac{1}{\sigma}e_1^2 + e_2^2 + e_3^2 \qquad (8)$$

with time-derivative along (7)

$$\dot{V}(e_1, e_2, e_3) = -2(e_1 - \frac{1}{2}e_2)^2 - \frac{3}{2}e_2^2 - 2be_3^2 \qquad (9)$$

showing that (e_1, e_2, e_3) asymptotically (and even exponentially!) converges to $(0, 0, 0)$. In other words, the receiver dynamics (6) asymptotically synchronizes with the chaotic transmitter (4) no matter how (4) and (6) are initialized.

In most of the synchronization literature (see e.g. [27]) the systems (1) and (2) are systems that are given beforehand, so that no synchronization will occur in general. This raises the question whether, given a system (1), a system (2) can be designed that synchronizes with (1). As was indicated in e.g. [25] this problem may be interpreted as viewing (2) as an observer for (1) given the output signal $y = h(x)$. So in those applications where one is able to design (2) freely, the synchronization problem may be tackled as an observer problem.

In its full generality the problem of observer design for nonlinear systems is unsolved. However, there are some important cases where a solution can be found. Some of them will be reviewed next. The natural way to approach the observer problem for (1) is to design another dynamical system driven by the measurements $y = h(x)$

$$\sigma z = f(z) + k(z, y) \tag{10}$$

where k in (10) should be such that $k(z, y) = 0$ if $h(z) = h(x) = y$. Recall that the dynamics (10) is an *observer* for (1) if $z(t)$ asymptotically converges to $x(t)$, for any pair of initial conditions $x(0)$ and $z(0)$. The structure of the observer (10) deserves some further attention. One may view (10) as an identical copy of (1) with an 'innovations' term $k(z, y)$ which vanishes when the estimated output $\hat{y} = h(z)$ coincides with $y = h(x)$. The latter could be phrased as: "we can not do better than our measurements allow for". In the Lorenz system (4,5) with receiver (6), it is easily checked that the system (6) indeed acts as an observer and can be put in the form (10):

$$\begin{cases} \dot{z}_1 = \sigma(z_2 - z_1) & +0 \\ \dot{z}_2 = rz_1 - z_2 - z_1 z_3 & +(r - z_3)(y - z_1) \\ \dot{z}_3 = -bz_3 + z_1 z_2 & +z_2(y - z_1) \end{cases} \tag{11}$$

Also, it may be worth noting that (10) is simply a computerized model and no hard-ware is required in building this system, even if a hardware realization of (1) is given.

In the following sections, we will explore some of the possibilities to solve the synchronization problem by making use of results from the theory of nonlinear observer design.

3   Observer Based Design of Synchronizing Systems

Probably the most successful approach to the observer design for nonlinear systems is to make use of so called Lur'e forms. A nonlinear discrete-time or continuous-time system in Lur'e form is a system of the form

$$\begin{cases} \sigma \xi = A\xi + \Phi(\tilde{y}) \\ \tilde{y} = C\xi \end{cases} \tag{12}$$

where $\xi \in \mathbf{R}^n$, $\tilde{y} \in \mathbf{R}$, A and C are matrices of appropriate dimensions, the pair (C, A) is observable, and Φ is smooth. As may be well known, for this kind of systems a synchronizing system is given by

$$\begin{cases} \sigma z = Az + \Phi(\tilde{y}) + K(\tilde{y} - \hat{y}) \\ \hat{y} = Cz \end{cases} \tag{13}$$

where the matrix K is such that the eigenvalues of $A - KC$ are in the open unit disc (for discrete-time systems) or in the open left half plane (for continuous-time systems).

From the above, it follows that synchronizing systems can also be designed quite easily for systems (1) that may be transformed into a Lur'e form by means of a change of coordinates $\xi = P(x)$ and an output transformation $\tilde{y} = p(y)$. Conditions in terms of f and h under which continuous-time systems can be transformed into Lur'e form are given in [20],[21], see also [25] where these conditions are given in the context of synchronization. For discrete-time systems, these conditions were given in [23] for the case without output transformations and in [13] for the case where also output transformations are allowed.

We illustrate the above with two examples.

Exemple 2: Consider the following discrete-time model of a bouncing ball ([35],[6],[18]):

$$\begin{cases} x_1(k+1) = x_1(k) + x_2(k) \\ x_2(k+1) = \alpha x_2(k) - \beta \cos(x_1(k) + x_2(k)) \end{cases} \tag{14}$$

In (14), $x_1(k)$ is the phase of the table at the k-th impact, $x_2(k)$ is proportional to the velocity of the ball at the k-th impact, α is the coefficient of restitution and $\beta = 2\omega^2(1 + \alpha)A/g$. Here ω is the angular frequency of the table oscillation, A is the corresponding amplitude, and g is the gravitational acceleration.

Suppose that only the phase x_1 is available for measurement, i.e., we have that

$$y = h(x) = x_1 \tag{15}$$

Defining new coordinates

$$\begin{cases} \xi_1 = x_1 \\ \xi_2 = -\alpha x_1 + x_2 + \beta \cos x_1 \end{cases} \tag{16}$$

we obtain the following Lur'e form:

$$\begin{cases} \xi_1(k+1) = \xi_2(k) + (1+\alpha)y(k) - \beta \cos y(k) \\ \xi_2(k+1) = -\alpha y(k) \\ \quad y(k) = \xi_1(k) \end{cases} \tag{17}$$

A synchronizing system for (17) is then given by

$$\begin{cases} z_1(k+1) = -\ell_1 z_1(k) + z_2(k) + \ell_1 y(k) + (1+\alpha)y(k) - \beta \cos y(k) \\ z_2(k+1) = -\ell_2 z_1(k) + \ell_2 y(k) - \alpha y(k) \end{cases} \tag{18}$$

where ℓ_1, ℓ_2 are chosen in such a way that the eigenvalues of the matrix $\begin{pmatrix} -\ell_1 & 1 \\ -\ell_2 & 0 \end{pmatrix}$ are in the open unit disc.

Example 3: Consider the hyperchaotic Rössler system, see [2],

$$\begin{cases} \dot{x}_1 = -x_2 + ax_1 \\ \dot{x}_i = x_{i-1} - x_{i+1} \qquad (i = 2, \cdots, n-1) \\ \dot{x}_n = \epsilon + bx_n(x_{n-1} - d) \\ y = x_n \end{cases} \tag{19}$$

where $a, b, d, \epsilon > 0$ and $n \geq 3$ arbitrary. The case $n = 3$ corresponds to the usual Rössler system, and when $n = 4$ the system has so called hyperchaotic flows and has two positive Lyapunov exponents. It is clear that the solutions of (19) with $x_n(0) > 0$ that have no finite escape time satisfy $x_n(t) > 0$ for all $t \geq 0$. Therefore, we may introduce new coordinates $\xi_i = x_i$ $(i = 1, \cdots, n-1)$, $\xi_n = \ln x_n$. When we also define a new output $\tilde{y} = \ln y$, the system in the new coordinates and with the new output has the form

$$\begin{cases} \dot{\xi}_1 = -\xi_2 + a\xi_1 \\ \dot{\xi}_i = \xi_{i-1} - \xi_{i+1} \qquad (i = 2, \cdots, n-2) \\ \dot{\xi}_{n-1} = \xi_{n-2} - \exp(\tilde{y}) \\ \dot{\xi}_n = b\xi_{n-1} - bd + \epsilon \exp(-\tilde{y}) \\ \tilde{y} = \xi_n \end{cases} \tag{20}$$

It is then clear that the system (20) is in Lur'e form. Hence a synchronizing system for (20) (and thus also for (19)) may be designed along the lines set out above.

The classes of continuous-time systems for which a successful observer design is possible, sofar almost all exploit Lur'e forms. There are however other cases where synchronization can be achieved without relying on a 'linearizability' assumption. To that end we return to the continuous-time system (1) and we introduce the following assumptions, see [11].

(i) The vector field f in (1) satisfies a global Lipschitz condition on its domain, need not to be $\mathbf{R}^n$.

(ii) The n functions $h(x), L_f h(x), L_f^2 h(x), \cdots, L_f^{n-1} h(x)$ define new coordinates (globally!). Here $L_f^i h(x)$ denotes the i-th iterated Lie-derivative of the function h in the direction of f.

If both (i) and (ii) hold an observer exists of the form

$$\dot{z} = f(z) + K(h(x) - h(z)) \qquad (21)$$

with K a constant suitable $(n, 1)$-vector (cf. [11]). Note that (21) obviously is of the form (10), though some of the entries in K may become very large (high-gain). An illustrative example of a system that fulfills (i) and (ii) is formed by the Lorenz-system (4,5), when this is restricted to a compact domain. Since it is known that (4) has an attractive compact box, the observer (21) is an interesting alternative for the observer (6).

Besides the above discussed cases for which a synchronizing system can be systematically designed we note that there exist further methods that may be applicable for other classes of systems, like bilinear systems. Also for certain mechanical systems 'physics-based' observers can be developed, and finally some systems admit a Kalman filter-like observer. *But*, no general method exists that works for all systems.

4   Static Synchronization and Extended Lur'e Forms for Discrete-time Systems

4.1   Static Synchronization

One of the methods that is extensively used in the synchronization literature for the design of synchronizing systems is based on the so called Takens-Aeyels-Sauer Reconstruction Theorem (see e.g. [33]). We will first give a brief statement of this theorem. Consider the discrete-time nonlinear system (1), and define the mapping $\tilde{\psi} : \mathbf{R}^n \to \mathbf{R}^{2n}$ by

$$\tilde{\psi}(x) := \begin{pmatrix} h(x) \\ \sigma h(x) \\ \vdots \\ \sigma^{2n-1} h(x) \end{pmatrix} = \begin{pmatrix} y \\ \sigma y \\ \vdots \\ \sigma^{2n-1} y \end{pmatrix} \qquad (22)$$

The Takens-Aeyels-Sauer Reconstruction Theorem ([34],[1],[30],[32]) then states that for *generic* f and h, the mapping $\tilde{\psi}$ is an embedding. In particular, this implies that there globally exists a mapping $\tilde{\psi}^+ : \mathbf{R}^{2n} \to \mathbf{R}^n$ such that

$$\tilde{\psi}^+(\tilde{\psi}(x)) = x$$

In other words, the mapping $\tilde{\psi}^+$ is a global left-inverse of $\tilde{\psi}$. This means that once the mapping $\tilde{\psi}^+$ is known, the state $x(k)$ may be reconstructed by first reconstructing $x(k - 2n + 1)$ by means of $x(k - 2n + 1) = \tilde{\psi}^+(y(k - 2n + 1), \cdots, y(k))$, which gives that $x(k) = f^{2n-1}(x(k - 2n + 1))$.

The static reconstruction approach set out above has the drawback that it is very sensitive to measurement errors, especially when chaotic systems (1) are being considered. In order to cope with this drawback, one would need to combine the static reconstruction with some kind of filtering in order to suppress the influence of measurement noise. One possibility to do this for discrete-time systems is by making use of so called *extended Lur'e forms*. This will be the topic of the next subsection.

Remark: When throughout "σ" is interpreted as the time-derivative, the Takens-Aeyels-Sauer Reconstruction Theorem also holds for continuous-time systems. Thus, in order to perform reconstruction based on the Takens-Aeyels-Sauer Reconstruction Theorem for continuous-time systems, one needs to repeatedly (numerically) differentiate the transmitted signal y. Especially when one is dealing with chaotic systems (1), this numerical differentiation is very unreliable. Besides the sensitivity to measurement errors, this forms a second drawback of static reconstruction methods for continuous-time systems.

4.2   Extended Lur'e Forms

The conditions for the existence of Lur'e forms for discrete-time systems given in [23],[13] are quite restrictive. For discrete-time systems however, there is a natural generalization of (12) that gives more freedom to design synchronizing systems. In this generalization, also past output measurements are employed in the synchronizing system, which takes the observer based design of synchronizing systems in the direction of the static reconstruction based on the Takens-Aeyels-Sauer Reconstruction Theorem.

More specifically, one considers discrete-time systems in so called extended Lur'e form with buffer N ($N = 1, \cdots, n - 1$). These are systems of the form

$$\begin{cases} \xi(k + 1) = A\xi(k) + \Phi(\tilde{y}(k), \tilde{y}(k - 1), \cdots, \tilde{y}(k - N)) \\ \quad \tilde{y}(k) = C\xi(k) \end{cases} \tag{23}$$

where $\xi \in \mathbf{R}^n$, $\tilde{y} \in \mathbf{R}$, A and C are matrices of appropriate dimensions, the pair (C, A) is observable, and Φ is smooth. Analogously to the previous subsection, for this kind of systems a synchronizing system is given by

$$\begin{cases} z(k+1) = Az(k) + \Phi(\tilde{y}(k), \tilde{y}(k-1), \cdots, \tilde{y}(k-N)) + \\ \qquad\qquad K(\tilde{y}(k) - \hat{y}(k)) \\ \hat{y}(k) = Cz(k) \end{cases} \tag{24}$$

where the matrix K is such that the eigenvalues of $A - KC$ are in the open unit disc.

As in the previous section, one is now interested in the question whether a given discrete-time system may be transformed into an extended Lur'e form with a buffer $N \in \{1, \cdots, n-1\}$. The transformations allowed here are so called extended coordinate changes parametrized by $y(k-1), \cdots, y(k-N)$ of the form $\xi(k) = P(x(k), y(k-1), \cdots, y(k-N))$, and possibly an output transformation $\tilde{y} = p(y)$. Conditions for the existence of an extended coordinate change and output transformation that transform (1) into an extended Lur'e form with a buffer $N \in \{1, \cdots, n-1\}$ are given in [13],[14],[15],[22]. A remarkable result in [13],[14],[15],[22] is that a strongly observable discrete-time system (see e.g. [13] for a definition of strong observability) *always* may be transformed into an extended Lur'e form with buffer $N = n - 1$.

Some remarks on the connection between static synchronization based on the Takens-Aeyels-Sauer Reconstruction Theorem and dynamic synchronization based on extended Lur'e forms with buffer $N = n - 1$ are in order here:

(i) As stated in the previous subsection, the static synchronization approach may be applied to generic discrete-time systems (1). As stated above, the dynamic reconstruction approach with buffer $N = n-1$ may be applied to strongly observable systems (1). As is well known, strong observability is a generic property for systems (1), and thus the dynamic synchronization approach with buffer $N = n - 1$ may also be applied to generic discrete-time systems (1). However, it remains an interesting question whether the generic class to which the static synchronization may be applied is the same as the generic class to which the dynamic reconstruction approach may be applied.

(ii) If there are no measurement errors, the static synchronization approach results in exact synchronization after $2n - 1$ time steps. In the dynamic synchronization approach with buffer $N = n - 1$, one can also achieve exact synchronization after $2n - 1$ time steps: $n - 1$ time steps are needed to initialize the receiver (24), and by an appropriate choice of the vector K in (24) one can obtain a dead-beat receiver which results in exact synchronization after another n time steps.

Remark: If one would like to follow the approach of extended Lur'e systems for the design of synchronizing systems for continuous-time systems, the extended Lur'e form as well as the synchronizing system would depend on time-derivatives of the output y instead of past output measurements. Conditions for the existence of extended Lur'e forms for continuous-time systems have been obtained in e.g. [12]. However, as said before, the fact that one needs to numerically differentiate y makes it questionable whether this approach would work for continuous-time systems.

5   Conclusions

We have tried to give a dynamical control view on synchronization. All in all, it is felt that nonlinear control may provide some useful tools to address certain synchronization problems. On the other hand, in many cases, a thorough study of certain time-varying dynamical systems is required and it may be concluded that further research along these lines requires knowledge from both dynamical systems and nonlinear control theory. The review as presented here gives only a partial view on synchronization. There are numerous variants of synchronization defined in the literature, of which one could mention, phase synchronization, partial synchronization and generalized synchronization, see [26], or [4] where a general definition of synchronization is proposed. In the study of synchronization several elements from control theory turn out to be relevant. This includes observers but also further aspects such as filtering (cf. [8],[31]), adaptive control (cf. [10]), robustness (cf. [28]), feedback control (cf. [17]), system inversion (cf. [9]), or system identification (cf. [16]). It should be clear that synchronization problems can be treated in other domains too. Even for transmitter/receiver dynamics described by partial differential equations one may expect some results along the lines set out in this paper, see e.g. [7] for a specific example of synchronizing PDE's. Likewise, synchronization with time-delayed feedback has also been studied in [7]. Synchronization has numerous potential applications running from coordination problems in robotics to mechanisms for secure communications. Precisely the latter area was mentioned in [27] as a potential field of application, although sofar much work remains to be done here.

References

1. Aeyels D. (1981) Generic Observability of Differentiable Systems. SIAM J Control Optimiz 19:595-603
2. Baier G., Sahle S. (1995) Design of Hyperchaotic Flows. Phys Rev E 51:R2712-R2714
3. Blekhman I.I., Landa P.S., Rosenblum M.G. (1995) Synchronization and Chaotization in Interacting Dynamical Systems. Appl Mech Rev 48:733-752, 1995
4. Blekhman I.I., Fradkov A.L., Nijmeijer H., Pogromsky A.Yu. (1997) On Self-synchronization and Controlled Synchronization. Syst Control Lett 31:299-305

5. Brunt M. (1998) Coordination of Redundant Systems. Ph.D. Thesis, Delft University of Technology, The Netherlands
6. Cao L., Judd K., Mees A. (1997) Targeting Using Global Models Built from Nonstationary Data. Phys Rev Lett 231:367-372
7. Chen G. (Ed.) (1999) Controlling Chaos and Bifurcations in Engineering Systems. CRC Press, Boca Raton
8. Cruz C., Nijmeijer H. (2000) Synchronization Through Filtering. Int J Bifurc Chaos: to appear
9. Feldman U., Hasler M., Schwarz W. (1996) Communication by Chaotic Signals: The Inverse System Approach. Int J Circ Theory Appl 24:551-579
10. Fradkov A.L., Nijmeijer H., Pogromsky A.Yu. (1999) Adaptive Observer Based Synchronization. In: [7], 417-438
11. Gauthier J., H. Hammouri H., Othman S. (1992) A Simple Observer for Nonlinear Systems, Application to Bioreactors. IEEE Trans Automat Control 37:875-880
12. Glumineau A., Moog C.H., Plestan F. (1996) New Algebro-geometric Conditions for the Linearization by Input-output Injection. IEEE Trans Automat Control 41:598-603
13. Huijberts H.J.C. (1999) On Existence of Extended Observer Forms for Nonlinear Discrete-time Systems. In: Nijmeijer H., Fossen T.I. (Eds.) New Directions in Nonlinear Observer Design. Springer, Berlin, 79-92
14. Huijberts H.J.C., Lilge T., Nijmeijer H. (1999) Control Perspective on Synchronization and the Takens-Aeyels-Sauer Reconstruction Theorem. Phys Rev E 59:4691-4694
15. Huijberts H.J.C., Lilge T., Nijmeijer H. (2000) Nonlinear Discrete-time Synchronization via Extended Observers. Preprint, submitted to Int J Bifurc Chaos
16. Huijberts H.J.C., Nijmeijer H., Willems R.M.A. (2000) System Identification in Communication with Chaotic Systems. IEEE Trans Circ Systems I: to appear
17. Huijberts H.J.C., Nijmeijer H., Willems R.M.A. (2000) Regulation and Controlled Synchronization for Complex Dynamical Systems. Int J Robust Nonl Control: to appear
18. Huijberts H.J.C., Nijmeijer H., Pogromsky A.Yu. (1999) Discrete-time Observers and Synchronization. In: [7], 439-455
19. Huijgens C. (1673) Horologium Oscilatorium. Paris, France
20. Krener A.J., Isidori A. (1983) Linearization by Output Injection and Nonlinear Observers. Syst Control Lett 3:47-52
21. Krener A.J., Respondek W. (1985) Nonlinear Observers with Linearizable Error Dynamics. SIAM J Control Optimiz 23:197-216
22. Lilge T. (1999) Zum Entwurf nichtlinearer zeitdiskreter Beobachter mittels Normalformen. VDI-Verlag, Düsseldorf
23. Lin W., Byrnes C.I. (1995) Remarks on Linearization of Discrete-time Autonomous Systems and Nonlinear Observer Design. Syst Control Lett 25:31-40
24. Liu Y.-H., Xu Y., Bergerman M. (1999) Cooperation Control of Multiple Manipulators with Passive Joints. IEEE Trans Robotics Automat 15:258-267
25. Nijmeijer H., Mareels I.M.Y. (1997) An Observer Looks at Synchronization. IEEE Trans Circ Systems I 44:882-890
26. Parlitz U., Kocarev L. (1999) Synchronization of Chaotic Systems. pp. 272-302 In: Schuster H.G. (Ed.) Handbook of Chaos Control, Wiley-VCH, New York, 272-302

27. Pecora L.M., Carroll T.L. (1990) Synchronization in Chaotic Systems. Phys Rev Lett 64:821-824
28. Pogromsky A.Yu., Nijmeijer H. (1998) Observer Based Robust Synchronization of Dynamical Systems. Int J Bifurc Chaos 8:2243-2254
29. Rayleigh J. (1945) The Theory of Sound. Dover Publ, New York
30. Sauer T., Yorke J.A., Casdagli M. (1991) Embeddology. J Stat Phys 65:579-616
31. Sobiski D.J., Thorp J.S. (1998) PDMA-1 Chaotic Communication via the Extended Kalman Filter. IEEE Trans Circ Syst I 46:841-850
32. Stark J. (1999) Delay Embedding for Forced Systems. I. Deterministic forcing. J Nonl Sci. 9:255-332
33. Stojanovski T., Parlitz U., Kocarev L., Harris R. (1997) Exploiting Delay Reconstruction for Chaos Synchronization. Physics Lett A 233:355-360
34. Takens F. (1981) Detecting Strange Attractors in Turbulence. In: D.A. Rand, Young L.S. (Eds.) Dynamical Systems and Turbulence, Springer, Berlin, 366-381
35. Tufillaro N.B., Abbott T., Reilly J. (1992) An Experimental Approach to Nonlinear Dynamics and Chaos. Addison-Wesley, Reading, MA
36. Van der Pol B. (1920) Theory of the Amplitude of Free and Forced Triod Vibration. Radio Rev 1:701-710

Regularity of the Sub-Riemannian Distance and Cut Locus

Sébastien Jacquet

Departamento de Matemtica
Universidade de Aveiro
Campus Santiago
3810 Aveiro, Portugal
jacquet@mat.ua.pt

Abstract. Sub-Riemannian distances are obtained by minimizing length of curves whose velocity is constrained to be tangent to a given sub-bundle of the tangent bundle. We study the regularity properties of the function $x \mapsto d(x_0, x)$ for a given sub-Riemannian distance d on a neighborhood of a point x_0 of a manifold M. We already know that this function is not C^1 on any neighborhood of x_0 (see [1]) and even if the data are analytic, the distance may fail to be subanalytic [8]. In this paper we make the link between the singular support of $x \mapsto d(x_0, x)$, the cut locus and the set of points reached from x_0 by singular minimizers.

1 Introduction.

In order to state the problem more precisely consider a smooth connected n dimensional manifold M, a smooth sub-bundle Δ of the tangent bundle TM and a smooth metric tensor g on Δ. We say that a H^1-curve $\gamma : [0, T] \longrightarrow M$ is admissible if its derivative is tangent to the distribution Δ almost everywhere i.e.

$$\dot{\gamma}(t) \in \Delta_{\gamma(t)}$$

for almost every t in $[0, T]$.

The length of an admissible curve γ is given by

$$L(\gamma) = \int_0^T \sqrt{g_{\gamma(t)}(\dot{\gamma}(t), \dot{\gamma}(t))} \, dt$$

and we can define the sub-Riemannian (or Carnot-Caratheodory) distance

$$d : M \times M \longrightarrow \mathbb{R} \cup \{+\infty\},$$

setting $d(x, y) = \inf\{L(\gamma) : \gamma$ admissible curve joining x to $y \}$. We make the Hörmander assumption on Δ that is:

$$\text{Lie}(\Delta)_x = T_x M, \quad \forall \, x \in M$$

where Lie(Δ) is the Lie algebra of smooth vector fields generated by $C^\infty(\Delta)$. Under the above assumption d is a distance and as a consequence of "Chow - Raschevski" Theorem, the metric space (M,d) induces on M the same topology as the original one.

The paper is devoted to study the value-function $f : x \longmapsto d(x_0,x)$ in a neighborhood U of x_0. The main difference between Riemannian (i.e. $\Delta = TM$) and sub-Riemannian case is that in any neighborhood of ¿x_0 there are points $x \neq x_0$ in which f is not continuously differentiable [1]. Results about subanalyticity of f have been established in [11,18,4,13,14]. The possible existence of "abnormal minimizers" make the study of the cut locus more complex. In this paper we give some results on the structure of the singularities of the value function and of the cut locus (see [3,10,2] for the study of the cut locus for contact distributions in $\mathbb{R}^3$ and [8] for the Martinet "flat" case).

The paper is organized as follows : we recall some basic facts about minimizers in section 2. The aim of the section 3 is to characterize the larger open set where the function f is of class C^1, the main results are stated in Theorem 1 and Corollary 1. In the last section we begin a description of the cut locus (Theorem 2).

2   Minimizers.

In this section we give a summary of known results on the subject and the definitions needed to state our main results. Since the results are local, without loss of generality we may assume that M is an open connected neighborhood of x_0 in $\mathbb{R}^n$ and the sub-Riemannian structure (Δ,g) is defined by an orthonormal frame $\{X_1, \cdots, X_m\}$. In this case the space $\mathcal{H}_T$ of all admissible curves defined on $[0,T]$ starting at x_0 can be identified with an open subset of $L_m^2([0,T])$ by means of

$$u \in L_m^2([0,T]) \longmapsto \gamma \in \mathcal{H}_T$$

where γ is the solution starting at x_0 of the control system

$$\dot{\gamma}(t) = \sum_{i=1}^m u_i(t) X_i(\gamma(t)) \qquad \text{a.e. in } [0,T], \tag{1}$$

Notice that, being the X_i's independent at each point, there is an unique $u \in L_m^2([0,T])$ corresponding to each γ in $\mathcal{H}_T$, such a u is called the control of γ.

Like in the case of Riemannian distance we associate to each curve a smooth "cost", called energy, defined by

$$E : \gamma \in \mathcal{H}_T \longmapsto \int_0^T g_{\gamma(t)}(\dot{\gamma}(t), \dot{\gamma}(t)) \, dt \in \mathbb{R}.$$

By our assumptions $E(\gamma) = ||u||^2$, where $||u||$ is the L^2-norm of the control of γ. The energy of a curve depends on the parametrization while the length does not and the following inequality holds

$$L(\gamma) \leq \sqrt{TE(\gamma)}$$

where the equality occurs if and only if the speed $\sqrt{g_{\gamma(t)}(\dot{\gamma}(t), \dot{\gamma}(t))}$ is constant almost everywhere. A curve that realizes the minimum of the energy is parametrized by constant speed and realizes also the minimum of the length. On the other side for every curve that realizes the minimum of the length there exists a reparametrization of this curve by constant speed so that it becomes a minimizer of the energy. We call minimizer an admissible curve that realizes the minimum of the energy among all admissible curves defined on the same interval joining two given points in M. According to our definition, minimizers are admissible curves parametrized on some $[0, T]$ with constant speed. In most of the proofs, it will be more convenient to consider minimizers parametrized by arc length, in this case, we get that $T = f(\gamma(1)) = \sqrt{E(\gamma)}$.

For $\varepsilon > 0$ small enough, every $u \in L^2_m([0, T])$ with $||u|| < 3T\varepsilon$ is the control of an admissible curve defined on $[0, T]$. For such an ε and for all $\delta \leq \varepsilon$ the sub-Riemannian ball

$$U_\delta = \{x \in M \ : \ f(x) < \delta\}$$

is an open set given by

$$U_\delta = \{\gamma(T) \ : \ E(\gamma) < T\delta^2\}.$$

Such an ε does not depend on T and **in all the paper U will be equal to such an U_ε and we denote again by $\mathcal{H}_T$ the set** $\{u \in L^2_m([0, T]) \ : \ ||u|| \leq T\varepsilon\}$. In the case when $T = 1$ we simply note $\mathcal{H}$. For ε sufficiently small, for all x in U_ε it always exists a minimizer going from x_0 to x and we have the following compactness result (see [1,14]).

Proposition 1. *Let K be a compact subset of U then the set of minimizers starting from x_0 with end-point in K is a compact subset of $\mathcal{H}_T$.*

Remark. Let V be an open subset of U that does not contain x_0 and suppose that you have a C^k optimal synthesis on V, i.e. a C^k map σ defined on V with values in $\mathcal{H}$ such that for all x in V, $\sigma(x)$ is a minimizer going from x_0 to x. Then f is given on V by $f^2_{|V} = E \circ \sigma$. and hence f is of class C^k on V. The main result of this paper will be to prove that this condition is in fact necessary for $k = 2$ and when the structure is analytic for $k = 1$.

Regularity of f is then related to the problem of finding smooth optimal synthesis and hence parametrization of minimizers. Let us recall now first order necessary optimal condition.

If $\gamma : [0, T] \to U$ is a minimizer then, by the Pontryagin Maximum Principal, there is $\lambda_0 \in \mathbb{R}$ and a nonzero solution $q : [0, T] \to (\mathbb{R}^n)^*$ of

$$\dot{q}(t) = -\sum_{i=1}^{m} u_i(t) \, {}^t DX_i(\gamma(t)) q(t) \tag{2}$$

such that

$$\langle q, X_i(\gamma) \rangle = 2\lambda_0 u_i, \quad i = 1, \cdots, m \tag{3}$$

where u is the control of γ. In fact if γ is an admissible curve and q satisfies the above conditions (2) and (3), $\lambda = (-q(T), \lambda_0)$ is a Lagrange multiplier for the map

$$\theta = e \times E : \gamma \in \mathcal{H}_T \longmapsto (\gamma(T), E(\gamma)) \in M \times \mathbb{R},$$

so that γ is an extremal for θ. The map $e : \mathcal{H}_T \to M$ is called the end point map. We call (γ, q, λ_0) an extremal and q an extremal lift of γ.

If $\lambda_0 = 0$ the extremal $(\gamma, q, 0)$ is called abnormal extremal and the corresponding curve γ is then a singularity of the end-point mapping. Such γ will be called singular.

If $\lambda_0 \neq 0$ the couple (q, λ_0) is defined up to multiplication by a scalar and we can choose $\lambda_0 = \frac{1}{2}$. In this case the control of γ can be reconstructed from (3) and (γ, q) is a solution of a Hamiltonian system. To be more precise, let us define $G_x : T_x^* M \to \Delta_x$ and $H : T^* M \to \mathbb{R}$ by

$$G_x \xi = \sum_{i=1}^{m} \langle \xi, X_i(x) \rangle X_i(x) \quad \text{and} \quad H(x, \xi) = \frac{1}{2} \langle \xi, G_x \xi \rangle.$$

It is not difficult to see that $(\gamma, q, \frac{1}{2})$ is an extremal iff (γ, q) is a solution of the Hamiltonian system associated to H, i. e. $(\dot{\gamma}, \dot{q}) = \boldsymbol{H}(\gamma, q)$. For (x, ξ) in $T^* M$ we shall denote by $\boldsymbol{H}(t)(x, \xi)$ the flow of $\boldsymbol{H}$ starting at (x, ξ). Let us remark that $G_x \xi$ is the only vector of Δ_x such that $\langle \xi, Y \rangle = g_x(G_x \xi, Y)$ for every Y in Δ_x and that the equality (3) is equivalent to $G_\gamma q = 2\lambda_0 \dot{\gamma}$.

Let $\mathcal{D}_{exp}$ be the open subset of points q_0 in $T_{x_0}^* M$ such that $\boldsymbol{H}(\cdot)(x_0, q_0)$ is defined on $[0, 1]$ and let $\rho : T^* M \to M$ be the canonical projection, we define the exponential map by:

$$\exp : q_0 \in \mathcal{D}_{exp} \longmapsto \rho \circ \boldsymbol{H}(1)(x_0, q_0) \in M.$$

If q_0 is in $\mathcal{D}_{exp}$ and (γ, q) is the corresponding integral curve of $\boldsymbol{H}$, then $\gamma(t) = \exp(tq_0)$ for every t in $[0, 1]$ and we will denote γ by $\exp_{q_0}$. The Hamiltonian H is constant along (γ, q) and we have $E(\gamma) = 2H(\gamma_0, q_0)$.

With the above notations the first order necessary condition for optimality can be reformulated as follows: every minimizer can be lifted to an extremal.

It can be easily checked that the space of abnormal lifts is a vector space and that the difference of two Hamiltonian lifts is an abnormal lift. In particular if ξ is an abnormal lift of $\exp_{q_0}$, then for each α in $\mathbb{R}$, $\exp(q_0 + \alpha \xi(0)) = \exp(q_0)$ and hence the restriction of exp to any neighborhood of q_0 is non injective. In particular, q_0 is a singularity of exp and $\gamma(1)$ is a singular value of both exp and e. On the other hand, if a minimizer is non singular then there exists q_0 such that $\gamma = \exp_{q_0}$ and this q_0 is unique.

In [19] every Hamiltonian curve is proved to be locally minimizing, the inverse statement is false in general : there exist sub-Riemannian structures with minimizers that cannot be lifted to a normal extremal that are so called strictly abnormal curves [16]. Until know, it is not known if a general strictly abnormal minimizer is smooth or not.

We call U_n the set of point that are reached by x_0 only by nonsingular minimizers. It comes directly from Proposition 1 that U_n is open and we have the following result.

Lemma 1. *Let K be a compact subset of U_n then the set of q in $T^*_{x_0} U$ such that exp_q is a minimizer with end point in K is compact.*

This result comes directly from Proposition 1. It is a particular case of compactness result obtained in [4,13] by using second order necessary condition for optimality proved in [5]. In this paper, first order necessary condition (maximum principle or Lagrange multiplier rule) will be enough to prove our results.

3 Regularity of f.

In this section we describe the set of point in which f is smooth.

Definition 1. *Let k be in $\mathbb{N} \cup \{+\infty, \omega\}$, the set $Reg_k(f)$ is defined by $x \in Reg_k(f)$ iff there exists a neighborhood of x such that the restriction of f to this neighborhood is of class C^k. We call $Sing_k(f)$ its complementary.*

We first state a sufficient condition for a point to be in $Reg_\infty(f)$.

Proposition 2. *Let x be in U such that*

- *there exists only one minimizer γ going from x_0 to x,*
- *γ is nonsingular,*
- *$\gamma = exp_{q_0}$ where q_0 is not a singularity of exp,*

then x belongs to $Reg_\infty(f)$ (resp. $Reg_\omega(f)$ if the structure is analytic).

Proof. Since q_0 is not a singularity of exp there exist a neighborhood V of x in U and a neighborhood W of q_0 in $T^*_{x_0} U$ such that $\exp_{|W}$ is a smooth diffeomorphism. Define the function f_1 on V by

$$f_1(y) = \sqrt{2H(x_0, (\exp_{|W})^{-1}(y))},$$

then f_1 is smooth (resp. analytic) and we will prove that we can choose V such that $f_{|V} = f_1$.

First remark that for y in V, $f_1(y)$ is the length of the curve $\exp_{q_y}$ where $q_y = (\exp_{|W})^{-1}(y)$ and then $f_{|V}(y) \le f_1(y)$ and the equality holds if and only if $\exp_{q_y}$ is a minimizer. Suppose that we cannot find V such that $f_{|V} = f_1$, then we can construct a sequence $\{y_\alpha\}$ that tends to x such that $\exp_{q_\alpha}$ is not minimizer where $q_\alpha = (\exp_{|W})^{-1}(y_\alpha)$.

For each α, let γ_α be a minimizer going from x_0 to y_α then using the fact that γ is the only minimizer going from x_0 to x and Proposition 1, $\{\gamma_\alpha\}$ tends to γ. In particular, for α big enough, γ_α is not singular since γ is not. Let ξ_α be unique such that $(\gamma_\alpha, \xi_\alpha, \frac{1}{2})$ is an extremal, then by extracting a subsequence we may assume that $(\gamma_\alpha, \xi_\alpha, \frac{1}{2})$ tends to some extremal $(\gamma, \xi, \frac{1}{2})$ (Lemma 1) and since γ is nonsingular, $\xi(0) = q_0$. Then, for α big enough, $\xi_\alpha(0)$ belongs to W and $\exp_{|W}$ being injective, $\xi_\alpha(0) = q_\alpha$ and gives us the contradiction.

$\square$

Remark. We keep the same notation of the proof of Proposition 2. We proved in fact that we can choose W and V such that for all q in W, $\exp_q$ is a minimizer going from x_0 to a point of V. In particular for $T > 1$ sufficiently close to 1, Tq_0 is in W and then $\exp_{Tq_0}$ is a minimizer. In other words, $t \mapsto \exp(tq_0)$ minimizes on some interval $[0, T]$ strictly bigger than $[0, 1]$ i.e. x is not a "cut point" (see definition below).

Definition 2. A point x in U is not in the cut locus (of x_0) if there is a minimizer γ which goes from x_0 to x and is the strict restriction of a minimizer starting from x_0. The cut locus will be denoted by $\mathcal{C}$ and points in $\mathcal{C}$ are called cut points.

Then a point x is in the cut locus only if at least one of the following condition is satisfied

- there are more than one minimizer going from x_0 to x,
- there is a singular minimizer going from x_0 to x,
- there is a minimizer $\exp_q$ going from x_0 to x where q is a singularity of $\exp$.

In Riemannian Geometry, singular curves does not exist and the two other conditions are sufficient condition for a point to be in the cut locus. The cut locus will be studied in the next sections, nevertheless we will need the following Lemma.

Lemma 2. *Suppose that the structure is analytic. Let x be in U such that*

- *there exists only one minimizer γ going from x_0 to x,*
- *γ is nonsingular*
- *$\gamma = exp_{q_0}$ where q_0 is a singularity of exp*

then x is in the cut locus.

Proof. Since γ is nonsingular, $\mathcal{H}_x := e^{-1}(x)$ is a submanifold. Let F be the restriction of E to $\mathcal{H}_x$ then since γ is a minimizer, it is a critical point of F and the Hessian F''_γ of F at γ (well defined since γ is a critical point of F) has to be positive. The negative index of F''_γ is computed in [6,7,13] where it is proved the following : for $T > 1$ sufficiently close to 1 the negative index of the Hessian at exp_{Tq_0} of the restriction of E to $e^{-1}(\{\exp(Tq_0)\})$ is equal to the dimension of the kernel of F''_γ that is also equal to the dimension of the kernel of the differential of exp at q_0. Then if $D\exp(x_0)$ has a non trivial kernel, exp_{Tq_0} fails to be a local minimizer.

□

Remark. Let q be in $T^*_{x_0}U$, a point $\exp(tq)$ is said to be conjugate to x_0 along $\exp_q$ if tq is a singularity of exp. The previous Lemma states that in the analytic case every nonsingular curve $\exp_q$ stop to minimize after its first conjugate point. Analyticity in this case is used to prove that each piece of a nonsingular analytic curve is non singular and then to prove that the set of conjugate points along a non singular curve $\exp_q$ is discrete. Up to now there is no counterexample to Lemma 2 in the non analytic case.

Remark. If we assume the structure to be analytic, we can replace the third sufficient condition of Proposition 2 by "x is not a cut point" by Lemma 2.

Let us now study necessary conditions for a point to be in $Reg_k(f)$. We shall use the following result that states that minimizers are integral curve of the "horizontal gradient", $x \mapsto G_x df(x)$ where it exists.

Proposition 3. *[12]The norm $|G_x df(x)|_\Delta$ is equal to one for all x in $Reg_1(f)$ and each restriction contained in $Reg_1(f)$ of a minimizer parametrized by arc length starting from x_0 is an integral curve of $x \mapsto G_x df(x)$.*

Proof. Let us suppose that there exists an open subset V of $Reg_1(f)$ such that $|G_x df(x)|_\Delta > 1$ for all x in V. Let x be in V and let T be small enough such that a solution on $[0, T]$ of

$$\dot{\gamma}(t) = G_{\gamma(t)} df(\gamma(t))$$
$$\gamma(T) = x$$

exists and remains in V. From one hand

$$f(\gamma(T)) - f(\gamma(0)) = \int_0^T \langle df(\gamma(t)), \dot{\gamma}(t) \rangle dt = \int_0^T |G_{\gamma(t)} df > (\gamma(t))|_\Delta^2 dt. \quad (4)$$

On the other hand triangular inequality gives us

$$f(\gamma(T)) \leq d(0, \gamma(0)) + d(\gamma(0), \gamma(T)) \leq f(\gamma(0)) + L(\gamma)$$

and then

$$f(\gamma(T)) - f(\gamma(0)) \leq \int_0^T |G_{\gamma(t)} df(\gamma(t))|_\Delta dt. \quad (5)$$

From (4) and (5) we obtain

$$\int_0^T |G_{\gamma(t)} df(\gamma(t))|_\Delta^2 dt \leq \int_0^T |G_{\gamma(t)} df(\gamma(t))|_\Delta dt$$

and contradicts the fact that $|G_x df(x)|_\Delta > 1$ on V. Then $|G_x df(x)|_\Delta \leq 1$.

Let us prove that the equality holds. indeed, if $|G_x df(x)|_\Delta < 1$ on some open subset V of $Reg_1(f)$ there exist a minimizer γ parametrized by arc length starting from x_0 and an open interval I such that $\gamma(t)$ belongs to V for all t in I. By derivating on I the equality $f(\gamma(t)) = d(x_0, \gamma(t)) = t$ we obtain

$$g_{\gamma(t)}(G_{\gamma(t)} df(\gamma(t)), \dot{\gamma}(t)) = \langle df(\gamma(t)), \dot{\gamma}(t) \rangle = 1$$

for almost t in I. Since $|\dot{\gamma}(t)|_\Delta$ is equal to one and $|G_{\gamma(t)} df(\gamma(t))|_\Delta < 1$, the previous equality cannot hold and gives us the contradiction. Then $|G_x df(x)|_\Delta = 1$ for all x in $Reg_1(f)$.

Let us now prove that a curve parametrized by arc length contained in $Reg_1(f)$ is a piece of a minimizer starting from x_0 if and only its is an integral of the horizontal gradient. Let γ be a minimizer parametrized by arc-length starting from x_0 and I and open interval such that $\gamma(I) \subset Reg_1(f)$, the equality

$$g_{\gamma(t)}(G_{\gamma(t)} df(\gamma(t)), \dot{\gamma}(t)) = 1$$

together with $|\dot{\gamma}(t)|_\Delta = 1$ implies that $\dot{\gamma}(t)$ is equal to $G_{\gamma(t)} df(\gamma(t))$ and then the restriction of γ to I is an integral curve of $x \mapsto G_x df(x)$.

$\square$

One could think that the above result implies that every $x \in Reg_1(f)$ can be reached from x_0 by an unique minimizer, but this last result need some more work. In fact a continuous vector field may have more than one integral curve, moreover, in principle, it could happen that two minimizers intersect themselves and continue together before reaching $Reg_1(f)$. Two minimizers may coincide in a subinterval if they are not analytic and this could be the case even if the structure is analytic because of the possible existence of strictly abnormal minimizers for which up to now no regularity results are proved.

Theorem 1. *Let x be in $Reg_1(f)$ then*

- *there exists only one minimizer γ going from x_0 to x,*
- *γ is nonsingular*
- *x is not in the cut locus.*

Proof. Let γ be a minimizer parametrized on $[0, 1]$ going from x_0 to x and let S_r be the sub-Riemannian sphere of radius $r = f(\gamma(1))$ centered at x_0. Since $E(\gamma) = r^2$, γ realizes the minimum of E among all the curves (parametrized on $[0, 1]$) going from x_0 to S_r. The Lagrange multiplier rule tells us that there exists $(\alpha, \lambda_0) \in \mathbb{R}^2$ nonzero such that

$$\alpha df(\gamma(1)) \cdot De_1(\gamma) + \lambda_0 \cdot DE(\gamma)$$

vanishes, i.e the adjoint lift q of γ such that $q(1) = -\alpha df(\gamma(1))$ satisfies

$$2\lambda_0 u_i(t) = \langle q(t), X_i(\gamma(t)) \rangle \qquad \text{for almost } t \text{ in } [0, 1]. \tag{6}$$

In a neighborhood of 1, $\dot{\gamma} = f(x) G_\gamma df(\gamma)$ from Proposition 3 and hence γ is of class C^1 on this neighborhood of 1 in $[0, 1]$. Then the previous equality holds for all t in a neighborhood of 1. Multiplying (6) by $u_i(1)$ and taking the sum from $i = 1$ to $i = m$ we get

$$2\lambda_0 |\dot{\gamma}(1)|_\Delta^2 = -\alpha \langle df(\gamma(1)), \dot{\gamma}(1) \rangle = -\alpha \langle df(x), f(x) G_x df(x) \rangle = -\alpha f(x)$$

by Proposition 3. Then λ_0 cannot be zero and for $\lambda_0 = \frac{1}{2}$, α is equal to $-f(x)$. Since γ is determined by $(x, f(x) df(x))$, it is the only minimizer that goes from x_0 to x.

Let us now prove that a minimizer going from x_0 to a point of $Reg_1(f)$ must be regular. Let Φ be the continuous map

$$\Phi : Reg_1(f) \longrightarrow T^* M,$$
$$x \longmapsto \Phi(x) = H(1)(x, -f(x) df(x)).$$

This map is injective and its image is contained in $T^*_{x_0} M$ since we just proved that the projection on M of the flow of H starting from $(x, -f(x) df(x))$ for x in $Reg_1(f)$ is the minimizer going from x to x_0. The restriction of the exponential map to Im Φ is continuous and its inverse mapping is Φ. Then Im Φ is a subset of $\mathbb{R}^n$ homeomorph to an open subset of $\mathbb{R}^n$ and then it has to be an open subset of $T^*_{x_0} M$. In particular, let γ be a minimizer going from x_0 to a point of $Reg_1(f)$ and let q_0 be such that $H(1)(x_0, q_0)$

is equal to $(\gamma(1), f(\gamma(1))df(\gamma(1)))$, the exponential mapping is injective in a neighborhood of q_0 and then γ is nonsingular.

Let us now prove that a point $x \in Reg_1(f)$ cannot belong to C. Let x be in $Reg_1(f)$ and $q_0 = \Phi(x)$, since Im Φ is open, for $\lambda > 1$ close enough to 1, λq_0 remains in Im Φ, that is $t \mapsto \exp(tq_0)$ minimize on $[0, \lambda]$ and then x does not belong to C.

□

Corollary 1. *We have that $Reg_2(f) = Reg_\infty(f)$ and if the structure is analytic $Reg_1(f) = Reg_\omega(f) = U_n \backslash C$.*

Proof. We keep the same notations of the proof of Theorem 1. The restriction of Φ to $Reg_2(f)$ is C^1 and then induces a C^1 diffeomorphism between $Reg_2(f)$ and its image by Φ. Then no point in $\Phi(Reg_2(f))$ is a singularity of exp. Then we just have to apply Theorem 1 and Proposition 2.

If the structure is analytic, every non singular minimizer stop to minimize after its first conjugate point (Lemma 2). Then, according to Theorem 1, no q in Im Φ is a singularity of the exponential map. Then we can apply Proposition 2 at a point of $Reg_1(f)$.

□

4   About the cut locus.

Points in the cut locus can be of different kind. We already saw that they are contained in one of the following three sets

- $U \backslash U_n$ that is the set of point that are reached from x_0 by at least one singular minimizer,
- C_m the set of x in U such that there exist more than one minimizer going from x_0 to x,
- C_s the set of x in U such that there exists a minimizer $\exp_q$ going from x_0 to x where q is a singularity of exp.

In all the section, we assume that U_n is dense in U. This assumption is equivalent to the fact that the set of points that can be reached from x_0 by strictly abnormal minimizer has an empty interior.

Theorem 2. *The cut locus is contained in the closure of C_m. Moreover*

$$Reg_1(f) = U_n \backslash \bar{C}_m$$

Proof. Remark that $Reg_1(f)$ is dense in $U_n \setminus C_m$. Indeed $Sing_1(f) \cap (U_n \setminus C_m)$ is contained in the set of singular values of the exponential mapping and then has an empty interior by Sard's Theorem.

Let x be in U_n but not the closure of C_m, let us prove that x does not belong to C. Let V be a neighborhood of x included in $U_n \setminus \bar{C}$. As we saw previously $Reg_1(f) \cap V$ is an open dense subset of V. Let us prove that the map $y \mapsto df(y)$ defined on $Reg_1(f) \cap V$ has a continuous continuation on V. Let us define the map Q on V as follows : for y in V there exists an unique minimizer φ that goes from x_0 to y (since y does not belongs to C_m) and an unique Hamiltonian lift q of φ (since y belongs to U_n), we set $Q(y) = q(1)$. This map is continuous by Lemma 1 and its restriction to $Reg_1(f) \cap V$ is fdf (see the proof of Theorem 1). Since f is continuous and never vanishes, $\frac{1}{f}Q$ is the continuous continuation on V of $y \mapsto df(y)$. It is then easy to check that f is in fact of class C^1 on V and then x is not a cut point by Theorem 1. In fact we have also proved that $Reg_1(f) = U_n \setminus \bar{C}_m$.

Let x be in U but not in the closure of C_m, let us prove that x is not in the cut locus. Let $\varepsilon < 1$ positive be such that the sub-Riemannian ball $B_{3\varepsilon}(x)$ of radius 3ε centered at x is contained in $U \setminus \bar{C}$.

We shall prove the following Lemma at the end of this proof.

Lemma 3. *For all y in $B_\varepsilon(x) \cap Reg_1(f) = B_\varepsilon(x) \cap U_n$ there exists a minimizer φ parametrized on $[0,1]$ such that $\varphi(\frac{f(y)}{f(y)+\varepsilon}) = y$.*

Consider now a sequence $\{y_\alpha\}$ contained in $B_\varepsilon(x) \cap U_n$ that tends to x (such a sequence exists since we made the assumption that U_n is dense in U) and , for each α, we call φ_α the minimizer defined on $[0,1]$ given by Lemma 3, i.e.

$$\varphi_\alpha\left(\frac{f(y_\alpha)}{f(y_\alpha)+\varepsilon}\right) = y_\alpha.$$

When α tends to infinity, $\{f_\alpha\}$ (or a subsequence) tends to some minimizer γ (Proposition 1) and

$$\gamma\left(\frac{f(x)}{f(x)+\varepsilon}\right) = x.$$

Since $\frac{f(x)}{f(x)+\varepsilon} < 1$, x cannot be in the cut locus.

$\square$

Proof of Lemma 3. We already proved that $Reg_1(f) = U_n \setminus \bar{C}_m$ and then $B_\varepsilon(x) \cap U_n = B_\varepsilon(x) \cap Reg_1(f)$. Let y be in $B_\varepsilon(x) \cap U_n$ then by Theorem 1 there is an unique q in $T^*_{x_0}M$ with $H(q) = \frac{1}{2}$ such that $t \mapsto \exp(tq)$ defined on $[0, f(y)]$ is the minimizer (nonsingular since $y \in Reg_1(f)$) joining x_0 to y and let us prove that it is still a minimizer on $[0, f(y) + \varepsilon]$. By contradiction, let T in $[f(y), f(y) + \varepsilon]$ be the smallest such that $t \mapsto \exp(tq)$ does not minimize on $[0, T']$ for any T' greater than T. We have that

$$d(x, \exp(Tq)) \le d(x, y) + d(y, \exp(Tq)) < \varepsilon + (T - f(y)) \le 2\varepsilon$$

and then $\exp(Tq)$ remains in $B_{2\varepsilon}(x) \subset U \backslash \bar{C}_m$, then $t \mapsto \exp(tq)$ defined on $[0, T]$ is a nonsingular minimizer (since its restriction to $[0, f(y)]$ is non singular) and there is no other minimizer going from x_0 to $\exp(Tq)$ (since $\exp(Tq)$ does not belong to C_m). Hence $\exp(Tq)$ belongs to $U_n \backslash \bar{C}_m$ and as we saw in the proof of Theorem 2, $\exp(Tq)$ is not a cut point. Then there exist $T' > T$ and a minimizer γ defined on $[0, T']$ whose restriction to $[0,T]$ is $t \mapsto \exp(tq)$ and by choosing T' small enough, we may assume that $\gamma(T)$ is in $U_n \backslash \bar{C}_m$. Then $\gamma = t \mapsto \exp(tq')$ where q' is unique and the equality $\exp(tq) = t \mapsto \exp(tq')$ for all t in $[0, T]$ together with the fact that $t \mapsto \exp(tq)$ is non singular implies that $q = q'$ and gives a contradiction. Then $t \mapsto \exp(tq)$ minimize on $[f(y), f(y) + \varepsilon]$ and the minimizer $t \mapsto \exp(t(f(y) + \varepsilon)q)$ parametrized on $[0, 1]$ is the one we were looking for.

$\square$

Acknowledgments: I would like to thank Professors A. Agrachev, A. Sarychev and G. Stefani for their encouragements and fruitful discussions.

This work was supported by the program "Breakthrough in the control of non linear systems" CEE.ERBFMRX CT 970137.

References

1. A.A. Agrachev "Compactness for Sub-Riemannian Length-minimizers and Subanalyticity." Preprint Ref. S.I.S.S.A. 26/99/M.
2. A.A. Agrachev, El-H. Ch. El-Alaoui, J.-P. Gauthier "Sub-Riemannian Metrics on $\mathbb{R}^3$." Canadian Mathematical Society, Conference Proceedings, Vol. 25 (1998).
3. A.A. Agrachev "Exponential Mappings for Contact Sub-Riemannian Structures." J. Dynam. Control Systems Vol. 2, No. 3, pp 321-358 (1996).
4. A.A. Agrachev, A.V. Sarychev, Sub-Riemannian metrics: minimality of abnormal geodesics versus subanalyticity. J. ESAIM: Control, Optimisation and Calculus of Variations, 1999, v.4.
5. A.A. Agrachev, A.V. Sarychev "Abnormal Sub-Riemannian Geodesics: Morse index and Rigidity",Ann. Inst. H. Poincaré Anal. Non Linéaire, vol 13 (1996).
6. A.A. Agrachev, R. V. Gamkrelidze, Symplectic methods for optimization and control, Geometry of Feedback and Optimal Control (New York)(B Jacubczyk and W Respondek, eds.), Marcel Dekker, 1997.
7. A.A. Agrachev, R.V. Gamkrelidze, Feedback–invariant optimal control theory and differential geometry, I. Regular extremals. J. Dynamical and Control Systems, 1997, v.3, 343-389.

8. A.A. Agrachev, B.Bonnard, M.Chyba, I. Kupka, "Sub-Riemannian spheres in the Martinet flat case", ESAIM Control Optim. Calc. Var., v.2, pp. 377-448, 1997.

9. W. L. Chow "Über Systeme von linearen partiellen Differentialgleichungen erster Ordnung", Math. Ann., v. 117, (1940) 98-105

10. El-H. Ch. El-Alaoui, J.-P. Gauthier, I. Kupka "Small Sub-Riemannian Balls on $\mathbb{R}^3$." J. Dynam. Control Systems Vol. 2, No. 3, pp 359-422 (1996).

11. Z. Ge "Horizontal path spaces and Carnot-Carathéodory metrics", Pacific J. Math., 1993, v.161, 255-286.

12. M. Grochowski, horizontal gradient and geodesics in sub-Riemannian geometry, preprint 593 Institue of Mathematics-Polish Academy of Sciences.

13. S.Jacquet, "Distance sous-riemannienne et sous analycité ", Thèse de doctorat (1998).

14. S. Jacquet, Subanalyticity of the Sub-Riemannian Distance, Journal of Dynamical and Control Systems, 5 (1999), 3: 303-328.

15. S. Jacquet, Regularity of sub-Riemannian distance and cut locus, Universita degli Studi di Firenze, Dip. di Matematica Appl. "G. Sansone", Preprint No. 35, May 1999.

16. R.Montgomery "Geodesics Which Do Not Satisfy the Geodesic Equations", SIAM J. Control Optim., v. 32, n.6, 1605-1620 (1994).

17. P. K. Rashevsky, Any two points of a totaly nonholonomic space may be connected by an admissible line. Uch. Zap. Ped. Inst. im. Liebknechta, Ser. Phys. Math.,2 (1938), 83-94.

18. H. J. Sussmann, "Subanalycity of the distance function for real analytic sub-Riemannian metrics on three dimensional manifolds", Report SYCON-91-05a, Rutgers, 1991.

19. W. S Liu, H. J. Sussmann, Shortest paths for sub-Riemannian metrics of rank two distributions, Mem. Amer. Math. Soc., 564, 104 p. (1995).

Industrial Sensorless Control of Induction Motors

Fabrice Jadot[1], Philippe Martin[2], and Pierre Rouchon[2]

[1] Schneider Electric SA
   33, avenue de Chatou
   92506 Rueil-Malmaison Cedex, France
[2] Centre Automatique et Systèmes
   École des Mines de Paris
   35 rue Saint-Honoré
   77305 Fontainebleau Cedex, France
   martin@cas.ensmp.fr

Abstract. In recent years, there has been an increased need in industry for high performance adjustable speed drives using an induction motor fed by a pulse width modulated voltage source. An important issue is the so-called "sensorless" control of the motor, i.e., the control without any mechanical (position or speed) measurements, relying only on the stator currents. This is a difficult control problem, both from a theoritical and practical point of view. We present experimental results achieved by the new Adjustable Frequency Controller ATV58-F of Schneider Electric SA. Its algorithm relies on "advanced" nonlinear control theory.

1 Introduction

The induction motor (also called asynchronous motor) has very good qualities –reliability, ruggedness, relatively low cost, etc– for industrial applications. The reason why is that there is no mechanical commutation: the rotor consists simply of closed windings in which currents are induced by a rotating magnetic field set up by the stator, hence creating a torque. The price to pay for these qualities is the need for a rather sophisticated control device, namely an Adjustable Frequency Controller (AFC) (also often referred to as Inverter, Adjustable Frequency Drive, Variable Frequency Drive or Power Converter).

This note aims at showing the design of a "sensorless" AFC leads to an interesting and difficult problem of nonlinear control theory. It is illustrated by experimental results of the Adjustable Frequency Controller ATV58-F of Schneider Electric SA (released April 2000), whose sensorless control algorithm was co-developed with the École des Mines de Paris.

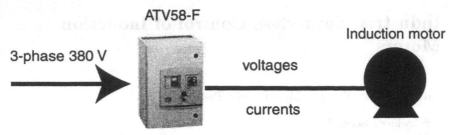

Fig. 1. Sensorless AFC driving a motor

2   What is "sensorless" control of induction motors?

The major role of an AFC is to deliver voltages to the motor so as to track a velocity reference supplied by the user; in other words, the device AFC + motor should behave as a "perfect velocity actuator". In the so-called sensorless problem, this must be achieved *without* any sensor "outside the box" (position, velocity, load,...); the AFC has to rely only on the currents, giving to the user the impression there are no sensors, and is simply plugged between the motor and the mains (figure 1).

It is worth emphasizing that the resulting control problem is much easier if the velocity can be measured. Nevertheless, adding a velocity sensor, when it is possible at all, makes the overall system less reliable from the industrial standpoint and less simple to use. This functionality is usually not offered (or optional) on most AFCs.

There are roughly three key points when manufacturing a "sensorless" AFC which bring the "added value" to the product:

1. The hardware (an AFC is more or less a microprocessor driving power electronics).
2. The "application software", i.e., the functionalities offered to the user.
3. The control algorithm. Many specifications of the product, as seen by the user, are expressed in terms of control performances.

The prominent role of the control algorithm makes the design of "sensorless" AFCs a very interesting and challenging problem of control theory. Since it is by no means easy and because the industrial market is very competitive, the problem has been widely studied for 15 years both in industry by AFC manufacturers (ABB, Alan Bradley, Mitsubishi, Schneider, Siemens, Toshiba, Yaskawa,...) and in academia (see, e.g., numerous papers in IEEE Transactions on Industry Applications,...). Surprisingly enough, though the control of the induction motor has aroused a lot of interest in the control community as an example of a nonlinear system, most authors have focused on control with velocity measurements.

3 A difficult control problem

Industrial sensorless control of induction motors is a difficult control problem. On the one hand, the hardware of a commercial AFC is for cost reasons far from perfect:

- The sensors are imperfect: besides the noise and quantization problems, there is a quite important bias which depends on the temperature inside the AFC.
- The actuators are imperfect: mainly because of the so-called dead times in the Pulse Width Modulation power electronics, the voltages actually delivered may differ from up to $20V$ from the desired voltages. This is a real problem at low velocities, where low voltages are required.
- The computing power is very limited: the algorithm must run on a fixed-precision 16-bit microcontroller. One multiplication takes about $1\mu s$, to be compared with a sampling period of about $250\mu s$.

On the other hand, there are several inherent theoretical difficulties, which are largely independent of the hardware:

- A high level of performance is required (static and dynamic).
- The system is multivariable and nonlinear: there are strong couplings and the operating range is very wide (for instance, the rotor velocity may vary from a fraction of Hertz to about $300Hz$, to be compared with a rated velocity of about $50Hz$).
- The measurements are "indirect": the velocity depends on the currents in a rather complicated way. Moreover, the observability of the system degenerates at first order on the "slip line" $\omega_s = 0$ (see the following section).
- Input constraints (voltages) and state constraints (currents) must be handled by the control algorithm. Voltage constraints appear at high velocities, current constraints at high torques (the current constraints are imposed to protect the power electronics).
- The stator and rotor resistances vary a lot with the temperature inside the motor.

Notice that the most severe problems occur for technological and theoretical reasons at low velocity around the first-order inobservable "slip line". No commercial sensorless AFC seems to be able to operate satisfactorily in this region without extra hardware.

4 Model of the induction motor

In the dq frame and using complex notations (time-varying phasors) for u_s, i_s, i_r, ψ_s and ψ_r, the standard two-phase equivalent machine representation of a symmetrical three-phase induction motor reads (see, e.g., [1, chapter 4] for a complete derivation of the equations)

Electromagnetism $\begin{cases} u_s = R_s i_s + \dfrac{d\psi_s}{dt} + j\omega_s \psi_s \\[2mm] 0 = R_r i_r + \dfrac{d\psi_r}{dt} + j(\omega_s - n_p\Omega)\psi_r \\[4mm] \psi_s = L_s i_s + L_m i_r \\[1mm] \psi_r = L_m i_s + L_r i_r \end{cases}$

Mechanics $\begin{cases} J\dfrac{d\Omega}{dt} = \tau_{em} - \tau_L \\[4mm] \tau_{em} = \dfrac{3n_p}{2}\,\Im(i_r^* \psi_r). \end{cases}$

Here u_s is the stator voltage; i_s and i_r the stator and rotor currents; ψ_s and ψ_r the stator and rotor fluxes; Ω the (mechanical) rotor velocity; ω_s the (electrical) stator velocity; L_s, L_r and L_m the stator, rotor and mutual inductances; R_s and R_r the stator and rotor resistances; n_p the number of pair of poles; J the moment of inertia; τ_{em} and τ_L the electromagnetic and load torques. Moreover, $j := \sqrt{-1}$, $\Im(z)$ denotes the imaginary part of the complex number z, and z^* the complex conjugate of z.

Eliminating the stator flux and rotor current and using only real variables, the system may be rewritten in the more familiar form

$$J\frac{d\Omega}{dt} = \frac{3n_p L_m}{2L_r}(i_{sq}\psi_{rd} - i_{sd}\psi_{rq}) - \tau_L$$

$$\frac{d\psi_{rd}}{dt} = -\frac{\psi_{rd}}{T_r} + (\omega_s - n_p\Omega)\psi_{rq} + \frac{L_m}{T_r}i_{sd}$$

$$\frac{d\psi_{rq}}{dt} = -\frac{\psi_{rq}}{T_r} - (\omega_s - n_p\Omega)\psi_{rd} + \frac{L_m}{T_r}i_{sq}$$

$$\sigma\frac{di_{sd}}{dt} = -\frac{i_{sd}}{T_s} + \sigma\omega_s i_{sq} + \frac{1-\sigma}{L_m}\left(\frac{\psi_{rd}}{T_r} + n_p\Omega\psi_{rq}\right) + \frac{u_{sd}}{L_s}$$

$$\sigma\frac{di_{sq}}{dt} = -\frac{i_{sq}}{T_s} - \sigma\omega_s i_{sd} + \frac{1-\sigma}{L_m}\left(\frac{\psi_{rq}}{T_r} - n_p\Omega\psi_{rd}\right) + \frac{u_{sq}}{L_s},$$

where $T_r := \frac{L_r}{R_r}$, $\frac{1}{T_s} := \frac{R_s}{L_s} + \frac{(1-\sigma)R_r}{L_r}$ and $\sigma := 1 - \frac{L_m^2}{L_s L_r}$.

There are 5 (real) states ($i_s = i_{sd} + ji_{sq}$, $\psi_r = \psi_{sd} + j\psi_{sq}$, Ω) and 3 (real) inputs ($u_s = u_{sd} + ju_{sq}$, ω_s). Notice the electrical velocity ω_s is in fact a dummy input; seeing it as an input is just a convenient way of taking into account the symmetries of the problem.

It is not difficult to see that, when only the stator currents are measured, the system becomes inobservable at first-order around every equilibrium point defined by $\omega_s = 0$. This corresponds to a straight line in the torque/velocity plane (the so-called "slip line").

5   Experimental results

The following experimental plots show the behavior of the ATV58-F of Schneider Electric driving a standard $22kW$ motor. They give an idea of the static and dynamic performances as well as the management of the current and voltage constraints that must be achieved by a good sensorless AFC.

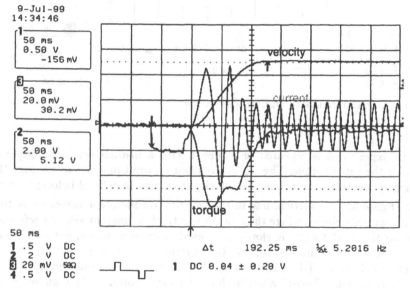

Fig. 2. Fast velocity transition ($0 \mapsto 50$ Hz), no load.

Figure 2 displays a fast velocity transition: initially the motor is at rest with zero flux and no load. It is first fluxed at zero velocity, then is required to track a constant $50Hz$ velocity reference. The overall sequence takes less than $0.2s$, which is fast (during the acceleration transient the current reaches its constraint).

On figure 3, a disturbance load torque corresponding to the rated torque is applied. The motor nearly perfectly tracks the desired $50Hz$ reference.

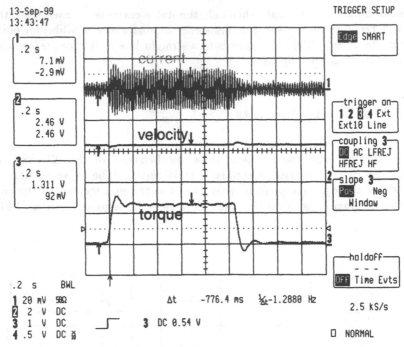

Fig. 3. Load torque step (τ_r), velocity 50 Hz.

The experiment is repeated on figure 4 with a disturbance load torque of twice the rated torque. The current reaches its limit in order to produce the largest possible torque, and the velocity decreases in a "well-behaved" way.

On figure 5, each vertical line in the velocity/torque plane corresponds to a different experiment where the motor must track a constant velocity reference while the load torque is slowly increased from zero to twice the nominal torque. At a certain load torque, the current constraint is reached, and the velocity decreases (the "dent" at the top of each vertical line corresponds to a short current "boost" allowing for brief extra torque). At high velocities (above the rated velocity), the voltage constraint is reached first; the allowable torque is smaller since constraining the voltage means reducing the flux, hence the torque.

Figure 6 illustrates the difficulty of controlling the motor at low velocity: the motor, loaded at the rated torque, must track a slow velocity ramp from $5Hz$ to $-5Hz$ and back (the ramp duration is $5s$ for the bottom plot and $10s$ for the top plot). The result should be a straight line in the velocity/torque plane: the "loops" around zero velocity correspond to the crossing of the slip line (the dissymmetry, due to the generator creating the load torque, is not relevant).

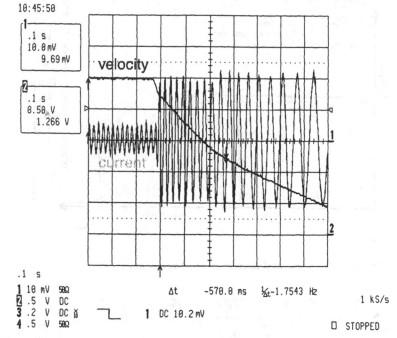

10:45:50

.1 s
10.0 mV
9.69 mV

velocity

.1 s
0.50 V
1.266 V

current

1

2

.1 s

1 10 mV 50Ω
2 .5 V   DC
3 .2 V   DC
4 .5 V   50Ω

Δt   -570.0 ms   ⅟Δt-1.7543 Hz

1 DC 10.2 mV

1 kS/s

□ STOPPED

Fig. 4. Load torque step ($2\tau_r$), velocity 50 Hz.

Figure 6 illustrates the difficulty of controlling the motor at low velocity: the motor, loaded at the rated torque, must track a slow velocity ramp from $5Hz$ to $-5Hz$ and back (the ramp duration is $5s$ for the bottom plot and $10s$ for the top plot). The result should be a straight line in the velocity/torque plane: the "loops" around zero velocity correspond to the crossing of the slip line (the dissymmetry, due to the generator creating the load torque, is not relevant).

6 Conclusion: from application to theory

Despite the important advances in nonlinear control theory, the sensorless control of induction motors does not seem to fit in any existing category of "solved problems". For instance, the control algorithm of the ATV58-F relies on ad hoc techniques using the many peculiarities of the system:

- A blend of oriented flux, flatness and symmetries for the "controller".
- An ad hoc nonlinear "observer" for velocity and flux.
- An ad hoc strategy for constraints.
- An ad hoc proof of convergence (away from $w_s = 0$).

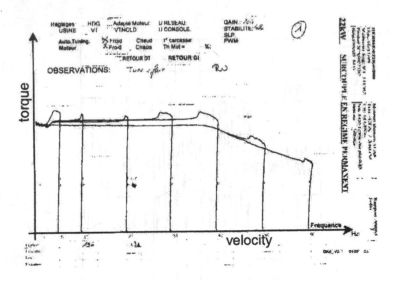

Fig. 5. Static performance and management of constraints.

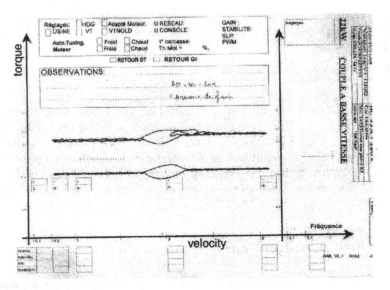

Fig. 6. Slow inversion at low velocity, load torque τ_r.

This problem can even be seen as a source of inspiration for control theory (stabilization under input and state constraints, role of symmetries, singularity of observability,...). In particular, the yet unsolved problem of operating the motor around the nonobservable "slip line", appears to be quite challenging.

References

1. P.C. Krause, O. Wasynczuk, and Sudhoff S.D. *Analysis of electric machinery.* IEEE Press, 1995.

Feedback Invariants, Critical Trajectories and Hamiltonian Formalism

Bronisław Jakubczyk*

Institute of Mathematics
Polish Academy of Sciences
Śniadeckich 8,
00-950 Warsaw, Poland
jakubczy@panim.impan.gov.pl

Abstract. We show that the Hamiltonian formalism, as used for optimal control problems, is a natural tool for studying the feedback equivalence problem and for constructing invariants. The most important invariants (covariants) are the critical curves or trajectories of the system, also called extremals or singular curves. They are obtained as the curves satisfying formally the necessary conditions of Pontryagin maximum principle. In particular, we show that for arbitrary scalar control systems the set of critical trajectories (if nonempty), together with a canonical involutive distribution, determine the system in neighbourhood of a regular point, up to feedback equivalence. We describe the structure of critical trajectories around regular points and show that the system can be decomposed into a feedback linearizable part and a fully nonlinear part. A complete set of invariants is given for fully nonlinear analytic systems.

1 Introduction

We consider control systems of the form

$$\Sigma : \qquad \dot{x} = f(x, u),$$

where $x(t) \in X$, $u(t) \in U$ and X, U are open subsets $X \subset \mathbb{R}^n$, $U \subset \mathbb{R}^m$, or differentiable manifolds (of class C^∞ or real-analytic). Further considerations will be in one of the two categories: smooth or real-analytic.

Given another system of the same form $\tilde{\Sigma} : \quad \dot{\tilde{x}} = \tilde{f}(\tilde{x}, \tilde{u})$ on $\tilde{X} \times \tilde{U}$, we say that both systems are *feedback equivalent* if one of them can be transformed into the other by smooth (respectively, analytic) invertible *feedback transformations*

$$(FT): \qquad x = \Phi(\tilde{x}), \qquad u = \Psi(\tilde{x}, \tilde{u}).$$

The two systems are called *pure feedback equivalent* if one can be obtained from the other by reparametrization of the control, i.e., by invertible transformations, called *pure feedback transformations*

$$(PFT): \qquad x = \tilde{x}, \qquad u = \Psi(\tilde{x}, \tilde{u}).$$

* Supported by Polish KBN grant 2P03A 035 16

In this paper we show how the Hamiltonian formalism of optimal control theory can be used for the feedback equivalence problem and for finding invariants and covariants.

The problem of constructing natural complete sets of feedback invariants (local, at least) is one of the most important in nonlinear control theory. So far three approaches were proposed or used.

The first one uses the language of vector fields and distributions spanned by vector fields and their properties related to the Lie bracket. Using this approach such problems were solved as characterization of feedback linearizable systems (cf. [11], [7] and the books [6], [18], [20]), classification of systems in the plane [12], classification of so called simple germs, i.e., system germs with structurally stable canonical forms [19], [23] which appear for control-affine systems with the number of controls $m = n - 1$. An approach using formal transformations in the jet spaces was developed in [13] which leads to formal normal forms for scalar-control systems [13], [22]. Several other properties of systems were characterized by using the language of vector fields, distributions, and Lie bracket. However, only the mentioned above classes of systems were classified in this way.

The second approach, proposed by Robert Gardner in [5], uses the Elie Cartan's method of moving frame for finding invariants of geometric structures. This method is very general and gives a procedure for finding abstract invariants, however, applying the method for particular classes of control systems creates serious difficulties. So far this method was carried out for partial construction of feedback invariants of scalar control regular systems [14], [15].

The third method, first proposed by B. Bonnard in [3], [4], uses the Hamiltonian formalism. An important advantage of this method is the fact that one can analyse the system microlocally (i.e., locally in the projectivised cotangent bundle T^*X), see [9], [10]. We will concentrate here on presenting partially the results obtained in [10] by using this method and by showing that it includes naturally some of the results obtained using the first method.

We will also present the following new results:

1. For a smooth system Σ fully nonlinear at (x_0, u_0) the critical trajectories determine Σ up to local (global) pure feedback equivalence (Section 3).

2. Any scalar control analytic system has a canonical decomposition around a generic point in $X \times U$ into a feedback linearizable part and a fully nonlinear part. The set of critical trajectories, if nonempty, and a canonical involutive distribution of the linearizable part determine the system, up to feedback equivalence, locally around a generic point (Section 4).

3. We show the structure of bi-critical curves which live on an explicitly constructed critical variety (a subset of T^*X).

We construct feedback invariants for the above situations by introducing symbols of the system, which are functions on the cotangent bundle, and defining momentum mapping (Sections 8 and 9). Complete proofs and a more comprehensive analysis of the problem will be presented in a forthcoming paper.

2   Critical and Bi-Critical Curves and Trajectories

Let T^*X denote the cotangent bundle of X. We denote the elements of T^*X by (p, x), where $x \in X$ and $p \in T_x^*X$ is a covector also called adjoint variable. We define the Hamiltonian of the system Σ as the function on $T^*X \times U$ given by

$$H(p, x, u) = pf(x, u) = \sum_{i=1}^{n} p_i f_i(x, u).$$

Here in the first formula we denote the duality product between a vector $v \in T_xX$ and a covector $p \in T^*X$ simply by pv (in coordinates $pv = \sum p_i v_i$). We will use the following terminology. *Bi-critical curves* are smooth curves

$$t \in I \to (p(t), x(t), u(t)) \in T^*X \times U, \quad p(t) \neq 0,$$

satisfying

$$\dot{x} = \frac{\partial H}{\partial p}(p, x, u), \qquad \dot{p} = -\frac{\partial H}{\partial x}(p, x, u), \qquad \frac{\partial H}{\partial u}(p, x, u) = 0, \qquad (1)$$

for all $t \in I$, where I is an open interval which depends on the curve. By "smooth" we mean "C^∞" though in most places C^1 would be enough.

Bi-critical trajectories are smooth curves

$$t \in I \to (p(t), x(t)) \in T^*X, \qquad p(t) \neq 0,$$

such that there exists a smooth control $t \in I \to u(t) \in U$ satisfying (1).

A *critical curve* is a smooth curve

$$t \in I \to (x(t), u(t)) \in X \times U,$$

for which there exists a smooth function $t \in I \to p(t) \in T_{x(t)}^*X$, $p(t) \neq 0$, such that $(p(\cdot), x(\cdot), u(\cdot))$ is a bi-critical curve.

A *critical trajectory* is a smooth curve

$$t \in I \to x(t) \in X,$$

for which there exist smooth functions $t \in I \to p(t) \in T_{x(t)}^*X$, $p(t) \neq 0$, and $t \in I \to u(t) \in U$ such that $(p(\cdot), x(\cdot), u(\cdot))$ is a bi-critical curve.

Example 1. For a dynamical system $\dot{x} = f(x)$ (where the control set U is a single point) the bi-critical trajectories are the trajectories of the Hamiltonian lift of the system,

$$\dot{x} = f(x), \qquad \dot{p} = -p\frac{\partial f}{\partial x}(x), \qquad p(t) \neq 0,$$

and the critical trajectories are all trajectories of the system (the critical equation $\partial H/\partial u = 0$ is satisfied automatically). Note that the critical trajectories determine the system in this case.

Our starting point is the following observation.

Proposition 1. *(a) Critical trajectories are invariants of pure feedback transformations (PFT). Under the feedback transformation (FT) applied to the system they are transformed by the state diffeomorphism:*

$$x(t) = \Phi(\tilde{x}(t)).$$

(b) Bi-critical trajectories are invariant under the pure feedback transformations (PFT) and they are transformed by the Hamiltonian lift of the state diffeomorphism Φ, under the feedback transformations (FT):

$$x(t) = \Phi(\tilde{x}(t)), \qquad p(t) = \tilde{p}(t)(d\Phi(\tilde{x}(t)))^{-1}.$$

The proof of the proposition follows trivially from the definitions of critical and bi-critical trajectories, the the formula

$$\frac{\partial}{\partial \tilde{u}} H(p, x, \Psi(x, \tilde{u})) = \frac{\partial H}{\partial u}(p, x, \Psi(x, \tilde{u}))\frac{\partial \Psi}{\partial \tilde{u}}(x, \tilde{u}),$$

and from invertibility of $\partial \Psi/\partial \tilde{u}$.

3   Fully nonlinear systems

Consider a scalar-control system Σ given by $\dot{x} = f(x, u)$.

We will call the system Σ *fully nonlinear* at (x_0, u_0) if

$$\frac{\partial f}{\partial u}(x_0, u_0), \qquad \frac{\partial^2 f}{\partial u^2}(x_0, u_0) \qquad \text{are linearly independent.}$$

We say that Σ is fully nonlinear (globally) if it is fully nonlinear at an open dense subset of $X \times U$.

We call the system *regular in control* if the set of velocities of the system

$$F = \bigcup_{x \in X} F(x), \qquad \text{where} \qquad F(x) = f(x, U) = \{f(x, u), u \in U\},$$

is a regular submanifold of $T_x X$, for any $x \in X$, and the map $u \in U \to F(x)$ is a diffeomorphism for any $x \in X$.

Systems regular in control are slightly easier to analyse since the control $u(\cdot)$ is determined by the trajectory $x(\cdot)$ (by inverting the map $u \to v = f(x, u)$, where $v = \dot{x}$). This property and the first claim in Proposition 1 imply that for systems regular in control we can eliminate the control when considering feedback invariants and, what it particulary important, we can eliminate the "unpleasant part" $\Psi(\tilde{x}, \tilde{u})$ in (FT) from our considerations.

The following observation is our second starting point.

Proposition 2. *A scalar-control system Σ which is fully nonlinear at a point in $X \times U$ is determined locally around this point by its critical trajectories, up to local pure feedback equivalence. If Σ is fully nonlinear globally, regular in control and has closed subset of velocities $F \subset TX$, then it is globally determined by its critical trajectories, up to pure feedback equivalence.*

The above means that two fully nonlinear systems having the same critical trajectories are pure feedback equivalent. This implies that any two fully nonlinear systems which have the critical trajectories equivalent by a state diffeomorphism are feedback equivalent. Therefore the feedback equivalence problem (local or global) reduces, for fully nonlinear systems, to the problem of classification of critical trajectories with the group of state diffeomorphisms.

In order to prove Proposition 2 let us define two subsets $\mathcal{D} \subset \mathcal{C} \subset T^*X \times U$,

$$\mathcal{C} = \left\{ (p, x, u) : \ \frac{\partial H}{\partial u}(p, x, u) = 0 \right\},$$

$$\mathcal{D} = \left\{ (p, x, u) : \ \frac{\partial H}{\partial u}(p, x, u) = 0, \ \frac{\partial^2 H}{\partial u^2}(p, x, u) = 0 \right\},$$

called, respectively, the *control critical set* and the *control discriminant set*. Let C and D be the projections of $\mathcal{C}$ and $\mathcal{D}$ onto T^*X, called the *critical set* and, respectively, the *discriminant set*.

The above proposition follows from the following trivial

Lemma 1. *Given a system Σ, through any point $(p, x) \in C \backslash D$ there passes a bi-critical trajectory and through any $(p, x, u) \in \mathcal{C} \backslash \mathcal{D}$ there passes a bi-critical curve.*

Proof. Let $(p, x) \in C \setminus D$, then there exists $u \in U$ such that $\frac{\partial H}{\partial u}(p, x, u) = 0$, $\frac{\partial^2 H}{\partial u^2}(p, x, u) \neq 0$. ¿From the implicit function theorem it follows that the critical equation $\frac{\partial H}{\partial u}(p, x, u) = 0$ has, locally, a unique smooth solution $u = u(p, x)$. Plugging this solution to the Hamiltonian equations we get the closed system of equations

$$\dot{x} = \frac{\partial H}{\partial p}(p, x, u(p, x)), \qquad \dot{p} = -\frac{\partial H}{\partial x}(p, x, u(p, x))$$

which has a unique, smooth in time solution through each point $(p, x) \in C \backslash \mathcal{D}$. Q.E.D.

Proof of Proposition 2. Consider a point at which Σ is fully nonlinear. At each (x, u) in a neighbourhood of this point the system is also fully nonlinear and there exists a $p \in T_x^* X$, $p \neq 0$, such that $(p, x, u) \in C \backslash \mathcal{D}$, i.e.,

$$\frac{\partial H}{\partial u}(p, x, u) = p\frac{\partial f}{\partial u}(x, u) = 0, \qquad \frac{\partial^2 H}{\partial u^2}(p, x, u) = p\frac{\partial^2 f}{\partial u^2}(x, u) \neq 0.$$

This means that for each such (x, u) there exist a p such that $(p, x, u) \in C \backslash \mathcal{D}$ and, by the lemma, there exists a bicritical curve $(p(\cdot), x(\cdot), u(\cdot))$ of Σ such that $p(0) = p$, $x(0) = x$, $u(0) = u$. Taking the tangent vector of $x(\cdot)$ at $t = 0$ gives $f(x, u)$. This means that the vector $f(x, u)$ is obtained as the tangent vector to a critical trajectory for any point (x, u) at which Σ is fully nonlinear. Since the set of points where Σ is fully nonlinear is dense in $X \times U$ (or in the neighbourhood of a given point, in the local case), the set of vectors V obtained as tangent vectors to critical trajectories is dense in the set F of the velocities of the system Σ. Taking the closure of V we obtain F, thus critical trajectories determine F. Since Σ is regular in u, the submanifold $F \subset TX$ determines the function $f : X \times U \to TX$ uniquely up to parametrization of control, i.e., up to pure feedback equivalence. Q.E.D.

Reduction to Control-Affine Systems

Let us replace the control system $\Sigma : \dot{x} = f(x, u)$ by the system

$$\Sigma_1 : \qquad \dot{x} = f(x, u), \quad \dot{u} = v,$$

called *affine extension* of Σ, where the new control v is in $\mathbb{R}$ and the new state is $(x, u) \in X \times U$. Consider the Hamiltonian of the affine extension

$$H_1(p, q, x, u, v) = pf(x, u) + qv,$$

where (p, q), $q \in \mathbb{R}$, is the new adjoint vector (covector). The equations for bi-critical curves of the new system are

$$\dot{x} = f(x, u), \quad \dot{u} = v, \quad \dot{p} = -p\frac{\partial f}{\partial x}(x, u), \quad \dot{q} = -p\frac{\partial f}{\partial u}(x, u),$$

and the equation $\partial H_1/\partial v = q = 0$. The last equation differentiated with respect to time along a trajectory of the Hamiltonian system gives the equation

$$p\frac{\partial f}{\partial u}(x, u) = 0.$$

We obtain as corollary the following claim.

Claim. The bi-critical curves (respectively, trajectories) of the system Σ are obtained from the bi-critical curves (respectively, trajectories) of the affine extension Σ_1 by the projection $(p, q, x, u, v) \to (p, x, u)$ (respectively, $(p, q, x, u) \to (p, x)$). Similarly, the critical curves (trajectories) of the system Σ are obtained from the critical curves (trajectories) of the affine extension Σ_1 by the projection $(x, u, v) \to (x, u)$ (respectively, $(x, u) \to x$).

We also have the following easy to prove fact (cf. [8]).

Claim. Two systems Σ are locally feedback equivalent if and only if their affine extensions are locally feedback equivalent.

The two observations imply that in analysing the feedback equivalence of systems using critical (bi-critical) curves and trajectories we may restrict ourselves to control-affine systems. This will slightly simplify our further considerations.

4   General scalar-control systems

General systems can be decomposed into a feedback linearizable part and a fully nonlinear part, as will be stated in Proposition 4. We will show that the critical trajectories, determine the fully nonlinear part and, together with an involutive distribution canonically assigned to the system, they determine the system but only up to feedback equivalence.

Consider the system

$$\Sigma: \qquad \dot{x} = f(x) + u g(x),$$

where $u \in \mathbb{R}$ is scalar. We will use the standard notation $[f, g] = ad_f(g)$ for the Lie bracket of f and g, and $ad_f^i(g) = [f, \ldots, [f, g] \ldots]$ for the i-th iterated Lie bracket. We define the sequence of distributions: $D_0(x) = \{0\}$ - the trivial 0-dimensional distribution, and

$$D_1(x) = span\{g(x)\}, \quad D_2(x) = span\{g(x), ad_f(g)(x)\}, \ldots$$

$$\cdots, D_n(x) = span\{g(x), ad_f(g)(x), \ldots, ad_f^{n-1}(g)(x)\}.$$

Fix a point x_0 and let k be the largest integer such that the distributions D_j, $j = 0, \ldots, k$ are involutive and of rank j at some points of an open subset $X_0 \subset X$ such that x_0 is in the closure of X_0. Then D_k is of rank k on X_0. Define the *maximal involutive distribution* of Σ at x_0 as

$$\mathcal{D} = D_k.$$

We call the point x_0 *regular* point of $\mathcal{D}$ (or *regular point of Σ*) if $x_0 \in X_0$, i.e., $\mathcal{D} = D_k$ is of rank k at x_0. Then all the distributions D_j, $j = 1, \ldots, k$ are involutive and of constant rank j in a neighbourhood of x_0.

Theorem 1. *For scalar control systems*

$$\Sigma : \qquad \dot{x} = f(x) + ug(x)$$

on X the maximal involutive distribution $\mathcal{D}$ and the critical trajectories determine the system (up to local feedback equivalence), locally around any regular point of $\mathcal{D}$ through which a critical trajectory passes.

The above statement means that, given two systems on X which have the same maximal involutive distribution $\mathcal{D}$ and the same nonempty sets of critical trajectories in a neighbourhood of a regular point $x_0 \in X$ of $\mathcal{D}$, then they are locally feedback equivalent at this point.

The above theorem can be deduced from Theorem 2 in Section 7 describing the set of bi-critical trajectories of the system. There we will present two nonequivalent scalar-control systems which have the same maximal involutive distributions and the same empty sets of critical trajectories at a regular point.

The following statement is a direct consequence of Theorem 1.

Corollary 1. *Two systems Σ, $\tilde{\Sigma}$ are locally feedback equivalent at their regular points iff their sets of critical trajectories (assumed to be nonempty) and their distributions $\mathcal{D}$, $\tilde{\mathcal{D}}$ are related by a local state diffeomorphism.*

The restriction to a generic (i.e. regular) point can not be removed, in general.

Example 2. Consider two scalar systems

$$\Sigma : \quad \dot{x} = x^2 u, \qquad \tilde{\Sigma} : \quad \dot{x} = x^3 u$$

at the point $x = 0$. It is easy to show that they are not locally feedback equivalent at 0, however, they have equal sets of critical trajectories since their equations are:

$$\dot{x} = x^r u, \quad \dot{p} = -rpx^{r-1}u, \quad px^r = 0,$$

where $r = 2$ or 3. Since $p \neq 0$ it follows that $x(t) \equiv 0$ and both systems have only constant critical trajectories $x \equiv 0$. Also their sets of bi-critical curves are equal and have the form: $(p(\cdot), x(\cdot), u(\cdot)) = (\text{const} \neq 0, 0, u(\cdot))$, where $u(\cdot)$ is an arbitrary smooth function with values in $U = \mathbb{R}$.

5   Linear Systems

Our further aim is to describe explicitly the structure of critical and bi-critical trajectories. As a preliminary exercise let us find the critical and bi-critical trajectories for a linear system

$$\Lambda : \qquad \dot{x} = Ax + Bu, \quad X = \mathbb{R}^n, \quad U = \mathbb{R}^m.$$

We have

$$H = pAx + pBu.$$

Denote $\partial H / \partial u = H_u = pB$. Then the conditions for critical curves in (1) read

$$\dot{x} = Ax, \quad \dot{p} = -pA, \quad H_u = pB = 0.$$

The third condition does not allow to compute u (as a function of p and x, instead it gives a condition on the function p, to be satisfied for all t. Differentiating this condition and using the system equations gives the equation $\frac{d}{dt}pB = -pAB = 0$ and, repeating the procedure $n - 1$ times,

$$pB = 0, \quad pAB = 0, \quad \ldots, \quad pA^{n-1}B = 0.$$

Further differentiation of the last equation does not give new independent conditions (by the Cayley-Hamilton theorem). This means that the conditions on bi-critical curves do not introduce any restriction on the control, neither on the state, and the following conclusion holds. Denote the subsets of $T^* \mathbb{R}^n$

$$C_k = \{(p, x) : \; pA^j b_i = 0, \; 0 \leq j \leq k - 1, \; 1 \leq i \leq m\}$$

and the annihilator of the controllability subspace

$$C = C_n = E^\perp, \quad E = \mathrm{Im}(B, AB, \ldots, A^{n-1}B).$$

Claim. If the system is controllable, then the set of bi-critical trajectories is empty. In the converse case through any point of the annihilator $C = E^\perp$ there passes a bi-critical trajectory. The bi-critical trajectories passing through such a point are parametrized by arbitrary smooth controls (in one-to-one manner, if $\mathrm{rank}\,B = m$).

The above conclusion implies that, in general, the bi-critical curves cannot distinguish between two nonequivalent controllable systems, in the multi-input case. (In the scalar control case this was possible in Theorem 1 since all controllable linear systems in $X = \mathbb{R}^n$ of fixed dimension n are equivalent.)

However, if we take into account the dimensions of the subspaces

$$C_k = (\mathrm{Im}(B, AB, \ldots, A^{k-1}B))^\perp \times \mathbb{R}^n$$

we see that they determine the Brunovsky controllability indices. This implies that *the critical subspaces $C_1, \ldots, C_n$ are complete feedback invariants in the class of controllable linear systems.* Splitting the system into controllable and uncontrollable part (which is a dynamical system) and using Example 1 we easily see that the following conclusion holds.

Claim. The critical subspaces $C_1, \ldots, C_n$ and the critical trajectories form a complete system of feedback invariants in the class of linear systems.

6   Critical Varieties

We shall use the canonical Poisson bracket defined, in canonical local coordinates, for two functions $h_1, h_2 : T^*X \to \mathbb{R}$, by

$$\{h_1, h_2\} = \sum_{i=1}^{n} \left(\frac{\partial h_1}{\partial p_i} \frac{\partial h_2}{\partial x_i} - \frac{\partial h_1}{\partial x_i} \frac{\partial h_2}{\partial p_i} \right)$$

and analogously for h_1, h_2 depending on u. We use the usual notation: $ad_h h_1 = \{h, h_1\}$, $ad_h^2 h_1 = \{h, \{h, h_1\}\}$ etc.

Let $H_{u_i} = \partial H / \partial u_i$, $i = 1, \ldots, m$. We define the families of functions on $T^*X \times U$

$$\mathcal{H}_1 = \{H_{u_i} : 1 \le i \le m\},$$

$$\vdots$$

$$\mathcal{H}_k = \{ad_H^j H_{u_i} : 0 \le j \le k-1, \ 1 \le i \le m\},$$

and

$$\mathcal{H} = \bigcup_{k \ge 1} \mathcal{H}_k.$$

We have

$$\mathcal{H}_1 \subset \mathcal{H}_2 \subset \cdots \subset \mathcal{H}_k \subset \cdots \subset \mathcal{H}.$$

We define the *control critical varieties* as the sets of zeros of $\mathcal{H}_k$ and $\mathcal{H}$,

$$\mathcal{C}_k = Z(\mathcal{H}_k) = \{(p, x, u) \in T^*X \times U : h(p, x, u) = 0 \ \forall h \in \mathcal{H}_k\},$$

and

$$\mathcal{C} = Z(\mathcal{H}) = \{(p, x, u) \in T^*X \times U : h(p, x, u) = 0 \ \forall h \in \mathcal{H} \}.$$

Their projections onto T^*X are called *critical varieties* and are denoted by

$$C_k = \{(p, x) \in T^*X \text{ such that } \exists u \in U \text{ for which } h(p, x, u) = 0 \ \forall h \in \mathcal{H}_k\}$$

and

$$C = \{(p, x) \in T^*X \text{ such that } \exists u \in U \text{ for which } h(p, x, u) = 0 \ \forall h \in \mathcal{H}\}.$$

We have

$$\mathcal{C}_1 \supset \mathcal{C}_2 \supset \mathcal{C}_3 \supset \cdots, \qquad C_1 \supset C_2 \supset C_3 \supset \cdots$$

and

$$\mathcal{C} = \bigcap_{i \ge 1} \mathcal{C}_i, \qquad C \subset \bigcap_{i \ge 1} C_i$$

(in the latter case we have equality if U is compact). In the analytic case there exists an $r \geq 0$ such that, for set germs at a given point, we have $C_r = C_{r+1} = C_{r+2} \ldots$. This follows from the fact that the ring of germs of real-analytic functions is noetherian.

Consider the case of scalar control affine systems $\dot{x} = f(x) + ug(x)$. Then we have a simpler stopping rule at regular (generic) points of Σ (which, by definition, are regular points of the maximal involutive distribution $\mathcal{D}$, see Section 4). Namely, it is easy to see that there exists an integer $r \geq 2$ such that either

(i) the ideal of function germs generated by the family $\mathcal{H}_{r+1}$ does not have generators independent of u, or

(ii) the ideal of function germs generated by the family $\mathcal{H}_{r+1}$ has generators independent of u but it is equal to the ideal generated by $\mathcal{H}_r$.

We choose the minimal r having the property (i) or (ii).

The structure of critical varieties around a regular point of an affine scalar-control system is described by the following

Proposition 3. *(a) The number r defined above is equal to $2k$ in the case (i) and it is equal to k in the case (ii), where k is the rank of the maximal involutive distribution defined in Section 4.*

(b) The critical varieties C_i are submanifolds of T^*X of codimension $\operatorname{codim} C_i = i$, for $i = 1, \ldots, r$ and they are anihilators of the involutive distributions D_i introduced in Section 4,

$$C_i = D_i^{\perp}, \qquad \text{for} \quad i = 1, \ldots, k,$$

where k is the rank of the maximal involutive distribution.

(c) In the case (i) the remaining critical varieties $C_{k+1}, \ldots, C_{2k}$ can be described as follows. There exist local coordinates $x_1, \ldots, x_n$ around the regular point such that, with the corresponding coordinates $p_1, \ldots, p_n$ on T^*X, we have $C_i = \{p_1 = \cdots = p_i = 0\}$, for $i = 1, \ldots, k$, and there exist functions $\psi_1, \ldots, \psi_k$ of the variables $q = (p_{k+1}, \ldots, p_n)$ and $y = (x_{k+1}, \ldots, x_n)$ such that locally around any non-discriminant point (i.e. a point satisfying $(\partial^2 \tilde{H} / \partial x_k^2)(q, y) \neq 0$) the varieties C_{k+i} have the form

$$C_{k+i} = \{p_1 = \cdots = p_k = 0, \quad x_k = \psi_k(q, y), \ldots, x_{k-i+1} = \psi_{k-i+1}(q, y)\},$$

for $i = 1, \ldots, k$.

Corollary 2. *The critical variety C of a scalar-control affine system is, in the coordinates of Proposition 3 locally around a regular point x and non-discriminant point (p, x), of the form*

$$C = C_{2k} = \{p_1 = \cdots = p_k = 0, \quad x_1 = \psi_1(q, y), \ldots, x_k = \psi_k(q, y)\},$$

in the case (i), and

$$C = C_k = \{p_1 = \cdots = p_k = 0\},$$

in the case (ii).

Remark 1. Under the assumptions of Proposition 3 the control critical varieties of the system are described as the subsets of $T^*X \times U$ given in the cases (i) and (ii), respectively, by the equations

$$C = \{p_1 = \cdots = p_k = 0,\ x_1 = \psi_1(q, y), \ldots, x_k = \psi_k(q, y), u = \psi_0(q, y)\},$$

$$C = \{p_1 = \cdots = p_k = 0\},$$

where ψ_0 is a smooth function.

We will see that the algorithm of computing the bi-critical trajectories of the system, via computing the critical varieties, discovers automatically a canonical decomposition of the system into a maximal feedback linearizable part and a nonlinear part.

7   Description of bi-critical trajectories

We are ready for describing explicitly the critical trajectories for scalar-control systems around regular points.

Theorem 2. *Any bi-critical trajectory of a system is contained in the critical variety C. Given a scalar-control affine system Σ and a point $(p, x) \in C$ such that x is a regular point $x \in X$ of Σ, the bi-critical trajectories fill a neighbourhood of (p, x) in the critical variety i.e. through any point of C in the neighbourhood there passes a bi-critical trajectory of Σ. In the case (i) through any point (p, x) of C, with $p \neq 0$, there passes a unique bi-critical trajectory, while in the case (ii) the bi-critical trajectories passing through (p, x) are parametrized in on-to-one manner by arbitrary smooth control functions.*

Example 3. For a given integer $r \geq 2$ we consider the system

$$\Sigma_r : \qquad \dot{x}_1 = u, \quad \dot{x}_2 = x_1, \quad \dot{x}_3 = x_2, \quad \dot{x}_4 = x_2^r$$

on $X = \mathbb{R}^4$ with the Hamiltonian

$$H = p_1 u + p_2 x_1 + p_3 x_2 + p_4 x_2^r$$

and the adjoint equations

$$\dot{p}_1 = -p_2, \quad \dot{p}_2 = -p_3 - r p_4 x_2^{r-1}, \quad \dot{p}_3 = 0, \quad \dot{p}_4 = 0.$$

We compute

$$H_u = p_1, \quad ad_H(H_u) = -p_2, \quad ad_H^2(H_u) = p_3 + rp_4x_2^{r-1},$$

$$ad_H^3(H_u) = r(r-1)p_4x_2^{r-2}x_1, \quad ad_H^4(H_u) = r(r-1)p_4x_2^{r-3}(x_2u + (r-2)x_1^2).$$

Thus, the first two critical varieties $C_1 = \{p_1 = 0\}$, $C_2 = \{p_1 = p_2 = 0\}$ are annihilators of involutive distributions. The number k defined in Section 4 is $k = 2$, the number r of Section 6 is $r = 4 = 2k$ and we have the case (i) in Proposition 3. Finally, the critical variety C and the control critical variety $\mathcal{C}$ are:

$$C = \{\ p_1 = p_2 = p_3 + rp_4x_2^{r-1} = p_4x_2^{r-2}x_1 = 0\ \},$$

$$\mathcal{C} = \{\ p_1 = p_2 = p_3 + rp_4x_2^{r-1} = p_4x_2^{r-2}x_1 = p_4x_2^{r-3}(x_2u + (r-2)x_1^2) = 0\ \}.$$

All points x are regular and all (p, x) with $p_4x_2^{r-2} \neq 0$ are non-discriminant. The critical control and the bi-critical trajectories are given by two alternative sets of equations

$$p_1(t) \equiv p_2(t) \equiv x_1(t) \equiv u(t) \equiv 0, \quad x_2(t) = \left(\frac{-p_3}{rp_4}\right)^{1/(r-1)} \neq 0, \quad p_4 \neq 0,$$

or

$$p_1(t) \equiv p_2(t) \equiv p_3(t) \equiv x_1(t) \equiv x_2(t) \equiv u(t) \equiv 0, \quad p_4 \neq 0,$$

together with the system equations $\dot{x}_3 = x_2$, $\dot{x}_4 = x_2^r$ where p_3, p_4 are constants which follows from the adjoint equations (the case $p_4 = 0$ leads to $p = 0$ and so it does not produce bi-critical curves).

Remark 2. Using the above example we can present two systems which locally have two identical maximal involutive distributions, empty sets of critical curves but are not locally feedback equivalent. Namely, taking Σ_{r_1} and Σ_{r_2} with nonequal $r_1, r_2 \geq 3$ we see that in a neighbourhood of a point $x = (x_1, \ldots, x_4)$ with $x_1 \neq 0$ they have empty sets of critical curves and the same maximal involutive distribution $\mathcal{D} = span\{\frac{\partial}{\partial x_1}, \frac{\partial}{\partial x_2}\}$. Suppose they are locally feedback equivalent around such a point. Since both systems are control-affine i.e., of the form $\dot{x} = f(x) + ug(x)$, the transformation establishing their equivalence must transform the distribution generated by the vector field $g = \frac{\partial}{\partial x_1}$ into itself, so it is of the form $x' = \Phi'(\tilde{x}')$, $x_1 = \Phi''(\tilde{x}', \tilde{x}_1)$, where $x' = (x_2, x_3, x_4)$. This transformation establishes local feedback equivalence of reduced systems on R^3, with the equation $\dot{x}_1 = u$ removed and the new control $v = x_1$. Repeating the operation for new systems we deduce that two systems on $\mathbb{R}^2$

$$\dot{x}_3 = u, \quad \dot{x}_4 = u^r,$$

with different $r = r_1, r_2$ are locally feedback equivalent. The set of velocities of such a system at a given point is of the form $F_r(x) = \{(u, u^r) : u \in U\} \subset \mathbb{R}^2$ and it is a piece of the zero level sets of the polynomial $v_1^r - v_2$. The sets of velocities are transformed one onto the other by a linear transformation (the Jacobian map of the state diffeomorphism establishing feedback equivalence), therefore the minimal degree of the polynomial giving $F(x)$ as the zero level set is invariant under this transformation. We deduce that equivalence is possible only if $r_1 = r_2$ which contradicts the supposition that the original systems with $r_1 \neq r_2$ were equivalent.

In order to prove Proposition 3 and Theorem 2 and show explicitly the structure of critical varieties and of bi-critical trajectories of a general system we will use a decomposition of the system into linear and non-linear parts.

Consider a system with scalar control

$$\Sigma: \quad \dot{x} = f(x, u), \quad x \in X, \quad u \in U.$$

Proposition 4. *If the system is real-analytic then there exists an open dense subset $G \subset X \times U$ such that locally around any point of G the system Σ is locally feedback equivalent to one of the following seminormal forms, written in a coordinate system $x = (x_1, \ldots, x_k, y)$ at the origin, with $y = (y_1, \ldots, y_{n-k})$,*

$$\dot{x}_1 = u, \quad \dot{x}_2 = x_1, \ldots, \dot{x}_k = x_{k-1},$$
$$(NF1): \qquad \dot{y} = \tilde{f}(y, x_k),$$

where the subsystem $\dot{y} = \tilde{f}(y, v)$ is fully nonlinear at $y = 0$ and $v = x_k = 0$, or to

$$\dot{x}_1 = u, \quad \dot{x}_2 = x_1, \ldots, \dot{x}_k = x_{k-1},$$
$$(NF2): \qquad \dot{y} = \tilde{f}(y).$$

Here $k = 0, 1, \ldots,$ or n and we denote $x_0 = u$ (if $k = 0$, then only the $\tilde{f}$ part appears in the canonical form).

The number k in the proposition is invariantly related to the system at its generic points and, in the case of control-affine systems, is given by the rank of the maximal involutive distribution defined in Section 4.

The proposition can be proved using standard arguments of feedback linearization (see [11], [16]) in the case of control-affine systems. The general case is reducible to control-affine as stated in Section 3.

Proof of Proposition 3. By Proposition 4 it is enough to compute the critical varieties for systems in the normal forms (NF1) and (NF2) and show that they have the required properties.

We can write the Hamiltonian of the system (NF1) in the form

$$H = p_1 u + p_2 x_1 + \cdots + p_k x_{k-1} + q\tilde{f}(y, x_k) = \sum_{j=1}^{k} p_j x_{j-1} + \tilde{H}(q, y, x_k),$$

where $x_0 = u$, $q = (q_1, \ldots, q_{n-k})$, and $\tilde{H} = q\tilde{f}(y, x_k)$. Then

$$H_u = p_1, \ ad_H(H_u) = -p_2, \ldots, ad_H^{k-1}(H_u) = (-1)^{k-1}p_k,$$

$$ad_H^k(H_u) = (-1)^k q \frac{\partial \tilde{f}}{\partial x_k}(y, x_k) = (-1)^k \tilde{H}_{x_k}(q, y, x_k)$$

and, for certain functions $h_1, \ldots, h_k$,

$$(-1)^k ad_H^{k+1}(H_u) = x_{k-1}\tilde{H}_{x_k x_k} + ad_{\tilde{H}} \tilde{H}_{x_k} = x_{k-1}\tilde{H}_{x_k x_k} + h_1(p_k, q, x_k, y),$$

$$(-1)^k ad_H^{k+2}(H_u) = x_{k-2}\tilde{H}_{x_k x_k} + h_2(p_{k-1}, p_k, q, x_{k-1}, x_k, y),$$

$$\vdots$$

$$(-1)^k ad_H^{2k}(H_u) = u\tilde{H}_{x_k x_k} + h_{2k}(p_1, \ldots, p_k, q, x_1, \ldots, x_k, y).$$

This means that
$$C_j = \{p_1 = 0, \ldots, p_j = 0\}$$
for $j = 1, \ldots, k$, and

$$C_{k+1} = \{p_1 = 0, \ldots, p_k = 0, \ \tilde{H}_{x_k}(q, y, x_k) = 0\}.$$

Since in our region $\tilde{H}_{x_k x_k}(q, y, x_k) \neq 0$, it follows by the implicit function theorem that the equation $\tilde{H}_{x_k}(q, y, x_k) = 0$ has a unique solution $x_k = \psi(q, y)$. Using the same inequality $\tilde{H}_{x_k x_k}(q, y, x_k) \neq 0$ we can compute x_{k-j-1} as a function of $(p_{k-j}, \ldots, p_k, q, x_{k-j}, \ldots, x_k, y)$ from the equation $ad_H^{k+j}H_u = 0$ for $j = 0, 1, \ldots, k$. In particular, from the last equation we compute u as a function of $p = (p_1, \ldots, p_k, q)$ and $x = (x_1, \ldots, x_k, y)$. We conclude that the control critical variety $\mathcal{C} = \mathcal{C}^{2k}$ is a submanifold of dimension $2(n-k)$ of $T^*X \times U$ of the form

$$\mathcal{C} = \{ p_1 = 0, \ldots, p_k = 0, \ u = \psi_0(q, y), \ x_1 = \psi_1(q, y) = 0, \ldots, x_k = \psi_k(q, y) \},$$

parametrized by the variables (q, y), and the critical variety C is

$$C = \{ p_1 = 0, \ldots, p_k = 0, \ x_1 = \psi_1(q, y) = 0, \ldots, x_k = \psi_k(q, y) \}.$$

The same but simpler calculation for the second normal form (NF2) and the Hamiltonian
$$H = p_1 u + p_2 x_1 + \cdots + p_k x_{k-1} + q\tilde{f}(y),$$

gives $ad_H^j(H_u) = p_{j+1}$, $j = 0, 1, \ldots, k-1$, and $ad_H^k(H_u) \equiv 0$. This means that
$$C_j = \{p_1 = 0, \ldots, p_j = 0\}$$
for $j = 1, \ldots, k$, and

$$C_k = C = \{p_1 = 0, \ldots, p_k = 0\}$$

as a submanifold of $T^*X \times U$. The critical variety C is given by the same equations in T^*X. This shows the structure of the critical varieties for systems in the normal forms (NF1) and (NF2) and completes the proof of Proposition 3.

Proof of Theorem 2 By Proposition 4 it is enough to compute the bi-critical trajectories for systems in the normal forms (NF1) and (NF2) and show that they have the required properties. This can be done in a straightforward way by following the procedure of computing the critical varieties of (NF1) and (NF2) in the proof of Proposition 4 and then using the argument in the proof of Proposition 2 (in the case (i)), or in Example 1 (in the case (ii)) for computation of bi-critical trajectories of fully nonlinear systems and of dynamical systems, respectively. We leave the details to the reader.

8   Symbols of the system

As explained in the preceeding section, any scalar-control analytic system can be canonically decomposed into a linear part and a nonlinear part, around a regular point. The feedback classification problem reduces then to classifying the nonlinear parts.

In the case (ii) the nonlinear part is a dynamical system. The corresponding problem of local classification of singular points of vector fields has wide literature which goes back to Poincaré (see [2]).

In the case (i) the problem reduces to classification of fully nonlinear systems. We know from Proposition 2 that the set of critical trajectories of the system is a complete covariant, i.e., the local feedback classification of fully nonlinear systems is equivalent to classification of critical trajectories under the group of state diffeomorphisms.

In this section we go further and construct a complete set of invariants of feedback equivalence for fully nonlinear systems. The result stated below is a special case of the results presented in [10].

Consider the analytic system

$$\Sigma : \quad \dot{x} = f(x, u)$$

on $I\!\!R^n$ and assume that it has *full span* at (x_0, u_0), namely the vectors

$$\frac{\partial f}{\partial u}(x_0, u_0), \quad \frac{\partial^2 f}{\partial u^2}(x_0, u_0), \quad \dots, \quad \frac{\partial^s f}{\partial u^s}(x_0, u_0)$$

span the tangent space $T_{x_0}X$ for some $s \geq n$. Let μ be the minimal s with this property, then we call $\mu = s - 1$ the *multiplicity* of Σ at (x_0, u_0). Let p_0 be a nonzero covector which annihilates the first $s - 1 = \mu$ vectors. Such a covector p_0 is defined uniquely, up to an invertible factor, and will be called *canonical covector.* Then Σ has finite multiplicity μ at (p_0, x_0, u_0) as defined in [10], Section 3.

The multiplicity describes the number of generic solutions of the critical equation (in the complex domain), as explained below. Consider the Hamiltonian $H(p, x, u) = pf(x, u)$. Since H is analytic with respect to all variables, in particular u, we can complexify it with respect to u, locally in a neighbourhood of (p_0, x_0, u_0). We denote the complexified Hamiltonian by $\hat{H}$, without changing notation for the complexified variable u. (For example, if H is polynomial with respect to u, the complexification is just considering the same polynomial as polynomial of the complex variable u. In the general case the same is done with the Taylor series of H at (p_0, x_0, u_0).) Consider the critical equation

$$\frac{\partial \hat{H}}{\partial u}(p, x, u) = 0$$

with respect to complex unknown u. From our assumptions it follows that:

$$\frac{\partial \hat{H}}{\partial u}(p_0, x_0, u_0) = \frac{\partial^2 \hat{H}}{\partial u^2}(p_0, x_0, u_0) = \cdots = \frac{\partial^\mu \hat{H}}{\partial u^\mu}(p_0, x_0, u_0) = 0,$$

$$\frac{\partial^{\mu+1} \hat{H}}{\partial u^{\mu+1}}(p_0, x_0, u_0) \neq 0.$$

From the Weierstrass preparation theorem we deduce that the critical equation has roots behaving like roots of a polynomial equation (with respect to u) of order μ, with coefficients depending analytically on (p, x) and vanishing at (p_0, x_0), except the highest coefficient. Let

$$u_1 = u_1(p, x), \cdots, u_\mu = u_\mu(p, x)$$

be those roots (complex, in general, possibly multiple), defined for (p, x) in a neighbourhood of (p_0, x_0). We call them *critical controls*. Let us plug the critical controls to the complexified Hamiltonian, then we obtain the functions

$$H_1(p, x) = \hat{H}(p, x, u_1(p, x)), \ldots, H_\mu(p, x) = \hat{H}(p, x, u_\mu(p, x)),$$

called *critical Hamiltonians*. The critical controls and the critical Hamiltonians are well defined, up to order, in the neighbourhood of (p_0, u_0) with the exlusion of the discriminant set

$$D = \{ (p, x) : \exists u \; \frac{\partial \hat{H}}{\partial u}(p, x, u) = 0, \quad \frac{\partial^2 \hat{H}}{\partial u^2}(p, x, u) = 0 \}.$$

In the regions where the critical controls are real the critical Hamiltonians define Hamiltonian vector fields on T^*X and their trajectories are bi-critical trajectories of Σ.

The critical Hamiltonians are complex-valued, in general, and do not have enough regularity for our purposes. Thus, we introduce the symmetric functions of the critical Hamiltonians, called *symbols of Σ*, as

$$S_1(p, x) = H_1(p, x) + \cdots + H_\mu(p, x),$$
$$S_2(p, x) = (H_1(p, x))^2 + \cdots + (H_\mu(p, x))^2,$$
$$\vdots$$
$$S_\mu(p, x) = (H_1(p, x))^\mu + \cdots + (H_\mu(p, x))^\mu.$$

Using the theory of functions of complex variables we can prove the following

Lemma 2. *The symbols $S_1, \ldots, S_\mu$ are well defined real-analytic functions in a neighbourhood of (p_0, x_0). They are homogeneous functions of p, namely, $S_j(tp, x) = t^j S_j(p, x)$.*

The critical Hamiltonians, as well as the symbols, describe the system Σ, up to pure feedback equivalence. In particular, we have the following

Theorem 3. *For regular in control analytic scalar-control systems Σ which have full span at a point (x_0, u_0) with multiplicity μ the symbols $S_1, \ldots, S_\mu$, as well as the critical Hamiltonians $H_1, \ldots, H_\mu$, determine the system locally around (x_0, u_0), up to pure feedback equivalence. In paricular, two such systems are locally feedback equivalent at (x_0, u_0) iff they have the same multiplicity μ and the same symbols.*

Corollary 3. *Under the assumptions of the above theorem two systems Σ and $\tilde{\Sigma}$ are locally feedback equivalent iff they have the same multiplicities $\mu = \tilde{\mu}$ and their symbols are related by the Hamiltonian lift of a state diffeomorphisms.*

Remark 3. The above theorem and corollary also hold for multi-input case and under a more general assumption on the system, namely the assumption of finite multiplicity. Moreover, the "direction" p_0 around which the two systems are compared can be chosen. This leads to microlocal equivalence and microlocal analysis of systems (see [9], [10] for more details).

The corollary shows that the problem of local feedback classification of systems having full span can be reduced to classification of a collection of analytic functions (the symbols) on the cotangent bundle T^*X, under the group of Hamiltonian lifts of state diffeomorphisms. This problem will be solved in the following section.

9   Invariant momentum

The following observation allows to construct the invariants. Given two functions $h_1, h_2 : T^*X \to I\!\!R$ and a local symplectomorphism $\chi : T^*X \to T^*X$

preserving a point $(p_0, x_0) \in T^*X$, we have $\{h_1 \circ \chi, h_2 \circ \chi\} = \{h_1, h_2\} \circ \chi$ and so the Poisson bracket $\{h_1, h_2\}(p_0, x_0)$ does not change when we transform h_1 and h_2 by χ. The same holds with iterated Poisson brackets. This remark together with the above corollary leads to the following result.

Let us fix a point $(x_0, u_0) \in X \times U$ and assume that the system Σ is analytic and has full span at this point. Then there exists a covector $p_0 \in T^*_{x_0}X$, $p_0 \neq 0$, defined uniquely up to a multiplicative constant, such that p_0 annihilates the first μ vectors $\partial^i f / \partial u^i (x_0, u_0)$, $i = 1, \ldots, \mu$, where μ is the multiplicity. Then the symbols $S_1, \ldots, S_\mu$ are well defined and analytic in a neighbourhood of (p_0, x_0), by Lemma 2.

Let

$$L = Lie\{1, \ldots, \mu\}$$

denote the free Lie algebra generated by the free generators $1, \ldots, \mu$ (here $1, 2, \ldots, \mu$ are treated as symbols or letters and not as numbers). The map

$$i \in \{1, \ldots, \mu\} \to S_i$$

extends to a unique Lie algebra homomorphism

$$\mathcal{M} : L \to C^\omega_{(p_0, x_0)}(T^*X),$$

where the target Lie algebra is the space of germs at (p_0, x_0) of real-analytic functions with the Poisson bracket as product. This homomorphism is the unique linear extension of the map

$$[i_1, \ldots, [i_{k-1}, i_k] \ldots] \to \{S_{i_1}, \ldots, \{S_{i_{k-1}}, S_{i_k}\} \ldots\},$$

where $[i_1, \ldots, [i_{k-1}, i_k] \ldots]$ denotes the formal Lie polynomial in L defined as the left iterated Lie product of the symbols $i_1, \ldots, i_k \in \{1, \ldots, \mu\}$ (for example, $[i_1, i_2] = i_1 i_2 - i_2 i_1$ and $[i_1, [i_2, i_3]] = i_1 i_2 i_3 - i_1 i_3 i_2 - i_2 i_3 i_1 + i_3 i_2 i_1$). Let L^* denote the dual to the linear space L. The homomorphism $\mathcal{M}$ defines the *momentum map*

$$M : T^*X \to L^*, \qquad (p, x) \to M(p, x),$$

(in fact, a map-germ at (p_0, x_0)) where

$$\langle M(p, x), W \rangle = (\mathcal{M}(W))(p, x) \qquad \text{for} \quad W \in L.$$

In particular, we define the linear functional $M(p_0, x_0)$ on L, called *invariant momentum*, by composing $\mathcal{M}$ with evaluation at (p_0, x_0), i.e.,

$$\langle M(p_0, x_0), W \rangle = (\mathcal{M}(W))(p_0, x_0), \qquad \text{for} \quad W \in L.$$

We have

$$\langle M(p_0, x_0), [i_1, \ldots, [i_{k-1}, i_k] \ldots] \rangle = \{S_{i_1}, \ldots, \{S_{i_{k-1}}, S_{i_k}\} \ldots\}(p_0, x_0).$$

The momentum $M(p_0, x_0)$ is an invariant of local feedback equivalence if the length of p_0 is chosen canonically (or in the same way for both systems).

Theorem 4. *Consider two regular in control analytic scalar-control systems* Σ, $\tilde{\Sigma}$ *which have full span at* (x_0, u_0) *and* $(\tilde{x}_0, \tilde{u}_0)$, *respectivaly. If* Σ *and* $\tilde{\Sigma}$ *are locally feedback equivalent at these points, then they satisfy the following equivalent conditions.*

(a) For any canonical covectors p_0, $\tilde{p}_0$ *of* Σ *and* $\tilde{\Sigma}$, *respectively, there exists a nonzero constant* $a \in \mathbb{R}$ *such that*

$$M(p_0, x_0) = \tilde{M}(a\tilde{p}_0, \tilde{x}_0).$$

(b) For any canonical covectors p_0, $\tilde{p}_0$ *there exists a nonzero constant* $a \in \mathbb{R}$ *such that*

$$\{S_{i_1}, \ldots, \{S_{i_{k-1}}, S_{i_k}\} \ldots\}(p_0, x_0) = a^{|I|}\{\tilde{S}_{i_1}, \ldots, \{\tilde{S}_{i_{k-1}}, \tilde{S}_{i_k}\} \ldots\}(\tilde{p}_0, \tilde{x}_0)$$

for all $I = (i_1, \ldots, i_k)$, $k \geq 1$, $1 \leq i_1, \ldots, i_k \leq \mu$, *where* $|I| = i_1 + \cdots + i_k$.

¿From the practical point of view we can always choose p_0 and $\tilde{p}_0$ so that $a = 1$ in the conditions (a) and (b), so that the normalization by the constant a is not needed. We will call such p_0 and $\tilde{p}_0$ *appropriately chosen.*

The momentum $M(p_0, x_0)$ itself is not a complete invariant of local feedback equivalence (it is a complete invariant of weaker microlocal feedback equivalence as stated in [10]). There is another invariant, independent of $M(p_0, x_0)$. A complete set of invariants can be described in several ways.

We introduce the following *rank condition*:

$$dim\ span\ \{\ dh(p_0, x_0)\ :\ h \in Lie\{S_1, \ldots, S_\mu\}\ \} = 2n.$$

Note that any $h \in Lie\{S_1, \ldots, S_\mu\}$ is of the form $h = \mathcal{M}(W)$, where $W \in L$. If the rank condition is satisfied then the map $M : T^*X \to L^*$ is the germ of an immersion at (p_0, x_0). Namely, there is a sequence $W = (W_1, \ldots, W_{2n})$ of elements of L such that the map $M_W : T^*X \to \mathbb{R}^{2n}$ given by

$$M_W = (\mathcal{M}(W_1), \ldots, \mathcal{M}(W_{2n}))$$

is the germ of a local diffeomorphism in a neighbourhood of (p_0, x_0). Then we say that the *rank condition is satisfied with* W. Choosing any linear basis $e_1, e_2, \ldots$ of L leads to a system of linear coordinates on L^*. We can identify L^* with $\mathbb{R}^\infty$ which can be endowed with the natural product topology. The momentum map $M : T^*X \to L^*$ can be identified with the map $T^*X \to \mathbb{R}^\infty$ given coordinatewise as

$$M = (\mathcal{M}(e_1), \mathcal{M}(e_2), \ldots).$$

If $e_1 = W_1, \ldots, e_{2n} = W_{2n}$, then the projection of the map M to its first r components, $r \geq 2n$, gives an immersion and it makes sense to speak of M being an immersion, as well as of its image $N \subset L^*$ being a submanifold. Let us denote by S the map germ $T^*X \to \mathbb{R}^\mu$ defined by the symbols

$$S(p, x) = (S_1(p, x), \ldots, S_\mu(p, x)).$$

Theorem 5. *Assume that two analytic systems Σ, $\tilde{\Sigma}$ have full span at (x_0, u_0) and $(\tilde{x}_0, \tilde{u}_0)$, respectively, and they satisfy the rank condition with the same sequence $\mathcal{W}$. Then the systems Σ and $\tilde{\Sigma}$ are locally feedback equivalent at (x_0, u_0) and $(\tilde{x}_0, \tilde{u}_0)$ if and only if the following conditions hold.*

(A) The multiplicities μ and $\tilde{\mu}$ are equal.

(B) The images of the momentum maps $M : T^*X \to L^*$ and $\tilde{M} : T^*\tilde{X} \to L^*$ coincide (as germs at $M(p_0, x_0) = \tilde{M}(\tilde{p}_0, \tilde{x}_0)$), with appropriately chosen canonical covectors p_0 and $\tilde{p}_0$).

(C) The images of the involutive distributions $\operatorname{span}\{\partial/\partial p_1, \ldots, \partial/\partial p_n\}$ and $\operatorname{span}\{\partial/\partial \tilde{p}_1, \ldots, \partial/\partial \tilde{p}_n\}$ under the momentum maps coincide.

Theorem 6. *The preceeding theorem holds with the conditions (B) and (C) changed for the following conditions.*

(B') The local images of the maps

$$\chi = (M_{\mathcal{W}}, S) : T^*X \to \mathbb{R}^{2n+\mu}, \qquad \chi = (\tilde{M}_{\mathcal{W}}, \tilde{S}) : T^*\tilde{X} \to \mathbb{R}^{2n+\mu}$$

coincide as germs at $\chi_{(p_0, x_0)} = \tilde{\chi}_{(\tilde{p}_0, \tilde{x}_0)}$, with appropriately chosen p_0 and $\tilde{p}_0$.

(C') The images of the involutive distributions $\operatorname{span}\{\partial/\partial p_1, \ldots, \partial/\partial p_n\}$ and $\operatorname{span}\{\partial/\partial \tilde{p}_1, \ldots, \partial/\partial \tilde{p}_n\}$ under the maps $M_{\mathcal{W}}$ and $\tilde{M}_{\mathcal{W}}$ coincide.

Remark 4. Since the momentum map is an immersion, under the rank assumption, its image $N \subset L^*$ is a submanifold of dimension $2n$ and the image of the distribution makes sense. The requirement of equality of images of the involutive distributions in (C) can be replaced by the equivalent condition that the non-vanishing contravariant n-tensor fields on $N \subset L^*$

$$T = \frac{\partial M}{\partial p_1} \wedge \cdots \wedge \frac{\partial M}{\partial p_n} \quad \text{and} \quad \tilde{T} = \frac{\partial \tilde{M}}{\partial \tilde{p}_1} \wedge \cdots \wedge \frac{\partial \tilde{M}}{\partial \tilde{p}_n}$$

are proportional (equal, up to a multiplicative nonvanishing function). Similarly, the corresponding replacement can be made in condition (C'), with M and $\tilde{M}$ replaced by $M_{\mathcal{W}}$ and $\tilde{M}_{\mathcal{W}}$, respectively. Instead of speaking of proportionality of these tensor fields we may normalize them by making one of the coefficients (the same for both tensors) equal to 1, and then speak of their equality. Equality of tensor fields on $N \subset L^*$ can be expressed as equality of their coefficients which are functions on N or, when composed with M, functions on T^*X. This idea will be realized below.

Let $\mathcal{W} = (W_1, \ldots, W_{2n})$ be a sequence of elements of L for which the rank condition is satisfied, i.e., $dM_1, \ldots, M_{2n}$ are linearly independent at (p_0, x_0), where we denote $M_1 = \mathcal{M}(W_1), \ldots, M_{2n} = \mathcal{M}(W_{2n})$. Consider the local diffeomorphism $M_{\mathcal{W}} = (M_1, \ldots, M_{2n}) : T^*X \to \mathbb{R}^{2n}$. We introduce the functions given as n-minors of the Jacobian matrix of $M_{\mathcal{W}}$ restricted to cotangent fibers,

$$T_{k_1 \cdots k_n} = Minor_{k_1 \cdots k_n}\left(\frac{\partial M_{\mathcal{W}}}{\partial p}\right),$$

where $1 \leq k_1 < \cdots < k_n \leq 2n$ are the indices of the rows taken when computing the n-minor. The rank assumption implies that at least one of them is nonvanishing, say $T_{\hat{K}}$, where $\hat{K} = \hat{k}_1 \cdots \hat{k}_n$. For $K = k_1 \cdots k_n$ we define

$$T_{\hat{K}}^{K} = T_{\hat{k}_1 \cdots \hat{k}_n}^{k_1 \cdots k_n} = \frac{T_{k_1 \cdots k_n}}{T_{\hat{k}_1 \cdots \hat{k}_n}}.$$

Let $\Gamma_{W,\hat{K}} : T^* X \to I\!\!R^N$ be the map-germ whose components are the functions $T_{\hat{K}}^{K}$ arranged in the lexicographic order of the multi-indicies $K = k_1 \cdots k_n$.

Theorem 7. *Theorem 5 holds with the conditions (B) and (C) replaced by the following condition.*

(D) The images of the maps

$$\eta = (M_W, S, \Gamma_{W,\hat{K}}) : T^* X \to I\!\!R^q, \qquad \tilde{\eta} = (\tilde{M}_W, \tilde{S}, \tilde{\Gamma}_{W,\hat{K}}) : T^* \tilde{X} \to I\!\!R^q$$

coincide as set-germs at $\eta(p_0, x_0) = \tilde{\eta}(\tilde{p}_0, x_0)$ for appropriately chosen $\hat{K}$ and canonical vectors p_0 and $\tilde{p}_0$, where $q = 2n + \mu + N$.

Replacing the functions $T_{\hat{K}}^{K}$ by their "Taylor coefficients" will lead to the second version of our result.

Let us choose a basis $e_1, e_2, \ldots$ of L (serving as a coordinate system on the dual L^*) and define the function germs $M_i : T^* X \to I\!\!R$ as $M_i = \mathcal{M}(e_i)$. For a multi-index $K = k_1 \cdots k_n$, with $1 \leq k_1 < \cdots < k_n$, we denote by T_K the determinant of the $n \times n$ matrix $\{\partial M_{k_i}/\partial p_j\}$. (This is a coefficient of the n-tensor field $T = \partial M/\partial p_1 \wedge \cdots \wedge \partial M/\partial p_n$ on the image submanifold $N \subset L^*$.) Under the rank assumption there exists a multi-index $\hat{K}$ such that the function germ $T_{\hat{K}}$ is nonvanishing. As before, we introduce the function germs $T_{\hat{K}}^{K} = T_K/T_{\hat{K}}$. Denote the iterated Poisson brackets of $T_{\hat{K}}^{K}$ with the symbols by

$$T_{\hat{K},I}^{K} = \{S_{i_k}, \ldots, \{S_{i_1}, T_{\hat{K}}^{K}\} \cdots\}, \quad \text{where} \quad I = i_1 \cdots i_k.$$

Let $\mathcal{T}_{\hat{K}}^{K}$ denote the collection of these function germs parametrized by all $I = i_1 \cdots i_k$ with $k \geq 0$ and $1 \leq i_1, \ldots, i_k \leq \mu$ (where for $k = 0$ we take $T_{\hat{K},I}^{K} = T_{\hat{K}}^{K}$). By $\mathcal{T}_{\hat{K}}^{K}(p_0, x_0)$ we denote the collection of values of the functions $T_{\hat{K},I}^{K}$ at (p_0, x_0), parametrized by the same multi-indices I.

Theorem 8. *Assume that two analytic systems Σ, $\tilde{\Sigma}$ have full span at points (x_0, u_0) and $(\tilde{x}_0, \tilde{u}_0)$, respectively, they satisfy the rank condition with the same sequence of words W, and they have a nonvanishing coefficient $T_{\hat{K}}$ with the same multi-index $\hat{K}$. Then the systems Σ and $\tilde{\Sigma}$ are locally feedback equivalent at (x_0, u_0) and $(\tilde{x}_0, \tilde{u}_0)$, respectively, if and only if the following conditions hold.*

(A) The multiplicities are equal, $\mu = \tilde{\mu}$.

(B) The invariant momenta are equal,

$$M_{(p_0, x_0)} = \tilde{M}_{(\tilde{p}_0, \tilde{x}_0)},$$

for an appropriate choice of p_0 and $\tilde{p}_0$.

(E) The following equality holds

$$\mathcal{T}_K^K(p_0, x_0) = \tilde{\mathcal{T}}_K^K(\tilde{p}_0, \tilde{x}_0)$$

for an appropriate choice of p_0 and $\tilde{p}_0$ and all multiindices K.

If the rank assumption is realized on the first $2n$ elements $W_1 = e_1, \ldots,$ $W_{2n} = e_{2n}$ of the basis of L, then it is enough that the condition (E) holds with the multiindices K satisfying $1 \leq k_1 < \cdots < k_n \leq 2n$.

Theorems 7 and 8 can be proved using Theorems 5 and 6.

References

1. Agrachev, A. (1995) Methods of Control Theory in Nonholonomic Geometry, in Proc. Int. Congress of Math., Zurich 1994, 1473-1483, Birkhäuser, Basel.

2. Anosov, D.V. and Arnold, V.I. (1988) Dynamical Systems I, Encyclopaedia of Mathematical Sciences, Springer Verlag, Heidelberg.

3. Bonnard, B. (1991) Feedback Equivalence for Nonlinear Systems and the Time Optimal Control Problem, SIAM J. Control and Optimiz. 29, 1300-1321.

4. Bonnard, B. (1992) Quadratic Control Systems, Mathematics of Control, Signals, and Systems 4, 139-160.

5. Gardner, R. (1989) The Method of Equivalence and Its Applications, CBMS Regional Conference Series in Applied Mathematics, CBMS 58, SIAM.

6. Isidori, A. (1989) Nonlinear Control Systems; An Introduction, Springer Verlag, New York 1989.

7. Hunt L.R., Su R. and Meyer G. (1983) Design for Muti-Input Nonlinear Systems, in "Differential Geometric Control Theory", R. Brockett, R. Millman, H. Sussmann eds. 268-298, Birkhäuser 1983.

8. Jakubczyk, B. (1990) Equivalence and Invariants of Nonlinear Control Systems, in "Nonlinear Controllability and Optimal Control", H. J. Sussmann (ed.), 177-218, Marcel Dekker, New York-Basel.

9. Jakubczyk, B (1992) Microlocal Feedback Invariants, preprint.

10. Jakubczyk, B. (1998) Critical Hamiltonians and Feedback Invariants, in "Geometry of Feedback and Optimal Control", B. Jakubczyk and W. Respondek (eds.), 219-256, Marcel Dekker, New York-Basel.

11. Jakubczyk, B. and Respondek, W. (1980) On linearization of control systems, Bull. Acad. Polon. Sci. Ser. Sci. Math. 28, 517-522.

12. Jakubczyk, B. and Respondek, W. (1991) Feedback Classification of Analytic Control Systems in the Plane, in "Analysis of Controlled Dynamical Systems", B. Bonnard et al. eds., 263-273, Progress in Systems and Control Theory 8, Birkhäuser, Boston-Basel.

13. Kang, W. and Krener, A.J. (1992) Extended Quadratic Controller Normal Form and Dynamic Feedback Linearization of Nonlinear Systyems, SIAM J. Control and Optimiz. 30, 1319–1337.
14. Kupka, I (1992) On Feedback Equivalence, Canadian Math. Society Conference Proceedings, Vol. 12, 1992, 105–117.
15. Kupka, I. (1994) Linear Equivalence of Curves, manuscript.
16. Marino, R., Boothby, W.M. and Elliott, D.L. (1985) Geometric Properties of Linearizable Control Systems, Math. Systems Theory 18, 97-123.
17. Montgomery, R. (1995) A Survey on Singular Curves in Sub-Riemannian Geometry, Journal of Dynamical and Control Systems 1, 49-90.
18. Nijmeijer H. and van der Schaft A.J. (1990), Nonlinear Dynamical Control Systems, Springer Verlag, New York.
19. Respondek W. and Zhitomirskii M. (1996) Feedback classification of Nonlinear Control Systems on 3-Manifolds, Mathematics of Control, Signals, and Systems 8, 299-333.
20. Sontag, E. (1990) Mathematical Control Theory: Deterministic Finite Dimensional Systems, Springer Verlag, New York.
21. Sussmann, H.J. (1974) An Extension of a Theorem of Nagano on Transitive Lie algebras, Proc. Am. Math. Soc. 45, 349-356.
22. Tall, I.A. and Respondek, W. (2000) Transforming Nonlinear Single-Input Control Systems to Normal Forms via Feedback, MTNS-2000.
23. Zhitomirskii M and Respondek W. (1998) Simple Germs of Corank 1 Affine Distributions, Banach Center Publications 44, Warsaw.

Paths in Sub-Riemannian Geometry

Frédéric Jean

Ecole Nationale Suprieure de Techniques Avances
Laboratoire de Mathématiques Appliquées
32, bd Victor
75739 Paris cedex 15, France
fjean@ensta.fr

Abstract. In sub-Riemannian geometry only horizontal paths – i.e. tangent to the distribution – can have finite length. The aim of this talk is to study non-horizontal paths, in particular to measure them and give their metric dimension. For that we introduce two metric invariants, the entropy and the complexity, and corresponding measures of the paths depending on a small parameter ε.

We give estimates for the entropy and the complexity, and a condition for these quantities to be equivalent. The estimates depend on a ε-norm on the tangent space, which tends to the sub-Riemannian metric as ε goes to zero. The results are based on an estimation of sub-Riemannian balls depending uniformly of their radius.

1 Length of a Path

Let M be a real analytic manifold and $X_1, \ldots, X_m$ analytic vector fields on M. We define a *sub-Riemannian metric* g on M by setting, for each $q \in M$ and $v \in T_q M$,

$$g_q(v) = \inf \left\{ u_1^2 + \cdots + u_m^2 \;\middle|\; \sum_{i=1}^{m} u_1 X_1(q) + \cdots + u_m X_m(q) = v \right\} .$$

The *length* of an absolutely continuous path $c(t)$ $(0 \leq t \leq \tau)$ is defined as

$$\text{length}(c) = \int_0^\tau \sqrt{g_{c(t)}(\dot{c}(t))} dt .$$

The *sub-Riemannian distance* is $d(p, q) = \inf \text{length}(c)$, where the infimum is taken on all the absolutely continuous paths joining p to q.

An *horizontal path* is an absolutely continuous path tangent almost everywhere to the distribution $\langle X_1, \ldots, X_m \rangle$. Only horizontal paths can have finite length. We are interested in the following questions: how to measure non-horizontal paths? how to compare the length of two such paths? More generally, which values can take metric invariants like Hausdorff dimension, entropy?

For an equiregular sub-Riemannian manifold, the Hausdorff dimension of submanifolds is known (Gromov [2]). However there are neither estimates of the entropy nor results in the general case.

In this talk, we will look at things from the point of view of approximating a path by finite sets. According to the chosen kind of finite set, a net or a chain, we introduce two metric invariants of a path, the entropy and the complexity, and the corresponding measures of the path.

We give computable estimates of these quantities (Theorems 2 and 3) and show in which case they are equivalent (Corollary 1). The result on the entropy allows in particular to compute Hausdorff and entropy dimensions. The estimates appear as integrals of some ε-norm of the tangent to the path. This ε-norm comes from a description of the shape of the sub-Riemannian balls depending uniformly of their radius (Theorem 1).

2   Nets and Chains

From now on we restrict the definition of path. What we call a path here is an analytic parameterized curve $c : [0, 1] \to M$.

Let c be a path in M. For $\varepsilon > 0$, the set $Z \subset M$ is called an ε-net for c if for any q in c there is $z \in Z$ with $d(q, z) \leq \varepsilon$. We define the *metric entropy* $e(c, \varepsilon)$ as the minimal number of the elements in ε-nets for c. It is the minimal number of closed balls of radius ε needed to cover c.

The notion of entropy has been introduced by Kolmogorov [4]. Notice that it is usually defined as the logarithm of $e(c, \varepsilon)$.

The asymptotic behavior of $e(c, \varepsilon)$ as ε tends to 0 reflects the geometry of c in M. This behavior is characterized essentially by the *entropy dimension*

$$\dim_e c = \varlimsup_{\varepsilon \to 0} \frac{\log e(c, \varepsilon)}{\log(\frac{1}{\varepsilon})} \ .$$

In other words $\dim_e c$ is the infimum of β for which $e(c, \varepsilon) \leq (1/\varepsilon)^{\beta}$ for ε small enough.

A maybe more usual characterization of the geometry of a space uses the Hausdorff dimension and measure. The entropy is however easier to evaluate. Moreover in our case the Hausdorff dimension can be deduced from the entropy dimension thanks to the following properties: $\dim_H c \leq \dim_e c$ and, if $e(c, \varepsilon) \sim \varepsilon^{-\beta_0}$, $\dim_H c = \dim_e c = \beta_0$.

For $\varepsilon > 0$, we call ε-*chain* for c a sequence of points $v_0 = c(0), v_1, \dots, v_k = c(1)$ in c where $d(v_i, v_{i+1}) \leq \varepsilon$ for $i = 0, \dots, k-1$. We define the *complexity* $\sigma(c, \varepsilon)$ as the minimum number of points in an ε-chain for c.

Finally, as suggested by Gromov [2, p. 278], we propose two definitions of an ε-length of a path:

$$\text{length}_\varepsilon^e(c) = \varepsilon \times e(c, \varepsilon) \qquad \text{and} \qquad \text{length}_\varepsilon^\sigma(c) = \varepsilon \times \sigma(c, \varepsilon) .$$

We want to compare these definitions. For that we use the notations $\succeq$, $\preceq$ to denote the corresponding inequalities $\geq$, $\leq$ up to multiplicative constants (uniform with respect to ε, for ε small enough). We say that f and g are equivalent, and we write $f \asymp g$, when $f \succeq g$ and $f \preceq g$.

Remark first that we can construct a 4ε-chain from an ε-net. This shows a first inequality

$$\sigma(c, \varepsilon) \preceq e(c, \varepsilon) .$$

To obtain a reverse inequality, an additional metric property of the path is needed. Let $\text{Tube}(c, \varepsilon)$ be the tube of radius ε centered at c. We say that a point q in the interior of c is a *cusp of c* if, for any constant $k \geq 1$, $\text{Tube}(c, \varepsilon)$ without the ball $B(q, k\varepsilon)$ is connected for ε small enough. We use the term cusp by analogy with curve's singularity. Notice however that here the path is smooth and that the property is a metric one.

Now, if c has no cusp, an ε-chain is a $k\varepsilon$-net for some k. This shows the following property.

Lemma 1. *For a path c without cusp, $\sigma(c, \varepsilon) \asymp e(c, \varepsilon)$.*

One can show that the cusps are isolated in c. We will see in Sect. 4 that we have to compare these points with another set of isolated points. Let us define it.

For $s \geq 1$ and $q \in M$, we denote by $L^s(q)$ the subspace of $T_q M$ spanned by values at q of the brackets of length $\leq s$ of vector fields $X_1, \ldots, X_m$. We say that $q \in c$ is a *regular point for c* if the integers $\dim L^s(q)$ $(s = 1, 2, \ldots)$ remain constant on c near q. Otherwise we say that q is a *singular point for c*. Singular points for c are isolated in c.

Notice that a cusp can be regular or singular for c and, conversely, that a singular point for c can be a cusp or not.

3   The ε-norm

Let $\Omega \subset M$ be a compact set (for instance a path). We denote by r the maximum of the degree of nonholonomy on Ω (recall that the degree of nonholonomy at q is the smaller s such that $\dim L^s(q) = n$).

Let $q \in \Omega$ and $\varepsilon > 0$. We consider the families of vector fields $(Y_1, \ldots, Y_n)$ such that each Y_j is a bracket $[[X_{i_1}, X_{i_2}], \ldots, X_{i_s}]$ of length $\ell(Y_j) = s \leq r$. On the set of these families, we have a function

$$\det \left(Y_1(q)\varepsilon^{\ell(Y_1)}, \ldots, Y_n(q)\varepsilon^{\ell(Y_n)} \right).$$

We say that the family $Y = (Y_1, \ldots, Y_n)$ is associated with (q, ε) if it achieves the maximum of this function. In particular the value at q of a family associated with (q, ε) forms a basis of $T_q M$.

With this definition we can describe the shape of the sub-Riemannian balls.

Theorem 1. *There exist constants k, K and $\varepsilon_0 > 0$ such that, for all $q \in \Omega$ and $\varepsilon \leq \varepsilon_0$, if $Y = (Y_1, \ldots, Y_n)$ is a family associated with (q, ε), then,*

$$B_Y(q, k\varepsilon) \subset B(q, \varepsilon) \subset B_Y(q, K\varepsilon), \tag{1}$$

where $B_Y(q, \varepsilon) = \{ q \exp(x_n Y_n) \cdots \exp(x_1 Y_1), \ |x_i| < \varepsilon^{\ell(Y_i)}, \ 1 \leq i \leq n \}$.

This theorem is proved in [3]. The proof is inspired by works of Bellaïche [1] and Nagel, Stein and Wainger [5].

Remark 1. This theorem extends the classical Ball-Box Theorem (Bellaïche [1], Gromov [2]). Indeed, q being fixed, for ε smaller than some $\varepsilon_1(q)$, the estimate (1) is equivalent to the one of Ball-Box Theorem. However $\varepsilon_1(q)$ can be infinitely small for q close to a singular point, though (1) holds for $\varepsilon \leq \varepsilon_0$, independent of q.

The theorem suggests to introduce the following notation. Let $u \in T_q M$ and $Y = (Y_1, \ldots, Y_n)$ be a family of vector fields such that $(Y_1(q), \ldots, Y_n(q))$ is a basis of $T_q M$. We denote by $(u_1^Y, \ldots, u_n^Y)$ the coordinates of u in this basis. We define then *the ε-norm on $T_q M$* as

$$\|u\|_{q,\varepsilon} = \max_{\substack{Y \text{ associated} \\ \text{with } (q,\varepsilon)}} \left(\left[\frac{u_1^Y}{\varepsilon^{\ell(Y_1)}} \right]^2 + \cdots + \left[\frac{u_n^Y}{\varepsilon^{\ell(Y_n)}} \right]^2 \right)^{1/2}.$$

Notice that, for a fixed $\varepsilon > 0$, the ε-norm induces a Riemannian metric. This metric, multiplied by ε, tends to the sub-Riemannian metric as ε goes to zero.

When u is the tangent $\dot{c}(t)$ to a path, the dependence with respect to $c(t)$ is implicit and we write the ε-norm as $\|\dot{c}(t)\|_\varepsilon$. It becomes then a function of t, which is piecewise continuous and so integrable on $[0, 1]$. Notice that in this case we choose $\Omega = c$.

4 Estimate of Entropy and Complexity

The theorems presented here are proved in [3].

Theorem 2. *For any path $c \subset M$,*

$$e(c, \varepsilon) \asymp \int_0^1 \|\dot{c}(t)\|_\varepsilon \, dt \ . \tag{2}$$

This estimate of the entropy allows to compute the Hausdorff dimension. For instance for a path c containing no singular points for c, $\dim_H c$ is equal to the smallest integer β such that $\dot{c}(t) \in L^\beta(c(t))$ for all $t \in [0,1]$.

More generally, $\dim_H c$ belongs to the interval $[\beta_{\text{reg}}, \beta_{\text{sing}}[$, where

- β_{reg} is the smallest integer β such that $\dot{c}(t) \in L^\beta(c(t))$ for all point $c(t)$ regular for c;
- β_{sing} is the smallest integer β such that $\dot{c}(t) \in L^\beta(c(t))$ for all $t \in [0,1]$.

The Hausdorff dimension can take not only the integer values belonging to the interval, but also rational ones (see [3] for examples).

Theorem 3. *Let $t_1 < \cdots < t_k$ be the parameters of the points which are both a cusp and singular for c $(0 < t_1$ and $t_k < 1)$. The complexity of c satisfies*

$$\int_0^1 \|\dot{c}(t)\|_\varepsilon \, dt - \sum_{i=1}^k \int_{t_i-\varepsilon}^{t_i+\varepsilon} \|\dot{c}(t)\|_\varepsilon \, dt \ \preceq \ \sigma(c, \varepsilon) \ \preceq \ \int_0^1 \|\dot{c}(t)\|_\varepsilon \, dt \ . \tag{3}$$

It results from these two theorems that, for a path without points both singular and cusps, complexity and entropy are equivalent. It is not always the case: we give in [3] an example where they are not equivalent. The inequality (3) provides however a sufficient condition on the integral of the ε-norm for complexity and entropy to be equivalent (we set $t_0 = 0$ and $t_{k+1} = 1$):

$$\text{if} \quad \int_{t_{i-1}}^{t_i} \|\dot{c}(t)\|_\varepsilon \, dt \asymp \int_{t_{i-1}+\varepsilon}^{t_i-\varepsilon} \|\dot{c}(t)\|_\varepsilon \, dt \quad \text{for } i = 1, \ldots, k+1 \ , \tag{4}$$

$$\text{then} \quad \sigma(c, \varepsilon) \asymp e(c, \varepsilon) \ .$$

More generally, a necessary and sufficient condition can be derived from the definition of a cusp. Let $q \in c$ and $\varepsilon > 0$. We denote by $\varrho = \varrho(q, \varepsilon)$ the biggest r such that $\text{Tube}(c, \varepsilon)$ without the open ball $B(q, r)$ is connected. By definition, q is a cusp if $\varrho(q, \varepsilon) \succeq \varepsilon$.

We set $c^\varrho = c \cap B(q, \varrho)$. The difference between the complexity and the entropy is that $\sigma(c, \varepsilon)$ is equivalent to $\sigma(c \setminus c^\varrho, \varepsilon)$ when $e(c, \varepsilon)$ is equivalent to $e(c \setminus c^\varrho, \varepsilon) + e(c^\varrho, \varepsilon)$. Notice that $c \setminus c^\varrho$ is a union of disjoined paths, so we defined its complexity (resp. entropy) as the sum of the complexities (resp. entropy) of each one of these paths.

It results from Lemma 1 that the complexity and the entropy of a path c are equivalent if and only if, for any cusp $q \in c$, $e(c^\ell, \varepsilon) \preceq e(c \setminus c^\ell, \varepsilon)$. A consequence of Theorem 3 is that this condition is always satisfied at a cusp regular for c.

Corollary 1. *We have $\sigma(c, \varepsilon) \asymp e(c, \varepsilon)$ if and only if, for any cusp singular for c, $e(c^\ell, \varepsilon) \preceq e(c \setminus c^\ell, \varepsilon)$.*

Remark 2. It is in general difficult to evaluate ϱ and to decide if a point is a cusp. On the other hand, the integral of $\|\dot{c}(t)\|_\varepsilon$ – and then the entropy – is computable as soon as $c(t)$ is known. Thus we use Condition (4) in the following way. Let us compute the two integrals of Condition (4) at each point $c(t_i)$ singular for c and compare them:

- if the integrals are equivalent at each point singular for c, then $\sigma(c, \varepsilon) \asymp e(c, \varepsilon)$;
- if the integrals at $c(t_i)$ are not equivalent, then $c(t_i)$ is a cusp.

For the ε-lengths, we see that the two definitions are not equivalent: the length of a ε-chain can be smaller than $\text{length}_\varepsilon^e$. Theorem 2 suggests a third definition, equivalent to $\text{length}_\varepsilon^e$:

$$\text{length}_\varepsilon(c) = \int_0^1 \varepsilon \|\dot{c}(t)\|_\varepsilon \, dt \ .$$

When ε goes to zero, the limit of $\text{length}_\varepsilon(c)$ is the length of c which is infinite if c is not horizontal. We can then compare the lengths of two non-horizontal paths by computing the limit of the ratio of their ε-lengths.

References

1. Bellaïche A. (1996) The Tangent Space in Sub-Riemannian Geometry. In: Bellaïche A., Risler J.-J. (Eds), Sub-Riemannian Geometry. Progress in Mathematics, Birkhäuser.
2. Gromov M. (1996) Carnot-Carathéodory Spaces Seen from Within. In: Bellaïche A., Risler J.-J. (Eds), Sub-Riemannian Geometry. Progress in Mathematics, Birkhäuser.
3. Jean F. (1999) Entropy and Complexity of a Path in Sub-Riemannian Geometry. Technical Report, ENSTA, Paris.
4. Kolmogorov A. N. (1956) On Certain Asymptotics Characteristics of some Completely Bounded Metric Spaces. Soviet Math. Dokl. **108**, 385–388.
5. Nagel A., Stein E. M., Wainger S. (1985) Metrics Defined by Vector Fields. Acta Math. **155**, 103–147.

Observability of C^∞-Systems for L^∞-Single-inputs and Some examples.

Philippe Jouan

AMS, UPRES-A CNRS 6085
Université de Rouen
Mathématiques, site Colbert
76821 Mont-Saint-Aignan Cedex, France
Philippe.Jouan@univ-rouen.fr

Abstract. For single-input multi-outputs C^∞-systems conditions under which observability for every C^∞ input implies observability for every almost everywhere continuous, bounded input (for every L^∞ input in the control-affine case) are stated. A normal system is then defined as a system whose only bad inputs are smooth on some open subset of definition. When the state space is compact normality turns out to be generic and enables to extend some results of genericity of observability to nonsmooth inputs.

1   Introduction and definitions.

In the paper [3] J.P. Gauthier and I. Kupka proved that a real analytic system, observable for any real analytic input, is observable for any L^∞-input. Herein observable for an input means that, for this input, any two different initial states give rise to different outputs. In the C^∞ case it is indeed natural to ask the question: *If a C^∞-system is observable for any C^∞-input, is it observable for any L^∞-input?*

In general the answer is negative as it is shown by the following examples, which prove the existence, for any $r \in \mathbb{N}$, of systems that are observable for all C^{r+1}-inputs but possess bad C^r-inputs.

Example 1.

Let Σ be the system on $\mathbb{R}^2$ whose dynamics is defined by

$$\begin{cases} \dot{x}_1 = 1 \\ \dot{x}_2 = u \end{cases}$$

If the output was $h(x_1, x_2) = x_1^2 + x_2$ the system would be observable for every input. We will see that it is possible to choose the output in such a way that the system is observable for all C^{r+1}-inputs but possesses a "bad" C^r-input. For this purpose let r be an integer and let v be a C^r-function on $[0,1]$, nowhere C^{r+1}.

Let H be a C^∞-function defined in the plane which vanishes exactly on the union of the graphs of the functions

$$x_2(x_1) = \int_0^{x_1} v(t)dt \qquad \text{and} \qquad \widetilde{x_2}(x_1) = 1 + \int_0^{x_1} v(t)dt \qquad (1)$$

defined on $[0,1]$.

The output is the $\mathbb{R}^2$-valued mapping:

$$\begin{cases} y_1 = (x_1^2 + x_2)H(x_1, x_2) \\ y_2 = H(x_1, x_2) \end{cases}$$

The system defined in this way is observable for all C^{r+1}-inputs because for such an input the trajectories are C^{r+2}-functions of x_1 and none of them can lie in the graphes of the functions (1). Therefore the following quantities are known:

$$x_1^2(t) + x_2(t)$$
$$2x_1(t) + u(t)$$
$$x_1(t) \text{ hence also } x_2(t)$$

Nevertheless for the input v previously defined the outputs do not distinguish between the initial states $(0,0)$ and $(0,1)$ because the function H vanishes identically along these trajectories.

Example 2.

The previous example can be modified in order to obtain a non locally observable system (thanks to Professor Respondek for this idea). Consider again the C^r-input v, nowhere C^{r+1}, and consider also the sequence of points $\{(0, \frac{1}{n}); n \geq 1\}$; let H be a C^∞-function which vanishes exactly on the union (which is closed) of the graphs of the functions

$$x_2(x_1) = \int_0^{x_1} v(t)dt \qquad \text{and} \qquad \chi_n(x_1) = \frac{1}{n} + \int_0^{x_1} v(t)dt \qquad n \geq 1 (2)$$

defined on $[0,1]$.

The output is again:

$$\begin{cases} y_1 = (x_1^2 + x_2)H(x_1, x_2) \\ y_2 = H(x_1, x_2) \end{cases}$$

The system is again observable for all C^{r+1}-inputs because for such an input the trajectories are C^{r+2}-functions of x_1 and none of them can lie in the graphes of the functions (2). However the system is not locally observable at the point $(0,0)$ for the input v because any neighbourhood of $(0,0)$ contains points of the type $(0, \frac{1}{n})$ that are not distinguished from $(0,0)$ by the outputs.

Despite theses examples one can expect that for a *generic* system observability for every C^∞-input implies observability for every L^∞-input.

For this purpose we consider single-input, multi-outputs, C^∞-systems of the form:

$$\Sigma = \begin{cases} \dot{x} = F(x, u) \\ y = h(x) \end{cases} \tag{3}$$

where x belongs to a C^∞-manifold X, u belongs to $\mathbb{R}$, and y belongs to $\mathbb{R}^p$ ($p \geq 1$).

The vector-field F and the output mapping $h = (h_1, h_2, \ldots, h_p)$, from X to $\mathbb{R}^p$, are C^∞.

An input $u \in L^\infty([0, T_u[)$ and an initial condition $x \in X$ being given, the solution of (3) such that $x(0) = x$ is denoted by $x(t)$, for $t \in [0, T[$ where T, $0 < T \leq T_U$, is the maximum time for which this solution exists.

Definition 1 *The system Σ is said to be observable for an input $u \in L^\infty([0, T_u[)$ if:*

$$\forall (x_1, x_2) \in X \times X, \quad x_1 \neq x_2 \qquad \forall T > 0$$

$$\exists \tau \in [0, T[\quad such\ that \quad h[x_1(\tau)] \neq h[x_2(\tau)].$$

An input $u \in L^\infty([0, T_u[)$ is said to be bad for $x_1 \neq x_2$ on $[0, T[$ where $0 < T \leq T_u$ if $x_1(.)$ and $x_2(.)$ are defined on $[0, T[$ and

$$\forall t \in [0, T[\quad h[x_1(t)] = h[x_2(t)]$$

This definition is not the usual one, that can be found for instance in ([4]), because the purpose is not here to find the pairs of points that are distinguished by at least one input but to find the universal inputs, i.e. the inputs for which any two points are distinguished by the outputs. Moreover in the previous definition two different initial states are required to be distinguished instantaneously. It would have been possible to choose the weaker definition:

$$\exists \tau \in [0, T_u[\quad such\ that \quad h[x_1(\tau)] \neq h[x_2(\tau)].$$

But when dealing with observability for a *class* of inputs both definitions are equivalent if the class is stable under restriction to a subinterval (and both definitions are also equivalent for analytic systems).

The inputs we will deal with belong to the following classes:

- The set IL^∞ of real valued functions that are defined and L^∞ on an interval $[0, T_u[$ where $0 < T_u \leq +\infty$.
- The set $\Re$ of real valued functions that are defined, almost everywhere continuous and bounded on an interval $[0, T_u[$ where $0 < T_u \leq +\infty$. For $T_u < +\infty$ these functions are the Riemann-integrable ones.

- The set IC^r of real valued functions that are defined and C^r on an interval $[0, T_u[$ where $0 < T_u \leq +\infty$ and $r \in \mathbb{N} \cup \{\infty\}$.

Definition 2 *The system Σ is said to be L^∞-observable (resp. $\Re$-observable), (resp. C^r-observable) if it is observable for any input belonging to IL^∞, (resp. $\Re$), (resp. IC^r).*

2   The control-affine case.

2.1   Main result.

We consider in this section systems whose vector-field is affine w.r.t. the control:

$$\Sigma = \begin{cases} \dot{x} = f(x) + ug(x) \\ y = h(x) \end{cases} \tag{4}$$

We denote by $\mathcal{G}^k$, for $k \in \mathbb{N}$, the sets of functions defined by the following induction:

$$\mathcal{G}^0 = \{h_1, h_2, \ldots, h_p\}$$
$$\mathcal{G}^{k+1} = L_f \mathcal{G}^k \cup L_g \mathcal{G}^k$$

where $L_f \mathcal{G}^k$ (resp. $L_g \mathcal{G}^k$) is the set of derivatives of the elements of $\mathcal{G}^k$ w.r.t. the vector-field f (resp. g). The set $\mathcal{G}$ stands for the union of all the $\mathcal{G}^k$'s.

Theorem 1 *Let Σ be a control-affine system. Σ is assumed to be C^∞-observable. If $u \in L^\infty([0, T_u[)$ is a "bad input" that does not distinguish between the two different initial states x_1 and x_2 on the interval $[0, T[$, where $0 < T \leq T_u$, then:*

$$\forall \varphi \in \mathcal{G}, \quad \forall t \in [0, T[, \quad \varphi(x_1(t)) = \varphi(x_2(t)).$$

Sketch of the proof : the key point is to show that if one function $\varphi \in \mathcal{G}$ distinguish between the two different initial states x_1 and x_2, then the "bad" input u must be of class C^∞.

Remark and example.

In case of the bad inputs of a system are of class C^∞ the conclusion of theorem 1 is false. For instance consider the system:

$$\begin{cases} \dot{x}_1 = x_1 + u x_2 \\ \dot{x}_2 = x_2 \\ \\ y = x_1 \end{cases}$$

Clearly $u \equiv 0$ is a bad input. However the set $\mathcal{G}$ separates the points because for this system $g = x_2 \frac{\partial}{\partial x_1}$ and therefore $y = h(x_1, x_2) = x_1$, $L_g h = x_2$.

2.2 Genericity results.

We assume now that the state space X is a compact manifold. For every positive integer p (the number of outputs) we consider the set $\mathcal{CAS}^p = (C^\infty(X))^p \times V^\infty(X) \times V^\infty(X)$ of all single-input, p-dimensional-output C^∞-systems. The set $\mathcal{CAS}^p$ is endowed with the product of the C^∞-topologies.

Definition 3 *A control-affine system Σ is said to be normal if it verifies the following property: let $u \in L^\infty([0, T_u[)$ be a bad input for Σ; then u is almost everywhere equal to a function that is C^∞ on an open and dense subset of $[0, T_u[$.*

The following lemma (see [2]) enables to state and prove two genericity theorems.

Lemma 1 *The set of systems Σ belonging to $\mathcal{CAS}^p$ such that the mapping*

$$\Phi_\Sigma : \qquad X \longrightarrow \mathbb{R}^{2d+1}$$

$$\Phi_\Sigma = (h_1, L_f h_1, L_f^2 h_1, ..., L_f^{2d} h_1)$$

where $d = \dim X$ and h_1 is the first component of the output, is an embedding, is open and dense in $\mathcal{CAS}^p$.

Theorem 2 *The state space X is assumed to be compact. For any $p \geq 1$ the set of normal control-affine systems contains an open and dense subset of $\mathcal{CAS}^p$.*

Theorem 3 *The state space X is assumed to be compact. For $p \geq 2$ the set of L^∞-observable control-affine systems is residual, hence dense, in $\mathcal{CAS}^p$.*

3 The general case.

3.1 Main result.

For general systems we denote by $\mathcal{H}^k$, for $k \in \mathbb{N}$, the sets:

$$\mathcal{H}^0 = \{h_1, h_2, \ldots, h_p\}$$

$$\mathcal{H}^{k+1} = L_F \mathcal{H}^k \cup \frac{\partial}{\partial u} \mathcal{H}^k$$

where $L_F \mathcal{H}^k$ (resp. $\frac{\partial}{\partial u} \mathcal{H}^k$) is the set of derivatives of the elements of $\mathcal{H}^k$ w.r.t. the vector-field F (resp. w.r.t. $\frac{\partial}{\partial u}$). The set $\mathcal{H}$ is the union of all the $\mathcal{H}^k$'s.

Theorem 4 *Let Σ be a C^∞-observable system. If u is an almost everywhere continuous and bounded "bad input" defined on $[0, T_u[$ that does not distinguish between the two different initial states x_1 and x_2 on the interval $[0, T[$, where $0 < T \leq T_u$, then:*

$$\forall \varphi \in \mathcal{H} \qquad \varphi(x_1(t), u(t)) = \varphi(x_2(t), u(t)), \qquad \text{for a.e. } t \in [0, T[.$$

Theorem 4 does not hold for L^∞-inputs because it is not true that under the assumptions of this theorem

$$\varphi(x_1(t), u(t)) = \varphi(x_2(t), u(t)) \quad \text{for a.e. } t \in [0, T[$$

implies

$$\frac{\partial}{\partial u} \varphi(x_1(t), u(t)) = \frac{\partial}{\partial u} \varphi(x_2(t), u(t)) \quad \text{for a.e. } t \in [0, T[.$$

Counter-example.

Let Σ be the system on $\mathbb{R}^3$ whose dynamics is defined by

$$\begin{cases} \dot{x}_1 = 1 \\ \dot{x}_2 = u \\ \dot{x}_3 = (x_1 - u)(1 - \phi(u))x_2 \end{cases}$$

where ϕ is an increasing C^∞-function whose value is 0 on $]-\infty, 1]$ and 1 on $[2, +\infty[$.

Let K be a compact subset of $[0, 1]$ with empty interior and positive Lebesgue measure. Let v be a continuous function on $[0, 1]$, nowhere derivable, and such that $\forall t \in [0, 1]$ $v(t) \geq 2$. We consider the function $\bar{u}$ defined on $[0, 1]$ by

$$\bar{u}(t) = t\chi_K(t) + v(t)\chi_{K^c}(t)$$

where χ_A stands for the characteristic function of the subset A of $\mathbb{R}$.

Let H be a C^∞-function defined in the (x_1, x_2)-plane which vanishes exactly on the union of the graphes of the functions

$$x_2(x_1) = \int_0^{x_1} u(t)dt \quad \text{and} \quad \widetilde{x_2}(x_1) = 1 + \int_0^{x_1} u(t)dt \tag{5}$$

defined on $[0, 1]$. The output of the system is the $\mathbb{R}^4$-valued mapping:

$$\begin{cases} y_1 = x_3 \\ y_2 = H(x_1, x_2) \\ y_3 = x_1 H(x_1, x_2) \\ y_4 = x_2 H(x_1, x_2) \end{cases}$$

The system defined in this way is C^∞-observable, the L^∞-input $\bar{u}$ is bad for the initial conditions $x = (0, 0, 0)$ and $\tilde{x} = (0, 1, 0)$ on $[0, 1]$ but does not verify the conclusion of theorem 4.

For these initial conditions

$$\frac{\partial}{\partial u} L_F h(x(t), \bar{u}(t)) - \frac{\partial}{\partial u} L_F h(\tilde{x}(t), \bar{u}(t))$$

does not vanish on K, therefore does not vanish everywhere almost everywhere since the Lebesgue measure of K is positive.

3.2 Genericity results.

The state space X is assumed to be compact.

A bound $B > 0$ is fixed, the inputs are restricted to the compact real interval $I_B = [-B, B]$, and for $r \in \mathbb{N} \cup \{+\infty\}$ we denote by $V_B^r(X)$ the set of C^r-vectorfields on X parametrized by $u \in I_B$.

In view of the forthcoming lemma 2 we first consider the set $C^r(X) \times V_B^r(X)$ of single-input C^r-systems. If $r \geq 2d$, where $d = \dim(X)$, and for $\Sigma = (h, F) \in C^r(X) \times V_B^r(X)$, we can define the mapping

$$\Psi_\Sigma \quad : \quad X \times I_B \longmapsto \mathbb{R}^{2d+3} \qquad \text{by}$$

$$\Psi_\Sigma(x, u) = \left(h(x), L_F h(x, u), L_F^2 h(x, u), ..., L_F^{2d} h(x, u), \frac{\partial}{\partial u} L_F h(x, u), u \right)$$

Lemma 2 *For r large enough the set of systems Σ belonging to $C^r(X) \times V_B^r(X)$ such that the mapping Ψ_Σ is an embedding is open and dense in $C^r(X) \times V_B^r(X)$.*

Let us denote by $S_B^p = (C^\infty(X))^p \times V_B^\infty(X)$ the set of all single-input, p-dimensional-output C^∞-systems on the compact $X \times I_B$. The set S_B^p is endowed with the product of the C^∞-topologies. $\Re_B$ is the set of I_B-valued elements of $\Re$.

Definition 4 *A system Σ is said to be $\Re_B$-normal if it verifies the following property: let $u \in \Re_B$ be a bad input for Σ defined on $[0, T_u[$; then u is C^∞ on an open and dense subset of $[0, T_u[$.*

Theorem 5 *For any $p \geq 1$ the set of $\Re_B$-normal systems contains an open and dense subset of S_B^p.*

Theorem 6 *For $p \geq 2$ the set of $\Re_B$-observable systems is residual, hence dense, in S_B^p.*

582   Philippe Jouan

References

1. M. Balde and Ph. Jouan, *Genericity of observability of control-affine systems*, Control Optimisation and Calculus of Variations, Vol. 3, 1998, 345-359.
2. J.P. Gauthier, H. Hammouri and I. Kupka, *Observers for nonlinear systems*, IEEE CDC conference, december 1991, 1483-1489.
3. J.P. Gauthier and I. Kupka, *Observability for systems with more outputs than inputs and asymptotic observers*, Mathematische Zeitschrift 223, 1996, 47-78.
4. R. Hermann and A.J. Krener, *Nonlinear controllability and observability*, IEEE Trans. Aut. Control, AC-22, 1977, 728-740.
5. Ph. Jouan, *Observability of real analytic vector-fields on compact manifolds*, Systems and Control Letters 26, 1995, 87-93.
6. Ph. Jouan, C^∞ and L^∞ *observability of single-input C^∞-Systems*, submitted to the Journal of Dynamical and Control Systems.
7. Ph. Jouan and J.P. Gauthier, *Finite singularities of nonlinear systems. Output stabilization, observability and observers*, Journal of Dynamical and Control Systems, vol.2, N° 2, 1996, 255-288.

Robust Control of a Synchronous Power Generator

Matei Kelemen, Aimé Francis Okou, Ouassima Akhrif, and
Louis-A. Dessaint

École de Technologie Supérieure GRÉPCI
Electrical Engineering Department
1100, rue Notre Dame Ouest
Montréal, Québec, H3C 1K3 Canada
mkelemen@ele.etsmtl.ca

Abstract. In this article we apply simple robust linear control to a synchronous power generator driven by a hydraulic turbine and connected to an infinite bus. The goal is to obtain good performance simultaneously for the terminal voltage and the rotor speed for short circuits and also when abrupt, large and permanent variations occur in the parameters of the infinite bus and transmission line. To achieve the goal we increase considerably the crossover frequencies of the terminal voltage and rotor speed feedback loops by taking into account some physical limitations on the controls. The proposed control - of Quantitative Feedback Theory type - is compared in simulations with a classical linear control and a modern nonlinear one.

1 Introduction

An important problem in the functioning of power systems is maintaining good stability and transient performance (of the terminal voltage, rotor speed, power transfer to the network) when confronted with significant disruptions from normal operating regimes. This requirement should be achieved by an adequate control of the system, in our case robust control due to the nature of the disruptions explained below.

In this article we address this problem for a single machine connected to an infinite bus (load) which may have short circuits and also sudden, large and permanent variations in the voltage and phase of the bus and the parameters of the transmission line. Simulations on the realistic dynamic model we have worked with [1] (nonlinear, 9 states, practical limitations on controls) showed that even in open loop the generator enjoys very good (bounded input bounded output) stability, but not robustness (the terminal voltage differed considerably of its desired value after a permanent change in the voltage of the infinite bus). The robustness problem was solved naturally in power engineering by using linear classical controllers based on (Taylor) linearization of the plant model around an operating point, ensuring good phase and gain margins [7], [9], [2], a fact perhaps not enough emphasized. The crossover frequency, however, was quite low which limited the transient performance of these compensators.

A significant improvement in the dynamic response of a single machine infinite bus system to a short circuit disturbance was achieved recently by using a multi input multi output (MIMO) nonlinear (partial feedback linearizing) control, [1]. Due to the exact cancellation of some of the plant nonlinearities this method had to renounce both the robustness and the simplicity of the control (the measurement of all 9 states was necessary to compute the nonlinear controller having at least the complexity of the plant).

In this article - building upon the good results of the mentioned nonlinear control - we show that it is possible to obtain the good robustness property of the classical control and the good dynamic response of the nonlinear one by using two linear compensators of order one each. They require two standard measurements, the terminal voltage and the speed, the control inputs being respectively the excitation and gate servomotor voltages. That is, our controllers are of automatic voltage regulator (AVR) and speed governor type. The compensators have almost the same performance at various operating points (power transfer levels), and the control effort is moderate. Technically we increase considerably the crossover frequencies of our designs: for the speed we use a new way to shape its feedback loop aiming to satisfy some physical constraints on the control input; for the voltage a high gain feedback is used. We note that this result was possible due to the good inherent properties of the power system (stable, minimum phase in the voltage, stable, robust in the speed). For design we have applied the Quantitative Feedback Design Theory (QFT) [5], a frequency domain graphical design methodology for robust control suitable for our purpose.

The results presented here qualify this type of linear robust compensators to the control of more complex power systems. We note that such a control was recently employed with good results in a realistic situation [8].

2   Mathematical Model

Our plant is a synchronous power generator driven by a hydraulic turbine. The generator is connected to an infinite bus by a transmission line. During the operation of this system two types of disruptions may occur: a symmetrical three-phase short circuit at the infinite bus, and abrupt, large and permanent changes in the resistance and inductance of the transmission line and the voltage and phase of the infinite bus, Fig. 1. The mathematical model of the plant we shall use is an early version of the one presented in some detail in [1] and based on formula (4.74) of [2, p. 101], therefore we shall not insist upon its derivation. The main difference is that we neglected the fast dynamics of the damper winding currents in the mathematical model but not

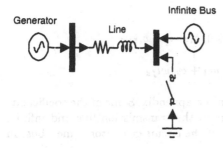

Fig. 1. Single machine infinite bus

in the simulation one, and worked with the gate dynamics given in Fig. 3. In state space form the resulting system is highly nonlinear not only in the state but in the input and output as well. The 8 state, 2 input and 2 output mathematical model has the following form

$$\dot{x} = f(x, u), \quad y = g(x, u),$$

where $x = [i_d \ i_q \ i_{fd} \ \delta \ w \ g \ \dot{g} \ q]^T$ is the state, $u = [u_f, u_g]^T$ is the control input and $y(x, u_f) = [(v_t(x, u_f), w(x))]$ is the output. The output is available for measurement and its behavior should be improved by a properly designed feedback.

Below we shall describe mathematically the model in Park coordinates and provide the physical significance of its elements. The coefficients and the parameters of the system are presented in the appendix. The parameters (typical for hydro power plants) were normalized to per unit - or p. u. -, except where the dimension is explicitly given. Thus

$$\frac{di_d}{dt} = A_{11}i_d + A_{12}i_{fd} + A_{13}wi_q + A_{14}cos(\delta - \alpha) + g_{11}u_f \tag{1}$$

$$\frac{di_q}{dt} = A_{21}wi_d + A_{22}wi_{fd} + A_{23}i_q + A_{24}sin(\alpha - \delta)$$

$$\frac{di_{fd}}{dt} = A_{31}i_d + A_{32}i_{fd} + A_{33}wi_q + A_{34}cos(\delta - \alpha) + g_{31}u_f$$

$$\frac{d\delta}{dt} = (w - 1)w_R$$

$$\frac{dw}{dt} = A_{51}i_q i_d + A_{52}i_q i_{fd} + A_{53}w + A_{54}T_m$$

$$\frac{dq}{dt} = \frac{1}{T_w}\left(1 - \frac{q^2}{G_A^2}\right), \quad G_A = A_t g,$$

where

$$T_m = P_m/w = q^3/(G_A^2 w)$$

$$v_t = (v_d^2 + v_q^2)^{1/2}$$

$$v_d = c_{11}cos(\delta - \alpha) + c_{12}i_d + c_{13}i_{fd} + c_{14}wi_q + c_{15}u_f$$

$$v_q = c_{21}sin(-\delta + \alpha) + c_{22}i_q + c_{23}wi_d + c_{24}wi_{fd}$$

and the constant factor A_t is defined in the appendix. Some of the coefficients from above depend on the parameters of the transmission line and infinite bus as a consequence of the equations of the circuit generator - line - bus. In (1)

t : time (s); i_d : direct axis current; i_q : quadrature axis current;

i_{fd} : field winding excitation current; δ : rotor (power) angle (rad);

w : rotor speed; g : opening of the turbine gate;

q : water flow rate to the turbine; T_m : turbine mechanical torque;

P_m : turbine mechanical power; u_f : field winding excitation voltage;

u_g : gate servomotor voltage; v_t : terminal (stator) voltage magnitude;

v_d : direct axis terminal voltage; v_q : quadrature axis terminal voltage.

Finally, there are also some practical limitations: on the control u_f - its magnitude, and on the gate opening g - its magnitude and speed. The actual values of the limits are listed in the appendix at the plant parameters.

Uncertain Parameters of the Environment

As mentioned, the environment of our power generator is an infinite bus connected to it by transmission lines. In this work we consider for control design that their parameters, V_{inf}, α, R_e, L_e, may have abrupt and permanent variation of up to 10 % of the nominal values given in the appendix. However, we allow variation up to 40 % to test our designs.

Also, a short circuit of up to 100 ms may occur at the infinite bus.

3   Feedback Control Design

Feedback Control Problem

To obtain good performance despite the disruptions to the normal operation of the power system described above we use a feedback structure with two

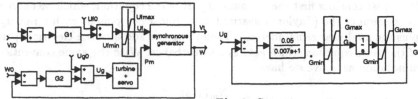

Fig. 3. Gate servomotor

Fig. 2. QFT control structure

loops: one electrical based on the measurement of v_t and actuated by u_f, and a mechanical one with the measurement of w and actuation by u_g, see Fig. 2 and 3. Our design method requires quantitative specifications to be achieved. These are provided by the time domain simulation obtained with a nonlinear controller for the same plant [1], see Fig. 6 and 7.

Open Loop Transfer Functions

We write our nonlinear MIMO plant in the following way. Let p be an arbitrary point in the set of uncertain parameters and associate the index 0 to all nominal values. Then consider a nominal operating point e_0, i.e. an equilibrium of (1) (7 states and 2 inputs) computed for p_0, and with $v_{t0} = 1$, $w_0 = 1$, and P_{m0} given. Working with displacements from the nominal values we denote: $v_{t1} = v_t - v_{t0}$, $w_1 = w - w_0$, $u_{f1} = u_f - u_{f0}$ and $u_{g1} = u_g - u_{g0}$.

Now in standard Laplace transform notations (a capital letter denotes the transform of a corresponding time domain function, and s is the transform argument) we have

$$V_{t1}(s, p) = P_1(s, e_0)U_{f1}(s) + N_{el}(s, p).$$

We obtained this formula by expressing v_{t1} as a sum of the (Taylor) linearized at e_0 linear component with input u_{f1} plus the nonlinear remainder n_{el} which depends on the uncertain parameters, and on the displacements from the nominal values of the states and inputs; for ease of notation we single out only the dependence on p. In n_{el} we absorb also the additive short circuit disturbance. The term n_{el} has not a useful formal expression, nevertheless, its effect will be taken into account via the open loop simulations from the beginning of voltage loop design.

We could apply the Laplace transform since we work with stable operating points. The transfer function P_1 is available numerically. For the chosen operating point, $P_{m0} = 0.7303$,

$$P_1(s) = \frac{0.1s^3 + 143s^2 + 24914s + 19688723}{s^3 + 75s^2 + 143554s + 73219}; \tag{2}$$

it is stable, minimum phase and proper.

For w_1 we consider first the mechanical part of the equation, which we express as the sum of the (Taylor) linearized at e_0 linear component with input u_{g1} plus the nonlinear remainder n_m. Then in n_m we add also the electrical part of the same equation which acts as a disturbance. With the same convention of notation as above we have

$$W_1(s,p) = P_2(s,e_0)U_{g1}(s) + \frac{N_m(s,p)}{s + A_{54}G_{A0}},$$

$$P_2(s,e_0) = \frac{A_t A_{54}}{s + A_{54}G_{A0}}, \tag{3}$$

where $G_{A0} = P_{m0}$ in steady state hence 0.7303 in our case. The following observations enable us to simplify the design.

Observation 1. We consider the nonlinear terms as disturbances to the plant (they contain the actual ones), see [5, section 11.8]. Hence to the extent we can attenuate them we may work with a decoupled linear MIMO plant.

Observation 2. For $P_2(s)$, if the gate movement was free, one would obtain a nonminimum phase property, i.e. a positive zero $z_0 = 1/(T_w A_t g_0)$ of the transfer function. This undesired property is removed (in design) by a slower movement of the gate, see Fig. 3, and by using the result of [6].

How fast this input may become will be determined *quantitatively* below.
Closed Loop Transfer Functions

From Fig. 2 and 3 we have

$$U_{f1}(s,p) = -G_1(s,e_0)V_{t1}(s,p),$$

$$U_{g1}(s,p) = -G_2(s,e_0)W_1(s,p). \tag{4}$$

Further simplifying the notation we shall keep only the Laplace argument; then we have the decoupled expressions (observation 1) of the two closed loops

$$V_{t1}(s) = \frac{1}{1 + G_1(s)P_1(s)}N_{el}(s), \tag{5}$$

$$W_1(s) = \frac{P_2(s)}{1 + G_2(s)P_2(s)}(N_m(s)/A_t A_{54}). \tag{6}$$

We perform the design in the frequency domain; ω (rad/s) denotes the frequency variable. For numerical computations we use the QFT Toolbox [3]. These are based on the Nichols chart (logarithmic complex plane) with axes the magnitude (dB) and phase (degree) of a complex number.

The transfer function $P_2(s)$ does not depend on the uncertain parameters while $P_1(s)$ has a large stability margin, being proper. Therefore the stability margin bound in the Nichols chart should be the same for all frequencies.

Design of the Speed Feedback Loop

In the design we have to take care of the observation 2 from above, i.e. we should produce a "slow" movement of the gate, via a "slow" control input, u_g, to the gate servomotor. One way to achieve this is to introduce a not very far from zero pole in G_2, and then determine the corresponding compensator gain. This process is iterative.

Our approach here is different: we extract frequency domain information from the time domain simulation in Fig. 7 (our specification) obtained with the nonlinear control [1], and for which all the constraints on the gate movement were in operation. Thus comparing this simulation with standard second order system responses we see that the damping ratio $\xi = 0.5$ and the natural frequency $\omega_n = 10$ rad/s are appropriate for it. In frequency domain this damping ratio corresponds, via formula (9.16) [4, p. 294], to the inequality

$$|\frac{G_2(s)P_2(s)}{1 + G_2(s)P_2(s)}|_{dB} \le 1.25 \ dB, \tag{7}$$

for all $s = j\omega$, $\omega \in R$, which limits the peak resonance of the closed loop. The peak appears at the frequency $\omega_m = 7$ rad/s according to formula (9.15) in the same reference. Then our frequency domain design goal is to attenuate the disturbance in (6) in the range of frequency [0 7] while satisfying the stability margin condition (7) as well. Now a straightforward loop shaping procedure led us to the compensator

$$G_2(s) = \frac{55}{s/9 + 1}.$$

In Fig. 4 the loop transmission $G_2(j\omega)P_2(j\omega)$ avoids from the left the closed bound (stability margin) to achieve also the Nyquist stability of the closed loop.

We note that the range of frequency on which the disturbance is attenuated translates, via (4) and (6), to the crossover frequency (bandwidth) of the transfer function from the (mechanical) disturbance to u_{g1}

$$\frac{-G_2(s)P_2(s)}{1 + G_2(s)P_2(s)}.$$

Remark:

• The usefulness of taking into account some practical constraints on the controls and using a loop shaping method aiming to achieve them can be seen in the *big increase* in the crossover frequency and a good phase margin compared

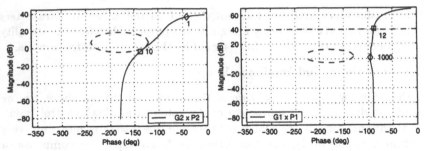

Fig. 4. Nichols chart: design of G_2 **Fig. 5.** Nichols chart: design of G_1

to classical linear control; this in turn leads to improved dynamical performance of the speed. Indeed, from the Fig. 4 we read the crossover frequency ≈ 7.2 rad/s and the phase margin ≈ 53 deg. On the other hand a straightforward computation shows that the transfer function of the (Taylor) linearized turbine is nonminimum phase having a positive zero $z_0 = 1/(T_w G_{A0})$ ≈ 0.5 rad/s, which restricts the crossover frequency at around $z_0/2 = 0.25$ rad/s for a phase margin of 40 deg, see [5, p. 235]. This result is compatible with [7, p. 402] and with [9, p. 175] Table 1 for smaller phase margins.

Design of the Voltage Feedback Loop

Time domain simulations in open loop showed that the maximal steady state error of v_t compared to v_{t0} obtained over the vertices of the uncertainty set (tested for abrupt and maximal variations from p_0) was about 10 %, and the oscillation of the response was around 4 rad/s. In view of the high accuracy and fast response required, see the nonlinear control result in Fig. 6, we attenuate the disturbance 100 times and over a larger frequency range than the open loop simulation indicates. Thus with (5) we get

$$\left|\frac{1}{1 + G_1(s)P_1(s)}\right|_{dB} \leq -40 \ dB,$$

for $s = j\omega$, $\omega \in [-12, 12]$. We satisfy also a standard stability margin condition

$$\left|\frac{G_1(s)P_1(s)}{1 + G_1(s)P_1(s)}\right|_{dB} \leq 2.3 \ dB,$$

for all $s = j\omega$, $\omega \in R$. By a usual rational function approximation to satisfy the two specifications from above and Nyquist stability (as before) we obtained the compensator, Fig. 5,

$$G_1(s) = \frac{10}{s/1000 + 1},$$

where the pole was introduced to have a strictly proper voltage loop (P_1 is proper) which gives a smoother solution.

Remarks

• The transfer function $P_1(s)$ is minimum phase therefore, in principle, one *can* obtain arbitrary good performance in v_t for arbitrary large disturbances if the constraint on u_f is removed. In reality, for good response at very large disturbances, the limited capability of the speed loop may call via the non-linear term N_{el} for higher and impractical control values of u_f, without being able to save the mechanical subsystem stability from the pressure of the N_m action. In our tests we obtained acceptable performance until the variation in V_{inf}, α, R_e and L_e was increased up to 35 %, however after 40 % the system (P_m) began to destabilize, i.e. we encountered a *bifurcation*.

• Our compensators display *almost similar* performance for a large class of stable operating points. This is due mainly to the good phase margin and crossover frequency they ensure (each in its own range) for the closed loops. This fact is illustrated in Fig. 12 to 15 for a worst case situation when P_{m0} is close to its maximal admissible value.

4 Simulation Results

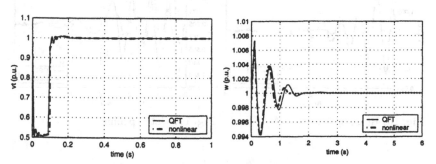

Fig. 6. Terminal voltage   **Fig. 7.** Rotor speed

A Simulink 9 state, nonlinear, dynamic model - with all the physical constraints on the control inputs in operation - of the power system (turbine, generator, transmission line, infinite bus) is used to test our linear robust compensators. For the operating point chosen, mechanical power $P_{m0} = 0.7303$, the steady state rotor angle is $\delta_0 = 17.1098$ deg. In the simulations presented below we compare the linear robust control - denoted QFT - to the best feature of the nonlinear control (dynamic performance) and linear classical control (robustness).

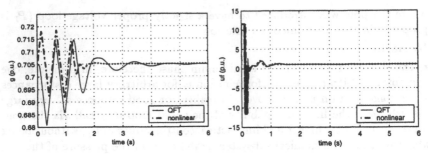

Fig. 8. Gate opening   **Fig. 9.** Excitation voltage

The simulations in Fig. 6 to 9 are for the nominal values of the parameters but with a symmetrical three-phase short circuit at the infinite bus of 100 ms at the moment $t = 0$, and the linear robust control is compared to the nonlinear one [1]. It may be seen that the results are very close, the speed of the nonlinear control stabilizing slightly ahead of the linear robust control. (The power angle is not represented being the integral of the speed variable.) In

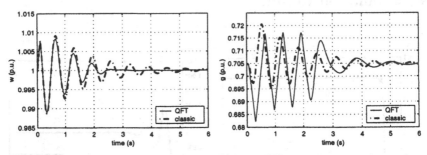

Fig. 10. Rotor speed   **Fig. 11.** Gate opening

Fig. 10, 11 in addition to the short circuit an abrupt and permanent variation from the nominal value of (worst case) -10 % in R_e, V_{inf}, α and $+10$ % in L_e at the moment $t = 0$ is simulated for the linear robust control and a linear classical control. For classical speed governor we consider a standard PID controller [10], while the AVR is our $G_1(s)$ (the simplest classical one is a pure gain). Therefore we represent only the mechanical variables, speed and gate movement; the electrical variables are presented in the context of Fig. 12 - 15. One can see that the classical control withstands all the disruptions as does the linear robust one, but has a slower dynamic response.

In Fig. 8, 9, 11 and 15 are represented the movement of the gate and the excitation voltage: u_f is a control input and g is in direct relation to the other control input u_g. We note (for all three control methods) that they saturate only for a short period of time - due to the limitations imposed on the controls - and do not oscillate much (especially the QFT and classical methods). Hence the control effort is moderate. Finally, in Fig. 12 to 15 we compare (in the

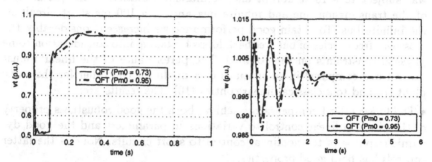

Fig. 12. Terminal voltage   **Fig. 13.** Rotor speed

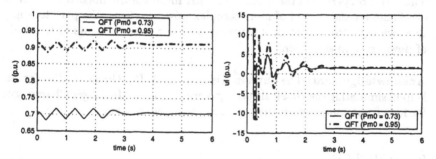

Fig. 14. Gate opening   **Fig. 15.** Excitation voltage

conditions of Fig. 10, 11) the performance of the QFT controllers at two "far" operating points: the initial $P_{m0} = 0.7303$ with $P_{m0} = 0.9459$. The results are indeed close.

The fragility of the nonlinear control not being surprising is not presented in simulation here. We mention only that for a permanent variation of $-10\ \%$ in R_e and $+10\ \%$ in L_e the mechanical power P_m destabilizes, and if a permanent variation of $-10\ \%$ in V_{inf} is added all the power system does not reach an acceptable steady state.

5 Conclusion

In this article we have solved the problem of simultaneous robust control of the terminal voltage and rotor speed of a synchronous power generator driven by a hydraulic turbine and connected to an infinite bus. The plant was subject to a 10 % abrupt and permanent variation in the parameters of the transmission line and infinite bus and to a 100 ms short circuit at the infinite bus. To obtain good performance we increased considerably the crossover frequencies of our feedback loops by taking into consideration some physical constraints on the controls. The performance of the linear robust control was compared in simulations to a nonlinear controller for the same plant [1], and to a classical linear control [10]. The results are:

• The linear robust compensators achieve both the good robustness property of the classical linear control to variations in parameters and the good dynamic response of the nonlinear control to short circuits. None of the latter controls has both good properties.

• The linear robust compensators are of order one each and require two standard measurements: the terminal voltage and the rotor speed. The compensators are of AVR and speed governor type. Moreover, they perform almost similarly at various operating points, and the control effort (input) is moderate.

The useful properties of the proposed robust linear control make it a plausible candidate for the control of more complex power systems. We plan to apply this technique to the control of networks of power generators and to the study of power system stabilizers.

Acknowledgement: The authors wish to thank Dr. Guchuan Zhu for his help in simulating the classical speed governor, and to Mr. Roger Champagne for his expert contribution with the computerized graphics.

6 Appendix

The Coefficients of the Mathematical Model

$$A_{11} = (R_s + R_e)L_{fd}w_R/a; \quad A_{12} = L_{md}R_{fd}w_R/a;$$

$$A_{13} = -L_{fd}(L_q + L_e)w_R/a; \quad A_{14} = L_{fd}V^\infty w_R/a;$$

$$A_{21} = -(L_d + L_e)w_R/b; \quad A_{22} = L_{md}w_R/b; \quad A_{23} = -(R_s + R_e)w_R/b;$$

$$A_{24} = -V^\infty w_R/b; \quad A_{31} = (R_s + R_e)L_{md}w_R/a;$$

$$A_{32} = R_{fd}(L_d + L_e)w_R/a; \quad A_{33} = -L_{md}(L_q + L_e)w_R/a;$$

$$A_{34} = L_{md}V^\infty w_R/a; \quad A_{51} = -(L_q - L_d)/(2H);$$

$$A_{52} = -L_{md}/(2H); \quad A_{53} = -F/(2H); \quad A_{54} = 1/(2H).$$

$$g_{11} = -L_{md}w_R/a; \quad g_{31} = -(L_d + L_e)w_R/a.$$

$$c_{11} = V^\infty + (L_e/w_R)A_{14}; \quad c_{12} = R_e + (L_e/w_R)A_{11}; \quad c_{13} = (L_e/w_R)A_{12};$$

$$c_{14} = -L_e + (L_e/w_R)A_{13}; \quad c_{15} = (L_e/w_R)g_{11};$$

$$c_{21} = V^\infty + (L_e/w_R)A_{24}; \quad c_{22} = R_e + (L_e/w_R)A_{23};$$

$$c_{23} = L_e + (L_e/w_R)A_{21}; \quad c_{24} = (L_e/w_R)A_{22}.$$

Here we have denoted

$$a = L_{md}^2 - L_{fd}(L_d + L_e); \quad b = L_q + L_e.$$

The Parameters of the Plant

$max|u_f| = 11.5:$ maximal excitation voltage;

$g_{min} = 0.01:$ gate minimal opening;

$g_{max} = 0.97518:$ gate maximal opening

$max|\dot{g}| = 0.1:$ gate maximal speed

$T_w = 2.67 \ (s):$ water time constant

$A_t = 1/(g_{max} - g_{min}):$ turbine gain

$1/k_a = (3/10) \ (s):$ gate servomotor time constant

$H = 3.19 \ (s):$ inertia coefficient of the rotor

$F \approx 0:$ friction coefficient of the rotor

$R_s = 0.003:$ stator winding resistance

$R_{fd} = 6.3581 \cdot 10^{-4}:$ rotor field resistance

$L_{fd} = 1.083:$ rotor self inductance

$L_d = 1.116:$ direct self inductance

$L_q = 0.416$:   quadrature self inductance

$L_{md} = 0.91763$:   direct magnetizing inductance

$L_{mq} = 0.21763$:   quadrature magnetizing inductance

$R_e = 0.0594$:   nominal transmission line resistance

$L_e = 0.2514$:   nominal transmission line inductance

$V^\infty = 0.99846$:   nominal magnitude of the voltage of the infinite bus

$\alpha = 78.72\ (deg)$:   nominal phase of the infinite bus

$w_R = (2\pi)60\ (rad/s)$:   reference speed of the rotor (1 p. u.).

The parameter L_{mq} does not appear explicitly in the coefficients of the model. However, it was used in evaluating the parameter L_q.

References

1. Akhrif O., Okou A. F., Dessaint L.-A., Champagne R. (1999) Application of a Multivariable Feedback Linearization Scheme for Rotor Angle Stability and Voltage Regulation of Power Systems. IEEE Trans. Power Systems, 14, 620 -629.
2. Anderson P. M., Fouad A. A. (1994) Power System Control and Stability. IEEE Press, New York.
3. Borghesani C., Chait Y., Yaniv Y. (1994) Quantitative Feedback Theory Toolbox. The Math Works Inc., Natick, Mass. 01760, USA.
4. D'Azzo J. J., Houpis C. H. (1995) Linear Control System Analysis and Design. McGraw-Hill, New-York.
5. Horowitz I. (1993) Quantitative Feedback Design Theory (QFT), 1. QFT Publications, Superior, Colorado 80027, USA.
6. Kelemen M. (1986) A Stability Property. IEEE Trans. Autom. Contr., 31 766-768.
7. Kundur P. (1994) Power System Stability and Control. McGraw Hill, Montreal.
8. Stanković A. M., Tadmor G., Sakharuk T. A. (1998) On Robust Control Analysis and Design for Load Frequency Regulation. IEEE Trans. Power Systems 13, 449 - 455.
9. IEEE Committee Report (1992) Hydraulic Turbine and Turbine Control Models for System Dynamic Studies. IEEE Trans. Power Systems, 7, 167 - 179.
10. Zhu G., Dessaint L.-A., Akhrif O. (1999) On the Damping of Local Mode Oscillations by Time-Scale Decomposition. Can. J. Elect. & Comp. Eng., 24, 175 -180.

Lecture Notes in Control and Information Sciences

Edited by M. Thoma

1997–2000 Published Titles: